ENVIRONMENTAL ENGINEERING

McGraw-Hill Series in Water Resources and Environmental Engineering

Rolf Eliassen, Paul H. King, and Ray K. Linsley
Consulting Editors

Bailey and Ollis: *Biochemical Engineering Fundamentals*
Bishop: *Marine Pollution and Its Control*
Biswas: *Models for Water Quality Management*
Bockrath: *Environmental Law for Engineers, Scientists, and Managers*
Bouwer: *Groundwater Hydrology*
Canter: *Environmental Impact Assessment*
Chanlett: *Environmental Protection*
Gaudy and Gaudy: *Microbiology for Environmental Scientists and Engineers*
Haimes: *Hierarchical Analysis of Water Resources Systems: Modelling and Optimization of Large-Scale Systems*
Hall and Dracup: *Water Resources Systems Engineering*
Linsley and Franzini: *Water Resources Engineering*
Linsley, Kohler, and Paulhus: *Hydrology for Engineers*
Metcalf & Eddy, Inc.: *Wastewater Engineering: Collection and Pumping of Wastewater*
Metcalf & Eddy, Inc.: *Wastewater Engineering: Treatment, Disposal, Reuse*
Peavy, Rowe, and Tchobanoglous: *Environmental Engineering*
Rich: *Low-Maintenance, Mechanically-Simple Wastewater Treatment Systems*
Sawyer and McCarty: *Chemistry for Environmental Engineering*
Steel and McGhee: *Water Supply and Sewerage*
Tchobanoglous, Theisen, and Eliassen: *Solid Wastes: Engineering Principles and Management Issues*

ENVIRONMENTAL ENGINEERING

Howard S. Peavy
Professor of Civil Engineering
Montana State University

Donald R. Rowe
Professor of Civil Engineering
King Saud University
Saudi Arabia

George Tchobanoglous
Professor of Civil Engineering
University of California, Davis

McGraw-Hill Book Company
New York St. Louis San Francisco Auckland Bogotá Hamburg
London Madrid Mexico Montreal New Dehli
Panama Paris São Paulo Singapore Sydney Tokyo Toronto

ENVIRONMENTAL ENGINEERING
INTERNATIONAL EDITION 1985

40 39 38 37 36 35 34
20 15 14 13
CTP COS

This book was set in Times Roman.
The editors were Kiran Verma and David A. Damstra.
The production supervisor was Leroy A. Young.

Library of Congress Cataloging in Publication Data

Peavy, Howard S.
Environmental engineering.

(McGraw-Hill series in water resources and environmental engineering)
Includes bibliographical references and indexes.
1. Environmental engineering. I. Rowe, Donald R.
II. Tchobanoglous, George. III. Title. IV. Series.
TD145.P43 1985 628 84-3854
ISBN 0-07-049134-8

When ordering this title use ISBN 0-07-100231-6

Printed in Singapore

CONTENTS

Part 3 Solid Waste

Appendixes

Indexes

PREFACE

Engineers and scientists from a number of related disciplines have been involved in the development of an academic basis for the understanding and management of the environment. The management of water quality has been dealt with in microbiology and sanitary engineering courses; air pollution problems have been covered in chemical and/or mechanical engineering courses; and the management of solid waste, long neglected by academicians, has been chiefly the purvey of those directly responsible for hauling and disposal operations.

During the last 10 to 13 years, schools of engineering have made considerable progress toward bringing the principles drawn from many related academic disciplines together and unifying them under the title *environmental engineering.* Not surprisingly, texts in this relatively new subject area have developed along classical, separatist lines. Thus, there have been a number of texts featuring in-depth treatment of one specific area (i.e., water, air, solid waste) and few texts attempting to treat the subject of environmental engineering as a whole.

The purpose of *Environmental Engineering* is to bring together and integrate in a single text the more general subject matter of the three principal areas of environmental engineering—water, air, and solid-waste management. And, as Chap. 1 indicates, this integration goes beyond binding three texts in a single cover.

Environmental Engineering introduces a unique approach to the overall concept of environmental engineering, an approach that emphasizes the relationship between the principles observed in natural purification processes and those employeed in engineered processes. First, the physical, chemical, mathematical, and biological principles of defining, quantifying, and measuring environmental quality are described. Next, the processes by which nature assimilates waste material are discussed and the natural purification processes that form the bases of engineered systems are detailed. Finally, the engineering principles and practices involved in the design and operation of conventional environmental engineering works are covered at length.

The breadth and depth of the material in this book precludes complete coverage in a one-semster or one-quarter course. However, the arrangement of the material lends itself to several different course formats.

1. For introductory engineering courses at the sophomore or junior level, Chaps. 1, 2, 3, 7, 8, and 10 provide an overview of the principles involved in environmental engineering systems. These chapters assume a basic knowledge of chemistry, biology, physics, and mathematics. However, because many engineering curricula contain few chemistry and even fewer biology or microbiology courses, the chapters review these subjects in detail. The introductory sections of Chaps. 4, 5, 9, 11, and 12 may be utilized to add relevance to the theoretical discussions. A course following this format will satisfy ABET requirements for engineering science.
2. If the first approach is used for an introductory course, the remaining chapters (4, 5, 6, 9, 11, and 12) can be used as a follow-up course in environmental engineering design. This course should be restricted to engineering students at the junior or senior level who have completed basic fluid mechanics. Such a course would meet ABET's engineering design criteria.
3. A more classical approach would be to use the first six chapters as a text for a one-semester or one-quarter course in water and wastewater engineering. A second one-semester/quarter course on air-pollution control and solid-waste management would use Chaps. 7 through 12. Designed for junior- or senior-level engineering students that have completed basic fluid mechanics, these two courses will meet ABET criteria for engineering design and science, or an approximate one to one ratio.
4. Chapters 1, 2, 3, 7, 8, and 10 can also be used for a companion course in environmental science for nonengineering students, provided allowance is made for the limited mathematical background of the students.

Whatever the approach used, the text should leave students with a clear understanding of the principles of all three of the major areas of enviromental engineering. User comments and suggestions concerning the effectiveness of this approach would be greatly appreciated.

The authors wish to acknowledge the fact that development and publication of *Environmental Engineering* would not have been possible without the help and inspiration of our former professors, the challenge and motivation of our students, the assistance and encouragement of our colleagues, the patience and forebearance of our editors, and the support and understanding of our families.

Howard S. Peavy
Donald R. Rowe
George Tchobanoglous

ENVIRONMENTAL ENGINEERING

CHAPTER

ONE

INTRODUCTION

Environmental engineering has been defined as the branch of engineering that is concerned with protecting the environment from the potentially deleterious effects of human activity, protecting human populations from the effects of adverse environmental factors, and improving environmental quality for human health and well-being. [1-2]

As the above definition implies, humans interact with their environment—sometimes adversely impacting the environment and sometimes being adversely impacted by pollutants in the environment. An understanding of the nature of the environment and of human interaction with it is a necessary prerequisite to understanding the work of the environmental engineer.

1-1 THE ENVIRONMENT

Simply stated, the environment can be defined as one's surroundings. In terms of the environmental engineer's involvement, however, a more specific definition is needed. To the environmental engineer, the word *environment* may take on global dimensions, may refer to a very localized area in which a specific problem must be addressed, or may, in the case of contained environments, refer to a small volume of liquid, gaseous, or solid materials within a treatment plant reactor.

The global environment consists of the atmosphere, the hydrosphere, and the lithosphere in which the life-sustaining resources of the earth are contained. The *atmosphere*, a mixture of gases extending outward from the surface of the earth, evolved from elements of the earth that were gasified during its formation and metamorphosis. The *hydrosphere* consists of the oceans, the lakes and streams, and the shallow groundwater bodies that interflow with the surface water. The *lithosphere* is the soil mantle that wraps the core of the earth.

The *biosphere*, a thin shell that encapsulates the earth, is made up of the atmosphere and lithosphere adjacent to the surface of the earth, together with the hydrosphere. It is within the biosphere that the life forms of earth, including humans, live. Life-sustaining materials in gaseous, liquid, and solid forms are cycled through the biosphere, providing sustenance to all living organisms.

Life-sustaining resources—air, food, and water—are withdrawn from the biosphere. It is also into the biosphere that waste products in gaseous, liquid, and solid forms are discharged. From the beginning of time, the biosphere has received and assimilated the wastes generated by plant and animal life. Natural systems have been ever active, dispersing smoke from forest fires, diluting animal wastes washed into streams and rivers, and converting debris of past generations of plant and animal life into soil rich enough to support future populations.

For every natural act of pollution, for every undesirable alteration in the physical, chemical, or biological characteristics of the environment, for every incident that eroded the quality of the immediate, or local, environment, there were natural actions that restored that quality. Only in recent years has it become apparent that the sustaining and assimilative capacity of the biosphere, though tremendous, is not, after all, infinite. Though the system has operated for millions of years, it has begun to show signs of stress, primarily because of the impact of humans upon the environment.

1-2 THE IMPACT OF HUMANS UPON THE ENVIRONMENT

In a natural state, earth's life forms live in equilibrium with their environment. The numbers and activities of each species are governed by the resources available to them. Species interaction is common, with the waste product of one species often forming the food supply of another. Humans alone have the ability to gather resources from beyond their immediate surroundings and process those resources into different, more versatile forms. These abilities have made it possible for human population to thrive and flourish beyond natural constraints. But the natural and manufactured wastes generated and released into the biosphere by these increased numbers of human beings have upset the natural equilibrium.

Anthropogenic, or human-induced, pollutants have overloaded the system. The overloading came relatively late in the course of human interaction with the environment, perhaps because early societies were primarily concerned with meeting natural needs, needs humans share in common with most of the higher mammals. These peoples had not yet begun to be concerned with meeting the acquired needs associated with more advanced civilizations.

Satisfying Natural Needs

Early humans used natural resources to satisfy their needs for air, water, food, and shelter. These natural, unprocessed resources were readily available in the biosphere, and the residues generated by the use of such resources were generally compatible with, or readily assimilated by, the environment. Primitive humans ate plant and animal foods without even disturbing the atmosphere with the smoke from a campfire. Even when use of fire became common, the relatively small amounts of smoke generated were easily and rapidly dispersed and assimilated by the atmosphere.

Early civilizations often drank from the same rivers in which they bathed and deposited their wastes, yet the impact of such use was relatively slight, as natural cleansing mechanisms easily restored water quality. These early humans used caves and other natural shelters or else fashioned their homes from wood, dirt, or animal skins. Often nomadic, early populations left behind few items that were not readily broken down and absorbed by the atmosphere, hydrosphere, or lithosphere. And those items that were not broken down with time were so few in number and so innocuous as to present no significant solid-waste problems.

Only as early peoples began to gather together in larger, more or less stable groupings did their impact upon their local environments begin to be significant. In 61 A.D., cooking and heating fires caused air pollution problems so severe that the Roman philosopher Seneca complained of "the stink of the smoky chimneys." By the late eighteenth century, the waters of the Rhine and the Thames had become too polluted to support game fish. From the Middle Ages the areas where food and human waste were dumped harbored rats, flies, and other pests.

Satisfying Acquired Needs

But these early evidences of pollution overload were merely the prelude to greater overloads to come. With the dawn of the industrial revolution, humans were better able than ever to satisfy their age-old needs of air, water, food, and shelter. Increasingly they turned their attention to other needs beyond those associated with survival. By the late nineteenth and early twentieth centuries, automobiles, appliances, and processed foods and beverages had become so popular as to seem necessities, and meeting these acquired needs had become a major thrust of modern industrial society.

Unlike the natural needs discussed earlier, acquired needs are usually met by items that must be processed or manufactured or refined, and the production, distribution, and use of such items usually results in more complex residuals, many of which are not compatible with or readily assimilated by the environment.

Take, for example, a familiar modern appliance—the toaster. The shell and the heating elements are likely to be made of steel, the handle of the lift lever of plastic. Copper wires and synthetic insulation may be used in the connecting cord, and rubber may be used on the plug. In assessing the pollutants generated by the manufacture and sale of this simple appliance, it would be necessary to include all the resources expended in the mining of the metals, the extracting and refining of the petroleum, the shipping of the various materials, then the manufacturing, shipping, and selling of the finished product. The potential impact of all of these activities upon air and water quality is significant. Furthermore, if the pollution potential involving the manufacture and use of the heavy equipment needed for the extraction and processing of the raw materials used in the various toaster components is considered, the list could go on ad nauseum. And the solid-waste disposal problems that arise when it is time to get rid of the toaster become a further factor.

As a rule, meeting the acquired needs of modern societies generates more residuals than meeting natural needs, and these residuals are likely to be less compatible with the environment and less likely to be readily assimilated into the biosphere. As societies ascend the socioeconomic ladder, the list of acquired needs, or luxuries, increases, as do the complexity of the production chain and the mass and complexity of the pollutants generated. Consequently, the impact of modern human populations upon the environment is of major concern to the environmental engineer.

1-3 THE IMPACT OF THE ENVIRONMENT UPON HUMANS

Though rivers become stagnant, skies smoke-shrouded, and dumping grounds odoriferous and unsightly, populations generally manage to ignore their impact on the environment until they begin to become aware of the ill effects that a polluted environment can have upon their own health and well-being. Though stagnant rivers, smoggy skies, and unsightly dumps were aesthetically displeasing to the citizens of overcrowded cities of earlier centuries, no attempt was made to reverse the negative impact humans had on their environment until it became evident that heavily polluted water, air, and soil could exert an equally negative impact on the health, the aesthetic and cultural pleasures, and the economic opportunities of humans.

Health Concerns

Elements of the air, the water, and the land may host harmful biological and chemical agents that impact the health of humans. A wide range of communicable diseases can be spread through elements of the environment by human and animal waste products. This is most clearly evidenced by the plagues of the Middle Ages when disease spread through rats that fed on contaminated solid and human waste and disease carried by waterborne parasites and bacteria ran rampant through the population of Europe.

It has only been in the last century that the correlation between waterborne biological agents and human diseases has been proved and effective preventive measures have been taken. Through immunization and environmental control programs, the major diseases transmitted via the environment have all but been eliminated in developed countries. No country, however, is totally immune from outbreaks of environmentally transmitted disease. The transmission of viruses and protozoa has proved particularly difficult to control, and lapses in good sanitary practice have resulted in minor epidemics of other waterborne diseases.

Pollution of the atmosphere has also posed severe health problems that are of great concern to environmental engineers. People in crowded cities have likely suffered from the ill effects of air pollution for centuries, but it is only in this century that increasingly heavy pollution has caused health problems so dramatic as to be easily attributed to air pollution. Several key incidents helped call attention to the potentially deadly effect of air pollution. Several killer smogs settled over London in the last quarter of the nineteenth century, but the true extent of the air-pollution problem in that city did not become apparent until 4000 deaths and countless illnesses were attributed to the London smog of 1952.

Though the 20 deaths caused by a smog over Donora, Pennsylvania, in 1948 raised some alarm, it was not until the New York inversion of 1963 claimed several hundred lives that this country began to take the fight against air pollution seriously. Monitoring of the sulfur dioxide, lead, and carbon monoxide levels in areas such as the smog-shrouded Los Angeles basin has revealed that the high levels of these and other contaminants pose direct and indirect threats to human health. These findings have made air-pollution control a top priority of the Environmental Protection Agency and a major concern of environmental engineers, who are now called upon to devise management programs designed to alter the pattern of air pollution begun centuries ago and continued until the present time.

Other environmentally related health problems also concern the environmental engineer. The widespread use of chemicals in agriculture and industry has introduced many new compounds into the environment. Some of these compounds have been diffused in small quantities throughout the environment, while others have been concentrated at disposal sites. Such chemicals may be spread through air, water, and soil, as well as through the food chain, and thus pose a potential threat to all humans.

The pesticide DDT was used extensively during the mid-century decades and has been instrumental in the elimination of malaria in many parts of the world. In addition, this pesticide was used extensively to control insect pests on food and fiber plants. Its beneficial use to humans was widely acclaimed, and its promoter, Paul Muller, was awarded a Nobel prize in 1958 for his contribution to public health. Subsequent research, however, has shown that DDT is a cumulative toxin that has adversely affected many nontarget species. Traces of DDT can be found in almost all living organisms throughout the world—including humans. Although the use of DDT is now banned in the United States and several other countries, the chemical is still being manufactured, primarily for use in several developing countries, particularly in tropical zones where its benefits are still considered to outweigh its liabilities.

A more recent example of chemical toxins that threaten health is the chemical dioxin. The formation of this chemical, the scientific name of which is 2,3,7,8-tetrachloro-dibenzoparadioxin, is an unintentional by-product of a manufacturing process used with some herbicides and wood-preserving compounds. It is also formed in the production of some disinfectants and industrial cleaning compounds. Dioxin is an extremely toxic substance, and its presence in excess of 1 ppb (part per billion) in the environmental elements becomes cause for concern. (One part per billion corresponds to one drop of water in a swimming pool measuring 15 ft wide, 30 ft long, and 6 ft deep.)

Chemicals containing dioxin residuals have been used on a widespread basis during the last few decades, and the level of this chemical in the general environment is not currently known. The discovery of dioxin residuals in waste-disposal sites and in soils that were contaminated through application of the parent material has caused great concern and has resulted in expensive cleaning efforts. The creation of a "superfund" in the Environmental Protection Agency, initially funded at several billion dollars, is but a start in the efforts to mitigate the hazards of chemicals in the environment.

Other Concerns

Clean air and water are an aesthetic delight, yet city dwellers have all but forgotten the smell of clean air, and clear, sparkling lakes, rivers, and streams are becoming increasingly rare. Littered streets and highways offend, rather than delight, and unfenced junkyards and uncontrolled dumps give further evidence of the aesthetically displeasing effect of improper solid-waste disposal techniques.

Our cultural as well as our aesthetic heritage is also being lost to pollution. The Parthenon in Athens, the Statue of Liberty in New York harbor, the statues and frescoes in Venice have withstood the onslaught of the elements for centuries, yet are in increasing danger of being destroyed by the constituents of a polluted atmosphere.

And pollution poses economic threats to human populations. Lake Erie once supported a thriving fishing industry and all the attendant processing and shipping facilities associated with that industry, yet the economic potential of the lake was nearly lost before serious cleanup efforts were begun. The silting in of rivers, harbors, and reservoirs due to uncontrolled erosion, often exacerbated by human activities, threatens to strengthen some industries at the expense of others.

Environmental engineers are committed to protecting humans from the threats a polluted environment pose to human health, aesthetic and cultural enjoyment, and economic well-being.

1-4 IMPROVEMENT OF ENVIRONMENTAL QUALITY

Vitally concerned with the improvement of environmental quality, the environmental engineer plays an important role in environmental management programs. Such programs might be said to involve two distinct aspects—environmental strategies and environmental tactics. [1-1] *Environmental strategies* are comprehensive plans that usually address a variety of problems that confront a single area. Typical environmental strategies might be a program to improve the quality of Lake Erie, to improve the air quality of the Los Angeles basin, or to collect and properly dispose of the solid waste from the city of Philadelphia.

Environmental strategies are usually worked out in public and political arenas. Considerations must include economic, social, and demographic factors. Historically, environmental engineers have not played a highly visible role in devising environmental strategies. Nevertheless, the environmental engineer should be an important member of a management team that includes persons drawn from a wide variety of disciplines. The input of the environmental engineer, especially in assessing the likely response of the environment to various levels of contaminant loading and in weighing the various technical solutions that may be proposed, is a necessary component of any environmental strategy.

Environmental engineers are usually more directly associated with the implementation of the *environmental tactics* that are the means for achieving the goals set forth in a specific portion of a given environmental strategy. The engineer's part in this implementation consists primarily of the design, construction, and operation of treatment facilities for water, air, and solid waste. For example, the environmental engineer would be involved directly in the addition of tertiary processes to remove phosphorus from the effluent of a wastewater-treatment facility emptying into Lake Erie, the installation of a hydrocarbon removal system at a gasoline refinery system in Los Angeles, or the design of a solid-waste processing plant in Philadelphia.

1-5 THE ROLE OF THE ENVIRONMENTAL ENGINEER

As pollutants enter air, water, or soil, natural processes such as dilution, biological conversions, and chemical reactions convert waste material to more acceptable forms and disperse them through a larger volume. Yet those natural processes can no longer perform the cleanup alone. The treatment facilities designed by the environmental engineer are based on the principles of self-cleansing observed in nature, but the engineered processes amplify and optimize the operations observed in nature to handle larger volumes of pollutants and to treat them more rapidly. Engineers adapt the principles of natural mechanisms to engineered systems for pollution control when they construct tall stacks to disperse and dilute air pollutants, design biological treatment facilities for the removal of organics from wastewater, use chemicals to oxidize and precipitate out the iron and manganese in drinking-water supplies, or bury solid wastes in controlled landfill operations.

Occasionally, the environmental engineer must design to reverse or counteract natural processes. For example, the containers used for disposal of hazardous wastes such as toxic chemicals and radioactive materials must isolate those materials from the environment in order to prevent the onset of the natural, but highly undesirable, processes of dilution and dispersion.

As will be demonstrated throughout this text, an understanding of natural and engineered purification processes requires an understanding of the biological and chemical reactions involved in these processes. Thus, in addition to being knowledgeable in the mathematical, physical, and engineering sciences, the environmental engineer must also be well grounded in the subject areas of chemistry and microbiology, subject areas not usually emphasized in engineering curricula. Indeed, an understanding of biological and chemical principles is as essential to the environmental engineer as the understanding of statics and strength of materials is to the structural engineer.

The environmental engineer's unique role is to build a bridge between biology and technology by applying all the techniques made available by modern engineering technology to the job of cleaning up the debris left in the wake of an indiscriminate use of that technology. The delicate balance of our biosphere has been disturbed, and the state in which we now find ourselves is a direct consequence of our having ignored the limits of the earth's ability to overcome heavy pollution loads, and of our having been ignorant of the constraints imposed by the limits of the self-cleansing mechanisms of our biosphere.

A keen awareness of these natural constraints plays an important role in the work of environmental engineers. For example, the laws of conservation of mass and energy prevent the destruction of pollutants, and the engineer is bound by these limits. The principles of waste treatment must therefore be to convert the objectionable material to other, less objectionable forms; to disperse the pollutants so that their concentrations are minimal; or to concentrate them for isolation from the environment.

In all instances, the end products of the treatment of polluted water or air or of the disposal of solid wastes must be compatible with the existing environmental resources and must not overtax the assimilative powers of hydrosphere, atmosphere, or lithosphere. In structural engineering, the engineer can simply specify a larger or stronger beam to carry a heavier load. The environmental engineer, on the other hand, must accept the carrying capacity of a stream, an airshed, or a landmass because these can seldom be changed.

It is the purpose of this text to demonstrate how the environmental engineer, working within these constraints, uses all available technological tools to design efficient control and treatment devices that are modeled after the natural processes that have so long preserved our biosphere. For only by bringing technology into harmony with the natural environment can the engineer hope to achieve the goals of the profession—the protection of the environment from the potentially deleterious effects of human activity, the protection of human populations from the effects of adverse environmental factors, and the improvement of environmental quality for human health and well-being.

REFERENCES

1-1 Bella, D. A., and W. S. Overton: "Environmental Planning and Ecological Possibilities," presented at the annual national environmental engineering meeting of ASCE, St. Louis, Mo., October 18–22, 1971.

1-2 "Guidelines for Environmental Engineering Visitors on ECPD Accreditation Teams," Engineers Council for Professional Development, United Engineering Center, 345 East 47th St., New York, October 1977.

PART ONE

WATER

CHAPTER

TWO

WATER QUALITY: DEFINITIONS, CHARACTERISTICS, AND PERSPECTIVES

The availability of a water supply adequate in terms of both quantity and quality is essential to human existence. Early people recognized the importance of water from a quantity viewpoint. Civilization developed around water bodies that could support agriculture and transportation as well as provide drinking water. Recognition of the importance of water quality developed more slowly. Early humans could judge water quality only through the physical senses of sight, taste, and smell. Not until the biological, chemical, and medical sciences developed were methods available to measure water quality and to determine its effects on human health and well-being.

It was not until the mid-nineteenth century that the relationship between human waste, drinking water, and disease was documented. Several more years intervened before the facts concerning this relationship became widely accepted and remedial action was taken. In 1854,* Dr. John Snow, a public-health worker in London, noted a high correlation between cholera cases and consumption of water from a well on Broad Street. Not only was cholera running rampant in the neighborhood around the well, but outbreaks of the disease in other parts of the city could be traced to individuals who had had occasion to drink from the Broad Street well. Although the proof was conclusive by modern epidemiology standards, the evidence was not accepted by Snow's contemporaries. It is alleged that he physically removed the pump handle to prevent use of the contaminated water, thus abating the epidemic. [2-21]

Advances in the germ theory of disease were made by Pasteur and others in the late nineteenth century, and by 1900 the concept of waterborne disease was well accepted. The development of the science of water chemistry roughly paralleled that of water microbiology. Many of the chemicals used in industrial processes

* This date is listed as 1849 in some publications.

and agriculture have been identified in water. However, the effort to identify other chemical compounds which may already be found in trace quantities in many water supplies and to determine their effect on human health was only recently begun. It is likely that new analytical techniques will be developed that will identify compounds not yet known to exist in water, and it is conceivable that these materials will also be linked to human health. Thus, the science of water quality will remain a challenge for engineers and scientists for years to come.

Like all sciences, the science of water quality has developed its own terminology and the means of quantifying these terms. The purpose of this chapter is to introduce the reader to the modern concepts of water quality. The means by which the nature and extent of contaminants in water are measured and expressed are presented along with the sources of various contaminants that find their way into water. An understanding of the material in this chapter will be essential in subsequent chapters dealing with water-quality changes in both natural and engineered systems.

2-1 THE HYDROLOGIC CYCLE AND WATER QUALITY

Water is one of the most abundant compounds found in nature, covering approximately three-fourths of the surface of the earth. In spite of this apparent abundance, several factors serve to limit the amount of water available for human use. As shown in Table 2-1, over 97 percent of the total water supply is contained in the oceans and other saline bodies of water and is not readily usable for most purposes. Of the remaining 3 percent, a little over 2 percent is tied up in ice caps and glaciers and, along with atmospheric and soil moisture, is inaccessible. [2-17] Thus, for their general livelihood and the support of their varied technical and agricultural activities, humans must depend upon the remaining 0.62 percent found in freshwater lakes, rivers, and groundwater supplies.

Table 2-1 World water distribution

Location	Volume, 10^{12} m^3	% of total
Land areas		
Freshwater lakes	125	0.009
Saline lakes and inland seas	104	0.008
Rivers (average instantaneous volume)	1.25	0.0001
Soil moisture	67	0.005
Groundwater (above depth of 4000 m)	8,350	0.61
Ice caps and glaciers	29,200	2.14
Total land area (rounded)	37,800	2.8
Atmosphere (water vapor)	13	0.001
Oceans	1,320,000	97.3
Total all locations (rounded)	1,360,000	100

Source: Adapted from Todd. [2-17]

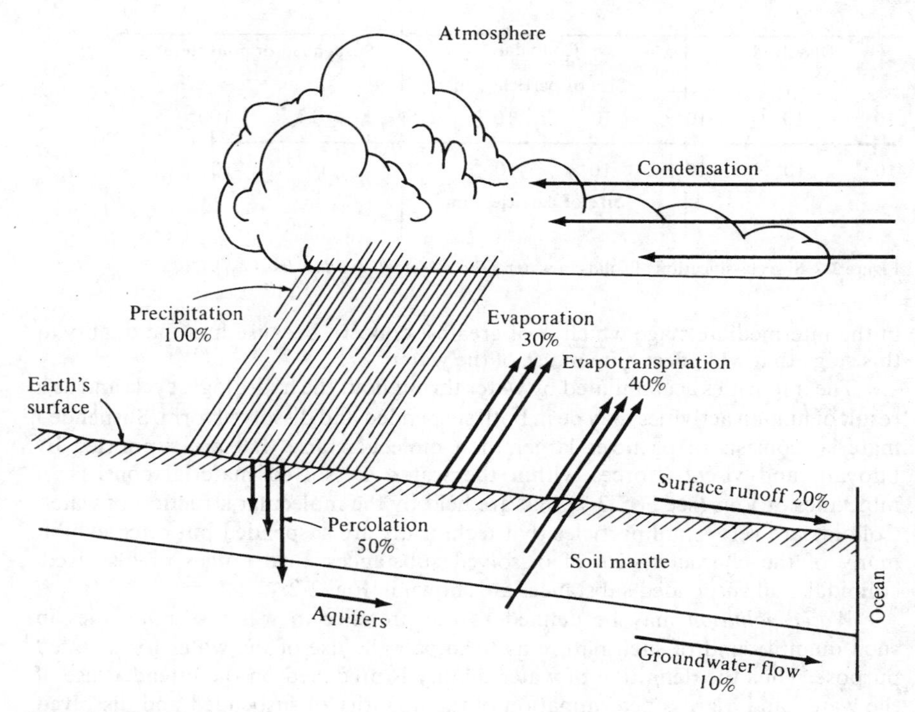

Figure 2-1 Hydrologic cycle.

Water is in a constant state of motion, as depicted in the hydrologic cycle shown in Fig. 2-1. Atmospheric water condenses and falls to the earth as rain, snow, or some other form of precipitation. Once on the earth's surface, water flows into streams, lakes, and eventually the oceans, or percolates through the soil and into aquifers that eventually discharge into surface waters. Through evaporation from surface waters or by evapotranspiration from plants, water molecules return to the atmosphere to repeat the cycle. Although the movement through some parts of the cycle may be relatively rapid, complete recycling of groundwater must often be measured in geologic time.

Water in nature is most nearly pure in its evaporation state. Because the very act of condensation usually requires a surface, or nuclei, water may acquire impurities at the very moment of condensation. Additional impurities are added as the liquid water travels through the remainder of the hydrologic cycle and comes into contact with materials in the air and on or beneath the surface of the earth. Human activities contribute further impurities in the form of industrial and domestic wastes, agricultural chemicals, and other, less obvious contaminants. Ultimately, these impure waters will complete the hydrologic cycle and return to the atmosphere as relatively pure water molecules. However, it is water quality

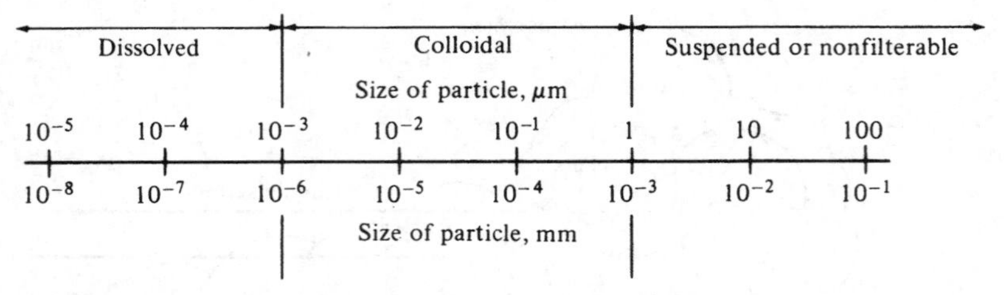

Figure 2-2 Size classification of solids in water. (*From Metcalf & Eddy, Inc.* [*2-8*].)

in the intermediate stage which is of greatest concern because it is the quality at this stage that will affect human use of the water.

The impurities accumulated by water throughout the hydrologic cycle and as a result of human activities may be in both suspended and dissolved form. Suspended material consists of particles larger than molecular size that are supported by buoyant and viscous forces within the water. Dissolved material consists of molecules or ions (see Sec. 2-7) that are held by the molecular structure of water. Colloids are very small particles that technically are suspended but often exhibit many of the characteristics of dissolved substances. Size ranges of dissolved, colloidal, and suspended substances are shown in Fig. 2-2.

Water pollution may be defined as the presence in water of impurities in such quantity and of such nature as to impair the use of the water for a stated purpose. Thus the definition of water quality is predicted on the intended use of the water, and a gross determination of the quantity of suspended and dissolved impurities, while useful in some cases, is not sufficient to completely define water quality. Many parameters have evolved that qualitatively reflect the impact that various impurities have on selected water uses. Analytical procedures have been developed that quantitatively measure these parameters. *Standard Methods for the Examination of Water and Wastewater* [2-15] has been the authoritative standard for test procedures for many years. For detailed coverage of the subject the interested reader is referred to this publication and to an Environmental Protection Agency publication that offers similar information. [2-9]

A knowledge of the parameters most commonly associated with water- and wastewater-treatment processes is essential to the environmental engineer. The remainder of this chapter will be devoted to a discussion of parameters used to assess the physical, chemical, and biological characteristics of water. Testing procedures described for each parameter are based on those described in *Standard Methods*. [2-15]

Physical Water-Quality Parameters

Physical parameters define those characteristics of water that respond to the senses of sight, touch, taste, or smell. Suspended solids, turbidity, color, taste and odor, and temperature fall into this category.

2-2 SUSPENDED SOLIDS

As noted earlier, solids can be dispersed in water in both suspended and dissolved forms. Although some dissolved solids may be perceived by the physical senses, they fall more appropriately under the category of chemical parameters and will be discussed more fully in a later section.

Sources

Solids suspended in water may consist of inorganic or organic particles or of immiscible liquids. Inorganic solids such as clay, silt, and other soil constituents are common in surface water. Organic material such as plant fibers and biological solids (algal cells, bacteria, etc.) are also common constituents of surface waters. These materials are often natural contaminants resulting from the erosive action of water flowing over surfaces. Because of the filtering capacity of the soil, suspended material is seldom a constituent of groundwater.

Other suspended material may result from human use of the water. Domestic wastewater usually contains large quantities of suspended solids that are mostly organic in nature. Industrial use of water may result in a wide variety of suspended impurities of either organic or inorganic nature. Immiscible liquids such as oils and greases are often constituents of wastewater.

Impacts

Suspended material may be objectionable in water for several reasons. It is aesthetically displeasing and provides adsorption sites for chemical and biological agents. Suspended organic solids may be degraded biologically, resulting in objectionable by-products. Biologically active (live) suspended solids may include disease-causing organisms as well as organisms such as toxin-producing strains of algae.

Measurement

There are several tests available for measuring solids. Most are gravimetric tests involving the mass of residues. The *total solids test* quantifies *all* the solids in the water, suspended and dissolved, organic and inorganic. This parameter is measured by evaporating a sample to dryness and weighing the residue. The total quantity of residue is expressed as milligrams per liter (mg/L) on a dry-mass-of-solids basis. A drying temperature slightly above boiling (104°C) is sufficient to drive off the liquid and the water adsorbed to the surface of the particles, while a temperature of about 180°C is necessary to evaporate the occluded water.

Most suspended solids can be removed from water by filtration. Thus, the suspended fraction of the solids in a water sample can be approximated by filtering the water, drying the residue and filter to a constant weight at 104°C ($\pm 1°$), and determining the mass of the residue retained on the filter. The results of this *suspended solids test* are also expressed as dry mass per volume (milligrams per liter). The amount of dissolved solids passing through the filters, also expressed as milligrams per liter, is the difference between the total-solids and suspended-solids content of a water sample.

It should be emphasized that filtration of a water sample does not exactly divide the solids into suspended and dissolved fractions according to the definitions presented earlier. Some colloids may pass through the filter and be measured along with the dissolved fraction while some of the dissolved solids adsorb to the filter material. The extent to which this occurs depends on the size and nature of the solids and on the pore size and surface characteristics of the filter material. For this reason, the terms *filterable residues* and *nonfilterable residues* are often used. Filterable residues pass through the filter along with the water and relate more closely to dissolved solids, while nonfilterable residues are retained on the filter and relate more closely to suspended solids. "Filterable residues" and "non-filterable residues" are terms more frequently used in laboratory analysis while the "dissolved solids" and "suspended solids" are terms more frequently used in water-quality-management practice. For most practical applications, the distinction between the two is not necessary.

Once samples have been dried and measured, the organic content of both total and suspended solids can be determined by firing the residues at 600°C for 1 h. The organic fraction of the residues will be converted to carbon dioxide, water vapor, and other gases and will escape. The remaining material will represent the inorganic, or fixed, residue. When organic suspended solids are being measured, a filter made of glass fiber or some other material that will not decompose at the elevated temperature must be used. The following example illustrates the calculations involved in suspended solids analysis.

Example 2-1: Determining the concentration of suspended solids: A filterable residue analysis is run on a sample of water as follows. Prior to filtering, the crucible and filter pad are kept overnight in the drying oven, cooled, and the dry mass (tare mass) of the pair determined to be 54.352 g. Two hundred and fifty milliliters of the sample is drawn through a filter pad contained in the porous-bottom crucible. The crucible and filter pad are then placed in a drying oven at 104°C and dried until a constant mass of 54.389 g is reached. Determine the suspended solids concentration of the sample.

SOLUTION

1. Determine the mass of solids removed.

$$\begin{array}{lr} \text{Tare mass + solids} & = 54.389 \text{ g} \\ -\text{Tare mass} & = 54.352 \text{ g} \\ \hline \text{Mass of solids} & = 0.037 \text{ g} \\ & = 37 \text{ mg} \end{array}$$

2. Determine the concentration of the solids.

$$\frac{\text{mg solids} \times 1000 \text{ mL/L}}{\text{mL of sample}} = \text{conc in mg/L}$$

$$\frac{37 \times 1000}{250} = 148 \text{ mg/L}$$

Use

Suspended solids, where such material is likely to be organic and/or biological in nature, are an important parameter of wastewater. The suspended-solids parameter is used to measure the quality of the wastewater influent, to monitor several treatment processes, and to measure the quality of the effluent. EPA has set a maximum suspended-solids standard of 30 mg/L for most treated wastewater discharges.

2-3 TURBIDITY

A direct measurement of suspended solids is not usually performed on samples from natural bodies of water or on potable (drinkable) water supplies. The nature of the solids in these waters and the secondary effects they produce are more important than the actual quantity. For such waters a test for turbidity is commonly used.

Turbidity is a measure of the extent to which light is either absorbed or scattered by suspended material in water. Because absorption and scattering are influenced by both size and surface characteristics of the suspended material, turbidity is not a direct quantitative measurement of suspended solids. For example, one small pebble in a glass of water would produce virtually no turbidity. If this pebble were crushed into thousands of particles of colloidal size, a measurable turbidity would result, even though the mass of solids had not changed.

Sources

Most turbidity in surface waters results from the erosion of colloidal material such as clay, silt, rock fragments, and metal oxides from the soil. Vegetable fibers and microorganisms may also contribute to turbidity. Household and industrial wastewaters may contain a wide variety of turbidity-producing material. Soaps, detergents, and emulsifying agents produce stable colloids that result in turbidity. Although turbidity measurements are not commonly run on wastewater, discharges of wastewaters may increase the turbidity of natural bodies of water.

Impacts

When turbid water in a small, transparent container, such as a drinking glass, is held up to the light, an aesthetically displeasing opaqueness or "milky" coloration is apparent. The colloidal material associated with turbidity provides adsorption sites for chemicals that may be harmful or cause undesirable tastes and odors and for biological organisms that may be harmful. Disinfection of turbid waters is difficult because of the adsorptive characteristics of some colloids and because the solids may partially shield organisms from the disinfectant.

In natural water bodies, turbidity may impart a brown or other color to water, depending on the light-absorbing properties of the solids, and may interfere with light penetration and photosynthetic reactions in streams and lakes. Accumulation of turbidity-causing particles in porous streambeds results in sediment deposits that can adversely affect the flora and fauna of the stream.

Measurement

Turbidity is measured photometrically by determining the percentage of light of a given intensity that is either absorbed or scattered. The original measuring apparatus, called a *Jackson turbidimeter*, was based on light absorption and employed a long tube and standardized candle. The candle was placed beneath the glass tube that was then housed in a black metal sheath so that the light from the candle could only be seen from above the apparatus. The water sample was then poured slowly into the tube until the lighted candle was no longer visible, i.e., complete absorption had occurred. The glass tube was calibrated with readings for turbidity produced by suspensions of silica dioxide (SiO_2), with one Jackson turbidity unit (JTU) being equal to the turbidity produced by 1 mg SiO_2 in 1 L of distilled water.

In recent years this awkward apparatus has been replaced by a turbidity meter in which a standardized electric bulb produces a light that is then directed through a small sample vial. In the absorption mode, a photometer measures the light intensity on the side of the vial opposite from the light source, while in the scattering mode, a photometer measures the light intensity at a 90° angle from the light source. Although most turbidity meters in use today work on the scattering principle, turbidity caused by dark substances that absorb rather than reflect light should be measured by the absorption technique. Formazin, a chemical compound, provides more reproducible standards than SiO_2 and has replaced it as a reference. Turbidity meter readings are now expressed as *formazin turbidity units*, or *FTUs*. The term *nephelometry turbidity units* (*NTU*) is often used to indicate that the test was run according to the scattering principle.

Use

Turbidity measurements are normally made on "clean" waters as opposed to wastewaters. Natural waters may have turbidities ranging from a few FTUs to several hundred. EPA drinking-water standards specify a maximum of 1 FTU, while the American Water Works Association has set 0.1 FTU as its goal for drinking water. [2-1]

2-4 COLOR

Pure water is colorless, but water in nature is often colored by foreign substances. Water whose color is partly due to suspended matter is said to have *apparent color*. Color contributed by dissolved solids that remain after removal of suspended matter is known as *true color*.

Sources

After contact with organic debris such as leaves, conifer needles, weeds, or wood, water picks up tannins, humic acid, and humates and takes on yellowish-brown hues. Iron oxides cause reddish water, and manganese oxides cause brown or blackish water. Industrial wastes from textile and dyeing operations, pulp and paper production, food processing, chemical production, and mining, refining, and slaughterhouse operations may add substantial coloration to water in receiving streams.

Impacts

Colored water is not aesthetically acceptable to the general public. In fact, given a choice, consumers tend to choose clear, noncolored water of otherwise poorer quality over treated potable water supplies with an objectionable color. Highly colored water is unsuitable for laundering, dyeing, papermaking, beverage manufacturing, dairy production and other food processing, and textile and plastic production. Thus, the color of water affects its marketability for both domestic and industrial use.

While true color is not usually considered unsanitary or unsafe, the organic compounds causing true color may exert a chlorine demand and thereby seriously reduce the effectiveness of chlorine as a disinfectant. Perhaps more important are the products formed by the combination of chlorine with some color-producing organics. Phenolic compounds, common constituents of vegetative decay products, produce very objectionable taste and odor compounds with chlorine. Additionally, some compounds of naturally occurring organic acids and chlorine are either known to be, or are suspected of being, carcinogens (cancer-causing agents).

Measurement

Although several methods of color measurement are available, methods involving comparison with standardized colored materials are most often used. Color-comparison tubes containing a series of standards may be used for direct comparison of water samples that have been filtered to remove apparent color. Results are expressed in true color units (TCUs) where one unit is equivalent to the color produced by 1 mg/L of platinum in the form of chlorplatinate ions. For colors other than yellowish-brown hues, especially for colored waters originating from industrial waste effluents, special spectrophotometric techniques are usually employed.

In fieldwork, instruments employing colored glass disks that are calibrated to the color standards are often used. Because biological and physical changes occurring during storage may affect color, samples should be tested within 72 h of collection.

Use

Color is not a parameter usually included in wastewater analysis. In potable water analysis, the common practice is to measure only the true color produced by organic acid resulting from decaying vegetation in the water. The resulting value can be taken as an indirect measurement of humic substances in the water.

2-5 TASTE AND ODOR

The terms *taste* and *odor* are themselves definitive of this parameter. Because the sensations of taste and smell are closely related and often confused, a wide variety of tastes and odors may be attributed to water by consumers. Substances that produce an odor in water will almost invariably impart a taste as well. The converse is not true, as there are many mineral substances that produce taste but no odor.

Sources

Many substances with which water comes into contact in nature or during human use may impart perceptible taste and odor. These include minerals, metals, and salts from the soil, end products from biological reactions, and constituents of wastewater. Inorganic substances are more likely to produce tastes unaccompanied by odor. Alkaline material imparts a bitter taste to water, while metallic salts may give a salty or bitter taste.

Organic material, on the other hand, is likely to produce both taste and odor. A multitude of organic chemicals may cause taste and odor problems in water, with petroleum-based products being prime offenders. Biological decomposition of organics may also result in taste- and odor-producing liquids and gases in water. Principal among these are the reduced products of sulfur that impart a "rotten egg" taste and odor. Also, certain species of algae secrete an oily substance that may result in both taste and odor. The combination of two or more substances, neither of which would produce taste or odor by itself, may sometimes result in taste and odor problems. This synergistic effect was noted earlier in the case of organics and chlorine.

Impacts

Consumers find taste and odor aesthetically displeasing for obvious reasons. Because water is thought of as tasteless and odorless, the consumer associates taste and odor with contamination and may prefer to use a tasteless, odorless water that might actually pose more of a health threat. And odors produced by organic substances may pose more than a problem of simple aesthetics, since some of those substances may be carcinogenic.

Measurement

Direct measurement of materials that produce tastes and odors can be made if the causative agents are known. Several types of analysis are available for measuring taste-producing inorganics. Measurement of taste- and odor-causing organics can be made using gas or liquid chromatography. Because chromatographic analysis is time-consuming and requires expensive equipment, it is not routinely performed on water samples, but should be done if problem organics are suspected. However, because of the synergism noted earlier, quantifying the sources does not necessarily quantify the nature or intensity of taste and odor.

Quantitative tests that employ the human senses of taste and smell can be used for this purpose. An example is the test for the *threshold odor number* (TON). Varying amounts of odorous water are poured into containers and diluted with enough odor-free distilled water to make a 200-mL mixture. An assembled panel of five to ten "noses" is used to determine the mixture in which the odor is just barely detectable to the sense of smell. The TON of that sample is then calculated, using the formula

$$\text{TON} = \frac{A + B}{A} \tag{2-1}$$

where A is the volume of odorous water (mL) and B is the volume of odor-free water required to produce a 200-mL mixture. Threshold odor numbers corresponding to various sample volumes are shown in Table 2-2. A similar test can be used to quantify taste, or the panel can simply rate the water qualitatively on an "acceptability" scale.

Table 2-2 Threshold odor numbers corresponding to sample volume diluted to 200 mL

Sample volume (A), mL	TON
200	1.0
175	1.1
150	1.3
125	1.6
100	2.0
75	2.7
67	3.0
50	4.0
40	5.0
25	8.0
10	20.0
2	100
1	200

Use

Although odors can be a problem with wastewater, the taste and odor parameter is only associated with potable water. EPA does not have a maximum standard for TON. A maximum TON of 3 has been recommended by the Public Health Service and serves as a guideline rather than a legal standard. [2-18]

2-6 TEMPERATURE

Temperature is not used to evaluate directly either potable water or wastewater. It is, however, one of the most important parameters in natural surface-water systems. The temperature of surface waters governs to a large extent the biological species present and their rates of activity. Temperature has an effect on most chemical reactions that occur in natural water systems. Temperature also has a pronounced effect on the solubilities of gases in water.

Sources

The temperature of natural water systems responds to many factors, the ambient temperature (temperature of the surrounding atmosphere) being the most universal. Generally, shallow bodies of water are more affected by ambient temperatures than are deeper bodies. The use of water for dissipation of waste heat in industry and the subsequent discharge of the heated water may result in dramatic, though perhaps localized, temperature changes in receiving streams. Removal of forest canopies and irrigation return flows can also result in increased stream temperature.

Impacts

Cooler waters usually have a wider diversity of biological species. At lower temperatures, the rate of biological activity, i.e., utilization of food supplies, growth, reproduction, etc., is slower. If the temperature is increased, biological activity increases. An increase of 10°C is usually sufficient to double the biological activity, if essential nutrients are present. At elevated temperatures and increased metabolic rates, organisms that are more efficient at food utilization and reproduction flourish, while other species decline and are perhaps eliminated altogether. Accelerated growth of algae often occurs in warm water and can become a problem when cells cluster into algae mats. Natural secretion of oils by algae in the mats and the decay products of dead algae cells can result in taste and odor problems. Higher-order species, such as fish, are affected dramatically by temperature and by dissolved oxygen levels, which are a function of temperature. Game fish generally require cooler temperatures and higher dissolved-oxygen levels.

Temperature changes affect the reaction rates and solubility levels of chemicals, a subject more fully explored in later sections of this chapter. Most chemical reactions involving dissolution of solids are accelerated by increased temperatures. The solubility of gases, on the other hand, decreases at elevated temperatures. Because biological oxidation of organics in streams and impoundments is dependent on an adequate supply of dissolved oxygen, decrease in oxygen solubility is undesirable. The relationship between temperature and dissolved oxygen levels is shown in Table C-3 of the appendix.

Temperature also affects other physical properties of water. The viscosity of water increases with decreasing temperature. The maximum density of water occurs at 4°C, and density decreases on either side of that temperature, a unique phenomenon among liquids. Both temperature and density have a subtle effect on planktonic microorganisms in natural water systems. The relationship of temperature and density to stratification of impoundments is discussed in Chap. 3.

Chemical Water-Quality Parameters

Water has been called the universal solvent, and chemical parameters are related to the solvent capabilities of water. Total dissolved solids, alkalinity, hardness, fluorides, metals, organics, and nutrients are chemical parameters of concern in water-quality management. The following review of some basic chemistry related to solutions should be helpful in understanding subsequent discussions of chemical parameters.

2-7 CHEMISTRY OF SOLUTIONS

An *atom* is the smallest unit of each of the elements. Atoms are building blocks from which *molecules* of elements and compounds are constructed. For instance, two hydrogen atoms combine to form a molecule of hydrogen gas:

$$H + H \longrightarrow H_2$$

Adding one atom of oxygen to the hydrogen molecule results in one molecule of the compound water:

$$H_2 + O \longrightarrow H_2O$$

A relative mass has been assigned to a single atom of each element based on a mass of 12 for carbon. The sum of the atomic mass of all the atoms in a molecule is the *molecular mass* of that molecule. The atomic mass of hydrogen is 1 and the atomic mass of oxygen is 16. Thus, the molecular mass of the hydrogen molecule is 2 and the molecular mass of water is 18. A *mole* of an element or compound is its molecular mass expressed in common mass units, usually grams. A mole of hydrogen is 2 g, while a mole of water is 18 g. One mole of a substance dissolved in sufficient water to make one liter of solution is called a one *molar solution*.

Bonding of elements into compounds is sometimes accomplished by electrical forces resulting from transferred electrons. When these compounds dissociate in water, they produce species with opposite charges. An example is sodium chloride:

$$NaCl \rightleftharpoons Na^+ + Cl^-$$

The charged species are called *ions*. Positively charged ions are called *cations*, and negatively charged ions are called *anions*. The number of positive charges must equal the number of negative charges to preserve electrical neutrality in a chemical compound. The number of charges on an ion is referred to as the *valence* of that ion. Thus, the valence of sodium (Na^+) is 1, while the valence of calcium (Ca^{2+}) is 2. Some compounds, called *radicals*, also possess charges. An example of a cationic radical is ammonium (NH_4^+), while carbonate (CO_3^{2-}) is an anionic radical.

When ions or radicals react with each other to form new compounds, the reactions may not always proceed on a one-to-one basis as was the case for sodium chloride. They do, however, proceed on an equivalence basis that can be related to electroneutrality. Technically, the *equivalence* of an element or radical is defined as the number of hydrogen atoms that element or radical can hold in combination or can replace in a reaction. In most cases, the equivalence of an ion is the same as the absolute value of its valence. An *equivalent* of an element or radical is its gram molecular mass divided by its equivalence. A *milliequivalent* is the molecular mass expressed in milligrams divided by the equivalence and is often more useful in water chemistry because concentrations of dissolved substances are more often in the milligrams per liter range. Compounds are formed by the combination of elements or radicals on a one-to-one equivalent basis. The calculation of equivalents is illustrated in Example 2-2.

Example 2-2: Calculating equivalents How many grams of calcium will be required to combine with 90 g of carbonate to form calcium carbonate?

SOLUTION

1. Carbonate (CO_3^{2-}) is a radical composed of carbon and oxygen. In this particular combination, carbon has an atomic mass of 12 and a valence of +4, while oxygen has an atomic mass of 16 and a valence of −2. Therefore, the radical has a total valence of −2 and an equivalence of 2. One equivalent of carbonate is

$$\frac{12 + 3(16)}{2} = 30 \text{ g/equiv}$$

2. The calcium ion has an atomic mass of 40 and a valence of +2; therefore, one equivalent of calcium is

$$\frac{40}{2} = 20 \text{ g/equiv}$$

3. The number of equivalents of calcium must equal the number of equivalents of carbonate, therefore

$$\frac{90 \text{ g}}{30 \text{ g/equiv}} = 3 \text{ equiv of carbonate}$$

Therefore, 3 equiv × 20 g/equiv = 60 g of calcium, and that amount will be required to react with 90 g of carbonate.

Equivalents are very important in water chemistry. In addition to being useful in calculating chemical quantities for desired reactions in water and wastewater treatment, equivalents also provide a means of expressing various constituents of dissolved solids in a common term. An equivalent of one substance is chemically equal to an equivalent of any other substance. Therefore, the concentration of substance A can be expressed as an equivalent concentration of substrate B by the following method.

$$\frac{(\text{g/L})\text{A}}{(\text{g/equiv})\text{A}} \times (\text{g/equiv})\text{B} = (\text{g/L})\text{A} \quad \text{expressed as B} \tag{2-2}$$

Historically, constituents of dissolved solids have been reported in terms of equivalent calcium carbonate concentrations. The following example illustrates this technique.

Example 2-3: Determining equivalent concentrations What is the equivalent calcium carbonate concentration of (*a*) 117 mg/L of NaCl and (*b*) 2×10^{-3} mol of NaCl?

SOLUTION

(*a*) 1. One equivalent of calcium carbonate is

$$\frac{40 + 12 + 3(16)}{2} = 50\ \text{g/equiv} = 50{,}000\ \text{mg/equiv} = 50\ \text{mg/mequiv}$$

2. One equivalent of sodium chloride is

$$\frac{23 + 35.5}{1} = 58.5\ \text{g/equiv} = 58.5\ \text{mg/mequiv}$$

3. By Eq. (2-2)

$$\frac{117\ \text{mg/L}}{58.5\ \text{mg/mequiv}} \times 50\ \text{mg/mequiv} = 100\ \text{mg/L of NaCl as CaCO}_3$$

(*b*) 1. One mole of a substance divided by its valence is one equivalent.

$$\frac{2 \times 10^{-3}\ \text{mol/L}}{1\ \text{mol/equiv}} = 2 \times 10^{-3}\ \text{equiv/L}$$

2. Thus, 2×10^{-3} equiv/L $\times$ 50,000 mg/equiv = 100 mg/L

Many solid substances, particularly those with crystalline structure, ionize readily in water. Water may or may not be a chemical reactant in the process. In Eq. (2-3), water is a reactant, while in Eq. (2-4) it is not.

$$\text{CaO} + \text{H}_2\text{O} \longrightarrow \text{Ca}^{2+} + 2\text{OH}^- \tag{2-3}$$

$$\text{NaCl} + \text{H}_2\text{O} \rightleftharpoons \text{Na}^+ + \text{Cl}^- + \text{H}_2\text{O} \tag{2-4}$$

When water is not a reactant, it is customary to omit it from the equation.

The double arrows in Eq. (2-4) indicate a reversible condition. That is, the solid form (NaCl) may be dissociating into its ionic components (*dissolution*), or the ionic components may be recombining into the solid form (*precipitation*). When the solid material is first contacted with water, the net reaction will be toward the ionic form. If a sufficient mass of solid is present, a condition of *dynamic equilibrium* will be reached in which the rate of dissolution and the rate of precipitation will be exactly equal. At this point, the water is saturated with the dissolved species.

Conditions of equilibrium can be expressed by the *mass action equation.* For the generalized reaction

$$\underset{\text{Solid compound}}{A_xB_y} \rightleftharpoons \underset{\text{Ionic components}}{xA + yB}$$

the mass action equation is

$$\frac{[A]^x[B]^y}{[A_xB_y]} = K \tag{2-5}$$

The brackets around the ionic and solid species indicate molar concentrations. The K value is an equilibrium constant for a given substance in pure water at a given temperature.

At equilibrium, the solid phase does not change concentrations because dissolution and precipitation are equal. Thus

$$[A_xB_y] = K_s = \text{const} \qquad \text{and} \qquad [A]^x[B]^y = KK_s = K_{sp} \tag{2-6}$$

The quantity K_{sp} is known as the *solubility product* for the ion pair. If the concentration of either or both of the ions is increased, the product of the ionic concentration will exceed the K_{sp} and precipitation will occur to maintain equilibrium conditions. The solubility products for several substances common to natural water systems are given in Table 2-3. Use of the solubility product to calculate ionic concentrations is illustrated in the following example.

Example 2-4: Determining equilibrium concentrations The solubility product for the dissociation of $Mg(OH)_2$ is shown in Table 2-3 as 9×10^{-12}. Determine the concentration of Mg^{2+} and OH^- at equilibrium, expressed as milligrams per liter of $CaCO_3$.

SOLUTION

1. Write the equation for the reaction

$$Mg(OH)_2 \rightleftharpoons Mg^{2+} + 2OH^-$$

2. The solubility product equation becomes

$$[Mg^{2+}][OH^-]^2 = 9 \times 10^{-12}$$

If x is the number of moles of Mg^{2+} resulting from the dissociation, then OH^- is equal to $2x$. Therefore,

$$[x][2x]^2 = 9 \times 10^{-12}$$

$$4x^3 = 9 \times 10^{-12}$$

$$x = 1.3 \times 10^{-4} \text{ mol/L} = \text{Mg}$$

$$2x = 2.6 \times 10^{-4} \text{ mol/L} = \text{OH}$$

3. $\dfrac{1.3 \times 10^{-4} \text{ mol/L}}{0.5 \text{ mol/equiv}} \times 50{,}000 \text{ mg/equiv} = 13.0 \text{ mg/L of Mg as } CaCO_3$

4. $\dfrac{2.6 \times 10^{-4} \text{ mol/L}}{1 \text{ mol/equiv}} \times 50{,}000 \text{ mg/equiv} = 13.0 \text{ mg/L of OH as } CaCO_3$

In addition to solid substances, many gases also dissolve in water. Elements from some of these gases may combine with water or with substances in the water to produce compounds or radicals that can be recovered in a solid form, thus becoming a part of the dissolved-solids load. An example is carbon dioxide.

$$CO_2 + H_2O \rightleftharpoons H_2CO_3 \rightleftharpoons H^+ + HCO_3^- \tag{2-7}$$

and

$$HCO_3^- \rightleftharpoons H^+ + CO_3^{2-} \tag{2-8}$$

Both the bicarbonate (HCO_3^-) and carbonate (CO_3^{2-}) are recoverable in solid form.

Table 2-3 Solubility products of selected ion pairs

Equilibrium equation	K_{sp} at 25°C	Significance in environmental engineering
$MgCO_3 \rightleftharpoons Mg^{2+} + CO_3^{2-}$	4×10^{-5}	Hardness removal, scaling
$Mg(OH)_2 \rightleftharpoons Mg^{2+} + 2OH^-$	9×10^{-12}	Hardness removal, scaling
$CaCO_3 \rightleftharpoons Ca^{2+} + CO_3^{2-}$	5×10^{-9}	Hardness removal, scaling
$Ca(OH)_2 \rightleftharpoons Ca^{2+} + 2OH^-$	8×10^{-6}	Hardness removal
$CaSO_4 \rightleftharpoons Ca^{2+} + SO_4^{2-}$	2×10^{-5}	Flue gas desulfurization
$Cu(OH)_2 \rightleftharpoons Cu^{2+} + 2OH^-$	2×10^{-19}	Heavy metal removal
$Zn(OH)_2 \rightleftharpoons Zn^{2+} + 2OH^-$	3×10^{-17}	Heavy metal removal
$Ni(OH)_2 \rightleftharpoons Ni^{2+} + 2OH^-$	2×10^{-16}	Heavy metal removal
$Cr(OH)_3 \rightleftharpoons Cr^{3+} + 3OH^-$	6×10^{-31}	Heavy metal removal
$Al(OH)_3 \rightleftharpoons Al^{3+} + 3OH^-$	1×10^{-32}	Coagulation
$Fe(OH)_3 \rightleftharpoons Fe^{3+} + 3OH^-$	6×10^{-36}	Coagulation, iron removal, corrosion
$Fe(OH)_2 \rightleftharpoons Fe^{2+} + 2OH^-$	5×10^{-15}	Coagulation, iron removal, corrosion
$Mn(OH)_3 \rightleftharpoons Mn^{3+} + 3OH^-$	1×10^{-36}	Manganese removal
$Mn(OH)_2 \rightleftharpoons Mn^{2+} + 2OH^-$	8×10^{-14}	Manganese removal
$Ca_3(PO_4)_2 \rightleftharpoons 3Ca^{2+} + 2PO_4^{3-}$	1×10^{-27}	Phosphate removal
$CaHPO_4 \rightleftharpoons Ca^{2+} + HPO_4^{3-}$	3×10^{-7}	Phosphate removal
$CaF_2 \rightleftharpoons Ca^{2+} + 2F^-$	3×10^{-11}	Fluoridation
$AgCl \rightleftharpoons Ag^+ + Cl^-$	3×10^{-10}	Chloride analysis
$BaSO_4 \rightleftharpoons Ba2^+ + SO_4^{2-}$	1×10^{-10}	Sulfate analysis

Source: Adapted from Sawyer and McCarty. [2-12].

2-8 TOTAL DISSOLVED SOLIDS

The material remaining in the water after filtration for the suspended-solids analysis is considered to be dissolved. This material is left as a solid residue upon evaporation of the water and constitutes a part of total solids discussed in Sec. 2-2.

Sources

Dissolved material results from the solvent action of water on solids, liquids, and gases. Like suspended material, dissolved substances may be organic or inorganic in nature. Inorganic substances which may be dissolved in water include minerals, metals, and gases. Water may come in contact with these substances in the atmosphere, on surfaces, and within the soil. Materials from the decay products of vegetation, from organic chemicals, and from the organic gases are common organic dissolved constituents of water. The solvent capability of water makes it an ideal means by which waste products can be carried away from industrial sites and homes.

Impacts

Many dissolved substances are undesirable in water. Dissolved minerals, gases, and organic constituents may produce aesthetically displeasing color, tastes, and odors. Some chemicals may be toxic, and some of the dissolved organic constituents have been shown to be carcinogenic. Quite often, two or more dissolved substances—especially organic substances and members of the halogen group—will combine to form a compound whose characteristics are more objectionable than those of either of the original materials.

Not all dissolved substances are undesirable in water. For example, essentially pure, distilled water has a flat taste. Additionally, water has an equilibrium state with respect to dissolved constituents. An undersaturated water will be "aggressive" and will more readily dissolve materials with which it comes in contact. Readily dissolvable material is sometimes added to a relatively pure water to reduce its tendency to dissolve pipes and plumbing.

Measurement

A direct measurement of total dissolved solids can be made by evaporating to dryness a sample of water which has been filtered to remove the suspended solids. The remaining residue is weighed and represents the *total dissolved solids* (*TDS*) in the water. The TDS is expressed as milligrams per liter on a dry-mass basis. The organic and inorganic fractions can be determined by firing the residue at 600°C as discussed in Sec. 2-2.

An approximate analysis for TDS is often made by determining the electrical conductivity of the water. The ability of a water to conduct electricity, known as the *specific conductance*, is a function of its ionic strength. Specific conductance is measured by a conductivity meter employing the Wheatstone bridge principle. The standard procedure is to measure the conductivity in a cubic-centimeter field at 25°C and express the results in millisiemens per meter (mS/m).

Unfortunately, specific conductance and concentration of TDS are not related on a one-to-one basis. Only ionized substances contribute to specific conductance. Organic molecules and compounds that dissolve without ionizing are not measured. Additionally, the magnitude of the specific conductance is influenced by the valence of the ions in solution, their mobility, and relative numbers. The temperature also has an important effect, with specific conductance increasing as the water temperature increases. Conversion of units to milligrams per liter or milliequivalents per liter must be made by use of an appropriate constant. A multiplier ranging from 0.055 to 0.09 is used to convert millisiemens to milligrams per liter. [2-15] To use specific conductance as a quantitative test, sufficient analysis for filterable residue must be run to determine the conversion factor. For this reason, specific conductance is most often used in a qualitative sense to monitor changes in TDS occurring in natural streams or treatment processes.

Use

Because no distinction among the constituents is made, the TDS parameter is included in the analysis of water and wastewater only as a gross measurement of the dissolved material. While this is often sufficient for wastewaters, it is frequently desirable to know more about the composition of the solids in water that is intended for use in potable supplies, agriculture, and some industrial processes. When this is the case, tests for several of the ionic constituents of TDS are made.

Ion Balance

The ions usually accounting for the vast majority of TDS in natural waters are listed in Table 2-4. Those listed under major constituents are often sufficient to

Table 2-4 Common ions in natural waters

Major constituents, 1.0–1000 mg/L	Secondary constituents, 0.01–10.0 mg/L
Sodium	Iron
Calcium	Strontium
Magnesium	Potassium
Bicarbonate	Carbonate
Sulfate	Nitrate
Chloride	Fluoride
	Boron
	Silica

Source: Adapted from Todd. [2-17]

characterize the dissolved-solids content of water. These are called *common ions* and are often measured individually and summed on an equivalent basis to represent the approximate TDS. As a check, the sum of the anions should equal the sum of the cations because electroneutrality must be preserved. A significant imbalance suggests that additional constituents are present or that an error has been made in the analysis of one or more of the ions. The following example illustrates the ion balance procedure.

Example 2-5: Testing for ion balance Tests for common ions are run on a sample of water and the results are shown below. If a 10 percent error in the balance is acceptable, should the analysis be considered complete?

Constituents

Ca^{2+} = 55 mg/L HCO_3^- = 250 mg/L
Mg^{2+} = 18 mg/L SO_4^{2-} = 60 mg/L
Na^+ = 98 mg/L Cl^- = 89 mg/L

SOLUTION

1. Convert the concentrations of cations and anions from milligrams per liter to milliequivalents per liter and sum them.

	Cations				Anions		
Ion	Conc, mg/L	Equiv, mg/mequiv	Equiv conc, meq/L	Ion	Conc, mg/L	Equiv, mg/mequiv	Equiv conc, mequiv/L
Ca^{2+}	55	40/2	2.75	HCO_3^-	250	61/1	4.10
Mg^{2+}	18	24.3/2	1.48	SO_4^{2-}	60	96/2	1.25
Na^+	98	23/1	4.26	Cl^-	89	35.5/1	2.51
Total ions			8.49				7.86

2. Calculate percent of error.

$$\frac{8.49 - 7.86}{7.86} 100 = 8\%$$

$$8\% < 10\%$$

Therefore, accept analysis.

A common ion balance can be displayed conveniently in the form of a bar diagram. A bar diagram for the water in Example 2-5 can be drawn as shown below.

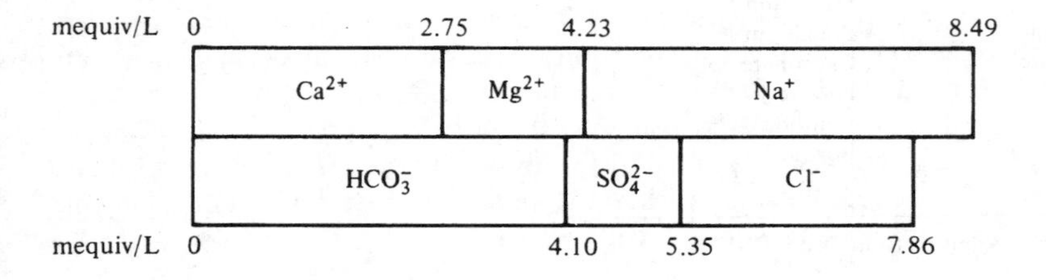

It is important to arrange the cations and anions in the order shown for convenience in determining types of hardness and the quantities of chemicals needed for softening, a subject more fully developed in a later chapter of this text.

Several of the constituents of dissolved solids have properties that necessitate special attention. These constituents include alkalinity, hardness, fluoride, metals, organics, and nutrients.

2-9 ALKALINITY

Alkalinity is defined as the quantity of ions in water that will react to neutralize hydrogen ions. Alkalinity is thus a measure of the ability of water to neutralize acids.

Sources

Constituents of alkalinity in natural water systems include CO_3^{2-}, HCO_3^-, OH^-, $HSiO_3^-$, $H_2BO_3^-$, HPO_4^{2-}, $H_2PO_4^-$, HS^-, and NH_3^0. [2-3] These compounds result from the dissolution of mineral substances in the soil and atmosphere. Phosphates may also originate from detergents in wastewater discharges and from fertilizers and insecticides from agricultural land. Hydrogen sulfide and ammonia may be products of microbial decomposition of organic material.

By far the most common constituents of alkalinity are bicarbonate (HCO_3^-), carbonate (CO_3^{2-}), and hydroxide (OH^-). In addition to their mineral origin, these substances can originate from carbon dioxide, a constituent of the atmosphere and a product of microbial decomposition of organic material. These reactions are as follows:

$$CO_2 + H_2O \rightleftharpoons H_2CO_3^* \quad \text{(dissolved } CO_2 \text{ and carbonic acid)} \tag{2-9}$$

$$H_2CO_3^* \rightleftharpoons H^+ + HCO_3^- \quad \text{(bicarbonate)} \tag{2-10}$$

$$HCO_3^- \rightleftharpoons H^+ + CO_3^{2-} \quad \text{(carbonate)} \tag{2-11}$$

$$CO_3^{2-} + H_2O \rightleftharpoons HCO_3^- + OH^- \quad \text{(hydroxide)} \tag{2-12}$$

The reaction represented by Eq. (2-12) is a weak reaction chemically. However, utilization of the bicarbonate ion as a carbon source by algae can drive the reaction to the right and result in substantial accumulation of OH^-. Water with heavy algal growths often has pH values as high as 9 to 10.

Because the reactions represented by the above equations involve hydrogen or hydroxide ions, the relative quantities of the alkalinity species are pH dependent. These relationships are shown graphically in Fig. 2-3.

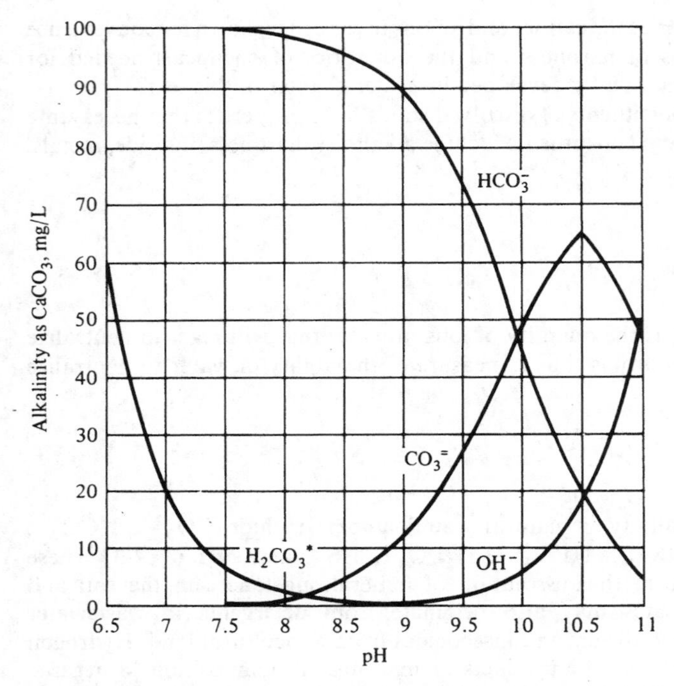

Figure 2-3 Alkalinity species vs. pH. Values are calculated for water at 25°C containing a total alkalinity of 100 mg/L as CaCO3. (*From Sawyer and McCarty* [*2-12*].)

Impacts

In large quantities, alkalinity imparts a bitter taste to water. The principal objection to alkaline water, however, is the reactions that can occur between alkalinity and certain cations in the water. The resultant precipitate can foul pipes and other water-systems appurtenances.

Measurement

Alkalinity measurements are made by titrating the water with an acid and determining the hydrogen equivalent. Alkalinity is then expressed as milligrams per liter of $CaCO_3$. If 0.02 *N* H_2SO_4 is used in the titration, then 1 mL of the acid will neutralize 1 mg of alkalinity as $CaCO_3$. Hydrogen ions from the acid react with the alkalinity according to the following equations:

$$H^+ + OH^- \rightleftharpoons H_2O \tag{2-13}$$

$$CO_3^{2-} + H^+ \rightleftharpoons HCO_3^- \tag{2-14}$$

$$HCO_3^- + H^+ \rightleftharpoons H_2CO_3 \tag{2-15}$$

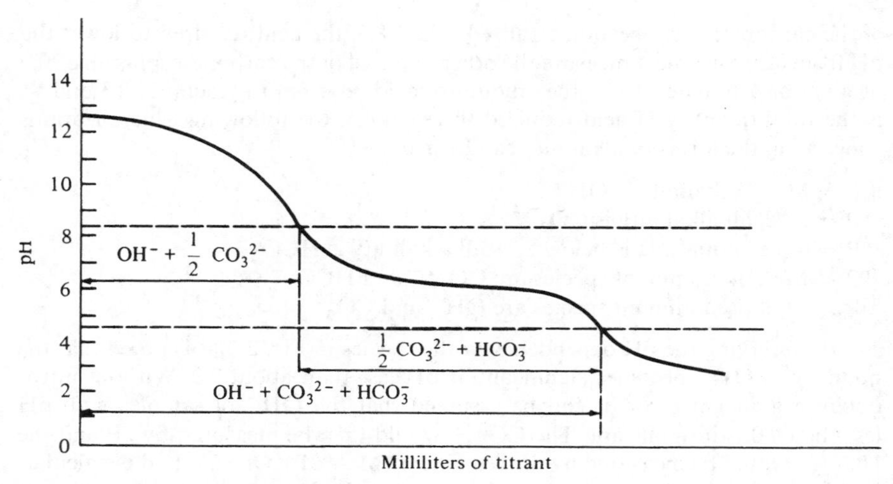

Figure 2-4 Alkalinity titration curve.

If acid is added slowly to water and the pH is recorded for each addition, a titration curve similar to that shown in Fig. 2-4 is obtained. Of particular significance are the inflection points in the curve that occur at approximately pH 8.3 and pH 4.5. The conversion of carbonate to bicarbonate [Eq. (2-14)] is essentially complete at pH 8.3. However, because bicarbonate is also an alkalinity species, an equal amount of acid must be added to complete the neutralization. Thus, the neutralization of carbonate is only one-half complete at pH 8.3. Because the conversion of hydroxide to water is virtually complete at pH 8.3 (see Fig. 2-3), all of the hydroxide and one-half of the carbonate have been measured at pH 8.3. At pH 4.5 all of the bicarbonate has been converted to carbonic acid [Eq. (2-15)], including the bicarbonate resulting from the reaction of the acid and carbonate [Eq. (2-14)]. Thus, the amount of acid required to titrate a sample to pH 4.5 is equivalent to the total alkalinity of the water. This point is illustrated in the following example.

Example 2-6: Determining total alkalinity A 200-mL sample of water has an initial pH of 10. Thirty milliliters of 0.02 *N* H_2SO_4 is required to titrate the sample to pH 4.5. What is the total alkalinity of the water in milligrams per liter as $CaCO_3$?

SOLUTION

Because each milligram of 0.02 *N* H_2SO_4 will neutralize 1 mg of alkalinity, there is 30 mg of alkalinity in the 200-mL sample. Therefore, the concentration of alkalinity expressed as milligrams per liter will be

$$\frac{30 \text{ mg}}{200 \text{ mL}} \times \frac{1000 \text{ mL}}{\text{L}} = 150 \text{ mg/L}$$

If the volume of acid needed to reach the 8.3 endpoint is known, the species of alkalinity can also be determined. Because all of the hydroxide and one-half of the carbonate have been neutralized at pH 8.3, the acid required to lower the pH from 8.3 to 4.5 must measure the other one-half of the carbonate, plus all of the original bicarbonate. If P is the amount of acid required to reach pH 8.3 and M is the total quantity of acid required to reach 4.5, the following generalizations concerning the forms of alkalinity can be made:

if $P = M$, all alkalinity is OH^-
$P = M/2$, all alkalinity is CO_3^{2-}
$P = 0$ (i.e., initial pH is below 8.3), all alkalinity is HCO_3^-
$P < M/2$, predominant species are CO_3^{2-} and HCO_3^-
$P > M/2$, predominant species are OH^- and CO_3^{2-}

In observing the pH dependency of the species in Fig. 2-3, it is noted that the quantity of OH^- becomes significant at pH less than about 9.0. Without introducing significant error, it can be assumed that the OH^- of samples with pH less than 9.0 is insignificant. The CO_3^{2-} would then be measured by 2P and the HCO_3^- would be measured by the remainder ($M - 2P$). One method of calculating the quantities of each species is illustrated in the following example.

Example 2-7: Determining alkalinity species Determine the species, and the quantity of each specie, of alkalinity in Example 2-6 if the 8.3 equivalence point is reached at 11 mL of acid.

SOLUTION

1. Because the initial pH is 10, the initial pOH of the water is 4. A determination of the OH^- concentration can be made as follows.

$$[OH^-] = \frac{10^{-4} \text{ mol } OH^-}{\text{L}} \times \frac{1 \text{ equiv}}{\text{mol } OH^-} \times \frac{50{,}000 \text{ mg } CaCO_3}{1 \text{ equiv}}$$
$$= 5 \text{ mg/L as } CaCO_3$$

2. Five milliliters of acid would be required to measure the OH^- in a 1-L sample. However, this sample is only 200 mL so the necessary volume of acid is:

$$5 \frac{200}{1000} = 1.0 \text{ mL}$$

3. If 1 mL of acid measures the OH^-, then 10 mL of acid measures one-half of the carbonate and 10 more will be required to measure the remaining one-half of the CO_3^{2-}, leaving 9 mL to measure the HCO_3^-. (See Fig. 2-4.) Thus, the quantity of each species is as follows:

$$OH^- \text{ (calculated from pH)} = 5 \text{ mg/L}$$

$$CO_3^{2-} = \frac{20 \text{ mg}}{200 \text{ mL}} \times \frac{1000 \text{ mL}}{\text{L}} = 100 \text{ mg/L}$$

$$HCO_3^- = \frac{9 \text{ mg}}{200 \text{ mL}} \times \frac{1000 \text{ mL}}{\text{L}} = 45 \text{ mg/L}$$

$$\text{Total alkalinity} = 150 \text{ mg/L}$$

Use

Alkalinity measurements are often included in the analysis of natural waters to determine their buffering capacity. It is also used frequently as a process control variable in water and wastewater treatment. Maximum levels of alkalinity have not been set by EPA for drinking water or for wastewater discharges.

2-10 HARDNESS

Hardness is defined as the concentration of multivalent metallic cations in solution. At supersaturated conditions, the hardness cations will react with anions in the water to form a solid precipitate. Hardness is classified as *carbonate hardness* and *noncarbonate hardness*, depending upon the anion with which it associates. The hardness that is equivalent to the alkalinity is termed carbonate hardness, with any remaining hardness being called noncarbonate hardness.

Carbonate hardness is sensitive to heat and precipitates readiiy at high temperatures.

$$Ca(HCO_3)_2 \xrightarrow{\Delta} CaCO_3 + CO_2 + H_2O \tag{2-16}$$

$$Mg(HCO_3)_2 \xrightarrow{\Delta} Mg(OH)_2 + 2CO_2 \tag{2-17}$$

Sources

The multivalent metallic ions most abundant in natural waters are calcium and magnesium. Others may include iron and manganese in their reduced states (Fe^{2+}, Mn^{2+}), strontium (Sr^{2+}), and aluminum (Al^{3+}). The latter are usually found in much smaller quantities than calcium and magnesium, and for all practical purposes, hardness may be represented by the sum of the calcium and magnesium ions.

Impacts

Soap consumption by hard waters represents an economic loss to the water user. Sodium soaps react with multivalent metallic cations to form a precipitate, thereby losing their surfactant properties. A typical divalent cation reaction is:

$$\underset{\text{Soap}}{2NaCO_2C_{17}H_{33}} + \text{cation}^{2+} \longrightarrow \underset{\text{Precipitate}}{\text{cation}^{2+}(CO_2C_{17}H_{33})_2} + 2Na^+ \tag{2-18}$$

Lathering does not occur until all of the hardness ions are precipitated, at which point the water has been "softened" by the soap. The precipitate formed by hardness and soap adheres to surfaces of tubs, sinks, and dishwashers and may stain clothing, dishes, and other items. Residues of the hardness-soap precipitate may remain in the pores, so that skin may feel rough and uncomfortable. In recent years these problems have been largely alleviated by the development of soaps and detergents that do not react with hardness.

Boiler scale, the result of the carbonate hardness precipitate may cause considerable economic loss through fouling of water heaters and hot-water pipes. Changes in pH in the water distribution systems may also result in deposits of precipitates. Bicarbonates begin to convert to the less soluble carbonates at pH values above 9.0.

Magnesium hardness, particularly associated with the sulfate ion, has a laxative effect on persons unaccustomed to it. Magnesium concentrations of less than 50 mg/L are desirable in potable waters, although many public water supplies exceed this amount. Calcium hardness presents no public health problem. In fact, hard water is apparently beneficial to the human cardiovascular system. [2-4]

Measurement

Hardness can be measured by using spectrophotometric techniques or chemical titration to determine the quantity of calcium and magnesium ions in a given sample. Hardness can be measured directly by titration with ethylenediamine tetraacetic acid (EDTA) using eriochrome black T (EBT) as an indicator. The EBT reacts with the divalent metallic cations, forming a complex that is red in color. The EDTA replaces the EBT in the complex, and when the replacement is complete, the solution changes from red to blue. If 0.01 *M* EDTA is used, 1.0 mL of the titrant measures 1.0 mg of hardness as $CaCO_3$.

Use

Analysis for hardness is commonly made on natural waters and on waters intended for potable supplies and for certain industrial uses. Hardness may range from practically zero to several hundred, or even several thousand, parts per million. Although acceptability levels vary according to a consumer's acclimation to hardness, a generally accepted classification is as follows:

Soft	<50 mg/L as $CaCO_3$
Moderately hard	50–150 mg/L as $CaCO_3$
Hard	150–300 mg/L as $CaCO_3$
Very hard	>300 mg/L as $CaCO_3$

The Public Health Service standards recommend a maximum of 500 mg/L of hardness in drinking water. [2-18] A maximum limit is not set by the EPA standards.

2-11 FLUORIDE

Generally associated in nature with a few types of sedimentary or igneous rocks, fluoride is seldom found in appreciable quantities in surface waters and appears in groundwater in only a few geographical regions. Fluoride is toxic to humans and other animals in large quantities, while small concentrations can be beneficial. Concentrations of approximately 1.0 mg/L in drinking water help to prevent dental cavities in children. During formation of permanent teeth, fluoride combines chemically with tooth enamel, resulting in harder, stronger teeth that are more resistant to decay. Fluoride is often added to drinking water supplies if sufficient quantities for good dental formation are not naturally present.

Excessive intakes of fluoride can result in discoloration of teeth. Noticeable discoloration, called *mottling*, is relatively common when fluoride concentrations in drinking water exceed 2.0 mg/L, but is rare when concentrations are less than 1.5 mg/L. Adult teeth are not affected by fluoride, although both the benefits and liabilities of fluoride during tooth-formation years carry over into adulthood. Excessive dosages of fluoride can also result in bone fluorosis and other skeletal abnormalities. Concentrations of less than 5 mg/L in drinking water are not likely to cause bone fluorosis or related problems, and some water supplies are known to have somewhat higher fluoride concentrations with no discernible problem other than severe mottling of teeth. On the assumption that people drink more water in warmer climates, EPA drinking-water standards base upper limits for fluoride on ambient temperatures. These standards are discussed more fully in Sec. 2-18.

2-12 METALS

All metals are soluble to some extent in water. While excessive amounts of any metal may present health hazards, only those metals that are harmful in relatively small amounts are commonly labeled toxic; other metals fall into the nontoxic group. Sources of metals in natural waters include dissolution from natural deposits and discharges of domestic, industrial, or agricultural wastewaters. Measurement of metals in water is usually made by atomic absorption spectrophotometry.

Nontoxic Metals

In addition to the hardness ions, calcium and magnesium, other nontoxic metals commonly found in water include sodium, iron, manganese, aluminum, copper, and zinc. Sodium, by far the most common nontoxic metal found in natural waters, is abundant in the earth's crust and is highly reactive with other elements. The salts of sodium are very soluble in water. Excessive concentrations cause a bitter taste in water and are a health hazard to cardiac and kidney patients. Sodium is also corrosive to metal surfaces and, in large concentrations, is toxic to plants.

Iron and manganese quite frequently occur together and present no health hazards at concentrations normally found in natural waters. As noted in Sec. 2-4, iron and manganese in very small quantities may cause color problems. Iron concentrations of 0.3 mg/L and manganese concentrations as low as 0.05 mg/L can cause color problems. Additionally, some bacteria use iron and manganese compounds for an energy source, and the resulting slime growth may produce taste and odor problems.

When significant quantities of iron are encountered in natural water systems, it is usually associated with chloride ($FeCl_2$), bicarbonate [$Fe(HCO_3)_2$], or sulfate [$Fe(SO_4)$] anions and exists in a reduced state. In the presence of oxygen, the ferrous (Fe^{2+}) ion is oxidized to the ferric (Fe^{3+}) ion and forms an insoluble compound with hydroxide [$Fe(OH)_3$]. Thus, significant quantities of iron will usually be found only in systems devoid of oxygen such as groundwaters or perhaps the bottom layers of stratified lakes. Similarly, manganese ions (Mn^{2+} and Mn^{4+}) associated with chloride, nitrates, and sulfates are soluble, while oxidized compounds (Mn^{3+} and Mn^{5+}) are virtually insoluble. It is possible, however, for organic acids derived from decomposing vegetation to chelate iron and manganese and prevent their oxidation and subsequent precipitation in natural waters.

The other nontoxic metals are generally found in very small quantities in natural water systems, and most would cause taste problems long before toxic levels were reached. However, copper and zinc are synergetic and when both are present, even in small quantities, may be toxic to many biological species.

Toxic Metals

As noted earlier, toxic metals are harmful to humans and other organisms in small quantities. Toxic metals that may be dissolved in water include arsenic, barium, cadmium, chromium, lead, mercury, and silver. Cumulative toxins such as arsenic, cadmium, lead, and mercury are particularly hazardous. These metals are concentrated by the food chain, thereby posing the greatest danger to organisms near the top of the chain.

Fortunately, toxic metals are present in only minute quantities in most natural water systems. Although natural sources of all the toxic metals exist, significant concentration in water can usually be traced to mining, industrial, or agricultural sources.

2-13 ORGANICS

Many organic materials are soluble in water. Organics in natural water systems may come from natural sources or may result from human activities. Most natural organics consist of the decay products of organic solids, while synthetic organics are usually the result of wastewater discharges or agricultural practices. Dissolved organics in water are usually divided into two broad categories: biodegradable and nonbiodegradable (refractory).

Biodegradable Organics

Biodegradable material consists of organics that can be utilized for food by naturally occurring microorganisms within a reasonable length of time. In dissolved form, these materials usually consist of starches, fats, proteins, alcohols, acids, aldehydes, and esters. They may be the end product of the initial microbial decomposition of plant or animal tissue, or they may result from domestic or industrial wastewater discharges. Although some of these materials can cause color, taste, and odor problems, the principal problem associated with biodegradable organics is a secondary effect resulting from the action of microorganisms on these substances.

Microbial utilization of dissolved organics can be accompanied by *oxidation* (addition of oxygen to, or the deletion of hydrogen from, elements of the organic molecule) or by *reduction* (addition of hydrogen to, or deletion of oxygen from, elements of the organic molecule). Although it is possible for the two processes to occur simultaneously, the oxidation process is by far more efficient and is predominant when oxygen is available. In *aerobic* (oxygen-present) environments, the end products of microbial decomposition of organics are stable and acceptable compounds. *Anaerobic* (oxygen-absent) decomposition results in unstable and objectionable end products. Should oxygen later become available, anaerobic end products will be oxidized to aerobic end products. The oxygen-demanding nature of biodegradable organics is of utmost importance in natural water systems. When oxygen utilization occurs more rapidly than oxygen can be replenished by transfer from the atmosphere, anaerobic conditions that severely affect the ecology of the system will result. This situation is covered in more detail in the next chapter.

The amount of oxygen consumed during microbial utilization of organics is called the *biochemical oxygen demand* (*BOD*). The BOD is measured by determining the oxygen consumed from a sample placed in an air-tight container and kept in a controlled environment for a preselected period of time. In the standard test, a 300-mL BOD bottle is used and the sample is incubated at 20°C for 5 days. Light must be excluded from the incubator to prevent algal growth that may produce oxygen in the bottle. Because the saturation concentration for oxygen in water at 20°C is approximately 9 mg/L, dilution of the sample with BOD-free, oxygen-saturated water is necessary to measure BOD values greater than just a few milligrams per liter.

The BOD of a diluted sample is calculated by

$$\text{BOD} = \frac{\text{DO}_I - \text{DO}_F}{\text{P}} \tag{2-19}$$

where DO_I and DO_F are the initial and final dissolved-oxygen concentrations (mg/L) and P is the decimal fraction of the sample in the 300-mL bottle.

Ranges of BOD covered by various dilutions are shown in Table 2-5. These values assume an initial dissolved-oxygen concentration of 9 mg/L in the mixture, with a minimum of 2 and a maximum of 7 mg/L of O_2 being consumed. Calculations of BOD_5 from this testing procedure are illustrated in the following example.

Example 2-8: Determining BOD_5 The BOD of a wastewater is suspected to range from 50 to 200 mg/L. Three dilutions are prepared to cover this range. The procedure is the same in each case. First the sample is placed in the standard BOD bottle and is then

Table 2-5 Ranges of BOD values covered by various dilutions

By using percent mixtures		By direct pipetting into 300-mL bottles	
% mixture	Range of BOD	mL	Range of BOD
0.01	20,000–70,000	0.02	30,000–105,000
0.02	10,000–35,000	0.05	12,000–42,000
0.05	4,000–14,000	0.10	6,000–21,000
0.1	2,000–7,000	0.20	3,000–10,500
0.2	1,000–3,500	0.50	1,200–4,200
0.5	400–1,400	1.0	600–2,100
1.0	200–700	2.0	300–1,050
2.0	100–350	5.0	120–420
5.0	40–140	10.0	60–210
10.0	20–70	20.0	30–105
20.0	10–35	50.0	12–42
50.0	4–14	100.0	6–21
100.0	0–7	300.0	0–7

Source: From Sawyer and McCarty. [2-12]

diluted to 300 mL with organic-free, oxygen-saturated water. The initial dissolved oxygen is determined and the bottles tightly stoppered and placed in the incubator at 20°C for 5 days, after which the dissolved oxygen is again determined.

Wastewater, mL	DO_I, mg/L	DO_5, mg/L	O_2 used, mg/L	P	BOD_5, mg/L
5	9.2	6.9	2.3	0.0167	138
10	9.1	4.4	4.7	0.033	142
20	8.9	1.5	7.4	0.067	110

If the third value is disregarded (the final DO being less than 2.0 mg/L), the average BOD of the wastewater is 140 mg/L.

Most natural water and municipal wastewaters will have a population of microorganisms that will consume the organics. In sterile waters, microorganisms must be added and the BOD of the material containing the organisms must be determined and subtracted from the total BOD of the mixture. The presence of toxic materials in the water will invalidate the BOD results.

The BOD_5 only represents the oxygen consumed in 5 days. The total BOD, or BOD for any other time period, can be determined provided additional information is known or obtained. The rate at which organics are utilized by microorganisms is assumed to be a first-order reaction; that is, the rate at which organics utilized is proportional to the amount available. Mathematically, this can be expressed as follows:

$$\frac{dL_t}{dt} = -kL_t \tag{2-20}$$

where L_t is the oxygen equivalent of the organics at time t, and k is a reaction constant. The units of L_t are milligrams per liter, and the units of k are d^{-1}.

Equation (2-20) can be rearranged and integrated as follows:

$$\frac{dL_t}{L_t} = -k\,dt$$

$$\int_{L_0}^{L} \frac{dL_t}{L_t} = -k \int_0^t dt$$

$$\ln \frac{L_t}{L_0} = -kt$$

$$L_t = L_0 e^{-kt} \tag{2-21}$$

The term L_0 in this equation represents the total oxygen equivalent of the organics at time 0, while L_t represents the amount remaining at time t, and decays exponentially with time, as shown in Fig. 2-5.

The oxygen equivalent remaining is not the parameter of primary importance. However, the amount of oxygen used in the consumption of the organics, the BOD_t, can be found from the L_t value. If L_0 is the oxygen equivalent of the total mass of

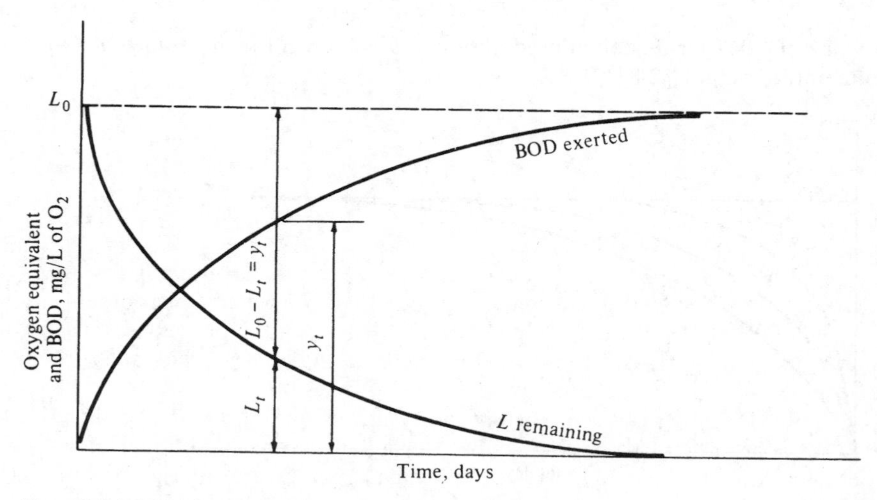

Figure 2-5 BOD and oxygen-equivalent relationships.

organics, then the difference between the value L_0 and L_t is the oxygen equivalent consumed, or the BOD exerted. Mathematically

$$y_t = L_0 - L_t$$

$$y_t = L_0 - L_0 e^{-kt}$$

$$y_t = L_0(1 - e^{-kt}) \qquad (2\text{-}22)$$

where y_t represents the BOD_t of the water. The value of y_t approaches L_0 asymptotically, indicating that the total, or ultimate, BOD (y_u) is equal to the initial oxygen equivalent of the water L_0. These relationships are shown in Fig. 2-5.

Equation (2-22) represents the BOD exerted by the carbon component of the organic compounds. Other components of organics, such as nitrogen and sulfur, may also be oxidized by microorganisms, resulting in an oxygen demand. Equations similar to Eq. (2-22) can be derived for these reactions.

The value of k determines the speed of the BOD reaction without influencing the magnitude of the ultimate BOD. This is shown graphically in Fig. 2-6. Numerical values of k range from about 0.1 to 0.5 d^{-1} depending on the nature of the organic molecules. Simple compounds such as sugars and starches are easily utilized by the microorganisms and have a high k rate, while complex molecules such as phenols are difficult to assimilate and have low k values. Some typical values of k are shown in Table 2-6.

The value of k for any given organic compound is temperature-dependent. Because microorganisms are more active at higher temperatures, the value of k increases with increasing temperatures. The change in k can be approximated by the van't Hoff-Arrhenius model:

$$k_T = k_{20}\theta^{T-20°} \qquad (2\text{-}23)$$

A value of 1.047 for θ is often used although θ is known to vary somewhat with temperature ranges. [2-8]

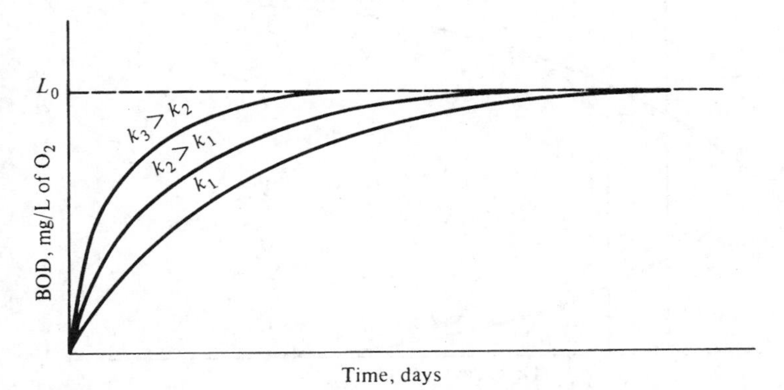

Figure 2-6 BOD exertion as a function of reaction constant k.

Table 2-6 Typical values of k and y_u for various waters

Water type	k, d^{-1} base e	y_u, mgL.
Tap water	<0.1	0–1
Surface waters	0.1–0.23	1–30
Weak municipal wastewater	0.35	150
Strong municipal wastewater	0.40	250
Treated effluent	0.12–0.23	10–30

The use of Eqs. (2-22) and (2-23) is illustrated in the following example.

Example 2-9: BOD conversions The BOD_5 of a wastewater is determined to be 150 mg/L at 20°C. The k value is known to be 0.23 per day. What would the BOD_8 be if the test were run at 15°C?

SOLUTION

1. Determine the ultimate BOD.

$$y_u = \frac{y_5}{1 - e^{-kt}}$$

$$= \frac{150}{1 - e^{-0.23 \times 5}}$$

$$= 220 \text{ mg/L}$$

2. Correct the k value for 15°C.

$$k_T = k_{20}\theta^{T-20}$$

$$k_{15} = 0.23(1.047^{-5})$$

$$= 0.18$$

3. Calculate y_8

$$y_t = y_u(1 - e^{-kt})$$

$$y_8 = 220\,(1 - e^{-0.18 \times 8})$$

$$= 168 \text{ mg/L}$$

Nonbiodegradable Organics

Some organic materials are resistant to biological degradation. Tannic and lignic acids, cellulose, and phenols are often found in natural water systems. These constituents of woody plants biodegrade so slowly that they are usually considered refractory. Molecules with exceptionally strong bonds (some of the polysaccharides) and ringed structures (benzene) are essentially nonbiodegradable.

An example is the detergent compound alkyl benzene sulfonate (ABS) which, with its benzene ring, does not biodegrade. Being a surfactant, ABS causes frothing and foaming in wastewater treatment plants and increases turbidity by stabilizing colloidal suspensions. This problem was largely alleviated when detergent manufacturers switched to a linear alkyl sulfonate (LAS) compound, which is biodegradable. Many of the organics associated with petroleum and with its refining and processing also contain benzene and are essentially nonbiodegradable.

Some organics are nonbiodegradable because they are toxic to organisms. These include the organic pesticides, some industrial chemicals, and hydrocarbon compounds that have combined with chlorine.

Pesticides, including insecticides and herbicides, have found wide-spread use in modern society in both urban and agricultural settings. Poor application practices and subsequent washoff by rainfall and runoff may result in contamination of surface streams. Organic insecticides are usually chlorinated hydrocarbons (i.e., aldrin, dieldrin, endrin, and lindane), while herbicides are usually chlorophenoxys (e.g., 2,4-dichlorophenoxyacetic acid and 2,4,5-trichlorophenoxypropionic acid). Many of the pesticides are cumulative toxins and cause severe problems at the higher end of the food chain. An example is the near-extinction of the brown pelican that feeds on fish and other macroaquatic species by the insecticide DDT, the use of which is now banned in the United States.

Measurement of nonbiodegradable organics is usually by the chemical oxygen demand (COD) test. Nonbiodegradable organics may also be estimated from a total organic carbon (TOC) analysis. Both COD and TOC measure the biodegradable fraction of the organics, so the BOD_u must be subtracted from the COD or TOC to quantify the nonbiodegradable organics. Specific organic compounds can be identified and quantified through analysis by gas chromatography.

2-14 NUTRIENTS

Nutrients are elements essential to the growth and reproduction of plants and animals, and aquatic species depend on the surrounding water to provide their nutrients. Although a wide variety of minerals and trace elements can be classified as nutrients, those required in most abundance by aquatic species are carbon, nitrogen, and phosphorus. Carbon is readily available from many sources. Carbon dioxide from the atmosphere, alkalinity, and decay products of organic matter all supply carbon to the aquatic system. In most cases, nitrogen and phosphorus are the nutrients that are the limiting factors in aquatic plant growth. A discussion of the consequences of overenrichment with nitrogen and phosphorus is presented in Chapter 3.

Nitrogen

Nitrogen gas (N_2) is the primary component of the earth's atmosphere and is extremely stable. It will react with oxygen under high-energy conditions (electrical discharges or flame incineration) to form nitrogen oxides. Although a few biological species are able to oxidize nitrogen gas, nitrogen in the aquatic environment is derived primarily from sources other than atmospheric nitrogen.

Nitrogen is a constituent of proteins, chlorophyll, and many other biological compounds. Upon the death of plants or animals, complex organic matter is broken down to simple forms by bacterial decomposition. Proteins, for instance, are converted to amino acids and further reduced to ammonia (NH_3). If oxygen is present, the ammonia is oxidized to nitrite (NO_2^-) and then to nitrate (NO_3^-). The nitrate can then be reconstituted into living organic matter by photosynthetic plants.

Other sources of nitrogen in aquatic systems include animal wastes, chemical (particularly chemical fertilizers), and wastewater discharges. Nitrogen from these sources may be discharged directly into streams or may enter waterways through surface runoff or groundwater discharge. Nitrogen compounds can be oxidized to nitrate by soil bacteria and may be carried into the groundwater by percolating water. Once in the aquifer, nitrates move freely with the groundwater flow. Groundwater contamination by nitrogen from animal feedlots and septic-tank drain fields has been recorded in numerous instances. [2-10, 2-11, 2-13]

In addition to the overenrichment problems alluded to earlier, nitrogen can have other serious consequences. Ammonia is a gas at temperatures and pressures normally found in natural water systems. The gas (NH_3) exists in equilibrium with the aqueous ionic form called ammonium (NH_4^+).

$$NH_3 + H_2O \rightleftharpoons NH_4^+ + OH^- \tag{2-24}$$

The hydroxyl ion concentration of the water, and thus the pH, controls the relative abundance of each species. Oxidation of NH_3 and NH_4^+ to nitrate and on to nitrate by aquatic microbes results in an additional biochemical oxygen demand as discussed in the preceding section.

Nitrate poisoning in infant animals, including humans, can cause serious problems and even death. Apparently, the lower acidity in an infant's intestinal tract permits growth of nitrate-reducing bacteria that convert the nitrate to nitrite, which is then absorbed into the bloodstream. Nitrite has a greater affinity for hemoglobin than does oxygen and thus replaces oxygen in the blood complex. The body is denied essential oxygen and, in extreme cases, the victim suffocates. Because oxygen starvation results in a bluish discoloration of the body, nitrate poisoning has been referred to as the "blue baby" syndrome, although the correct term is *methemoglobinemia*. Once the flora of the intestinal tract has fully developed, usually after the age of 6 months, nitrate conversion to nitrite and subsequent methemoglobinemia from drinking water is seldom a problem. Fortunately, the natural oxidation of nitrite to nitrate occurs quickly so that significant quantities of nitrites are not found in natural water.

Tests for nitrogen forms in water commonly include analysis for ammonia (including both ammonia and ammonium), nitrate, and organic nitrogen. The results of the analyses are usually expressed as milligrams per liter of the particular species as nitrogen. Tests for ammonium and organic nitrogen are more common on wastewater and other polluted waters, while the test for nitrate is the most common on clean-water samples and treated wastewaters.

Phosphorus

Phosphorus appears exclusively as phosphate (PO_4^{3-}) in aquatic environments. There are several forms of phosphate, however, including orthophosphate, condensed phosphates (pyro-, meta-, and polyphosphates), and organically bound phosphates. These may be in soluble or particulate form or may be constituents of plant or animal tissue. Like nitrogen, phosphates pass through the cycles of decomposition and photosynthesis.

Phosphate is a constituent of soils and is used extensively in fertilizer to replace and/or supplement natural quantities on agricultural lands. Phosphate is also a constituent of animal waste and may become incorporated into the soil in grazing and feeding areas. Runoff from agricultural areas is a major contributor to phosphate in surface waters. The tendency for phosphate to adsorb to soil particles limits its movement in soil moisture and groundwater, but results in its transport into surface waters by erosion.

Municipal wastewater is another major source of phosphate in surface water. Condensed phosphates are used extensively as builders in detergents, and organic phosphates are constituents of body waste and food residue. Other sources include industrial waste in which phosphate compounds are used for such purposes as boiler-water conditioning.

While phosphates are not toxic and do not represent a direct health threat to human or other organisms, they do represent a serious indirect threat to water quality. As noted earlier, phosphate is often the limiting nutrient in surface waters. When the available supply is increased, rapid growth of aquatic plants usually results, with severe consequences. Phosphate can also interfere with water-treatment processes. Concentrations as low as 0.2 mg/L interfere with the chemical coagulation of turbidity. [2-20]

Phosphates are measured colorimetrically. Orthophosphates can be measured directly, while condensed forms must be converted to orthophosphate by acid hydrolyzation and organic phosphates must be converted to orthophosphates by acid digestion. Results of the analysis are reported as milligrams per liter of phosphate as phosphorus. Careful handling of samples prior to analysis is crucial. For example, acid-washed glass bottles should be used for sampling, as bottles washed in phosphate detergent may contaminate samples.

Biological Water-Quality Parameters

Water may serve as a medium in which literally thousands of biological species spend part, if not all, of their life cycles. Aquatic organisms range in size and complexity from the smallest single-cell microorganism to the largest fish. All members of the biological community are, to some extent, water-quality parameters, because their presence or absence may indicate in general terms the characteristics of a given body of water. As an example, the general quality of water in a trout stream would be expected to exceed that of a stream in which the predominant species of fish is carp. Similarly, abundant algal populations are associated with a water rich in nutrients.

Biologists often use a species-diversity index (related to the number of species and the relative abundance of organisms in each species) as a qualitative parameter for streams and lakes. A body of water hosting large numbers of species with well-balanced numbers of individuals is considered to be a healthy system. Based on their known tolerance for a given pollutant, certain organisms can be used as indicators of the presence of pollutants. A more detailed coverage of this topic is given in Chap. 3.

2-15 PATHOGENS

From the perspective of human use and consumption, the most important biological organisms in water are pathogens, those organisms capable of infecting, or of transmitting diseases to, humans. These organisms are not native to aquatic systems and usually require an animal host for growth and reproduction. They can, however, be transported by natural water systems, thus becoming a temporary member of the aquatic community. Many species of pathogens are able to survive in water and maintain their infectious capabilities for significant periods of time. These waterborne pathogens include species of bacteria, viruses, protozoa, and helminths (parasitic worms). The characteristics of the primary waterborne pathogens are listed in Table 2-7.

Bacteria

The word *bacteria* comes from the Greek word meaning "rod" or "staff," a shape characteristic of most bacteria. Bacteria are single-cell microorganisms, usually colorless, and are the lowest form of life capable of synthesizing protoplasm from the surrounding environment. In addition to the rod shape (bacilli) mentioned above, bacteria may also be spherical (cocci) or spiral-shaped (spirilla). Gastrointestinal disorders are common symptoms of most diseases transmitted by waterborne pathogenic bacteria.

Cholera, the disease that ravaged Europe during the eighteenth and nineteenth centuries, is transmitted by *Vibrio comma*. Among the most violent of the waterborne bacterial diseases, cholera causes vomiting and diarrhea that, without treatment, result in dehydration and death. Symptoms of typhoid, a disease transmitted by the waterborne pathogen, *Salmonella typhosa*, include gastrointestinal disorders, high fever, ulceration of the intestines, and possible nerve damage. Although immunization of individuals and disinfection of water supplies have eliminated cholera and typhoid in most parts of the world, areas of developing countries where overcrowding and poor sanitary conditions prevail still

Table 2-7 Common waterborne pathogens

Organism	Disease
Bacteria	
Francisella tularensis	Tularemia (deer fly fever)
Leptospirae	Leptospirosis (Weil's disease, swineherd's disease, hemorrhagic jaundice)
Salmonella paratyphi (A,B,C)	Paratyphoid (enteric fever)
Salmonella typhi	Typhoid fever, enteric fever
Shigella (*S. flexneri, S. sonnei, S. dysenteriae, S. boydii*)	Shigellosis (bacillary dysentery)
Vibrio comma (*Vibrio cholerae*)	Cholera (Asiatic, Indian, El Tor)
Viruses	
Enteric cytopathogenic human orphan (ECHO) (ECHO)	Aseptic meningitis, epidemic exanthem, infantile diarrhea
Poliomyelitis (3 types)	Acute anterior poliomyelitis, infantile paralysis
Unknown viruses	Infectious hepatitis
Protozoa	
Entamoeba histolytica	Amebiasis (amebic dysentery, amebic enteritis, amebic colitis)
Giardia lamblia	Giardiasis (Giardia enteritis, lambliasis)
Helminths (parasitic worms)	
Dracunculus medinensis	Dracontiasis (dracunculiasis; dracunculosis; medina; serpent, dragon, or guinea-worm infection)
Echinococcus	Echinococcosis (hydatidosis; granulosus; dog tapeworm)
Schistosoma (*S. mansoni, S. japonicum, S. haematobium*)	Schistosomiasis (bilharziasis or "Bill Harris" or "blood fluke" disease)

experience occasional outbreaks of these two diseases. Temporary lapses in good sanitary practices sometimes result in outbreaks of gastroenteritis caused by some of the other bacterial pathogens listed in Table 2-7.

Viruses

Viruses are the smallest biological structures known to contain all the genetic information necessary for their own reproduction. So small that they can only be "seen" with the aid of an electron microscope, viruses are obligate parasites that require a host in which to live. Symptoms associated with waterborne viral infections usually involve disorders of the nervous system rather than of the gastrointestinal tract. Waterborne viral pathogens are known to cause poliomyelitis and infectious hepatitis, and several other viruses are known to be, or suspected of being, waterborne.

Immunization of individuals has reduced the incidence of polio to a few isolated cases each year in developed nations. Outbreaks of hepatitis are more common, with around 60,000 cases reported in the United States each year. Most of the hepatitis cases result from persons eating shellfish contaminated by viruses from polluted waters, [2-2] although an occasional outbreak will occur at campgrounds or other facilities where crowds gather and where water-supply protection and sanitary facilities are poor.

Although standard disinfection practices are known to kill viruses, confirmation of effective viral disinfection is difficult, owing to the small size of the organism and the lack of quick and conclusive tests for viable virus organisms. The uncertainty of viral disinfection is a major obstacle to direct recycling of wastewater and is a cause of concern regarding the increasing practice of land application of wastewater.

Protozoa

The lowest form of animal life, protozoa are unicellular organisms more complex in their functional activity than bacteria or viruses. They are complete, self-contained organisms that can be free-living or parasitic, pathogenic or nonpathogenic, microscopic or macroscopic. Highly adaptable, protozoa are widely distributed in natural waters, although only a few aquatic protozoa are pathogenic.

Protozoal infections are usually characterized by gastrointestinal disorders of a milder order than those associated with the bacterial infections discussed earlier. Protozoal infections can be serious nonetheless, as illustrated by an epidemic in Chicago in 1933 in which over 1400 people were affected and 98 deaths resulted when drinking water was contaminated by sewage containing *Entamoeba histolytica.* [2-14]. Many cases of giardiasis, or backpackers disease, have been reported in recent years among persons that drank untreated water from surface streams. This infection is caused by *Giardia lamblia,* a protozoan that may be carried by wild animals living in or near natural water systems.

Under adverse environmental circumstance, aquatic protozoa form cysts that are difficult to deactivate by disinfection. Usually complete treatment, including filtration, is necessary to remove protozoal cysts.

Helminths

The life cycles of helminths, or parasitic worms, often involve two or more animal hosts, one of which can be human, and water contamination may result from human or animal waste that contains helminths. Contamination may also be via aquatic species of other hosts, such as snails or insects. While aquatic systems can be the vehicle for transmitting helminthal pathogens, modern water-treatment methods are very effective in destroying these organisms. Thus, helminths pose hazards primarily to those persons who come into direct contact with untreated water. Sewage plant operators, swimmers in recreational lakes polluted by sewage or stormwater runoff from cattle feedlots, and farm laborers employed in agricultural irrigation operations are at particular risk. [2-5]

2-16 PATHOGEN INDICATORS

Analysis of water for all the known pathogens would be a very time-consuming and expensive proposition. Tests for specific pathogens are usually made only when there is a reason to suspect that those particular organisms are present. At other times, the purity of water is checked using indicator organisms.

An *indicator organism* is one whose presence presumes that contamination has occurred and suggests the nature and extent of the contaminant(s). The ideal pathogen indicator would (1) be applicable to all types of water, (2) always be present when pathogens are present, (3) always be absent when pathogens are absent, (4) lend itself to routine quantitative testing procedures without interference from or confusion of results because of extraneous organisms, and (5) for the safety of laboratory personnel, not be a pathogen itself. [2-6]

Most of the waterborne pathogens are introduced through fecal contamination of water. Thus, any organism native to the intestinal tract of humans and meeting the above criteria would be a good indicator organism. The organisms most nearly meeting these requirements belong to the fecal coliform group. Composed of several strains of bacteria, principal of which is *Escherichia coli*, these organisms are found exclusively in the intestinal tract of warm-blooded animals and are excreted in large numbers with feces. Fecal coliform organisms are nonpathogenic and are believed to have a longer survival time outside the animal body than do most pathogens. Because the die-off rate of fecal coliforms is logarithmic, the number of surviving organisms may be an indication of the time lapse since contamination.

There are other coliform groups which flourish outside the intestinal tract of animals. These organisms are native to the soil and decaying vegetation and are often found in water that was in recent contact with these materials. Because the life cycles of some pathogens (particularly helminths) may include periods in the soil, this group of coliform organisms also serves as an indicator of pathogens.

It is the usual practice in the United States to use the total coliform group (those of both fecal and nonfecal origin) as indicators of the sanitary quality of drinking water, while the indicator of choice for wastewater effluents is the fecal coliform group. Relatively simple tests have been devised to determine the presence of coliform bacteria in water and to enumerate the quantity. The tests for total coliform organisms employ slightly different culture media and lower incubation temperatures than those used to identify fecal coliform organisms.

The membrane-filter technique, a technique popular with environmental engineers, gives a direct count of coliform bacteria. In this test, a portion of the sample is filtered through a membrane, the pores of which do not exceed 0.45 μm. Bacteria are retained on the filter that is then placed on selective media to promote growth of coliform bacteria while inhibiting growth of other species. The membrane and media are incubated at the appropriate temperature for 24 h, allowing coliform bacteria to grow into visible colonies that are then counted. The results are reported in number of organisms per 100 mL of water.

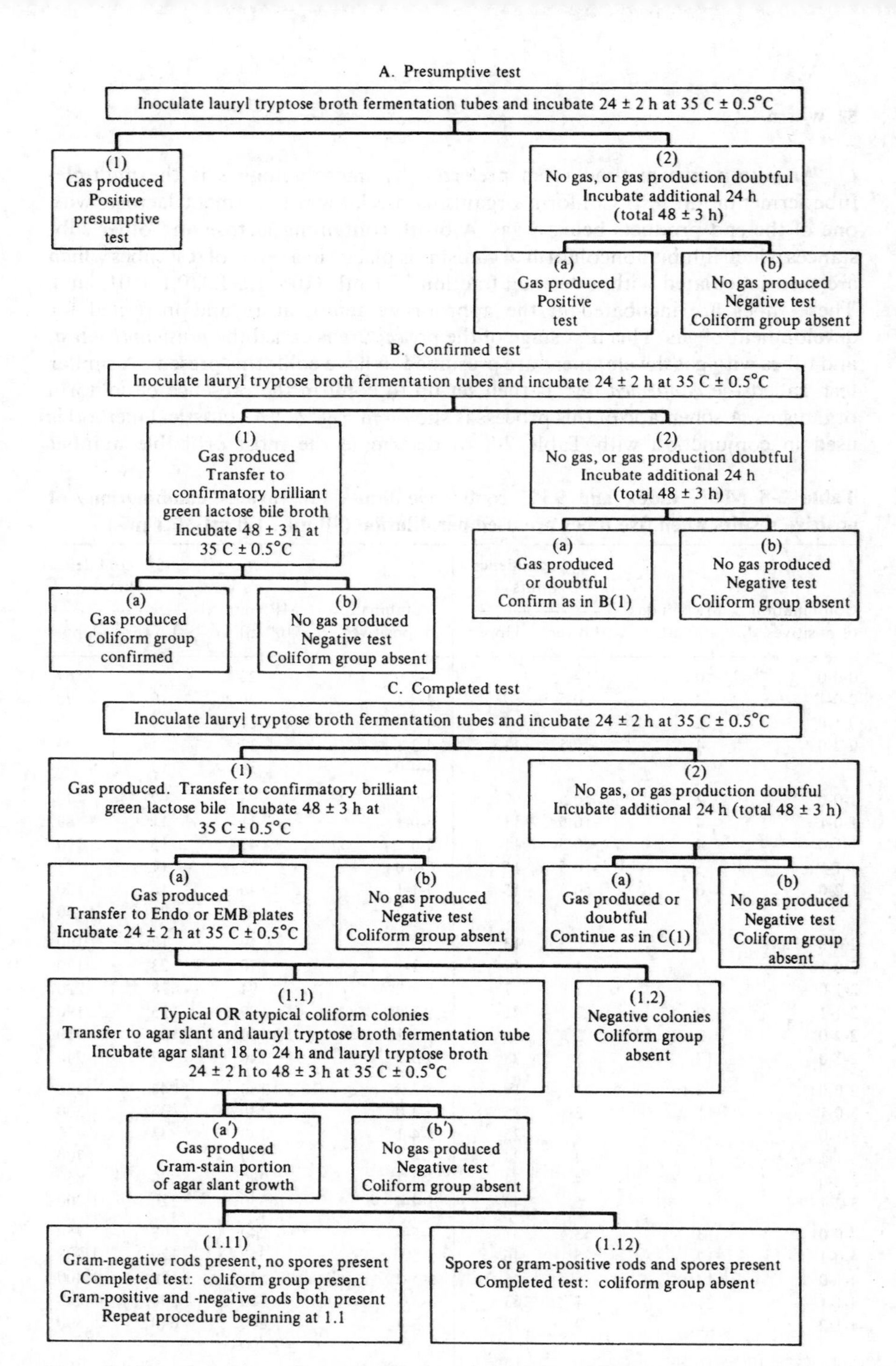

Figure 2-7 Procedure for running total coliform analysis by the multiple-tube fermentation method. (*From Standard Methods* [2-15].)

An alternative method often preferred by microbiologists is the multiple-tube fermentation test. Coliform organisms are known to ferment lactose, with one of the end products being a gas. A broth containing lactose and other substances which inhibit noncoliform organisms is placed in a series of test tubes which are then inoculated with a decimal fraction of 1 mL (100, 10, 1.0, 0.1, 0.01, etc.). These tubes are incubated at the appropriate temperature and inspected for development of gas. This first stage of the procedure is called the *presumptive test*, and tubes with gas development are presumed to have coliforms present. A similar test, called the *confirmed test*, is then set up to confirm the presence of coliform organisms. A schematic of this process is shown in Fig. 2-7. A statistical method is used in conjunction with Table 2-8 to determine the most probable number

Table 2-8 MPN index and 95% confidence limits for various combinations of positive results when five tubes are used per dilution (10 mL, 1.0 mL, 0.1 mL)

Combination of positives	MPN index /100 mL	95% confidence limits Lower	95% confidence limits Upper	Combination of positives	MPN index /100 mL	95% confidence limits Lower	95% confidence limits Upper
0-0-0	<0	—	—	4-2-0	22	7	67
0-0-1	2	<0.5	7	4-2-1	26	9	78
0-1-0	2	<0.5	7	4-3-0	27	9	80
0-2-0	4	<0.5	11	4-3-1	33	11	93
				4-4-0	34	12	93
1-0-0	2	<0.5	7	5-0-0	23	7	70
1-0-1	4	<0.5	11	5-0-1	31	11	89
1-1-0	4	<0.5	11	5-0-2	43	15	110
1-1-1	6	<0.5	15	5-1-0	33	11	93
1-2-0	6	<0.5	15	5-1-1	46	16	120
				5-1-2	63	21	150
2-0-0	5	<0.5	13	5-2-0	49	17	130
2-0-1	7	1	17	5-2-1	70	23	170
2-1-0	7	1	17	5-2-2	94	28	220
2-1-1	9	2	21	5-3-0	79	25	190
2-2-0	9	2	21	5-3-1	110	31	250
2-3-0	12	3	28	5-3-2	140	37	340
3-0-0	8	1	19	5-3-3	180	44	500
3-0-1	11	2	25	5-4-0	130	35	300
3-1-0	11	2	25	5-4-1	170	43	490
3-1-1	14	4	34	5-4-2	220	57	700
3-2-0	14	4	34	5-4-3	280	90	850
3-2-1	17	5	46	5-4-4	350	120	1,000
4-0-0	13	3	31	5-5-0	240	68	750
4-0-1	17	5	46	5-5-1	350	120	1,000
4-1-0	17	5	46	5-5-2	540	180	1,400
4-1-1	21	7	63	5-5-3	920	300	3,200
4-1-2	26	9	78	5-5-4	1,600	640	5,800
				5-5-5	≥2,400	—	—

Source: From Smith. [2-14]

(MPN) of coliform bacteria in 100 mL of the water sample. This method is illustrated in the following example.

Example 2-10: Determining the most probable number of coliforms A standard multiple-tube fermentation test is run on a sample of water from a surface stream. The results of the analysis for the confirmed test are shown below.

Size of sample, mL	No. positive	No. negative
10	4	1
1	2	3
0.1	1	4
0.01	1	4
0.001	0	5

Determine the most probable number of coliform organisms.

SOLUTION

1. Select a series where three tubes each have positive results (not necessary, but recommended); use sample sizes 10, 1, and 0.1.
2. Enter Table 2-8 with the number of positive tubes out of five (4, 2, 1). The corresponding MPN is 26 with a range of 9–78 organisms per 100 mL possible at a 95 percent confidence level.

ALTERNATE SOLUTION

1. Select sample sizes 1.0, 0.1, and 0.01.
2. From Table 2-8, the corresponding MPN is 9 and the 95 percent confidence range is 2 through 21.
3. Because the series of samples used is one-tenth of the 10, 1, and 0.1 sample sizes used in the table, multiply the values by 10. Therefore, the MPN of the sample is 90 organisms per 100 mL and the 95 percent confidence range is 20 to 210.

Sampling techniques and subsequent handling of the samples are extremely important because samples can easily be contaminated. Detailed procedures for sampling, sample preparation, and sterilization of laboratory equipment for both the membrane filter technique and the multiple-tube technique are presented in *Standard Methods.* [2-15]

It should be emphasized again that pathogens are not identified by the coliform test. The presence of coliform organisms in water does, however, indicate that some portion of the water has recently contacted soil or decaying vegetation or has been through the intestinal tract of a warm-blooded animal. The assumption must then be made that pathogens may have accompanied the coliform bacteria.

Water-Quality Requirements

Water-quality requirements vary according to the proposed use of the water. Detailed descriptions of the quality criteria for agricultural use, fish and wildlife propagation, specific industrial and recreational uses, and power generation are presented elsewhere, [2-7, 2-6, 2-17], and such descriptions are beyond the scope of this text. Water unsuitable for one use may be quite satisfactory for another, and water may be deemed acceptable for a particular use if water of better quality is not available.

Water-quality requirements should not be confused with water-quality standards. Set by the potential user, *water-quality requirements* represent a known or assumed need and are based on the prior experience of the water user. *Water-quality standards* are set by a governmental agency and represent a statutory requirement. For example, a farmer may know from prior experience that highly saline water will damage the crops, but there are no official water-quality standards that say such water cannot be used for irrigation purposes.

In the United States, standards have been promulgated for streams and lakes, for public water supplies, and for wastewater discharges. Standards for surface waters, potable water, and wastewaters are discussed in the following section.

2-17 IN-STREAM STANDARDS

For reasons of aesthetics as well as health, it is generally considered desirable to maintain natural water systems at as high a quality level as possible. All 50 of the states have set minimum quality standards for all surface waters within their boundaries. These standards often reflect the beneficial use made of the stream. That is, more stringent standards are applied to a stream used as a source of water for municipal purposes than to streams used for other purposes.

Many factors affect stream quality. Wastewater discharges and other human activities often have significant impact on in-stream water quality. These activities may lend themselves to control by legislation and government regulations. For example, consistent with what it believed to be public opinion, the United States Congress passed the Water Pollution Control Act of 1972 (Public Law 92-500) and, with minor modifications, renewed it in 1977. A stated goal of this legislation is that all surface waters in the United States be maintained at "fishable, swimmable" quality. Attainment of this goal should result in a quality sufficient for most water uses and simplify treatment processes for waters intended for potable and industrial use.

While legislation such as Public Law 92-500 can control some aspects of water pollution, the geology of a watershed, coupled with other natural phenomena, is often the controlling factor in water quality. This fact must be considered if in-stream water-quality standards are to be realistic.

2-18 POTABLE-WATER STANDARDS

Standards for drinking water have evolved over the years as knowledge of the nature and effects of various contaminants has grown. Currently, it is considered desirable that drinking water be free of suspended solids and turbidity, that it be tasteless and odorless, that dissolved inorganic solids be in moderate quantities, and that organics, toxic substances, and pathogens be absent. As more is learned about the constituents of water, additional requirements will probably be added to this list, making drinking-water requirements even more stringent.

The World Health Organization has established minimum criteria for drinking water that all nations are urged to meet. These standards are listed in Table 2-9. Countries with more advanced technology generally have standards that exceed this quality.

Table 2-9 Drinking-water standards of the World Health Organization

	Concentrations in milligrams per liter				
	WHO International (1958)			WHO European (1961)	
Chemical constituent	Permissible limit	Excessive limit	Maximum limit	Recommended limit	Tolerance limit
Ammonia (NH_4)	—	—	—	0.5	—
Arsenic	—	—	0.2	—	0.2
Cadmium	—	—	—	—	0.05
Calcium	75	200	—	—	—
Chloride	200	600	—	350	—
Chromium (hexavalent)	—	—	0.05	—	0.05
Copper	1.0	1.5	—	3.0*	—
Cyanide	—	—	0.01	—	0.01
Fluoride	—	—	—	1.5	—
Iron	0.3	1.0	—	0.1	—
Lead	—	—	0.1	—	0.1
Magnesium	50	150	—	125†	—
Magnesium + sodium sulfates	500	1000	—	—	—
Manganese	0.1	0.5	—	0.1	—
Nitrate (as NO_3)	—	—	—	50	—
Oxygen, dissolved (minimum)	—	—	—	5.0	—
Phenolic compounds (as phenols)	0.001	0.002	—	0.001	—
Selenium	—	—	0.05	—	0.05
Sulfate	200	400	—	250	—
Total solids	500	1500	—	—	—
Zinc	5.0	15	—	5.0	—

* After 16 h contact with new pipes; but water entering a distribution system should have less than 0.05 mg/L of copper.

† If there is 250 mg/L of sulfate present, magnesium should not exceed 30 mg/L.

Source: Adapted from Todd. [2-17]

The Safe Drinking Water Act of 1974 (Public Law 93-523) mandated the U.S. Environmental Protection Agency to establish drinking-water standards for all public water systems serving 25 or more people or having 15 or more connections. Pursuant to this mandate, EPA has established maximum contaminant levels for drinking water delivered through public water supply distribution systems. These standards were published in 1975 under Title 40, Subchapter D, Part 141 of the Code of Federal Regulations. [2-19] The maximum contaminant level (MCL) of inorganics, organic chemicals, turbidity, and microbiological contaminants are shown in Tables D-1 through D-5 of the appendix. EPA has also issued proposed regulations to serve as guidelines to the states with regard to the so-called secondary drinking-water standards. [2-16] These appear in Table D-6 of the appendix.

2-19 WASTEWATER EFFLUENT STANDARDS

The water Pollution Control Act of 1972 (Public Law 92-500) mandated the Environmental Protection Agency to establish standards for wastewater discharges. Current standards require that as a minimum all municipal wastewater be treated to "secondary" standards shown in Table D-7 of the appendix. More stringent standards may be imposed where necessary, and in some cases less stringent standards may be permitted for small flows. Industrial dischargers are required to treat their wastewater to the level obtainable by the "best available technology" for wastewater treatment in that particular type of industry. If industry discharges to a municipal wastewater collection system, the industrial waste must be pretreated so as to be compatible with the untreated municipal wastewater.

The EPA regulations define receiving streams as either "effluent-limited" or "water-quality-limited." An *effluent-limited stream* is a stream which will meet its in-stream standards if all discharges to that stream meet the secondary-treatment and best-available-technology standards. Municipalities and industries discharging to effluent-limited streams are assigned discharge permits under the National Pollution Discharge Elimination System (NPDES) which reflects the secondary-treatment and best-available-technology standards.

A *water-quality-limited-stream* would *not* meet its proposed in-stream standards even if all discharges met secondary-treatment and best-available-technology criteria. Discharges to these streams may be required to meet effluent conditions more stringent than secondary-treatment and best-available-technology. These discharge limits are established on a case-by-case basis.

Effluent standards, potable-water standards, and in-stream standards are obviously interrelated. Enforcement of effluent standards, along with the control of nonpoint sources of pollution, should result in the attainment of in-stream standards. The improvement of in-stream quality should result in a better raw water for potable supplies. However, it is impractical to expect surface waters, even if in-stream standards are met, to meet all of the maximum contaminant levels imposed by the potable-water standards. Treatment of surface water for potable use will always be required, the nature and level of treatment depending on the in-stream quality of the water source.

DISCUSSION TOPICS AND PROBLEMS

2-1 Name the physical water-quality parameters of concern to environmental engineers.

2-2 Discuss the sources and impacts of suspended solids.

2-3 How are suspended solids measured?

2-4 An analysis for suspended solids is run as follows: (1) A fiberglass filter is dried to a constant mass of 0.137 g; (2) 100 mL of a sample is drawn through the filter; and (3) the filter and residue are placed in a drying over at 104°C until a constant mass of 0.183 g is reached. Determine the suspended-solids concentration in milligrams per liter.

2-5 One hundred milliliters of the filtrate from a suspended-solids analysis is placed in an evaporation dish whose tare mass has been determined to be 327.485 g. The contents of the dish are evaporated to dryness, and the total mass of the dish and solids is found to be 327.517 g. Determine the quantity of filterable residue (in milligrams per liter).

2-6 The crucible, filter pad, and solids of Example 2-1 are placed in a muffle furnace at 600°C for 1 h. After cooling, the mass is determined to be 54.367 g. Determine the concentration of the volatile (organic suspended solids).

2-7 The evaporation dish and residue from Prob. 2-5 is placed in a muffle furnace at 600°C for 1 h. After cooling, its mass is found to be 327.498 g. Determine the filterable volatile (organic) solids of the sample in milligrams per liter.

2-8 Discuss the sources and impacts of tastes and odors in water supplies.

2-9 How are tastes and odors measured?

2-10 What are the sources of temperature increases in water bodies? What are the impacts of elevated temperatures?

2-11 Name the chemical parameters of concern in water-quality management.

2-12 The reaction of soda ash (Na_2CO_3) with calcium sulfate in water is represented by the following chemical statement:

$$Na_2CO_3 + CaSO_4 \longrightarrow CaCO_3 + Na_2SO_4$$

Assuming that this reaction is complete and that there is 153 mg/L of $CaSO_4$ initially present, what is the mass of soda ash that must be added to (*a*) 1 L of the water and (*b*) 10,000 m^2 of the water to complete the reactions.

2-13 How many grams of CaO are required to be the chemical equivalent of 246 g of $Mg(HCO_3)_2$?

2-14 Express the following concentrations of elements and compounds as milligrams per liter of $CaCO_3$.

95 mg/L Ca^{2+}	420 mg/L Mg SO_4
87 mg/L Mg^{2+}	189 mg/L $NaHCO_3$
125 mg/L Na^{+}	221 mg/L $Ca(HCO_3)_2$

2-15 Express the following molar concentrations of elements and compounds as milligrams per liter of $CaCO_3$.

1×10^{-2} mol/L Al^{3+}	1.8×10^{-3} mol/L $CaSO_4$
1.5×10^{-3} mol/L SO_4^{2-}	2.1×10^{-3} mol/L $Mg(Cl)_2$
3.2×10^{-3} mol/L Cl^-	3.5×10^{-3} mol/L NaOH

2-16 Determine the concentration of the following ions in solution at equilibrium with the solid at 25°C.

$$Ca^{2+} + 2OH^- \rightleftharpoons Ca(OH)_2$$

$$Mg^{2+} + CO_3^{2-} \rightleftharpoons MgCO_3$$

$$Ca^{2+} + SO_4^{2-} \rightleftharpoons CaSO_4$$

2-17 A wastewater containing $Fe(HCO_3)_2$ is discharged to a surface pond. Assuming complete oxidation of the Fe^{2+} to Fe^{3+} and sufficient OH for the following reaction to occur

$$Fe^{3+} + 3OH^- \longrightarrow Fe(OH)_3s$$

determine the concentration (mg/L as $CaCO_3$) of the Fe^{3+} remaining dissolved in the pond water.

2-18 A sample of water from a surface stream is analyzed for the common ions with the following resuls:

Ca^{2+} = 98 mg/L

Cl^- = 89 mg/L

HCO_3^- = 317 mg/L

Mg^{2+} = 22 mg/L

Na^+ = 71 mg/L

SO_4^{2-} = 125 mg/L

(*a*) What is the percent error in the cation-anion balance?
(*b*) Draw a bar diagram for the water.

2-19 A sample of water was analyzed for common ions with the result shown below.

HCO_3^- = 300 mg/L

Na^+ = 115 mg/L

SO_4^{2-} = 240 mg/L

Mg^{2+} = 36.6 mg/L

Cl^- = 71.0 mg/L

Ca^{2+} = 100 mg/L

Construct a bar diagram in milliequivalents per liter for this water.

2-20 Draw a milliequivalent-per-liter bar diagram for the water with the following common ion concentrations.

$$Ca^{2+} = 70 \text{ mg/L} \qquad Mg^{2+} = 28 \text{ mg/L} \qquad Na^{+} = 124 \text{ mg/L}$$

$$HCO_3^{-} = 165 \text{ mg/L} \qquad SO_4^{2-} = 173 \text{ mg/L} \qquad Cl^{-} = 202 \text{ mg/L}$$

Determine the error in the ion balance.

2-21 An analysis of water from a surface stream yields the following results.

$$Ca^{2+} = 60 \text{ mg/L} \qquad HCO_3^{-} = 115 \text{ mg/L}$$

$$Mg^{2+} = 10 \text{ mg/L} \qquad SO_4^{2-} = 96 \text{ mg/L}$$

$$Na^{+} = 7 \text{ mg/L} \qquad NO_3^{-} = 10 \text{ mg/L}$$

$$K^{+} = 20 \text{ mg/L} \qquad Cl^{-} = 11 \text{ mg/L}$$

If an error of 10 percent is acceptable, should the analysis be considered complete?

2-22 What are the sources and impacts of dissolved solids in water supplies?

2-23 How are dissolved solids measured? How are TDS measurements expressed?

2-24 A solids analysis is to be conducted on a sample of wastewater. The procedure is as follows:

1. A Goch crucible and filter pad are dried to a constant mass of 25.439 g.
2. Two hundred milliliters of a well-shaken sample of the wastewater is passed through the filter.
3. The crucible, filter, pad, and removed solids are dried to a constant mass of 25.645 g.
4. One hundred milliliters of the filtrate [water passing through the filter in (2) above] is placed in an evaporation dish that had been preweighed at 275.419 g.
5. The sample in (4) is evaporated to dryness and the dish and residue are weighed at 276.227 g.
6. Both the crucible from (3) and the evaporation dish from (5) are placed in a muffle furnace at 600°C for an hour. After cooling, the mass of the crucible is 25.501 g and the mass of the dish is 275.944 g.

Determine the following:

(*a*) The filterable solids (mg/L)
(*b*) The nonfilterable solids (mg/L)
(*c*) The total solids (mg/L)
(*d*) The organic fraction of the filterable solids (mg/L)
(*e*) The organic fraction of the nonfilterable solids (mg/L)

2-25 What are the most common constituents of alkalinity, and what are their sources and impacts?

2-26 How is alkalinity measured?

2-27 Determine the alkalinity of the waters described in Probs. 2-18 to 2-21.

2-28 A 100-mL sample of water is titrated with 0.02 N H_2SO_4. The initial pH is 9.5, and 6.2 mL of acid is required to reach the pH 8.3 endpoint. An *additional* 9.8 mL is required to reach the 4.5 endpoint. Determine the species of alkalinity present and the concentration of each species.

2-29 A 200-mL sample of water with an initial pH of 10.6 is titrated with 0.02 N H_2SO_4. The sample reaches pH 8.3 after an addition of 8.8 mL of the acid, and an additional 5.5 mL is required to bring the sample to pH 4.5. Identify the species of alkalinity present and determine the concentrations (mg/L) of each.

2-30 The initial pH of a water sample is 7.5. A 200-mL sample is titrated with 0.01 N H_2SO_4. The pH 4.5 endpoint is reached after the addition of 15 mL of the acid. Determine the species of alkalinity present and the concentration (mg/L) of each.

2-31 Define "hardness" of water, note the two broad classifications of hardness, and discuss the sources and impacts of hardness.

2-32 Would hard water be acceptable in most drinking-water supplies? Why or why not? Would hard water be an acceptable coolant for an industrial plant? Why or why not?

2-33 How is hardness measured?

2-34 Determine the carbonate hardness, noncarbonate hardness, and total hardness of the water described in Probs. 2-19 through 2-21.

2-35 Discuss the sources and impacts of fluorides in drinking-water supplies.

2-36 Name the most common nontoxic metals found in water supplies, identify their sources, and discuss their impacts.

2-37 Name toxic metals that may be dissolved in water, identify their principal sources, and discuss their impacts.

2-38 Define biodegradable organics. Give examples, discuss sources, and assess the impact of biodegradable organics in water.

2-39 Define biochemical oxygen demand (BOD) and outline the steps in the standard 5-day BOD test.

2-40 The 5-day BOD of a wastewater is 190 mg/L. Determine the ultimate oxygen demand. Assume $k_1 = 0.25\ d^{-1}$.

2-41 In a BOD determination, 6 mL of wastewater containing no dissolved oxygen is mixed with 294 mL of dilution water containing 8.6 mg/L of dissolved oxygen. After a 5-day incubation at 20°C, the dissolved oxygen content of the mixture is 5.4 mg/L. Calculate the BOD of the wastewater.

2-42 The 5-d 20°C BOD of a wastewater is 210 mg/L. What will be the ultimate BOD? What will be the 10-day BOD? If the sample had been incubated at 30°C what would the 5-day BOD have been ($k_1 = 0.23\ d^{-1}$)?

2-43 An analysis for BOD_5 is to be run on a sample of wastewater. The BOD is expected to range from 50 to 350, and the dilutions are prepared accordingly. In each case, a standard 300-mL BOD bottle is used. The data are recorded below.

Bottle no.	Wastewater, mL	DO_I	DO_5
1	20	8.9	1.5
2	10	9.1	2.5
3	5	9.2	5.8
4	2	9.2	7.5

(*a*) Determine the BOD_5 of the wastewater.

(*b*) If you know that the oxygen utilization rate is 0.21 per day at 20°C, what will be the BOD_3 if the test is run at 30°C?

2-44 A BOD analysis is begun on Monday. Thirty (30) milliliters of waste with a DO of zero is mixed with 270 mL of dilution water with a DO of 10 mg/L. The sample is then put in the incubator. Since the fifth day falls on Saturday and lab personnel do not work on Saturday, the final DO does not get measured until Monday, the seventh day. The final DO is measured at 4.0 mg/L. However, it is discovered that the incubator was set at 30°C. Assume a k_1 of 0.2 at 20°C and $k_T = k_{20} 1.05^{T-20}$. Determine the 5-day, 20°C BOD of the sample.

2-45 Define nonbiodegradable organics. Give examples, discuss sources, and assess the impact of nonbiodegradable organics in water.

2-46 Define chemical oxygen demand (COD) and total organic carbon (TOC), and discuss how these and other tests are used to quantify nonbiodegradable organics in water.

2-47 Name the nutrients required in greatest abundance by aquatic species.

2-48 Discuss the sources and impacts of nitrogen and phosphorus in water bodies.

2-49 How are nitrogen and phosphorus measured?

2-50 Define methemoglobinemia and discuss it as a water-related illness.

2-51 Pathogens are not always bacteria. Name two pathogenic bacteria, two viruses, and one protozoan sometimes found in water supplies.

2-52 With which waterborne pathogens are the following diseases associated?

(*a*) Cholera
(*b*) Swineherd's disease
(*c*) Amebic dysentery
(*d*) Giardiasis
(*e*) Bacillary dysentery
(*f*) Typhoid fever
(*g*) Paratyphoid
(*h*) Infantile paralysis
(*i*) Infectious hepatitis.

2-53 What is an indicator organism? Discuss the characteristics of the ideal pathogen indicator and indicate which organisms most nearly exhibit these characteristics.

2-54 Discuss the use of total coliform and fecal coliform tests in the measurement of pathogens. Discuss the membrane-filter technique and explain how test results are reported when this technique is used.

2-55 Discuss the multiple-tube fermentation test. What is a presumptive test? A confirmed test? How are results expressed?

2-56 A sample of wastewater is analyzed for coliform organisms by the multiple-tube fermentation method. The results of the confirmed test are as follows:

Sample size, mL	Number of positive results out of 5 tubes	Number of negative results out of 5 tubes
0.01	5	0
0.001	4	1
0.0001	2	3
0.00001	2	3
0.000001	0	1

Determine the most probable number and range of coliform organisms per 100 mL at the 95 percent confidence level.

2-57 Discuss in-stream standards, effluent standards, and potable-water standards. Who sets these standards in the United States? Elsewhere?

2-58 What is an effluent-limited stream?

REFERENCES

2-1 American Water Works Association: "Quality Goals for Public Water—Statement of Policy," *J AWWA*, **60**: 1317 (1968).
2-2 Berg, Gerald: *Transmission of Viruses by the Water Route*, Wiley, New York, 1965.
2-3 Camp, T. R.: *Water and Its Impurities*, Reinhold, New York, 1973.
2-4 *Environmental Quality, the Eighth Annual Report of the Council of Environmental Quality*. U.S. Gov. Printing Office, Washington, D.C., December 1977.
2-5 Geldreich, Edwin E.: "Water Borne Pathogens," in Ralph Mitchell (ed.), *Water Pollution Microbiology*, Wiley, New York, 1972.
2-6 Hahn, Roy W., Jr.: *Fundamental Aspects of Water Quality Management*, Technomic, Westport, Conn., 1972.
2-7 McKee, J. E., and H. W. Wolf: *Water Quality Criteria*, publ. no. 3-A, State Water Resources Control Board, Sacramento, Calif., 1971.
2-8 Metcalf & Eddy, Inc.: *Wastewater Engineering: Treatment, Disposal, Reuse*, 2d ed., McGraw-Hill, New York. 1979.
2-9 *Methods for Chemical Analysis of Water and Waste*, EPA 600/4-79-020, U.S. EPA, Cincinnati, March 1979.
2-10 Miller, J. C.: *Nitrate Contamination of the Water Table Aquifer of Delaware*, Report of Investigations no. 20, Delaware Geological Survey, University of Delaware, 1972.
2-11 ———, P. S. Hackenberry, and F. A. DeLucca: *Groundwater Pollution Problems in the Southeastern United States*, EPA 600/3-77-012, U.S. EPA, 1977.
2-12 Sawyer, C. N., and P. L. McCarty: *Chemistry for Environmental Engineers*, 3d ed., McGraw-Hill, New York, 1978.
2-13 Schmidt, K. D.: "Nitrate in Groundwater of the Fresno-Clovis Metropolitan Area, California," *Groundwater*, **10**: 50 (1972).
2-14 Smith, Alice: *Microbiology and Pathology*, Mosby, St. Louis, 1976.
2-15 *Standard Methods for the Examination of Water and Wastewater*, 15th ed., American Public Health Association, Washington, D.C., 1981.
2-16 Steele, E. W., and T. J. McGhee: *Water Supply and Sewerage*, 5th ed., McGraw-Hill, New York, 1979.
2-17 Todd, D. K.: *The Water Encyclopedia*, Water Information Center, Port Washington, New York, 1970.
2-18 U.S. Department of Health, Education, and Welfare: *Drinking Water Standards*, PHS bulletin no. 956, Public Health Service, 1962.
2-19 U.S. Environmental Protection Agency: "National Interim Primary Drinking Water Regulations," *Federal Register*, pt. IV, December 24, 1975.
2-20 Walker, Rodger: *Water Supply, Treatment, and Distribution*, Prentice-Hall, Englewood Cliffs, N. J., 1978.
2-21 Vesilind, P. Aarne: *Environmental Pollution and Control*, Ann Arbor Science, Ann Arbor, Mich., 1975.

CHAPTER

THREE

WATER PURIFICATION PROCESSES IN NATURAL SYSTEMS

Natural forms of pollutants have always been present in surface waters. Long before the dawn of civilization, many of the impurities discussed in the previous chapter were washed from the air, eroded from land surfaces, or leached from the soil and ultimately found their way into surface water. With few exceptions, natural purification processes were able to remove or otherwise render these materials harmless. Indeed, without these self-cleaning processes, the water-dependent life on earth could not have developed as it did.

As civilization evolved, human activity increased the amount and changed the nature of pollutants entering watercourses. As settlements grew into villages, villages into towns, and towns into cities, the quantity of waste products increased until the self-purification capacity of local water bodies was exceeded. Smaller streams were affected first, with larger streams and lakes ultimately becoming polluted. Only in recent decades have pollution control programs been initiated in an attempt to reduce the contaminants discharged to these water bodies to the level that the natural purification processes can once again assimilate them.

The self-purification mechanisms of natural water systems include physical, chemical, and biological processes. The speed and completeness with which these processes occur depend on many variables that are system-specific. Hydraulic characteristics such as volume, rate, and turbulence of flow, physical characteristics of bottom and bank material, variations in sunlight and temperature, as well as the chemical nature of the natural water, are all system variables that have an influence on the natural purification processes. In natural waters, these system variables are set by nature and can seldom be altered.

The same physical, chemical, and biological processes that serve to purify natural water systems also work in engineered systems. In water- and wastewater-treatment plants, the rate and extent of these processes are managed by controlling the system variables. A thorough knowledge of the natural purification processes is thus essential to the understanding of both the assimilative capacity of surface waters and the operation of engineered systems. The self-purification of natural water systems is discussed in this chapter, while water purification in engineered systems is covered in Chaps. 4 and 5.

Physical Processes

The major physical processes involved in self-purification of watercourses are dilution, sedimentation and resuspension, filtration, gas transfer, and heat transfer. These processes are not only important in and of themselves, but are also of significance in their relation to certain chemical and biochemical self-purification processes.

3-1 DILUTION

Through the first decades of the present century, wastewater disposal practices were based on the premise that "the solution to pollution is dilution." Dilution was considered the most economical means of wastewater disposal and as such was considered good engineering practice. [3-5, 3-25] Early workers in the field devised mixing-zone concepts based on the lateral, vertical, and longitudinal dispersion characteristics of the receiving waters. Formulas predicting space and time requirements for diluting certain pollutants to preselected concentrations were developed. Highly polluted water in the immediate vicinity of the discharge was tolerated as inevitable, and little thought was given to the low levels of material transported downstream.

Although dilution is a powerful adjunct to self-cleaning mechanisms of surface waters, its success depends upon discharging relatively small quantities of waste into large bodies of water. Growth in population and industrial activity, with attendant increases in water demand and wastewater quantities, precludes the use of many streams for dilution of raw or poorly treated wastewaters. In the United States, legal constraints further limit use of water bodies for wastewater dilution. Under present regulations, maximum allowable loads are set independently of dilution capacity. Only when the standard maximum loads result in violation of in-stream water-quality standards is the dilution capacity considered, and then only to determine the increment of treatment necessary.

The dilution capacity of a stream can be calculated using the principles of mass balance. If the volumetric flow rate and the concentration of a given material are known in both the stream and waste discharge, the concentration after mixing can be calculated as follows.

$$C_s Q_s + C_w Q_w = Q_m C_m \qquad (3\text{-}1)$$

where C represents the concentration (mass/volume) of the selected material, Q is the volumeric flow rate (volume/time), and the subscripts s, w, and m designate stream, waste, and mixture conditions. The following example illustrates the use of this formula.

Example 3-1: Measuring dilution in streams A treated wastewater enters a stream as shown in the accompanying figure. The concentration of sodium in the stream at point A is 10 mg/L, and the flow rate is 20 m^3/s. The concentration of sodium in the waste stream is 250 mg/L, and the flow rate is 1.5 m^3/s. Determine the concentration of sodium at point B assuming that complete mixing has occurred.

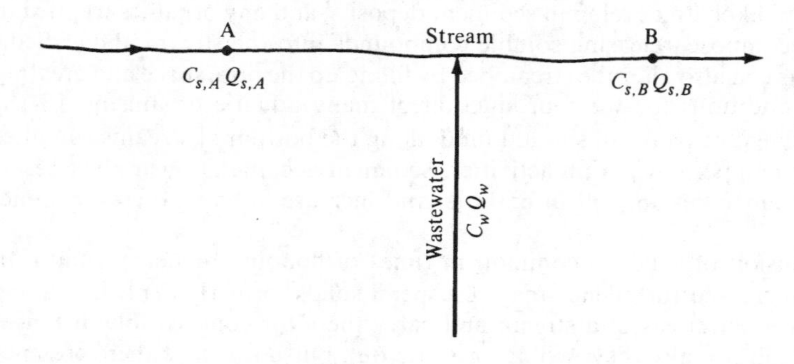

SOLUTION

1. Write a mass balance between points A and B

$$\text{Mass in} = \text{Mass out}$$

$$C_{s,B}Q_{s,B} = C_{s,A}Q_{s,A} + C_wQ_w$$

Since $Q_{s,B}$ is the sum of the other two flows

$$C_{s,B} = \frac{C_{s,A}Q_{s,A} + C_wQ_w}{Q_{s,A} + Q_w}$$

2. Insert numerical values and solve for $C_{s,B}$

$$C_{s,B} = \frac{10 \times 20 + 250 \times 1.5}{20 + 1.5}$$

$$C_{s,B} = 26.7 \text{ mg/L}$$

3-2 SEDIMENTATION AND RESUSPENSION

Sources of suspended solids, one of the most common water pollutants, include domestic and industrial wastewater and runoff from agricultural, urban, or silvicultural activities. As discussed in Sec. 2-2, these solids may be inorganic or organic materials and/or live organisms, and they may vary in size from large organic particles to tiny, almost invisible, colloids. In suspension, solids increase turbidity (see Sec. 2-3), and the reduced light penetration may restrict the photosynthetic activity of plants, inhibit the vision of aquatic animals, interfere with feeding of aquatic animals that obtain food by filtration, and be abrasive to respiratory structures such as gills of fish. [3-27]

Settling out, or sedimentation, is nature's method of removing suspended particles from a watercourse, and most large solids will settle out readily in quiescent water. Particles in the colloidal size range can stay in suspension for long periods of time, though eventually most of these will also settle out.

This natural sedimentation process is not without its drawbacks. Anaerobic conditions are likely to develop in sediment deposits, and any organics trapped in them will decompose, releasing soluble compounds into the stream above. Sediment deposits can also alter the streambed by filling up the pore space and creating unsuitable conditions for the reproduction of many aquatic organisms. [3-18] The development of banks of silt and mud along the bottom of streams can alter its course or hamper navigation activities. Sediment accumulations reduce reservoir storage capacities and silt in harbors, and increase flooding due to channel fill-in.

Resuspension of solids is common in times of flooding or heavy runoff. In such cases, increased turbulence may resuspend solids formerly deposited along normally quiescent areas of a stream and carry them for considerable distances downstream. Eventually they will again settle out, but not before their presence has increased the turbidity of the waters into which they have been introduced.

3-3 FILTRATION

As large bits of debris wash along a streambed, they often lodge on reeds or stones where they remain caught until high waters wash them into the mainstream again. Small bits of organic matter or inorganic clays and other sediments may be filtered out by pebbles or rocks along the streambed. As water percolates from the surface downward into groundwater aquifers, filtration of a much more sophisticated type occurs, and, if the soil layers are deep enough and fine enough, removal of suspended material is essentially complete by the time water enters the aquifer. Many streams interchange freely with the alluvial aquifers underneath them, so the filtered water may reenter the stream at some point downstream.

3-4 GAS TRANSFER

The transfer of gases into and out of water is an important part of the natural purification process. The replenishment of oxygen lost to bacterial degradation or organic waste is accomplished by the transfer of oxygen from the air into the water. Conversely, gases evolved in the water by chemical and biological processes may be transferred from the water to the atmosphere. A knowledge of the principles of gas transfer is essential to understanding these natural processes.

Consider the simple system shown in Fig. 3-1 in which a container of liquid is sealed with a gas above it. If the liquid is initially pure with respect to the gas, molecules of gas will migrate across the gas-liquid interface and become dissolved in the liquid. Although some molecules of gas will begin leaving the liquid and returning to gas phase, the net reaction will be toward the liquid until a state of

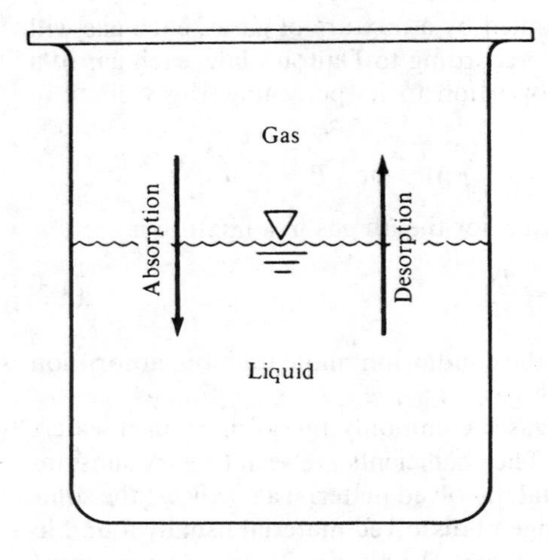

Figure 3-1 Gas-liquid contact with gas transfer between the phases.

equilibrium is reached. At this point, the number of molecules leaving the liquid is equal to the number of molecules entering it again, and the liquid is said to be *saturated* with the gas. Equilibrium in this case implies a dynamic steady state, not a static state in which all movement of gas molecules would stop once saturation occurred.

Two characteristics of the above process that are important in water are (1) *solubility*, or the extent to which the gas is soluble in the water (i.e., the concentration of gas in the water at equilibrium), and (2) *transfer rate*, or the rate at which dissolution or release occurs.

Solubility

The solubility of a gas in equilibrium with a liquid is quantified by Henry's law and is expressed mathematically by

$$x = \frac{P}{H} \tag{3-2}$$

in which x is the equilibrium mole fraction of the dissolved gas at 1 atm or

$$x = \frac{\text{moles of gas } (n_g)}{\text{moles of gas } (n_g) + \text{moles of liquid } (n_1)}$$

H is the coefficient of absorption (*Henry's coefficient*, which is unique for each gas-liquid system), and P is the pressure of the gas above the liquid. Other factors that affect x are temperature (the solubility increases as temperature decreases) and the concentration of other dissolved gases and solids (the solubility decreases as other dissolved material in the liquid increases).

If the space above the liquid is occupied by a mixture of gases, each gas will have its own equilibrium mole fraction. According to Dalton's law, each gas in a mixture exerts a partial pressure in proportion to its percentage by volume in the mixture; that is,

$$PV = (p_1 + p_2 + p_3 + \cdots + p_n)V \quad \text{or} \quad P = \sum p_i$$

Substituting into Henry's law, we see that x for the ith gas in a mixture is

$$x_i = \frac{p_i}{H_i} \tag{3-3}$$

in which x_i, H_i, and p_i are, respectively, the equilibrium mole fraction, absorption coefficient, and partial pressure of the ith gas.

Absorption coefficients for several gases commonly found in natural waters are given in Table C–2 of the appendix. The coefficients are seen to vary substantially with temperature. Although the total dissolved material also affects the solubility, the effect is insignificant in the range of dissolved material usually found in fresh water. To be precisely accurate, the partial pressure of water vapor must be accounted for in Eqs. (3-2) and (3-3). Conversion of the equilibrium mole fraction x to an equilibrium concentration C_s is illustrated in the following example.

Example 3-2: Calculating the solubility of air in water Calculate the solubility of air in water at 0°C and 1 atm pressure. Assume other dissolved material is negligible.

SOLUTION

1. From Table C-2 in the appendix, Henry's constant for air at 0°C is

$$H = 4.32 \times 10^4 \text{ atm/mol fraction}$$

at 1 atm pressure. The mole fraction of air in water is found by Eq. (3-2).

$$x_{\text{air}} = \frac{P}{H}$$

$$= \frac{1.0 \text{ atm}}{4.32 \times 10^4 \text{ atm/mol fraction}}$$

$$= 2.31 \times 10^{-5} \text{ mol fraction}$$

2. One liter of water contains

$$\frac{1000 \text{ g/L}}{18 \text{ g/mol}} = 55.6 \text{ g} \cdot \text{mol/L}$$

and

$$2.31 \times 10^{-5} = \frac{n_g}{n_g + 55.6}$$

$$n_g - (2.31 \times 10^{-5} n_g) = 2.31 \times 10^{-5} \times 55.6$$

$$n_g = 1.287 \times 10^{-3} \text{ g} \cdot \text{mol/L}$$

3. The saturation concentration is

$$C_s = 1.287 \times 10^{-3}\ \text{g·mol/L} \times 28.9\ \text{g/g·mol} \times 10^3\ \text{mg/g}$$
$$= 37.2\ \text{mg/L}$$

The solubility of air can also be found by using its components and Dalton's law. The components of air by volume are approximately as follows:

$$N_2 \cong 79\%$$
$$O_2 \cong 21\%$$
$$CO_2 \cong 0.03\%$$

4. The molecular mass of nitrogen is 28 g/mol and H from Table C-2 in the appendix is 5.29×10^4.

$$x_{N_2} = \frac{0.79}{5.29 \times 10^4} = 1.49 \times 10^{-5}$$

$$n_{N_2} - 1.49 \times 10^{-5}\ n_{N_2} = 1.49 \times 10^{-5} \times 55.6$$
$$= 8.3 \times 10^{-4}\ \text{mol/L}$$

$$C_s = 8.3 \times 10^{-4}\ \text{g·mol/L} \times 28\ \text{g·mol} \times 10^3\ \text{mg/g}$$
$$= 23.25\ \text{mg/L}$$

5. The equilibrium concentrations for O_2 and CO_2 can be found similarly and are 16.65 and 0.02 mg/L, respectively.

The equilibrium concentration of air is

$$23.25 + 16.65 + 0.02 = 39.92\ \text{mg/L}$$

The discrepancy is accounted for by the rounding off of the percentage of N_2, O_2, and CO_2 in air.

Transfer Rate

The rate of gas transfer is an important parameter in aeration. The rate of transfer is governed by several factors and is mathematically expressed as

$$dC/dt = (C_s - C)k_a$$

where dC/dt is the instantaneous rate of change of the concentration of gas in the liquid, C_s and C are the saturation concentration and the actual concentration, respectively, and k_a is a constant related to given physical conditions. It should be noted that desorption of the gas occurs when C is greater than C_s. The magnitude of k_a is known to depend upon the temperature of the system, the interfacial area available for gas transfer, and resistance to movement from one phase to the other.

While the effect of temperature can be predicted by the van't Hoff-Arrhenius rule, the other variables are system-specific. The interfacial area available for gas transfer is measured by the total contact surface between the gas and liquid. Larger interfacial area per given volume will result in greater opportunity for gas transfer.

The resistance to movement between the phases is most often explained by the two-film theory of mass transfer initially postulated by Lewis and Whitman in 1924. According to this theory, the interface is composed of two distinct films, one on the gas side and one on the liquid side, that serve as a barrier between the bulk phases. This system is shown graphically in Fig. 3-2*a*. In order for a molecule of gas somewhere in the interior of the gas phase to be transferred into the interior of the liquid phase, it must move through the bulk gas to the interface, across the gas film, across the liquid film, and, finally, away from the interface and into the bulk liquid. In systems where the liquid is supersaturated with respect to the gas, movement of the gas molecule will be in the reverse direction (Fig. 3-2*b*).

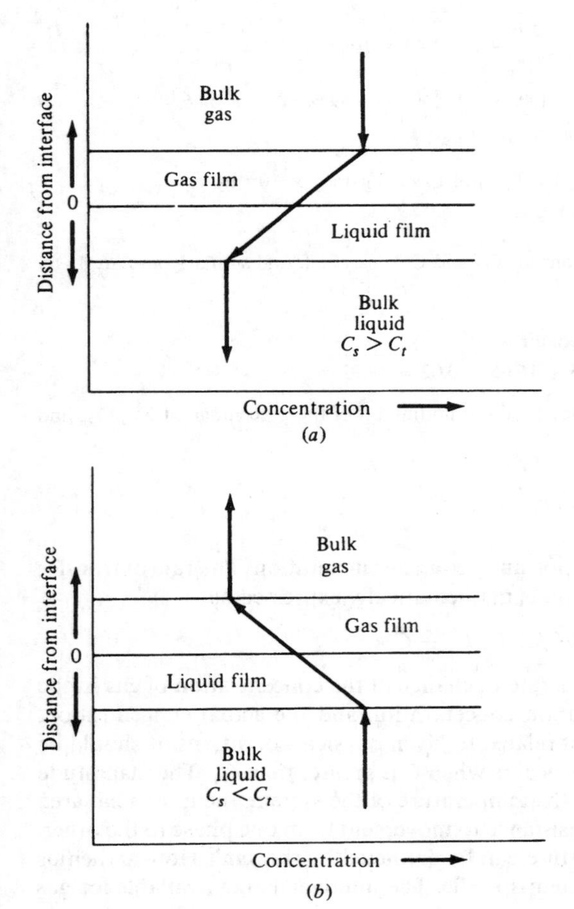

Figure 3-2 Two-film model of the interface between gas and liquid: (*a*) absorption mode and (*b*) desorption mode.

The driving force causing mass transfer is the concentration gradient, $C_s - C$. Resistance to mass transfer must be overcome for the process to occur, and each one of the steps listed above is likely to exhibit a different level of resistance. The step which offers the most resistance to the movement of gas molecules becomes the rate-limiting step. In stagnant situations (i.e., no internal movement of the bulk phases), movement of gas molecules to and away from the interface depends totally upon diffusion, and the process is very slow. However, if internal movement of the bulk phases occurs, molecules of gas are transferred to and away from the interface by turbulence and eddy diffusion, and the rate of mass transfer is most likely to be governed by one or both of the films.

In most natural water, sufficient agitation of the bulk phases exists, and the films become the limiting factors. In general, gases that are highly soluble in water, such as ammonia, encounter more resistance in passing through the gas film and the process is said to be *gas-film-controlled.* Conversely, slightly soluble gases such as oxygen and nitrogen encounter more resistance in the liquid film, and the system is *liquid-film-controlled* with respect to these gases. Gases of intermediate solubility, such as hydrogen sulfide, encounter approximately equal resistance through the two films and the system is said to be *mixed-film-controlled.*

3-5 HEAT TRANSFER

Bodies of water lose and gain heat much more slowly than do land or air masses, and under most circumstances, water temperature is fairly constant and changes gradually with the seasons. Consequently, aquatic plants and animals have not developed sufficient adaptability to deal with abrupt changes in temperature, and only the most hardy species survive such changes. Thus, heat increases tend to decrease the number of species of aquatic plants and animals. [3-17] Furthermore, increases in water temperature affect ionic strength, conductivity, dissociation constants, solubility, and corrosion potential, all factors associated with water quality.

Given constant meteorologic conditions, water theoretically requires an infinite time of exposure to attain equilibrium after a heat load. Furthermore, an infinite surface area would be required to cool warm water introduced into a river or basin to the equilibrium temperature. However, because temperature decline is nearly logarithmic, equilibrium can be closely approached within practical limitations of time and surface area. Many meteorological variables—plus other factors such as channel characteristics (depth, width, surface area), channel volume, etc.—affect the rate of heat transfer in bodies of water. For streams heated by solar radiation over several miles of heat-load area, cooling begins only in shaded areas or at night and may proceed much more slowly than cooling in streams which receive their heat load in one discharge.

In temperate zones, heat transfer in reservoirs and lakes where the influence of turbulence and current is negligible is controlled by a phenomenon known as *thermal statification.* Fresh waters reach their maximum density at 4°C (39°F), with density declining as water moves toward the freezing point or grows warmer. (See Fig. 3-3.) Thus, during warm seasons and in impoundments of sufficient depth, water divides into an upper layer of warm, circulating water known as the *epilimnion* and a lower layer of cool, relatively undisturbed water known as the *hypolimnion.* These two layers are separated by the *thermocline*, or *metalimnion*, a region of sharp thermal gradient. This stratification is shown in Fig. 3-4*a*.

Stratification is usually interrupted in autumn (Fig. 3-4*b*) as surface waters cool and begin to sink. Wind action can then cause circulation throughout the entire body of water so that turnover in the lake's strata occurs, stratification disappears, and the body of water reverts to a uniform temperature throughout its depth. In cold regions, surface waters freeze over as winter sets in. Waters at 2°C (36°F), being denser than the colder waters above, form a layer along the bottom, a layer in which the aquatic ecosystem survives as long as sufficient oxygen is available, despite the freezing of the lake's surface (see Fig. 3-4*c*). In spring, the process is reversed as ice melts and turnover occurs (Fig. 3-4*d*), and summer stratification begins as surface waters are warmed by increased solar radiation. [3-27]

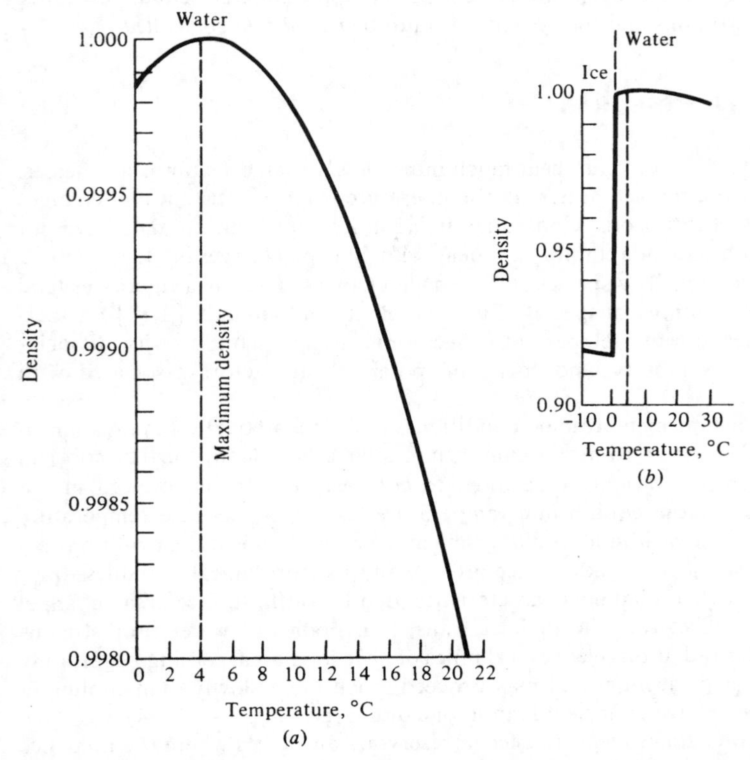

Figure 3-3 Changes in the density of (*a*) water and (*b*) ice with changes in temperature. (*From Warren [3-27].*)

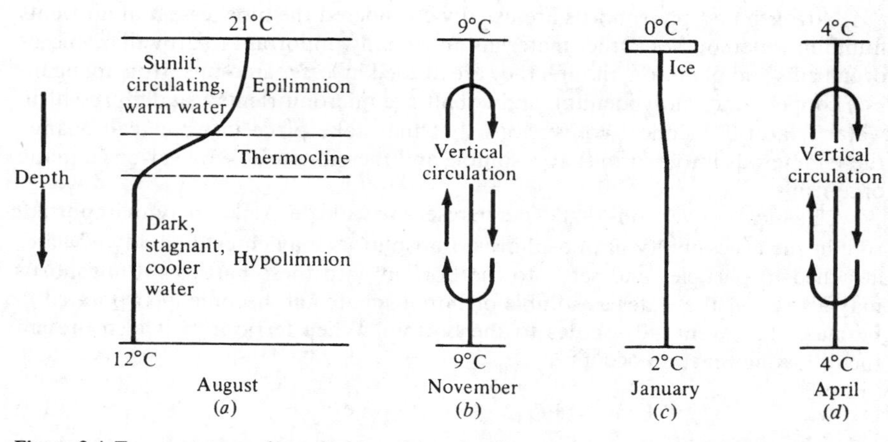

Figure 3-4 Temperature profiles of a deep lake, showing (*a*) thermal stratification; (*b*) autumnal circulation; (*c*) winter stagnation; and (*d*) spring overturn. (*Adapted from Hammer [3-8]*.)

The nature and extent of stratification varies, depending upon the size, depth, configuration, and terrain of the body of water, area-volume-stage relations, orientation of prevailing winds, and hydrologic (or induced) inflow and outflow characteristics, as well as with seasonal variations in temperature. [3-25]

Chemical Processes

Natural watercourses contain many dissolved minerals and gases that interact chemically with one another in complex and varied ways. Oxidation-reduction, dissolution-precipitation, and other chemical conversions may alternately aid or obstruct natural purification processes of natural water systems.

3-6 CHEMICAL CONVERSIONS

Strictly speaking, most of the oxidation-reduction conversions that play a part in self-purification of watercourses are biochemically mediated and will therefore be discussed in subsequent sections of this chapter. Because the solids dissolved in water are essential to the metabolic and reproductive activities of the microorganisms that degrade and stabilize organic wastes, many of those processes are directly or indirectly influenced by the dissolution-precipitation conversions that occur in the watercourses. As certain minerals pass into and out of solution, they become more or less readily available to the microorganisms that rely upon them for the successful completion of their life processes.

Nitrogen and phosphorus are usually considered the most essential nutrients found in watercourses. Other materials are equally important to growth of microorganisms and plankton, though they are needed in lesser amounts. Iron, manganese, copper, zinc, molybdenum, and cobalt are micronutrients usually present in water. Natural chemical conversions that may take place in water can change these materials into a form that is soluble and therefore usable by various aquatic organisms.

Chemical conversions that occur in reservoirs and deep lakes play an important role in the accessibility of phosphorus. Phosphorus may enter the body of water attached to particles and settle to the bottom with these particles. Phosphorus may also enter the water as soluble orthophosphate and become incorporated in biomass that eventually settles to the bottom. When ferric iron is also present the following reaction occurs:

$$Fe^{3+} + PO_4{}^{3-} \longrightarrow FePO_4 \downarrow \tag{3-4}$$

The insoluble ferric phosphate is precipitated and settles to the bottom. There, in the relative absence of oxygen, the iron is reduced to the ferrous form and the ions go into solution. During spring or fall turnover, the phosphorus is mixed throughout the entire depth of the lake, with some of it being used by plant life and some of it recombining with ferric iron and reforming the insoluble ferric phosphate compound, with that precipitate again settling to the bottom to await reduction. [3-19]

Chemical conversions that take place in streams and lakes can help to stabilize the pH of those bodies of water. For example, limestone and other forms of calcium carbonate ($CaCO_3$) dissolve readily in water containing CO_2. [3-28]

$$H_2O + CO_2 \rightleftharpoons H_2CO_3{}^*$$

$$H_2CO_3{}^* \rightleftharpoons H^+ + HCO_3^-$$

The hydrogen ions thus formed react with slightly soluble calcium carbonate to yield highly soluble calcium and more bicarbonate ions.

$$CaCO_3 + H^+ \rightleftharpoons Ca^{2+} + HCO_3^-$$

The bicarbonate acts as a buffer to protect a stream from pH fluctuations that can be harmful to aquatic systems.

Biochemical Processes

Many of the chemical reactions involved in the self-purification process must be biologically mediated. These chemical reactions are not spontaneous but require an external source of energy for initiation. In the case of biodegradable organics and other nutrients, this activation energy can be supplied by microorganisms that utilize these materials for food and energy. The sum total of the processes by which living organisms assimilate and use food for subsistence, growth, and reproduction is called *metabolism*. The metabolic processes and the organisms involved are a vital part of the self-purification process of natural water systems.

3-7 METABOLIC PROCESSES

The biochemical reactions involved in metabolism are extremely complicated and are not yet completely understood. It is known, however, that two types of processes, each involving many steps, must occur simultaneously. One process, called *catabolism*, provides the energy for the synthesis of new cells, as well as for the maintenance of other cell functions. The other process, called *anabolism*, provides the material necessary for cell growth. When an external food source is interrupted, the organisms will use stored food for maintenance energy in a process called *endogenous catabolism*. Each type of microorganism has its own metabolic pathway, from specific reactants to specific end products. A generalized concept of metabolic pathways of importance in natural water systems is shown in Fig. 3-5.

Enzymes play a major role in biochemical reactions. *Enzymes* may be considered as organic catalysts that influence reactions without becoming a reactant themselves. In biochemical processes, enzymes lower the activation energy necessary to initiate reactions. The enzyme then reverts to its original form for reuse. A model of enzyme-substrate (food) reactions is shown in Fig. 3-6. Enzymes are complex protein compounds and are very specific in terms of the reactions that they support. A microorganism thus needs specific enzymes for each reaction in its

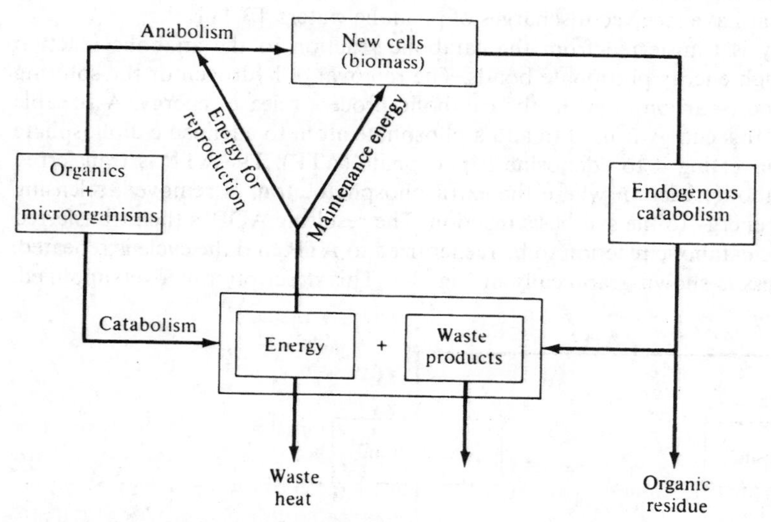

Figure 3-5 Generalized metabolic pathway.

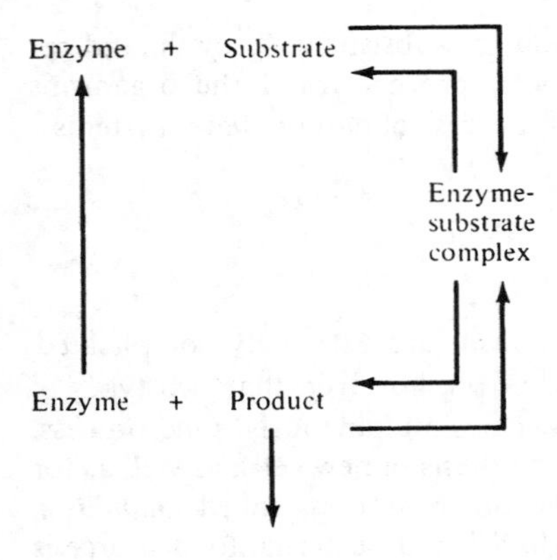

Figure 3-6 Enzyme reaction model.

metabolic pathway. The fact that enzymes are not used up in the metabolic processes is indeed fortunate, as this frees the microorganism to devote its energies and resources to the building of new cellular material rather than to the constant rebuilding of enzymes.

Microorganisms are equipped with enzymes that are especially well suited to the use of particular types of organic matter. When these enzymes are a normal part of a particular microorganism, they are called *constitutive*. Cells produce special enzymes, called *adaptive enzymes*, when they are exposed to unusual, even toxic, substances. This acclimation occurs naturally, though at a relatively slow rate. In many cases, the continued presence of a toxic substance will lead to the gradual development of a specific bacteria capable of decomposing and utilizing that toxic compound. For example, phenol-splitting bacteria are often found in streams that have received discharges of phenolic waters. [3-11]

Energy is transferred from the catabolic reaction to the anabolic reaction through high-energy phosphate bonds. The removal of hydrogen or the splitting of the carbon–carbon bond in the catabolic process releases energy. A sizeable fraction of this energy is used to add a phosphate atom to adenosine diphosphate (ADP), converting it to adenosine triphosphate (ATP). The ATP is transferred to the anabolic reaction where the extra phosphate atom is removed, releasing the stored energy to the synthesis reaction. The resulting ADP is then transferred back to the catabolic reaction to be reenergized to ATP, and the cycle is repeated. This process is shown graphically in Fig. 3-7. This description is oversimplified,

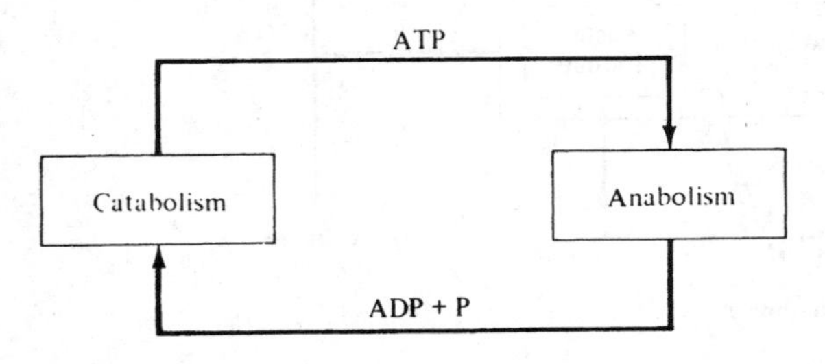

Figure 3-7 Energy transfer model.

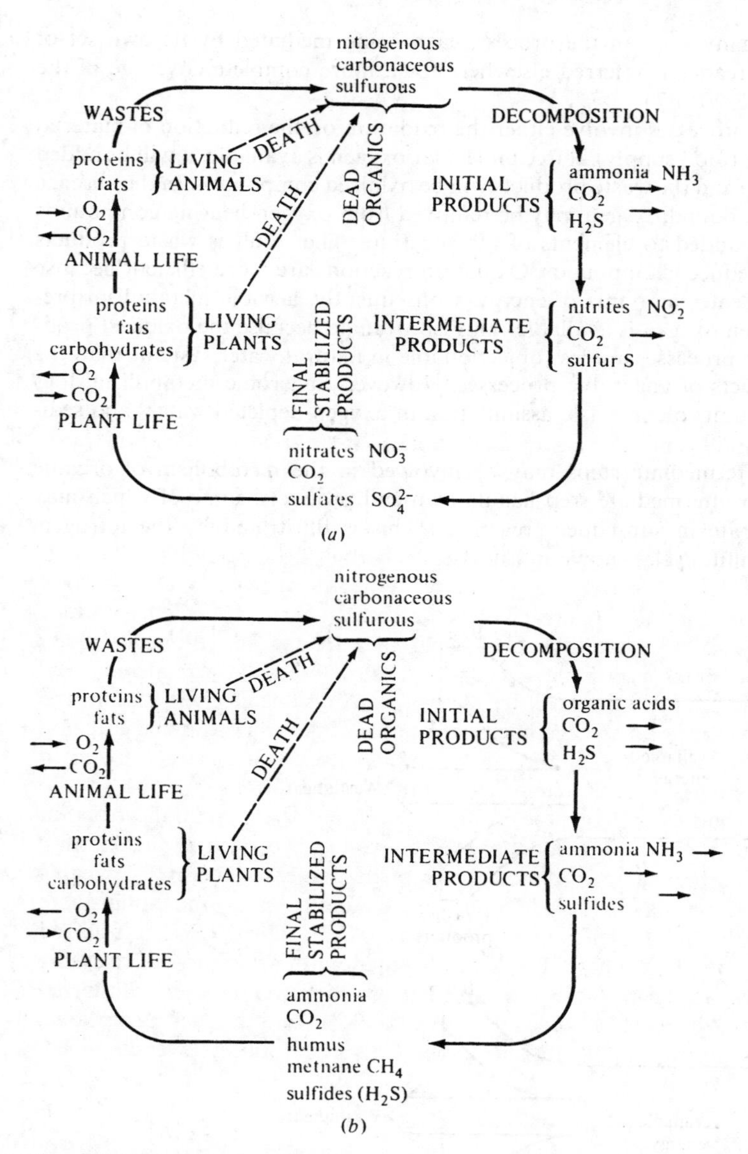

Figure 3-8 Nitrogen, carbon, and sulfur cycles: (*a*) aerobic and (*b*) anaerobic. (*From Vesilind* [3-26].)

as there are many steps in the process, each being mediated by its own set of enzymes. The reader is referred elsewhere for a more complete coverage of the subject. [3-14, 3-20, 3-7]

Catabolic processes involve either the oxidation or the reduction of material in the *substrate* (food supply). If free molecular oxygen is available, it will be added to the substrate and the waste products will be oxidized compounds. In the absence of free oxygen, bound oxygen may be removed from oxygen-bearing compounds and hydrogen added to elements of the substrate. The result is waste products composed of reduced compounds. Oxidation reactions are more efficient because they release greater amounts of energy. Consequently, aerobic metabolism predominates when oxygen is available. This is fortunate because the oxidized products of aerobic processes are less objectionable in natural water systems than the reduced products of anaerobic processes. However, anaerobic metabolism does play an important role in waste assimilation in oxygen-depleted waters and sediment.

Several intermediate steps may be involved in the metabolism of organic material. Each intermediate step has its own end products, some of which may become substrate in subsequent reactions. This is illustrated by the nitrogen, carbon, and sulfur cycles shown in Fig. 3-8.

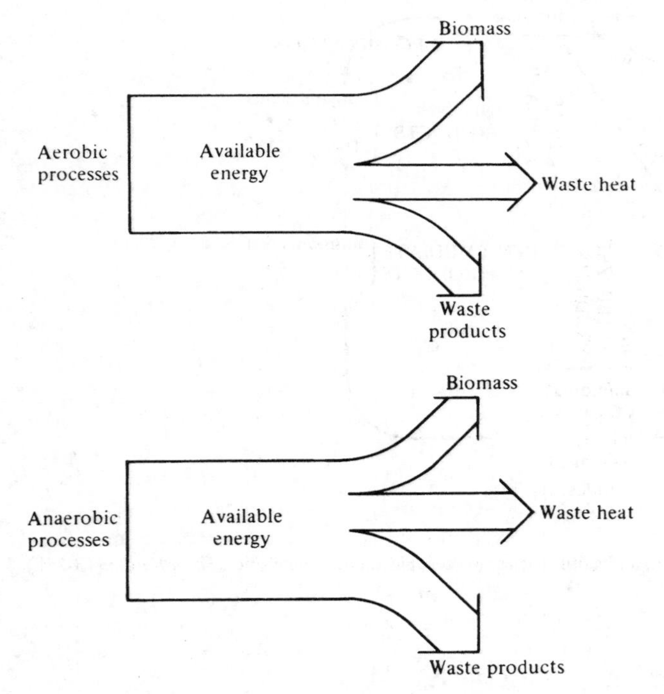

Figure 3-9 Energy balance in metabolism. (*After Steele and McGhee* [3-22].)

Like matter, energy can be neither created nor destroyed. Energy released in the catabolic process is transferred to the cellular material synthesized in the anabolic process, stored in the waste products of catabolism, or released as heat or mechanical energy. The relative quantities dispersed in these ways depend upon the nature of the reaction, as depicted in Fig. 3-9. The end products of aerobic catabolism are low-energy, stable compounds, with most of the energy being stored in the cellular material. By contrast, most of the energy released in anaerobic catabolism remains in the waste products.

3-8 MICROORGANISMS IN NATURAL WATER SYSTEMS

Classical nomenclature divides living organisms into two major subdivisions or kingdoms, plants and animals. The term *protista* is often used to classify organisms in which there is no cell specialization; that is, each cell is capable of carrying out all of the functions of that organism. Members of the protista group are called *protists* and may belong to either the plant or animal kingdom under the classical nomenclature. Most of the organisms of significance in natural purification processes—bacteria, algae, and protozoa—are protists.

Bacteria

Bacteria are the primary decomposers of organic material. Bacteria are single-cell protists that utilize soluble food. Although bacteria may link together into chains or clusters, each cell is an independent organism capable of carrying out all the necessary life functions. [3-14] The structures of bacterial cells typical of natural water systems are illustrated in Fig. 3-10*a*. A listing of the relative abundance of the elements comprising the cell is presented in Table 3-1. The chemical formula for bacterial cells is assumed to be $C_5H_7O_2N$. [3-15]

Energy for bacterial growth and reproduction may be derived from the biochemical oxidation of inorganic or organic compounds, or from the reduction of these compounds. A few bacteria are able to utilize ultraviolet energy from sunlight. Material sources can be derived from either organic or inorganic compounds.

Bacteria are often classified according to the energy and material sources that they require. Organisms that derive both energy and material from inorganic sources are called *autotrophs*, while bacteria that obtain both energy and material from organic compounds are called *heterotrophs*. *Phototrophs*, bacteria which utilize sunlight for an energy source and inorganic substances for a material source, play an insignificant role in the natural water purification processes.

Heterotrophic bacteria are the most important species in the degradation of organic material. *Aerobic heterotrophs* require oxygen in their metabolic processes while *anaerobic heterotrophs* utilize organics in the absence of oxygen. A third group, called *facultative heterotrophs*, function as aerobes when oxygen is present but switch to anaerobic processes when oxygen becomes unavailable. A major

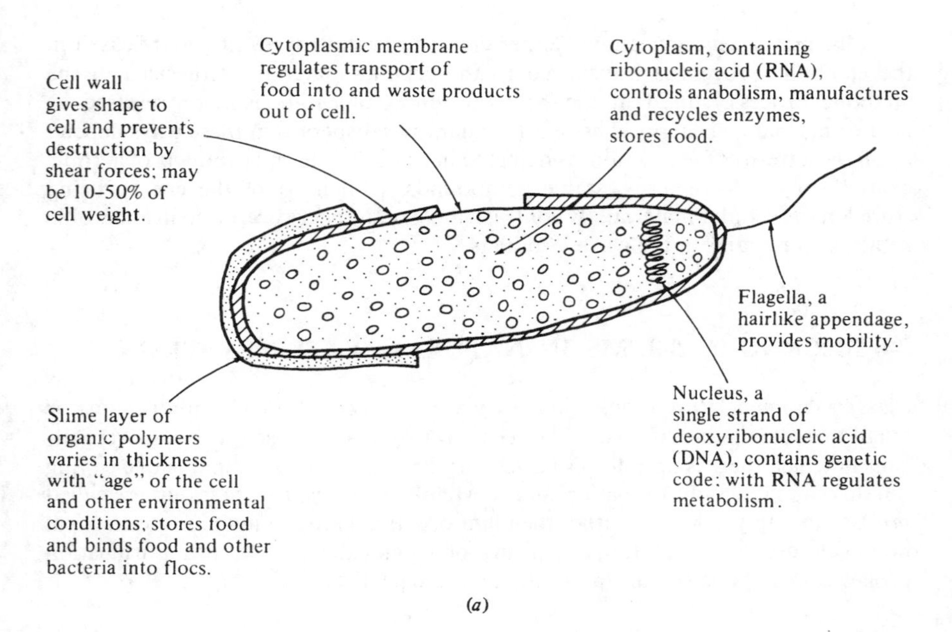

Cytoplasmic membrane regulates transport of food into and waste products out of cell.
Cytoplasm, containing ribonucleic acid (RNA), controls anabolism, manufactures and recycles enzymes, stores food.
Cell wall gives shape to cell and prevents destruction by shear forces; may be 10–50% of cell weight.
Flagella, a hairlike appendage, provides mobility.
Nucleus, a single strand of deoxyribonucleic acid (DNA), contains genetic code; with RNA regulates metabolism.
Slime layer of organic polymers varies in thickness with "age" of the cell and other environmental conditions; stores food and binds food and other bacteria into flocs.

(a)

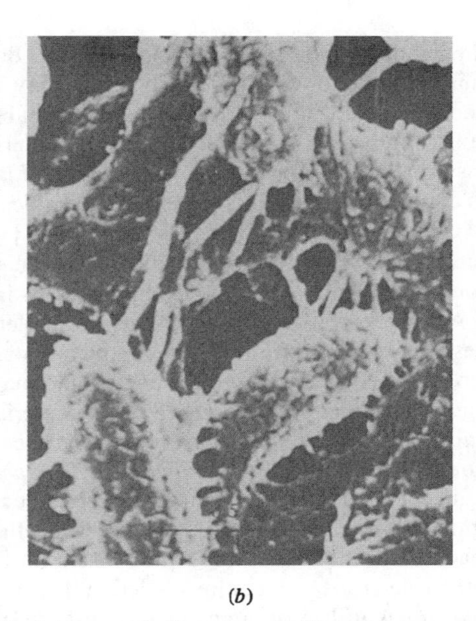

(b)

Figure 3-10 (*a*) Generalized structure of a bacterial cell; (*b*) photomicrograph of freshwater bacteria attached to a surface. Threadlike materials are extracellular polymers that bind the organisms together and to the surface (*photo courtesy of W. G. Characklis*).

Table 3-1 Bacterial cell composition

Element	Dry weight, %
Carbon	50
Oxygen	20
Nitrogen	14
Hydrogen	8
Phosphorus	3
Sulfur	1
Potassium	1
Sodium	1
Calcium	0.5
Magnesium	0.5
Chlorine	0.5
Iron	0.2
All others	0.3

Source: From Gaudy and Gaudy. [3-7]

function of autotrophic bacteria is the oxidation of nitrogen and sulfur compounds to stable end products.

Protozoa

Like bacteria, protozoa are single-cell organisms that reproduce by binary fission. Unlike bacteria, protozoa ingest solid organics for food. Since protozoa are one to two orders of magnitude larger than bacteria, the protozoa diet often includes bacterial cells as well as colloidal organics. There are many aquatic species of protozoa, most of which are strict aerobes. Like heterotrophic bacteria, they obtain both energy and material for growth and reproduction from the same organic food source.

The most important protozoal group in natural water systems is the *ciliata*. These organisms are characterized by hairlike appendages called *cilia* and may be either free-swimming or stalked (attached to a solid particle), as illustrated in Fig. 3-11. The free-swimming protozoa use a rapid movement of their cilia to propel themselves through the water in search of food. The stalked protozoa use their cilia to bring food in from the surrounding water. Protozoa are voracious consumers of organic material and are important members of the aquatic community.

Algae

Algae are autotrophic, photosynthetic organisms and, even though they do not utilize organic compounds directly, play a significant role in the natural purification process. In the presence of sunlight, algae metabolize the waste products of heterotrophic bacteria (CO_2, NO_3^-, PO_4^{3-}, etc.) while obtaining energy from sunlight.

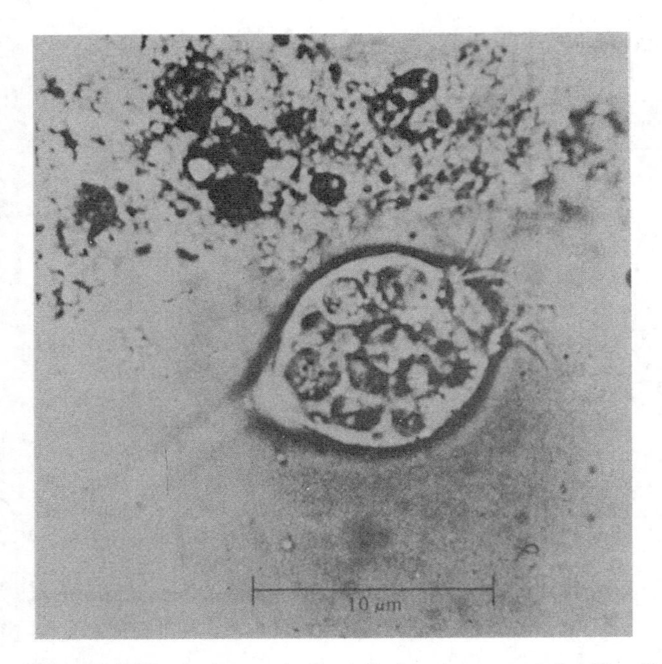

Figure 3-11 Photomicrograph of a stalked protozoan: note hairlike cilia used in the food-gathering process.

One of the waste products of this reaction is oxygen. When sunlight is not available, the algae catabolize stored food for energy and use oxygen in the process. This diurnal nature of algal catabolism is an important factor in the oxygen balance of natural waters that are nutrient-rich.

There are literally thousands of species of algae of various sizes, shapes, and colors. Algal cells may be found in clusters, in long filaments attached to banks or bottom material, or may remain as single cells suspended in water. Some species of algae can have negative effects on water quality because they produce oily substances that cause taste and odor problems.

Other Organisms

Other microorganisms may also play important roles in the natural purification process. *Rotifers* and *crustacea* are lower-order animals that prey on bacteria, protozoa, and algae. They help to maintain a balance in the populations of primary producers and serve as an important link in the chain by which organic materials are passed on to higher-order animals. *Sludge worms* such as tubifex and bloodworms, as well as other helminths and insect larvae, feed on sludge deposits and help to break down and solubilize the particulate organics.

Response of Streams To Biodegradable Organic Waste

The self-purification of natural water systems is a complex process that often involves physical, chemical, and biological processes working simultaneously. Chemical and biochemical reactions are conversion processes rather than removal processes. The nature, and perhaps phase, of the waste may be changed, but the products remain in the water until physical processes remove them from suspension by sedimentation or by transfer to the atmosphere. This is illustrated by the reaction in Eq. (3-4). Here chemical processes combine iron and phosphate into solid form, and the physical process of sedimentation removes it from suspension. Another example is the metabolism of organics by microorganisms. Biochemical reactions convert the organics to biological solids and other end products that may be recycled several times (Fig. 3-8) before ultimately being incorporated into bottom sediments or released as gases to the atmosphere, both by physical processes.

The self-purification processes can be modeled, provided the waste characteristics and the system variables of the water body are known. The modeling process is complicated in lakes and estuaries by dilution and dispersion characteristics that are variable with time. A complete discussion of water-quality modeling is beyond the scope of this text. However, some examples will be used to illustrate the interaction of the physical, chemical, and biochemical processes described earlier. The examples chosen relate to the assimilation of organic material by streams and the resulting effects on the oxygen balance and the ecosystem. Although these topics are the ones most frequently covered in the literature on natural purification processes, the reader should be aware that other self-purification processes, with respect to other contaminants, may be of equal importance.

3-9 DISSOLVED-OXYGEN BALANCE

Dissolved oxygen is one of the most important constituents of natural water systems. Fish and other aquatic animal species require oxygen, and a stream must have a minimum of about 2 mg/L of dissolved oxygen to maintain higher life forms. At least 4 mg/L of dissolved oxygen is required for game fish and some species may require more. In addition to this life-sustaining aspect, oxygen is important because the end products of chemical and biochemical reactions in anaerobic systems often produce aesthetically displeasing colors, tastes, and odors in water.

When biodegradable organics are discharged to a stream containing dissolved oxygen, microorganisms begin the metabolic processes that convert the organics, along with the dissolved oxygen, into new cells and oxidized waste products. The quantity of oxygen required for this conversion is the biochemical oxygen demand discussed in Sec. 2-13. The rate at which the dissolved oxygen is used will depend on the quantity of the organics, the ease with which they are bic-degraded, and the dilution capacity of the stream.

The dissolved oxygen that is used from the stream must be replaced or anaerobic conditions will develop. Two mechanisms are known to contribute oxygen to surface waters: (1) dissolution of oxygen from the atmosphere, often called reaeration, and (2) production of oxygen by algal photosynthesis.

Reaeration

The principles of equilibrium between water and gas in contact with each other are described in Sec. 3-4. Equilibrium concentrations of oxygen in water at various temperatures and salinity values are given in Table C-3 of the appendix. When concentrations of dissolved oxygen drop below the equilibrium value, the net movement of oxygen will be from the atmosphere into the water. The difference between the equilibrium concentration and the actual concentration is called the *oxygen deficit* and is represented mathematically by

$$D = C_s - C \tag{3-5}$$

where D is the dissolved oxygen deficit and C_s and C are the equilibrium concentration and actual oxygen concentration, respectively. The units of all the terms are milligrams per liter of oxygen. For constant equilibrium conditions, i.e., C_s does not change, the rate of change in the deficit is

$$\frac{dD}{dt} = -\frac{dC}{dt} \tag{3-6}$$

The deficit thus increases at the same rate that the oxygen is used up.

The dissolved oxygen deficit is the driving force for reaeration. The greater the deficit, the greater the rate of reaeration. It follows, then, from Eq. (3-6) that the rate of reaeration increases as the concentration of dissolved oxygen decreases.

Algal Photosynthesis

In the presence of sunlight, algae metabolize inorganic compounds, with one of the waste products being oxygen. The following formula is a simplified representation of this reaction.

$$CO_2 + 2H_2O \xrightarrow{\text{Light}} \underbrace{CH_2O}_{\substack{\text{New}\\ \text{algal}\\ \text{cells}}} + O_2 + H_2O \tag{3-7}$$

The oxygen thus released is immediately available to replenish the dissolved oxygen in the water. In the presence of excessive nutrients and bright sunlight, algal metabolism may produce so much oxygen that the water becomes supersaturated. That is, $C > C_s$ and the deficit has a negative value.

Adverse factors associated with excessive algal growths often outweigh the benefits of the oxygen they produce. Because algae use the waste products from bacterial metabolism, major algal activity usually occurs downstream from, rather than within, the area of greatest bacterial activity where the oxygen is needed the most. Also, in the absence of light, algae obtain energy from endogenous catabolism represented by the following reaction.

$$CH_2O + O_2 \longrightarrow CO_2 + H_2O \tag{3-8}$$

This reaction contributes to the oxygen demand rather than to the oxygen supply of the stream.

The difference in algal catabolism during light and dark periods results in diurnal variations in the dissolved oxygen in streams with heavy algal growths. The dissolved-oxygen concentration often peaks around 2 to 4 P.M., with the lowest levels occurring just before sunrise. Unfortunately, the excess oxygen generated during the day cannot be stored for use during the night, as it is expelled to the atmosphere to maintain equilibrium. In cases where the algal growth is heavy, the endogenous catabolism may deplete the dissolved oxygen to the point where fish kills occur.

Because of the variability of photosynthetically produced oxygen, reaeration is considered the most dependable source of dissolved oxygen. It may be necessary however, to include photosynthetic oxygen in a dissolved-oxygen model for waters where algal growths are heavy.

3-10 DISSOLVED-OXYGEN MODEL

Most all of the dissolved-oxygen models in current use relate in some way to the model developed by Streeter and Phelps in 1925. This model predicts changes in the deficit as a function of BOD exertion and stream reaeration.

Rate of Oxygen Removal

The rate at which dissolved oxygen disappears from the stream coincides with the rate of BOD exertion. Therefore

$$\frac{dy}{dt} = -\frac{dC}{dt} \tag{3-9}$$

Substituting into Eq. (3-6)

$$\frac{dy}{dt} = \frac{dD}{dt} \tag{3-10}$$

confirming that an increase in the rate of BOD exertion results in an increase in the rate of change of oxygen deficit. In Sec. 2-13, it was shown that

$$y = L_0 - L_t$$

Because L_0 is the ultimate BOD and therefore a fixed value,

$$\frac{dy}{dt} = -\frac{dL_t}{dt} \tag{3-11}$$

Recalling Eq. (2-20)

$$\frac{dL_t}{dt} = -kL_t$$

and making appropriate substitutions in Eqs. (3-11) and (3-10), the following relationship is obtained

$$\frac{dD}{dt} = kL_t \tag{3-12}$$

which states that the rate of change in the dissolved oxygen deficit at time t due to the BOD is a first-order reaction proportional to the oxygen equivalent of the remaining organics. A more convenient form of Eq. (3-12) is

$$r_D = k_1 L_t \tag{3-13}$$

where r_D replaces the differential form as the rate of change in the oxygen deficit due to oxygen utilization. The reaction rate constant k_1 is the same parameter described in Sec. 2-13 and is derived from laboratory tests on the wastewater. The rate constant is adjusted for temperature changes, but is not usually adjusted for other effects of dilution with the stream water.

Rate of Oxygen Addition

As noted in Sec. 3-9, the rate of reaeration is a first-order reaction with respect to the magnitude of the oxygen deficit. This is expressed mathematically by

$$r_R = -k_2 D \tag{3-14}$$

where r_R is the rate at which oxygen becomes dissolved from the atmosphere, D is the oxygen deficit defined by Eq. (3-5), and k_2 is a reaeration rate constant that is system-specific. The negative sign reflects the fact that an increase in the oxygen supply due to reaeration reduces the oxygen deficit. Factors affecting k_2 include stream turbulence (a function of velocity and channel characteristics), surface area, water depth, and temperature. Temperature corrections are made by Eq. (2-23) with a value of 1.016 for θ being most common. Several models are available for determining numerical values for k_2, [3-16, 3-4], the development of which is beyond the scope of this text. A range of values typically found applicable to various flow regimes is given in Table 3-2.

The Oxygen Sag Curve

The oxygen deficit in a stream is a function of both oxygen utilization and reaeration. Inspection of Eqs. (3-13) and (3-14) shows that these two processes have

Table 3-2 Reaeration constants

Water body	Ranges of k_2 at 20°C, base *e*
Small ponds and backwaters	0.1–0.23
Sluggish streams and large lakes	0.23–0.35
Large streams of low velocity	0.35–0.46
Large streams of normal velocity	0.46–0.69
Swift streams	0.69–1.15
Rapids and waterfalls	Greater than 1.15

Source: After Metcalf & Eddy, Inc. [3-15]

opposite effects on the deficit. This is shown graphically in Fig. 3-12. The rate of change in the deficit is the sum of the two reactions

$$\frac{dD}{dt} = r_D + r_R$$

$$= k_1 L_t - k_2 D \qquad (3\text{-}15)$$

The actual oxygen concentration $(C_s - D_t)$ has a characteristic dip as shown in Fig. 3-12, resulting in the term *oxygen sag curve*, commonly used to describe the process.

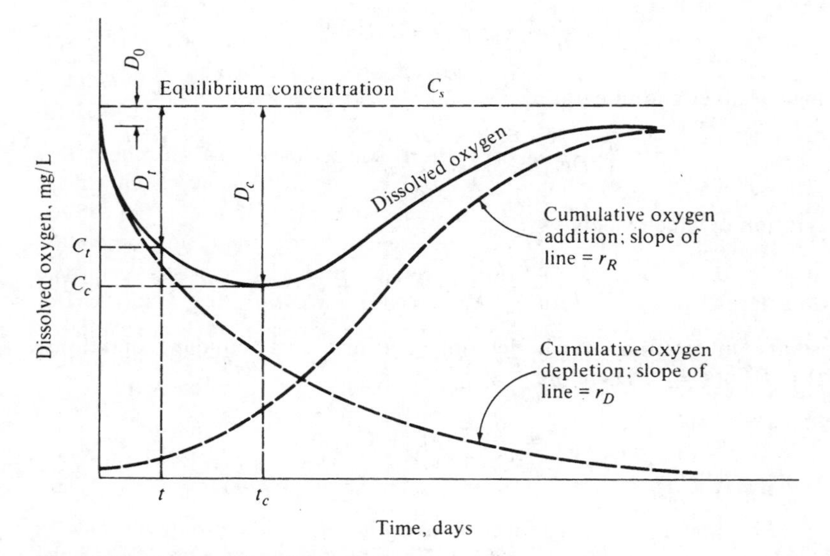

Figure 3-12 Characteristics of the oxygen sag curve.

The oxygen deficit, and therefore the oxygen concentration, at any point in time after the discharge can be determined by integrating Eq. (3-15). This is not, however, a straightforward operation. Recalling from Eq. (2-21) that

$$L_t = L_0 e^{-k_1 t}$$

and rearranging Eq. (3-15), the following equation is obtained

$$\frac{dD}{dt} + k_2 D = k_1 L_0 e^{-k_1 t} \tag{3-16}$$

which is a first-order differential equation of the form

$$\frac{dy}{dx} + Py = Q$$

where P and Q are functions of x. [3-1] The use of the integrating factor $\exp(\int P\,dx)$ is necessary for the solution of this type equation. For Eq. (3-16), the integrating factor is

$$e^{\int k_2 dt} = e^{k_2 t} \tag{3-17}$$

Multiplying both sides of Eq. (3-16) by the right side of Eq. (3-17) yields

$$e^{k_2 t}\frac{dD}{dt} + k_2 D e^{k_2 t} = k_1 L_0 e^{(k_2 - k_1)t} \tag{3-18}$$

The left side of this equation can be factored as follows

$$e^{k_2 t}\frac{dD}{dt} + k_2 D e^{k_2 t} = \frac{d}{dt} D e^{k_2 t}$$

Separating variables and integrating

$$\int dDe^{k_2 t} = k_1 L_0 \int e^{(k_2 - k_1)t}\,dt$$

The integration of which yields

$$De^{k_2 t} = \frac{k_1 L_0}{k_2 - k_1}(e^{(k_2 - k_1)t}) + C \tag{3-19}$$

The constant of integration C can be determined from known boundary conditions, that is, $D = D_0$ at $t = 0$. Therefore

$$D_0 = \frac{k_1 L_0}{k_2 - k_1} 1 + C$$

and

$$C = D_0 - \frac{k_1 L_0}{k_2 - k_1}$$

The final solution becomes

$$De^{k_2 t} = \frac{k_1 L_0}{k_2 - k_1}\left(e^{(k_2 - k_1)t} + D_0 - \frac{k_1 L_0}{k_2 - k_1}\right)$$

or

$$D = \frac{k_1 L_0}{k_2 - k_1}\left(\frac{e^{(k_2 - k_1)t}}{e^{k_2 t}}\right) - \frac{k_1 L_0}{(k_2 - k_1)e^{k_2 t}} + \frac{D_0}{e^{k_2 t}}$$

and in final form

$$D = \frac{k_1 L_0}{k_2 - k_1}(e^{-k_1 t} - e^{-k_2 t}) + D_0 e^{-k_2 t} \tag{3-20}$$

In this equation, t represents the time of travel in the stream from the point of discharge and is the only independent variable in the equation. The time of travel from the point of discharge to any given downstream point is:

$$t = \frac{x}{u} \tag{3-21}$$

where x is the distance along the stream and u is the stream velocity. The units of t must always be days. Substituting values for t, or x/u, into Eq. (3-20), will result in a value of D for that point in the stream.

The most important point on the oxygen sag curve is often the point of lowest concentration because this point represents the maximum impact on the dissolved oxygen due to wastewater discharge. This point is called the *critical deficit* D_c, and the time of travel to this point is termed the *critical time* t_c. Recognizing that the rate of change of the deficit is zero at the maximum deficit, an expression for D_c can be found from Eq. (3-16).

$$0 + k_2 D_c = k_1 L_0 e^{-k_1 t_c}$$

or

$$k_2 D_c = k_1 L_0 e^{-k_1 t_c}$$

and

$$D_c = \frac{k_1}{k_2} L_0 e^{-k_1 t_c} \tag{3-22}$$

The solution of this equation depends on a numerical value for t_c, which is somewhat more difficult to obtain. First, Eq. (3-20) is differentiated and set equal to zero, again because D_c is a maximum at t_c.

$$0 = \frac{k_1 L_0}{k_2 - k_1}(-k_1 e^{-k_1 t_c} + k_2 e^{-k_2 t_c}) - k_2 D_0 e^{-k_2 t_c}$$

Dividing through by $e^{-k_2 t_c}$

$$0 = \frac{k_1 L_0}{k_2 - k_1}(-k_1 e^{(k_2 - k_1)t_c} + k_2) - k_2 D_0$$

rearranging

$$k_2 D_0 \frac{k_2 - k_1}{k_1 L_0} = k_2 - k_1 e^{(k_2 - k_1)t_c}$$

and

$$k_1 e^{(k_2 - k_1)t_c} = k_2 - D_0 \frac{k_2}{k_1} \frac{k_2 - k_1}{L_0}$$

dividing through by k_1 and taking the logarithm of both sides

$$(k_2 - k_1)t_c = \ln \left(\frac{k_2}{k_1} - D_0 \frac{k_2}{k_1^2} \frac{k_2 - k_1}{L_0} \right)$$

or in more conventional form

$$t_c = \frac{1}{k_2 - k_1} \ln \left[\frac{k_2}{k_1} \left(1 - D_0 \frac{k_2 - k_1}{k_1 L_0} \right) \right] \tag{3-23}$$

Equations (3-22) and (3-23) can be used to determine the critical oxygen level in the stream and the position at which it occurs. The procedure is illustrated in the following example.

Example 3-3: Applying the BOD sag curve A municipal wastewater-treatment plant discharges secondary effluent to a surface stream. The worst conditions are known to occur in the summer months when stream flow is low and water temperature is high. Under these conditions, measurements are made in the laboratory and in the field to determine the characteristics of the wastewater and stream flows. The wastewater is found to have a maximum flow rate of 15,000 m^3/day, a BOD_5 of 40 mg/L, a dissolved oxygen concentration of 2 mg/L, and a temperature of 25°C. The stream (upstream from the point of wastewater discharge) is found to have a minimum flow rate of 0.5 m^3/s, a BOD_5 of 3 mg/L, a dissolved oxygen concentration of 8 mg/L, and a temperature of 22°C. Complete mixing of the wastewater and stream is almost instantaneous, and the velocity of the mixture is 0.2 m/s. From the flow regime, the reaeration constant is estimated to be 0.4 day^{-1} for 20°C conditions.

Sketch the dissolved oxygen profile a 100-km reach of the stream below the discharge.

SOLUTION

1. Determine characteristics of wastewater-stream mixture.

a.

$$Q_w = 15{,}000 \text{ m}^3/\text{d} \times \frac{1 \text{ d}}{24 \text{ h}} \times \frac{1 \text{ h}}{60 \text{ min}} \times \frac{1 \text{ min}}{60 \text{ s}}$$

$$= 0.17 \text{ m}^3/\text{s}$$

$$Q_{mix} = 0.17 + 0.5 = 0.67 \text{ m}^3/\text{s}$$

b. BOD [Eq. (3-1)]:

$$y_{mix} = \frac{y_s Q_s + y_w Q_w}{Q_s + Q_w}$$

$$= \frac{3.0 \times 0.5 + 40 \times 0.17}{0.67}$$

$$= 12.4 \text{ mg/L}$$

Convert to ultimate BOD. (Assume $k_1 = 0.23$ for mixture.)

$$y_u = L_0 = \frac{y}{1 - e^{-k_1 t}}$$

$$= \frac{12.4}{1 - e^{-0.23 \times 5}}$$

$$= 18.2 \text{ mg/L}$$

c. Dissolved oxygen:

$$DO_{mix} = \frac{8.0 \times 0.5 + 2.0 \times 0.17}{0.67}$$

$$= 6.5 \text{ mg/L}$$

d. Temperature:

$$T_{mix} = \frac{22 \times 0.5 + 25 \times 0.17}{0.67}$$

$$= 22.8°\text{C}$$

2. Correct reaction constants for temperature.
   a. BOD reaction rate [Eq. (2-23)]:

$$k_{22.8} = k_{20}(1.047^{22.8-20})$$

$$= 0.23 \times 1.14$$

$$k_{22.8} = 0.26 \text{ d}^{-1} = k_1$$

   b. Stream reaeration rate

$$k_{22.8} = k_{20}(1.016^{22.8-20})$$

$$= 0.4 \times 1.05$$

$$k_{22.8} = 0.42 \text{ d}^{-1} = k_2$$

3. Determine initial oxygen deficit D_0.
   a. At $T = 22.8$, the equilibrium concentration of oxygen in fresh water is 8.7; therefore

$$D_0 = 8.7 - 6.5 = 2.2 \text{ mg/L}$$

4. Determine the critical deficit and its location.

a.
$$t_c = \frac{1}{k_2 - k_1} \ln\left[\frac{k_2}{k_1}\left(1 - D_0 \frac{k_2 - k_1}{k_1 L_0}\right)\right] \tag{3-23}$$

$$= \frac{1}{0.42 - 0.26} \ln\left[\frac{0.42}{0.26}\left(1 - 2.2 \frac{0.42 - 0.26}{0.26 \times 18.2}\right)\right]$$

$$t_c' = 2.5 \text{ d}$$

b.
$$D_c = \frac{k_1}{k_2} L_0 e^{-k_1 t_c}$$

$$= \frac{0.26}{0.42} 18.2 e^{-0.26 \times 2.5}$$

$$= 5.9 \text{ mg/L}$$

c. This condition will occur at a distance of

$$x = 0.2 \text{ m/s} \times 86{,}400 \text{ s/d} \times 2.5 \text{ d}$$
$$= 43.2 \text{ km downstream from point of discharge}$$

5. Determine the deficit at points 20, 75. and 100 km from the point of discharge.

a.
$$t = \frac{x \text{ km}}{u \text{ km/d}}$$

$$u = 0.2 \text{ m/s} \times \frac{86{,}400 \text{ s}}{\text{d}} \times \frac{1 \text{ km}}{1000 \text{ m}} = 17.3 \text{ km/d}$$

$$t_{20} = 20/17.3 = 1.16 \text{ d}$$

$$t_{75} = 75/17.3 = 4.3 \text{ d}$$

$$t_{100} = 100/17.3 = 5.8 \text{ d}$$

b. The deficits at these times are:

$$D = \frac{k_1 L_0}{k_2 - k_1}(e^{-k_1 t} - e^{-k_2 t}) + D_0 e^{-k_2 t} \tag{3-20}$$

$$D_{20} = \frac{0.26 \times 18.2}{0.42 - 0.26}(e^{-0.26 \times 1.16} - e^{-0.42 \times 1.16}) + 2.2 e^{-0.42 \times 1.16}$$

$$= 5.1 \text{ mg/L}$$

$$D_{75} = 5.2 \text{ mg/L}$$

$$D_{100} = 4.1 \text{ mg/L}$$

6. The dissolved-oxygen concentrations at each point are found to be:

$$C_{20} = 8.8 - 5.1 = 3.4 \text{ mg/L}$$

$$C_{43.2} = 2.8 \text{ mg/L}$$

$$C_{75} = 3.5 \text{ mg/L}$$

$$C_{100} = 4.1 \text{ mg/L}$$

These points are connected by a smooth curve as shown in the accompanying figure to yield the desired oxygen profile of the stream.

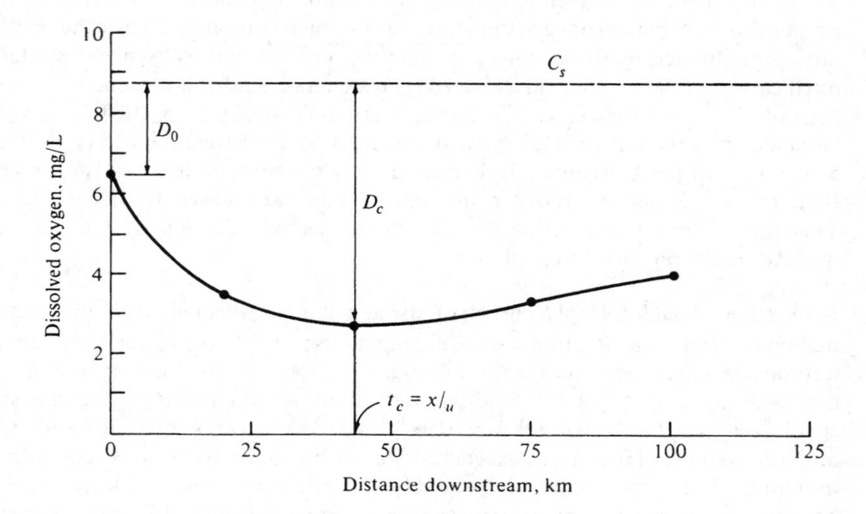

Both the position and magnitude of the critical deficit are related to the system variables (k_1, k_2, L_0, D_0, and u). The time of travel to the critical deficit (t_c) is influenced more strongly by the values of k_1 and k_2, while the magnitude of the deficit is most affected by the L_0 value. Not only do heavier loads result in greater deficits, but they extend the influence of the waste farther downstream. Heavy loads of organics may result in the development of anaerobic conditions. Under these conditions, oxygen is transferred in at a high rate [Eq. (3-14)] but is used up by facultative organisms that may also be utilizing the organic material produced by anaerobic metabolism. In a deep stream, true anaerobic organisms may flourish near the bottom. Only after the strength of the waste has been sufficiently reduced will aerobic conditions be restored. Since anaerobic metabolism is a slow process, recovery of an overloaded stream will be slow and the oxygen sag will extend far downstream.

Limitations of the Oxygen Sag Curve

The limitations of the oxygen sag curve should be at once apparent. The rate of deoxygenation and the rate of reaeration are each affected by many variables for which the model makes no allowance.

BOD variables The equation is based on the assumption that there is one source of BOD when there may actually be several different point or nonpoint sources of BOD. Additional discharges can be taken into consideration by subdividing a river into short reaches, each fed by a single point source. If tributaries empty into the mainstream, any discharge they may have received must also be taken into consideration, as well as the increase in flow of the receiving stream.

Even when care is given to consider all organic loads introduced at discharge points, the biochemical oxygen demand of a stream may be affected by other factors not approximated by the k_1 constant. Algal respiration in the absence of sunlight, nitrification processes that increase oxygen demand, and the presence of sludge deposits in pool areas can all increase a stream's BOD. [3-5, 3-8] In shallow streams, masses of microbial growth attached to the streambed may be more efficient at utilizing organics, and consequently consume more dissolved oxygen than the suspended microorganisms used in the laboratory BOD test. [3-21] This fact, though recognized and valued in engineered treatment systems, is often ignored in self-purification studies.

Reaeration variables Replacement of oxygen is also affected by many factors not taken into consideration by the formulas used to derive oxygen sag curves, notably the reaeration contribution of algae photosynthesis. Further, the mathematics assumes steady-state conditions all along a river channel. Because such steady-state conditions would indeed be rare, most streams must be subdivided and a k_2 value assigned to each reach. Even with subdivision into reaches, determination of the k_2 constant is probably the one area most prone to error in oxygen-sag-curve work, because no theoretical assumption of flow characteristics —channel formation, obstacles, pools, effects of impoundments, and other such variables—is likely to fit any one particular stream perfectly.

Additions have been made to the basic Streeter–Phelps model that incorporate the diurnal effect of algal photosynthesis, the nitrification process, and the sedimentation-resuspension of organic material. These models are presented elsewhere in the literature [3-22] and require a much more sophisticated data base for use.

Confirmation of the Oxygen Sag Curve

The dissolved-oxygen profile obtained from mathematical models should be confirmed by actual field measurements. Ideally, there should be a comprehensive sampling under conditions of known waste loads and river hydrology. A period of warm weather and low flows is desirable, and daily sampling for 1 month for all parameters is preferred. Once the DO deficit and the time to the critical O_2 concentration have been verified by a detailed water-quality survey, oxygen sag curves can be used to forecast stream conditions that can be expected for given waste loads and stream flows.

3-11 ORGANIC DISCHARGE AND STREAM ECOLOGY

In addition to variations in the oxygen concentrations, many other physical, chemical, and biological changes occur in streams after the discharge of biodegradable organic material. Together with the oxygen supply, these processes and their products greatly influence the *ecology* (the relationship between living organisms and their environment) of the stream. Like the oxygen balance, the ecological balance of a stream receiving a biodegradable organic discharge can be modeled. Most of the models assume that the organic waste is composed primarily of municipal wastewater and does not contain significant quantities of materials that would be toxic to the flora and fauna of the stream.

Ecological modeling usually involves dividing the stream into reaches, or zones, in which certain species or certain processes predominate. The model most commonly used in the United States is the one devised by Whipple, Fair, and Whipple. This model divides the stream into four zones labeled the zones of *degradation, active decomposition, recovery,* and *clean water.* A summary of the physical, chemical, and biological characteristics of each zone is presented in Table 3-3.

Many of the physical characteristics described in Table 3-3 may be noted by the casual observer, but the chemical characteristics (with the exception of the presence of highly odorous H_2S) can be determined only through sampling and laboratory testing. Biological species and numbers are markedly different from zone to zone, and species diversity is a primary means of establishing zone boundaries. The change in species and numbers of organisms in each species is illustrated in Fig. 3-13.

The food supply is a primary factor in determining the type of organisms that predominate. Near the point of discharge, bacteria, protozoa, and molds predominate. Bacteria find an abundant food supply in the form of carbohydrates, proteins, and fats. As these microorganisms decompose organic wastes, they convert them into nutrient materials such as nitrates, phosphates, and carbon dioxide. The bacterial populations flourish until dissolved oxygen and/or the food supply is exhausted. Because bacteria provide food for protozoa, ciliates, rotifers, and crustaceans, these higher forms of life diminish as bacteria die off.

The abundant supply of nutrient materials made available by the bacterial decomposition of organic matter brings about still further changes. About midway through the zone of active decomposition, where mineral nutrients (notably nitrates) abound, algae begin a rapid increase. Blue-green (*Phormidium, Lyngbya,* and *Oscillatoria*) and green algae (*Spirogyra* and *Stigeoclonium*), and diatoms (*Gomphonema* and *Nitzschia*) may be present in this zone. [3-3]

In the zone of recovery, algae growth peaks, then declines, with algal populations in the clean zone beginning to approximate those found in the predischarge portions of the stream. Blue-green (*Microcystis* and *Anabaena*), pigmented flagellates (*Euglena* and *Pandorina*), green algae (*Cladophora* and *Ankistrodesmus*), and diatoms (*Meridion* and *Cyclotella*) are species found in the zone of recovery. [3-3]

As nutrient loads decline, BOD decreases, and DO levels return to their predischarge levels, algae and bacteria populations return to their clean-water status, and clean-water invertebrate and vertebrate fauna again populate the stream. At this point, the stream's natural self-purification process has essentially been completed, but only insofar as biodegradable organic wastes are concerned.

Table 3-3 Whipple, Fair, and Whipple model for zones of stream self-purification

Zone	Physical characteristics	Chemical characteristics	Biological characteristics
Degradation (Zone 2 in Fig. 3-13)	The water is turbid; there are sludge deposits and floating debris	Oxygen is reduced to about 40% of saturation.	Fish and green algae are declining; littoral forms of green and blue-green algae are trailing from frequently wetted stones. These include *Stigeoclonium*, *Oscillatoria*, and *Ulothrix*. Bottom forms in sludge include reddish worms (Tubificidae) similar to earthworms, such as *Tubifex* and *Limnodrilus*. Water fungi are typically white, olive green, putty gray, rusty brown. *Sphaerotilus natans*, *Leptomitus*, and *Achlya* appear, as do ciliated protozoa or ciliata such as *Carchesium*, *Epistylis*, and *Vorticella*.
Active decomposition (Zone 3 in Fig. 3-13)	Water is grayish and darker than in degradation zone; scum may form; septic conditions may have set in.	Oxygen level moves between 40% of saturation and zero; then as active decomposition diminishes, oxygen content rises. Methane, hydrogen, and sulfide are given off.	Bacteria flora flourish; anaerobes displace aerobes, which eappear toward the lower end of the zone. Protozoa follow course of aerobic bacteria, first diminishing and then reappearing. Fungi follow a similar course, disappearing under true septic conditions and then reappearing. Organisms are threadlike and develop pink, cream, and grayish tints. Algae are present to a very slight extent at the lower end of the zone. *Tubifex* are present only at the upper and lower ends of the zone. *Psychoda* (sewage fly) larvae are present in all but the most septic stage. Rattail maggots (*Eristabis*) and mosquito larvae (*Culex*) are found. There is no fish life.

(*continued*)

Table 3-3 (continued)

Zone	Physical characteristics	Chemical characteristics	Biological characteristics
Recovery (Zone 4 in Fig. 3-13)	Water is clearer.	Dissolved oxygen content moves upward from 40% of saturation; nitrates are present.	Protozoa, rotifers, and crustaceans appear. Fungi are present to a limited degree. Algae appear in the following order: *Cyanophycaea*, *Chlorophycaea*, and diatoms. Large plants (sponges, bryozoans) appear. Bottom organisms include *Tubifex*, mussels, snails, and insect larvae. Carp, suckers, and more resistant forms of fish occur.
Clean water (Zones 1 and 5 in Fig. 3-13)	Natural stream conditions are restored.	Dissolved oxygen is close to saturation.	Mayflies (*Ephemeropteria*), stone flies (*Plecoptera*), caddis flies (Trichoptera), and gamefish are found.

Source: Adapted from Babbitt. [3-2]

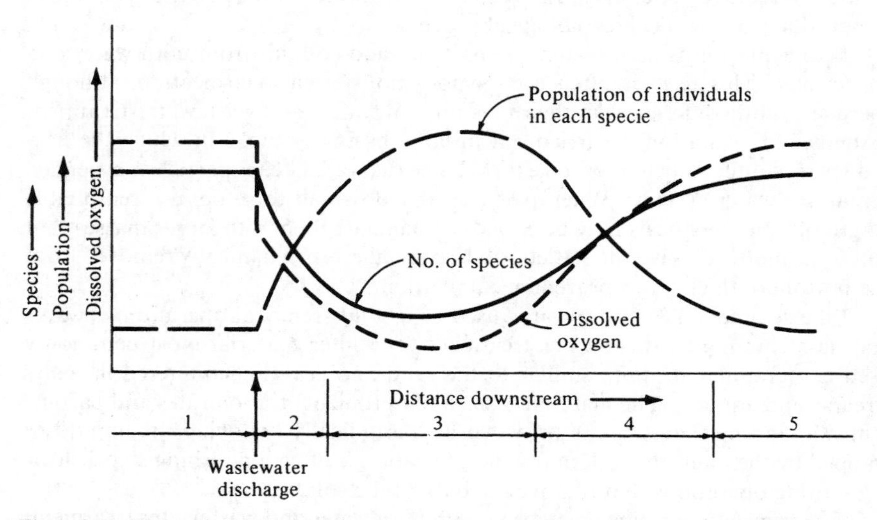

Figure 3-13 Changes in population of macroorganisms caused by waste discharge into a clean stream. (*From Kemmer*.) [3-10]

Should anaerobic conditions develop in the zone of active decomposition, a drastic change in both materials and flora and fauna would be observed. Reduced compounds, rather than oxidized end products, would appear, and aerobic species would give way to anaerobic and facultative organisms that, without competition from the aerobes, would flourish in great numbers.

Application of Natural Processes in Engineered Systems

Many of the physical, chemical, and biological processes that function in natural water systems have been incorporated into engineered systems for water and wastewater treatment. By carefully controlling the system variables, the rate at which the processes occur is maximized and the time required for purification is minimized. Reactions may thus be carried to completion in engineered systems in a fraction of the time and space required for similar efficiencies in natural water systems. The following section gives an overview of the application of natural processes in engineered systems while Chaps. 4 and 5 provide a more complete coverage of the engineered systems.

3-12 PHYSICAL PROCESSES

The physical processes frequently used in engineered systems include sedimentation, filtration, and gas transfer. These are the basic removal processes and may be used to remove materials in raw water or wastewater or may be used to remove the products of chemical or biological processes.

Sedimentation is used to remove particles and colloids from both water and wastewater. This term is often used synonymously with clarification, although there are subtle differences in their meaning. In water- and wastewater-treatment systems, sedimentation is carried out in large basins or tanks in which the flow is dispersed uniformly to minimize turbulence that often keeps particles suspended in natural water systems. When particles are too small to settle in a reasonable length of time, chemicals may be added to coagulate them into larger masses that will settle more quickly. The settled solids, or *sludge*, is mechanically removed from the bottom of the tank to prevent accumulation.

Like sedimentation, filtration is used as a solids-removal operation in water and, less commonly, wastewater treatment. The filter material most commonly used is a granular medium similar to the sand and gravel encountered in many streams and aquifers. The material is sized to optimize filtration rates and particle removal, and mechanisms are provided for periodically removing the impurities trapped by the filter. In modern practice, filtration is often a polishing step following settling operations that remove the bulk of the solids.

Gas-transfer operations may be used in both water and wastewater treatment. Depending on the treatment objectives, gases may be removed from or added to the liquid. Removal of gases that are in low concentrations in the atmosphere is enhanced by maximizing contact between the water and air, an operation often used to strip undesirable gases from water intended for potable use. Oxygen, a major constituent of the atmosphere, may be added to wastewaters by much the same principle. The addition of gases such as carbon dioxide and chlorine to meet specific treatment objectives (recarbonation and disinfection, respectively) is usually accomplished in closed pressurized systems.

3-13 CHEMICAL PROCESSES

Chemicals are used in many water- and wastewater-treatment processes. Chemicals may be added to alter equilibrium conditions and cause precipitation of undesirable species. An example is the addition of lime to precipitate hardness in potable water treatment and to precipitate phosphate in wastewater treatment. Often the chemical adjustment of pH is necessary to effect the desired precipitation. Oxidizing agents may be used if reduced compounds are to be removed. For example, potassium permanganate may be added to oxidize soluble forms of iron and manganese to forms that precipitate. Chlorine is sometimes used as an oxidizing agent as well as a disinfectant in both water and wastewater treatment. Chemical coagulation, often used as an adjunct to sedimentation or filtration, conditions small particles and colloids so that they form large, settleable flocs.

In addition to the above, many other chemicals may be used for special purposes in water and wastewater treatment. Again, it should be kept in mind that chemical processes are conversion processes and that actual removal is accomplished by physically separating the solid, liquid, or gaseous products of the chemical reactions.

3-14 BIOLOGICAL PROCESSES

Biological processes have found little use in the treatment of potable water supplies because of the low levels of biodegradable organics in the raw water. However, biological processes are used extensively in wastewater treatment to convert biodegradable organics and other nutrients into a more manageable form. Biological processes form the basis for secondary treatment in which dissolved and colloidal organics are converted into biomass that is subsequently separated from the liquid stream. Secondary treatment systems are designed to optimize contact between microorganisms and organics under the most favorable environmental conditions.

Once separated, the biomass becomes a concentrated waste stream that must be dealt with promptly. Biological treatment of this and other organic wastewater sludges, called *sludge digestion*, is one of the most important, and most difficult, processes in wastewater treatment.

DISCUSSION TOPICS AND PROBLEMS

3-1 Name and briefly describe the major physical processes involved in self-purification of watercourses.

3-2 Two streams converge as shown in the sketch below. Determine the flow, temperature, and dissolved oxygen in the merged streams at point C.

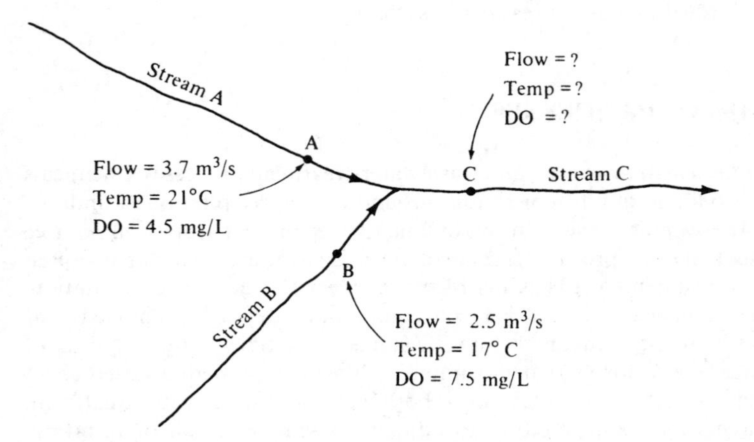

3-3 Effluent from a wastewater-treatment plant is discharged to a surface stream. The characteristics of the effluent and stream are as follows:

Effluent	Stream
Flow = 8640 m^3/d	Flow = 1.2 m^3/s
BOD_5 = 25 mg/L	BOD_5 = 2.1 mg/L
Ammonia = 7 mg/L	Ammonia = 0 mg/L
Nitrate = 10 mg/L	Nitrate = 3.0 mg/L
Chloride = 15 mg/L	Chloride = 5.0 mg/L

Determine the stream characteristics after mixing with the waste has occurred.

3-4 Cooling tower blowdown from a power plant is discharged to a surface stream. The characteristics of each are given as:

Stream	Cooling water
Flow = 10 m^3/s	Flow = 40 m^3/min
Temperature = 15°C	Temperature = 28°C
TDS = 125 mg/L	TDS = 2520 mg/L
Chromate = 0	Chromate = 0.9 mg/L

Determine the characteristics of the stream after mixing.

3-5 An industrial wastewater is discharged into a municipal wastewater sewer. The characteristics of the two wastes are as follows:

Industrial	Municipal
Flow = 3500 m^3/d	Flow = 17,400 m^3/d
BOD_5 = 1200 mg/L	BOD_5 = 210 mg/L
PO_4^{3-} = 140 mg/L	PO_4^{3-} = 2.3 mg/L

Determine the characteristics of the mixture.

3-6 Discuss thermal stratification and its importance in temperature of streams and lakes.

3-7 What are three major chemical processes that may alternately aid or obstruct natural purification processes of water systems?

3-8 Calculate the solubility of hydrogen sulfide in water at 20°C.

3-9 Determine the solubility of the components of air in water at 20°C and 1.5 atm pressure.

3-10 What is the solubility of methane in water at 20°C?

3-11 Define (*a*) metabolism; (*b*) catabolism; (*c*) anabolism; and (*d*) endogenous catabolism.

3-12 What are adaptive enzymes? What role do they play in natural purification processes of bodies of water?

3-13 Define (*a*) autotrophs, (*b*) heterotrophs, (*c*) phototrophs, (*d*) aerobic heterotrophs, (*e*) anaerobic heterotrophs, and (*f*) facultative heterotrophs.

3-14 Explain the role of rotifers, crustacea, and sludge worms in natural purification processes of bodies of water.

3-15 What are the two mechanisms known to contribute oxygen to surface waters?

3-16 What is the oxygen deficit of a stream and how is this deficit represented mathematically?

3-17 Write a simplified formula for the photosynthetic process by which algae populations may replenish oxygen in a body of water.

3-18 Write a formula for the endogenous catabolism by which algae populations may contribute to oxygen demand.

3-19 A wastewater-treatment plant disposes of its effluent in a surface stream. Characteristics of the stream and effluent are shown below.

	Wastewater	Stream
Flow, m^3/s	0.2	5.0
Dissolved oxygen, mg/L	1.0	8.0
Temperature, °C	15	20.2
BOD_5 at 20°C, mg/L	100 mg/L	2.0 mg/L
K_1 at 20°C, d^{-1}	0.2	—
K_2 at 20°C, d^{-1}	—	0.3

(*a*) What will be the dissolved oxygen concentration in the stream after 2.0 d?

(*b*) What will be the lowest dissolved oxygen concentration as a result of the waste discharge?

3-20 A municipal wastewater-treatment plant discharges 18,925 m^3/d of treated wastewater to a stream. The wastewater has a BOD_5 of 30 mg/L with a k_1 of 0.23 d^{-1}. The temperature of the wastewater is 27°C, and the dissolved oxygen is 2.0 mg/L. The stream just above the point of wastewater discharge flows at 0.65 m^3/s, has a BOD_5 of 5.0 mg/L, and is 90 percent saturated with oxygen. The temperature of the stream is 23°C. After mixing, the stream and wastewater flows at a velocity 0.5 m/s and the reaeration constant is 0.45 d^{-1}.

(*a*) What is the oxygen level of the stream after 2 d?

(*b*) What is the critical oxygen level in the stream and how far downstream will it occur?

3-21 A wastewater-treatment plant discharges to a small stream. The characteristics of the wastewater and the characteristics of stream are given below.

Stream	Waste
Flow = 0.4 m^3/s	Flow = 10,000 m^3/d
BOD = 2.0 mg/L	DO = 0 ms/L
DO = 90% saturation	Temperature = 21°C
Temperature = 24°C	k_1 = 0.23 d^{-1}
k_2 = 0.45 d^{-1}	

Determine the maximum BOD_5 (20°C) that can be discharged if a minimum of 4.0 mg/L of oxygen must be maintained in the stream.

3-22 A milk-products industry discharges a wastewater to a stream. Characteristics of the wastewater and the stream are shown below.

(*a*) If no treatment at all is given to the wastewater, what will be the lowest oxygen level in the stream as a result of the discharge?

Parameter	Wastewater	Stream
Flow	1000 m^3/d	19,000 m^3/d
BOD_5 at 20°C	1250 mg/L	2.0 mg/L
DO	0 mg/L	10.0 mg/L
Temperature, °C	50	10
k_1 at 20°C	0.35 d^{-1}	—
k_2 at 20°C	—	0.55 d^{-1}

(*b*) If the stream is a trout fishery and the stream standards require a minimum DO of 5.0 mg/L, what is the maximum BOD_5 (20°C) that can be discharged by the industry?

3-23 Write a computer program to model the Streeter–Phelps equation. Repeat Probs. 3-19, 3-20, and 3-21 using the computer.

3-24 What are the four zones in the Whipple, Fair, and Whipple model? Define the zones by explaining what happens in each.

3-25 Most of the natural purification processes discussed in this chapter have their counterparts in engineered processes for the treatment of potable water supplies or the treatment of wastewaters. Discuss briefly the ways in which the following natural processes are utilized in engineering systems: (*a*) sedimentation, (*b*) filtration, (*c*) gas transfer, (*d*) precipitation, and (*e*) microbial action.

REFERENCES

3-1 Amirtharajah, A.: by personal communication, 1982.

3-2 Babbitt, H. E., and E. R. Baumann: *Sewerage and Sewage Treatment*, 8th ed., Wiley, New York, 1952.

3-3 Bartsch, A. F., and W. M. Ingram: *Biology of Water Pollution*, U.S. Dept. of Interior, Water Pollution Control Administration, 1967.

3-4 Churchill, M. A., H. L. Elmore, and R. A. Buckingham: "The Prediction of Stream Reaeration Rates," *Water Pollution Research*, vol. 1, Pergamon, London, 1964.

3-5 Clark, John W., Warren Viessman, Jr., and Mark J. Hammer: *Water Supply and Pollution Control*, 3d ed., Harper & Row, New York, 1977.

3-6 Foin, Theodore C., Jr.: *Ecological Systems and the Environment*, Houghton Mifflin, Boston, 1976.

3-7 Gaudy, A. F., Jr., and E. T. Gaudy: *Microbiology for Environmental Scientists and Engineers*, McGraw-Hill, New York, 1980.

3-8 Hammer, Mark J.: *Water and Waste-Water Technology*, Wiley, New York, 1975.

3-9 Hynes, H. B. N.: *The Biology of Polluted Water*, Liverpool University Press, Liverpool, 1960.

3-10 Kemmer, Frank N.: *The NALCO Water Handbook*, McGraw-Hill, New York, 1979.

3-11 Klein, Louis: *River Pollution II. Causes and Effects*, Butterworth, London, 1962.

3-12 Lewis, W. K., and W. G. Whitman: "Principles of Gas Absorption," *Ind. Eng. Chem.*, **16**:1215 (1924).

3-13 Lindsley, R. K., and J. B. Franzini: *Water Resources Engineering*, 3d ed., McGraw-Hill, New York, 1979.

3-14 McKinney, R. E.: *Microbiology for Sanitary Engineers*, McGraw-Hill, New York, 1962.

3-15 Metcalf & Eddy, Inc.: *Wastewater Engineering: Treatment and Disposal*, 2d ed., McGraw-Hill, New York, 1979.

3-16 O'Connor, D. J., and W. E. Dobbins: "The Mechanisms of Reaeration in Natural Streams," *J San Eng Div*, A.S.C.E., **82**:SA6 (1956).

3-17 Parker, F. L., and P. A. Krenkel: *Thermal Pollution: Status of the Art*, Dept. Environmental and Resources Engineering, Vanderbilt University, Nashville, December 1969.

3-18 Patrick, R.: "Effect of Suspended Solids, Organic Matter and Toxic Materials on Aquatic Life in Rivers," *Water and Sewage Works*, February 1968, p. 90.

3-19 Ruttner, Franz: *Fundamentals of Limnology*, 3d ed., D. G. Frey and F. E. J. Fry (trans.), University of Toronto Press, Toronto, 1963.

3-20 Sawyer, C. N., and P. L. McCarty: *Chemistry for Environmental Engineers* 3d ed., McGraw-Hill, New York, 1978.

3-21 Srinanthakumar, S., and A. Amirtharajah: "Organic Carbon Decay in a Stream with Biofilm Kinetics," *J. Inv. Eng., ASCE*, **109**(1): 102 (February 1983).

3-22 Steel, E. W., and T. J. McGhee: *Water Supply and Sewerage*, 9th ed., McGraw-Hill, New York, 1979.

3-23 Streeter, H. W., and E. B. Phelps: U.S. Pub. Health bulletin no. 146, 1925.

3-24 Tsivoglou, E. C.: *Tracer Measurement of Stream Reaeration*, U.S. Dept. of Interior, Water Pollution Control Administration, Washington, D.C., June 1976.

3-25 Velz, Clarence, J.: *Applied Stream Sanitation*, Wiley Interscience, New York, 1970.

3-26 Vesilind, P. Aarne: *Environmental Pollution and Control*, Ann Arbor Science, Ann Arbor, Mich., 1975.

3-27 Warren, Charles E.: *Biology and Water Pollution Control*, Saunders, Philadelphia, 1971.

3-28 Whipple, C. C.: *The Microscopy of Drinking Water*, 4th ed., rev. by G. M. Fair and M. C. Whipple, Wiley, New York, 1927.

CHAPTER

FOUR

ENGINEERED SYSTEMS FOR WATER PURIFICATION

An adequate supply of pure water is absolutely essential to human existence. The consequences of a contaminated water supply can be illustrated by conditions prevalent during the industrial revolution in Europe when large numbers of peasants were attracted to the cities where they crowded together with little or no sanitary facilities. Human waste, or "night soil" as it was called, was tossed into the streets or emptied into pits in common courtyards, often near the shallow wells that served as the neighborhood water supply. Seepage into these wells and runoff into nearby streams provided a direct link in the infection cycle, and once an outbreak of disease occurred it usually spread rapidly through the community. The resulting loss of life and suffering left scarcely a family untouched during several centuries prior to the 1900s.

The development of effective water-treatment methods has virtually eliminated major waterborne epidemics in developed countries. This is not to suggest, however, that the problem of waterborne diseases has been eliminated. Developing nations, where treated water is not available to all the population, still experience occasional epidemics of cholera and typhoid, as well as many outbreaks of less severe disease. Even highly developed countries, including the United States, where public water supplies are almost universally treated, are not totally immune from an occasional outbreak of gastrointestinal illnesses traceable to biologically contaminated water supplies.

Chemical contamination of water supplies has become a concern in more recent times. Industrial facilities in developed countries produce and use literally thousands of chemical compounds. Along with an abundant array of household and agricultural chemicals, these materials often find their way into water supplies. While some of these chemical compounds are known toxicants, mutagents, or carcinogens, the health effects of many others are not presently known. Sufficient data are not presently available to predict the consequences of ingesting small quantities of chemicals over long periods of time. It is ironic that the high standard of living that allows industrialized nations to provide biologically pure water to the majority of their populations also results in the discharge of chemical waste that may eventually have more deleterious effects on human health than the domestic waste that helped spread the plagues of past centuries.

4-1 HISTORICAL OVERVIEW OF WATER TREATMENT

The treatment of water intended for human consumption is a very old practice. Baker [4-6] reports references in Sanskrit literature dating back to 2000 B.C. to such practices as the boiling and filtering of drinking water. Wick siphons that transferred water from one vessel to another, filtering out the suspended impurities in the process, were pictured in Egyptian drawings of the thirteenth century B.C. and were referred to in early Greek and Roman literature. The fact that these practices were recorded in the medical documents of the times indicates that the connection between water and health had been observed. In fact, Hippocrates (460–354 B.C.), considered to be the father of modern medicine, wrote that "...whosoever wishes to investigate medicine properly should —consider the water that the inhabitants use—for water contributes much to health." [4-6]

These early water-treatment devices were used in individual households; there is no indication of community water supplies being treated until around the first century. Some of the Roman aqueducts had settling basins at the headworks and incorporated "pebble catchers" in the aqueduct channel. These aqueducts supplied a few private taps and provided fountains or reservoirs for use by the general public. The city of Venice, situated on islands with no freshwater resource, channeled rainwater from roofs and courtyards into elaborate cisterns through sand filters surrounding the reservoir. The first of these cisterns was built around the fifth century A.D. and provided private and public water supplies for about 13 centuries. [4-6]

Water-treatment practice apparently lagged during the Middle Ages, with a renewed interest emerging in the eighteenth century. Several patents were issued for filtering devices, primarily in France and England. As in ancient times, however, these devices were for use in private households, institutions, ships, etc. It was not until the beginning of the nineteenth century that the treatment of public water supplies was attempted on a large scale. The city of Paisley, Scotland, is generally credited with being the first city with a treated water supply. That system consisted of settling operations followed by filtration and was put in service in 1804. [4-6] This practice slowly spread through Europe and by the end of the century, most major municipal water supplies were filtered. These filters were the "slow sand" type described in Sec. 4-8.

The development of water treatment in America lagged behind the European practice. The first attempt at filtration was made at Richmond, Virginia, in 1932. This project was a failure, and several years intervened before another significant effort was made. [4-6] After the Civil War, other attempts were made to follow the sand filtration practice of Europe, few of which were successful. Apparently the nature of the suspended solids in American streams was significantly different from that of the solids in European streams, and the slow sand process was not as effective. The development of the hydraulically cleaned rapid sand filter during the latter part of the nineteenth century provided a more workable process, and by the end of the century its use was widespread.

During the first two-thirds of the nineteenth century, filtration was practiced to improve the aesthetic quality of the drinking water. An unknown benefit was the removal of microorganisms, including pathogens, which made the water more wholesome as well. The acceptance of this fact in the last quarter of the century spurred the construction of the filter plants throughout Europe and America. At the turn of the century, filtration was the primary defense against waterborne disease.

Acceptance of the germ theory of disease transmittal led to the disinfection of public water supplies. First used on a temporary basis, disinfection with bleach powders and hypochlorites was used in isolated cases in the eighteen-nineties. The first permanent installation for chlorinating water was made in Belgium in 1902. The production of liquid chlorine began in 1909 and was first used for water disinfection in Philadelphia in 1913. [4-6] Other means of disinfection, notably ozonation, were developed simultaneously but did not find widespread use. The drastic reduction in deaths due to waterborne diseases as a result of disinfection led to the widespread chlorination of public water supplies.

Other water-treatment processes developed more slowly and less dramatically. Coagulation as an adjunct to settling was developed along with the rapid sand filter in America. Softening of hard waters was demonstrated in Europe during the nineteenth century but did not find widespread use in public water supplies until well into the twentieth century. The capacity of charcoal to remove dissolved organics was observed by early experimenters in filtration but did not find application in public water supplies. The improvement of this material into "activated carbon" and its use in water-treatment plants is a recent occurrence, as is the use of synthetic membranes for hyperfiltration to remove dissolved inorganic material.

More progress has been made in water purification in the last century than in all of the previously recorded history. With few exceptions, treatment processes developed in the absence of scientific knowledge concerning the basic principles upon which they operate, and often with little means to quantitatively assess their effectiveness. Only within the last 30 to 40 years has the body of scientific knowledge caught up with the practice of water purification. It is interesting to note that the development of a theory base has resulted in few changes in the basic processes of water purification. Understanding of scientific principles has, however, led to refinements of processes, development of better equipment, and an overall increase in operating efficiencies in water treatment. The following section gives an overview of modern water-treatment processes, while the remaining sections of the chapter contain a detailed description of the individual processes.

4-2 WATER-TREATMENT PROCESSES

Past practices in America have often been to obtain the purest possible source, even at the expense of transporting water over long distances, and to deliver it to the consumer with little or no treatment. Some cities still own large tracts of land near the headwaters of stream and restrict activities on these watersheds to minimize contamination. Although the benefits of source protection are recognized as a "first line of defense" in preserving water quality, all natural waters will require some degree of treatment in order to meet modern drinking-water standards. The nature and extent of treatment will, of course, depend upon the nature and extent of impurities.

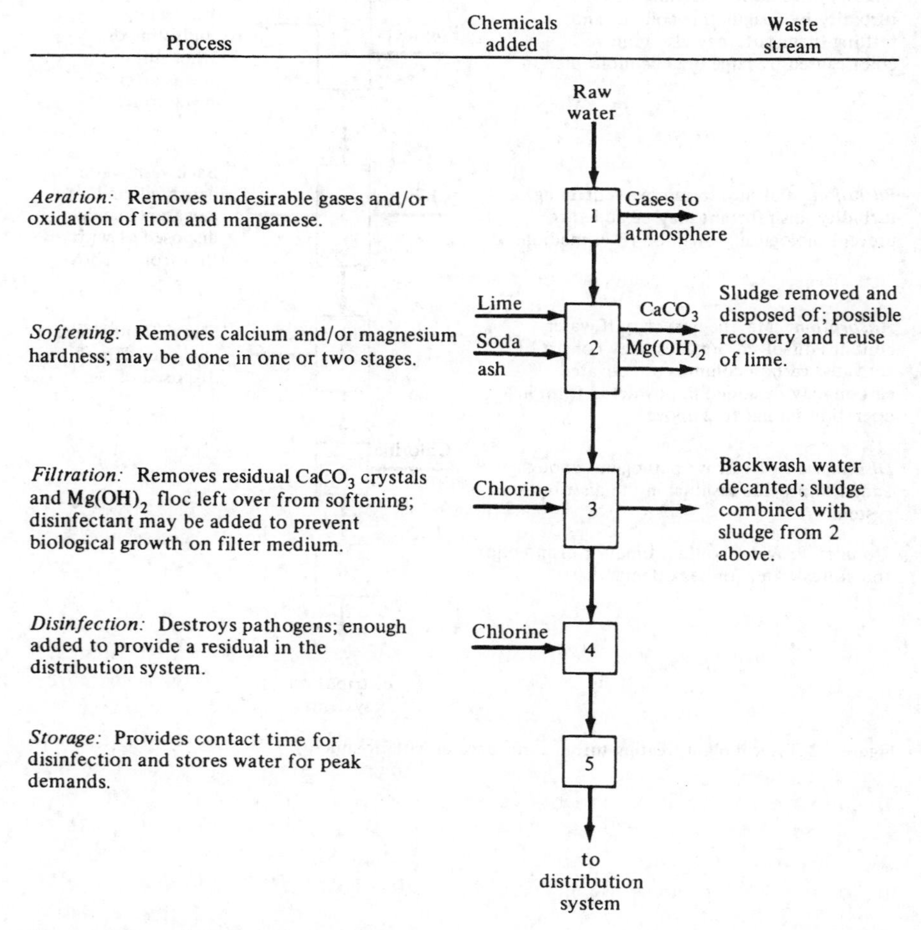

Figure 4-1 Typical plant treating hard groundwater.

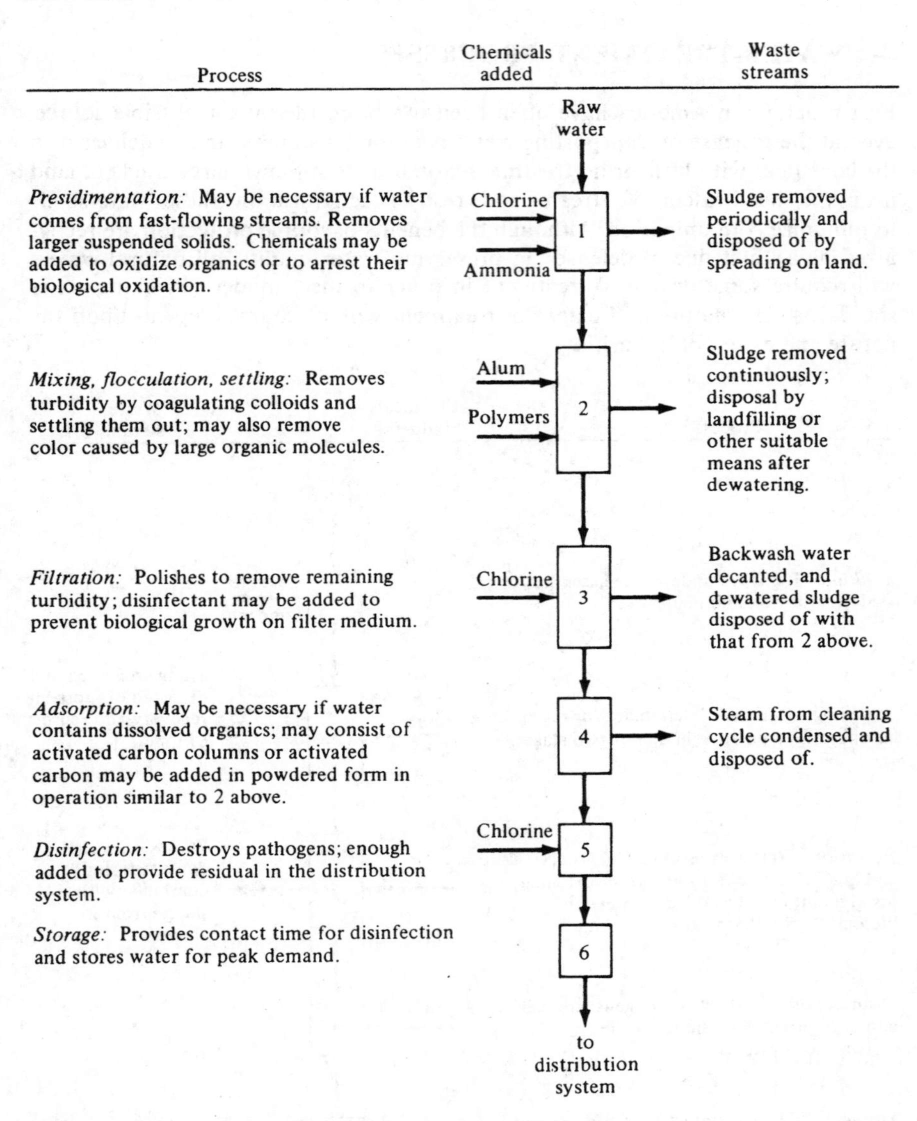

Process
Chemicals added
Waste streams
Raw water
Presidimentation: May be necessary if water comes from fast-flowing streams. Removes larger suspended solids. Chemicals may be added to oxidize organics or to arrest their biological oxidation.
Chlorine
Ammonia
1
Sludge removed periodically and disposed of by spreading on land.
Mixing, flocculation, settling: Removes turbidity by coagulating colloids and settling them out; may also remove color caused by large organic molecules.
Alum
Polymers
2
Sludge removed continuously; disposal by landfilling or other suitable means after dewatering.
Filtration: Polishes to remove remaining turbidity; disinfectant may be added to prevent biological growth on filter medium.
Chlorine
3
Backwash water decanted, and dewatered sludge disposed of with that from 2 above.
Adsorption: May be necessary if water contains dissolved organics; may consist of activated carbon columns or activated carbon may be added in powdered form in operation similar to 2 above.
4
Steam from cleaning cycle condensed and disposed of.
Disinfection: Destroys pathogens; enough added to provide residual in the distribution system.
Chlorine
5
Storage: Provides contact time for disinfection and stores water for peak demand.
6
to distribution system

Figure 4-2 Typical plant treating turbid surface water with organics.

The processes selected for the treatment of potable water depend on the quality of the raw water supply. Most groundwaters are clear and pathogen-free and do not contain significant amounts of organic materials. Such waters may often be used in potable systems with a minimal dose of chlorine to prevent contamination in the distribution system. Other groundwaters may contain large quantities of dissolved solids or gases. When these include excessive amounts of iron, manganese, or hardness, chemical and physical treatment processes may be required. Treatment systems commonly used to prepare potable water from groundwater are shown in Fig. 4-1.

Surface waters often contain a wider variety of contaminants than groundwater, and treatment processes may be more complex. Most surface waters contain turbidity in excess of drinking-water standards. Although fast-moving streams may carry larger material in suspension, most of the solids will be colloidal in size and will require chemical coagulation for removal. Depending on the geology of the watershed, hardness may or may not be a problem in surface waters. If low levels of color and other organic material are present, adsorption onto surface-active material, a process not significant in natural water systems, may be necessary. A wide variety of microorganisms, some of which may be pathogenic, are also common constituents of surface waters. Treatment systems commonly used in treating surface waters are shown in Fig. 4-2.

Water-Treatment Processes: Theory and Application

It is generally convenient to group human use of water into two broad categories depending upon the location of the use relative to the source. In-place use of water includes navigation, recreation, wildlife propagation, and the dilution, assimilation, and transportation of wastewater. Although hydroelectric power generation requires brief diversion of water through turbine penstocks, this use is also considered an in-place use. Quantitatively, in-place use is a nonconsumptive use and will not be covered in this text.

For irrigation and industrial use, and for individual and public domestic supplies, water must be withdrawn from streams, lakes, or aquifers in the natural hydrologic cycle. The pollutants most deleterious to crops (inorganic salts and metals) are difficult and expensive to remove. The vast quantity of irrigation water used and the low margin of profit associated with farming virtually preclude any treatment of this water. Water not suited for irrigation is simply abandoned, and available capital is used instead to secure an alternate source of acceptable quality. Many industries with needs for small amounts of essentially potable water obtain their supplies from public systems. Some industrial water supplies, such as boiler-feed water, may require a chemical purity an order of magnitude greater than potable water. Engineering design for treatment of other types of industrial water supplies may also be necessary. Cooling water, particularly that used only once and discharged back to nature, has few quality constraints. Individual domestic supplies are usually drawn from wells or springs of acceptable quality and serve individual homes or farmsteads. Such systems are seldom engineered but are installed and operated by the home owners, perhaps with the advice of the well-driller and the distributor of home water-treatment units.

Public water supplies, while only a fraction of the total water use, require by far the largest amount of effort expended by environmental engineers in the water-treatment field. The remainder of this chapter will be devoted to the principles of water purification for potable supplies. The processes involved are discussed first from a theoretical standpoint and then from an applications standpoint.

4-3 AERATION

Aeration is a process sometimes used in preparing potable water. It may be used to remove undesirable gases dissolved in water (*degasification*) or to add oxygen to water to convert undesirable substances to a more manageable form (*oxidation*). Aeration is more often used to treat groundwater, as most surface waters have been in contact with the atmosphere for a sufficient period of time for gas transfer to occur naturally.

Groundwater may contain appreciable quantities of gases such as carbon dioxide (CO_2) and hydrogen sulfide (H_2S). These gases are biological waste products from bacterial decomposition of organic matter in the soil or by-products of reduction of sulfur from mineral deposits. Excessive carbon dioxide concentration results in a corrosive water. High carbon dioxide levels may also interfere with other treatment processes. Hydrogen sulfide imparts an unpleasant taste and odor to water, even in small concentrations. Although these gases are only slightly soluble at atmospheric conditions, groundwater may contain considerably higher concentrations under pressures commonly found in deep aquifers. Aeration of water supersaturated with these gases serves to speed the release toward equilibrium conditions.

Although volatile liquids such as humic acids and phenols can be removed from water by aeration, the removal rates are too slow for the process to be practical except in extreme cases where excessive quantities must be reduced to more manageable levels.

Iron and manganese are common elements widely distributed in nature. In the absence of oxidizing agents, both of these elements are soluble in water. Forming compounds with other soluble ions, both iron and manganese are soluble in significant quantities only in the +2 oxidation state, i.e., Fe^{2+} and Mn^{2+}. Upon contact with oxygen, or any other oxidizing agents, both ferrous iron and manganese are oxidized to higher valances, forming new ionic complexes that are not soluble to any appreciable extent. Thus, the iron and manganese may be removed as a precipitate after aeration. Chemically, these reactions may be written as follows.

$$4Fe^{2+} + O_2 + 10H_2O \longrightarrow 4Fe(OH)_3 \downarrow + 8H^+ \qquad (4\text{-}1)$$

$$2Mn^{2+} + O_2 + 2H_2O \longrightarrow 2MnO_2 \downarrow + 4H^+ \qquad (4\text{-}2)$$

In Eq. (4-1), Fe goes from the +2 to the +3 oxidation state and in Eq. (4-2) Mn goes from the +2 to the +4 oxidation state. In both equations the free oxygen (O_2) is reduced, and the anion originally associated with the ferrous and manganous ions recombines with other cations in the solution. In both cases, the pH of the solution is lowered by the production of hydrogen ions.

Iron and manganese are found in appreciable amounts only in groundwater and in water from the hypolimnion of stratified lakes where anaerobic conditions exist. Aeration of this water provides the oxygen necessary to convert both elements to the insoluble form. Chemical oxidants, such as potassium permanganate, can also be used for this purpose. They are sometimes used in connection with aeration to speed up the process. When aeration is used to precipitate iron and manganese, additional treatment will be required to remove the precipitated solids.

Both degasification and oxidation are governed by the principles of gas transfer that were presented in Sec. 3-4. Subtle differences in liquid-gas contact systems can have a pronounced effect on the overall gas-transfer process. An understanding of gas-transfer principles is essential in aerator design, and the student is encouraged to reread Sec. 3-4 before proceeding into the following discussion.

Liquid-Gas Contact Systems

Liquid-gas contact systems are designed to drive the water-gas mixture toward equilibrium as quickly as possible for degasification purposes and to provide supersaturation of oxygen for oxidation purposes. These goals may be accomplished by either dispersing the water into the air or by dispersing the air into the water.

When water is dispersed into the air, as depicted in Fig. 4-3, the interfacial area per volume of water is maximized by minimizing the drop size. This will increase the desorption rate for supersaturated solutions (Fig. 4-3*a*) or increase the absorption rate for undersaturated solutions (Fig. 4-3*b*). In general, this approach works better for desorbing gases than for absorbing oxygen, although the latter can be accomplished for undersaturated waters.

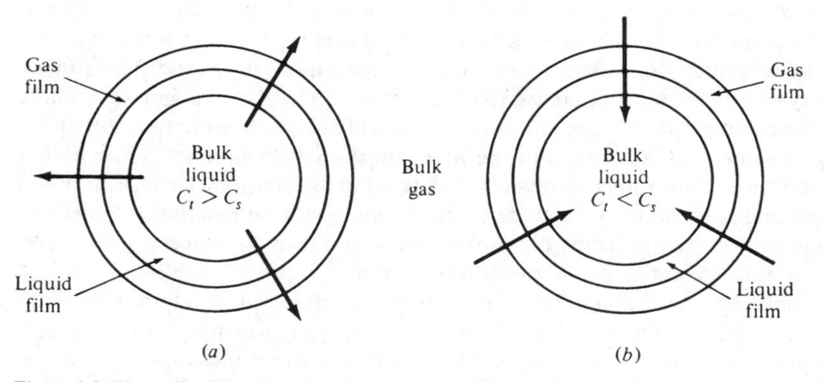

Figure 4-3 Water dispersed in air: (*a*) desorption and (*b*) absorption.

In water purification plants, water-in-air systems may consist of fountains, cascade towers, or tray towers. *Fountains* consist of a piping grid suspended over a catch basin. Nozzles located at the intersection of the pipes are fixed to direct the flow of water upward. Once its kinetic energy is dissipated, the water falls back into the catch basin where it is recovered, portions of the flow perhaps being recycled. The height of the spray, and therefore the water-air contact time, is determined by the pressure in the pipes, while the dispersion pattern is determined by the nozzle characteristics. Nozzle size may vary from 2 to 4 cm in diameter. While smaller nozzles result in finer sprays, which yield greater surface-to-volume ratios, frequent clogging of small nozzles can result in high maintenance cost.

Design parameters for spray aerators include system pressure, nozzle spacing, and flow rates per nozzle. Pressures of around 70 kPa (10 lb/in^2) are common and produce flow rates of from 5 to 10 L/s through each nozzle. Grid spacing may vary from 0.6 to 3.5 m depending on the distance necessary to prevent extensive overlap of nozzle discharges. A typical design may consist of 2.54-cm nozzles on 1.25-m centers operating at 70-kPa pressure, resulting in an area requirement of approximately 10 m^2/(50 L/s) of water treated (or about 100 ft^2/(Mgal/d)).

Cascade towers consist of a series of waterfalls that drop into small pools. In this case the water is not dispersed as droplets but is exposed to the atmosphere in thin sheets as it cascades down each step. Each step in a cascade tower is usually about 0.3 m in height, and as many as 10 steps may be employed. The number of steps determines the contact time between the water and the air. Head loss through the system is simply the height of the topmost step. The cascades may be arranged longitudinally like stair steps or may be arranged in a circle, with the steps extending concentrically outward from top to bottom. Area requirement for cascade aerators ranges from 4 to 9 m^2/(50 L/s) (40–90 ft^2/(Mgal/d)), depending upon the number of steps used. [4-50]

Tray towers are similar in nature to cascade towers in that the water is lifted and allowed to fall to a lower elevation. Instead of being intercepted in pools, tray towers intercept the flow with solid surfaces over which the water must pass in its downward journey. The solid surfaces may be a series of redwood slat trays which break the flow of the water or a series of porous-bottom trays containing stones, ceramic spheres, or other porous packing. In any case, tray material provides large surface areas over which the flow is spread in thin films. Porosity of the system must be sufficient to ensure circulation of air around the surfaces.

Tray towers are most often used for oxidation of iron and manganese. Usually the tray packing will be large chunks of coke which have been precoated with a strong oxidant such as potassium permanganate to help initiate the oxidation process. Films of iron and manganese solids are deposited on the surface of the medium, and these films serve to catalyze the precipitation reaction. Manganese precipitates very slowly below a pH of about 9, and it may be necessary to raise the pH to this level in order to speed the reaction.

In addition to the above operations, there are many proprietary devices on the market which make use of one or more of the basic principles just discussed. Information on these devices may be obtained from current literature or from the manufacturers or distributors.

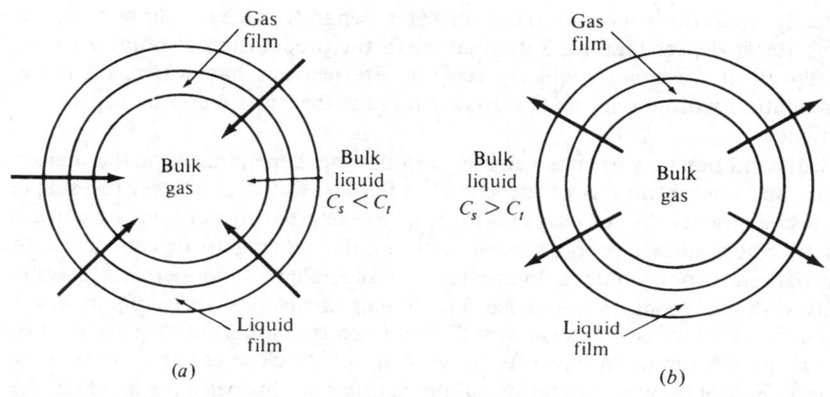

Figure 4-4 Air dispersed in water: (*a*) desorption and (*b*) absorption.

Another method of aerating water is to disperse the air into the water. Again, both absorption and desorption are enhanced by maximizing the interfacial area, in this case by minimizing the size of the air bubble. Figure 4-4*a* can be used to illustrate the situation for a supersaturated water (desorption), and the process for an undersaturated water (absorption) is shown in Fig. 4-4*b*. In general, this approach works better for absorption than for desorption.

Air-in-water systems most often consist of tanks from 2.5 to 5.0 m deep through which the water flows. Air is then injected through a porous bottom or through spargers near the bottom. Since the energy for this system is expended on the air, not the water, smaller, less complicated equipment is required. Blower capacity need only be sufficient to deliver the required air volume at the pressure determined by head loss through the distributing mechanism, plus the depth of the water. This type of aeration device has found greater use in wastewater treatment than in potable water treatment.

Several variations of this process may be employed. Carrying out the process in an enclosed tank with a positive pressure in the atmosphere above the liquid will speed the absorption rate, although it will also decrease the desorption rate. An impeller placed just above the point of air injection will break the air flow into smaller bubbles and enhance mixing patterns. As in the case of water-in-air systems, there are several proprietary devices which make unique application of the basic principles discussed here.

All aeration operations must be well ventilated to prevent the buildup of gases which may be toxicants or asphixants.

4-4 SOLIDS SEPARATION

The terms *sedimentation* and *clarification* are commonly used interchangeably with regard to preparation of potable water. Although there are some subtle differences in the connotations of the two words, they both convey the idea of physically separating solid material from water. Separation may occur by flotation if the water is denser than the solid matter. In the preparation of potable water, virtually all of the solids requiring removal are heavier than water; therefore, sedimentation with gravity as the driving force is the most common separation technique.

Sedimentation may be classified into various types depending upon the characteristics and concentrations of suspended materials. Particles whose size, shape, and specific gravity do not change with time are referred to as *discrete particles.* Particles whose surface properties are such that they aggregate, or coalesce, with other particles upon contact, thus changing size, shape, and perhaps specific gravity with each contact, are called *flocculating particles.* Suspensions in which the concentration of particles is not sufficient to cause significant displacement of water as they settle or in which particles will not be close enough for velocity field interference to occur are termed *dilute suspensions.* Suspensions in which the concentration of particles is too great to meet these conditions are termed *concentrated suspensions.* These differences result in significantly different settling patterns and require separate analysis. Settling in dilute suspensions is discussed below. Since concentrated suspensions are most often encountered in wastewater treatment, that discussion is presented in Chap. 5.

Type-1 Settling

Discrete particles in dilute suspension, type-1 settling, is the easiest situation to analyze. If a particle is suspended in water, it initially has two forces acting upon it: (1) the force of gravity

$$f_g = \rho_p g V_p$$

in which ρ_p is the density of the particle, g is the gravitational constant, and V_p is the volume of the particle; and (2) the buoyant force quantified by Archimedes as

$$f_b = \rho_w g V_p$$

where ρ_w is the density of the water. Since these forces are in opposite directions, there will be no net force when $\rho_p = \rho_w$, and no acceleration of the particle in relation to the water will occur. If, however, the density of the particle differs from that of the water, a net force is exerted and the particle is accelerated in the direction of the force:

$$f_{\text{net}} = (\rho_p - \rho_w) g V_p$$

This net force becomes the driving force for acceleration.

Once motion has been initiated, a third force is created due to viscous friction. This force, called the *drag force*, is quantified by

$$f_d = C_D A_p \rho_w \frac{v^2}{2}$$

where C_D is the coefficient of drag, A_p is the cross-sectional area of the particle perpendicular to the direction of movement, and v is the velocity of the particle. Because the drag force acts in the opposite direction to the driving force and increases as the square of the velocity, acceleration occurs at a decreasing rate until a steady velocity is reached at a point where the drag force equals the driving force:

$$(\rho_p - \rho_w)gV_p = C_D A_p \rho_w \frac{v^2}{2} \tag{4-3}$$

For spherical particles,

$$\frac{V_p}{A_p} = \frac{\frac{4}{3}\pi(d/2)^3}{\pi(d/2)^2} = \frac{2}{3}d$$

Substituting into Eq. (4-3)

$$v_t^2 = \frac{4}{3}g\frac{(\rho_p - \rho_w)d}{C_D \rho_w} \tag{4-4}$$

Expressions for C_D change with characteristics of different flow regimes. For laminar, transitional, and turbulent flow, the values of C_D are:

$$C_D = \frac{24}{\text{Re}} \text{ (laminar)} \tag{4-5}$$

$$= \frac{24}{\text{Re}} + \frac{3}{\text{Re}^{1/2}} + 0.34 \text{ (transitional)} \tag{4-6}$$

$$= 0.4 \text{ (turbulent)} \tag{4-7}$$

where Re is the Reynolds number

$$\text{Re} = \frac{\phi v_t \rho_w d}{\mu} \tag{4-8}$$

Reynolds numbers less than 1.0 indicate laminar flow, while values greater than 10^4 indicate turbulent flow. Intermediate values indicate transitional flow. The shape factor ϕ is added to correct for lack of spherosity. For perfect spheres, the value of ϕ is 1.0. For laminar flow, substitution of Eq. (4-5) into Eq. (4-4) yields:

$$v_t = \frac{g(\rho_p - \rho_w)d^2}{18\,\mu} \tag{4-9}$$

which is known as the *Stokes equation*. Terminal settling velocities for the transitional flow involve simultaneous solutions of Eqs. (4-6) and (4-4). Use of the above equations in determining the terminal settling velocities of discrete particles in dilute suspensions is illustrated in the following example.

Example 4-1: Finding the terminal settling velocity of a sphere in water Find the terminal settling velocity of a spherical particle with diameter 0.5 mm and specific gravity of 2.65 settling through water at 20°C.

SOLUTION

1. Assume laminar flow; from Eq. (4-8) with $\rho_w = 998.2\ \text{kg/m}^3$ and $\mu = 1.002 \times 10^{-3}$ $\text{N}\cdot\text{s/m}^2$ at 20°C

$$v_t = \frac{9.81\ \text{m/s}(2650 - 998.2)\ \text{kg/m}^3 \times (5.0 \times 10^{-4})^2\ \text{m}^2}{18 \times 1.002 \times 10^{-3}\ \text{N}\cdot\text{s/m}^2}$$

(Recall that the units of N are $\text{kg}\cdot\text{m/s}^2$.)

$$v_t = 0.22\ \text{m/s}$$

2. Check Reynolds number:

$$\text{Re} = \frac{0.22\ \text{m/s} \times 5 \times 10^{-4}\ \text{m} \times 998.2\ \text{kg/m}^3}{1.002 \times 10^{-3}\ \text{N}\cdot\text{s/m}^2}$$

$$= 112, \text{ which indicates transitional flow}$$

3.
$$C_D = \frac{24}{112} + \frac{3}{112^{1/2}} + 0.34$$
$$= 0.84$$

4.
$$v_t^2 = \frac{4}{3} \times 9.81 \times \frac{(2650 - 998.2)}{0.84 \times 998.2} 5 \times 10^{-4}$$
$$v_t = 0.11\ \text{m/s}$$

5. With $v_t = 0.11$, repeat steps 2, 3, and 4.

$\text{Re} = 55$

$C_D = 1.18$

$v_t = 0.10\ \text{m/s} \simeq 0.11\ \text{m/s}$ (see step 4)

Direct application of Eqs. (4-4) through (4-9) is seldom possible in water treatment because the size of particles must be known and a correction factor to account for departure from sphericity has to be determined. An indirect method of measuring settling velocities of discrete particles in dilute suspensions, and of determining settling characteristics of a suspension, was devised by Camp. [4-11] A settling column is constructed as shown in Fig. 4-5. The suspension to be tested is placed in the column and is mixed completely to ensure uniform distribution of the particles. The suspension is then allowed to settle quiescently.

Suppose that a particle is just at the surface at time equal zero and its settling velocity is such that it arrives at the sampling port at a later time, say $t = t_0$. Now, the averaging settling velocity of this particle can be calculated as

$$v_0 = \frac{\text{distance traveled}}{\text{time of travel}} = \frac{Z_0}{t_0}$$

Suppose also that another particle is initially suspended at a distance Z_p above the sampling port and that its terminal settling velocity, less than v_0, is such that it arrives at the port at the same time as the previous particle. Its settling velocity can be calculated as

$$v_p = \frac{\text{distance traveled}}{\text{time of travel}} = \frac{Z_p}{t_0}$$

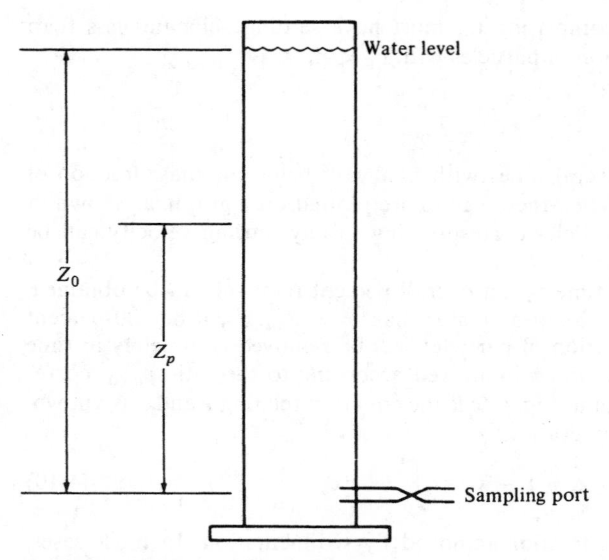

Figure 4-5 Settling column for analyzing type-1 suspension.

Observing that the time of travel is equal for the two particles, it follows that

$$t_0 = \frac{Z_0}{v_0} = \frac{Z_p}{v_p} \quad \text{and} \quad \frac{v_p}{v_0} = \frac{Z_p}{Z_0}$$

Some generalized statements can be made based on the above equation.

1. All particles with diameters equal to or greater than d_0, such that their settling velocities equal or exceed v_0, will arrive at or pass the sampling port in time t_0.
2. A particle with diameter $d_p < d_0$ will have a terminal settling velocity $v_p < v_0$ and will arrive at or pass the sampling port in time t_0, provided its original position was at, or below, a point Z_p.
3. If the suspension is mixed uniformly (i.e., all particle sizes are randomly distributed from top to bottom of the column), then the fraction of particles of size d_p with settling velocity v_p which will arrive at or pass the sampling port in time t_0 will be $Z_p/Z_0 = v_p/v_0$. Thus, the removal efficiency of any size particle from suspension is the ratio of the settling velocity of that particle to the settling velocity v_0 defined by Z_0/t_0.

These principles can be used to determine the settleability of any given suspension. An apparatus similar to that shown in Fig. 4-5 is filled with the suspension to be tested. Theoretically, the depth of the water column is not a factor in the analysis, but practical considerations dictate a depth of about 2 m. The suspension is mixed completely to ensure an initially uniform distribution of particles. A suspended-solids test is run on a sample of the completely mixed suspension, and an initial concentration C_0 is determined. After the suspension is allowed to settle for a time t_1 a second sample is then drawn off and another concentration C_1 is determined. All particles comprising C_1 must have settling velocities less than Z_0/t_1. Thus, the mass fraction of particles with $v_1 < Z_0/t_1$ is

$$x = \frac{C_1}{C_0}$$

The process is repeated several times with x_i always being the mass fraction of particles with $v_i < Z_0/t_i$. When these values are plotted on a graph, as shown in Fig. 4-6, the fraction of particles corresponding to any settling velocity can be obtained.

For a given detention time t_0, an overall percent removal can be obtained. All particles with settling velocities greater than $v_0 = Z_0/t_0$ will be 100 percent removed. Thus, $1 - x_0$ fraction of particles will be removed completely in time t_0. The remaining particles will be removed according to the ratio v_i/v_0, corresponding to the shaded area in Fig. 4-6. If the equation relating v and x is known, the area can be found by integration:

$$X = 1 - x_0 + \int_0^{x_0} \frac{v_i}{v_0}\,dx \tag{4-10}$$

where X is the total mass fraction removed by sedimentation. In most cases, it is simpler to integrate by finite intervals as demonstrated in the following example.

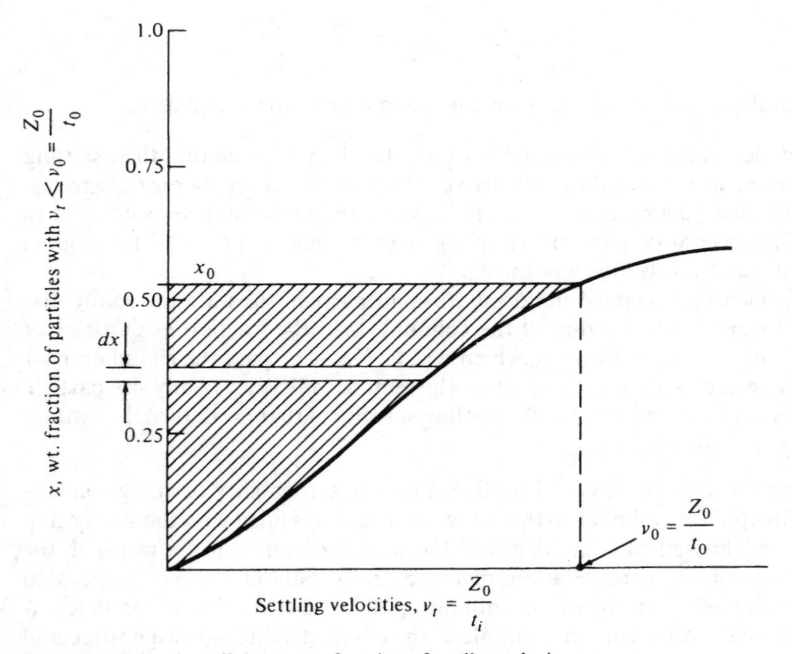

Figure 4-6 Collection efficiency as a function of settling velocity.

Example 4-2: Settling column analysis of type-1 suspension A settling analysis is run on a type-1 suspension. The column is 1.8 m deep, and data are shown below.

Time, min	0	60	80	100	130	200	240	420
Conc, mg/L	300	189	180	168	156	111	78	27

What will be the theoretical removal efficiency in a settling basin with a loading rate of 25 m^3/m^2-d (25 m/d)?

SOLUTION

1. Calculate mass fraction remaining and corresponding settling rates.

Time, min	60	80	100	130	200	240	420
mass fraction remaining	0.63	0.60	0.56	0.52	0.37	0.26	0.09
$v_t \times 10^2$, m/min	3.0	2.5	2.0	1.55	1.0	0.83	0.48

2. Plot mass fraction remaining vs. settling velocity.

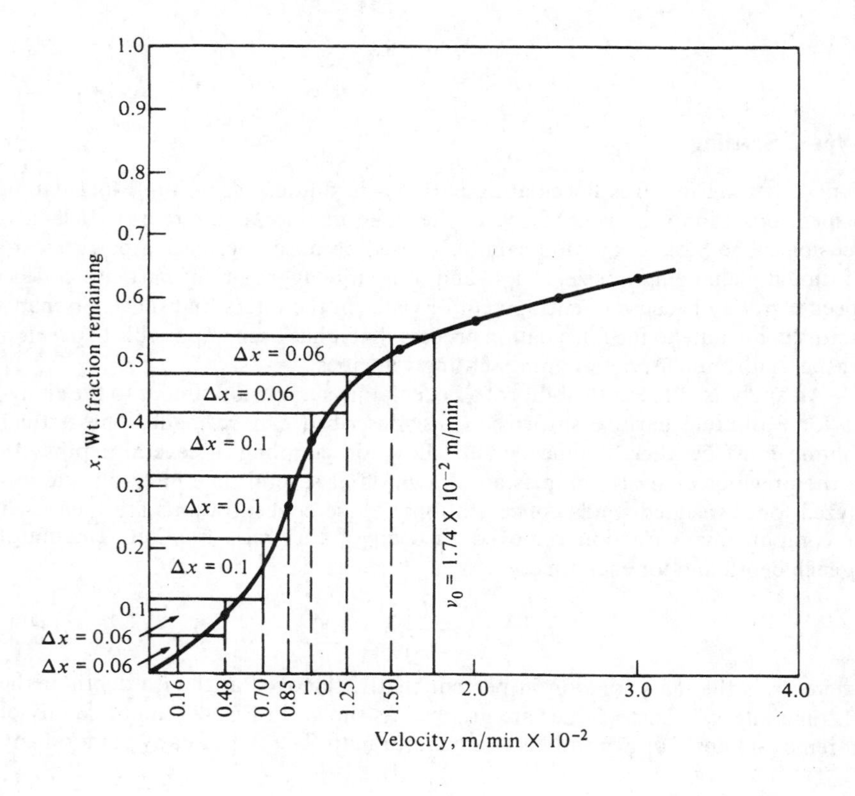

3. Determine $v_0 = 25\ m^3/m^2 \cdot d = 1.74 \times 10^{-2}$ m/min.
4. Determine $x_0 = 54$ percent.
5. Determine $\Delta x \cdot v_t$ by graphical integration.

Δx	v_t	$\Delta x \cdot v_t$
0.06	1.50	0.09
0.06	1.22	0.07
0.1	1.00	0.10
0.1	0.85	0.09
0.1	0.70	0.07
0.06	0.48	0.03
0.06	0.16	0.01

$\Sigma \Delta x \cdot v_t = 0.46$

6. Determine overall removal efficiency.

$$X = 1 - x_0 + \sum \frac{\Delta x \cdot v_t}{v_0}$$

$$= 0.54 + \frac{0.46}{1.74}$$

$$= 0.46 + 0.26$$

$$= 72\%$$

Type-2 Settling

Type-2 settling involves flocculating particles in dilute suspension. Flocculating suspensions cannot be generalized in the same manner as discrete particle suspensions. The Stokes equation cannot be used because flocculating particles are continually changing in size, shape, and, if a large aggregate of particles collect, specific gravity because of entrapment of water in the interstitial space. So many factors contribute to the flocculation process that it has been impossible to develop a general formula for determining settling velocities.

An analysis of the settleability of a flocculating suspension similar to the analysis for a discrete particle suspension just described can be made. The settling column must be altered somewhat to allow for sampling at several depths. As in the previous analysis, samples are drawn off at several time intervals and analyzed for suspended-solids concentrations. These concentrations are then used to compute mass fraction removed (instead of the mass fraction remaining) at each depth and for each time.

$$x_{ij} = \left(1 - \frac{C_{ij}}{C_0}\right) \times 100 \tag{4-11}$$

where x_{ij} is the mass fraction in percent that is removed at the ith depth at the jth time interval. These values are graphed as shown in Fig. 4-7, and a family of isoremoval lines is drawn similar to a contour map. The slope at any point on any

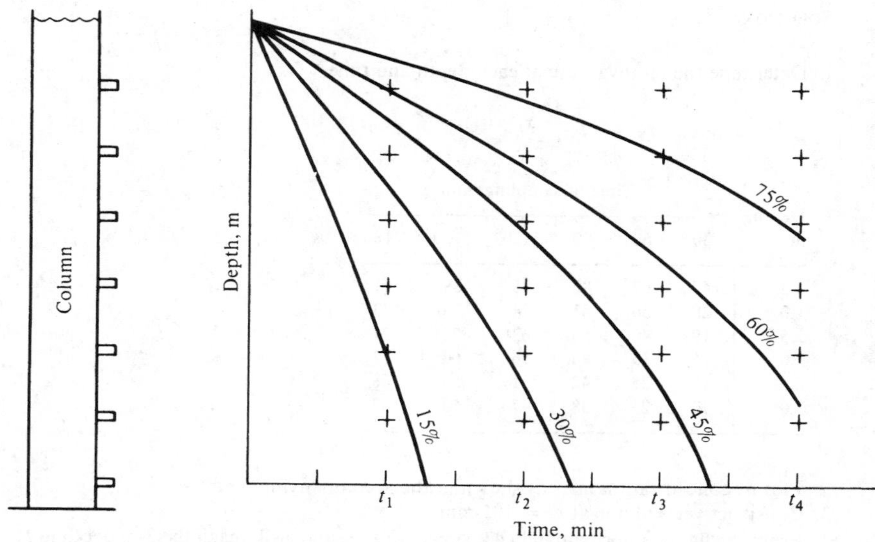

Figure 4-7 Isoremoval lines from settling analysis.

given isoremoval line is the instantaneous velocity of the fraction of particles represented by that line.

It should be noted that the velocity becomes greater (i.e., the slope of the isoremoval lines becomes steeper) at greater depth. This common characteristic of flocculating suspensions reflects the increase in particle size and subsequent increase in settling velocity because of continued collision and aggregation with other particles.

For any predetermined detention time, the overall percentage removed can be obtained as illustrated in the example below.

Example 4-3: Settling column analysis of flocculating particles A column analysis of a flocculating suspension is run in the apparatus shown below. The initial solids concentration is 250 mg/L. The resulting matrix is shown below. What will be the overall removal efficiency of a settling basin which is 3 m deep with a detention time of 1 h and 45 min?

Depth,	Time of sampling, min					
m	30	60	90	120	150	180
0.5	133*	83	50	38	30	23
1.0	180	125	93	65	55	43
1.5	203	150	118	93	70	58
2.0	213	168	135	110	90	70
2.5	220	180	145	123	103	80
3.0	225	188	155	133	113	95

* Results of suspended solids test on sample $C_i \cdot$ mg/L

SOLUTION

1. Determine the removal rate at each depth and time.

$$x_{ij} = (1 - C_{ij}/C_0) \times 100$$

Depth, m	Time of sampling, min 30	60	90	120	150	180
0.5	47	67	80	85	88	91
1.0	28	50	63	74	78	83
1.5	19	40	53	63	72	77
2.0	15	33	46	56	64	72
2.5	12	28	42	51	59	68
3.0	10	25	38	47	55	62

2. Plot isoconcentration lines as shown in the accompanying figure.
3. Construct vertical line at $t_0 = 105$ min.
4. From the figure, approximately 43 percent of the solids will reach the 3-m depth in t_0. They will be 100 percent removed. Some percentage of the remaining particles will be removed. Working upward along the t_0 line, determine increments of removal and depths to the midpoint of these increments.

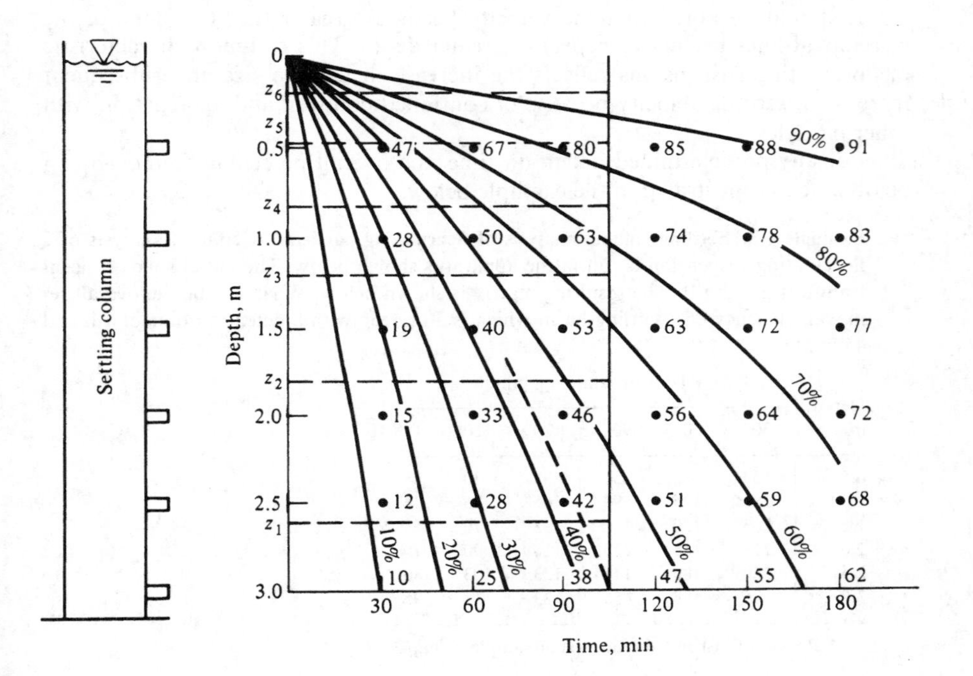

Δr	Z_i	$\Delta r \cdot Z_i$
0.07	2.6	0.18
0.1	1.8	0.18
0.1	1.2	0.12
0.1	0.8	0.08
0.1	0.45	0.04
0.1	0.15	0.01
	$\Sigma \Delta r Z_i =$	0.61

5. Determine the removal efficiency, $R = r_0 + \Sigma \dfrac{\Delta r \cdot Z_i}{Z_0}$

$$= 0.43 + \frac{0.61}{3.0}$$

$$= 63\%$$

4-5 SETTLING OPERATIONS

The sedimentation process has many applications in the preparation of potable water. Materials that may be removed by sedimentation include suspended solids originally present in the water or dissolved solids which have been precipitated in the course of other treatment processes. Suspensions in water-treatment plants are assumed to be dilute, although some zone settling may occur near the bottom of settling basins.

Criteria for design of settling basins have evolved as much from practice as from theory. Settling basins employed for solids removal in water-treatment plants are classified as either long-rectangular, circular, or solids-contact clarifiers. Although these are all continuous-flow systems, the settling theory for batch analysis discussed in the previous section can be applied.

Long-Rectangular Basins

Long-rectangular basins are commonly used in treatment plants processing large flows. This type of basin is hydraulically more stable, and flow control through large volumes is easier with this configuration. A typical long-rectangular tank is shown in Fig. 4-8. Typical designs consist of basins whose length ranges from 2 to 4 times their width and from 10 to 20 times their depth. The bottom is slightly sloped to facilitate sludge scraping. A slow-moving mechanical sludge scraper, usually redwood slats on a chain drive, continuously pulls the settled material into a sludge hopper where it is pumped out periodically.

A long-rectangular settling tank can be divided into four different functional zones:

1. The inlet zone in which baffles intercept the incoming water and spread the flow uniformly both horizontally and vertically across the tank

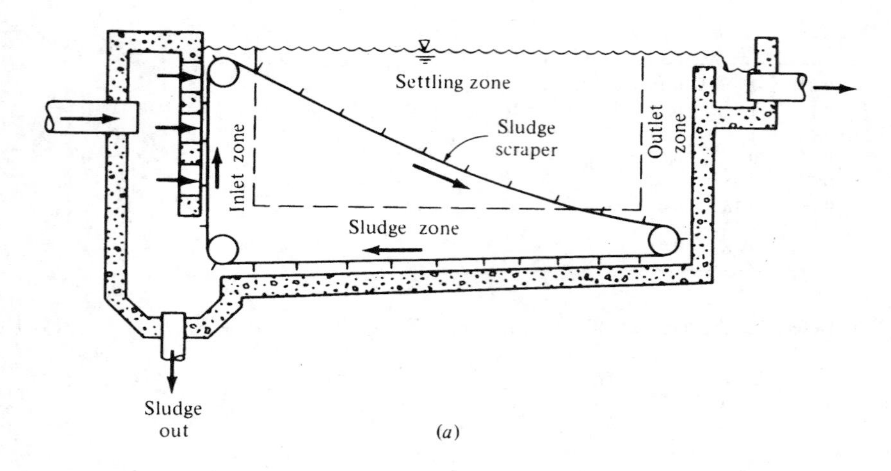

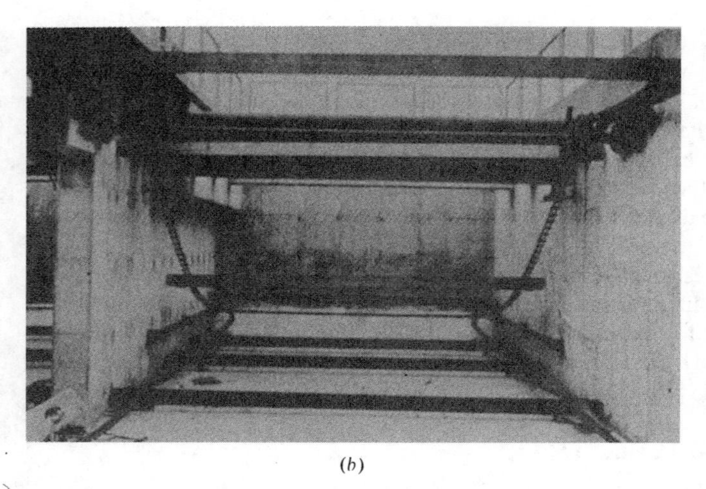

(b)

Figure 4-8 Long-rectangular settling basin: (*a*) diagrammatic sketch and (*b*) view of interior showing sludge scraper arrangement. (*Photo courtesy of Envirex Inc., a Rexnord Company.*)

2. The outlet zone in which water flows upward and over the outlet weir
3. The sludge zone, which extends from the bottom of the tank to just above the scraper mechanism
4. The settling zone, which occupies the remaining volume of the tank

Although all four zones must function properly for efficient solids removal, primary attention here will be focused on the settling zone. Assume that the settling column in Fig. 4-5 is suspended in the flow of the settling zone as shown in Fig. 4-9. The column travels with the flow across the settling zone. Discrete

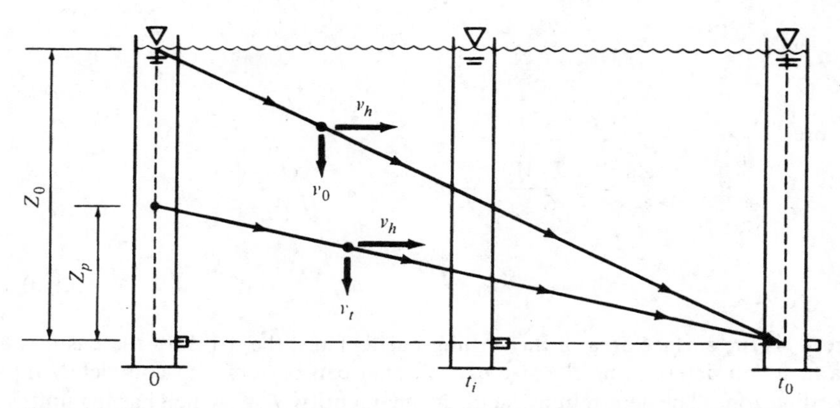

Figure 4-9 Discrete particle removal in the settling zone of a long-rectangular settling basin.

particles falling through the settling column will now have two components of velocity; the vertical component,

$$v_t = g\frac{(\rho_w - \rho_p)d_p^2}{18\,\mu}$$

and the horizontal component, that is, the velocity of flow through the settling zone,

$$v_h = \frac{Q}{A}$$

where A is the cross-sectional area, or the width times the depth. The sum of these velocities is the absolute velocity of the particles.

Now consider the particle in the batch analysis for type-1 settling which was initially at the surface and settled through the depth of the column Z_0, in the time t_0. If t_0 also corresponds to the time required for the column to be carried horizontally across the settling zone, then the particle will fall into the sludge zone and be removed from suspension at the point at which the column reaches the end of the settling zone. As in the batch analysis, all particles with $v_t > v_0$ will be removed from suspension at some point along the settling zone. Now consider the particle with settling velocity $< v_0$. If the initial depth of this particle was such that $Z_p/v_t = t_0$, this particle will also be removed, as shown in the batch analysis. Therefore, the removal of suspended particles passing through the settling zone will be in proportion to the ratio of the individual settling velocities to the settling velocity v_0.

Another point can be made by this analysis. The time t_0 corresponds to the retention time in the settling zone.

$$t_0 = \frac{V}{Q} = \frac{LZ_0W}{Q}$$

Also

$$t_0 = \frac{Z_0}{v_0}$$

therefore

$$\frac{Z_0}{v_0} = \frac{LZ_0 W}{Q} \qquad \text{and} \qquad v_0 = \frac{Q}{LW}$$

or

$$v_0 = \frac{Q}{A_s} \tag{4-12}$$

where A_s is the surface area of the settling basin. Thus, the depth of the basin is not a factor in determining the size particle that can be removed completely in the settling zone. The determining factor is the quantity Q/A_s, which has the units of velocity and is referred to as the overflow rate q_0. This overflow rate, expressed as cubic meters per square-meter hour (or gallons per square-foot day), is the design factor for settling basins and corresponds to the terminal settling velocity of the particle that is 100 percent removed.

If a similar comparison of flocculating particles is made between batch settling and continuous settling in long-rectangular tanks, the path of the falling particles will not be a straight line. As determined in the batch analysis, average velocities of flocculating particles increase with depth. Since the paths of particles tend to curve downward as illustrated in Fig. 4-7, depth *is* a factor in flocculant settling. Therefore, the batch analysis must be performed in a column of the same depth as the basin which it is to model.

Settling basins designed for discrete particles are usually from 2.5 to 3 m deep, while those for flocculating particles are usually 3 to 4 m deep. [4-44] From a practical standpoint, widths in excess of about 12 m create problems with sludge removal equipment; thus lengths are usually kept to less than 48 m. Multiple units in parallel are used to obtain the volume and retention times needed for large flows. In fact, it is always good practice to have at least two units so one can continue functioning while the other is down for repairs or routine maintenance.

For dilute suspensions, overflow rates for discrete particle settling usually range from 1.0 to 2.5 m/h (0.4 to 1.0 gal/ft^2 · min), while overflow rates for flocculating suspensions range from 0.6 to 1.0 m/h (0.25 to 0.4 gal/ft^2 · min). Detention times range from 2 to 4 h for discrete particles and from 4 to 6 h for flocculating suspensions. [4-44]

Although selection of the overflow rate and the detention time determine the size of the basin, other parameters also have to be considered. These include the horizontal velocity v_h and the weir overflow rate q_w.

The motion of the sludge scraper may momentarily resuspend lighter particles and flocs a few centimeters above the scraper blades. Since excessive horizontal velocities would move this material progressively toward the outlet zone where it would be lost in the overflow, horizontal flow velocity should not exceed 9.0 m/h (0.5 ft/min) for light flocculent suspensions or about 36 m/h (2 ft/min) for heavier, discrete-particle suspensions. [4-44]

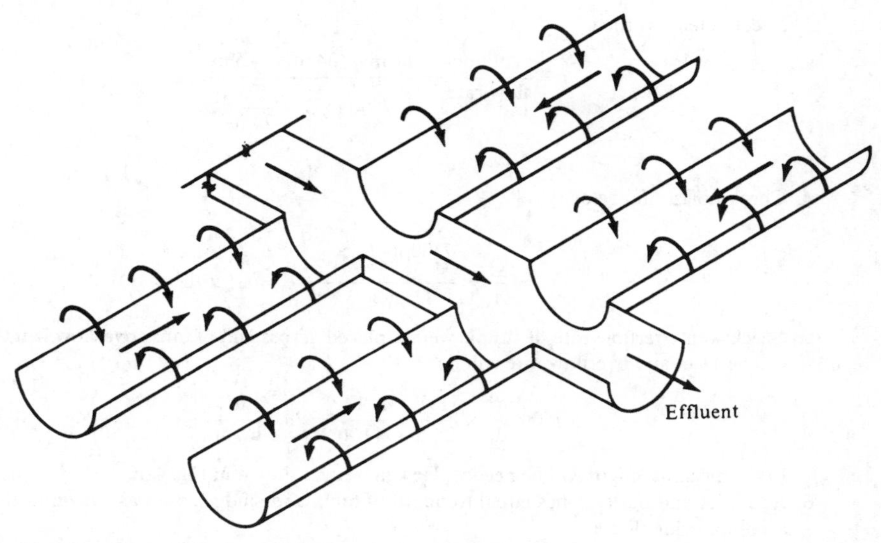

Figure 4-10 Inboard weir arrangement to increase weir length.

Large weir overflow rates result in excessive velocities at the outlet. These velocities extend backward into the settling zone, causing particles and flocs which would otherwise be removed as sludge to be drawn into the outlet. Overflow rates ranging from 6 m^3/h per meter of weir for light flocs to about 14 m^3/h per meter of weir for heavier discrete-particle suspensions are commonly used. [4-44] It may be necessary to provide special inboard weir designs such as the one in Fig. 4-10 to accommodate the lower weir overflow rates.

The design of long-rectangular settling basins is illustrated in the following example.

Example 4-4: Designing a long-rectangular settling basin for type-2 settling A city must treat about 15,000 m^3/d (4 Mgal/d) of water. Flocculating particles are produced by coagulation, and a column analysis indicates that an overflow rate of 20 m/d will produce satisfactory removal at a depth of 3.5 m. Determine the size of the required settling tank.

SOLUTION

1. Compute surface area (provide two tanks at 7500 m^3/d each).

$$Q = q_0 A_s$$

$$7500 \text{ m}^3/\text{d} = A_s \times 20 \text{ m/d}$$

$$A_s = 7500/20 = 375 \text{ m}^2$$

2. Selecting a length-to-width ratio of 3/1, calculate surface dimensions.

$$w \times 3w = 375 \text{ m}^2$$

$$\text{Width} = 11.18, \text{ say } 11 \text{ m}$$

$$\text{Length} = 33.54, \text{ say } 34 \text{ m}$$

3. Check retention time.

$$t = \frac{\text{volume}}{\text{flow rate}} = \frac{11\text{ m} \times 34\text{ m} \times 3.5\text{ m}}{7500\text{ m/d} \times \dfrac{1\text{ d}}{24\text{ h}}}$$

$$= 4.19\text{ h}$$

4. Check horizontal velocity.

$$v_h = \frac{Q}{A_x} = \frac{7500\text{ m}^3\text{/d} \times \dfrac{\text{d}}{24\text{ h}}}{11\text{ m} \times 3.5\text{ m}} = 8.1\text{ m/h}$$

5. Check weir overflow rate. If simple weir is placed across end of tank, overflow length will be 11 m and overflow rate will be

$$7500\,\frac{\text{m}^3}{\text{d}} \times \frac{1\text{ d}}{24\text{ h}} \times \frac{1}{11\text{ m}} = 28.4\,\frac{\text{m}^3}{\text{h} \cdot \text{m}}$$

Five times this length will be needed. Design weir as shown in Fig. 4-10.

6. Add inlet and outlet zones equal to depth of tank, and sludge zones as shown in the accompanying figure.

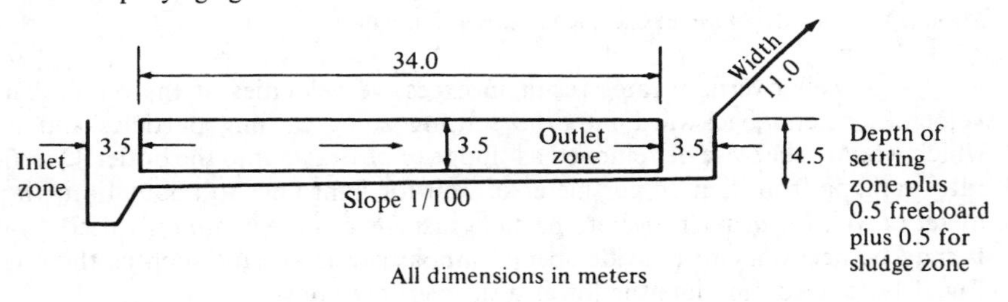

Circular Basins

Circular settling basins have the same functional zones as the long-rectangular basin, but the flow regime is quite different. When the flow enters at the center and is baffled to flow radially toward the perimeter, the horizontal velocity of the water is continually decreasing as the distance from the center increases. Thus, a discrete particle with a settling velocity v_0 is continually undergoing a change in its absolute velocity due to the decrease in horizontal velocity. Thus, the particle path in a circular basin is a parabola as opposed to the straight particle path line in the long-rectangular tank.

Circular tanks have certain advantages. Sludge removal mechanisms are simpler and require less maintenance. [4-57] Excessive weir overflow should never be a problem because the entire circumference is used for overflow. In fact, to prevent extremely thin sheets of water from being drawn off, overflow weirs on circular tanks usually consist of V-notched metal plates which reduce the effective overflow area. These strips are bolted onto the collection trough and can be adjusted to correct for differential settling of the basin after construction.

It is essential that the weir plates be precisely level, since a very slight difference in elevation will result in considerable short circuiting (direct channeling from influent to effluent). Uneven distribution and wind currents can also cause short circuiting. These factors make flow control more difficult in circular basins than in long-rectangular ones. Because flow-control problems become more difficult to control as tank size increases, it is usually advisable to limit circular tank diameters to 30 m or less.

Design of circular settling basins is based on overflow rates and detention times. The limits presented for long-rectangular tanks are applicable to circular tanks. For obvious reasons, neither horizontal velocity nor weir overflow rates are a consideration in the design of circular settling basins.

The following example illustrates the design of circular settling basins.

Example 4-5: Designing a circular settling basin Using the data in Example 4-4, determine the diameter required for circular settling basins.

SOLUTION

1. Again providing two tanks, the surface area is calculated as before.

$$A_s = 375 \text{ m}^2$$

2. The diameter is calculated by

$$\pi d^2/4 = 375 \text{ m}^2$$
$$d = 21.85, \text{ say } 22 \text{ m}$$

3. Inlet, outlet, and sludge zones are provided as shown in the accompanying figure.

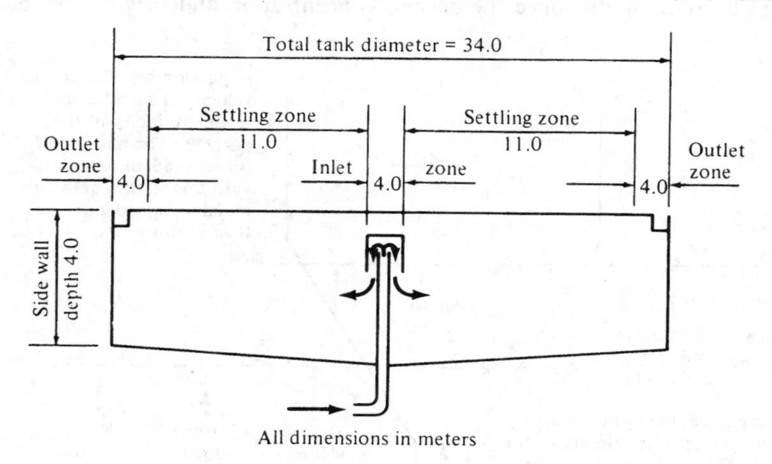

All dimensions in meters

Solids-Contact Basins

Although solids-contact basins differ considerably from either circular or long-rectangular basins with respect to flow regimes, they still make use of the principles of sedimentation previously discussed. The upward velocity of the flow in solids-contact basins works contrary to the settling velocity of the suspended particles. Referring to Fig. 4-11 the vector sum of velocities for any given particle is its terminal settling velocity minus the upward velocity of the flow at the level of the particle. At high upward velocities (near the bottom of the cone), virtually all particles and flocs are swept upward with the flow. As the cross-sectional area of the cone increases, the upward velocity of the water decreases, and the vector sum of the velocities reaches zero; the particle stops and is suspended at that height in the cone. At this point, a particle is considered "removed" from the water. Obviously, larger particles with greater settling velocities become suspended nearer the bottom of the cone where upflow velocities are higher. Particles whose settling velocity is exceeded by the upflow velocity at the top of the cone are swept on upward and into the overflow of the tank.

Particles collect at positions dictated by their settling velocities until a solids blanket is formed, creating a concentrated suspension, even if the original suspension was dilute. The creation of the solids blanket is an important part of upflow clarification. Very small particles, which would normally be swept out of the clarifier, must first pass through the concentrated zone. In flocculating suspensions, the chances are excellent that enmeshment in the sludge blanket will occur so that even very small particles or flocs will be removed. Thus, the blanket acts similarly to a filter for solids removal.

The elevation of the sludge blanket in the cone is determined by the concentration of the solids in the blanket, higher concentration resulting in a greater rise. The concentration, and thus the elevation, of the solids blanket is controlled by drawing off excess sludge once the desired concentration and height have been reached.

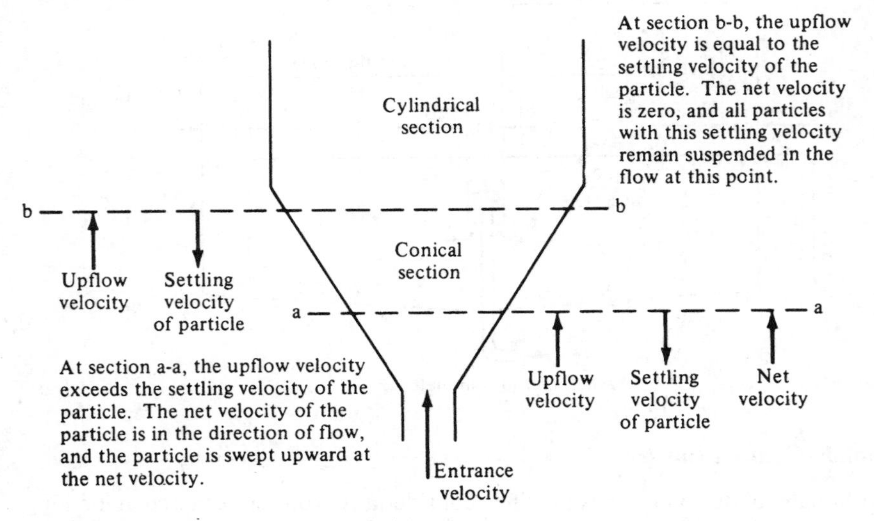

Figure 4-11 Principles of upflow clarification.

As a factor of safety, the unit is usually sized so that the upflow velocity at a point 1.5 m below the top of the cone is one-half of the settling velocity of the particle that is to be removed. [4-1] Generally speaking, upflow velocities of about 1 m/h at the liquid-solids interface for flocculent sludges and about 2 m/h for nonflocculated slurries are adequate. Solids concentration of approximately 3 percent by weight for flocculated sludge blankets and approximately 5 percent by weight for nonflocculated slurries should be maintained in the solids blanket. Since horizontal velocities are to be avoided in this process, overflow weirs should be arranged to ensure essentially vertical flow. Horizontal flow distances should not exceed the depth of the clarified zone by more than a factor of 3. [4-44]

4-6 COAGULATION

Virtually all surface water sources contain perceptible turbidity. Some particle sizes common to most surface waters are listed in Table 4-1, along with their terminal settling velocity (assuming quiescent conditions and specific gravity of 2.65). From these values it is obvious that plain sedimentation will not be very efficient for the smaller suspended particles. Under conditions normally encountered in settling basins, efficient removal of particles less than 50 μm in diameter cannot be expected.

Agglomeration of particles into groups, increasing the effective size and therefore the settling velocities, is possible in some instances. Particles in the colloidal size range, however, possess certain properties that prevent agglomeration. Surface waters with turbidity resulting from colloidal particles cannot be clarified without special treatment. A knowledge of the nature of colloidal suspensions is essential to this removal.

Colloidal Stability

Colloidal suspensions that do not agglomerate naturally are called *stable*. The most important factor contributing to the stability of colloidal suspensions is the excessively large surface-to-volume ratio resulting from their very small size.

Table 4-1 Settling velocities of various size particles*

Particle diameter mm	Size typical of	Settling velocity
10	Pebble	0.73 m/s
1	Coarse sand	0.23 m/s
0.1	Fine sand	1.0×10^{-2} m/s (0.6 m/min)
0.01	Silt	1.0×10^{-4} m/s (8.6 m/d)
0.0001	Large colloid	1.0×10^{-8} m/s (0.3 m/yr)
0.000001	Small colloid	1.0×10^{-13} m/s (3 m/million yr)

* Spheres with specific gravity of 2.65 in water at 20°C.

Surface phenomena predominate over mass phenomena. The most important surface phenomenon is the accumulation of electrical charges at the particle surface. Molecular arrangement within crystals, loss of atoms due to abrasion of the surfaces, or other factors may result in the surfaces being charged. In most surface waters, colloidal surfaces are negatively charged.

Ions contained in the water near the colloid will be affected by the charged surface. A negatively charged colloid with a possible configuration of ions around it is shown in Fig. 4-12. The first layer of cations attracted to the negatively charged surface is "bound" to the colloid and will travel with it, should displacement of the colloid relative to the water occur. Other ions in the vicinity of the colloid arrange themselves as shown, with greater concentrations of positive, or counter, ions being closer to the colloidal surface. This arrangement produces a net charge that is strongest at the bound layer and decreases exponentially with distance from the colloid.

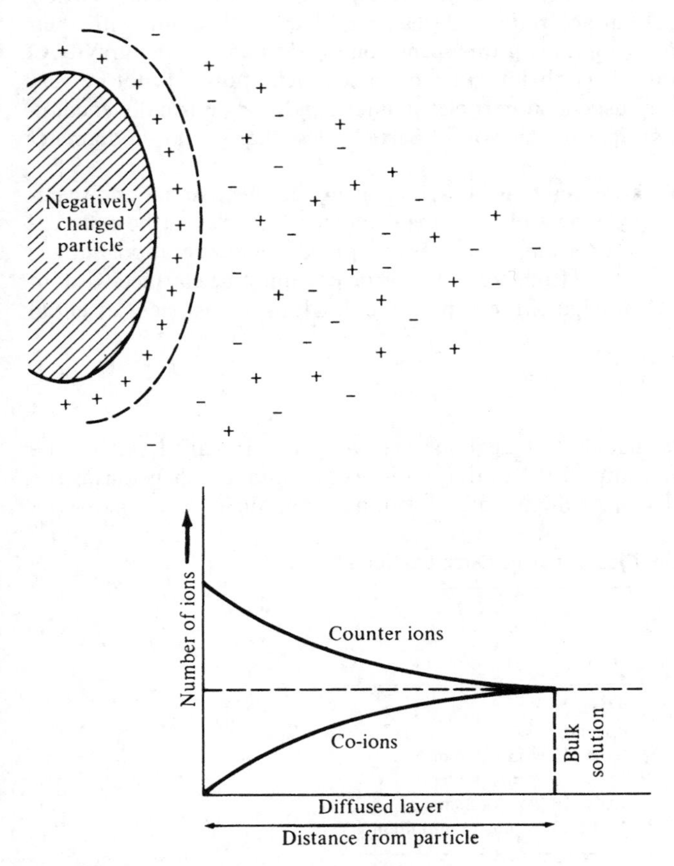

Figure 4-12 Charge system in a colloidal suspension.

When two colloids come in close proximity there are two forces acting on them. The electrostatic potential created by the "halo" of counter ions surrounding each colloid reacts to repel the particles, thus preventing contact. The second force, an attraction force called the *van der Waals force*, supports contact. This force is inversely proportional to the sixth power of the distance between the particles and also decays exponentially with distance. It decreases more rapidly than the electrostatic potential, but is a stronger force at close distances. The sum of the two forces as they relate to one colloid in close proximity to another is illustrated in Fig. 4-13. As noted in the figure, the net force is repulsive at greater distances and becomes attractive only after passing through a maximum net repulsive force, called the *energy barrier*, at some distance between colloids. Once the force becomes attractive, contact between the particles takes place.

A means of overcoming the energy barrier must be available before agglomeration of particles can occur. Brownian movement, the random movement of smaller colloids because of molecular bombardment, may produce enough momentum for particles to overcome the energy barrier and thus collide. Mechanical agitation of the water may impart enough momentum to larger particles to move them across the energy barrier. These processes are too slow, however, to be efficient

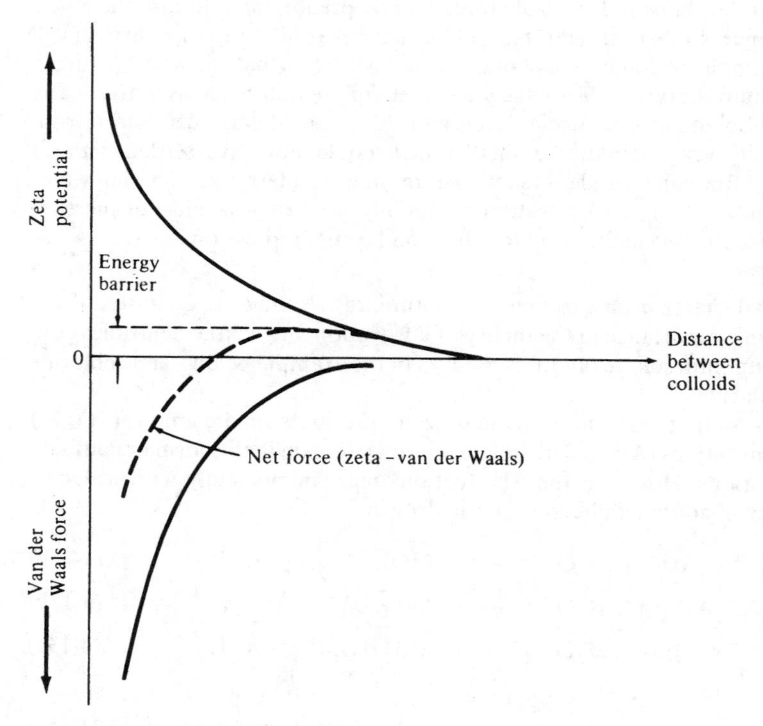

Figure 4-13 Force fields between colloids of like charge.

in water purification, and neither results in collisions of medium-sized colloids. Thus, other means of agglomeration must be used. In water purification this is generally accomplished by chemically coagulating the colloids into clusters, or flocs, which are large enough to be removed by gravity settling.

Coagulation Theory

In water-treatment plants, chemical coagulation is usually accomplished by the addition of trivalent metallic salts such as $AL_2(SO_4)_3$ (aluminum sulfate) or $FeCl_3$ (ferric chloride). Although the exact method by which coagulation is accomplished cannot be determined, four mechanisms are thought to occur. These include ionic layer compression, adsorption and charge neutralization, entrapment in a flocculent mass, and adsorption and interparticle bridging.

Ionic layer compression The quantity of ions in the water surrounding a colloid has an effect on the decay function of the electrostatic potential. As illustrated in Fig. 4-14 a high ionic concentration compresses the layers composed predominantly of counter ions toward the surface of the colloid. If this layer is sufficiently compressed, then the van der Waals force will be predominant across the entire area of influence, so that the net force will be attractive and no energy barriers will exist. An example of ionic layer compression occurs in nature when a turbid stream flows into the ocean. There the ion content of the water increases drastically and coagulation and settling occur. Eventually, deposits (deltas) are formed from material which was originally so small that it could not have settled without coagulation. Although coagulants such as aluminum and ferric salts used in water treatment ionize, at the concentration commonly used they would not increase the ionic concentration sufficiently to affect ion layer compression.

Adsorption and charge neutralization The nature, rather than the quantity, of the ions is of prime importance in the theory of adsorption and charge neutralization. Although aluminum sulfate (alum) is used, as in the example below, ferric chloride behaves similarly.

The ionization of aluminum sulfate in water produces sulfate anions (SO_4^{2-}) and aluminum cations (Al^{3+}). The sulfate ions may remain in this form or combine with other cations. However, the Al^{3+} cations react immediately with water to form a variety of aquometallic ions and hydrogen.

$$Al^{3+} + H_2O \longrightarrow AlOH^{2+} + H^+ \tag{4-13a}$$

$$Al^{3+} + 2H_2O \longrightarrow Al(OH_2)^+ + 2H^+ \tag{4-13b}$$

$$7Al^{3+} + 17H_2O \longrightarrow Al_7(OH)_{17}{}^{4+} + 17H^+ \tag{4-13c}$$

$$\vdots$$

$$Al^{3+} + 3H_2O \longrightarrow Al(OH)_3 + 3H^+ \tag{4-13n}$$

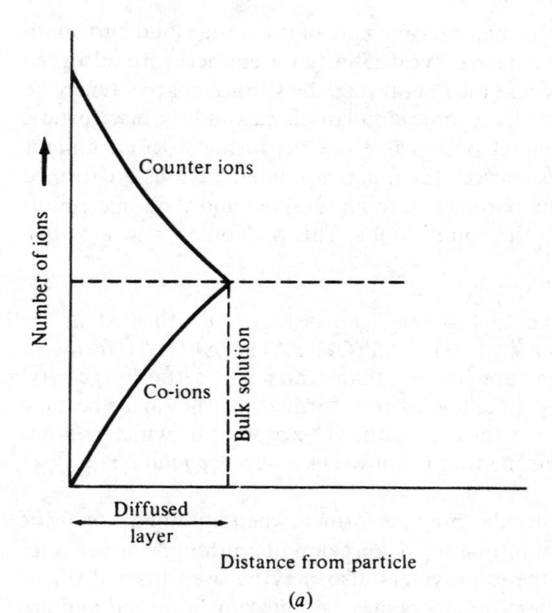

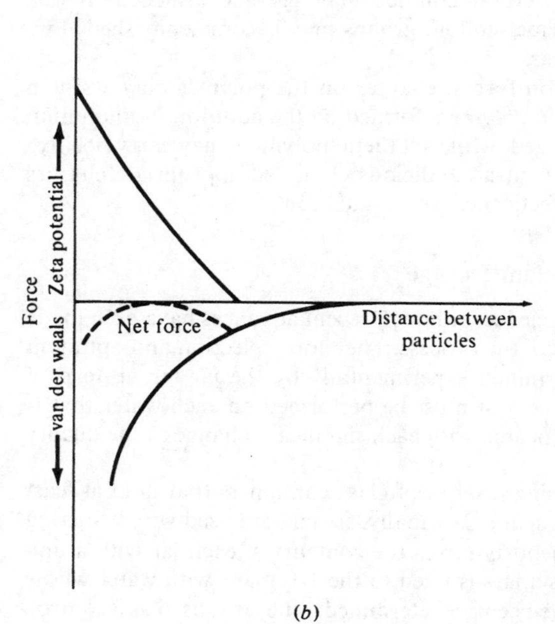

Figure 4-14 Ionic compression: (*a*) reduction of thickness in diffused layer; (*b*) reduction of net force.

The aquometallic ions thus formed become part of the ionic cloud surrounding the colloid and, because they have a great affinity for surfaces, are adsorbed onto the surface of the colloid where they neutralize the surface charge. Once the surface charge has been neutralized, the ionic cloud dissipates and the electrostatic potential disappears so that contact occurs freely. Overdosing with coagulants can result in restabilizing the suspension. If enough aquometallic ions are formed and adsorbed, the charges on the particles become reversed and the ionic clouds reform, with negative ions being the counter ions. This phenomenon will be discussed more fully in a later section.

Sweep coagulation According to Eq. (4-13*n*), the last product formed in the hydrolysis of alum is aluminum hydroxide, $Al(OH)_3$. The $Al(OH)_3$ forms in amorphous, gelatinous flocs that are heavier than water and settle by gravity. Colloids may become entrapped in a floc as it is formed, or they may become enmeshed by its "sticky" surface as the flocs settle. The process by which colloids are swept from suspension in this manner is known as *sweep coagulation.*

Interparticle bridging Large molecules may be formed when aluminum or ferric salts dissociate in water. Equation (4-13*c*) is an example, although larger ones are probably formed also. Synthetic polymers also may be used instead of, or in addition to, metallic salts. These polymers may be linear or branched and are highly surface reactive. Thus, several colloids may become attached to one polymer and several of the polymer-colloid groups may become enmeshed (Fig. 4-15), resulting in a settleable mass.

In addition to the adsorption forces, charges on the polymer may assist in the coagulation process. Metallic polymers formed by the addition of aluminum or ferric salts are positively charged, while synthetic polymers may carry positive or negative charges or may be neutral. Judicious choice of appropriate charges may do much to enhance the effectiveness of coagulation.

Jar Tests for Optimum Coagulant Dosage

Coagulation is not yet an exact science, although recent advances have been made in understanding the mechanics of the process. Therefore, selection and optimum dosages of coagulants are determined experimentally by the jar test instead of quantitatively by formula. The jar test must be performed on each water that is to be coagulated and must be repeated with each significant change in the quality of a given water.

The jar test is performed using a series of glass containers that hold at least 1 L and are of uniform size and shape. Normally, six jars are used with a stirring device (Fig. 4-16) that simultaneously mixes the contents of each jar with a uniform power input. Each of the six jars is filled to the 1-L mark with water whose turbidity, pH, and alkalinity have been predetermined. One jar is used as a control, while the remaining five are dosed with different amounts of coagulant(s) at different pH values until the minimum values of residual turbidity are obtained.

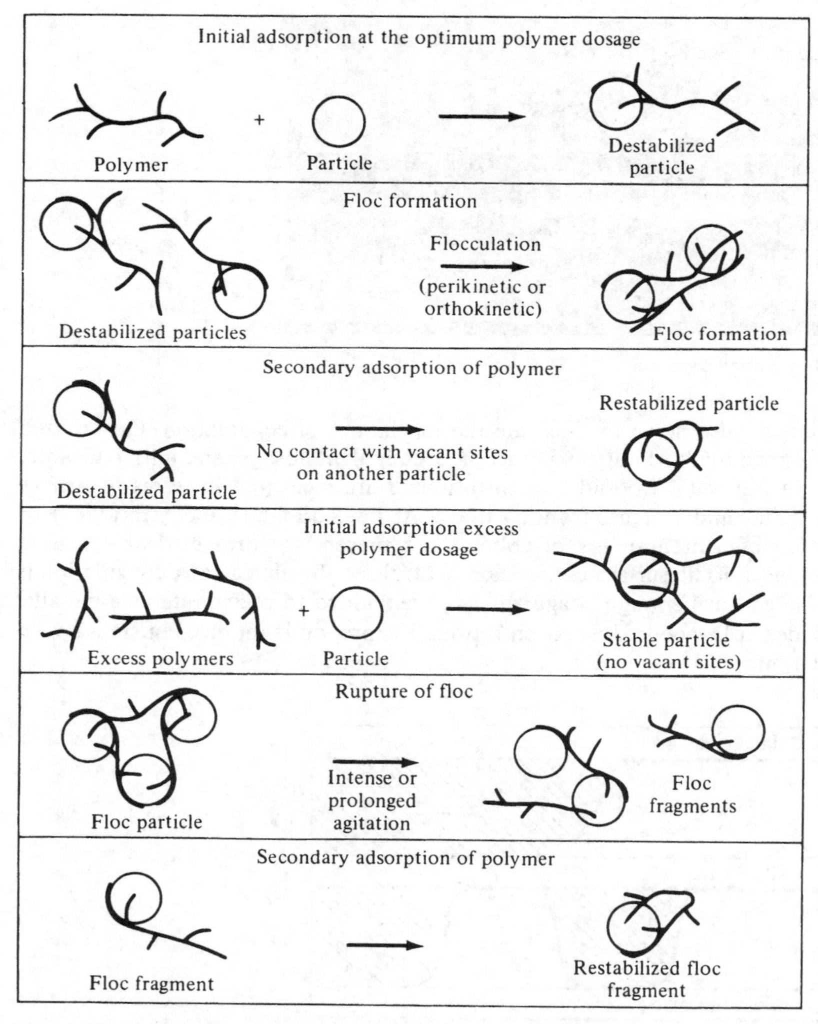

Figure 4-15 Interparticle bridging with polymers. (*After O'Melia* [4-41].)

After chemical addition, the water is mixed rapidly for about 1 min to ensure complete dispersion of the chemicals, then mixed slowly for 15 to 20 min to aid in the formation of flocs. The water is next allowed to settle for approximately 30 min, or until clarification has occurred. Portions of the settled water are then removed and tested to determine the remaining turbidity. Test results are used to calculate the type and quantity of coagulant to be used in the water-treatment plant.

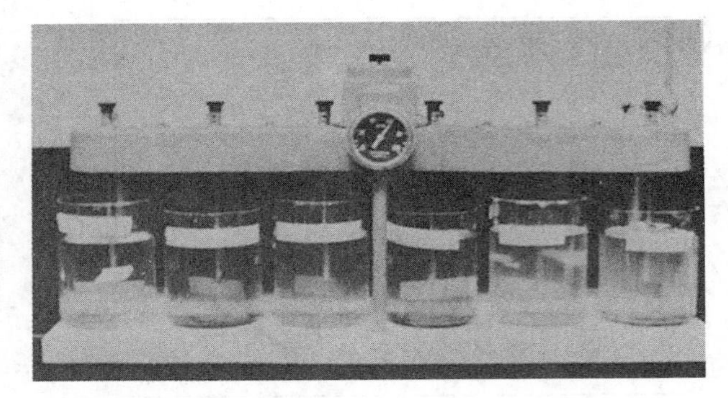

Figure 4-16 Jar test apparatus.

Jar tests also serve to illustrate the mechanics of coagulation. Generalized curves for residual turbidity as a function of coagulant dosages and initial turbidity appear in Fig. 4-17. Colloid concentrations S are measured in terms of surface area per liter and increase from S_1 to S_4. At low colloidal concentrations (Fig. 4-17*a*), insufficient numbers of colloids are present to form settleable masses (zone 1), even if the surface charges are neutralized. In such a case, coagulation is not initiated until enough coagulant has been added to precipitate as a metallic hydroxide. At low colloidal concentrations, the predominant mechanism is sweep coagulation.

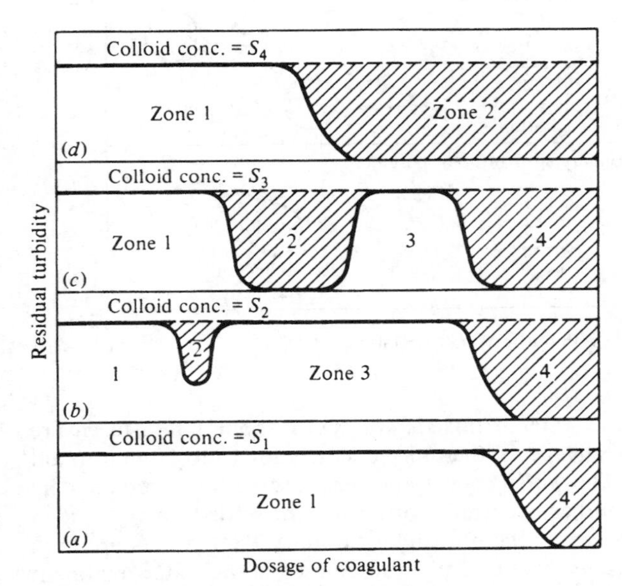

Figure 4-17 Results of jar tests at increasing turbidities. (*After O'Melia* [4-41].)

At higher colloidal concentrations (Fig. 4-17*b*), destabilization by adsorption and charge neutralization occurs early (zone 2), but continued addition of the coagulant results in charge reversal and restabilization (zone 3). Still greater colloidal concentrations (Fig. 4-17*c*) result in more chances of collisions, and thus better coagulation over a wider range of concentration. Continued addition of coagulants results first in restabilization and eventually in hydroxide floc formation and sweep coagulation (zone 4).

Extremely large colloidal concentrations (Fig. 4-17*d*) theoretically provide enough colloids to result in coagulation by adsorption and charge neutralization (zone 2), although it is probable that polymer bridging and sweep coagulation also occur. Restabilization of highly turbid waters is seldom a problem.

Information from many curves similar to those in Fig. 4-17 can be summarized as shown in Fig. 4-18. Two very useful observations can be made from this figure. First, coagulation by adsorption and charge neutralization (zone 2) is impractical unless coagulant dosage can be very carefully controlled. As illustrated by the coagulation region for the colloid concentrations S_2, a very slight overdose results in restabilization. Second, highly turbid waters may require a lesser amount of coagulant for good coagulation than waters with slight turbidity. For this reason it is sometimes advantageous to *add* turbidity to relatively clear water. For example, if water with an initial turbidity corresponding to S_1 were made more turbid,

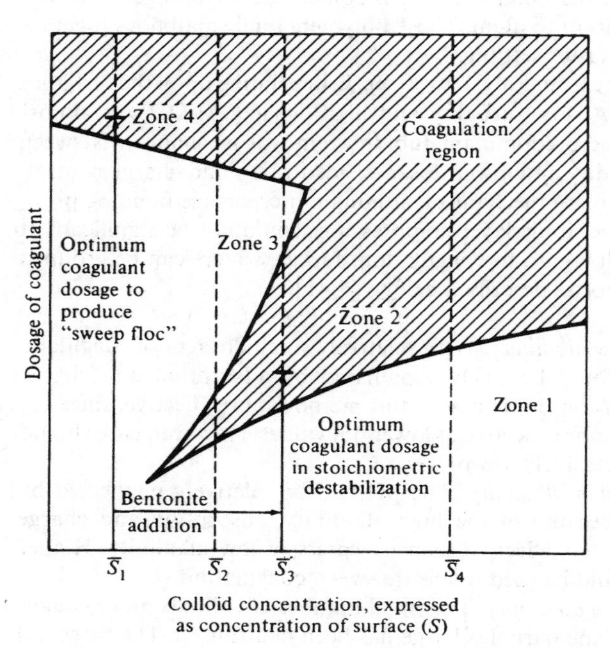

Figure 4-18 Coagulant dosage as a function of turbidity. (*After O'Melia* [*4-41*].)

say to S_3 or beyond, savings in coagulants could be realized. Bentonite clay is generally used for this purpose.

Alkalinity-Coagulation Relationships

As noted in Eq. (4-13), the coagulation of metallic salts releases hydrogen ions as well as coagulant species. These hydrogen ions neutralize alkalinity. Hydrogen resulting from the addition of 1 mg/L of alum will neutralize 0.5 mg/L of alkalinity. If the initial alkalinity of a water is low, further reduction will destroy its buffering capacity and the pH will drop rapidly. Since optimum pH values must be maintained for best coagulation and since alkalinity must be present for hydroxide floc formation, low alkalinity waters must be artificially buffered. This is usually accomplished by the addition of lime [$Ca(OH)_2$] or soda ash (Na_2CO_3).

Coagulation Practice

Aluminum sulfate, the most commonly used coagulant in water purification, is most effective between pH ranges of 5.0 and 7.5. Ferric chloride, effective down to pH 4.5, and ferrous sulfate, effective only above pH 9.5, are sometimes used. [4-54] Although less expensive than alum, these coagulants can cause color problems if the precipitate is not removed completely. It is sometimes advantageous to use synthetic polymers in addition to alum. These polymers bind small flocs together to make larger masses for faster settling.

Alum dosage may range from 5 mg/L to 50 mg/L, depending upon the turbidity and nature of the water. At low turbidity and high dosage, $Al(OH)_3$ is almost certain to form so that the predominant tubidity-removal mechanism is sweep coagulation. At high turbidity and lower dosages, adsorption and charge neutralization will be the predominant mechanism, although interparticle bonding probably plays a significant role. Ionic layer compression would not be significant at these concentrations. With regard to coagulation, surface waters can be grouped into the four general categories described below. [4-41]

Group 1: *High turbidity-low alkalinity.* With relatively small dosages of coagulant, water of this type should be easily coagulated by adsorption and charge neutralization. Depression of pH makes this method more effective, since the aquometallic ions are more effective at lower pH values. However, care should be used to prevent excessively low pH.

Group 2: *High turbidity-high alkalinity.* The pH will be relatively unaffected by coagulant addition. Because of the high alkalinity, adsorption and charge neutralization will be less effective than in waters of low alkalinity. Higher coagulant dosage should be used to ensure sweep coagulation.

Group 3: *Low turbidity-high alkalinity.* The small number of colloids make coagulation difficult, even if the particle charge has been neutralized. The principal coagulation mechanism is sweep coagulation with moderate coagulant dosage. Addition of some turbidity may decrease the amount of coagulant needed.

Group 4: *Low turbidity-low alkalinity*. Again, the small number of colloids make coagulation difficult, and low alkalinity prevents effective $Al(OH)_3$ formation. Additional turbidity can be added to convert this water to that of group 1, or additional alkalinity can be added to convert it to a Group 3 type. It may be advantageous to add both turbidity and alkalinity.

It should be recognized that the above cases are generalizations. Optimum treatment of any water can only be determined by careful analysis using the jar test or other pilot-plant procedures.

The laboratory coagulation jar test is modeled after the coagulation process used in most water-treatment plants. In both cases, the coagulant is first added to the water, and the water is then violently agitated to ensure uniform mixing. This operation is termed *rapid mixing*. The water is then gently stirred to keep all the solids in suspension and to promote collisions between destabilized particles and between particles and flocs. This operation is called *flocculation*. Finally, the water is passed through a settling basin where the flocculated solids are removed by type-2 settling. These operations are illustrated in Fig. 4-19, and detailed descriptions are given in the following sections.

Rapid mixing Thorough mixing is essential if uniform coagulation is to occur. Consequently, careful attention must be paid to the design of rapid-mix units. Design parameters for rapid-mix units are mixing time t and velocity gradient G. The velocity gradient is a measure of the relative velocity of two particles of fluid and the distance between. As an example, two water particles moving 1 m/s relative to each other at a distance 0.1 m apart would have a velocity gradient of:

$$\frac{1.0 \text{ m/s}}{0.1 \text{ m}} = 10 \text{ s}^{-1}$$

A more useful concept of velocity gradients, however, is given in terms of power dissipation per volume. [4-12]

$$G = \left(\frac{P}{V\mu}\right)^{1/2} \tag{4-14}$$

where G = velocity gradient, s^{-1}
P = power input, W (N · m/s)
V = volume of mixing basin, m^3
μ = viscosity, $N \cdot s/m^2$

Rapid mixing can be accomplished in numerous ways, including injection of coagulants at the suction side of pumps, upstream from hydraulic jumps, or in

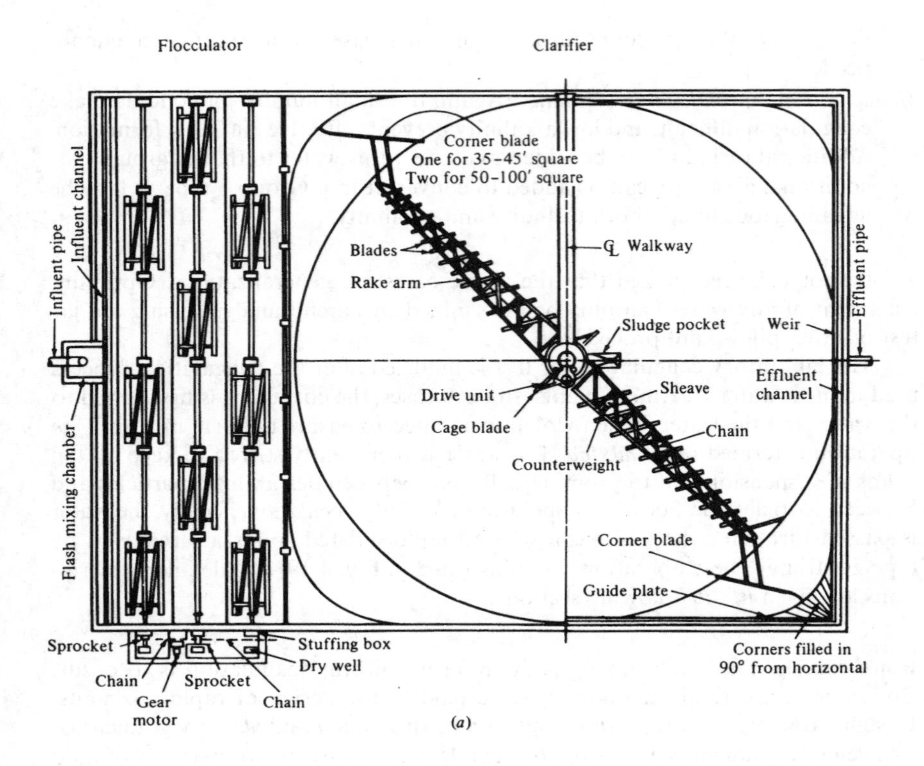

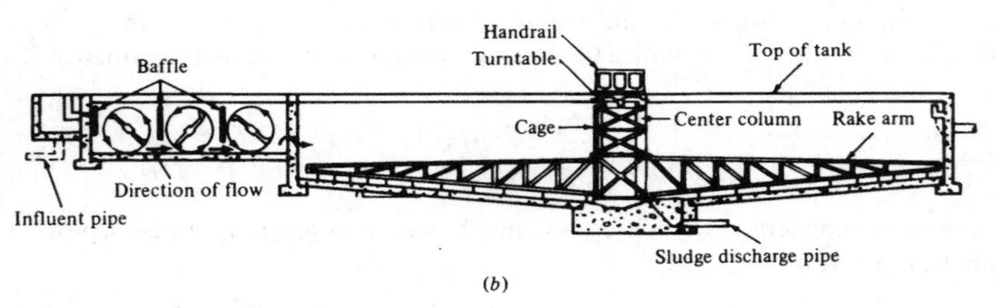

Figure 4-19 Rapid mixing and flocculation followed by a square settling basin: (*a*) plan and (*b*) sectional elevation (*courtesy of Dorr-Oliver, Inc.*).

flow-through basins where head loss around baffles provides power input. Most modern designs, however, use either mixing tanks with back-mix impellers or in-line flash mixers. In-line flash mixers may have mechanically driven impellers or may rely on head loss created by static constrictions in the pipe.

Rapid-mixing tanks operate best at G values from 700 to 1000, with detention times of approximately 2 min. [4-10, 4-36] Numerous configurations of tanks and impellers are used, with the most popular units being square tanks with back-mix

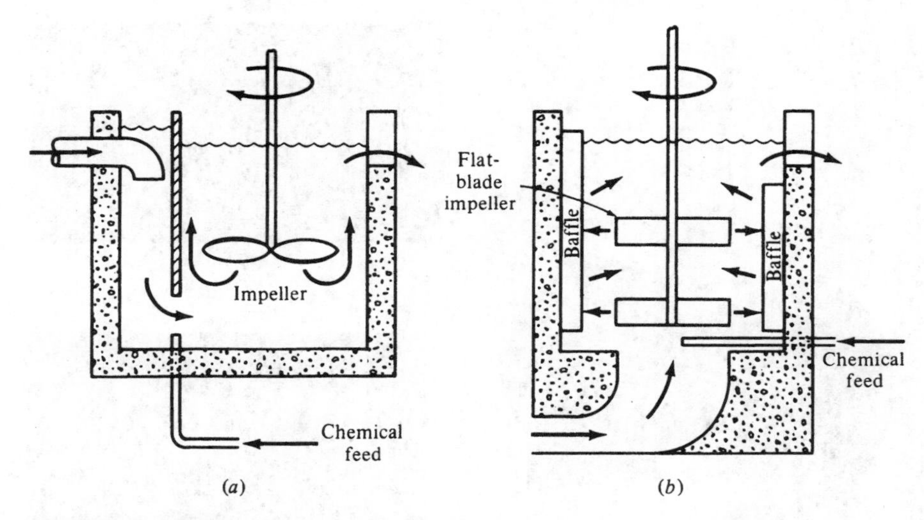

Figure 4-20 Typical rapid-mixing tanks: (*a*) back-mix impeller and (*b*) flat-blade impeller.

impellers (Fig. 4-20*a*). A more effective unit might be a square tank with baffles and flat blade impellers as shown in Fig. 4-20*b*. [4-4]

In-line blenders are designed for complete mixing in less than 1 s. Values of G for in-line blenders, calculated from flow rate and power input or head loss, range from 3000 to 50000 s^{-1}. [4-32] Several configurations of in-line blenders are available; two models are shown in Fig. 4-21.

Flocculation The flocculation process relies on turbulence to promote collisions. Velocity gradients are also a convenient way of measuring this turbulence. Time is an important factor, and the design parameter for flocculation is Gt, a dimensionless number. Values of Gt from 10^4 to 10^5 are commonly used, with t ranging from 10 to 30 min. [4-28] Large G values with short times tend to produce small, dense flocs, while low G values and long times produce larger, lighter flocs. Since large, dense flocs are more easily removed in the settling basin, it may be advantageous to vary the G values over the length of the flocculation basin. The small, dense flocs produced at high G values subsequently combine into larger flocs at the lower G values. Reduction in G values by a factor of 2 from the influent end to the effluent end of the flocculator has been shown to be effective. [4-33]

Traditional flocculator design is illustrated in Fig. 4-22. These units consist of long-rectangular basis equipped with mechanically operated paddles to provide power input. The paddles are usually constructed of redwood or aluminum slats and may operate either transverse or parallel to the longitudinal axis of the basin.

More recent design tends toward units which combine rapid mixing, flocculation, and settling in one tank. Such a unit is shown in Fig. 4-23. The principles of mixing and flocculation in this unit are the same as those for the long-rectangular

(*a*)

(*b*)

(*c*)

Figure 4-21 In-line blenders: (*a*) power-driven (*courtesy of Walker Process Corp.*); (*b*) static mixer pipe section (*courtesy of Komax Systems, Inc.*); (*c*) static mixer in 2.5 × 2.5-m-square channel section. This mixer processes 530,000 m^3/d at the Val Vista water-treatment plant in Phoenix, Arizona. Note chemical feed lines in front of vanes (*courtesy of Komax Systems, Inc.*)

basin, although the method of operation is somewhat different. While these units are limited in the quantity of flow that they can handle, multiple units can be provided in parallel to meet any demand.

The G value for mechanically driven flocculators is calculated as follows. First, the power input is determined by

$$P = Dv_p \tag{4-15}$$

where P = power input, W (N · m/s)
D = drag force on paddles, N
v_p = velocity of paddles, m/s

(*a*)

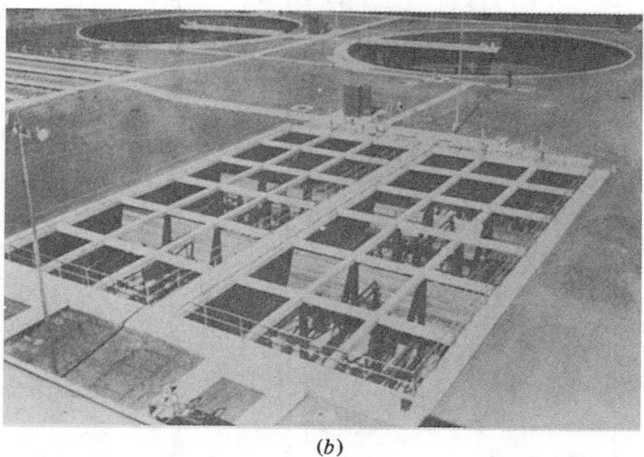

(*b*)

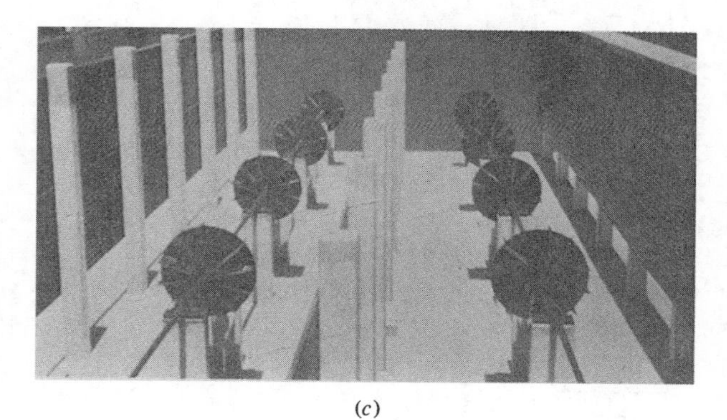

(*c*)

Figure 4-22 Flocculator units: (*a*) paddle arrangement in long tank (*photo courtesy of Envirex Inc., a Rexnord Company*); (*b*) multistage units (*photo courtesy of Walker Process Corp.*); (*c*) turbine-type units (*photo courtesy of Envirex, Inc., a Rexnord Company*).

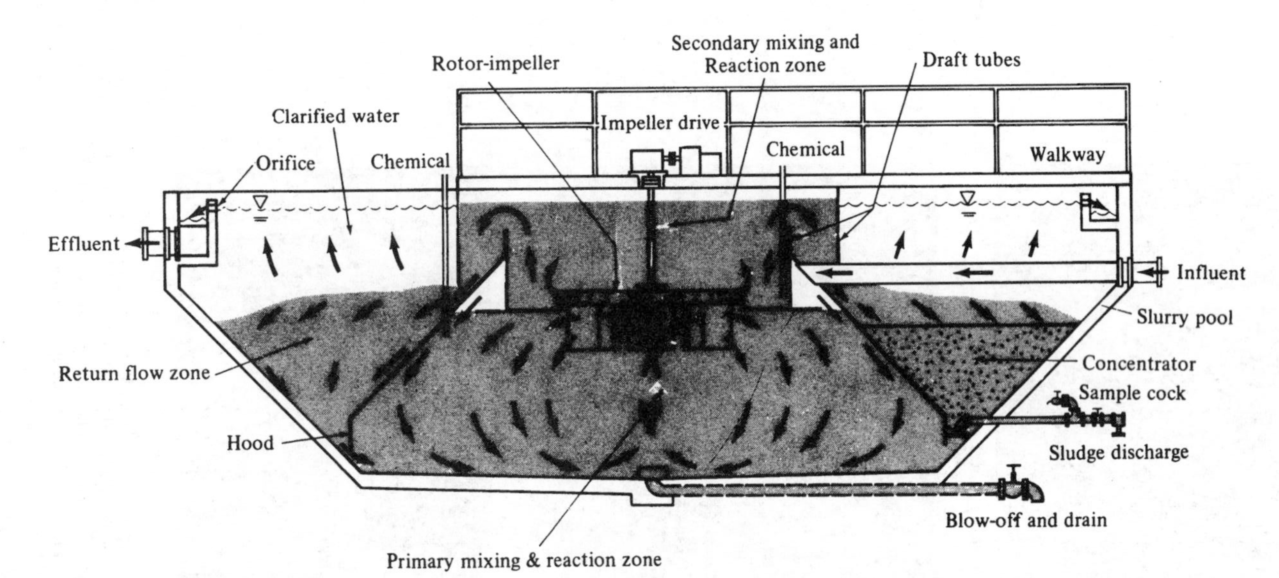

Figure 4-23 Solids-contact clarifier with mixing and flocculation included. (*Courtesy of Infilco Degremont, Inc.*)

The drag force on the paddle is given by

$$D = C_D A_p \rho \frac{v_p^2}{2} \tag{4-16}$$

where C_D = dimensionless coefficient of drag, 1.8 for flat blades
A_p = area of paddle blades, m^2
ρ = density of water, kg/m^3

Equation (4-15) now becomes

$$P = \frac{C_D A_p \rho v_p^3}{2}$$

Substituting into Eq. (4-14)

$$G = \left(\frac{C_D A_p \rho v_p^3}{2V\mu}\right)^{1/2}$$

The area of the paddle A_p refers to the combined area of the slats that are perpendicular to the cylinder of rotation. This area should not exceed 40 percent of the total area encompassed by the paddle. The velocity of the paddle tip v_p is the velocity relative to the water and is about 75 percent of the actual paddle speed. Paddle velocity should be less than 1 m/s, and a minimum distance of 0.3 m should be maintained between paddle tips and all other structures in the flocculator to prevent local areas of excessive velocity gradients.

It should be noted that in transferring water from the flocculator basin to the settling basin, extreme care must be exercised to avoid turbulence that can break up the floc. This is usually not a problem in units in which mixing, flocculation, and settling are combined. In the long-rectangular units, the settling basin is often constructed adjacent to the flocculator, with the common wall omitted. Necessary baffles are designed for low G values.

Design of flocculation units is illustrated in the following example.

Example 4-6: Designing a flocculator A water-treatment plant is being designed to process 50,000 m^3/d of water. Jar testing and pilot-plant analysis indicate that an alum dosage of 40 mg/L with flocculation at a Gt value of 4.0×10^4 produces optimal results at the expected water temperatures of 15°C. Determine:

1. The monthly alum requirement.
2. The flocculation basin dimensions if three cross-flow horizontal paddles are to be used. The flocculator should be a maximum of 12 m wide and 5 m deep in order to connect appropriately with the settling basin.
3. The power requirement.
4. The paddle configuration.

SOLUTION

1. Monthly alum requirements:

$$40 \text{ mg/L} = 0.04 \text{ kg/m}^3$$

and

$$\frac{0.04 \text{ kg}}{\text{m}^3} \times 50{,}000 \frac{\text{m}^3}{\text{d}} \times 30 \text{ d/mo} = 60{,}000 \text{ kg/mo}$$

2. Basin dimensions:

a. Assume an average G value of 30 s^{-1}

$$Gt = 4.0 \times 10^4$$

$$t = \frac{4.0 \times 10^4}{30} \frac{1 \text{ min}}{60 \text{ s}}$$

$$t = 22.22 \text{ min}$$

b. Volume of the tank is

$$V = Qt = 50{,}000 \text{ m}^3/\text{d} \times 22.22 \text{ m/m} \times 1 \text{ d}/1440 \text{ min}$$
$$= 771.5 \text{ m}^3$$

c. The tank will contain three cross-flow paddles, so its length will be divided into three compartments. For equal distribution of velocity gradients, the end area of each compartment should be square, i.e., depth equals $\frac{1}{3}$ length. Assuming maximum depth of 5 m, length is

$$3 \times 5 = 15 \text{ m}$$

and width is

$$5 \times 15 \times w = 771.5$$

$$w = 10.3 \text{ m}$$

d. The configuration of the tanks and paddles should be as follows:

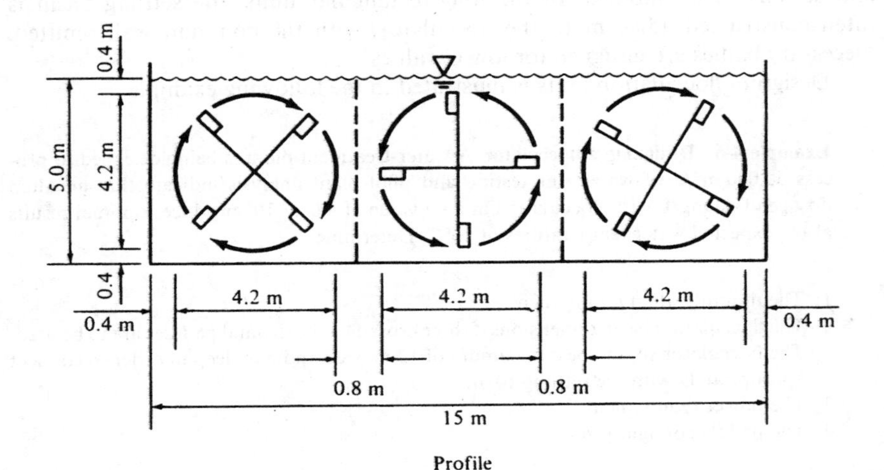

Profile

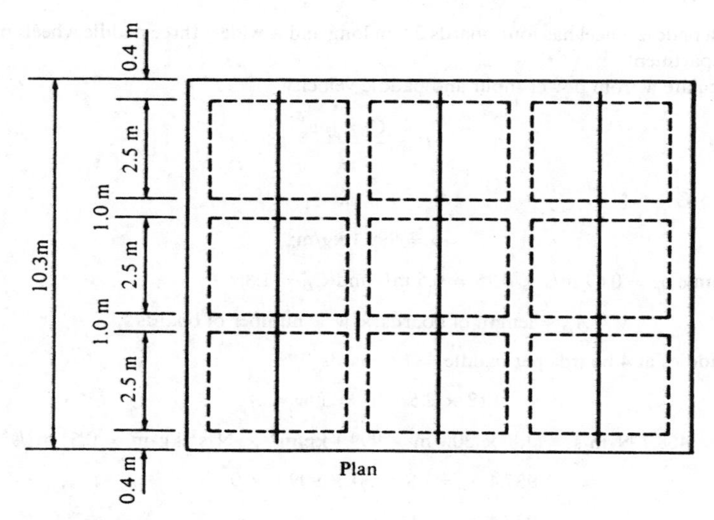

3. Power requirements:
   a. Assume G value tapered as follows.

$$\text{First compartment, } G = 40\ \text{s}^{-1}$$
$$\text{Second compartment, } G = 30\ \text{s}^{-1}$$
$$\text{Third compartment, } G = 20\ \text{s}^{-1}$$

   b. Calculate power requirements for compartments 1, 2, and 3:

$$P = G^2 V \mu$$
$$V = 771.5\ \text{m}^3/3 = 257.2\ \text{m}^3$$

At 15°C

$$\mu = 1.139 \times 10^{-3}\ \text{N}\cdot\text{s/m}^2$$
$$P_1 = 40^2/\text{s}^2 \times 257.2\ \text{m}^3 \times 1.139 \times 10^{-3}\ \text{N}\cdot\text{s/m}^2$$
$$= 468.7\ \text{N}\cdot\text{m/s} \times 10^{-3}\ \text{kW/N}\cdot\text{m/s} = 0.47\ \text{kW}$$
$$P_2 = 30^2 \times 257.2 \times 1.139 \times 10^{-3} \times 10^{-3} = 0.26\ \text{kW}$$
$$P_3 = 20^2 \times 257.2 \times 1.139 \times 10^{-3} \times 10^{-3} = 0.12\ \text{kW}$$

4. Paddle configuration
   a. Assume paddle design as shown below.

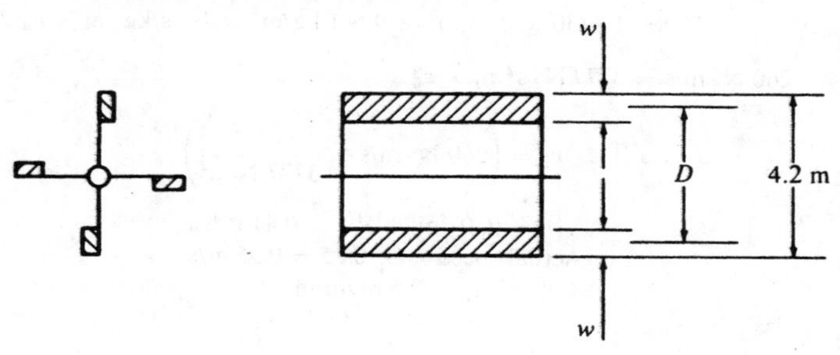

Each paddle wheel has four boards 2.5 m long and w wide—three paddle wheels per compartment.

b. Calculate w from power input and paddle velocity.

$$P = \frac{C_D A_p \rho v_p^3}{2}$$

At 15°C

$$\rho = 999.1 \text{ kg/m}^3$$

Assume $v_p = 0.67 \text{ m/s} \times 0.75 = 0.5 \text{ m/s}$ and $C_D = 1.8$.

$$A_p = \text{length of boards} \times w \times \text{number of boards}$$

3 paddles at 4 boards per paddle = 12 boards

$$12 \times 2.5 \times w = 30w = A_p$$

$$P_1 = 468.7 \text{ N·m/s} = (1.8 \times 30w \text{ m} \times 999.1 \text{ kg/m}^3 \times \text{N·s}^2\text{/kg·m} \times 0.5^3 \text{ m}^3\text{/s}^3)/2$$

$$937.4 \text{ m} = 1.8 \times 30 \times 999.1 \times 0.5^3 w$$

$$937.4 \text{ m} = 6744w$$

$$w = 0.14 \text{ m}$$

c. Calculate rotational speed of paddles.

First compartment:

$$v_p = \pi D \frac{\text{m}}{\text{rev}} \times \omega$$

$$= \pi 4.2 \frac{\text{m}}{\text{rev}} \times \omega$$

$$0.67 \text{ m/s} \times \frac{\text{rev}}{4.2\pi \text{ m}} \times \frac{60 \text{ s}}{\text{min}} = 3.05 \text{ rev/min} = \omega$$

Second compartment:

$$P = 0.26 \text{ kW} \times 10^3 \frac{\text{N} \cdot \text{m/s}}{\text{kW}}$$

$$= 260 \text{ N} \cdot \text{m/s} = \frac{C_D A_p \rho v_p^3}{2}$$

$$= 1.8\,(30 \times 0.14)\text{m}^2 \times 999.1 \text{ kg/m}^3 \times \text{N} \cdot \text{s/kg} \cdot \text{m} \times v_p^3/2$$

$$260 \text{ N} \cdot \text{m/s} = 3777 \text{ N} \cdot \text{s}^2\text{/m} \times v_p^3$$

$$v_p = \left(260 \text{ N} \cdot \text{m/s} \times \frac{1 \text{ m}^2}{3777 \text{ N} \cdot \text{s}^2}\right)^{1/3}$$

$$= (0.07 \text{ m}^3\text{/s}^3)^{1/3} = 0.41 \text{ m/s}$$

$$\text{Actual speed} = v_p/0.75 = 0.55 \text{ m/s}$$

$$\omega = 2.5 \text{ rev/min}$$

Third compartment:

$$P_3 = 120\ \text{N}\cdot\text{m/s} = 3777\ \text{N}\cdot\text{s}^2/\text{m}^2 \times v_p^3$$

$$v_p = \left(120\ \text{N}\cdot\text{m/s} \times \frac{1\ \text{m}^2}{3777\ \text{N}\cdot\text{s}^2}\right)^{1/3}$$

$$= (0.03\ \text{m}^3/\text{s}^3)^{1/3} = 0.32\ \text{m/s}$$

$$v_p\ \text{actual} = 0.32\ \text{m/s} \times \frac{1}{0.75} = 0.42\ \text{m/s}$$

$$\omega = 1.91\ \text{rev/min}$$

4-7 SOFTENING

Hardness as a water-quality parameter was discussed in Sec. 2-10. The reduction of hardness, or *softening*, is a process commonly practiced in water treatment. Softening may be done by the water utility at the treatment plant or by the consumer at the point of use, depending on the economics of the situation and the public desire for soft water. Generally, softening of moderately hard water (50 to 150 mg/L hardness) is best left to the consumer, while harder water should be softened at the water-treatment plant. Softening processes commonly used are chemical precipitation and ion exchange, either of which may be employed at the utility-owned treatment plant. Home-use softeners are almost exclusively ion-exchange units.

Chemical Precipitation

The different species of hardness have different solubility limits, as shown in Table 4-2. The least soluble forms are calcium carbonate and magnesium hydroxide. Chemical precipitation is accomplished by converting calcium hardness to calcium carbonate and magnesium hardness to magnesium hydroxide. This can be accomplished by the lime–soda ash process or by the caustic soda process.

Lime–soda ash All forms of carbonate hardness as well as magnesium noncarbonate hardness can be converted to the precipitating species by the addition of lime (CaO). In the following equations, the symbol s is used to indicate that a solid precipitate forms and that it is sufficiently dense to settle by gravity.

$$Ca^{2+} + 2(HCO_3)^- + CaO + H_2O \longrightarrow 2CaCO_3\$ + 2H_2O \tag{4-17}$$

$$Mg^{2+} + 2(HCO_3)^- + CaO + H_2O \longrightarrow CaCO_3\$ + Mg^{2+} + CO_3^{\,2-} \tag{4-18}$$

$$Mg^{2+} + CO_3^{\,2-} + CaO + H_2O \longrightarrow CaCO_3\$ + Mg(OH)_2\$ \tag{4-19}$$

$$Mg^{2+} + \left\{\begin{matrix} SO_4^{\,2-} \\ 2Cl^- \\ 2NO_3^{\,-} \end{matrix}\right\} + CaO + H_2O \longrightarrow Ca^{2+} + \left\{\begin{matrix} SO_4^{\,2-} \\ Cl_2^{\,-} \\ 2NO_3^{\,-} \end{matrix}\right\} + Mg(OH)_2\$ \tag{4-20}$$

Table 4-2 Equilibrium of solid and dissolved species of common ions

Mineral	Formula	Solubility, mg/L $CaCO_3$ at 0°C
Calcium bicarbonate	$Ca(HCO_3)_2$	1,620
Calcium carbonate	$CaCO_3$	15
Calcium chloride	$CaCl_2$	336,000
Calcium sulfate	$CaSO_4$	1,290
Calcium hydroxide	$Ca(OH)_2$	2,390
Magnesium bicarbonate	$Mg(HCO_3)_2$	37,100
Magnesium carbonate	$MgCO_3$	101
Magnesium chloride	$MgCl_2$	362,000
Magnesium hydroxide	$Mg(OH)_2$	17
Magnesium sulfate	$MgSO_4$	170,000
Sodium bicarbonate	$NaHCO_3$	38,700
Sodium carbonate	Na_2CO_3	61,400
Sodium chloride	$NaCl$	225,000
Sodium hydroxide	$NaOH$	370,000
Sodium sulfate	Na_2SO_4	33,600

Source: Adapted from Loewenthal and Marais. [4-38]

As seen in Eq. (4-20), removal of magnesium noncarbonate hardness results in the formation of calcium noncarbonate hardness. This calcium noncarbonate hardness, as well as any initially present in the water, can be removed by the addition of soda ash (Na_2CO_3):

$$Ca^{2+} + \begin{Bmatrix} SO_4^{\ 2-} \\ Cl_2^{\ -} \\ 2NO_3^{\ -} \end{Bmatrix} + Na_2 + CO_3 \longrightarrow CaCO_3\$ + 2Na^+ + \begin{Bmatrix} SO_4^{\ 2-} \\ 2Cl^- \\ 2NO_3^{\ -} \end{Bmatrix} \tag{4-21}$$

The sodium in Eq. (4-21) is soluble and, unless excessive amounts are added, is permissible in potable water.

The precipitation of $CaCO_3$ and $Mg(OH)_2$ is pH-dependent, as is illustrated in Fig. 4-24. The optimum pH for $CaCO_3$ precipitation by line addition is from 9 to 9.5, while effective precipitation of $Mg(OH)_2$ under water-treatment plant conditions requires a pH of about 11.0. Since most natural waters have a pH considerably below these values, it is often necessary to artificially raise the pH. This can be accomplished by the addition of an excess amount of lime:

$$CaO + H_2O \longrightarrow Ca^{2+} + 2OH^- \tag{4-22}$$

The addition of about 1.25 mequiv/L of lime is sufficient to raise the pH to 11.0.

If dissolved carbon dioxide is present in water it will also react with lime.

$$CO_2 + CaO \longrightarrow CaCO_3\$ \tag{4-23}$$

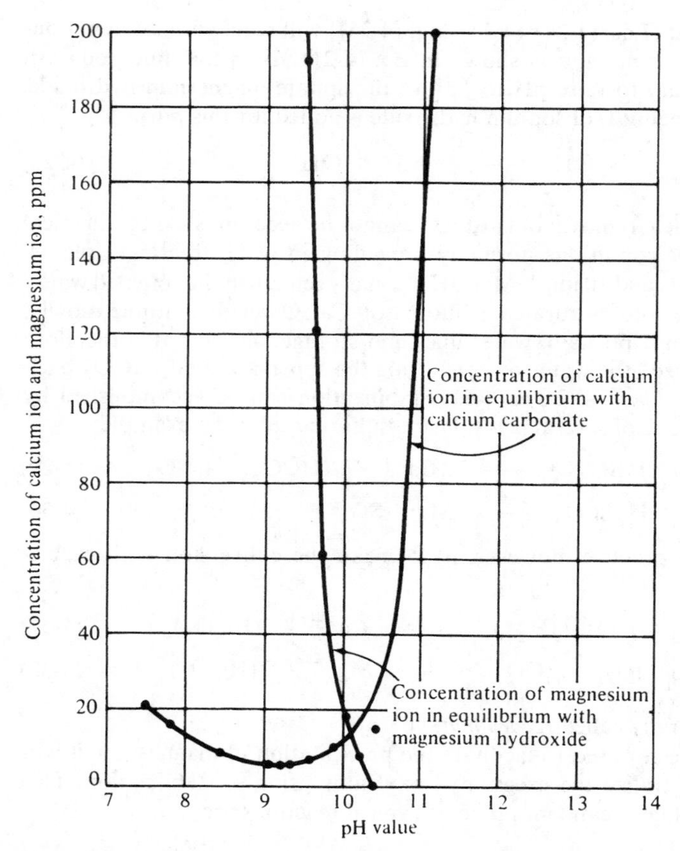

Figure 4-24 Equilibrium concentration of calcium and magnesium ions as a function of pH. (*From Powell* [*4-43*].)

Although this reaction does not reduce hardness, it does consume lime. Removal of supersaturated CO_2 by aeration is often practiced to reduce lime requirements. If CO_2 exceeds 10 mg/L it may be economically advantageous to remove it prior to softening.

Caustic soda. All forms of hardness can also be converted to the precipitating species by the addition of caustic soda (NaOH).

$$CO_2 + 2NaOH \longrightarrow 2Na^+ + CO_3^{2-} + H_2O \tag{4-24}$$

$$Ca^{2+} + 2(HCO_3)^- + 2NaOH \longrightarrow CaCO_3\downarrow + 2Na^+ + CO_3^{2-} + 2H_2O \tag{4-25}$$

$$Mg^{2+} + 2(HCO_3)^- + 4NaOH \longrightarrow Mg(OH)_2\downarrow + 4Na^+ + 2CO_3^{2-} + 2H_2O \tag{4-26}$$

$$Mg^{2+} + SO_4^{2-} + 2NaOH \longrightarrow Mg(OH)_2\downarrow + 2Na^+ + SO_4^{2-} \tag{4-27}$$

The soda ash formed [Eqs. (4-24), (4-25), and (4-26)] will react with calcium non-carbonate hardness as previously shown in Eq. (4-21). As in the lime–soda ash process, it is necessary to raise pH to 11.0 to precipitate magnesium hydroxide. An excess of 1.25 mequiv/L of sodium hydroxide is added for this purpose.

$$NaOH \longrightarrow Na^+ + OH^- \tag{4-28}$$

Stabilization Complete removal of hardness cannot be accomplished by chemical precipitation. Under conditions normally prevailing in water-treatment plants, up to 40 mg/L $CaCO_3$ and 10 mg/L $Mg(OH)_2$ usually remain in the softened water. Precipitation of the supersaturated solution of $CaCO_3$ will continue slowly, however, resulting in deposits in water lines and storage facilities. It is therefore necessary to "stabilize" the water by converting the supersaturated $CaCO_3$ back to the soluble form, $Ca^{2+} + 2(HCO_3)^-$. Stabilization can be accomplished by the addition of any one of several acids. Using sulfuric acid as an example:

$$2CaCO_3 + H_2SO_4 \longrightarrow 2Ca^{2+} + 2(HCO_3)^- + SO_4{}^{2-} \tag{4-29}$$

$$Mg(OH)_2 + H_2SO \longrightarrow Mg^{2+}\ SO_4{}^{2-} + 2H_2O \tag{4-30}$$

The most common practice, however, is to make the conversion with carbon dioxide:

$$CaCO_3 + CO_2 + H_2O \longrightarrow Ca^{2+} + 2(HCO_3)^- \tag{4-31}$$

$$Mg(OH)_2 + 2CO_2 \longrightarrow Mg^{2+} + 2(HCO_3) \tag{4-32}$$

This process is generally called *recarbonation.*

If the pH has been raised to facilitate the precipitation of magnesium, it will be necessary to neutralize the excess hydroxyl ions prior to stabilization. This necessitates a two-stage treatment process. Typical reactions are:

With sulfuric acid

$$Ca^{2+} + 2OH^- + H_2SO_4 \longrightarrow Ca^{2+} + SO_4{}^{2-} + 2H_2O \tag{4-33}$$

$$2Na^+ + 2OH^- + H_2SO_4 \longrightarrow 2Na^+ + SO_4{}^{2-} + 2H_2O \tag{4-34}$$

With carbon dioxide

$$Ca^{2+} + 2OH^- + 2CO_2 \longrightarrow CaCO_3\$ + H_2O \tag{4-35}$$

$$2Na^+ + 2OH^2 + CO_2 \longrightarrow 2Na^+ + CO_3{}^{2-} + H_2O \tag{4-36}$$

The pH must be lowered to approximately 9.5 before significant stabilization occurs.

Chemical requirement The quantity of chemicals to soften water can be calculated using the appropriate formulas from Eqs. (4-17) through (4-36). These calculations are illustrated in Examples 4-7 and 4-8.

Example 4-7: Single-stage softening A water with the ionic characteristics shown in the bar diagram below is to be softened to the minimum calcium hardness by the lime–soda ash process. Magnesium removal is not deemed necessary.

1. Calculate the chemical requirements and solids produced in milliequivalents per liter.
2. Draw a bar diagram for the finished water.
3. For a flow of 25,000 m^3/d, calculate the daily chemical requirement and the mass of solids produced. Assume that the lime used is 90 percent pure and the soda ash is 85 percent pure.

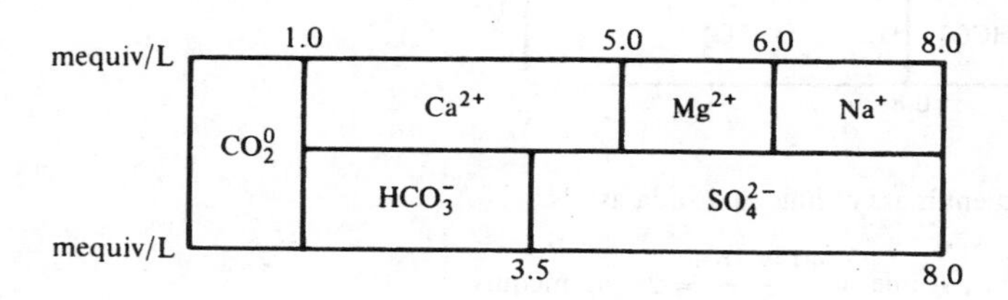

SOLUTION

The following treatment scheme will be used.

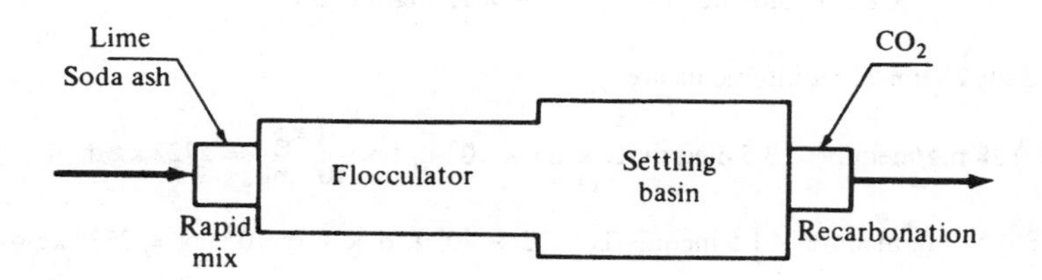

1. Calculate chemical requirements using appropriate formulas.

$$1.0CO_2 + 1.0CaO \longrightarrow 1.0CaCO_3\downarrow$$

$$2.5(Ca^{2+} + 2HCO_3) + 2.5CaO \longrightarrow 5.0CaCO_3\downarrow + 5.0H_2O$$

$$1.5(Ca^{2+} + SO_4{}^{2-}) + 1.5Na_2CO_3 \longrightarrow 1.5CaCO_3\downarrow + 1.5(2Na^+ + SO_4{}^{2-})$$

Second-stage recarbonation will be required to stabilize the water. Assuming a $CaCO_3$ concentration of 40 mg/L in the effluent from the settling basin, 25 mg/L should be converted to reach the equilibrium of 15 mg/L of $CaCO_3$.

$$0.5CaCO_3 + 0.5CO_2 + 0.5H_2O \longrightarrow 0.5Ca(HCO_3)_2$$

Total chemical requirements are:

Lime = 1.0 + 2.5 = 3.5 mequiv/L

Soda ash = 1.5 = 1.5 mequiv/L

CO_2 = 0.5 mequiv/L

Solids produced are

$$CaCO_3 = 1.0 + 5.0 + 1.5 - 0.8 = 6.7 \text{ mequiv/L}$$

2. The bar diagram for the finished water is

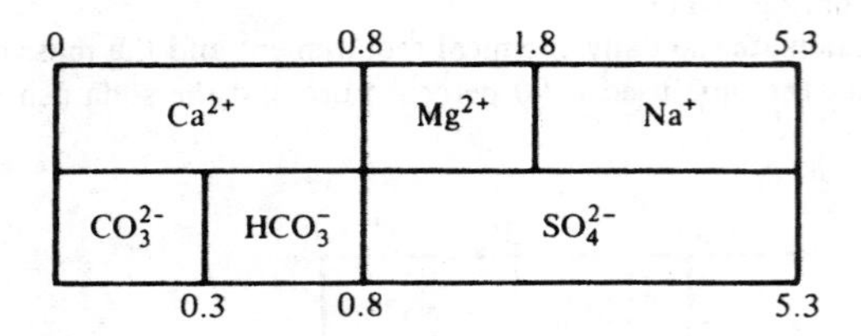

3. The equivalent mass of lime and soda ash is

$$\text{Lime} = \frac{40 + 16}{2} = 28 \text{ mg/mequiv}$$

$$\text{Soda ash} = \frac{2(23) + 12 + 3(16)}{2} = 53 \text{ mg/mequiv}$$

$$\text{Carbon dioxide} = \frac{12 + 2(16)}{2} = 22 \text{ mg/mequiv}$$

The daily chemical requirements are:

$$(1/0.9)\ 28 \text{ mg/mequiv} \times 3.5 \text{ mequiv/L} \times 25 \times 10^6 \text{ L/d} \times \frac{1 \text{ kg}}{10^6 \text{ mg}} = 2722 \text{ kg/d}$$

$$(1/0.85)\ 53 \text{ mg/mequiv} \times 1.5 \text{ mequiv/L} \times 25 \times 10^6 \text{ L/d} \times 1 \text{ kg}/10^6 \text{ mg} = 2338 \text{ kg/d}$$

$$22 \text{ mg/mequiv} \times 0.5 \text{ mequiv/L} \times 25 \times 10^6 \text{ L/d} \times 1 \text{ kg}/10^6 \text{ mg} = 275 \text{ kg/d}$$

The mass of dry solids produced per day is

$$50 \text{ mg/mequiv} \times 6.7 \text{ mequiv/L} \times 25 \times 10^6 \text{ L/d} \times 1 \text{ kg}/10^6 \text{ mg} = 8375 \text{ kg/d}$$

Example 4-8: Two-stage softening A water with the ionic characteristics shown below is to be softened to the minimum possible hardness by the lime–soda-ash–excess-lime process. Calculate the required chemical quantities in milliequivalents per liter. Draw a bar diagram of the finished water.

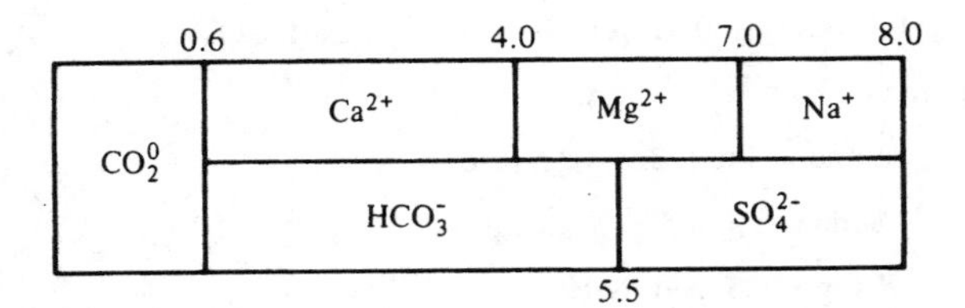

SOLUTION The following treatment scheme will be used.

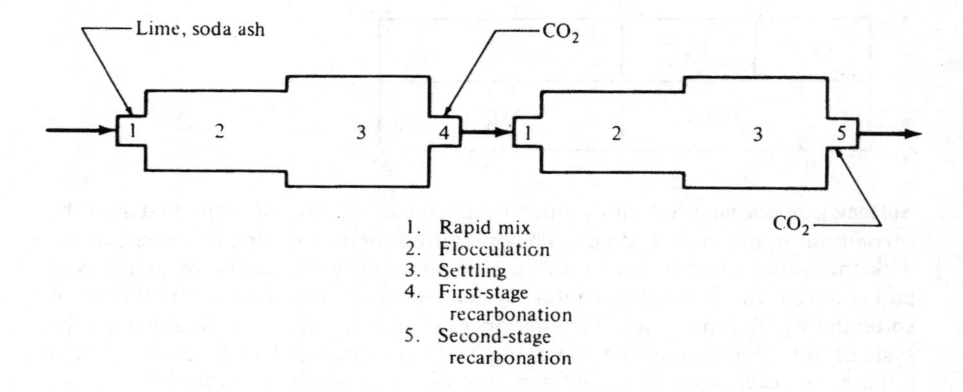

1. Calculate chemical quantities using appropriate formulas.

$$0.6\,CO_2 + 0.6\,CaO \longrightarrow 0.6\,CaCO_3 \downarrow$$

$$3.4(Ca^{2+} + 2\,HCO_3^-) + 3.4\,CaO \longrightarrow 6.8\,CaCO_3 \downarrow + 6.8\,H_2O$$

$$1.5(Mg^{2+} + 2\,HCO_3^-) + 3.0\,CaO \longrightarrow 1.5\,Mg(OH)_2 \downarrow + 3.0\,CaCO_3 \downarrow$$

$$1.5(Mg^{2+} + SO_4^{2-}) + 1.5\,CaO + 1.5\,H_2O \longrightarrow 1.5\,Mg(OH)_2 \downarrow + 1.5(Ca^{2+} + SO_4^{2-})$$

$$1.5(Ca^{2+} + SO_4^{2-}) + 1.5\,Na_2CO_3 \longrightarrow 1.5\,CaCO_3 \downarrow + 1.5(2\,Na^+ + SO_4^{2-})$$

Excess lime = 1.25 mequiv/L

For first-stage recarbonation, use CO_2 to neutralize excess lime.

$$1.25(Ca^{2+} + 2\,OH^-) + 1.25\,CO_2 \longrightarrow 1.25\,CaCO_3 \downarrow + 1.25\,H_2O$$

Assuming 40 mg/L $CaCO_3$ and 10 mg/L $Mg(OH)_2$ remaining in solution after second-stage settling,

$$0.2\,Mg(OH)_2 + 0.4\,CO_2 \longrightarrow 0.2\,Mg(HCO_3)_2$$

$$0.5\,CaCO_3 + 0.5\,CO_2 + 0.5\,H_2O \longrightarrow 0.5\,Ca(HCO_3)_2$$

Total chemical quantities are

$$\text{Lime} = 0.6 + 3.4 + 3.0 + 1.5 + 1.25 = 9.75$$

$$\text{Soda ash} = 1.5$$

$$CO_2 = 1.25 + 0.4 + 0.5 = 2.15$$

2. Bar diagram of final water:

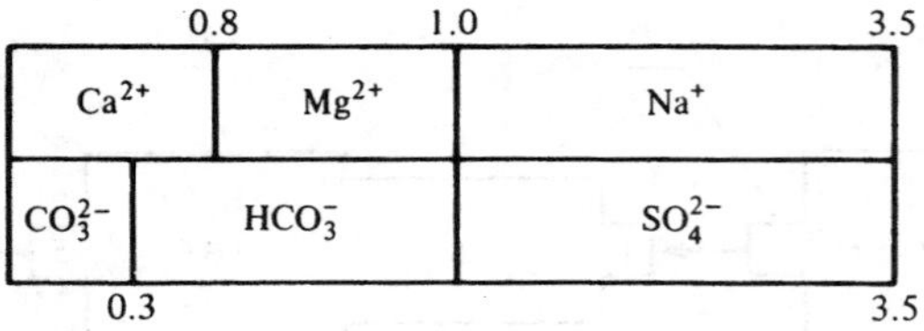

0.8	1.0	3.5
Ca^{2+}	Mg^{2+}	Na^{+}

0.3		3.5
CO_3^{2-}	HCO_3^-	SO_4^{2-}

Softening operations Softening operations consist of several steps and may be carried out in one or two stages. The operations include mixing of the chemicals with the water, flocculation to aid in precipitate growth, settling of precipitate, and stabilization. The solids-contact system shown in Fig. 4-23 is often used for softening operations. These systems operate in much the same manner as the systems for coagulating and removing turbidity discussed in Sec. 4-6. Design criteria, however, are slightly different and are summarized in Table 4-3.

Table 4-3 Typical design criteria for softening systems

Parameter	Mixer	Flocculator	Settling basin	Solids-contact basin
Detention time*	5 min	30–50 min	2–4 h	1–4 h
Velocity gradient, s^{-1}	700	10–100	NA	†
Flow-through velocity, ft/s	NA	0.15–0.45	0.15–0.45	NA
Overflow rate, gal/min/ft^2	NA	NA	0.85–1.71	4.27‡

* This should be confirmed by pilot-plant analysis for each water.

† Velocity gradient in mixer and flocculator component should be approximately the same as in flow-through units.

‡ At slurry blanket-clarifier water interface.

Source: Adapted from *Recommended Standards*. [4-44]

Water with high magnesium hardness is often softened by a process called *split treatment*. This process bypasses the first-stage softening unit with a part of the incoming water. Excess lime is added to facilitate the removal of magnesium in the first stage and, instead of being neutralized thereafter, is used to precipitate the calcium hardness in the bypassed water in the second stage. Since no magnesium is removed in the bypassed water, the initial magnesium hardness and the allowable magnesium hardness in the finished water govern the quantity that may be bypassed:

$$Q_x = \frac{\mathrm{Mg}_f - \mathrm{Mg}_1}{\mathrm{Mg}_r - \mathrm{Mg}_1} \tag{4-37}$$

where Q_x = fraction of the total flow bypassed

Mg_f = Magnesium concentration in the finished water, 40–50 mg/L (as $CaCO_3$) usually acceptable

Mg_r = magnesium concentration in the raw water, mg/L
Mg_1 = magnesium concentration remaining in the fraction of the water receiving first-stage treatment. [As previously stated, practical limits are 10 mg/L $Mg(OH)_2$ (as $CaCO_3$).]

A typical split-treatment system for removing magnesium is shown in Fig. 4-25. The quantity of softening chemicals saved by this system is illustrated in the following example.

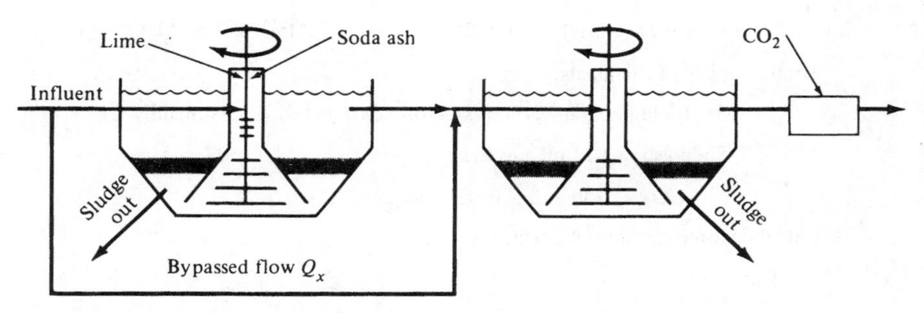

Figure 4-25 Flow diagram for softening by split treatment.

Example 4-9: Softening by split treatment Use split treatment to soften the water with ionic strength given in Example 4-8. Assume that a final hardness of less than 100 mg/L is acceptable, provided the magnesium is less than 45 mg/L. Calculate the chemical requirements and draw a bar diagram of the finished water.

SOLUTION

The treatment scheme shown in Fig. 4-25 will be used.

1. Calculate the bypass fraction:

$$Q_x = \frac{Mg_f - Mg_1}{Mg_r - Mg_1}$$

$$= \frac{0.9 - 0.2}{3.0 - 0.2}$$

$$= 0.25$$

2. Calculate the quantity of chemicals added to first stage:

$$0.6CO_2 + 0.6CaO \longrightarrow CaCO_2 \downarrow$$

$$3.4(Ca^{2+} + 2HCO_3^-) + 3.4CaO \longrightarrow .8CaCO_3 \downarrow + 6.8H_2O$$

$$(1.0 - 0.25)(1.5)(Mg^{2+} + 2HCO_3^-) + (1.0 - 0.25)3.0CaO$$
$$\longrightarrow 1.13Mg(OH)_2 \downarrow + 2.25CaCO_3 \downarrow$$

$$(1.0 - 0.25)(1.5)(Mg^{2+} + SO_4^{2-}) + (1.0 - 0.25)1.5CaO$$
$$\longrightarrow 1.13Mg(OH)_2 + 1.13(Ca^{2+} + SO_4^{2-})$$

$$1.13(Ca^{2+} + SO_4^{2-}) + 1.13(Na_2CO_3)$$
$$\longrightarrow 1.13CaCO_3 \downarrow + 1.13(2Na^+ + SO_4^{2-})$$

Check to make sure extra lime is enough to provide 1.25 mequiv/L:

$$\frac{(0.6 + 3.4)0.25}{0.75} = 1.33$$

1.33 > 1.25, so acceptable

For second-stage recarbonation:

$$0.5CaCO_3 + 0.5CO_2 + 0.5H_2O \longrightarrow 0.5(Ca^{2+} + 2HCO_3^-)$$

$$0.75 \times 0.2Mg(OH)_2 + 0.30CO_2 \longrightarrow 0.15(Mg^{2+} + 2HCO_3^-)$$

Total quantity of chemicals:

$$\text{Lime} = 0.6 + 3.4 + (1.0 - 0.25)(3.0 + 1.5) = 7.38 \text{ mequiv/L}$$

$$\text{Soda ash} = 1.13 \text{ mequiv/L}$$

$$CO_2 = 0.30 + 0.5 = 0.80 \text{ mequiv/L}$$

3. Calculate ionic strength of finished water:

$$Ca^{2+} = 0.8$$

$$Mg^{2+} = 0.75 \times 0.2 \text{ (first stage)} + 0.25 \times 3.0 \text{ (in bypass)} = 0.9$$

$$Na^+ = 1.0 + 1.13 = 2.13$$

$$CO_3^{2-} = 0.3$$

$$HCO_3^- = 0.5(\text{conversion of } CaCO_3) + 0.15(\text{conversion of } Mg(OH)_2) + 0.25 \times 1.5 \text{ (associated with by passed Mg)} = 1.03$$

$$SO_4^{2-} = 2.5$$

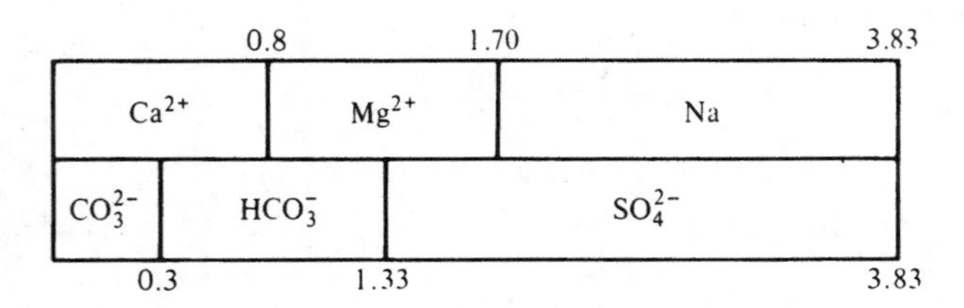

For a more complete description of split treatment, the reader is referred to Cleasby and Dellingham. [4-20]

Recarbonation Recarbonation for pH reduction and stabilization takes place in a closed reactor. Carbon dioxide is added under pressure and dissolved according to gas-transfer principles previously discussed. Figure 4-26 shows a typical recarbonation process.

Typical recarbonation units consist of two chambers, one for mixing the CO_2 and one in which the reactions occur. Detention time in the mixing chamber should be from 3 to 5 min, with a total detention time of at least 20 min. [4-44]

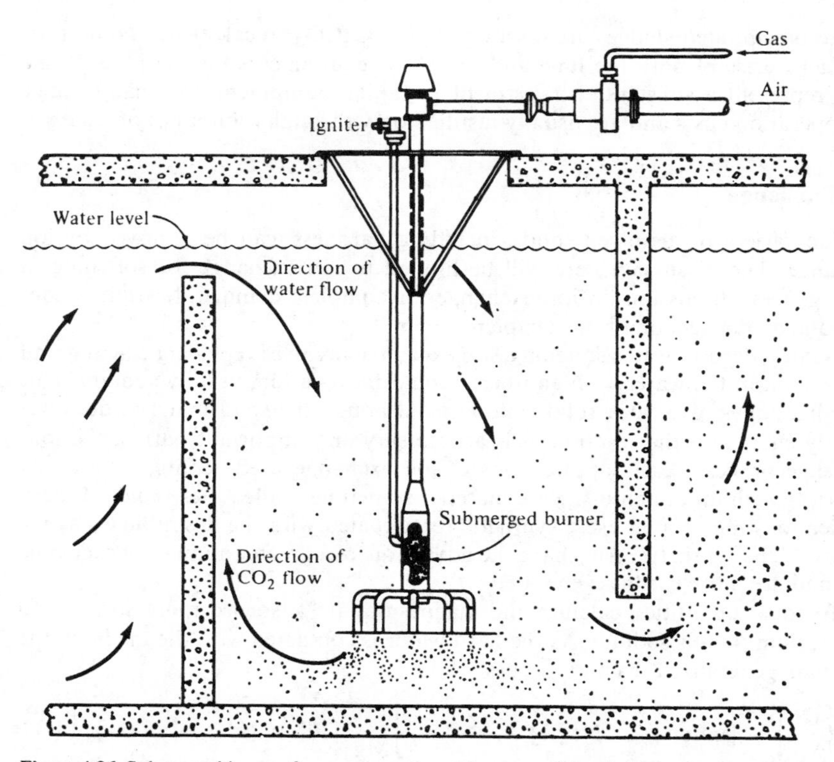

Figure 4-26 Submerged burner for recarbonation. (*Courtesy of Ozark-Mahoning Company.*)

Where split treatment is employed it may be necessary to follow the recarbonation unit with a settling chamber if the influent to the units still contains an excess of lime. [4-45] All recarbonation units should have provisions for periodic cleaning as some precipitate will accumulate.

The source of CO_2 may be the exhaust from combustion of natural gas ($CH_4 + 2O_2 \rightarrow CO_2 + H_2O$) or CO_2 which has been purified and shipped to the plant in containers. Walker [4-58] suggests that the stoichiometric quantity of CO_2 be multiplied by a factor of 2 to compensate for inefficiency of CO_2 transfer from the exhaust gases if submerged burners are used. Liquified CO_2 that is essentially pure (99.5 percent) can be obtained; this greatly enhances the efficiency of the recarbonation process. Storage of liquid CO_2 presents a problem since it gasifies at 31°C, resulting in extremely high vapor pressure. The usual procedure is to store liquid CO_2 at around −20°C and 2000 kPa. This necessitates strong tanks and refrigeration equipment.

Large water-treatment plants often find it economically advantageous to recalcify the $CaCO_3$ sludge, recovering both lime and carbon dioxide.

$$CaCO_3 \xrightarrow{\text{Heat}} CaO + CO_2 \qquad (4\text{-}38)$$

Where precipitated sludges are essentially pure $CaCO_3$, recalcifying should produce an excess of both the lime and the CO_2 requirements for the plant. Lime kilns represent a substantial investment in capital equipment and maintenance and operation costs and are usually justified only through economies of scale.

Ion Exchange

A wide variety of dissolved solids, including hardness, can be removed by ion exchange. The discussion here will be limited to ion exchange for softening; a more general discussion on ion exchange for complete demineralization is contained in a later section of this chapter.

As practiced in water softening, ion exchange involves replacing calcium and magnesium in the water with another, nonhardness cation, usually sodium. This exchange takes place at a solids interface. Although the solid material does not directly enter into the reaction, it is a necessary and important part of the ion-exchange process. Early applications of ion exchange used zeolite, a naturally occurring sodium alumino-silicate material sometimes called *greensand*. Modern applications more often use a synthetic resin coated with the desirable exchange material. The synthetic resins have the advantage of a greater number of exchange sites and are more easily regenerated.

In equal quantities, calcium and magnesium are adsorbed more strongly to the medium than is sodium. As the hard water is contacted with the medium, the following generalized reaction occurs.

$$\left\{\begin{matrix} \mathrm{Ca} \\ \mathrm{Mg} \end{matrix}\right\} + [\mathrm{anion}] + 2\,\mathrm{Na}[\mathrm{R}] \longrightarrow \left\{\begin{matrix} \mathrm{Ca} \\ \mathrm{Mg} \end{matrix}\right\}[\mathrm{R}] + 2\,\mathrm{Na} + [\mathrm{anion}] \quad (4\text{-}39)$$

The reaction is virtually instantaneous and complete as long as exchange sites are available. The process is depicted graphically in Fig. 4-27.

When all of the exchange sites have been utilized, hardness begins to appear in the effluent. Referred to as *breakthrough*, this necessitates the regeneration of the medium by contacting it with a strong sodium-chloride solution. The strength of the solution overrides the selectivity of the adsorption site, and calcium and magnesium are removed and replaced by the sodium.

$$\left\{\begin{matrix} \mathrm{Ca} \\ \mathrm{Mg} \end{matrix}\right\}[\mathrm{R}] + 2\,\mathrm{NaCl}\,(\mathrm{excess}) \longrightarrow \left\{\begin{matrix} \mathrm{Ca} \\ \mathrm{Mg} \end{matrix}\right\} 2\,\mathrm{Cl} + 2\,\mathrm{Na}[\mathrm{R}] \quad (4\text{-}40)$$

The system can again function as a softener according to Eq. (4-39).

The capacity and efficiency of ion-exchange softeners vary with many factors, including type of solid medium, type of exchange material used for coating, quantity of regeneration materials, and regeneration contact time. The overall quality of the water to be softened is also an important factor. A complete discussion of these factors is beyond the scope of this text and the reader is referred to Refs. [4-47] and [4-53] for greater details. Generally, the capacity of ion-exchange materials ranges from 2 to 10 mequiv/g or about 15 to 100 kg/m^3. Regeneration

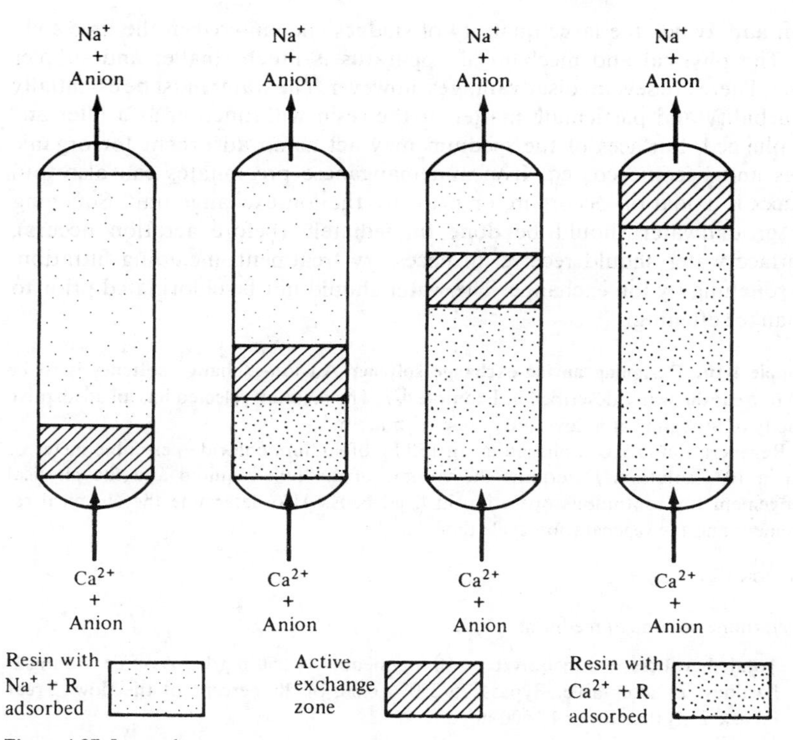

Figure 4-27 Ion-exchange process.

is accomplished using from 80 to 160 kg of sodium chloride per cubic meter of resin in 5 to 20% solution at a flow rate of about 40 L/min · m².

The effluent from the regeneration cycle will contain the hardness accumulated during the softening cycle as well as excess sodium chloride. After regeneration, the medium should be flushed with softened water to remove the excess sodium chloride. These highly mineralized waters constitute a waste stream that must be disposed of properly.

Ion-exchange operations are usually conducted in enclosed structures containing the medium. Water is forced through the material under pressure at up to 0.4 m³/min · m². Single or multiple units may be used and the medium may be contained in either a fixed or a moving bed. Where continuous operation is necessary, multiple units or moving beds are used. Single-stage fixed beds can be used when the flow of treated water can be interrupted for regeneration. Most treatment-plant operations are of the continuous type, while home softeners are serviced intermittently.

Ion-exchange softening at water-treatment plants is becoming more commonplace as more efficient resins are developed and as the process is better understood by design engineers. Ion exchange produces a softer water than chemical precipitation and avoids the large quantity of sludges encountered in the lime-soda process. The physical and mechanical apparatus is much smaller and simpler to operate. There are several disadvantages, however. The water must be essentially free of turbidity and particulate matter or the resin will function as a filter and become plugged. Surfaces of the medium may act as an adsorbent for organic molecules and become coated. Iron and manganese precipitates can also foul the surfaces if oxidation occurs in, or prior to, the ion-exchange unit. Softening of clear groundwater should be done immediately (before aeration occurs), while surface water should receive all necessary treatment, including filtration, prior to softening by ion exchange. The water should not be chlorinated prior to ion-exchange softening.

Example 4-10: Designing an ion-exchange softener An ion-exchange softener is to be used to treat the water described in Example 4-7. The medium selected has an adsorptive capacity of 90 kg/m^3 at a flow rate of 0.4 m^3/min · m^2.

Regeneration is accomplished using 150 kg of sodium chloride per cubic meter of resin in 10% solution. Determine the volume of medium required and the physical arrangement for continuous operation in fixed beds. Also determine the chemical requirement and the regeneration cycle time.

SOLUTION

1. Determine volume of medium.

   a. Total hardness = 6 mequiv/L × 50 mg/mequiv = 250 mg/L. Assume 75 mg/L hardness is acceptable. Bypass 75/250 = 0.30, or 30 percent of the flow. Treat 0.70 × 25,000 m^3/d, or 17,500 m^3/d.

   b. Hardness to be removed:

$$5.0 \text{ equiv/m}^3 \times 17{,}500 \text{ m}^3\text{/d} \times 50 \text{ gm/equiv} \times 1 \text{ kg}/10^3 \text{ gm} = 4375 \text{ kg/d}$$

   c. Volume of medium for 1-d operation:

$$4375 \text{ kg/d} \times 1 \text{ m}^3\text{/90 kg} = 48.6 \text{ m}^3 \text{ medium/d operation}$$

2. Determine surface area and height of medium.

   a. 17,500 m^3/d × d/1440 min = 12.15 m^3/min

$$\text{Area} = 12.15 \text{ m}^3\text{/min} \times \text{min}/0.4 \text{ m} = 30.38 \text{ m}^2$$

   b. Use tanks 2.0 m in diameter.

$$A = \pi d^2/4 = 3.14 \text{ m}^2$$

$$\text{No. of tanks} = \frac{30.38}{3.14} = 9.67\text{; use 9 tanks.}$$

   c.

$$\text{Height of medium} = \frac{\text{total volume}}{\text{total area}}$$

$$= \frac{48.6 \text{ m}^3}{9 \times 3.14 \text{ m}^2}$$

$$= 1.72 \text{ m, say 2 m}$$

d. Add three extra tanks for use during regeneration cycle. Total volume of exchange resin is:

$$V = \text{No. of tanks} \times \text{area} \times \text{height}$$
$$= 12 \times 3.14\ \text{m}^2 \times 2.0\ \text{m}$$
$$= 75.4\ \text{m}$$

3. Determine chemical requirements for regenerations.

a. Volume of one unit

$$V = 3.14 \times 2.0 = 6.28\ \text{m}^3$$

b. Salt requirement

$$150\ \text{kg/m}^3 \times 6.28\ \text{m}^3 = 942\ \text{kg}$$

Regenerating 9 units/d will require 9 × 942 = 8,478 kg/d of NaCl.

c. Using a 10% solution, the volume of regenerate liquid is 942 kg/0.1 = 9,420 kg, or approximately 9 m^3 for each unit.

d. At a loading rate of 0.04 $\text{m}^3/\text{m}^2 \cdot \text{min}$, the regeneration time is

$$t = 9.0\ \text{m}^3/(0.04\ \text{m}^3/\text{m}^2 \cdot \text{min} \times 3.14\ \text{m}^2)$$
$$= 72\ \text{min}$$

Assuming a total of 2 h for all operations necessary to regenerate units in groups of three, all 12 units can be regenerated in an 8-h workday.

4-8 FILTRATION

As practiced in modern water-treatment plants, filtration is most often a polishing step to remove small flocs or precipitant particles not removed in the settling of coagulated or softened waters. Under certain conditions, filtration may serve as the primary turbidity-removal process, e.g., in direct filtration of raw water. Although filtration removes many pathogenic organisms from water, filtration should not be relied upon for complete health protection.

Precoat filtration, a process in which a thin sheet of diatomaceous earth, or other very fine media, combine with solids in the water to form a "cake" on a microscreen, may have advantages under certain circumstances. A discussion of precoat filtration is beyond the scope of this text and the reader is referred to Baumann [4-7] for a thorough discussion of the subject.

The most commonly used filtration process involves passing the water through a stationary bed of granular medium. Solids in the water are retained by the filter medium. Several modes of operation are possible in granular medium filtration. These include upflow, biflow, pressure, and vacuum filtration. While any of these may find application under specialized conditions, the most common practice is gravity filtration in a downward mode, with the weight of the water column above the filter providing the driving force. The above operations are depicted graphically in Fig. 4-28.

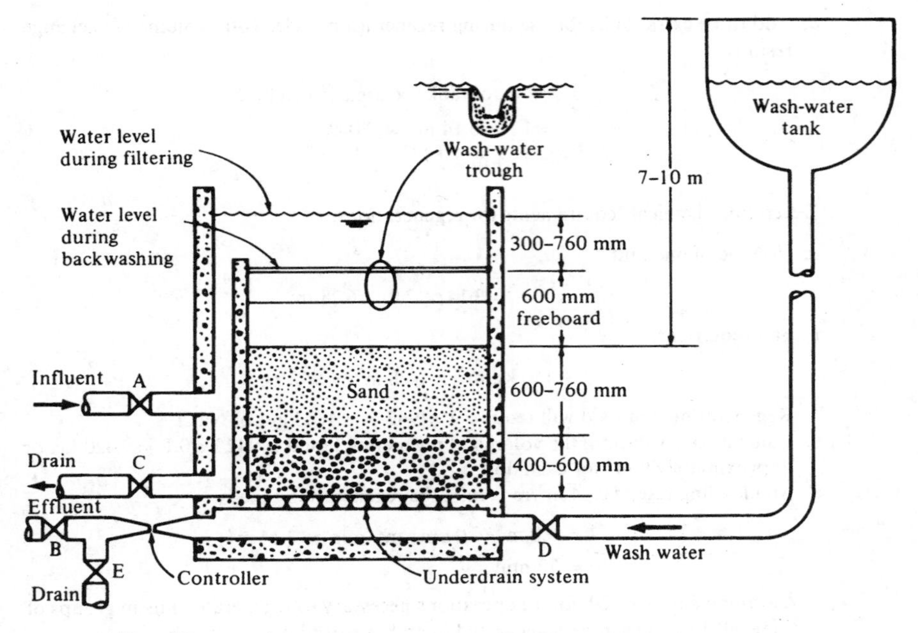

How filter operates

1. Open valve A. (This allows influent to flow to filter.)
2. Open valve B. (This allows water to flow through filter.)
3. During filter operation all other valves are closed.

How filter is backwashed

1. Close valve A.
2. Close valve B when water in filter drops down to top of overflow.
3. Open valves C and D. (This allows water from wash-water tank to flow up through the filtering medium, loosening up the sand and washing the accumulated solids from the surface of the sand, out of the filter. Filter backwash water is returned to head end of treatment plant.

How to filter to waste (if used)

1. Open valves A and E. All other valves closed. Effluent is sometimes filtered to waste for a few minutes after filter has been washed to condition the filter before it is put into service.

Figure 4-28 Typical gravity flow filter operation. (*From Metcalf & Eddy, Inc.* [*4-40*].)

The solids-removal operation with granular-medium filters involves several complicated processes. The most obvious process is the physical straining of particles too large to pass between filter grains. Other processes are also important, since most of the solid material contained in settled water is too small to be removed by straining. Removal of particles and flocs in the filter bed depends on mechanisms that transport the solids through the water to the surface of the filter grains, and on retention of the solids by the medium once contact has occurred. Transport mechanisms include settling (pore openings act as miniature settling basins), inertial impaction, diffusion of colloids into areas of lower concentrations and/or lower hydraulic shear, [4-42] and, to a lesser extent, Brownian movement and van der Waals forces. Retention of solids once contact has occurred can be attributed primarily to electrochemical forces, van der Waals force, and physical adsorption.

With chemical preconditioning of the water, a well-designed and operated filter should remove virtually all solids down to the submicron size. Removal begins in the top portion of the filter. As pore openings are filled by the filtered material, increased hydraulic shear sweeps particles farther into the bed. When the storage capacity of bed has thus become exhausted, the filter must be cleaned. Modern practice is to clean the filter by hydraulic backwashing. Backwash water containing the accumulated solids is disposed of and the filter returned to service.

Many variables influence the performance of granular media filters. An understanding of filter hydraulics, media characteristics, and operating procedures is necessary for the design of effective granular medium filters.

Filter Hydraulics

Filter hydraulics falls into two separate categories, the actual filtration process by which the water is cleaned and the backwashing operation by which the filter is cleaned. These operations are equally important in the overall filtration process. Flow through the packed bed can be analyzed by classic hydraulic theory. Carmen [4-14] modified the Darcy–Wiesbach equations for head loss in a pipe to reflect conditions in a bed of porous media of uniform size. Development of this equation is presented in several texts (Refs. [4-16], [4-29], [4-53]) and will not be repeated here. The resulting equation, known as the Carmen–Kozeny equation, is:

$$h_f = \frac{f'L(1 - e)V_s^2}{e^3 g d_p} \tag{4-41}$$

where h_f = friction loss through bed of particles of uniform size d_p, m
L = depth of the filter, m
e = porosity of bed
V_s = filtering velocity, i.e., the velocity of the water just above the bed (total flow Q to the filter divided by the area of the filter), m/s
g = gravitational acceleration, m/s^2
d_p = diameter of filter media grains, m

The remaining term f' is a friction factor related to the coefficient of drag around the particles. In the usual range of filter velocities (laminar flow) this can be calculated by

$$f' = 150\frac{(1 - e)}{\text{Re}} + 1.75 \tag{4-42}$$

where

$$\text{Re} = \text{Reynolds number} = \frac{\phi \rho_w V_s d}{\mu} \tag{4-43}$$

and ρ_w and μ are the density and dynamic viscosity, respectively, of the water. The units of ρ_w are kilograms per cubic meter, and the units of μ are newton-seconds per square meter. The shape factor ϕ ranges from 0.75 to 0.85 for most filter material.

Equation (4-41) can be modified for a bed of nonuniform medium. From a sieve analysis of the medium, the weight fraction x_{ij} between adjacent sieve sizes is determined. The average particle size d_{ij} is assumed to be halfway between the sieve sizes. The depth of the particles between adjacent sieve sizes can be taken as $x_{ij}L$ and Eq. (4-41) can be rewritten as follows:

$$h_f = \frac{L(1 - e)V_s^2}{e^3 g} \sum \frac{f_{ij} x_{ij}}{d_{ij}} \tag{4-44}$$

Equation (4-44) assumes that the bed is stratified by size and that the porosity is uniform throughout. Calculation of head loss across a uniform and a stratified media is illustrated in the following examples.

Example 4-11: Determining head loss across a bed of uniform-size particles Clean water at 20°C is passed through a bed of uniform sand at a filtering velocity of 5.0 m/h (1.39×10^{-3} m/s). The sand grains are 0.4 mm in diameter with a shape factor of 0.85 and a specific gravity of 2.65. The depth of the bed is 0.67 m and the porosity is 0.4. Determine the head loss through the bed.

SOLUTION

1. Calculate the Reynolds number by Eq. (4-43).

At 20°C $\qquad \rho = 998.2 \text{ kg/m}^3$

$$\mu = 1.002 \times 10^{-3} \text{ N} \cdot \text{s/m}^2 \times \frac{\text{kg} \cdot \text{m}}{\text{s}^2 \cdot \text{N}}$$

$$= 1.002 \times 10^{-3} \frac{\text{kg}}{\text{m} \cdot \text{s}}$$

$$\text{Re} = 0.85 \frac{998.2 \text{ kg/m}^3 \times 4.0 \times 10^{-4} \text{ m} \times 1.39 \times 10^{-3} \text{ m/s}}{1.002 \times 10^{-3} \text{ kg/m} \cdot \text{s}}$$

$$= 0.47 < 1.0 \text{ (laminar flow confirmed)}$$

2. Calculate f' by Eq. (4-42).

$$f' = 150 \frac{(1 - 0.4)}{0.47} + 1.75$$

$$= 193.24$$

3. Calculate head loss by Eq. (4-41).

$$h_f = \frac{193.24 \times 0.67 \text{ m}(1 - 0.4) \times (1.39 \times 10^{-3})^2 \text{ m}^2/\text{s}^2}{0.4^3 \times 9.81 \text{ m/s}^2 \times 4.0 \times 10^{-4} \text{ m}}$$

$$= 0.60 \text{ m}$$

Example 4-12: Determining head loss across a bed of nonuniform, stratified particles Water at 20°C is passed through a filter bed at 1.2×10^{-3} m/s (4.32 m/h). The bed is 0.75 m deep and is composed of nonuniform sand (specific gravity of 2.65) stratified so that the smallest particles are on top and the largest on bottom. The porosity and shape factors are 0.4 and 0.85 throughout the depth of the bed. The size distribution of the granules is given in the table below. Determine the head loss for clean water flow through the bed.

Sieve analysis

U.S. sieve no.		Particle size range, mm		Average size	Mass fract. in size range
Passing	Retained	Passing	Retained	d_{ij}, mm	x_{ij}
	14		1.41	1.41	0.01
14	20	1.41	0.84	1.13	0.11
20	25	0.84	0.71	0.78	0.20
25	30	0.71	0.60	0.66	0.32
30	35	0.60	0.50	0.55	0.21
35	40	0.50	0.42	0.46	0.13
40		0.42		0.42	0.02

SOLUTION

1. From Eq. (4-43):

$$\text{Re} = \frac{0.85 \times 998.2 \times 1.2 \times 10^{-3} \text{ m/s}}{1.002 \times 10^{-3} \text{ kg/m} \cdot \text{s}} d_{ij} \text{ m}$$

$$= 1016 d_{ij}$$

2. From Eq. (4-42):

$$f'_{ij} = \frac{150(1 - 0.4)}{1016 d_{ij}} + 1.75$$

$$= \frac{0.09}{d_{ij}} + 1.75$$

3. Determine $\Sigma f_{ij} \frac{x_{ij}}{d_{ij}}$ as follows:

Particle size, d_{ij} m × 10^3	x_{ij}	f_{ij}	$f_{ij} \frac{x_{ij}}{d_{ij}}$ 1/m
1.41	0.01	65.6	465
1.13	0.11	81.4	7,924
0.78	0.20	117.1	30,026
0.66	0.32	138.1	66,958
0.55	0.21	165.4	63,153
0.46	0.13	197.4	55,787
0.42	0.02	216.0	10,286
		$\Sigma f'_{ij} \frac{x_{ij}}{d_{ij}}$ =	234,599

4. Calculate h_f from Eq. (4-44):

$$h_f = \frac{0.73 \text{ m} \times (1 - 0.4) \times (1.2 \times 10^{-3})^2 \text{ m}^2/\text{s}^2}{0.4^3 \times 9.81 \text{ m/s}^2} \times 234{,}599 \text{ 1/m}$$

$$= 0.24 \text{ m}$$

It should be noted that Eqs. (4-41) and (4-44) are applicable only to clean filter beds. Once solids begin to accumulate, the porosity of the bed changes. As the porosity decreases, the head loss increases. The rate at which solids accumulate in the filter, and therefore the rate of head-loss change, is a function of the nature of the suspension, the characteristics of the media, and filter operation.

Although attempts to formulate a mathematical expression of a general nature to quantify changes in head loss with solids removal have not been very successful, some general observations can be made. To maintain a constant filtering velocity V_s, an increment in driving force must be applied to match each increment in head loss resulting from decreased porosity. Conversely, if a constant driving force is applied, the filtering velocity will diminish as the porosity decreases. In filter operations, a run is terminated when sufficient solids have accumulated to (1) use up the available driving force; (2) cause the filtering velocity to drop below a predetermined level; or (3) exhaust the storage capacity of the bed so that solids begin to "break through" into the effluent. At this point, the filter must be backwashed.

Backwashing of granular-medium filters is accomplished by reversing the flow and forcing clean water upward through the media. To clean the interior of the bed, it is necessary to expand it so that the granules are no longer in contact with each other, thus exposing all surfaces for cleaning. To hydraulically expand a porous bed, the head loss must be at least equal to the buoyant weight of the particles in the fluid. For a unit area of filter this is expressed by

$$h_{fb} = L(1 - e)\frac{\rho_m - \rho_w}{\rho_w} \tag{4-45}$$

where h_{fb} = head loss required to initiate expansion, m
L = bed depth, m
$1 - e$ = fraction of the packed bed composed of medium
ρ_m = density of the medium, kg/m^3
ρ_w = density of the water, kg/m^3

The head loss through an expanded bed is essentially unchanged because the total buoyant weight of the bed is constant. Therefore:

$$\text{Weight of packed bed} = \text{weight of bed fluidized}$$

$$L(1 - e)\frac{\rho_m - \rho_w}{\rho_w} = L_{fb}(1 - e_{fb})\frac{\rho_m - \rho_w}{\rho_w}$$

or

$$L_{fb} = L\frac{(1 - e)}{(1 - e_{fb})} \tag{4-46}$$

where L_{fb} = the depth of the fluidized bed
e_{fb} = the porosity of the fluidized bed

The quantity e_{fb} is a function of the terminal settling velocity of the particles and the backwash velocity. An increase in the backwash velocity will result in a greater expansion of the bed. The expression commonly used to relate the bed expansion to backwash velocity and particle settling velocity is [4-28]:

$$e_{fb} = \left(\frac{V_B}{v_t}\right)^{0.22} \tag{4-47}$$

where V_B is the backwash velocity (backwash flow Q divided by the total filter area). The depth of the fluidized bed and the backwash velocity for a given size medium (with known v_t) can now be related as follows:

$$L_{fb} = \frac{L(1 - e)}{1 - \left(\dfrac{V_B}{v_t}\right)^{0.22}} \tag{4-48}$$

This equation can also be modified for a stratified bed of nonuniform particles where

$$x_{ij} = \frac{L_{ij}}{L}$$

Again x_{ij} is the weight fraction between adjacent sieve sizes. Assuming uniform porosity in the packed bed, L_{ij} will be the depth of the layer of media represented by x_{ij}. The expansion of this layer is represented by

$$L_{fb,ij} = x_{ij} \frac{L(1-e)}{1 - \left(\dfrac{V_B}{v_{t,ij}}\right)^{0.22}}$$

The total expansion is the sum of the individual layers

$$L_{fb} = L(1-e) \sum \frac{x_{ij}}{1 - \left(\dfrac{V_B}{v_{t,ij}}\right)^{0.22}} \qquad (4\text{-}49)$$

Total expanded depth should range from 120 to 155 percent of the unexpanded depth. [4-7] Amirtharajah [4-5] has shown that the optimum expansion for hydraulic backwashing occurs at expanded porosities of from 0.65 to 0.70.

Example 4-13: Finding the expanded depth of a uniform medium The filter medium described in Example 4-11 is to be expanded to a porosity of 0.7 by hydraulic backwash. Determine the required backwash velocity and the resulting expanded depth.

SOLUTION

1. The terminal settling velocity for the medium is first calculated from Stokes' law [Eq. (4-9)].

$$v_t = \frac{9.81 \text{ m/s } (2650 - 998.2) \text{ kg/m}^3 \times (4 \times 10^{-4} \text{ m})^2}{18 \times 1.002 \times 10^{-3} \text{ N}\cdot\text{s/m}^2}$$

$$= 0.14 \text{ m/s (rounded)}$$

2. Check Reynolds number [Eq. (4-43)].

$$\text{Re} = 0.85 \times \frac{0.14 \text{ m/s} \times 4 \times 10^{-4} \text{ m} \times 998.2 \text{ kg/m}^3}{1.002 \times 10^{-3} \text{ N}\cdot\text{s/m}^2}$$

$$= 47.4 \text{ (transitional flow)}$$

3.
$$C_D = \frac{24}{47.4} + \frac{3}{47.4^{1/2}} + 0.34$$

$$= 1.28$$

4.
$$v_t^2 = \frac{4/3 \times 9.81 \text{ m/s}^2 (2650 - 998.2) \text{ kg/m}^3 \times 4.0 \times 10^{-4} \text{ m}}{1.28 \times 998.2 \text{ kg/m}^3}$$

$$v_t = 0.08 \text{ m/s}$$

5. Repeat steps 2, 3, and 4.

$$\text{Re} = 26.6$$

$$C_D = 1.85$$

$$v_t = 0.07 \text{ m/s}$$

6. From Eq. (4-47):

$$0.7 = \left(\frac{V_B}{0.07}\right)^{0.22}$$

$$V_B = 0.7^{4.55} \times 0.07 \text{ m/s}$$

$$= 1.4 \times 10^{-3} \text{ m/s}$$

7. From Eq. (4-46):

$$L_{fb} = \frac{0.67 \text{ m } (1 - 0.4)}{1 - 0.7}$$

$$= 1.34 \text{ m}$$

Example 4-14: Finding the expanded depth of a nonuniform stratified bed The filter bed described in Example 4-12 is to be backwashed at a velocity of 1.5×10^{-2} m/s. Determine the depth of the expanded bed.

SOLUTION

Each "layer" of particles defined by the sieve analysis of Example 4-12 must be treated separately and the results summed.

For the bottom layer, $d_{ij} = 1.41$ and $x_{ij} = 0.01$.

1. Estimate an initial velocity assuming turbulent flow [Eq. (4.4) with $C_D = 0.4$].

a.

$$v_t = \left(\frac{4}{3} \times \frac{9.81 \text{ m/s}^2}{0.4} \frac{\rho_m - \rho_w}{\rho_w} \times d_{ij} \text{ mm} \times \frac{10^{-3} \text{ m}}{\text{mm}}\right)^{1/2}$$

$$= (5.4 \times 10^{-2} \text{ m}^2/\text{s}^2\, d_{ij})^{1/2} \quad *$$

$$= (5.4 \times 10^{-2} \times 1.41)^{1/2} \text{ m/s}$$

$$= 0.28 \text{ m/s}$$

b.

$$\text{Re} = \phi v_t d\rho/\mu = \frac{0.85 \times 998.2 \text{ kg/m}^3 \times v_t \text{ m/s} \times d_{ij} \times 10^{-3} \text{ m/mm}}{1.002 \times 10^{-3} \text{ N}\cdot\text{s/m}^2}$$

$$= 847 \times v_t \times d_{ij} \quad *$$

$$= 847 \times 0.28 \times 1.41$$

$$= 329 \text{ (transitional flow)}$$

c.

$$C_D = \frac{24}{329} + \frac{3}{329^{1/2}} + 0.34$$

$$= 0.58$$

d.

$$v_t = \frac{4}{3} \times 9.81 \text{ m/s}^2 \frac{\rho_m - \rho_w}{\rho_w} \frac{d_{ij}}{C_D} \text{ mm} \times \frac{10^{-3}}{\text{mm}}$$

$$= \left(2.158 \times 10^{-3} \frac{\text{m}^2}{\text{s}^2 \cdot \text{mm}} \frac{d_{ij} \text{ mm}}{C_D}\right)^{1/2} \quad *$$

$$= 0.23 \text{ m/s}$$

e. Repeat steps *b*, *c*, and *d* using * expansions. Final solution is:

$$\mathrm{Re} = 274$$

$$C_D = 0.61$$

$$v_t = 0.22 \text{ m/s}$$

f. Determine expanded porosity of layer by Eq. (4-47).

$$e_{fb} = \left(\frac{V_B}{v_{t,ij}}\right)^{0.22} = \left(\frac{1.5 \times 10^{-2} \text{ m/s}}{v_{t,ij}}\right)^{0.22}$$

$$= 0.55$$

g. Find $x_{ij}/(1 - e_{fb})$.

$$x_{ij}\Bigg/\left(1 - \left(\frac{V_B}{v_{t,ij}}\right)^{0.22}\right) = \frac{0.01}{1 - 0.55} = 0.02$$

2. Repeat all preceding steps for each layer of particles. Again, * expression can be used directly with proper values inserted. The results are tabulated below

Average particle	v_{tij}, m/s	$\left(\frac{V_B}{v_{t,ij}}\right)^{0.22}$	$1 - \left(\frac{V_B}{v_{t,ij}}\right)^{0.22}$	x_{ij}	$\frac{x_{ij}}{1 - \left(\frac{V_B}{v_{t,ij}}\right)^{0.22}}$
1.41	0.22	0.55	0.45	0.01	0.02
1.13	0.19	0.57	0.43	0.11	0.26
0.78	0.14	0.61	0.39	0.20	0.51
0.66	0.12	0.63	0.37	0.32	0.86
0.55	0.10	0.66	0.34	0.21	0.62
0.46	0.08	0.69	0.31	0.13	0.42
0.42	0.07	0.71	0.29	0.02	0.07

$$\sum \frac{x_{ij}}{1 - \left(\frac{V_B}{v_{t,ij}}\right)^{0.22}} = 2.76$$

3. The expanded bed depth is found by Eq. (4-49).

$$L_{fb} = L(1 - e) \times \frac{x_{ij}}{1 - \left(\frac{V_B}{v_{t,ij}}\right)^{0.22}}$$

$$= 1.24 \text{ m}$$

and

$$\frac{1.24}{0.75} \times 100 = 165\% \text{ of original bed depth}$$

The repetitive nature of the above example suggests solution by computer.

The principal cleaning mechanism in backwashing filters is hydrodynamic shear, which tears adhered material away from medium grains. While increased backwash velocity might increase this shear, the resulting expansion could result in several undesirable effects. Jets of water aimed at the surface of the filter and/or mechanically powered rakes are often employed to create turbulence in the expanded bed during backwash. In addition to increasing the shear forces without increasing backwash velocity, these operations also promote collision of media grains, with the inherent abrasion assisting in the cleaning process.

Another technique, air scour, is also useful in increasing shear forces in backwashing filters. Air is introduced along with the backwash water and creates additional turbulence without substantially increasing expansion. Cleasby et al. [4-22, 4-19, 4-21] and Amirtharajah [4-5] have shown that air scour at subfluidizing water flows may provide more effective cleaning of granular-medium filters.

Filter Components

A typical granular-medium filter system used in water treatment was shown in Fig. 4-28. Filter components include the containment structure (filter box), an underdrain system, and filtering media. Additionally, piping systems, pumps, valves, backwash troughs, and other appurtenances for controlling the flow of water to and from the filter are necessary.

Filter box Containment structures for filters are usually constructed of reinforced concrete, although corrosion-resistant steel or other suitable material may be used. Structurally, the filter box must be strong enough to support the weight of the underdrain system, filter medium, and water column. Additionally, the structure must be watertight at pressures corresponding to the height of the maximum water column expected.

Usually square or rectangular in shape, filter boxes are arranged facing each other across an access corridor containing common piping and other appurtenances. If more than two filters are necessary, a series of multiples of two provides the economy of common walls and minimized piping. These *filter galleries*, as they are commonly called, are usually enclosed in suitable housing with the controls located for central operations.

Underdrain systems The purpose of the underdrain system is to collect and remove the filtered water and to disperse the backwash water. Underdrain systems in filters may consist of built-in-place main and lateral pipe arrangements or of proprietary units manufactured elsewhere and assembled on site. Figure 4-29 shows several types of underdrain systems. Many systems of this type require a graded gravel packing to prevent loss of filter media into the underdrain system. Figure 4-29*a* is illustrative of the sizing of the gravel. No gravel packing is required for underdrain systems such as the ones shown in Fig. 4-29*c* and 29*d*. These

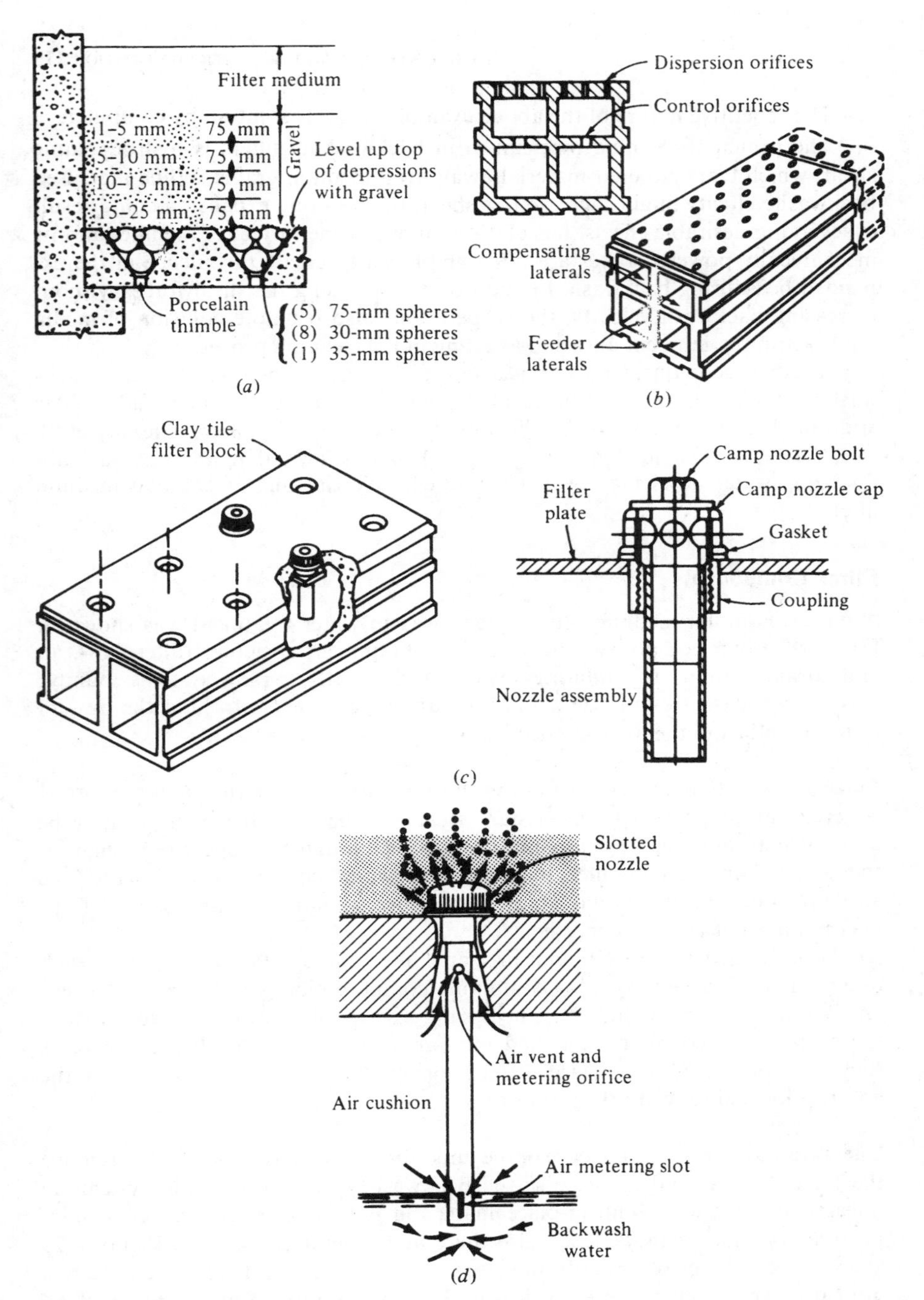

Figure 4-29 Proprietary filter underdrains: (*a*) BIF, Unit of General Signal Corp.; (*b*) F. B. Leopold Company, subsidiary of Moeller Company; (*c*) Walker Process Corp.; (*d*) Infilco Degremont, Inc.

systems have slit openings which are too narrow for grains of filter media to pass through.

Underdrain systems of all types contribute significantly to head loss due to friction during filter runs and during backwash. Hydraulically, underdrain systems must be designed to handle backwash flow rates, which usually exceed filtering rates by at least a factor of 2. An excellent discussion of underdrain design is presented by Cleasby. [4-17]

Filter media Traditionally, silica sand has been the medium most commonly used in granular-medium filters. Modern filter applications often make use of anthracite coal and garnet sand in place of, or in combination with, silica sand. The important properties of these materials are size, size distribution, and density.

The smaller the size of granular media, the smaller the pore openings through which the water must pass. Small pore openings increase filtration efficiency not only because of straining but also because of other removal mechanisms. However, as size of pore openings decreases, head loss through the medium increases, resulting in a diminished flow rate. Larger media increase pore size, reduce head loss, and increase flow rate, but at a sacrifice of filtration efficiency.

Since large quantities of filter medium of any uniform size would be difficult to obtain and therefore quite expensive, filter media vary in diameter within a selected size range. In modern filtration practice, the effect of varying size ranges becomes important because of stratification during backwashing operations. When the bed is expanded, small grains are lifted farther than larger grains and settle more slowly once the wash cycle is ended. Thus, a bed of nonuniform medium will stratify with smaller particles, and therefore smaller pore openings, at the top, an inefficient arrangement because most of the removal and most of the head loss during the filtration cycle will occur at the surface.

Another factor which influences the height of expansion during backwash and the rate of settling after backwash is the density of the medium. When two or more materials with different specific gravities are used, the lighter material is located above the denser material of the same size.

The choice of size, size distribution, and density of the filter medium is an important aspect of filter design. Through these variables, the engineer attempts to match the filter to the characteristics of the water to be filtered and to the desired quality and quantity of the output. Examples of engineered filter designs are given in the following paragraphs.

Slow sand filter The first filters to be used on a widespread basis for water purification were slow sand filters. These filters were constructed of fine sand with an effective size of about 0.2 mm. The *effective size* is the size of the openings of the sieve that retains 10 percent of the medium. This small size resulted in virtually all of the suspended material being removed at the filter surface. Additionally, a mat of biological organisms was allowed to develop at the water-sand interface, which aided in the filtration process. The resulting high head loss produced very low flow rates (0.12 to 0.32 m/h), necessitating the construction of very large filters. Cleaning was accomplished by periodically (usually no more frequently than once a month) draining the filters and mechanically removing the top few centimeters of sand, along with the accumulated solids and the biological mat.

Slow sand filters have large space requirement and are capital-intensive. Additionally, they do not function well with highly turbid water since the surface plugs quickly, requiring frequent cleaning. The rapid sand filter was developed in the mid-1800s to alleviate these difficulties.

Rapid sand filter The rapid sand filter utilizes a bed of silica sand ranging from 0.6 to 0.75 m in depth. Sizes may range from 0.35 to 1.0 mm or even larger, with effective sizes from 0.45 to 0.55 mm. A uniformity coefficient (60 percent less than size/10 percent size) of 1.65 is commonly specified. These larger sizes, coupled with frequent cleaning and the absence of a biological mat, result in a rate of filtration an order of magnitude larger than that of the slow sand filter. Common filtration rates in rapid sand filters range from 2.5 to 5.0 m/h.

An important feature of the rapid sand filter is that it is cleaned by hydraulic backwashing with resulting stratification of the medium. Filtration of relatively clean water presents few problems. However, filtration of turbid water necessitates frequent backwashing. Coagulated water with large, strong flocs causes binding at the fine-grained surface and results in a rapid buildup of head loss, necessitating frequent cleaning. This situation could be alleviated if the gradation of the filter could be reversed so that larger grains were deposited on top with media of progressively decreasing size below, so that the smallest grains were on the bottom. Such an arrangement would mean the large pores on top would retain mostly larger suspended material, while subsequently smaller pores would retain succeedingly finer material. Thus, the entire depth of the filter would function efficiently, and larger volumes of suspended solids could be retained between backwashes. The overall result would be longer filter runs, less head loss, and greater filtering rates.

By careful selection of medium with regard to size and density, it is possible to approximate this reverse gradation. Dual-media filters do this to some extent, and mixed-media filters essentially approximate reverse gradation.

Dual-media filters Dual-media filters are usually constructed of silica sand and anthracite coal. The depth of the sand may range from 0.15 to 0.4 m, with the coal depth ranging from 0.3 to 0.6 m. Size and uniformity coefficients of the two media can be selected to produce either a distinct separation or a given degree of mixing after backwashing. These conditions are illustrated in Fig. 4-30. As an example, the following material would produce a filter 0.6 m deep with approximately 0.15 m of intermixing. [4-17]

	Sand	Coal
Depth, m	0.3	0.3
Specific gravity	2.65	1.4–1.6
Effective size, mm	0.5–0.55	0.9–1.0
Uniformity coefficient	<1.65	<1.8

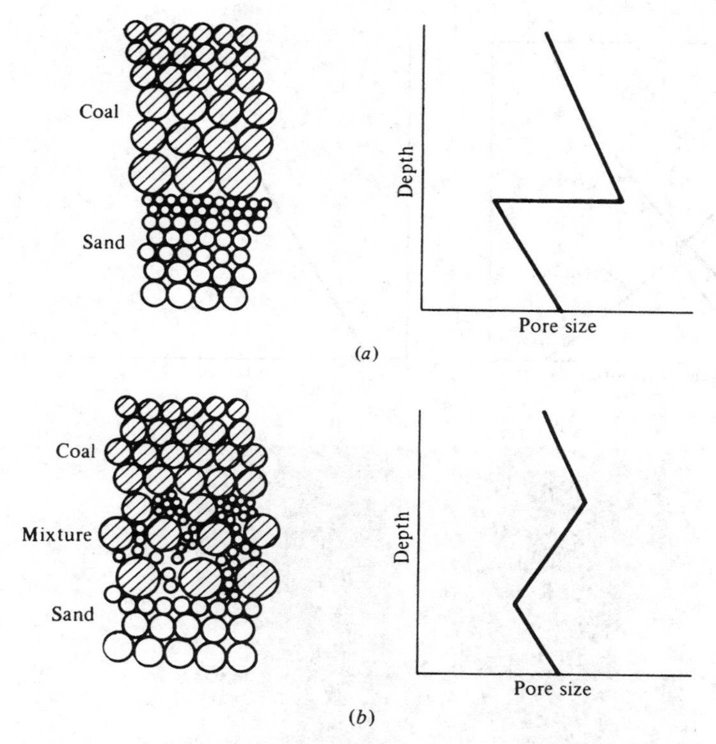

Figure 4-30 Size gradation in dual-media filters: (*a*) sharp gradation and (*b*) partial mixing.

The large pores in the anthracite layer remove large particles and flocs, while most of the smaller material penetrates to the sand layer before it is removed. Dual-media filters thus have the advantage of more effectively utilizing pore space for storage. This results in longer filter runs and greater filtration rates because of lower head losses. A disadvantage of dual-media filters is that the filtered material is held rather loosely in the anthracite layer. Any sudden increase in hydraulic loading dislodges the material and transports it to the surface of the sand layer, resulting in rapid binding at this level.

Mixed-media filters As noted earlier, the ideal filter would consist of a medium graded evenly from large at the top to small at the bottom. This can be accomplished by using three or more types of media with carefully selected size, density, and uniformity coefficients. A typical installation might consist of a 0.75-m bed with 60 percent anthracite, 30 percent silica sand, and 10 percent garnet sand, with specific gravities of 1.6, 2.6, and 4.2, respectively. Effective sizes ranging from a maximum of 1.0 mm for the anthracite to a minimum of 0.15 mm for the garnet, coupled with carefully selected uniformity coefficients, will produce intermixing and result in a pore-size gradation as shown in Fig. 4-31. [4-24]

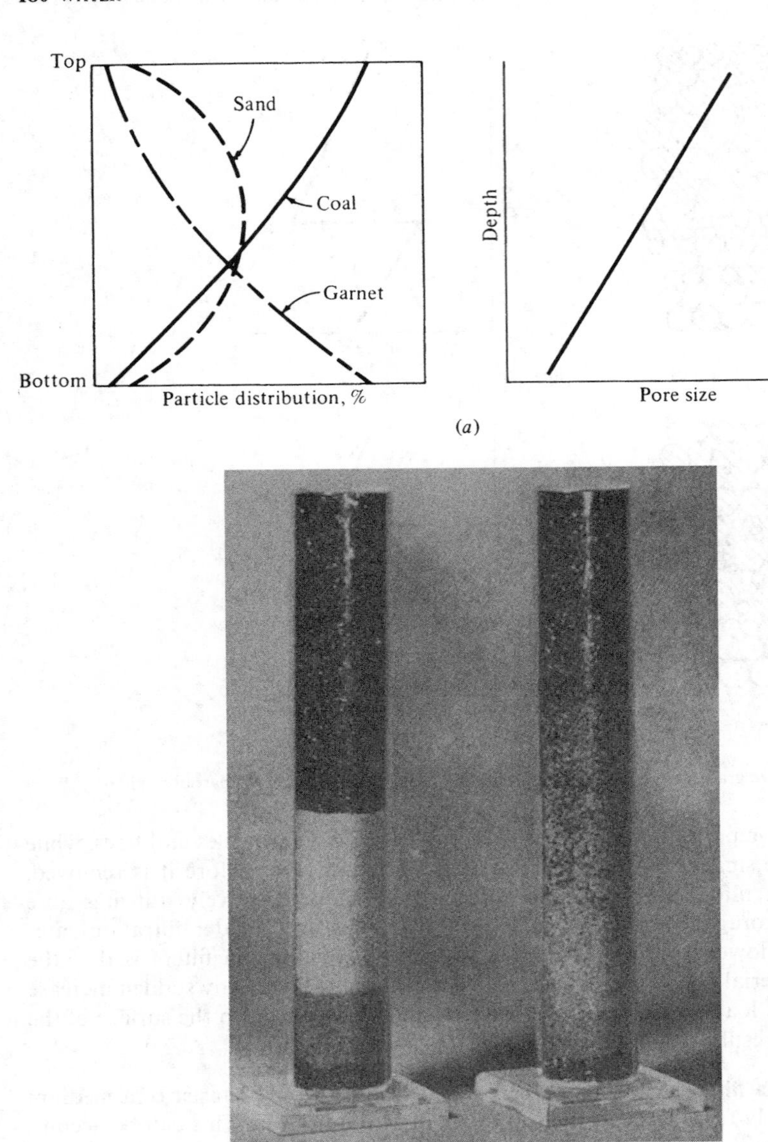

Figure 4-31 Size gradation of mixed-media filter: (*a*) particle distribution and pore size and (*b*) multi-media segregated and mixed by backwashing (*courtesy of Neptune Microfloc, Inc.*).

Thus, the mixed-media filter (perhaps "mixed-up" media is a more descriptive term) approaches an ideal filter. Filtration rates range from 10 to 20 m/h, considerably higher than rapid sand filters and about the same as for dual-media filters. The reverse gradation avoids the major problems of each of these media, however.

Dual- and mixed-media filters make possible the direct filtration of water of low turbidity without settling operations. Coagulating chemicals are often added to the influent of the filter to produce small, strong flocs to enhance turbidity removal.

Filter Operation

The two basic modes of operating granular-medium filters are (1) constant head–variable flow and (2) constant flow–variable head. These two modes are often modified to obtain better results.

In the constant head–variable flow mode, the water level above the filter is kept at a preselected level. Since a clean filter bed presents limited head loss, the flow rate will be quite large. As the filter becomes clogged, the head loss increases and the flow rate diminishes. When the flow reaches the design minimums, the filter must be backwashed.

Because very rapid flow through a clean filter results in poor efficiency, throttling the flow from the filter with a flow-control valve may be necessary. This valve is designed to provide additional head loss in the underdrain system and decrease the flow rate to an acceptable level. When the filter medium is clean, the valve operates with a small opening to produce a large supplemental head loss. As the head loss increases due to medium plugging, the valve gradually opens to decrease supplemental head loss and to maintain a more or less constant head loss across the entire system. The result is essentially a constant head–constant flow filter.

If water is introduced into a clean filter at a constant rate, an equilibrium will be established between the height of the water column and the application rate. At first, the head will be low due to the minimal head loss in the medium. As the medium becomes plugged and greater head loss occurs, the height of the water column must increase to provide the needed driving force. When the water column reaches a predetermined level, the filter is backwashed and the cycle repeats itself.

Filters that operate in this way must be designed to prevent dewatering of the bed during the initial filter cycle. A minimum depth of water above the bed can be assured by elevating the entrance to the clear well above the surface of the filter media.

More recent design of larger filter plants usually makes use of a combination of the above modes of operation. A constant flow is delivered to a bank of several filters through a common header and is allowed to distribute itself according to the operating rate of each individual filter. The height of the water column is the same above all the filter units, with the cleanest filter accepting the greatest flow. When the flow rate through any one unit decreases to a predetermined level, that filter is taken off-line and backwashed. Removal of one filter results in an increase in flow to the remaining filters, with a subsequent increase in head and flow rate through each filter. When backwashing is completed, the newly cleaned filter is returned to service and will accommodate a larger flow rate. Water level will therefore drop slightly in all the filters, resulting in a decrease in flow through each filter.

Regardless of the operating mode, a uniform flow rate is essential to the best performance of a granular-medium filter. Any increases in flow rate must occur gradually, or the quality of the effluent will deteriorate. Large changes occurring quickly produce the greatest degree of deterioration.

When automatic control valves are used to regulate filter output, they must be maintained to ensure that they produce gradual changes in the orifice opening. Otherwise, a rapid change in flow rate will occur, with significant deterioration of the effluent quality. As noted earlier, fluctuations in filtering rates occur in variable-declining-rate filtration each time a filter is taken off-line for backwashing. The magnitude of the fluctuations increases as the number of filter units in the system decreases. To prevent significant disruptions in filter quality, a minimum of four filters should be used in this mode of operation. [4-18]

4-9 DISINFECTION

As practiced in water treatment, disinfection refers to operations aimed at killing, or rendering harmless, pathogenic microorganisms. Sterilization, the complete destruction of all living matter, is not usually the objective of disinfection. The effect of disinfection on the reduction of waterborne disease is quite dramatic, as evidenced in Fig. 4-32. [4-56]

Other water-treatment processes assist in removing pathogens. In excess of 90 percent of the bacteria and viruses should be removed by coagulation, settling, and filtration. Excess-lime softening is an effective disinfectant due to the high pH involved. However, to meet the EPA's standard of one coliform organism per 100 mL and to provide protection against regrowth, additional disinfection must be practiced.

A good disinfectant must be toxic to microorganisms at concentrations well below the toxic thresholds to humans and higher animals. Additionally, it should have a fast rate of kill and should be persistent enough to prevent regrowth of organisms in the distribution system. The rate of kill is often postulated as a first-order reaction:

$$\frac{dN}{dt} = -kN$$

$$\ln\frac{N_t}{N_0} = -kt$$

$$N_t = N_0 e^{-kt} \qquad (4\text{-}50)$$

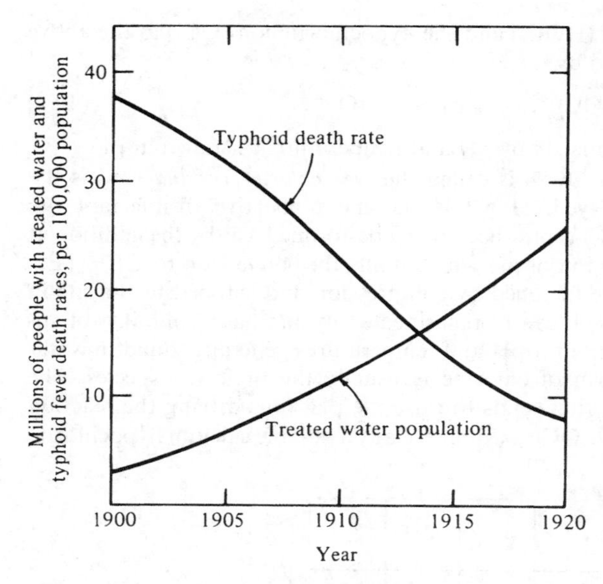

Figure 4-32 Typhoid fever and treated water supplies during two decades. (*From Vesilind* [*4-56*].)

Complete disinfection cannot be accomplished because N_t, the number of organisms remaining at time t, will only approach zero asymptotically as time gets excessively large. However, since the number of organisms initially present (N_0) should be small, 99.9 percent kill can be affected in a reasonable time. The value of the constant k must be determined experimentally.

Factors which militate against effective disinfection are turbidity and resistant organisms. Turbidity-producing colloids offer sanctuary to organisms, thus shielding them from the full action of the disinfectant. Particulate matter may adsorb the disinfectant.

Viruses, cysts, and ova are more resistant to disinfectants than are bacteria. Additional exposure time and higher concentrations of the disinfectant will be required for an effective kill of these organisms.

Disinfectants include chemical agents such as the halogen group, ozone, or silver; irradiation with gamma waves or ultraviolet light; and sonification, electrocution, heating, or other physical means. In America, disinfection and chlorination have become synonymous terms, while ozonation has been practiced more widely in Europe.

Chlorination

Chlorine may be applied to water in gaseous form (Cl_2) or as an ionized product of solids [$Ca(OCl)_2$, $NaOCl$]. The reactions in water are as follows:

$$Cl_2 + H_2O \longrightarrow H^+ + HOCl \tag{4-51}$$

$$Ca(OCl)_2 \longrightarrow Ca^{2+} + 2OCl^- \tag{4-52}$$

$$NaOCl \longrightarrow Na + OCl^- \tag{4-53}$$

The hypochlorous acid (HOCl) and the hypochlorite ion (OCl) in the above equations are further related by

$$HOCl \rightleftharpoons H^+ + OCl^- \tag{4-54}$$

a relationship governed primarily by pH and temperature, as shown in Fig. 4-33.

The sum of HOCl and OCl^- is called the *free chlorine residual* and is the primary disinfectant employed. HOCl is the more effective disinfectant. As indicated in Eq. (4-51), HOCl is produced on a one-to-one basis by the addition of Cl_2 gas, along with a reduction of pH which limits the conversion to OCl^- [Eq. (4-54)]. Chlorine gas can be liquefied by compression and shipped to the site in compact containers. Because it can be regasified easily and has a solubility of approximately 700 mg/L in water at pH and temperatures generally found in water purification plants, this form of chlorine is usually the preferred species. The application of the hypochlorites tends to raise the pH, thus driving the reaction more toward the less effective OCl^-. Commercially available calcium hypochlorite

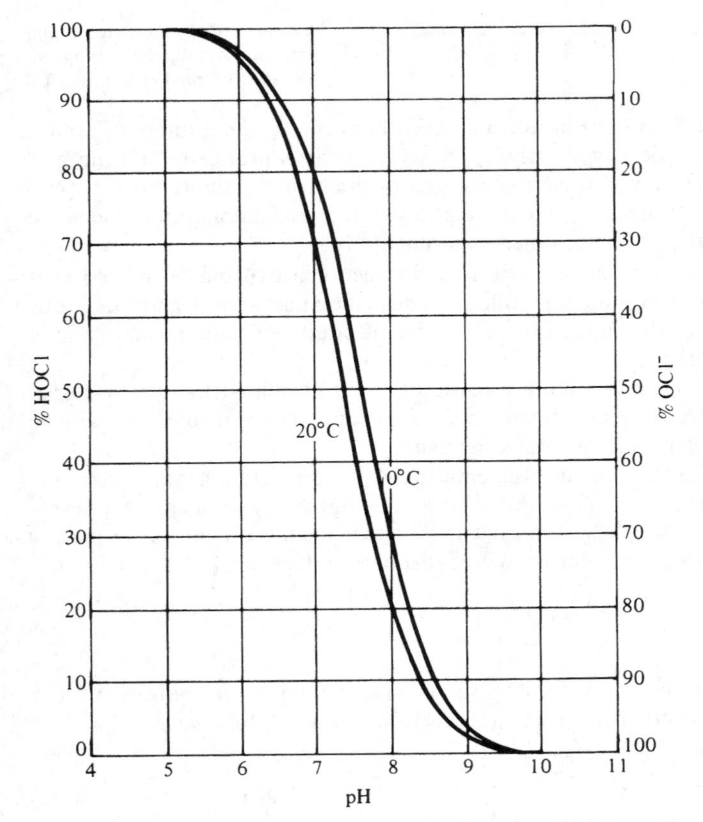

Figure 4-33 Distribution of HOCL and OCL^- as a function of pH. (*From Sawyer and McCarty* [*4-48*].)

contains approximately 70 to 80 percent available chlorine, while NaOCl contains only 3 to 15 percent available chlorine. [4-53] Some practical difficulty is involved in dissolving $Ca(OCl)_2$, and both hypochlorites are more expensive on an equivalence basis than liquefied Cl_2.

There are other considerations, however, which sometimes dictate the use of hypochlorites. Chlorine gas is a very strong oxidant that is toxic to humans. Since it is heavier than air, it spreads slowly at ground level. Therefore, extreme care must be exercised in its manufacture, shipping, and use. Accounts of evacuations of populated areas because of rail or barge accidents involving chlorine gas have become common news items. The use of hypochlorites is often mandated where large quantities of chlorine are needed in treatment plants located in highly populated areas.

At low concentrations, chlorine probably kills microorganisms by penetrating the cell and reacting with the enzymes and protoplasm. At higher concentrations, oxidation of the cell wall will destroy the organism. Factors affecting the process are

1. Form of chlorine
2. pH
3. Concentration
4. Contact time
5. Type of organism
6. Temperature

Hypochlorous acid is more effective than the hypochlorite ion by approximately two orders of magnitude. Because the free-chlorine species is related to pH, one would expect a relationship between efficiency and pH. Empirically, it has been found that chlorine dosages must be increased to compensate for higher pH.

Chlorine concentration and contact time relationship is often expressed by

$$C^n t_p = k \tag{4-55}$$

where C = concentration of chlorine, mg/L
t_p = time required for given percent kill, min
n, k = experimental derived constants for a given system

An example of this relationship was reported by Berg and is shown in Fig. 4-34. [4-49]

The effects of temperature variations can be modeled by the following equation derived from the van't Hoff–Arrhenius equation. [4-40]

$$\ln \frac{t_1}{t_2} = \frac{E'(T_2 - T_1)}{R_{T1,T2}} \tag{4-56}$$

where t_1, t_2 = time required for given kills
$T_1 T_2$ = temperature corresponding to t_1 and t_2, K
R = gas constant, 1.0 cal/K-mol
E' = activation energy, related to pH (as shown in Table 4-4)

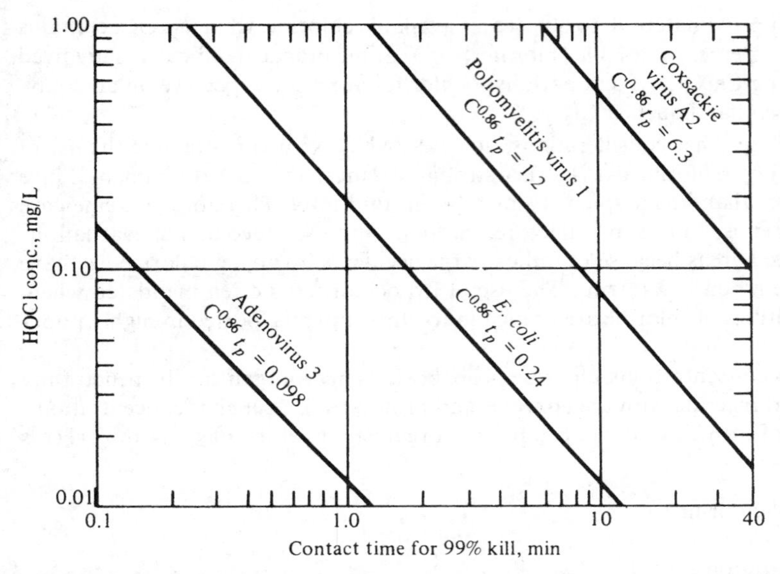

Figure 4-34 Concentration of free residual chlorine and contact time necessary for 99 percent kill at 0 and 6°C. (*From Schroeder* [4-49].)

Being a strong oxidant, chlorine will react with almost any material that is in a reduced state. In water, this usually consists of Fe^{2+}, Mn^{2+}, H_2S, and organics. Ammonia (NH_3) is sometimes present in small quantities or may be added for purposes to be presently discussed. These oxidizable materials will consume chlorine before it has a chance to act as a disinfectant. The amount of chlorine required for this purpose must be determined experimentally, since the exact nature and quantity of oxidizable material in water is seldom known. A typical titration curve is shown in Fig. 4-35.

The products of organics oxidized by chlorine are often undesirable. Organic acids (humic, fulvic) form chlorinated hydrocarbon compounds that are suspected of being carcinogenic. Minute quantities of phenolic compounds react with chlor-

Table 4-4 Activation energies for aqueous chlorine

pH	E', cal
7.0	8,200
8.5	6,400
9.8	12,000
10.7	15,000

Source: From Fair et al. [4-30]

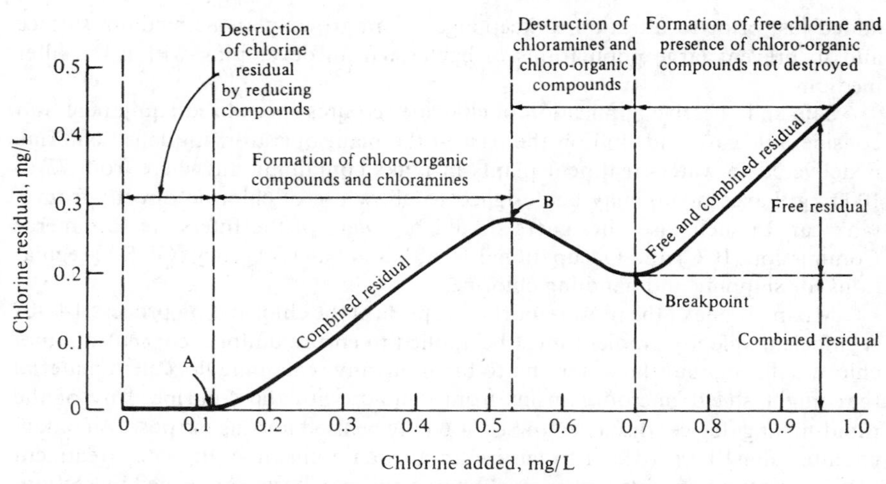

Figure 4-35 Generalized curve obtained during breakpoint chlorination. (*From Metcalf & Eddy, Inc.* [*4-40*].)

ine to form severe taste and odor problems. The original organics must be removed before chlorination, and undesirable compounds must be removed after chlorination, or the compounds must be prevented from forming. The compounds can be removed by adsorption onto activated carbon, or their formation can be prevented by the substitution of chloramines, which do not react with the organics or phenols, for free chlorine. Chloramines can be formed by first adding a small quantity of ammonia to the water, then adding chlorine. The reactions of chlorine with ammonia are as follows:

$$NH_3 + HOCl \longrightarrow NH_2Cl \text{ (monochloramine)} + H_2O \tag{4-57}$$

$$NH_2Cl + HOCl \longrightarrow NHCl_2 \text{ (dichloramine)} + H_2O \tag{4-58}$$

$$NHCl_2 + HOCl \longrightarrow NCl_3 \text{ (nitrogen trichloride)} + H_2O \tag{4-59}$$

These reactions are dependent on several factors, the most important of which are pH, temperature, and reactant quantities. At pH greater than 6.5 monochloramine will be the predominant species. [4-53] Since combined residuals are less effective as a disinfectant, concentration of 2 to 3 mg/L with contact time in excess of 30 min is often required. Chloramines are persistent and provide continued protection against regrowth in the distribution system.

Provisions may be made for application of chlorine at several points within the water-treatment process. When treating raw water of good quality, no early applications may be necessary, yet it is advisable to design a plant to allow for easy addition of early applications later, should future conditions require them.

Chlorine may be added to the incoming flow (prechlorination) to assist with the oxidation of inorganics or to arrest biological action that may produce undesirable gases in the sludge at the bottom of clarifiers. Chlorine is frequently added just prior to filtration to keep algae from growing at the medium surface and to prevent large populations of bacteria from developing within the filter medium.

Safe and effective application of chlorine requires specialized equipment and considerable care and skill on the part of the plant operator. Liquefied chlorine is delivered to water-treatment plants in tanks containing anywhere from 75 to 1000 kg. Large plants may be designed to allow use of chlorine directly from a tank car. In such cases, designers should be aware of the Interstate Commerce Commission (ICC) and Occupational Health and Safety Agency (OHSA) regulations for shipping and handling chlorine.

Mixing is one of the most important aspects of the chlorination process. [4-40] A sufficient velocity gradient must be applied to ensure uniform concentration of chlorine throughout the water and to break up any remaining flocculent material that might shield microorganisms from contact with the chlorine. Any of the rapid-mixing devices discussed in Sec. 4-6 may be used for this purpose. A contact chamber must be provided to ensure an adequate kill time. In water-treatment plant operations, mixing and contact operations may be accomplished by sectioning off part of the clear well.

Safety considerations mandate storing of chlorine tanks in a separate room. Storage and operating rooms should not be directly connected, nor should they be directly connected to other enclosed areas of the treatment plant. All doors to these facilities should be open to the outside, and windows should be provided for visual inspection from the outside. Safety equipment, including masks with air tanks, chlorine detection devices, and emergency repair equipment, should be provided in strategic locations.

Other Means of Disinfection

Given the problems associated with chlorination, it is not surprising that a search for a substitute means of disinfection has been in progress for years. However, the list of candidates for replacement remains quite small, with chlorine dioxide and ozone being the leading contenders. Although both of these are effective in destroying pathogens, ozone does not leave a disinfecting residual that can guard against pathogen regrowth in the distribution system, and both are more expensive than chlorine and have practical problems associated with their use.

Ozone Ozone, the allotropic form of oxygen, can be produced in a high-strength electrical field from oxygen in pure form or from the ionization of clean, dry air.

$$O_2 \xrightarrow{\text{high voltage}} O + O$$

$$O + O_2 \rightleftharpoons O_3 \qquad (4\text{-}60)$$

Ozone is a powerful oxidant which reacts with reduced inorganic compounds and with organic material. The difference, however, is that an oxygen atom, instead of a chloride atom, is added to the organics, the end result being an environmentally acceptable compound. Once this ozone demand has been met, the ozone reacts vigorously with bacteria and viruses. It is reported to be more effective than chlorine in inactivating resistant strains of bacteria and viruses. [4-58]

Because ozone is chemically unstable it must be produced on-site and used immediately. Typical dosages range from 1.0 to 5.3 kg/1000 m^3 [4-58], with power consumptions of from 10 to 20 kW · h/kg of ozone. [4-53] Cost of ozonation is two to three times higher than the cost of chlorination. Since no residual remains, it will be necessary to use a small amount of chlorine *after* ozonation to provide continued protection against regrowth in the distribution system.

Because ozone has a low solubility in water, it must be mixed thoroughly with the water to ensure adequate contact. This can be a problem when air is used as the oxygen source, since large volumes of nitrogen must also be handled.

In spite of these problems, ozone is widely used in Europe for disinfecting water containing color and organic compounds. In the United States, ozonation has been limited, primarily because chlorination has been more economical. However, ozonation will no doubt come into wider use as a result of recently adopted standards on chlorinated hydrocarbon compounds. When raw water is known to contain precursors of haloform compounds, use of ozone should be seriously considered.

Chlorine dioxide Chlorine dioxide (ClO_2) has many of the same properties as ozone. A strong oxidant which forms neither chloroforms nor chloramines, it is particularly effective in oxidizing phenolic compounds. Although highly soluble, chlorine dioxide does not react chemically with water. Contact with the atmosphere will result in loss of ClO_2 by gas transfer, and the presence of light results in photooxidation. Chlorine dioxide must therefore be generated on-site, in aqueous form, usually by the chlorination of sodium chlorite at low pH. [4-25]

The disinfectant properties of chlorine dioxide are similar to those of chlorine, and its use results in a measurable residual. Although its principal application has been in wastewater disinfection, chlorine dioxide has had limited use in potable water treatment for oxidizing iron and manganese and for removal of taste and odor compounds. Its possible reduction to chlorate, a substance which may be toxic to humans, makes questionable its use in potable water.

Other disinfectants Irradiation with ultraviolet light is a promising method of disinfection. Although it provides no residual, this method is effective in inactivating both bacteria and viruses. [4-58] Ultraviolet light spans the wavelengths of 2000 to 3900 Å (angstrums). The most effective band for disinfection is in the shorter range of 2000 to 3000 Å. Light with this wavelength can be generated with low-pressure mercury vapor lamps. A power input of 30 $\mu W/cm^2$ applied to thin sheets of turbidity-free water should be sufficient. A means of keeping the glass surface clear of deposits must be provided.

A variety of other disinfection methods may be used in special circumstances. These include other halogens (iodine, bromine), metals (copper, silver), other oxidants ($KMnO_4$), sonification, electrical current, and gamma-ray irradiation. It is unlikely, however, that any of these processes will find widespread use in disinfecting public water supplies in the foreseeable future.

Other Water-Treatment Processes

The water treatment processes discussed in the previous sections of this chapter are sufficient to render most natural surface water or groundwater potable. In some instances, however, the water supply may contain materials that are not removed by the conventional water-treatment processes. Examples are groundwater with excessive dissolved solids and surface waters that contain organic compounds from domestic or industrial wastewaters or naturally occurring organics such as humic and fulvic acids or products of algal blooms. Processes are available for removing these contaminants. These processes involve sophisticated equipment, require highly skilled operators, and are therefore quite expensive. Their use in potable water preparation should be considered only when a better-quality water supply is not available.

The following sections will discuss processes for removing inorganic and organic dissolved solids from water intended for potable use. These same processes may act as tertiary treatment for wastewater with some modification. The discussion is arranged according to target contaminants rather than process type.

4-10 DISSOLVED-SOLIDS REMOVAL

Target contaminants in dissolved-solids removal processes may be inorganic minerals or refractory organic compounds. Several processes are available for reducing the levels of these compounds in water intended for potable use, and process selection must be based on economics and dependability.

Inorganic Material

Demineralization and *desalinization* are synonymous terms applied to the removal of inorganic mineral substances from water. This is most often accomplished by selective, staged ion-exchange units or by processes employing the use of semi-permeable membranes. Both procedures require virtually complete removal of suspended solids prior to their application.

Ion exchange The principles of the ion-exchange process were described in Sec. 4-7 as related to water softening. In that process, sodium ions were exchanged for calcium and magnesium ions on an equivalence basis and there was no net decrease in dissolved solids. For demineralization, however, the exchanged ions must not contribute dissolved solids to the effluent. This is accomplished by exchanging hydrogen for the dissolved cations and hydroxide for the dissolved anions. The two then combine in equal amounts to form H_2O, leaving no residual and not affecting the pH. The resins are regenerated with acids and bases, respectively.

The ion-exchange process must be carried out in two or more steps. Generally, the cations are removed first, followed by the anions. The process and related chemical reactions are shown in Fig. 4-36. Because completely demineralized water is undesirable, a portion of the water is bypassed and blended with the process effluent to provide a stable water.

Microporous membranes Demineralization of water can be accomplished using thin, microporous membranes. There are two basic modes of operation in use. One system uses pressure to drive water through the membrane against the force of osmotic pressure and is called *reverse osmosis*, even though the pressure applied is several orders of magnitude in excess of the natural osmotic pressure. The other process, called *electrodialysis*, uses electrical forces to drive ions through ion-selective membranes.

The membrane commonly used in reverse osmosis is composed of cellulose acetate and is about 100 μm thick. Special techniques of casting result in an asymmetric arrangement, with one side of the membrane having a thin (0.2 μm), dense film, while the remainder is more porous. The film contains microscopic openings that allow water molecules to pass through but reject dissolved solids by either molecular sieving or by some other mechanisms not yet completely understood. [4-25] The process results in a concentrated solution of the ions on the pressure side of the membrane and a product water which is relatively free of ions.

Three basic membrane configurations are used in reverse-osmosis systems. These are the spiral-wound system (Fig. 4-37*a*), in which membranes and support material are placed in alternate layers, rolled into a cylindrical shape, and placed in tubes of suitable material. The support material is porous and serves as a transport medium for the liquid streams. Separation of the product water and concentrate is accomplished by internal arrangement within the containment tube. Tubular systems (Fig. 4-37*b*) are available in which the membrane and its porous support system are formed to fit inside a containment tube of up to 125 mm in diameter. Product water is withdrawn from the porous support medium, while the concentrate passes through the core of the membrane. Hollow-fiber membranes (Fig. 4-37*c*) are extremely small tubes, diameters of 1.0 μm or less being common. The large wall-thickness-to-diameter ratio provides a good radial strength, and the fibers can be suspended in the fluid without the use of the support medium. The feed water is usually on the outside of the fiber, while the product water is withdrawn through the center.

The spiral-wound and hollow-fiber systems generally provide higher flow rates but are more susceptible to fouling than are the tubular systems and are more often used for demineralizing potable water. Tubular units are better suited for wastewater treatment because membrane fouling can be minimized by increasing the flow rate through the tube.

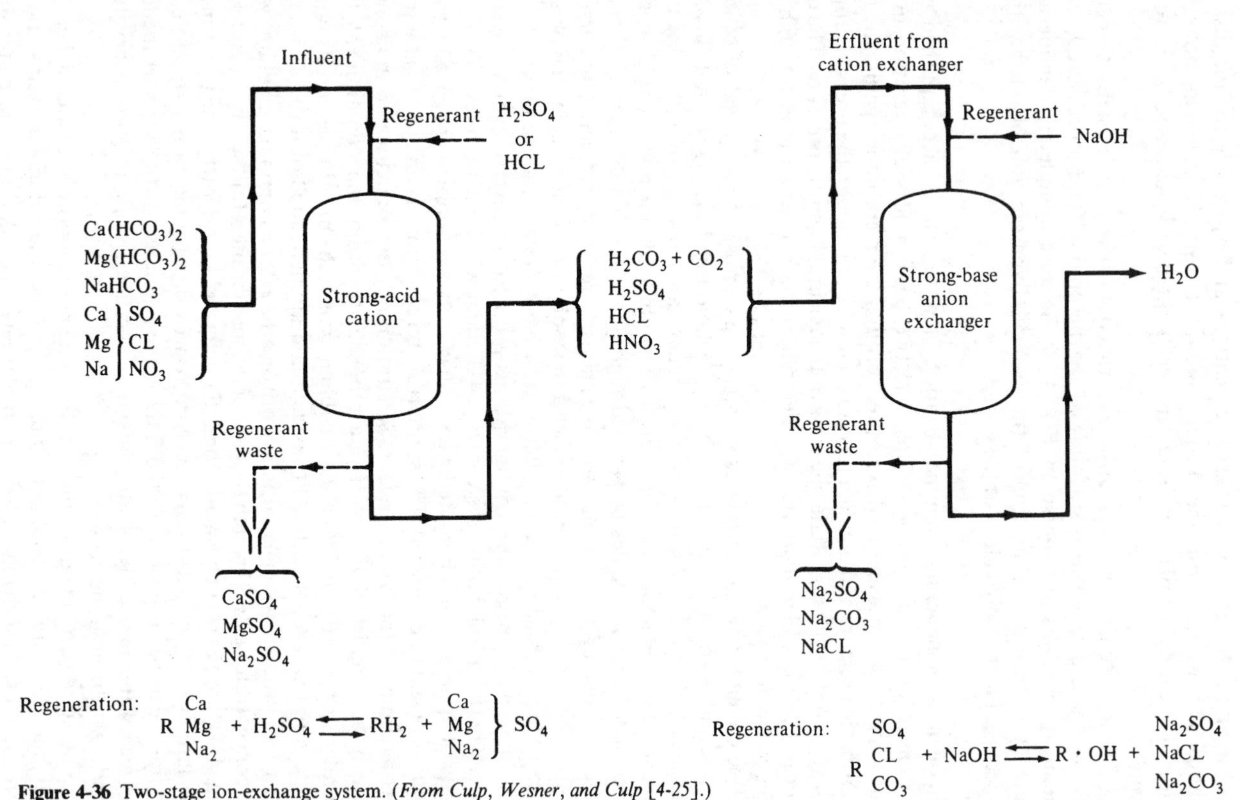

Figure 4-36 Two-stage ion-exchange system. (*From Culp, Wesner, and Culp* [4-25].)

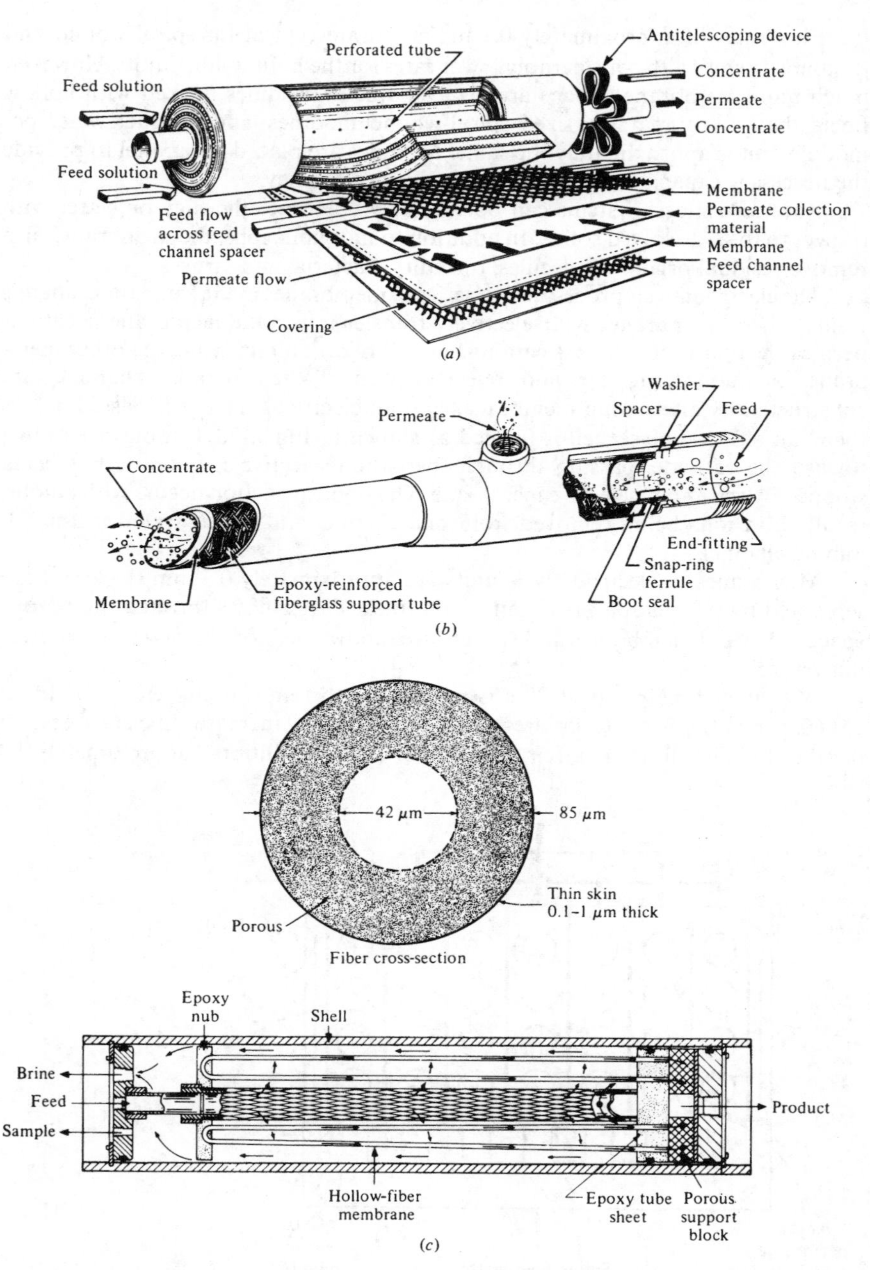

Figure 4-37 Reverse osmosis modules: (*a*) spiral-wound ultrafiltration model (*courtesy of Abcor, Inc.*); (b) tubular model (*courtesy of Abcor, Inc.*); (c) hollow-fiber model (*courtesy of Dupont*).

Flux rates of approximately 0.1 $m^3/m^2 \cdot d$ are typical for spiral-wound and tubular systems, with considerably lower rates for the hollow-fiber units. However, much more membrane surface area is available in modules packed with hollow fibers than in comparably sized spiral-wound modules, and product water per module unit is approximately the same. Modules are placed in parallel to provide the necessary capacity and in series to increase efficiency.

Reverse osmosis systems can operate at 90 percent efficiency or better with respect to total dissolved solids. In addition to inorganic ions, the membranes also remove residual organic molecules, turbidity, bacteria, and viruses.

The electrodialysis process uses a series of membranes made from ion-exchange resins. These membranes will selectively transfer ions. One membrane is cation-permeable, that is, it will pass cations but will reject anions, while the other membrane is anion-permeable and rejects cations. When parallel channels are constructed by alternating membranes and an electrical current is passed across them, an electrodialysis cell is formed as shown in Fig. 4-38. Cations are drawn toward the cathode, passing through the cation-selective membrane but being stopped by the anion-selective membrane. The opposite action occurs with anions, resulting in ions being removed from one channel and concentrated in the adjoining channel.

Membranes in electrodialysis units are approximately 0.5 mm thick and are separated by porous spacers about 1 mm thick. Water flows through the porous spacers. Several membranes and spacers are sandwiched together into one electrodialysis cell.

A contact time of 10 to 20 s is required with removal efficiencies of about 25–60 percent. [4-25] Cells are placed in series to increase efficiency and in parallel to meet flow requirements. Under ideal conditions, approximately 90

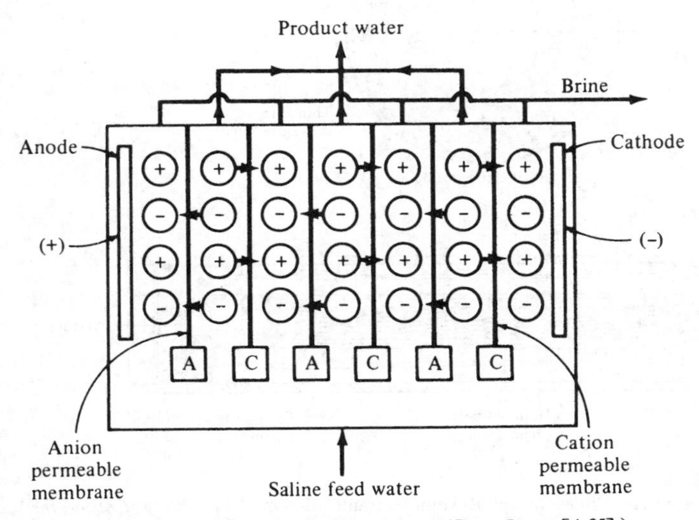

Figure 4-38 Schematic of electrodialysis process. (*From Lacey* [4-35].)

percent of the feed water is deionized, with the ions being concentrated in the remaining 10 percent.

Both reverse osmosis and electrodialysis require a high degree of treatment prior to their application. Suspended solids removal is absolutely necessary, and dissolved organics should be removed to prevent fouling. Adjustment of pH to the slightly acidic range may be necessary to prevent inorganic precipitation. Reverse osmosis generally produces a higher-quality effluent than does electrodialysis, although at a higher cost. Design parameters for demineralization processes are given in Culp et al. [4-25]

Both reverse osmosis and electrodialysis produce a waste stream that may range from 10 to 25 percent of the feed water. In potable water supply systems, an additional volume of water must be processed to offset this loss. In both water- and wastewater-treatment systems the concentrated wastewater streams must be disposed of properly.

Organic Material

Refractory organics can be removed from water and wastewater by adsorption processes or by chemical oxidation. The processes are essentially the same for both water and wastewater treatment, although the applications may differ somewhat.

Adsorption Adsorption can be defined as the accumulation of substances at the interface between two phases. [4-53] In water and wastewater treatment, the interface is between the liquid and solid surfaces that are artificially provided. The material removed from the liquid phase is called the *adsorbate*, and the material providing the solid surfaces is called the *adsorbent*.

The adsorbent most commonly used in water and wastewater treatment is activated carbon. Activated carbon is manufactured from carbonaceous material such as wood, coal, petroleum residues, etc. A char is made by burning the material in the absence of air. The char is then oxidized at higher temperatures to create a very porous structure. This "activation" step provides irregular channels and pores in the solid mass, resulting in a very large surface-area-per-mass ratio. Surface areas ranging from 500 to 1500 m^2/g have been reported [4-53], with all but a small fraction of the surface area being associated with the pores.

Once formed, activated carbon is crushed into granules ranging from 0.1 to 2 mm in diameter or is pulverized to a very fine powder. Dissolved organic material adsorbs to both exterior and interior surfaces of the carbon. When these surfaces become covered, the carbon must be regenerated. Although adsorption properties and mechanisms are essentially the same, application techniques for granular activated carbon and powdered activated carbon are considerably different.

The contact system for granular activated carbon (GAC) consists of a cylindrical tank which contains a bed of the material (Fig. 4-39). The water is passed through the bed with sufficient residence time allowed for completion of the adsorption process. The system may be operated in either a fixed-bed or moving-bed

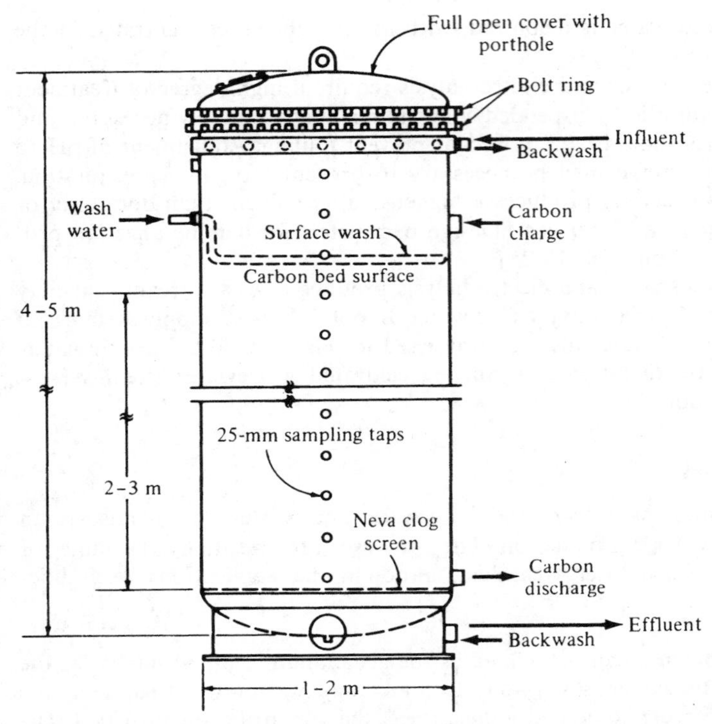

Figure 4-39 Typical activated-carbon adsorption column. (*From Metcalf & Eddy, Inc.* [*4-40*].)

mode. Fixed-bed systems are batch operations that are taken off the line when the adsorptive capacity of the carbon is used up.

Although fixed granular carbon beds can be cleaned in a place with superheated steam, the most common practice is to remove the carbon for cleaning in a furnace. The regeneration process is essentially the same as the original activation process. The adsorbed organics are first burned at about 800°C in the absence of oxygen. An oxidizing agent, usually steam, is then applied at slightly higher temperatures to remove the residue and reactivate the carbon.

In a moving-bed system, spent carbon is continuously removed from the bottom of the bed, with regenerated carbon being replaced at the top. Most modern applications use the moving-bed system with a countercurrent flow; that is, the water is introduced at the bottom of the bed and moves upward against the flow of carbon.

The major problem associated with granular-activated-carbon-contact systems is plugging of the bed by suspended solids in the water. Provisions may be made in the design of the vessel for backwashing the bed in a fashion similar to filter backwashing. Other designs avoid plugging by operating with the bed in a fluidized state. Sufficient upflow velocity is provided to maintain the bed at about

10 percent expansion at all times so suspended solids in the influent can pass through. This mode of operation has an added advantage in that adsorbed organics increase the density of the carbon, and the spent carbon migrates to the bottom of the fluidized bed for removal to the regeneration process.

Design of granular-activated-carbon systems is based on flow rates and contact times. Flow rates of 0.08 to 0.4 $m^3/m^2 \cdot min$ and contact times of 10 to 50 min, based on empty-tank cross section and volume, are common practice. Downtime of up to 40 percent should be included in the plant capacity, with 5 to 10 percent makeup carbon being provided after each regeneration cycle. The interested reader is referred to Culp, Wesner, and Culp [4-25] for a more detailed discussion of design.

Carbon columns can be arranged in parallel to increase the capacity and in series to increase the contact time. To approximate the countercurrent approach in a series of fixed-bed columns, water proceeds from the column which has been used the longest to the one in use for the shortest time.

Powdered activated carbon (PAC) cannot be used in a fixed-bed arrangement because of its small size and the subsequent high head loss that would result from passing water through it. Powdered activated carbon is contacted with the water in open vessels where it is maintained in suspension for the necessary contact time and then removed by conventional solids-removal processes. Flocculation equipment described in Sec. 4-6 is sufficient for this purpose.

Powdered activated carbon is much more difficult to regenerate than granular. Most systems employ a fluidized bed arrangement in which a mixture of steam and other hot gases holds the carbon in suspension while the regeneration processes occur. In some cases, sand is fluidized along with the carbon to help hold heat in the system.

In wastewater treatment, powdered activated carbon can be added to the aeration basin and removed with the biological solids in the secondary clarifier. In this case, both refractory and biodegradable organics are adsorbed. Biomass growth on the carbon surface utilizes the biodegradable fraction. Removal efficiency for biodegradable organics may be improved by this process, but usually at the expense of refractory organic removal efficiency. Use of powdered activated carbon in secondary wastewater systems results in an inseparable mixture of biological solids and carbon. Thermal regeneration of the carbon also results in destruction of the biomass, eliminating the need for other sludge processing and disposal techniques, but increasing the size of the carbon regeneration system.

In current practices, most systems treating potable water use powdered activated carbon, while advanced wastewater systems use moving-bed, granular activated carbon. Better regeneration procedures would greatly enhance the use of powdered activated carbon in wastewater treatment, particularly in the secondary processes.

Chemical oxidation Chemical oxidation of refractory organic compounds can be used as an alternative to the adsorption process in both potable water– and wastewater-treatment systems. Large, complex organic molecules, ring-structured detergents, and phenolic and humic compounds can be broken into simpler compounds by strong oxidants such as ozone or chlorine. Further oxidation by chemical or biological means may result in stable end products. Added advantages of this process may include ammonia removal, oxidation of inorganic substances such as iron and manganese, and disinfection. The discovery that chlorine reacts with some organic compounds to form undesirable haloforms makes its use as a chemical oxidant questionable.

The application of ozone for both disinfection and chemical oxidation in potable water treatment has been a long-standing practice in Europe. The destruction of taste and odor compounds and color-producing organics by ozonation is quite effective. Application of ozone to wastewater organics is less efficient. Some of the biologically resistant compounds in secondary effluents are also chemically resistant. Generally, a 3-1 ratio of ozone to organics on a mass basis is sufficient to reduce the COD by approximately 70 percent. [4-40] Better efficiencies can be obtained only by significantly increasing the dosage.

Application of ozone for chemical oxidation and disinfection is a simultaneous operation in potable water systems. In wastewater systems, chemical oxidation is more cost-effective when applied after secondary treatment, or after tertiary processes if these are included in the system. The specific characteristics of ozone and systems for contacting it with water were discussed in Sec. 4-9 and will not be repeated here.

DISCUSSION TOPICS AND PROBLEMS

4-1 Briefly discuss the differences in the quality of water obtained from groundwater sources and water obtained from surface water sources. What kinds of treatment processes might be needed for groundwater intended for potable use? For water taken from a mountain reservoir and intended for potable use? For water taken from the Mississippi River and intended for potable use?

4-2 What kind of treatment is needed for well water intended for agricultural use? Surface waters intended for agricultural use?

4-3 Would ordinary tap water from a city water supply be adequate, without further treatment, for all industrial uses? Why or why not?

4-4 Why is aeration used in water-treatment plants? Is it more commonly used with groundwater or surface water? Why?

4-5 Name and describe three commonly used water-in-air systems found in water purification plants.

4-6 Describe an air-in-water system commonly used in water purification plants.

4-7 Define (*a*) discrete particles, (*b*) flocculating particles, (*c*) dilute suspension, and (*d*) concentrated suspension.

4-8 What is a type-1 suspension?

4-9 Determine the settling velocity of a spherical particle with a diameter of 100 μm and a specific gravity of 2.3 in water at 25°C.

4-10 A particle with a diameter of 1.0 mm and a specific gravity of 3.0 is released in water at 30°C. How long will it take the particle to travel 2 m?

4-11 A particle with a diameter of 0.5 mm and a specific gravity of 2.5 is released in water with a temperature of 25°C. How far does the particle travel in 3 s?

4-12 Two particles are released in water at the same time. Particle A has a diameter d_A of 0.4 mm. Particle B has a diameter d_B, of 0.9 mm. What is the ratio of the settling velocity of particle A to that of particle B? Assume equal densities.

4-13 Suppose that a column is filled with water containing a uniform suspension of particles A and B as described in Prob. 4-12. Particle B is removed with 100 percent efficiency in exactly 10 s. What is the percent removal of particle A?

4-14 Name three types of settling basins employed for solids removal in water-treatment plants.

4-15 Describe the four functional zones of a long-rectangular settling tank.

4-16 A settling column analysis is run on a type-1 suspension. The settling column is 2 m tall, and the initial concentration of the well-mixed sample is 650 mg/L. Results of the analysis are shown below.

Time, min	0	58	77	91	114	154	250
Conc remaining, mg/L	650	560	415	325	215	130	52

What is the theoretical efficiency of the settling basins that receive this suspension if the loading rate is 2.4×10^{-2} m/min?

4-17 Using the data from Prob. 4-16, determine the theoretical efficiency of a settling basin with a loading rate of 3.0×10^{-2} m/min.

4-18 Using the data from Prob. 4-16, determine the theoretical efficiency of a settling basin with a surface area of 500 m^2 and an inflow of 14,400 m^3/d.

4-19 Determine the theoretical efficiency of the settling basin in Prob. 4-18 if $v_i = 0.04x$.

4-20 A settling column analysis is run on a type-2 suspension with the following results. (Entries are suspended-solids concentrations at stated times.)

	Time, min							
Depth, m	0	40	80	120	160	200	240	280
0.5	820	369	238	164	107	66	41	33
1.0	820	442	369	279	213	164	115	90
1.5	820	631	476	361	287	230	180	148
2.0	820	672	558	426	353	287	238	187
2.5	820	713	590	492	402	344	262	230
3.0	820	722	615	533	460	394	320	262
3.5	820	738	656	574	492	418	360	303

Determine the theoretical efficiency of a settling basin with a depth of 3.5 m, a volume of 1400 m^3, and an inflow of 11,200 m^3/d.

4-21 Using the data from Prob. 4-20, determine the theoretical efficiency of a settling basin with a depth of 2.5 m, a volume of 2,200 m^3, and an inflow of 13,200 m/d^3.

4-22 A settling basin processing 14,400 m^3/d of water has a depth of 4.0 m and a volume of 1200 m^3. Using the data from Prob. 4-20, determine the theoretical efficiency.

4-23 What is a type-2 suspension?

4-24 A water-treatment plant is to process 19,000 m^3/d. A settling basin for a type-2 suspension is to operate at 0.75 m/h. Determine the dimension of the basin for (*a*) a long-rectangular unit and (*b*) a circular unit. Check detention times, horizontal velocities, and weir overflow rates.

4-25 Determine the appropriate number of units and dimensions for settling basins to treat 75,000 m^3/d at an overflow rate of 0.8 m/h.

4-26 Assume that the settling basins in Prob. 4-25 will be constructed of reinforced concrete and that the cost of forming and pouring circular walls is 1.25 times the cost of forming and pouring straight walls. What will be the relative costs of using circular tanks compared to rectangular tanks utilizing common walls where possible?

4-27 Chemical coagulation in water-treatment plants is accomplished by the addition of trivalent metallic salts. Name two of these.

4-28 Name and discuss the four mechanisms thought to occur during coagulation.

4-29 Explain the importance of the jar test in coagulation operations and describe the test.

4-30 Under what conditions might it be desirable to add turbidity to water in a treatment plant?

4-31 Under what circumstances are lime and/or soda ash added to waters during coagulation operations?

4-32 Name and describe the four general categories into which surface waters are grouped with regard to coagulation.

4-33 Define (*a*) rapid mixing and (*b*) flocculation.

4-34 A water-treatment plant is to process 30,000 m^3/d. The rapid mixing tank will blend 35 mg/L of alum with the flow and is to have a detention time of 2 min. The tank is to have a square cross section with vertical baffles and a flat blade impeller similar to Fig. 4-20*b*. Determine the following:

(*a*) Quantity (kilograms per day) of alum added
(*b*) Dimensions of the tank
(*c*) Power input (kilowatts) necessary for a G value of 900 s^{-1}. The water temperature is 22°C.

4-35 The flow described in Prob. 4-34 is to be flocculated in a basin having four flocculators with transverse paddle units. (See Fig. 4-22*a*.) The basin may be a maximum of 10 m wide and 4 m deep to connect to the settling basin. Determine:

(*a*) Basin dimensions
(*b*) Power requirements
(*c*) Paddle configuration and rotational speed

The best Gt value for this system has been found to be 3.5×10^4.

4-36 The flow through a flocculator processes 16,800 m^3/d of water at 17°C. The paddles are arranged longitudinally. The optimum Gt value has been found from jar tests to be 4.5×10^4. Determine

(*a*) Basin dimensions
(*b*) Power applied to the water
(*c*) Paddle configuration and rotational speed

4-37 A flocculator paddle of the design and dimensions shown below is rotated through water at 20°C with an angular speed of 4.0 r/min.

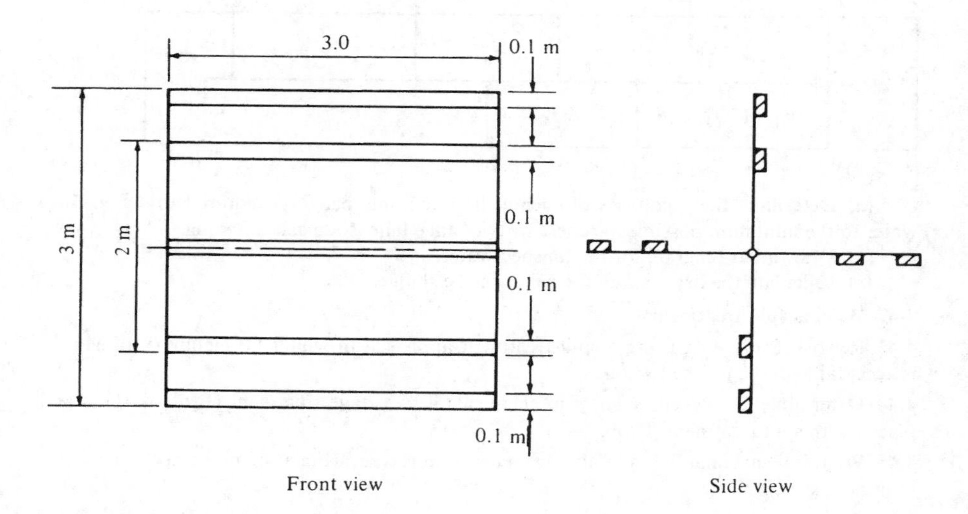

(*a*) How much power is dissipated into the water?

(*b*) If the tank in which this paddle is rotating has the dimensions of 4 × 4 × 4 m and the flow through the tank is 5000 m^3/d, determine the Gt value for the flocculator.

4-38 Softening of hard water may be done at a water utility treatment plant or by the consumer. As a general rule of thumb, what hardness level indicates the need to soften at the treatment plant?

4-39 Differentiate between single-stage and two-stage softening processes.

4-40 A water has the following ionic constituents (mequiv/L):

Ca^{2+} = 4.7 HCO_3^- = 2.5
Mg^{2+} = 1.0 SO_4^{2-} = 2.9
Na^+ = 2.2 Cl^- = 2.5

CO_2 = 0.6

(*a*) Calculate the chemical requirements (milliequivalents per liter) required to remove as much of the calcium as possible and to restabilize the water. (No Mg^{2+} removal is required.)

(*b*) Draw a bar diagram of the finished water.

(*c*) Calculate the daily quantity (kilograms per day) of lime and soda ash (assume a purity of 92 percent for the lime and 90 percent for the soda ash) to treat 17,500 m^3/d of this water.

(*d*) Determine the dry mass (kilograms per day) of the sludges produced.

4-41 A water-treatment plant processes 24,500 m^3/d of water with the following ionic concentration:

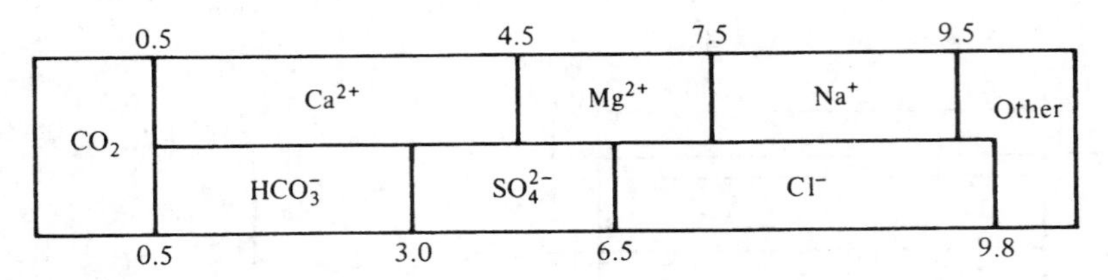

(*a*) Determine the quantities of chemicals (kilograms per day) required to soften this water to the minimum possible hardness by two-stage lime–soda ash softening.

(*b*) Draw a bar diagram for the finished water.

(*c*) Calculate the dry mass of the solids in the sludge.

4-42 What is split treatment?

4-43 Rework Prob. 4-41 using a split-treatment approach in which 1.0 mequiv/L of Mg^{2+} is acceptable in the finished water.

4-44 Determine the percent savings in chemicals if two-stage treatment (Prob. 4-41) is replaced with split treatment (Prob. 4-43).

4-45 What is recarbonation and under what conditions is recarbonation necessary in a water-treatment system?

4-46 An ion-exchange system is to be used to soften the water described in Prob. 4-40. The resin has an exchange capacity of 95 kg/m^3 when operated at a flow rate of 0.35 $m^3/m^2 \cdot min$. Determine the volume of resin needed and a tank configuration to allow continuous operation if the regeneration time is 2 h.

4-47 Determine the chemical requirement for regeneration of the ion-exchange system in Prob. 4-46 if regeneration is accomplished using 140 kg of sodium chloride per cubic meter of resin. What volume of backwash fluid must be disposed of if the salt used is in 10% solution?

4-48 Define "breakthrough" as it relates to treatment of hardness and discuss what steps must be taken after breakthrough point is reached.

4-49 A bed of filter sand 0.75 m deep is composed of uniform particles with diameter 0.5 mm, specific gravity 2.64, and shape factor 0.9. The porosity of the packed bed is 0.45. Plot a curve for head loss vs. filtering velocity over the filter velocity range of 2.0 to 7.0 m/h at a water temperature of 13°C.

4-50 Discuss filtration as a means of water treatment. What is precoat filtration?

4-51 What is the principal cleaning mechanism in backwashing filters?

4-52 Differentiate between slow sand filters and rapid sand filters.

4-53 What are dual-media filters? What are their advantages and disadvantages?

4-54 What are mixed-media filters? What are their advantages and disadvantages?

4-55 A hydrostatic head of 2 m is maintained above a 0.6-m-deep bed of filter sand. The sand is uniformly sized with diameter 0.4 mm, specific gravity 2.65, and shape factor 0.85. Determine the flow rate through the bed if the water temperature is 15°C.

4-56 An experimental filter consists of a 2-m depth of uniform sand with a diameter of 0.85 mm and a shape factor of 0.7. The porosity of the bed is 0.35 and the specific gravity of the sand is 2.65. Determine the head (meters of water column and kilopascals) to maintain a flow of water through the bed at a flow rate of 10 m/h. The water temperature is 15°C.

4-57 A rapid sand filter has a bed depth of 0.7 m. It is composed of sand grains that have a specific gravity of 2.65 and a shape factor of 0.82. The porosity of the bed is 0.45 throughout. The sieve analysis of the sand is shown below.

Sieve no.	Mass retained, %	Average particle size, mm
14–20	0.87	1.0
20–28	8.63	0.71
28–32	21.30	0.54
32–35	28.10	0.46
35–42	23.64	0.38
42–48	7.09	0.32
48–60	3.19	0.27
60–65	2.16	0.23
65–100	1.02	0.18

Determine the head loss through the bed if the flow rate is 5.0 m/s and the water temperature is 17°C.

4-58 A constant head of 2.5 m is maintained above the filter bed described in Prob. 4-57. Determine the flow rate through the filter.

4-59 Write a computer program and rework Prob. 4-57 by computer (or a hand-held programmable calculator).

4-60 Determine the backwash velocity (V_s) at which the filter bed in Prob. 4-49 will just begin to fluidize.

4-61 Determine the backwash velocity at which the filter bed in Prob. 4-55 will just begin to fluidize.

4-62 The filter bed in Prob. 4-49 is to be expanded to 1.5 times its original depth during backwash. Determine the required backwash rate (m/h) assuming that the backwash water is 18°C.

4-63 The filter bed in Prob. 4-55 is to be expanded to 1.6 times its original depth during backwash. Determine the required backwash rate (m/h) assuming a water temperature of 20°C.

4-64 The filter bed described in Prob. 4-57 is to be completely fluidized during backwash.

(*a*) Determine the backwash velocity that will just fluidize the largest particles in the bed.

(*b*) For the velocity determined in (*a*), calculate the expanded bed depth.

4-65 Write a program and solve Problem 4-64*b* by computer or hand-held programmable calculator.

4-66 A filter plant is to be constructed to process 75,700 m^3/d. Pilot-plant analysis on mixed media indicates that a filtration rate of 15 m/h will be acceptable. Assuming a surface configuration of approximately 5 × 8 m, how many filter units will be required? Allow one unit out of service for backwashing.

4-67 The backwash velocity required to expand the filters in Prob. 4-66 is 36 m/h. Each backwash period requires 20 min, and the water is wasted for the first 10 min of each filter run. Determine the net production of each filter if it is backwashed once a day.

4-68 Describe the characteristics of a good disinfectant.

4-69 Name several commonly used disinfectants and discuss the advantages and disadvantages posed by each. Which is the most commonly used in the United States? In Europe?

4-70 What factors militate against effective disinfection?

4-71 What methods are commonly used for desalinization of water?

4-72 How are refractory organics removed from water and wastewater?

4-73 Why is powdered activated carbon (PAC) unsuitable for use in a fixed-bed adsorption arrangement?

4-74 A city draws its water supply from a large reservoir. The water has consistent quality throughout the year. It has a turbidity ranging from 20 to 50 units, and its maximum hardness is less than 100 mg/L as $CaCO_3$. Refractory organics are not a problem and the TDS is low. Draw a schematic diagram of a treatment plant that may be used to render this water potable. Identify each unit and briefly state its purpose. Show points of chemical additions and identify the chemicals.

4-75 A city water supply is obtained from a deep aquifer. The water has uniform quality. It is clear and free of organics; hardness is in excess of 300 mg/L and consists of both calcium and magnesium. Dissolved CO_2 is approximately 15 mg/L and iron (Fe^{2+}) is about 1.0 mg/L. Other dissolved constituents are below problem levels. Draw a schematic diagram of a treatment plant that will render this water potable. Identify each unit and briefly state its purpose. Show points of chemical addition and identify the chemicals.

4-76 A large stream flowing through a highly industrialized area must serve as a raw water supply for a community. The water is consistently turbid, has hardness in excess of 300 mg/L, and has refractory organics that are known precursors of trihalomethanes. Draw a schematic diagram of a treatment plant that should render this water potable. Identify all units, state their purpose, and show points of chemical addition. Identify all chemicals.

REFERENCES

4-1 American Society of Civil Engineers, American Water Works Association, and Conference of State Sanitary Engineers: *Water Treatment Plant Design*, AWWA, New York, 1969.

4-2 American Water Works Association: *Water Quality and Treatment, A Handbook of Public Water Supplies*, 3d ed., McGraw-Hill, New York, 1971.

4-3 Amirtharajah, A.: "Design of Flocculation Systems," in R. L. Sanks (ed.), *Water Treatment Plant Design*, Ann Arbor Science, Ann Arbor, Mich., 1978.

4-4 ———: "Design of Rapid Mix Units," in R. L. Sanks (ed.), *Water Treatment Plant Design*, Ann Arbor Science, Ann Arbor, Mich., 1978.

4-5 ———: "Optimum Backwashing of Sand Filters," *J Env Eng Div ASCE*, **104**(EE5):917 (October 1978).

4-6 Baker, M. N.: *The Quest for Pure Water*, AWWA, New York, 1948.

4-7 Baumann, E. R.: "Granular Media Deep Bed Filtration," in R. L. Sanks (ed.), *Water Treatment Plant Design*, Ann Arbor Science, Ann Arbor, Mich., 1978.

4-8 ———: "Precoat Filtration," in R. L. Sanks (ed.), *Water Treatment Plant Design*, Ann Arbor Science, Ann Arbor, Mich., 1978.

4-9 Bernado, L. D., and J. R. Cleasby: "Declining-Rate vs. Constant-Rate Filtration," *J Env Eng Div, ASCE*, **106**(EE6):1023 (December 1980).

4-10 Camp, T. R.: "Floc Volume Concentration," *J AWWA*, **60**(6):656 (1968).

4-11 ———: "Velocity Gradients and Internal Work in Fluid Motion," *J. Boston Society of Civil Engineering*, **30**:219 (1943).

4-12 ——— and P. C. Stein: "Sedimentation and Design of Settling Tanks," *Trans ASCE*, **111**:895 (1946).

4-13 Carl, K. J., R. A. Young, and G. C. Anderson: "Guidelines for Determining Fire Flows," *J AWWA*, **65**(5):335 (1973).

4-14 Carmen, P. C.: "Fluid Flow Through Granular Beds," *Trans Inst Chem Eng* (London), **15**:150 (1937).

4-15 Chanlett, E. T.: *Environmental Protection*, 2d ed., McGraw-Hill, New York, 1979.
4-16 Clark, J. W., W. Viessman, Jr., and M. J. Hammer: *Water Supply and Pollution Control*, 3d ed., Harper & Row, New York, 1977.
4-17 Cleasby, J. L.: "Filtration" in W. J. Weber, Jr. (ed.), *Physiochemical Processes for Water Quality Control*, Wiley Interscience, New York, 1972.
4-18 ———: by personal communication, January 1980.
4-19 ———, J. Arboleda, D. E. Burns, P. W. Prendiville, and E. S. Savage: "Backwashing of Granular Filters," *J AWWA*, **69**: 115 (February 1977).
4-20 ———, and J. H. Dellingham: "Rational Aspects of Split Treatment," *Proc ASCE, J San Eng Div*, **92**(SA2): 1 (1966).
4-21 ——— and J. C. Lorence: "Effectiveness of Backwashing for Wastewater Filters," *J Env Eng Div, ASCE*, **104**(EE4): 749 (August 1978).
4-22 ———, E. W. Stangl, and G. H. Rice: "Developments in Backwashing of Granular Filters," *J Env Eng Div, ASCE*, **101**(EE5): 713 (October 1975).
4-23 Cohen, J. M., and S. H. Hannah: "Coagulation and Flocculation," in *Water Quality and Treatment*, 3d ed., McGraw-Hill, New York, 1971.
4-24 Conley, W. R.: "Water Pollution Technology Report," *Neptune McroFloc Inc.*, **2**(1), February 1968.
4-25 Culp, R. L., G. M. Wesner, and G. L. Culp: *Handbook of Advanced Wastewater Treatment*, Van Nostrand Reinhold, New York, 1978.
4-26 Davis, S. N., and R. J. M. DeWiest: *Hydrogeology*, Wiley, New York, 1966.
4-27 *Design Manual for Suspended Solids Removal*, U.S. Environmental Protection Agency, 1975.
4-28 Fair, G. M., and J. C. Geyer: *Water Supply and Wastewater Disposal*, Wiley, New York, 1961.
4-29 ———, ———, and D. A. Okun: *Elements of Water Supply and Wastewater Disposal*, 2d ed., Wiley, New York, 1971.
4-30 ———, F. C. Morris, S. L. Chang, I. Weil, and R. A. Burden: "The Behavior of Chlorine as a Water Disinfectant," *J AWWA*, **40**: 1051 (1948).
4-31 Gehm, H. W., and J. I. Bregman (eds.): *Handbook of Water Resources and Pollution Control*, Van Nostrand Reinhold, New York, 1976.
4-32 Hudson, H. E., and J. P. Wolfner: "Design of Mixing and Flocculation Basins," *J AWWA*, **59**: 1257 (1967).
4-33 Ives, J. K., and A. C. Bhole: "Theory of Flocculation for Continuous Flow Systems," *J Env Eng Div ASCE*, **99**: 17 (1973).
4-34 Kammerer, J. C.: "Water Quantity Requirements for Public Supplies and Other Uses," in H. W. Gehm and J. I. Bregman (eds.), *Handbook of Water Resources and Pollution Control*, Van Nostrand Reinhold, New York, 1976.
4-35 Lacey, R. E.: "Membrane Separation Processes," *Chem Eng*, **4**: 56 (September 1972).
4-36 Letterman, R. D., J. E. Quan, and R. S. Gemmell: "Influence of Rapid-Mix Parameters on Flocculation," *J AWWA*, **65**: 716 (1973).
4-37 Lindsley, R. K., and J. B. Franzini: *Water Resources Engineering*, 3d ed., McGraw-Hill, New York, 1979.
4-38 Lowenthal, R. E., and B. v. R. Marais: *Carbonate Chemistry of Aquatic Systems, Theory and Applications*, Ann Arbor Science, Ann Arbor, Mich., 1976.
4-39 McWhimie, R. C., and P. R. Johnson: "Water Storage and Distribution," in A. W. Gehm and J. I. Bregman (eds.), *Handbook of Water Resources and Pollution Control*, Van Nostrand Reinhold, New York, 1976.
4-40 Metcalf & Eddy, Inc.: *Wastewater Engineering: Treatment, Disposal, Reuse*, 2d ed., McGraw-Hill, New York, 1979.
4-41 O'Melia, C. R.: "Coagulation and Flocculation," in W. J. Weber, Jr. (ed.), *Physiochemical Processes for Water Quality*, Wiley Interscience, New York, 1972.
4-42 ——— and D. K. Crapps: "Some Chemical Aspects of Rapid Sand Filtration," *J AWWA*, **56**(10): 1326 (October 1964).
4-43 Powell, S. T.: *Water Conditioning for Industry*, McGraw-Hill, New York, 1954.
4-44 *Recommended Standards for Water Works*, Health Education Service, Albany, N.Y., 1976.

4-45 Reh, Carl W.: "Lime-Soda Softening Processes," in R. L. Sanks (ed.), *Water Treatment Plant Design*, Ann Arbor Science, Ann Arbor, Mich., 1978.
4-46 Rich, L. G.: *Environmental Systems Engineering*, McGraw-Hill, New York, 1973.
4-47 Sanks, R. L.: "Ion Exchange," in R. L. Sanks (ed.), *Water Treatment Plant Design*, Ann Arbor Science, Ann Arbor, Mich., 1978.
4-48 Sawyer, C. N., and P. L. McCarty: *Chemistry for Environmental Engineers*, 3d ed., McGraw-Hill, New York, 1978.
4-49 Schroeder, E. D.: *Watér and Wastewater Treatment*, McGraw-Hill, New York, 1977.
4-50 Scott, G. R.: "Aeration," in *Water Quality and Treatment*, 3d ed., McGraw-Hill, New York, 1971.
4-51 Steel, E. W., and T. J. McGhee: *Water Supply and Sewerage*, 9th ed., McGraw-Hill, 1979.
4-52 Stumm, W., and C. R. O'Melia: "Stoichiometry of Coagulation," *J AWWA*, **60**:514 (1968).
4-53 Sunderstron, D. W., and H. E. Klei: *Wastewater Treatment*, Prentice-Hall, Englewood Cliffs, N.J., 1979.
4-54 Tebbutt, T. H. Y.: *Principles of Water Quality Control*, 2d ed., Pergamon, Oxford, England, 1977.
4-55 Todd, D. K.: *Groundwater Hydrology*, Wiley, New York, 1960.
4-56 Vesilind, P. A.: *Environmental Pollution and Control*, Ann Arbor Science, Ann Arbor, Mich., 1975.
4-57 Walker, D. J.: "Sedimentation," in R. L. Sanks (ed.), *Water Treatment Plant Design*, Ann Arbor Science, Ann Arbor, Mich., 1978.
4-58 Walker, Rodger: *Water Supply Treatment and Distribution*, Prentice-Hall, Englewood Cliffs, N.J., 1978.
4-59 Weber, W. J. (ed.): *Physiochemical Processes for Water Quality Control*, Wiley Interscience, New York, 1972.

CHAPTER
FIVE

ENGINEERED SYSTEMS FOR WASTEWATER TREATMENT AND DISPOSAL

In modern societies proper management of wastewater is a necessity, not an option. The public health consequences of poor wastewater management have been discussed in previous chapters. Historically, the practice of collecting and treating wastewater prior to disposal is a relatively recent undertaking. Although remains of sewers have been found in ancient cities, the extent of their use for wastewater carriage is not known. The elaborate drainage system of ancient Rome was not used for waste disposal, and wastes were specifically excluded from the sewerage systems of London, Paris, and Boston until well after the turn of the nineteenth century.

Prior to this time, city residents placed "night soil" in buckets along the streets and workers emptied the waste into "honeywagon" tanks. The waste was transported to rural areas for disposal over agricultural lands. The invention of the flush toilet in the nineteenth century drastically changed waste-disposal practices. Existing systems for transporting urban wastes for disposal on agricultural lands were not adequate to handle the large volume of liquid generated by the flush toilets. Faced with this transportation problem, cities began to use natural drainage systems and storm sewers for wastewater carriage, against the advice of such men as Edwin Chadwick, who in 1842 recommended "rain to the river and sewage to the soil." [5-21] Construction of combined sewers was commonplace in large cities during the latter half of the nineteenth century. Since storm drain systems naturally ended at watercourses, waterborne wastes were discharged directly to streams, lakes, and estuaries without treatment. Gross pollution often resulted, and health problems were transferred from the sewered community to downstream users of the water.

The first "modern" sewerage system for wastewater carriage was built in Hamburg, Germany, in 1842 by an innovative English engineer named Lindley. Lindley's system included many of the principles that are still in use today. [5-10] Most of the improvements in wastewater collection systems over the last 100 years have consisted of improved materials and the inclusion of manholes, pumping stations, and other appurtenances.

The treatment of wastewater lagged considerably behind its collection. Treatment was considered necessary only after the self-purification capacity of the receiving waters was exceeded and nuisance conditions became intolerable. Various treatment processes were tried in the late 1800s and early 1900s, and by the 1920s, wastewater treatment had evolved to those processes in common use today. Design of wastewater-treatment facilities remained empirical, however, until midcentury. In the last 30 to 40 years, great advances have been made in understanding wastewater treatment, and the original processes have been formulated and quantified. The science of wastewater treatment is far from static, however. Advanced wastewater-treatment processes are currently being developed that will produce potable water from domestic wastewater. Problems associated with wastewater reuse will no doubt challenge the imagination of engineers for many years to come.

Philosophies concerning the ultimate disposal of wastewater have also evolved over the years. As previously mentioned, the practice of land disposal was replaced by the convenience of the water carriage system with direct discharge to surface waters. Operating under the assumption that the "solution to pollution is dilution," the assimilative capacity of streams was utilized before treatment was deemed necessary. For many years, little, if any, treatment was required of small communities located on large streams, while a high level of treatment was required by large cities discharging to small streams. In more recent times, the policy has shifted to require a minimum level of treatment of all waste discharges, regardless of the capacity of the receiving stream. Under current practice in the United States, all dischargers are given a permit stating the maximum amount of each pollutant that they are allowed to discharge. Discharge permits are no longer intended to just prevent discharges that exceed the self-purification capacity of the streams, but are concerned with obtaining the "fishable, swimmable" goals mentioned in Sec. 2-17.

Where extensive treatment of wastewater is necessary to meet stringent discharge permits, the quality of the treated effluent often approaches that of the receiving stream. These effluents should be considered a valuable water resource, particularly where water is scarce. Regulatory agencies encourage utilization of these wastewaters for irrigation, non-body-contact recreational activities, groundwater recharge, some industrial processes, and other nonpotable uses.

5-1 WASTEWATER CHARACTERISTICS

Wastewaters are usually classified as industrial wastewater or municipal wastewater. Industrial wastewater with characteristics compatible with municipal wastewater is often discharged to the municipal sewers. Many industrial wastewaters require pretreatment to remove noncompatible substances prior to discharge into the municipal system. Characteristics of industrial wastewater vary greatly from industry to industry, and, consequently, treatment processes for industrial wastewater also vary, although many of the processes used to treat municipal wastewater are also used in industrial wastewater treatment. A complete coverage of industrial wastewater treatment is beyond the scope of this text, and the interested reader is referred to other texts on the subject. See Refs. [5-7, 5-18, and 5-38].

Water collected in municipal wastewater systems, having been put to a wide variety of uses, contains a wide variety of contaminants. A list of contaminants commonly found in municipal wastewater along with their sources and their environmental consequences is given in Table 5-1.

Quantitatively, constituents of wastewater may vary significantly, depending upon the percentage and type of industrial waste present and the amount of dilution from infiltration/inflow into the collection system. Results of an analysis of a typical wastewater from a municipal collection system are given in Table 5-2.

The composition of wastewater from a given collection system may change slightly on a seasonal basis, reflecting different water uses. Additionally, daily fluctuations in quality are also observable and correlate well with flow conditions as noted in Fig. 5-1. Generally, smaller systems with more homogeneous uses produce greater fluctuations in wastewater composition.

The most significant components of wastewater are usually suspended solids, biodegradable organics, and pathogens. *Suspended solids* are primarily organic in nature and are composed of some of the more objectionable material in sewage.

Table 5-1 Important wastewater contaminants

Contaminant	Source	Environmental significance
Suspended solids	Domestic use, industrial wastes, erosion by infiltration/inflow	Cause sludge deposits and anaerobic conditions in aquatic environment
Biodegradable organics	Domestic and industrial waste	Cause biological degradation, which may use up oxygen in receiving water and result in undesirable conditions
Pathogens	Domestic waste	Transmit communicable diseases
Nutrients	Domestic and industrial waste	May cause eutrophication
Refractory organics	Industrial waste	May cause taste and odor problems, may be toxic or carcinogenic
Heavy metals	Industrial waste, mining, etc.	Are toxic, may interfere with effluent reuse
Dissolved inorganic solids	Increases above level in water supply by domestic and/or industrial use	May interfere with effluent reuse

Table 5-2 Typical analysis of municipal wastewater

Constituent, mg/L*	Concentration		
	Strong	Medium	Weak
Solids, total:	1200	720	350
Dissolved, total	850	500	250
Fixed	525	300	145
Volatile	325	200	105
Suspended, total	350	220	100
Fixed	75	55	20
Volatile	275	165	80
Settleable solids, mL/L	20	10	5
Biochemical oxygen demand, 5-day, 20°C (BOD_5)	400	220	110
Total organic carbon (TOC)	290	160	80
Chemical oxygen demand (COD)	1000	500	250
Nitrogen (total as N):	85	40	20
Organic	35	15	8
Free ammonia	50	25	12
Nitrites	0	0	0
Nitrates	0	0	0
Phosphorus (total as P):	15	8	4
Organic	5	3	1
Inorganic	10	5	3
Chlorides	100	50	30
Alkalinity (as $CaCO_3$)	200	100	50
Grease	150	100	50

* Unless otherwise noted.
Source: From Metcalf & Eddy, Inc. [5-36]

Body wastes, food waste, paper, rags, and biological cells form the bulk of suspended solids in wastewater. Even inert materials such as soil particles become fouled by adsorbing organics to their surface. Removal of suspended solids is essential prior to discharge or reuse of wastewater.

Although suspended organic solids are biodegradable through hydrolysis, biodegradable material in wastewater is usually considered to be soluble organics. *Soluble organics* in domestic wastewater are composed chiefly of proteins (40 to 60 percent), carbohydrates (25 to 50 percent), and lipids (approximately 10 percent). [5-52] Proteins are chiefly amino acids, while carbohydrates are compounds such as sugars, starches, and cellulose. Lipids include fats, oil, and grease. All of these materials contain carbon that can be converted to carbon dioxide biologically, thus exerting an oxygen demand as discussed in Sec. 2-13. Proteins also contain nitrogen, and thus a nitrogenous oxygen demand is also exerted. The biochemical oxygen demand test is therefore used to quantify biodegradable organics.

All forms of *waterborne pathogens* may be found in domestic wastewater. As discussed in Sec. 2-15, these include bacteria, viruses, protozoa, and helminths. These organisms are discharged by persons who are infected with the disease.

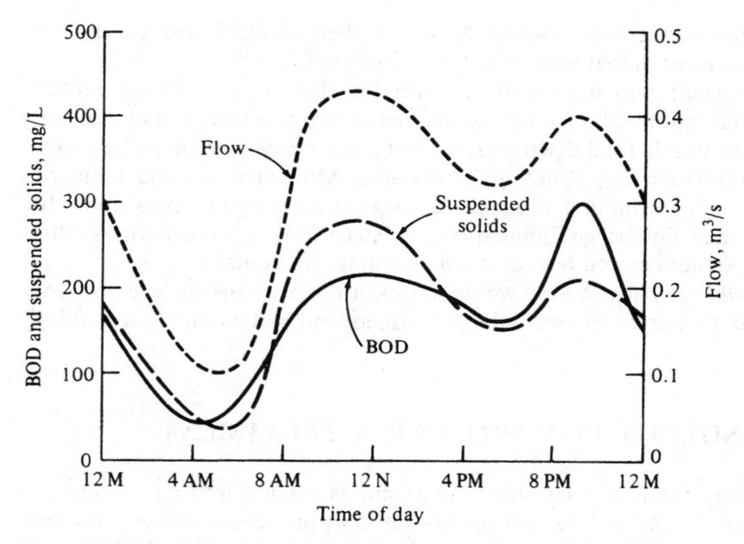

Figure 5-1 Typical variation in flow, suspended solids, and BOD_5 in municipal wastewater. (*From Metcalf & Eddy, Inc.* [*5-36*].)

Although pathogens causing some of the more exotic diseases may rarely be present, it is a safe assumption that a sufficient number of pathogens are present in all untreated wastewater to represent a substantial health hazard. Fortunately, few of the pathogens survive wastewater treatment in a viable state.

Traditional wastewater-treatment processes are designed to reduce suspended solids, biodegradable organics, and pathogens to acceptable levels prior to disposal. Additional wastewater-treatment processes may be required to reduce levels of nutrients if the wastewater is to be discharged to a delicate ecosystem. Processes to remove refractory organics and heavy metals and to reduce the level of inorganic dissolved solids are required where wastewater reuse is anticipated.

5-2 EFFLUENT STANDARDS

The Water Pollution Control Act of 1972 (Public Law 92-500) mandated the Environmental Protection Agency to establish standards for wastewater discharges. Current standards require that municipal wastewater be given secondary treatment and that most effluents meet the conditions shown in Table D-7 of the appendix. Secondary treatment of municipal wastewater is generally assumed to include settling, biological treatment, and disinfection, along with sludge treatment and disposal. Thus, the principal components of municipal wastewater, suspended solids, biodegradable material, and pathogens should be reduced to acceptable levels through secondary treatment. Industrial dischargers are required to treat their wastewater to the level obtainable by the "best available technology" for wastewater treatment in that particular type of industry.

The EPA regulations further define receiving streams as "effluent-limited" and "water-quality-limited". An *effluent-limited stream* is a stream that will meet its in-stream standards if all discharges to that stream meet the secondary-treatment and "best-available-technology" standards. Municipalities and industries discharging to effluent-limited streams are assigned discharge permits under the National Pollution Discharge Elimination System (NPDES); these permits reflect the secondary treatment and best-available-technology standards.

A *water-quality-limited stream* would *not* meet the proposed in-stream standards, even if all discharges met secondary-treatment and best-available-technology levels.

5-3 TERMINOLOGY IN WASTEWATER TREATMENT

The terminology used in wastewater treatment is often confusing to the uninitiated person. Terms such as unit operations, unit processes, reactors, systems, and primary, secondary, and tertiary treatment frequently appear in the literature, and their usage is not always consistent. The meanings of these terms, as used in this text, are discussed in the following paragraphs.

Methods used for treating municipal wastewaters are often referred to as either unit operations or unit processes. Generally, *unit operations* involve contaminant removal by physical forces, while *unit processes* involve biological and/or chemical reactions.

The term *reactor* refers to the vessel, or containment structure, along with all of its appurtenances, in which the unit operation or unit process takes place. Although unit operations and processes are natural phenomena, they may be initiated, enhanced, or otherwise controlled by altering the environment in the reactor. Reactor design is a very important aspect of wastewater treatment and requires a thorough understanding of the unit processes and unit operations involved.

A *wastewater-treatment system* is composed of a combination of unit operations and unit processes designed to reduce certain constituents of wastewater to an acceptable level. Many different combinations are possible. Although practically all wastewater-treatment systems are unique in some respects, a general grouping of unit operations and unit processes according to target contaminants has evolved over the years. Unit operations and processes commonly used in wastewater treatment are listed in Table 5-3 and are arranged according to conventional grouping. Actually, only a few wastewater-treatment methods fall completely into one category. Thus the usefulness of this classification system is somewhat compromised.

Municipal wastewater-treatment systems are often divided into primary, secondary, and tertiary subsystems. The purpose of *primary treatment* is to remove solid materials from the incoming wastewater. Large debris may be removed by

Table 5-3 Unit operations, unit processes, and systems for wastewater treatment

Contaminant	Unit operation, unit process, or treatment system
Suspended solids	Sedimentation
	Screening and comminution
	Filtration variations
	Flotation
	Chemical-polymer addition
	Coagulation/sedimentation
	Land treatment systems
Biodegradable organics	Activated-sludge variations
	Fixed-film: trickling filters
	Fixed-film: rotating biological contactors
	Lagoon and oxidation pond variations
	Intermittent sand filtration
	Land treatment systems
	Physical-chemical systems
Pathogens	Chlorination
	Hypochlorination
	Ozonation
	Land treatment systems
Nutrients:	
Nitrogen	Suspended-growth nitrification and denitrification variations
	Fixed-film nitrification and denitrification variations
	Ammonia stripping
	Ion exchange
	Breakpoint chlorination
	Land treatment systems
Phosphorus	Metal-salt addition
	Lime coagulation/sedimentation
	Biological-chemical phosphorus removal
	Land treatment systems
Refractory organics	Carbon adsorption
	Tertiary ozonation
	Land treatment systems
Heavy metals	Chemical precipitation
	Ion exchange
	Land treatment systems
Dissolved inorganic solids	Ion exchange
	Reverse osmosis
	Electrodialysis

Source: From Metcalf & Eddy, Inc. [5-36]

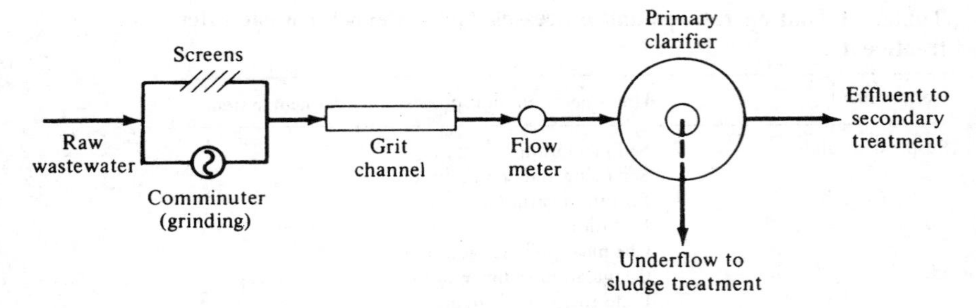

Figure 5-2 Typical primary treatment system.

screens or may be reduced in size by grinding devices. Inorganic solids are removed in grit channels, and much of the organic suspended solids is removed by sedimentation. A typical primary treatment system (Fig. 5-2) should remove approximately one-half of the suspended solids in the incoming wastewater. The BOD associated with these solids accounts for about 30 percent of the influent BOD.

Secondary treatment usually consists of biological conversion of dissolved and colloidal organics into biomass that can subsequently be removed by sedimentation. Contact between microorganisms and the organics is optimized by suspending the biomass in the wastewater or by passing the wastewater over a film of biomass attached to solid surfaces. The most common suspended biomass system is the activated-sludge process shown in Fig. 5-3*a*. Recirculating a portion of the biomass maintains a large number of organisms in contact with the wastewater and speeds up the conversion process. The classical attached-biomass system is the trickling filter shown in Fig. 5-3*b*. Stones or other solid media are used to increase the surface area for biofilm growth. Mature biofilms peel off the surface and are washed out to the settling basin with the liquid underflow. Part of the liquid effluent may be recycled through the system for additional treatment and to maintain optimal hydraulic flow rates.

Secondary systems produce excess biomass that is biodegradable through endogenous catabolism and by other microorganisms. Secondary sludges are usually combined with primary sludge for further treatment by anaerobic biological processes as shown in Fig. 5-4. The results are gaseous end products, principally methane (CH_4) and carbon dioxide (CO_2), and liquids and inert solids. The methane has significant heating value and may be used to meet part of the power requirements of the treatment plant. The liquids contain large concentrations of organic compounds and are recycled through the treatment plant. The solid residue has a high mineral content and may be used as a soil conditioner and fertilizer on agricultural lands. Other means of solids disposal may be by incineration or by landfilling.

Sometimes primary and secondary treatment can be accomplished together, as shown in Fig. 5-5. The oxidation pond (Fig. 5-5*a*) most nearly approximates

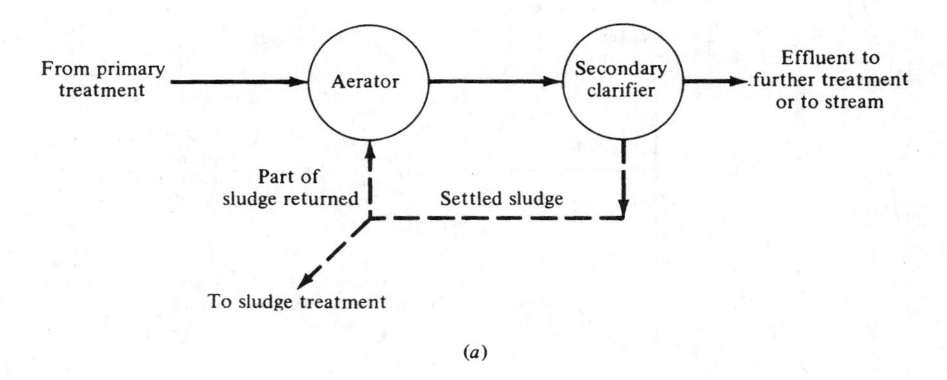

(a)

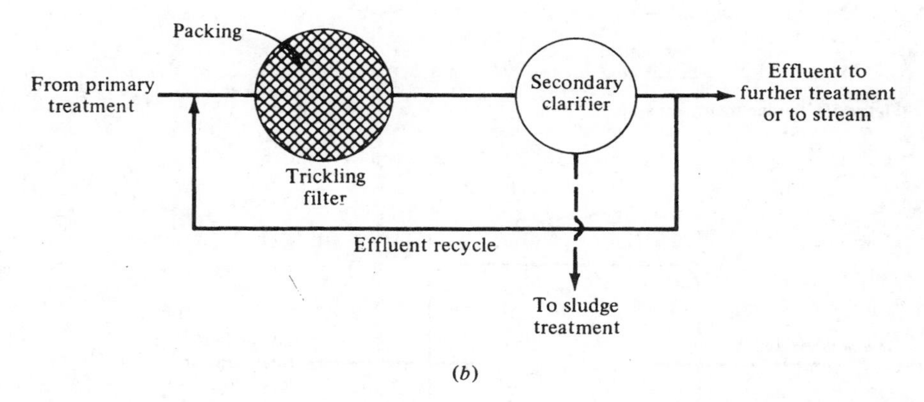

(b)

Figure 5-3 Secondary treatment system: (*a*) activated sludge system and (*b*) trickling filter system.

natural systems, with oxygen being supplied by algal photosynthesis and surface reaeration. This oxygen seldom penetrates to the bottom of the pond, and the solids that settle are decomposed anaerobically. In the aerated lagoon system (Fig. 5-5*b*) oxygen is supplied by mechanical aeration, and the entire depth of the pond is aerobic. Decomposition of the biomass occurs by aerobic endogenous catabolism. The small quantity of excess sludge that is produced is retained in the bottom sediments.

In most cases, secondary treatment of municipal wastewater is sufficient to meet effluent standards. In some instances, however, additional treatment may be required. *Tertiary treatment* most often involves further removal of suspended solids and/or the removal of nutrients. Solids removal may be accomplished by filtration, and phosphorus and nitrogen compounds may be removed by combinations of physical, chemical, and biological processes.

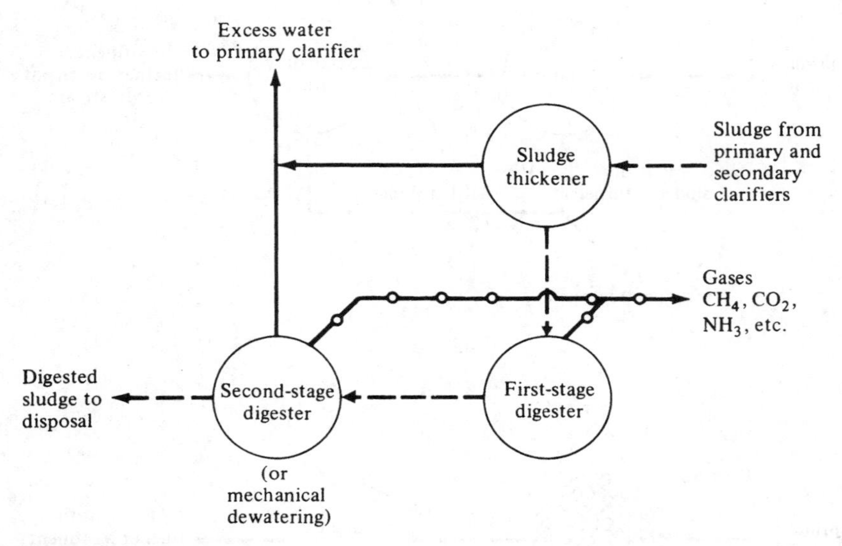

Figure 5-4 Sludge treatment system.

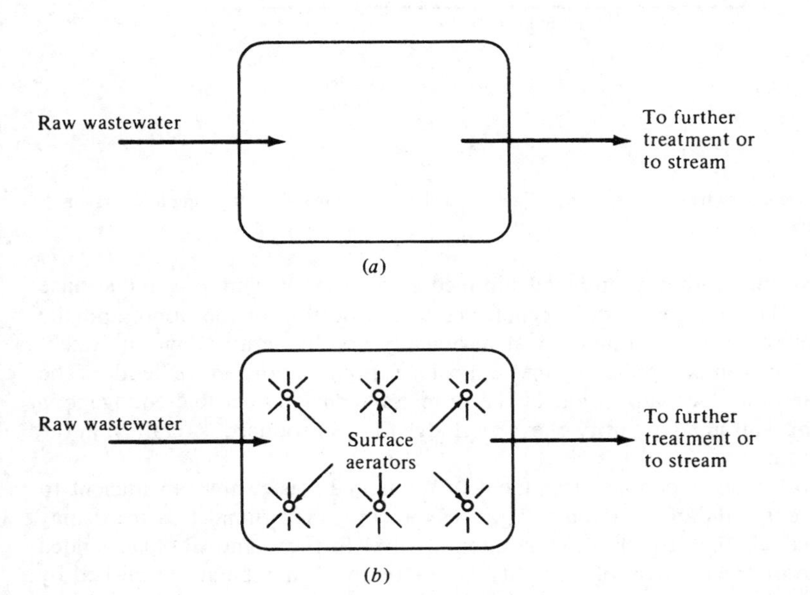

Figure 5-5 Primary and secondary water treatment in combination: (*a*) oxidation pond and (*b*) aerated lagoon.

A careful inspection of Figs. 5-2 through 5-5 leads to an interesting observation. The "removal" processes in wastewater treatment are essentially concentrating, or thickening, processes. Suspended solids are removed as sludges, and dissolved solids are converted to suspended solids and subsequently become removable sludges. Hammer [5-25] states that primary and secondary treatment, followed by sludge thickening, may concentrate organic material represented by 250 mg/L of suspended solids and 200 mg/L BOD in 375 L of municipal wastewater (the average per capita contribution) to 2.0 L of sludge containing 50,000 mg/L of solids. Most of the objectionable material initially in the wastewater is concentrated in the sludges and must be disposed of in a safe and environmentally acceptable manner. Vesilind [5-55] notes that a majority of the expenses, effort, and problems of wastewater treatment and disposal are associated with the sludges.

Design of wastewater-treatment systems is an important part of an environmental engineer's work. A thorough understanding of the unit operations and processes is necessary before the reactors can be designed. The following sections of this chapter are devoted to the various unit operations and processes commonly used in treating municipal wastewater. Many of these are similar, if not identical, to those used in preparing potable water. With the exception of nutrient removal, tertiary treatment operations for wastewater involve essentially the same principles used in preparing water of poor chemical quality for a potable supply. Where material would be duplicated, the reader is referred back to Chap. 4.

Primary Treatment

Wastewater contains a wide variety of solids of various shapes, sizes, and densities. Effective removal of these solids may require a combination of unit operations such as screening, grinding, and settling. Although no material is removed by the process, flow-measurement devices are essential for the operation of wastewater-treatment plants and are generally included in the primary system. Operations to eliminate large objects and grit, along with flow measurement, often referred to as *preliminary treatment*, are an integral part of primary treatment. Operations common to primary systems in most wastewater-treatment plants are described in the following paragraphs.

5-4 SCREENING

Screening devices are used to remove coarse solids from wastewater. Coarse solids consist of sticks, rags, boards, and other large objects that often and, inexplicably, find their way into wastewater collection systems. Because the primary purpose of screens is to protect pumps and other mechanical equipment and to prevent clogging of valves and other appurtenances in the wastewater plant, screening is normally the first operation performed on the incoming wastewater.

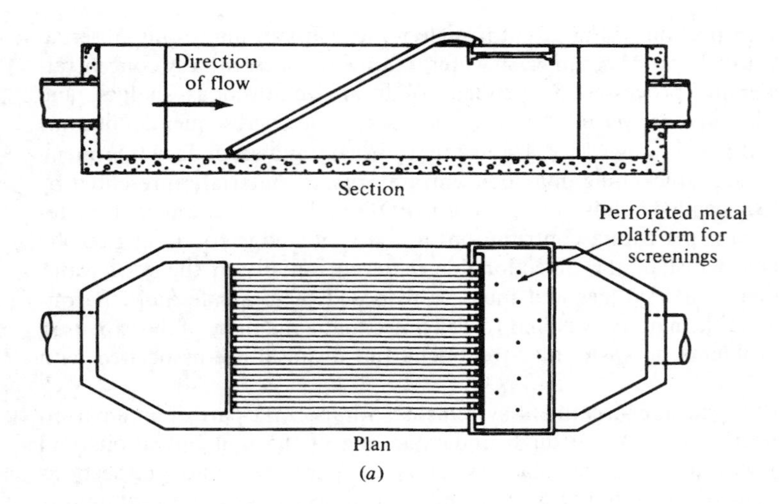

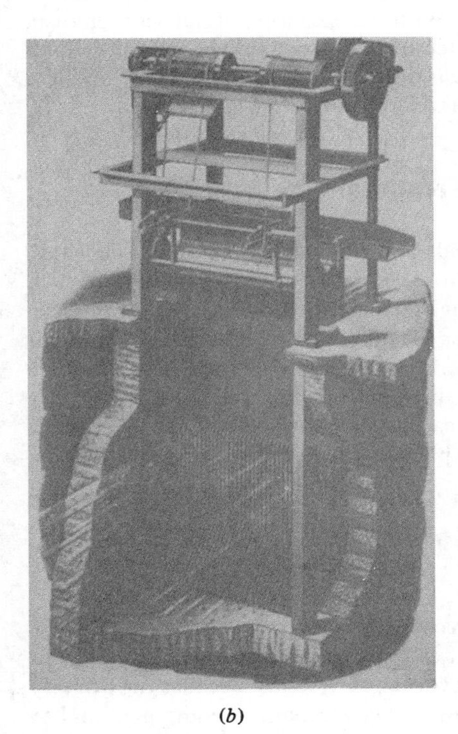

(b)

Figure 5-6 Screening devices used in wastewater treatment: (a) manually cleaned bar rack (*from Steele and McGhee* [*5-50*] (b) mechanically cleaned bar screen (*courtesy of Envirex Inc., a Rexnord Company*).

Wastewater screens are classified as fine or coarse, depending on their construction. Coarse screens usually consist of vertical bars spaced 1 or more centimeters apart and inclined away from the incoming flow. Solids retained by the bars are usually removed by manual raking in small plants, while mechanically cleaned units are used in larger plants. Fine screens usually consist of woven-wire cloth or perforated plates mounted on a rotating disk or drum partially submerged in the flow, or on a traveling belt. Fine screens should be mechanically cleaned on a continual basis. Typical screening devices are shown in Fig. 5-6.

Screening devices are contained in rectangular channels that receive the flow from the collection system. Manually cleaned devices should be readily accessible for cleaning, and mechanically cleaned systems should be enclosed in suitable housing. Proper ventilation must be provided to prevent accumulation of explosive gases. A straight channel section should be provided a few meters ahead of the screen to ensure good distribution of flow across the screen. Hydraulically, flow velocity should not exceed 1.0 m/s (3.3 ft/s) in the channel, with 0.3 m/s (1 ft/s) considered good design. Head loss across the screen will depend on the degree of clogging. Clean bars and screens result in a head loss of less than 0.1 m. Provisions should be made for a head loss of up to 0.3 m for manually cleaned or for manually operated, mechanically cleaned screens.

The quantity of solids removed by screening depends primarily on screen-opening size. The quantity of screenings removed from a typical municipal wastewater as a function of the screen size is illustrated in Fig. 5-7. Screened solids are coated with organic material of a very objectionable nature and should be promptly disposed of to prevent a health hazard and/or nuisance condition. Disposal in a sanitary landfill, grinding and returning to the wastewater flow, and incineration are the most common disposal practices.

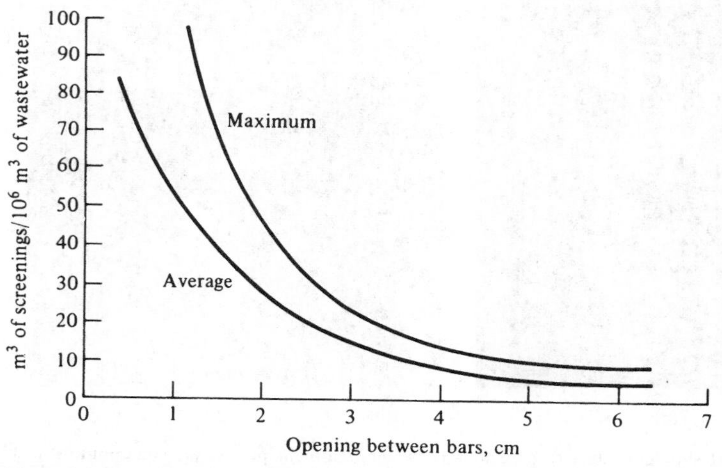

Figure 5-7 Quantity of screening from municipal wastewater as a function of bar spacing using mechanically cleaned bar screens. (*From Metcalf & Eddy, Inc.* [5-36].)

5-5 COMMINUTING

As mentioned above, screenings are sometimes shredded and returned to the wastewater flow. A hammermill device is most often used for this purpose. More often, a shredding device called a *comminutor* is located across the flow path and intercepts the coarse solids and shreds them to approximately 8 mm ($\frac{1}{4}$ in) in size. These solids remain in the wastewater.

Many kinds of comminutors are available. Basic parts include a screen and cutting teeth. The screen may be a slotted drum that rotates in the vertical plane. Stationary teeth then shred material that is intercepted by the screen. Other types use a stationary semicircular screen and rotating or oscillating cutting teeth. Another device, called a *barminutor*, uses a vertical bar screen with a cutting head that travels up and down the rack of bars, shredding the intercepted material. Typical shredding units are shown in Fig. 5-8.

Channel design for comminutors is similar to that for screens. Since material does not accumulate on the device, head loss rarely exceeds 10 cm (4 in). Comminutors are high-maintenance items, and provisions should be made to bypass the unit when repairs are needed. In small plants, bypass through a bar screen is usually provided. Larger plants may operate several comminutors in parallel so that flow from one or more disabled units may be proportioned through the remaining units.

(*a*)

(*b*)

Figure 5-8 Typical shredding devices used in wastewater treatment plants: (*a*) comminutors with rotating teeth behind the stationary slotted drum; (*b*) barminutor with traveling cutting head (*courtesy of Clow Corporation*).

Shredding devices should be located ahead of pumping facilities at the treatment plant. Grit removal ahead of the shredder will save wear on the cutting head. Usually, however, grit chambers are located at or above ground level to facilitate grit handling, and pumps may be necessary to lift the sewage to them. In this case, shredding is done ahead of the pumps and cutter wear must be tolerated.

5-6 GRIT REMOVAL

Municipal wastewater contains a wide assortment of inorganic solids such as pebbles, sand, silt, egg shells, glass, and metal fragments. Operations to remove these inorganics will also remove some of the larger, heavier organics such as bone chips, seeds, and coffee and tea grounds. Together, these compose the material known as *grit* in wastewater treatment systems.

Most of the substances in grit are abrasive in nature and will cause accelerated wear on pumps and sludge-handling equipment with which it comes in contact. Grit deposits in areas of low hydraulic shear in pipes, sumps, and clarifiers may absorb grease and solidify. Additionally, these materials are not biodegradable and occupy valuable space in sludge digesters. It is therefore desirable to separate them from the organic suspended solids.

Because infiltration is a major source of inorganics, the quantity of grit varies with the type, age, and condition of the pipe in the collection system. The type and quantity of industrial waste and the prevalence of domestic garbage grinders are also contributing factors. Quantities ranging from 4 to 200 $m^3/10^6$ m^3 have been reported, with a typical value of around 15 $m^3/10^6$ m^3 of wastewater. [5-36]

Grit removal facilities basically consist of an enlarged channel area where reduced flow velocities allow grit to settle out. Many configurations of grit tanks are available, with the most recent installations usually being channel-type or aerated rectangular basins such as those shown in Fig. 5-9. The deposited grit is removed by mechanical scrapers.

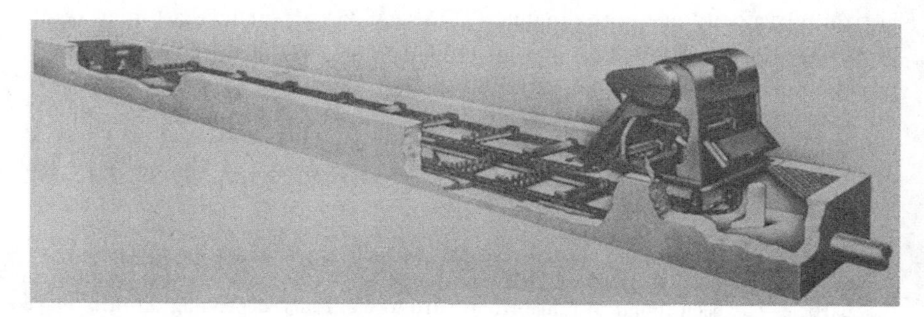

Figure 5-9 Typical grit removal equipment: channel-type chain and bucket grit chamber (*courtesy of Envirex Inc., a Rexnord Company*).

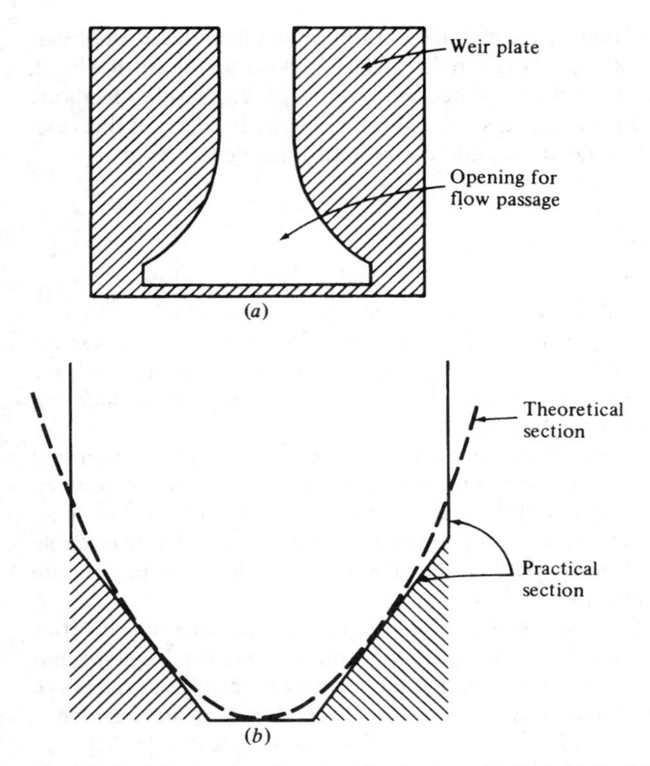

Figure 5-10 Velocity control sections for horizontal grit channels: (*a*) proportioning weir and (*b*) parabolic channel section. (*Adapted from Steele and McGhee* [*5-50*].)

Hydraulically, grit chambers are designed to remove, by type-1 settling, discrete particles with diameters of 0.2 mm and specific gravity of 2.65. In channel-type, horizontal-flow grit chambers (Fig. 5-9), it is important to maintain the horizontal velocity at approximately 0.3 m/s. A 25-percent increase may result in washout of grit, while a 25-percent reduction may result in retention of non-target organics. Since a wide variation in flow rates may be encountered, the horizontal velocity must be artificially controlled. A proportioning weir on the effluent end of the tank (Fig. 5-10*a*) or a parabolic tank section (Fig. 5-10*b*) is often used to maintain steady flow at 0.3 m/s. The design of channel-type grit chambers is illustrated in Example 5-1.

Example 5-1: Designing a channel-type grit chamber A grit chamber is designed to remove particles with a diameter of 0.2 mm, specific gravity 2.65. Settling velocity for these particles has been found to range from 0.016 to 0.022 m/s, depending on their shape factor. A flow-through velocity of 0.3 m/s will be maintained by a proportioning weir. Determine the channel dimensions for a maximum wastewater flow of 10,000 m^3/d.

SOLUTION

1. Assume a rectangular cross section with depth 1.5 × width at maximum flow

$$A_x = w \times 1.5w = 1.5w^2$$

$$= \frac{Q}{v_h} = 10{,}000\,\frac{\text{m}^3}{\text{d}} \times \frac{\text{s}}{0.3\text{ m}} \times \frac{\text{d}}{1440\text{ min}} \times \frac{\text{min}}{60\text{ s}}$$

$$= 0.39\text{ m}^2$$

$$w = 0.51\text{ m}$$

$$D = 0.76\text{ m}$$

2. Assuming a settling velocity of 0.02 m/s, the detention time is

$$t_d = D/v_t$$
$$= 0.76\text{ m}/(0.02\text{ m/s})$$
$$= 38\text{ s}$$

3. Determine length.

$$L = t_d v_h$$
$$= 38\text{ s} \times 0.3\text{ m/s}$$
$$= 11.4\text{ m}$$

Tank dimensions are therefore

$$w = 0.51\text{ m}$$

$$D = 0.76\text{ m}$$

$$L = 11.4\text{ m}$$

In larger treatment plants, the trend is toward aerated grit chambers. Turbulence created by the injection of compressed air keeps lighter organic material in suspension while the heavier grit falls to the bottom. Since roll velocity, rather than horizontal velocity, serves to separate the nontarget organics from the grit, artificial control of the horizontal velocity is not necessary. Adjustment of air quantities provides settling control. The design of aerated grit chambers is based on detention time at peak flow. Typical design parameters are shown in Table 5-4.

Aerated grit chambers may serve another useful purpose. If the sewage is anaerobic when it arrives at the plant, aeration serves to strip noxious gases from the liquid and to restore it immediately to an aerobic condition, which allows for better treatment. When an aerated grit chamber is used for this purpose, the aeration period is usually extended from 15 to 20 min.

Grit, particularly from channel-type grit chambers, may contain a sizable fraction of biodegradable organics that must be removed by washing, or must be disposed of quickly to avoid nuisance problems. Grit containing organics must either be placed in a sanitary landfill or incinerated, along with screenings, to a sterile ash for disposal.

Table 5-4 Design parameter for aerated grit chambers

Item	Value	
	Range	Typical
Dimensions:		
Depth, m	2–5	
Length, m	7.5–20	
Width, m	2.5–7.0	
Width-depth ratio	1 : 1–5 : 1	2 : 1
Detention time at peak flow, min	2–5	3
Air supply, $m^3/min \cdot m$ of length	0.15–0.45	0.3
Grit and scum quantities:		
Grit, $m^3/10^3\ m^3$	0.004–0.200	0.015

Source: From Metcalf & Eddy, Inc. [5-36]

5-7 FLOW MEASUREMENT

Although the measurement of wastewater flows does not in itself result in removal of contaminants, it is an important adjunct to wastewater treatment. A knowledge of hydraulic loading rates is necessary for the operation of many of the reactors in a wastewater-treatment plant. Chemical additives, air volume, recirculation rates, and many other operating parameters depend upon the hydraulic flow rate. Additionally, records of flows should be kept to establish trends in flow quantities for evaluation of infiltration/inflow quantities and to estimate future capacity needs.

The most common devices used for measuring flows in a wastewater-treatment plant are Parshal flumes and Palmer–Bowlus flumes. These devices, essentially open-channel venturi meters, have an established flow-head relationship from which the flow is determined by simply measuring the water elevation at a given point. Continuous-stage recording devices can be installed to provide flow records. The hydraulic design of flow-measuring devices is beyond the scope of this text, and the interested reader is referred to other texts. See for instance, Refs [5-50, 5-30, 5-48].

5-8 PRIMARY SEDIMENTATION

Primary sedimentation is a unit operation designed to concentrate and remove suspended organic solids from the wastewater. When primary treatment was considered sufficient as the total treatment, primary settling was the most important operation in the plant. Its design and operation were critical in reducing waste loads to receiving streams. With the current universal requirement for secondary treatment, primary sedimentation plays a lesser role. Indeed, many of the secondary wastewater-treatment unit processes are capable of handling the organic solids if good grit and scum removal are provided for in preliminary treatment. The theory and practice of primary settling operations in wastewater are essentially the same as those for clarifying water for potable supplies, and the reader should review Secs. 4-4 and 4-5 before proceeding.

Most of the suspended solids in wastewater are "sticky" in nature and flocculate naturally. Primary settling operations proceed essentially as type-2 settling without the addition of chemical coagulants and mechanical mixing and flocculation operations. The organic material is slightly heavier than water and settles slowly, usually in the range of from 1.0 to 2.5 m/h. Lighter materials, primarily oils and grease, float to the surface and must be skimmed off.

Primary sedimentation is accomplished in either long-rectangular tanks or circular tanks similar to those described in Sec. 4-5. Scum removal in rectangular tanks is accomplished by having the sludge scrapers penetrate through the surface as they return to the effluent end of the tank. Floating material is carried to a collection point some distance behind the effluent weirs where it is removed over a scum weir or by a transverse scum scraper. Circular tanks have a skimmer arm attached to the sludge-scraper drive mechanisms. The scum is wiped up an inclined apron and into a scum trough for removal. In both cases, a scum baffle should be located between scum removal facilities and the effluent weir. The modifications necessary for scum removal are shown in Fig. 5-11. Separated scum is usually disposed of with screenings, unwashed grit, or digested sludge.

Design criteria for primary sedimentation tanks are presented in Table 5-5.

Table 5-5 Design criteria for primary sedimentation tanks

	Value	
Parameter	Range	Typical
Detention time, h	1.5–2.5	2.0
Overflow rate, $m^3/m^2 \cdot d$		
Average flow	32–48	
Peak flow	80–120	100
Weir loading, $m^3/m \cdot d$	125–500	250
Dimensions, m		
Rectangular		
Depth	3–5	3.6
Length	15–90	25–40
Width*	3–24	6–10
Sludge scraper speed, m/min	0.6–1.2	1.0
Circular		
Depth	3–5	4.5
Diameter	3.6–60	12–45
Bottom slope, mm/m	60–160	80
Sludge scraper speed, r/min	0.02–0.05	0.03

* Must divide into bays of not greater than 6.0 m wide for mechanical sludge removal equipment.

Source: From Metcalf & Eddy, Inc. [5-36]

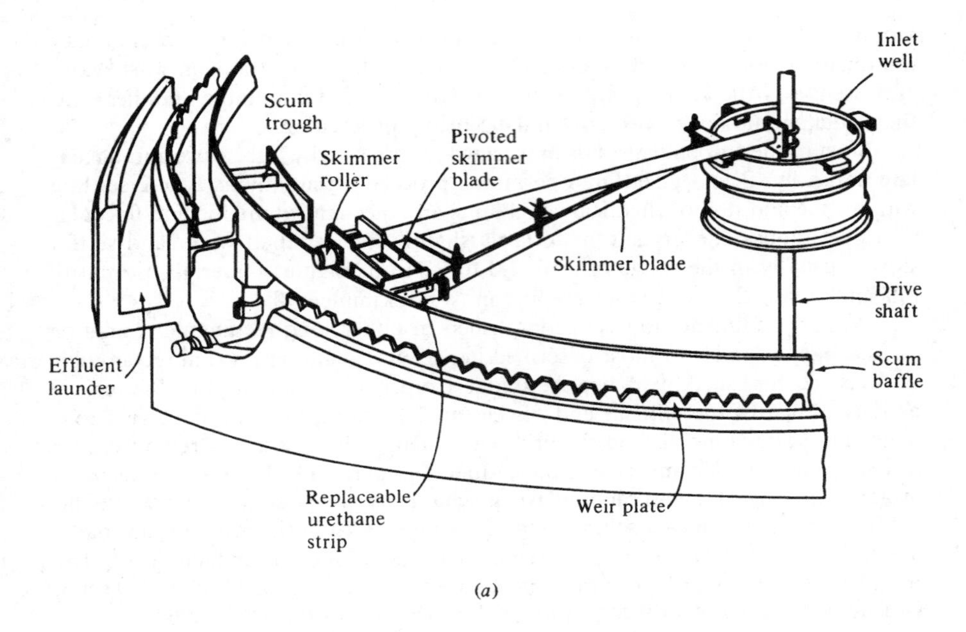

(*a*)

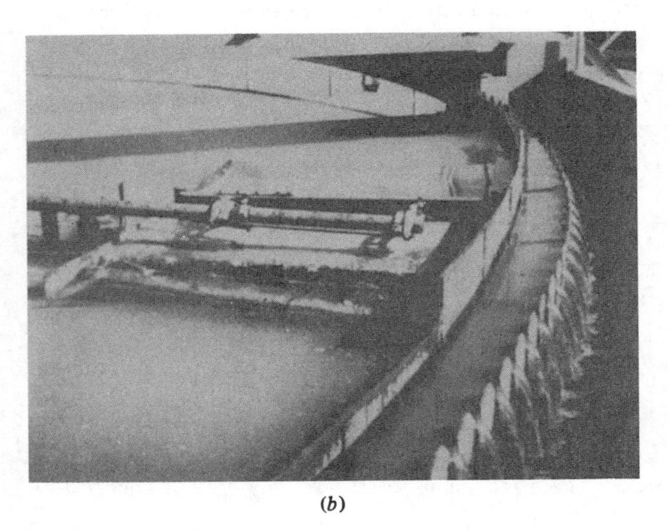

(*b*)

Figure 5-11 Scum removal from surface of primary clarifier: (*a*) diagram of skimming device (*courtesy of Infilco Degremont, Inc.*); (*b*) scum trough arrangement (*photo courtesy of Montana Department of Health and Environmental Sciences*).

In large plants, the use of several rectangular tanks with common walls reduces construction costs and space requirements. Smaller plants tend to use circular tanks because of the simplicity of sludge removal. Some settling basin arrangements commonly used in primary treatment are shown in Fig. 5-12.

Sludge should be removed from the primary sedimentation tank before anaerobic conditions develop. If the sludge begins to decompose anaerobically,

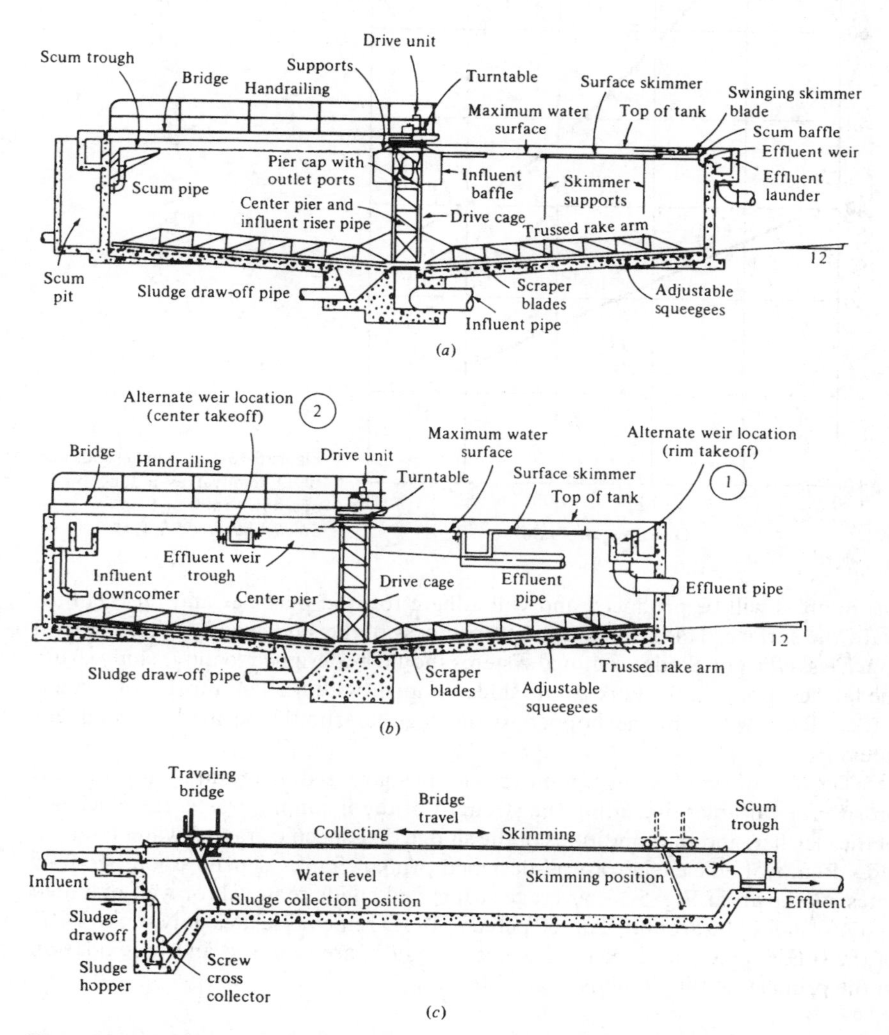

Figure 5-12 Typical primary clarifiers: (*a*) circular basin, center feed (*from Metcalf & Eddy, Inc.* [*5-36*]); (*b*) circular basin, rim feed (*from Metcalf & Eddy* [*5-36*]); (*c*) long-rectangular basin with traveling bridge sludge scraper (*courtesy of FMC Corporation, Material Handling Systems Division*).

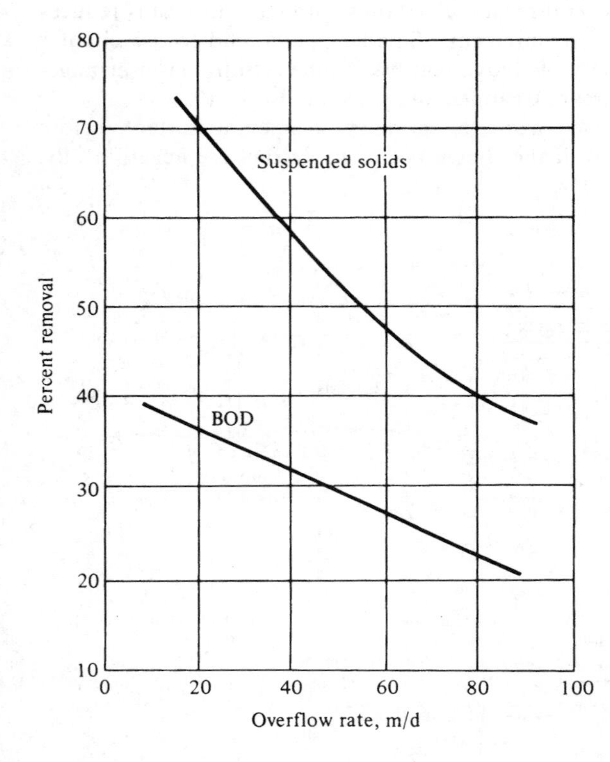

Figure 5-13 Suspended solids and BOD removal as a function of overflow rate. (*Adapted from Steele and McGhee* [5-50].)

gas bubbles will be produced and will adhere to solid particles and lift them toward the surface. This reduces the compactness of the sludge and makes removal much less efficient. Sludge removal systems should be designed to move sludge from the farthest point in the tank to the sludge hopper within 30 min to 1 h of when it settles. Removal from the hopper to the digester should be made at frequent intervals.

The quantity of sludge removed in primary sedimentation may depend on several variables, including the strength of the incoming waste, the efficiency of the clarifier, and the conditions of the sludge (i.e., specific gravity, water content, etc.). Removal efficiencies of well-designed primary tanks depend upon overflow rates, as shown in Fig. 5-13. Average suspended-solids removal for well-operated systems should be around 50 to 60 percent. BOD removal relates only to the BOD of the solids removed, since no dissolved organics are removed and biooxidation in the primary settling tank is negligible.

Example 5-2: Designing a primary settling basin A municipal wastewater-treatment plant processes an average flow of 5000 m^3/d, with peak flows as high as 12,500 m^3/d. Design a primary clarifier to remove approximately 60 percent of the suspended solids at average flow.

SOLUTION

1. From Fig. 5-13, an overflow rate of 35 $m^3/m^2 \cdot d$ should yield a suspended-solids removal efficiency of about 60 percent. Required surface area is

$$\frac{5000\ m^3/d}{35\ m^3/m^2 \cdot d} = 143\ m^2$$

2. Using a circular tank, the diameter is

$$d = \left(\frac{4A}{\pi}\right)^{1/2} = \left(\frac{4 \times 143}{3.14}\right)^{1/2} = 13.5\ m$$

3. Assuming a sidewall depth of 3 m, volume of tank is approximately

$$143 \times 3 = 429\ m^3$$

and the detention time at average flow is

$$\frac{429\ m^3}{5000\ m^3/d} = 0.09\ d = 2.06\ h$$

4. At peak flow conditions, the overflow rate is

$$\frac{12{,}500\ m^3/d}{143\ m^2} = 87\ m/d$$

and the detention time is approximately 50 min (a little low). From Fig. 5-13, the suspended-solids removal efficiency drops to about 38 percent for peak flow conditions.

Secondary Treatment

The effluent from primary treatment still contains 40 to 50 percent of the original suspended solids and virtually all of the original dissolved organics and inorganics. To meet the minimum EPA standards for discharge, the organic fraction, both suspended and dissolved, must be significantly reduced. This organic removal, referred to as secondary treatment, may consist of chemical-physical processes or biological processes. Combinations of chemical-physical operations such as coagulation, microscreening, filtration, chemical oxidation, carbon adsorption, and other processes can be used to remove the solids and reduce the BOD to acceptable levels. Currently, these operations represent a high-cost option with respect to both capital and operating expenses, and thus are not commonly used. Biological processes are used in practically all municipal wastewater-treatment systems where secondary treatment is employed.

In biological treatment, microorganisms use the organics in wastewater as a food supply and convert them into biological cells, or *biomass*. Because wastewater contains a wide variety of organics, a wide variety of organisms, or a *mixed culture*, is required for complete treatment. Each type of organism in the mixed culture utilizes the food source most suitable to its metabolism. Most mixed cultures will also contain grazers, or organisms that prey on other species. The newly created biomass must be removed from the wastewater to complete the treatment process.

The microorganisms involved in wastewater treatment are essentially the same as those that degrade organic material in natural freshwater systems. These organisms and their metabolic pathways were described in Secs. 3-7 and 3-8. The processes are not allowed to proceed in their natural fashion, however, but are controlled in carefully engineered reactors to optimize both the rate and completeness of organic removal. Removal efficiencies that would be effected over a period of days in natural systems are accomplished in a period of hours in engineered systems. Design of biological systems requires an understanding of the biological principles, kinetics of metabolism, principles of mass balance, and physical operations necessary to control the environment in the reactors. Basic biological principles were discussed in Secs. 3-7 and 3-8 and should be reviewed by the reader before proceeding. The following sections describe the kinetics of biological growth and substrate utilization and the principles of reactor design.

5-9 GROWTH AND FOOD UTILIZATION

The relationship of cell growth and food utilization can be illustrated by a simple batch reactor such as a stoppered bottle. A given quantity of a food containing all the necessary nutrients is placed in the bottle and inoculated with a mixed culture of microorganisms. If S represents the quantity of soluble food (in milligrams per liter) and X represents the quantity of biomass (in milligrams per liter), the rate of utilization of food dS/dt and the rate of biomass growth dX/dt can be represented by curves as shown in Fig. 5-14.

There are several distinct segments in the biomass curve that warrant further examination. The microorganisms must first become acclimated to their surrounding environment and to the food provided. The acclimation period, called the *lag phase*, is represented by segment 1 on the curve and will vary in length, depending on the history of the seed organisms. If the organisms have been accustomed to a similar environment and similar food, the lag phase will be very brief. Once growth has been initiated, it will proceed quite rapidly. Bacterial cells reproduce by binary fission; that is, cells divide into segments that separate to become two new independent cells. The regeneration time, or the time required for a cell to mature and separate, depends on environmental factors and food supply and may be as short as 20 min. When maximum growth is occurring, the rate of reproduction is exponential according to the equation

$$N = 2^{n-1} \tag{5-1}$$

where N is the number of organisms produced from one individual after n regeneration times. Maximum growth thus occurs at a logarithmic rate, and segment 2 on the growth curve is called the *log-growth phase*.

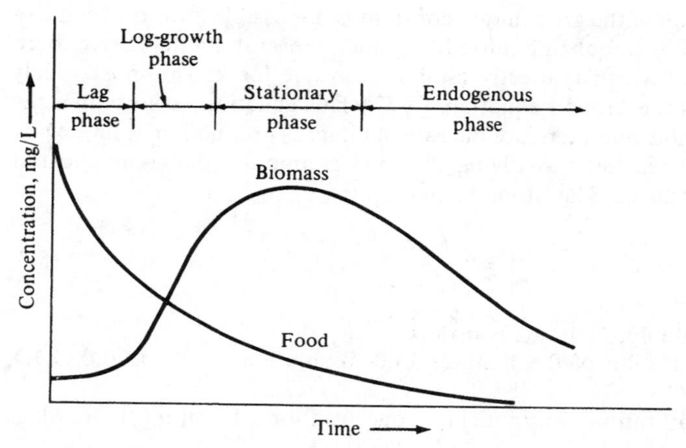

Figure 5-14 Biomass growth and food utilization.

Maximum growth cannot continue indefinitely. The food supply may become limiting, environmental conditions may change (i.e., overcrowding, waste-product buildup, etc.), and a population of grazers may develop. Cells that are unable to obtain food from external sources begin endogenous catabolism, or the catabolizing of stored protoplasm for maintenance energy. Other cells die and lyse, or break open, releasing their protoplasm, which adds to the available food. Segment 3 of the curve, the *stationary phase* represents the time during which the production of new cellular material is roughly offset by death and endogenous respiration.

Although some reproduction continues beyond the stationary phase, endogenous respiration and death predominate in segment 4 of the curve. In this final *endogenous phase*, biomass slowly decreases, approaching zero asymptotically after a very long time.

The most common method of quantifying biomass is the suspended-solids test. When the wastewater contains only soluble organic material, this test should be fairly representative, although it does not distinguish between living and dead cells. The volatile suspended-solids test is a better test when the wastewater contains a sizable fraction of suspended inorganics. Neither test will differentiate between biological solids and organic particles originally in the wastewater.

In the log-growth phase, the biomass increases according to

$$\frac{dX}{dt} = kX \tag{5-2}$$

where $\frac{dX}{dt}$ = the growth rate of the biomass mg/L$\cdot t$

X = the concentration of biomass, mg/L

k = the growth rate constant, t^{-1}

Direct evaluation of the growth rate constant is impossible for mixed cultures of microorganisms metabolizing mixed organics. Several models have been developed, however, which indirectly establish a value for k. The most widely accepted of these is the Monod equation. [5-37] This equation assumes that the rate of food utilization, and therefore the rate of biomass production, is limited by the rate of enzyme reactions involving the food compound that is in shortest supply relative to its need. The Monod equation is

$$k = \frac{k_0 S}{K_s + S} \tag{5-3}$$

where k_0 = maximum growth rate constant, t^{-1}
S = concentration of the limiting food in solution, mg/L, BOD, COD, or TOC
K_s = half saturation constant, i.e., concentration of limiting food when $k = \frac{1}{2}k_0$, mg/L

The growth rate of biomass is therefore a hyperbolic function of the food concentration, as shown in Fig. 5-15.

Several observations can be made relative to Eq. (5-3). When there is an excess of the limiting food, i.e., $S \gg K_s$, then the growth rate constant k is approximately equal to the maximum growth rate k_0 in Eq. (5-3), and the system is enzyme-limited. Since the enzymes are supplied by the microbial mass, the system is essentially biomass-limited, and the equation

$$\frac{dX}{dt} = r_x = k_0 X$$

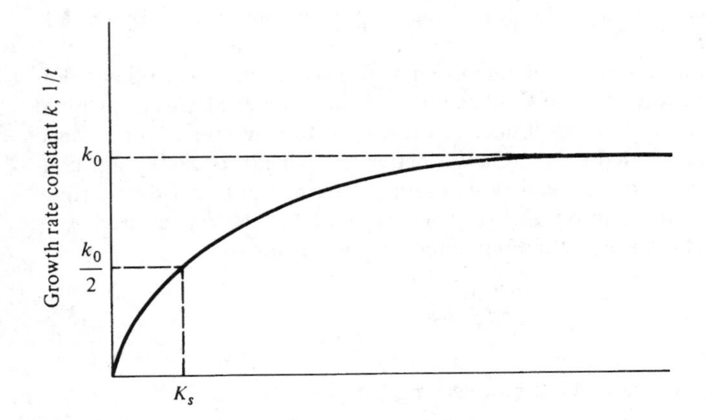

Figure 5-15 Monod growth rate constant as a function of limiting food concentration.

is a first-order equation in biomass; that is, the growth rate r_x is proportional to the first power of the biomass present. When $S \ll K_s$, the system is food-limited. In this case

$$r_x = \text{constant}$$

and the growth rate is zero order in biomass; that is, the growth rate is independent of the biomass present. When $S = K_s$, the growth rate constant is one-half the maximum as per the definition of K_s.

Substituting Eq. (5-3) into Eq. (5-2), the rate of biomass production becomes

$$r_x = \frac{dX}{dt} = \frac{k_0 SX}{K_s + S} \tag{5-4}$$

If all of the food were converted to biomass, then the rate of food utilization would equal the rate of biomass production in Eq. (5-4). Because catabolism converts part of the food into waste products, the rate of food utilization will be greater than the rate of biomass production.

$$r_x = -Y\frac{dS}{dt} = -Yr_s$$

or

$$r_s = -\frac{r_x}{Y} = -\frac{k_0 SX}{Y(K_s + S)} \tag{5-5}$$

where Y = decimal fraction of food mass converted to biomass

$$= \frac{\text{mg/L biomass}}{\text{mg/L food utilized}}$$

$$r_s = \frac{dS}{dt} = \text{rate of food utilization, mg/L}\cdot t$$

The factor Y varies depending on the metabolic pathway used in the conversion process. Aerobic processes are more efficient than anaerobic processes with respect to biomass conversion and thus have a greater value for Y. Typical values of Y for aerobic reactions are about 0.4 to 0.8 kg biomass per kilogram of BOD_5, while anaerobic reactions range from 0.08 to 0.2 kg biomass per kilogram of BOD_5.

Equation (5-4) is incomplete without an expression to account for depletion of biomass through endogenous respiration. Endogeneous decay is also taken to be first order in biomass concentration.

$$\frac{dX}{dt}(\text{end}) = -k_d X \tag{5-6}$$

where k_d = endogeneous decay rate constant t^{-1}. Incorporation of Eq. (5-6) into Eq. (5-4) results in

$$\frac{dX}{dt} = \frac{k_0 SX}{K_s + S} - k_d X \tag{5-7}$$

Endogeneous decay has very little effect on the overall growth rate in the initial phases of the growth curve in Fig. 5-14. In the stationary phase, however, endogeneous decay is equal to the growth rate and becomes predominant in the endogeneous phase.

Several external factors may affect the rate of biomass production and food utilization. These include temperature, pH, and toxins. Rate constants increase with increasing temperatures within the range of 0 to 55°C, with a corresponding increase in biomass production and food utilization. Increases in reaction rates approximately follow the van't Hoff-Arrhenius rule of doubling with every 10°C increase in temperature [5-47] up to a maximum temperature. Excessive heat denatures the enzymes and can destroy the organism.

The pH of the surrounding microorganism is also important. Enzyme systems have a fairly narrow range of tolerance. Microorganisms that degrade wastewater organics function best near neutral pH, with a tolerance range of from about pH 6 to pH 9.

Other factors such as toxicants, salt concentration, and oxidants influence biomass growth. Toxicants poison the microorganism, salt concentrations interfere with internal-external pressure relationships, and oxidants destroy enzyme and cell materials. Microorganisms are capable of adjusting to a wide range of most environmental factors, provided changes occur gradually. Sudden changes, such as a rapid drop in pH or a slug of salt, may do irreparable damage to the culture.

Several types of reactors may be used in biological treatment of wastewater. Although batch reactors may be useful in a few applications, those considered here will be continuous-flow systems. Reactors may contain suspended cultures or attached cultures. In *suspended cultures*, the microorganisms are suspended in the wastewater either as single cells or as clusters of cells called *flocs*. They are thus surrounded by the wastewater which contains their food and other essential elements. *Attached cultures* consist of masses of organisms adhered to inert surfaces with wastewater passing over the microbial film.

5-10 SUSPENDED-CULTURE SYSTEMS

Suspended-culture reactors may be of three basic types: (1) completely mixed without sludge recycle, (2) completely mixed with sludge recycle, and (3) plug-flow with sludge recycle. Recycling sludge, which consists primarily of microorganisms, increases the biomass in the reactor and therefore directly affects the biomass production and food utilization rates described by Eqs. (5-5) and (5-7).

5-11 ACTIVATED SLUDGE

The activated-sludge process is a suspended-culture system that has been in use since the early 1900s. The process derives its name from the fact that settled sludge containing living, or active, microorganisms is returned to the reactor to

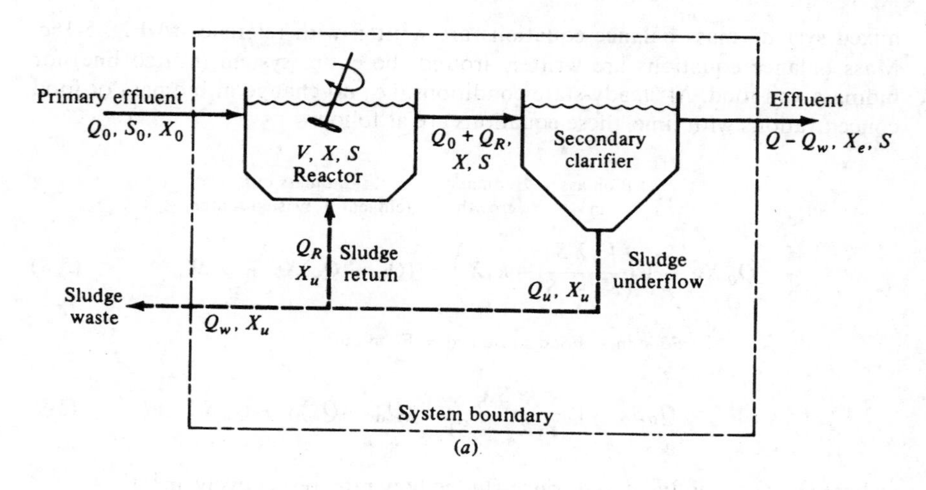

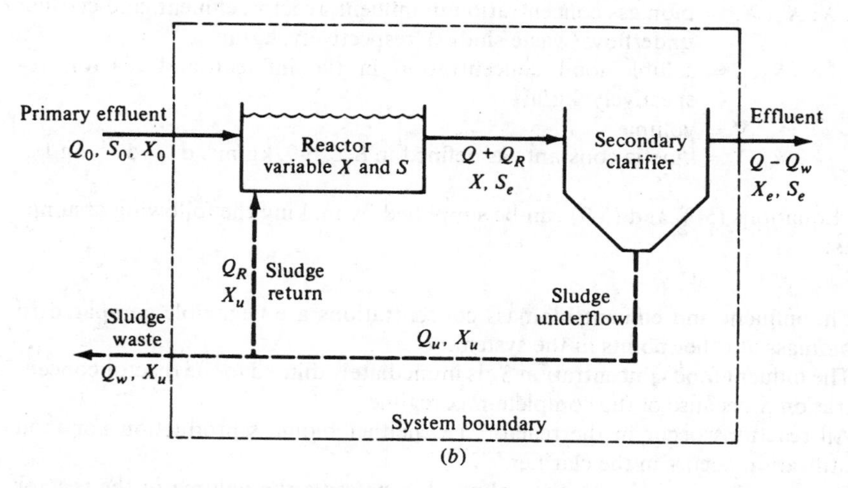

Figure 5-16 Typical activated-sludge systems: (*a*) completely mixed reactor and (*b*) plug-flow reactor.

increase the available biomass and speed up the reactions. The activated-sludge process is thus a suspended-culture process with sludge return and may be either a completely mixed or a plug-flow process, as depicted in Fig. 5-16*a* and *b*. The process is aerobic, with oxygen being supplied by dissolution from entrained air.

Completely Mixed Reactors

Equations (5-5) and (5-7) serve as a starting point for activated-sludge analysis. Reaction rate equations are coupled with system variables, and, for a completely mixed system, mass balance equations are written with reference to Fig. 5-16*a*. Mass balance equations are written around the entire system (dotted line) for biomass and food. At steady-state conditions, i.e., no change in biomass or food concentrations with time, these equations are as follows:

$$\text{Biomass in} + \text{Biomass growth} = \text{Biomass out (effluent + wasted sludge)}$$

$$Q_0 X_0 + V\left(\frac{k_0 XS}{K_s + S} - k_d X\right) = (Q_0 - Q_w)X_e + Q_w X_u \tag{5-8}$$

$$\text{Food in} - \text{Food consumed} = \text{Food out}$$

$$Q_0 S_0 - V\frac{k_0 SX}{Y(K_s + S)} = (Q_0 - Q_w)S + Q_w S \tag{5-9}$$

where Q_0, Q_w = influent and waste-sludge flow rate, respectively, m^3/d
X_0, X, X_e, X_u = biomass concentration in influent, reactor, effluent, and clarifier underflow (waste sludge), respectively, kg/m^3
S_0, S = soluble food concentration in the influent and reactor, respectively, kg/m^3
V = volume
K_s, k_0, k_d, Y = kinetic constants as defined in Sec. 5-9, kg/m^3, d^{-1}, d^{-1}, kg/kg

Equations (5-8) and (5-9) can be simplified by making the following assumptions:

1. The influent and effluent biomass concentrations are negligible compared to biomass at other points in the system.
2. The influent food concentration S_0 is immediately diluted to the reactor concentration S because of the complete-mix regime.
3. All reactions occur in the reactor; i.e., neither biomass production nor food utilization occurs in the clarifier.
4. Because of assumption 3, the volume V represents the volume of the reactor only.

With these assumptions, Eqs. (5-8) and (5-9) are rearranged as follows:

$$\frac{k_0 S}{K_s + S} = \frac{Q_w X_u}{VX} + k_d \tag{5-10}$$

$$\frac{k_0 S}{K_s + S} = \frac{Q_0}{V}\frac{Y}{X}(S_0 - S) \tag{5-11}$$

Combining these equations gives

$$\frac{Q_w X_u}{VX} = \frac{Q_0}{V}\frac{Y}{X}(S_0 - S) - k_d \tag{5-12}$$

The inverse of the expressions $Q_w X_u/VX$ and Q_0/V have unique physical significance in the activated-sludge system modeled in Fig. 5-16*a*. The quantity

$$V/Q_0 = \theta \tag{5-13}$$

is the hydraulic detention time in the reactor based on influent flow. The ratio of the total biomass in the reactor to the biomass wasted per given time

$$\frac{VX}{Q_w X_u} = \theta_c \tag{5-14}$$

represents the average time that microorganisms spend in the reactor. This parameter, called the *mean cell-residence time* will be greater than the hydraulic detention time since most of the sludge from the clarifier is returned to the reactor. Substituting Eqs. (5-13) and (5-14) into Eq. (5-12):

$$\frac{1}{\theta_c} = \frac{Y(S_0 - S)}{\theta X} - k_d \tag{5-15}$$

The concentration of biomass, or *mixed-liquor suspended solids* (*MLSS*) as it is more often called, in the reactor is related to the mean cell-residence time and the hydraulic detention time, and is found by solving Eq. (5-15) for X.

$$X = \frac{\theta_c Y(S_0 - S)}{\theta(1 + k_d \theta_c)} \tag{5-16}$$

Although this equation indicates that shortening the hydraulic detention time increases the MLSS when the other variables are held constant, there is a limit beyond which this is not true. When the hydraulic detention time approaches the regeneration time for the microorganisms, cells are washed out of the reactor before growth can occur. Consequently, X decreases and S approaches S_0, meaning that no treatment is occurring.

Plug-Flow Reactors

The plug flow with sludge recycle reactor (Fig. 5-16*b*) is often used in the activated-sludge process. Assuming complete mixing in the transverse plane but minimal mixing in the direction of flow, the mixture of wastewater and returned sludge travels as a unit through the reactor. Reaction kinetics for biomass production is similar to the batch process (Sec. 5-9), with the exception of an initially higher biomass concentration and lower food concentration because of sludge return. Lawrence and McCarty [5-29] derived expressions for an average MLSS and food utilization as follows:

$$\bar{X} = \frac{\theta_c Y(S_0 - S)}{\theta(1 + k_d \theta_c)} \tag{5-17}$$

and

$$r_s = -\frac{k_0}{Y}\frac{S\bar{X}}{K_s + S} \tag{5-18}$$

where $\bar{X}$ = average biomass concentration in the reactor (milligrams per liter). These equations are applicable only when $\theta_c/\theta \geq 5$.

Integrating Eq. (5-18) over the detention time in the reactor and substituting the appropriate boundary conditions and recycle factor yields the following equation:

$$\frac{1}{\theta_c} = \frac{k_0(S_0 - S)}{(S_0 - S) + (1 - \alpha)(K_s \ln S_i/S)} - k_d \tag{5-19}$$

where α = recycle factor, Q/Q_r

S_i = concentration of substrate after mixing with recycled sludge, mg/L

$$S_i = \frac{S_0 + \alpha S}{1 + \alpha}$$

Process Variations

In practice, several variations of the completely mixed and plug-flow systems are often used. Some involve subtle differences, such as rates and points of air or wastewater applications, detention times, reactor shapes, and methods of introducing air. Others involve more drastic differences, such as sorption and settling prior to biological oxidation and the use of pure oxygen rather than air. The most common of these variations are identified in Fig. 5-17.

Design Considerations

There are many factors which must be considered in the design of activated-sludge systems. Combinations of the process variations and reactor types that are compatible with the wastewater characteristics and environmental constraints must be selected. External factors such as construction costs, operation and maintenance difficulties and cost, and space limitations must also be considered. Usually, the engineer selects for detailed analysis several of the schemes that appear most promising. Biological constants associated with the wastewater and the reactor are determined, and the operating parameters that will produce the desired degree of treatment are quantified. A preliminary design of each alternative is made, and the one proving the most cost-effective is selected for the more detailed design necessary for its construction.

Although few absolutes apply to process and reactor selection, some general observations can be made in light of recent experiences. Because of required reactor volume, extended aeration systems are often limited to flows of 7500 m^3/d (2 Mgal/d) or less. High-rate processes, except for the pure oxygen system, produce a hard-to-settle sludge and are not usually used where a high-quality effluent is required. Complete-mix reactors are superior to plug-flow reactors where wide fluctuations in flow rates occur. Instantaneous dilution in the aerator "dampens" out shock loads that would carry through plug-flow systems and result in variable effluent characteristics. Where loading is reasonably constant, plug-flow systems produce a more mature sludge with excellent settling characteristics.

One factor in activated-sludge design that should be stressed is the interdependence of the biological reactor and the secondary clarifier. High biomass concentrations and short aeration periods may produce good treatment efficiencies with respect to soluble BOD. The savings in aeration tank volume is offset, however, by the large secondary clarifier required to clarify the effluent and thicken the sludge. Because of thickening limitations, it is the secondary clarifier that usually sets the upper limits on the biomass concentrations in the reactor.

Design variables for activated sludge reactors have included (1) volumetric loading rates, (2) food-to-mass ratios, and (3) mean cell-residence times. The volumetric loading rate V_L is the mass of BOD in the influent divided by the volume of the reactor, or

$$V_L = \frac{QS_0}{V} \tag{5-20}$$

the units of which are kilograms of BOD per cubic meter-day. The food-to-mass ratio F/M is the mass of BOD removed divided by the biomass in the reactor, or

$$F/M = \frac{Q(S_0 - S)}{VX} \tag{5-21}$$

the units being kilograms of BOD per kilogram of biomass · day.

The mean cell residence time, θ_c in Eqs. (5-14), (5-15), and (5-19), is currently the most commonly used design parameter. Both the F/M-ratio and θ_c approach to design allow for trade-off between reactor volume and concentration of MLSS in the reactor.

Typical design parameters for activated sludge systems are given in Table 5-6. The design of a completely mixed activated-sludge reactor is illustrated in the example on page 243.

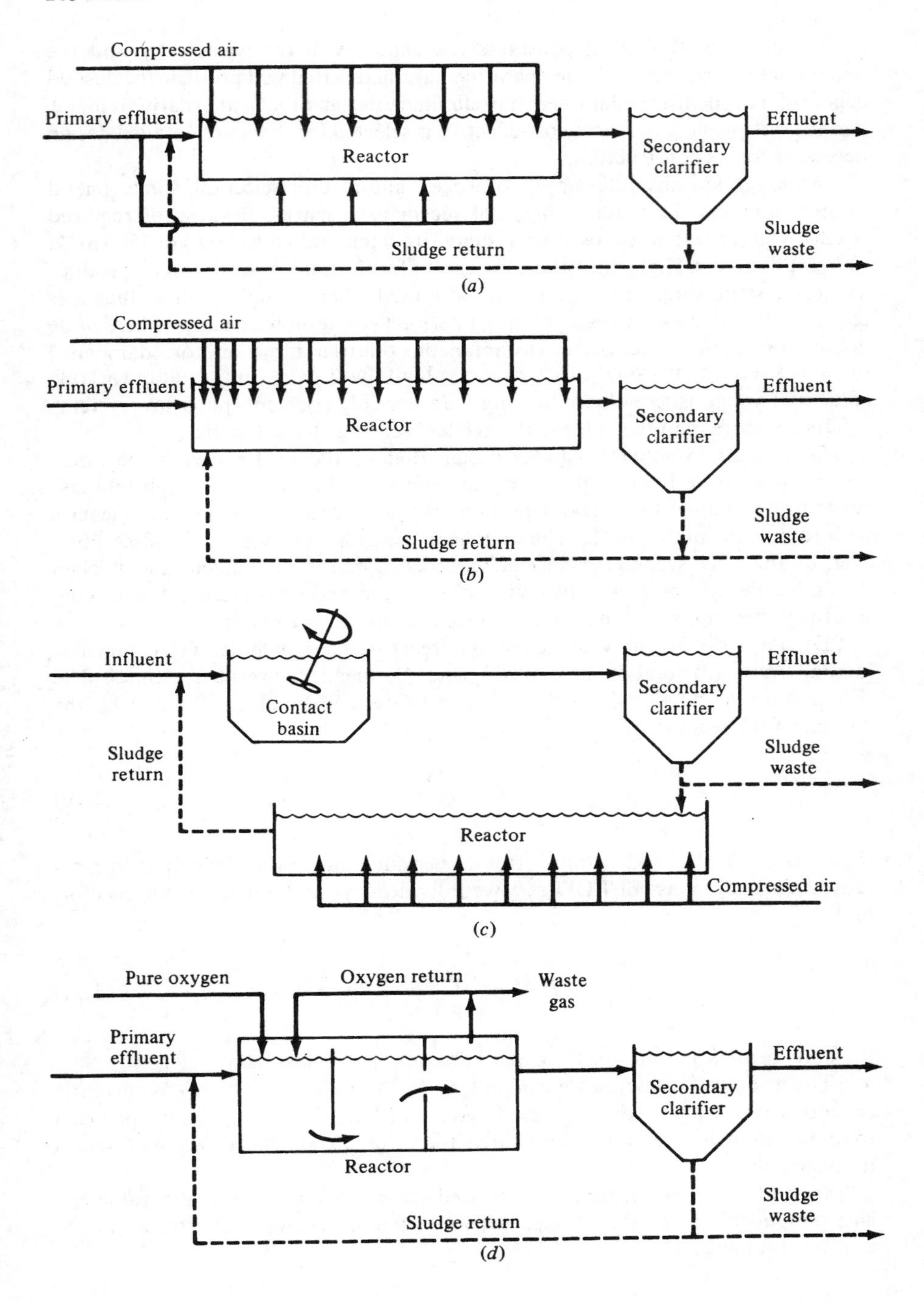
Compressed air
Primary effluent
Reactor
Secondary clarifier
Effluent
Sludge waste
Sludge return
(a)
Compressed air
Primary effluent
Reactor
Secondary clarifier
Effluent
Sludge waste
Sludge return
(b)
Influent
Contact basin
Secondary clarifier
Effluent
Sludge return
Sludge waste
Reactor
Compressed air
(c)
Pure oxygen
Oxygen return
Waste gas
Primary effluent
Reactor
Secondary clarifier
Effluent
Sludge waste
Sludge return
(d)

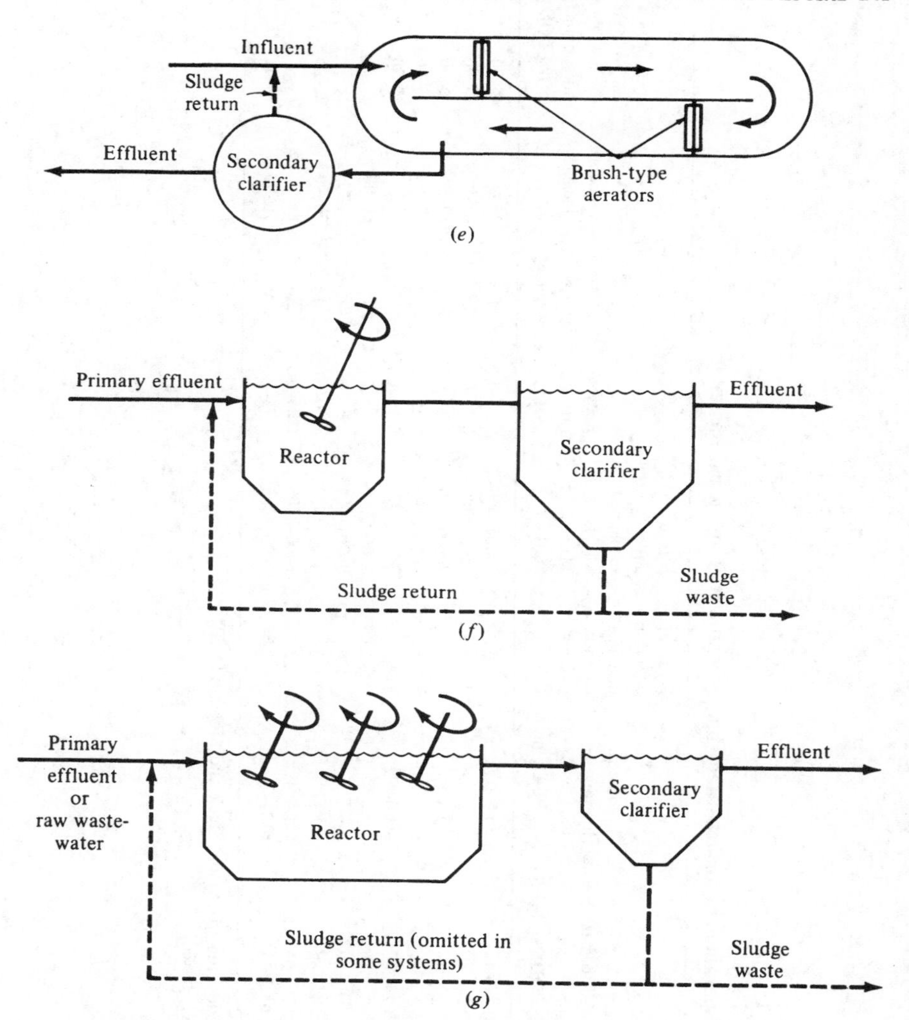

Figure 5-17 Common variations of the activated-sludge process. (*a*) Step aeration: influent addition at intermediate points provides more uniform BOD removal throughout tank. (*b*) Tapered aeration: air is added in proportion to BOD exerted. (*c*) Contact stabilization: biomass adsorbs organics in contact basin and settles out in secondary clarifier; the thickened sludge is aerated before being returned to the contact basin. (*d*) Pure-oxygen activated sludge: oxygen added under pressure keeps dissolved oxygen level high. (*e*) Oxidation ditch, plan view. (*f*) High rate: short detention time and high food/mass ratio in aerator to maintain culture in log-growth phase. (*g*) Extended aeration: long detention time and low food/mass ratio in aerator to maintain culture in endogenous phase.

Table 5-6 Design and operational parameters for activated-sludge treatment of municipal wastewater

Type of process	Mean cell residence time, θ d	Food-to-mass ratio kg BOD_5/kg MLSS	Volumetric loading, V_L kg BOD_5/m^3	Hydraulic retention time in aeration basin, θ, h	Mixed liquor suspended solids, MLSS, mg/L	Recycle ratio, Q_r/Q	Flow regime*	BOD_5 removal efficiency, %	Air supplied, m^3/kg BOD_5
Tapered aeration	5–15	0.2–0.4	0.3–0.6	4–8	1500–3000	0.25–0.5	PF	85–95	45–90
Conventional	4–15	0.2–0.4	0.3–0.6	4–8	1500–3000	0.25–0.5	PF	85–95	45–90
Step aeration	4–15	0.2–0.4	0.6–1.0	3–5	2000–3500	0.25–0.75	PF	85–95	45–90
Completely mixed	4–15	0.2–0.4	0.8–2.0	3–5	3000–6000	0.25–1.0	CM	85–95	45–90
Contact stabilization	4–15	0.2–0.6	1.0–1.2			0.25–1.0			45–90
Contact basin				0.5–1.0	1000–3000		PF	80–90	
Stabilization basin				4–6	4000–10000		PF		
High-rate aeration	4–15	0.4–1.5	1.6–16	0.5–2.0	4000–10000	1.0–5.0	CM	75–90	25–45
Pure oxygen	8–20	0.2–1.0	1.6–4	1–3	6000–8000	0.25–0.5	CM	85–95	—
Extended aeration	20–30	0.05–0.15	0.16–0.40	18–24	3000–6000	0.75–1.50	CM	75–90	90–125

Source: Adapted from Metcalf & Eddy, Inc. [5-36] and Steele and McGhee [5-50].

* PF = plug flow, CM = completely mixed.

Example 5-3: Designing an activated-sludge reactor An activated-sludge system is to be used for secondary treatment of 10,000 m^3/d of municipal wastewater. After primary clarification, the BOD is 150 mg/L, and it is desired to have not more than 5 mg/L of soluble BOD in the effluent. A completely mixed reactor is to be used, and pilot-plant analysis has established the following kinetic values: $Y = 0.5$ kg/kg, $k_d = 0.05$ d^{-1}. Assuming an MLSS concentration of 3000 mg/L and an underflow concentration of 10,000 mg/L from the secondary clarifier, determine (1) the volume of the reactor, (2) the mass and volume of solids that must be wasted each day, and (3) the recycle ratio.

SOLUTION A schematic of the system is shown in the accompanying figure.

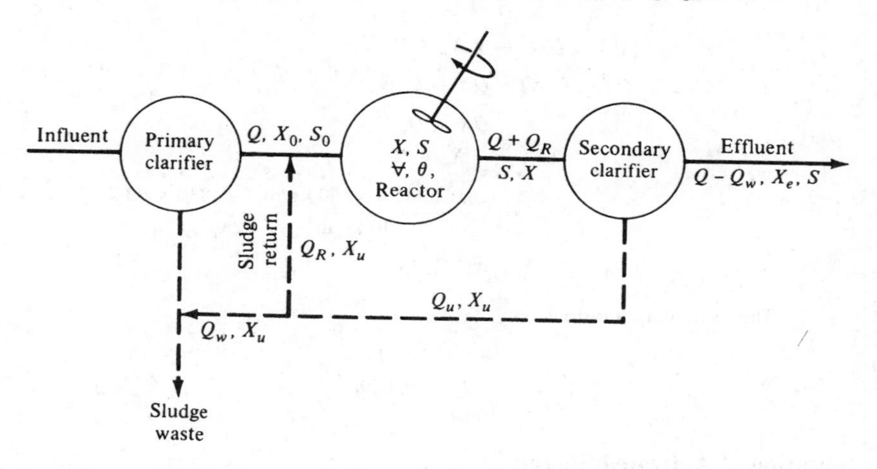

1. Select $\theta_c = 10$ d, solve Eq. (5-15) with $\theta = V/Q$.

$$\frac{1}{\theta_c} = \frac{QY(S_0 - S)}{VX} - k_d$$

$$0.1\ \text{d}^{-1} = \frac{10{,}000\ \text{m}^3/\text{d}\ 0.5\ (0.15\ \text{kg/m}^3 - 0.005\ \text{kg/m}^3)}{V\ 3.0\ \text{kg/m}^3} - 0.05\ \text{d}^{-1}$$

$$0.15\ \text{d}^{-1} = \frac{241.67\ \text{m}^3/\text{d}}{V}$$

$$V = 1611\ \text{m}^3$$

2. At equilibrium conditions, Eq. (5-14) applies.

$$\theta_c = \frac{\text{mass of solids in reactor}}{\text{mass of solids wasted}} = \frac{VX}{Q_w X_u}$$

$$Q_w X_u = \frac{VX}{\theta_c}$$

$$= \frac{1611\ \text{m}^3 \times 3.0\ \text{kg/m}^3}{10\ \text{d}}$$

$$Q_w X_u = 483.3\ \text{kg/d}$$

If the concentration of solids in the underflow is 10,000 mg/L

$$Q_w = \frac{483.3 \text{ kg/d}}{10 \text{ kg/m}^3} = 48.3 \text{ m}^3\text{/d}$$

3. A mass balance around the secondary clarifier can be written as follows:

$$(Q + Q_r)X = (Q + Q_r - Q_w)X_e + (Q_r + Q_w)X_u$$

Assuming that the solids in the effluent are negligible compared to the influent and underflow,

$$QX + Q_rX = Q_rX_u + Q_wX_u$$

$$Q_r(X_u - X) = QX - Q_wX_u$$

$$Q_r = \frac{QX - Q_wX_u}{X_u - X}$$

$$= \frac{10{,}000 \text{ m}^3\text{/d} \times 3.0 \text{ kg/m}^3 - 483.3 \text{ kg/d}}{10 \text{ kg/m}^3 - 3 \text{ kg/m}^3}$$

$$Q_r = 4217 \text{ m}^3\text{/d}$$

The recirculation ratio is

$$\frac{Q_r}{Q} = \frac{4{,}217}{10{,}000} = 0.42$$

Aeration of Activated Sludge

The rate at which oxygen is consumed by the microorganism in the biological reactor is called the *oxygen utilization rate.* For the activated-sludge processes, the oxygen utilization rate will always exceed the rate of natural replenishment, thus some artificial means of adding oxygen must be used. With the exception of the pure oxygen system, oxygen is supplied by aerating the mixed liquor in the biological reactor.

The oxygen utilization rate is a function of the characteristics of both the wastewater and the reactor. Treatment of ordinary municipal wastewater by extended aeration usually results in an oxygen utilization rate of approximately 10 mg/L · h. Treatment of the same waste by conventional activated-sludge processes results in an oxygen utilization rate of about 30 mg/L · h, and up to 100 mg/L · h if treatment is by the high-rate process. [5-25] Oxygen addition should be sufficient to match the oxygen utilization rate and still maintain a small excess in the mixed liquor at all times to ensure aerobic metabolism.

Aeration techniques consist of using air diffusers to inject compressed air into the biological reactor and/or using mechanical mixers to stir the contents violently enough to entrain and distribute air through the liquid. It is common practice to use diffused air in plug-flow systems and mechanical aerators in completely mixed systems, although there are exceptions in both cases.

Air diffusers Many types of air diffusers are available from manufacturers. Fine-bubble diffusers produce many bubbles of approximately 2.0 to 2.5 mm in diameter, while coarse-bubble diffusers inject fewer bubbles of a larger (up to 25-mm diameter) size. Both types have advantages and disadvantages. With respect to oxygen transfer, the fine-bubble diffuser is more efficient because of the larger surface area per volume of air. However, head loss through the small pores necessitates greater compression of the air and thus greater energy requirements, and compressed air must be filtered to remove all particulates that would plug the tiny diffuser openings. Coarse-bubble diffusers offer less maintenance and lower head loss, but poorer oxygen transfer efficiencies. One compromise is to locate a mechanical turbine just above a coarse-bubble diffuser so that the shearing action of the blade at high rotational speed breaks the large bubbles into smaller ones and disperses them through the liquid. Typical installations of air diffuser systems are illustrated in Fig. 5-18.

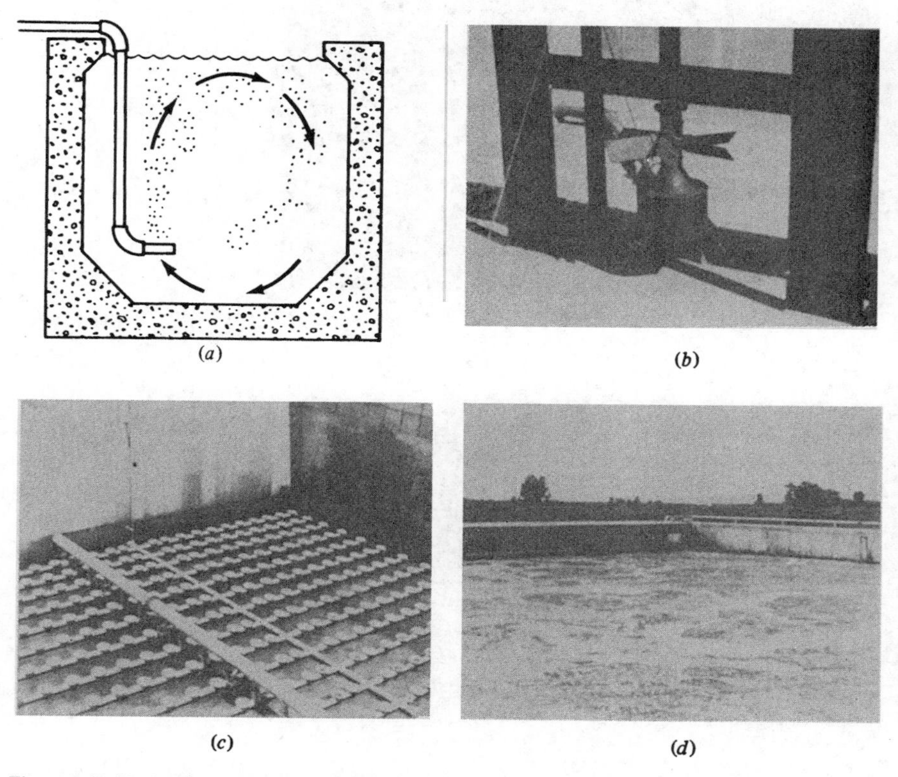

Figure 5-18 Typical application of diffused aeration systems. (*a*) Mixing action by diffusers mounted on side of long, narrow tank; (*b*) mechanical turbine above a coarse-bubble diffuser (*courtesy of Montana State Department of Health*); (*c*) porous diffusers at bottom of aeration basin; (*d*) bubble action resulting from installation in (*c*).

Mechanical aerators Mechanical aerators produce turbulence at the air-liquid interface, and this turbulence entrains air into the liquid. Mechanical aerators may have high-speed impellers that add large quantities of air to relatively small quantities of water. This aerated water is then mixed with the reactor contents through velocity gradients. Large impellers driven at slow speeds agitate larger quantities of water less violently. Typical units of both types are shown in Fig. 5-19. Use of the smaller, high-speed units is common in extended aeration systems, while the slow-speed units are more common in conventional activated-sludge systems. Brush-type aerators are used to provide both aeration and momentum to wastewater in the oxidation-ditch variation of the activated-sludge process. Their use is illustrated in Fig. 5-20.

(*a*)

(*b*)

Figure 5-19 Typical mechanical aerators used in activated-sludge processes. (*a*) Low-speed mechanical aerator mounted on fixed platform (*photo courtesy of Envirex Inc., a Rexnord Company*); (*b*) high-speed floating aerator (*photo courtesy of Envirex Inc., a Rexnord Company*).

(a)

(b)

(c)

Figure 5-20 Oxidation ditch aeration. (*a*) Rotor aerator brush used in oxidation ditch (*photo courtesy of Kathleen Miller-Hoard*); (*b*) brush action in an oxidation ditch (*photo courtesy of Lakeside Equipment Corp.*); (*c*) typical arrangement of aerators in oxidation ditch (*photo courtesy of Lakeside Equipment Corp.*).

5-12 PONDS AND LAGOONS

In addition to the activated-sludge processes, other suspended-culture biological systems are available for treating wastewater, the most common being ponds and lagoons. A wastewater pond, alternatively known as a *stabilization pond*, *oxidation pond*, and *sewage lagoon*, consists of a large, shallow earthen basin in which wastewater is retained long enough for natural purification processes to provide the necessary degree of treatment. At least part of the system must be aerobic to produce an acceptable effluent. Although some oxygen is provided by diffusion from the air, the bulk of the oxygen in ponds is provided by photosynthesis. Lagoons are distinguished from ponds in that oxygen for lagoons is provided by artificial aeration. There are several varieties of ponds and lagoons, each uniquely suited to specific applications.

Shallow ponds in which dissolved oxygen is present at all depths are called *aerobic ponds*. Most frequently used as additional treatment processes, aerobic ponds are often referred to as *polishing* or "*tertiary*" *ponds*. Deep ponds in which oxygen is absent except for a relatively thin surface layer are called *anaerobic ponds*. Anaerobic ponds can be used for partial treatment of a strong organic wastewater but must be followed by some form of aerobic treatment to produce acceptable end products. Under favorable conditions *facultative ponds* in which both aerobic and anaerobic zones exist may be used as the total treatment system for municipal wastewater.

Lagoons are classified by the degree of mechanical mixing provided. When sufficient energy is supplied to keep the entire contents, including the sewage solids, mixed and aerated, the reactor is called an *aerobic lagoon*. The effluent from an aerobic lagoon requires solids removal in order to meet suspended-solids effluent standards. When only enough energy is supplied to mix the liquid portion of the lagoon, solids settle to the bottom in areas of low velocity gradients and proceed to degrade anaerobically. This facility is called a *facultative lagoon*, and the process differs from that in the facultative pond only in the method by which oxygen is supplied.

The majority of ponds and lagoons serving municipalities are of the facultative type. The remainder of this discussion will relate to the facultative processes, the interested reader being referred elsewhere for more information on the other systems. See especially Refs. [5-6] and [5-36].

Facultative ponds and lagoons are assumed to be completely mixed reactors without biomass recycle. Raw wastewater is transported into the reactor and is released near the bottom. Wastewater solids settle near the influent while biological solids and flocculated colloids form a thin sludge blanket over the rest of the bottom. Outlets are located so as to minimize short circuiting.

System Biology

A generalized diagram of the processes that occur in facultative ponds is shown in Fig. 5-21. Aerobic conditions are maintained in the upper portions of the pond

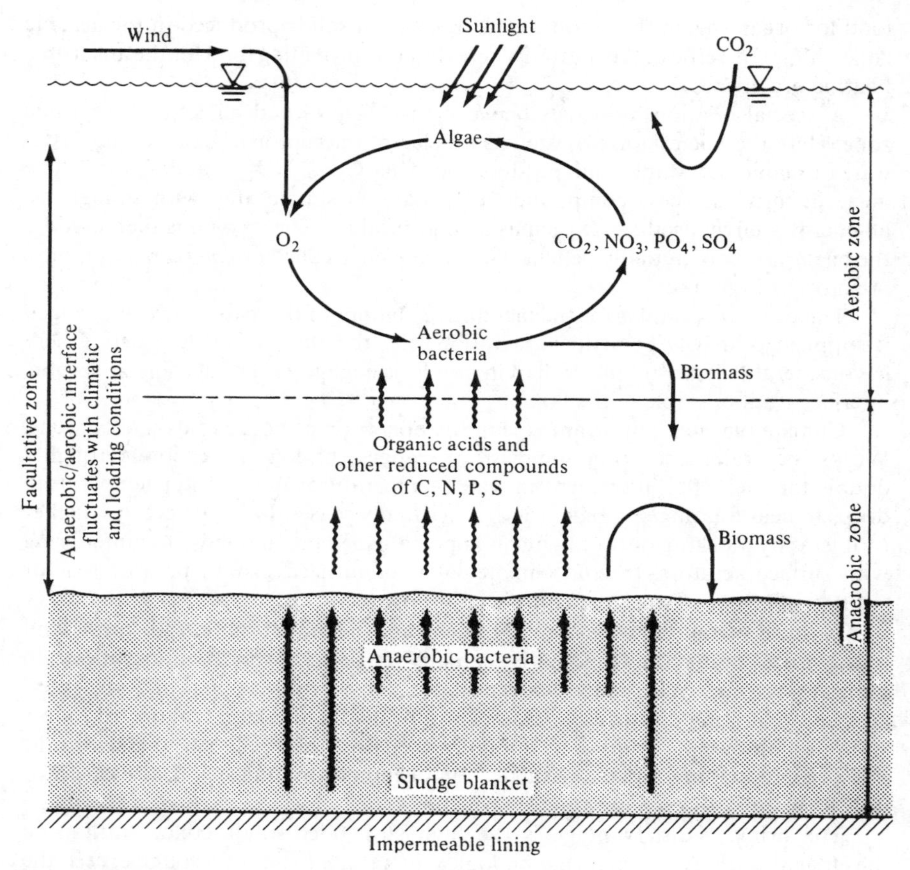

Figure 5-21 Generalized diagram of facultative pond reaction.

by oxygen generated by algae, and, to a lesser extent by penetration of atmospheric oxygen. Stagnant conditions in the sludge along the bottom prevent oxygen transfer to that region and anaerobic conditions prevail there. The boundary between the aerobic and anaerobic zones is not stationary. Mixing by wind action and penetration by sunlight may extend the aerobic area downward. Conversely, calm waters and weak lighting result in the anaerobic layer rising toward the surface. Diurnal changes in light conditions may result in diurnal fluctuations in the aerobic-anaerobic interface. The volume through which the presence of dissolved oxygen fluctuates is called the *facultative zone*, because organisms in this zone must be capable of adjusting their metabolism to the change in oxygen conditions.

Considerable interaction exists between the zones. Organic acids and gases, products of decomposition in the anaerobic zone, are released and become soluble food for organisms in the aerobic zone. Biological solids produced in the aerobic zone ultimately settle to the bottom where they die, providing food for the anaerobic benthic organisms.

A special relationship exists between the bacteria and algae in the aerobic zone. Here the bacteria use oxygen as an electron acceptor to oxidize the wastewater organics to stable end products such as CO_2, NO_3^-, and PO_4^{3-}. The algae in turn use these compounds as a material source and, with sunlight as an energy source, produce oxygen as an end product. The oxygen is then used by the bacteria. Such mutually beneficial arrangements, called *symbiotic relationships*, often occur in nature.

The process is similar for the facultative lagoon. In this case, however, oxygen is supplied primarily by artificial aeration, and the effect of algae, existing here in considerably lesser numbers than in ponds, is negligible. The aerobic-anaerobic interface is more stable in the lagoon.

Climate plays an important role in the operation of both ponds and lagoons. Within naturally occurring temperature ranges, biological reactions roughly double for each 10°C increment in water temperature. When water temperature drops to near freezing, biological activity virtually ceases. Ice cover creates further problems by blocking out sunlight, an important element in ponds, and interfering with surface aerators. In cold climates, it is often necessary to provide reactor volume sufficient to store the entire winter flow.

Because of the large land requirement and incompatibility with most land uses, the use of ponds and lagoons is generally restricted to small communities with semirural settings. Where their use is possible, considerable savings in both capital cost and operating cost can be realized. Additionally, the large volume-to-inflow ratio provides ample dilution to minimize the effects of highly variable hydraulic and organic loading typical of small, homogenous communities. Little operator skill is required to keep the system operating.

The primary disadvantage is the high suspended-solids concentrations in the effluent. Although principally biological in nature, these solids often exceed the secondary-treatment standards. This problem has been relieved somewhat by reevaluation of the discharge standards by EPA to allow up to 75 mg/L of biological solids in the effluent from ponds and lagoons. An additional disadvantage for ponds may be odor problems during the spring if poorly treated wastewater has been stored during the winter season.

Design of Ponds and Lagoons

Several approaches to the design of ponds and lagoons have been proposed. The model most commonly assumed is the completely mixed reactor without solids recycle. In the case of facultative systems, complete mixing is assumed to apply only to the liquid portion of the reactor. Wastewater solids and biological solids that fall to the bottom are not resuspended. Because the rate at which solids are removed by sedimentation is not quantifiable, a mass balance for solids cannot be written. A mass balance for the soluble food can be written, because soluble food is assumed to be uniformly distributed throughout the reactor by mixing of the liquid. If the conversion rate is assumed to be first order in food concentration, then mass balance can be written as follows:

$$\text{BOD in} = \text{BOD out} + \text{BOD consumed}$$

$$QS_0 = QS + V(kS) \tag{5-22}$$

Upon rearranging, Eq. (5-22) becomes

$$\frac{S}{S_0} = \frac{1}{1 + kV/Q} = \frac{1}{1 + k\theta} \tag{5-23}$$

where S/S_0 = fraction of soluble BOD remaining
k = reaction rate coefficient, d^{-1}
θ = hydraulic detention time, d
V = reactor volume, m^3
Q = flow rate, m^3/d

If several reactors are arranged in series, the effluent of one pond becomes the influent to the next. A substrate balance written across a series of n reactors results in the following equation:

$$\frac{S_n}{S_0} = \frac{1}{(1 + k\theta/n)^n} \tag{5-24}$$

When facultative ponds are used to treat municipal wastewater, it is common practice to use at least three ponds in series to minimize short circuiting. Marais [5-35] and Mara [5-34] demonstrated that maximum efficiency occurs when ponds in series are of the same approximate size. When this is the case, the first pond, called the *primary pond*, will retain most of the sewage solids and will thus be the most heavily loaded. It may be necessary to provide aeration in the primary pond to prevent complete anaerobic conditions with their attendant odor problems. The result is one facultative lagoon followed by two or more facultative ponds.

Although the above models are useful for visualizing the pond and lagoon processes, it is impractical to expect instantaneous mixing of influent with such large reactor volumes. In practice, a wide range of dispersion occurs because of reactor shape and size, mixing by wind action or aerators, and influent and effluent arrangements. Thirumurthi [5-53] developed graphical relationships between food removal and values for $k\theta$ for dispersion factors ranging from infinity for completely mixed reactors to zero for plug-flow reactors. These relationships, shown in Fig. 5-22, can be used for design, provided values of k are known or assumed.

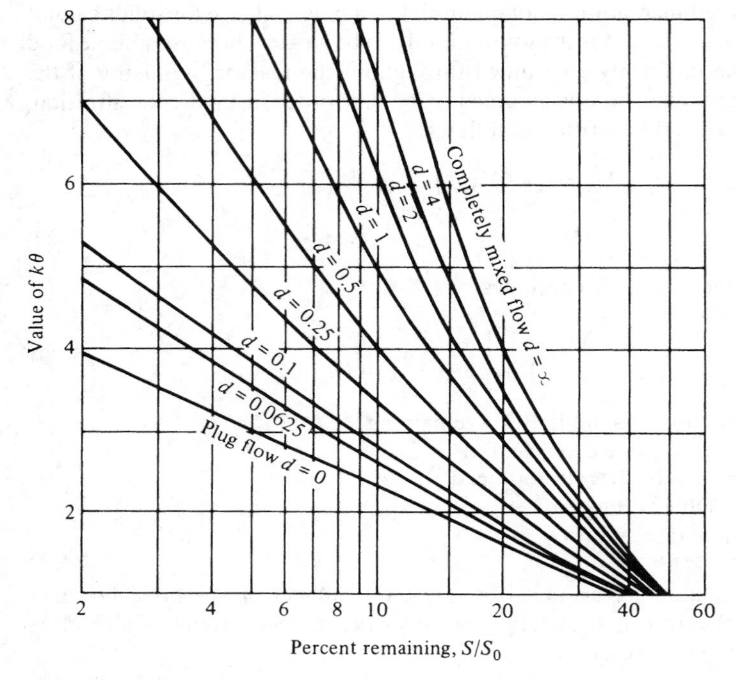

Figure 5-22 Graphic relationship between S/S_0 and $k\theta$ in Eq. (5-24). (*From Thirumurthi* [5-53].)

A wide range of values for k is encountered in the literature. Although many variables relating to both the reactor and wastewater affect k, water temperature appears to be the most significant. Equations of the form

$$\frac{k_T}{k_{20}} = \phi^{T-20}$$

are commonly used. Values frequently used for k_{20} range from about 0.2 to 1.0, while values of the temperature coefficient ϕ may range from 1.03 to 1.12. These values must often be determined experimentally for a given pond system. Because of the complexity of accurately assessing the k constant, design of ponds and lagoons is often based on loading factors and other empirically derived parameters. Parameters and values frequently used are given in Table 5-7.

Although some photosynthesis undoubtedly occurs in facultative lagoons, oxygen requirements are assumed to be met by aeration. Two kilograms of oxygen should be supplied for each kilogram of BOD_5 in the influent to ensure adequate oxygen for the soluble BOD released from the anaerobic zone as well as for the BOD in the raw wastewater. The rate of oxygen transfer is a function of water temperature, oxygen deficit, and aerator characteristics. In conditions normally encountered in wastewater lagoons, the rate of oxygen transfer correlates well

Table 5-7 Typical design parameters for facultative ponds and lagoons

Parameter	Facultative pond	Facultative lagoon
Flow regime		Mixed surface layer
Pond size, ha	1–4 multiples	1–4 multiples
Operation*	Series or parallel	Series or parallel
Detention time, d*	7–30	7–20
Depth, m	1–2	1–2.5
pH	6.5–9.0	6.5–8.5
Temperature range, °C	0–50	0–50
Optimum temperature, °C	20	20
BOD_5 loading, kg/ha · d†	15–18	50–200
BOD_5 conversion	80–95	80–95
Principal conversion products	Algae, CO_2, CH_4, bacterial cell tissue	Algae, CO_2, CH_4, bacterial cell tissue
Algal concentration, mg/L	20–80	5–20
Effluent suspended solids, mg/L‡	40–100	40–60

* Depends on climatic conditions.

† Typical values (much higher values have been applied at various locations). Loading values are often specified by state pollution-control agencies.

‡ Includes algae, microorganisms, and residual influent suspended solids. Values are based on an influent soluble BOD_5 of 200 mg/L and an influent suspended-solids concentration of 200 mg/L.

Source: From Metcalf & Eddy, Inc. [5-36]

with the energy dissipated by the aerator, and transfer rates ranging from 0.3 to 2.0 kg/kW · h are common. More exact figures can be obtained from the equipment manufacturers when operation conditions are known. The design of facultative ponds and lagoons is illustrated by the following example.

Example 5-4: Designing facultative ponds and lagoons Wastewater flow from a small community averages 3000 m^3/d during the winter and 5000 m^3/d during the summer. The average temperature of the coldest month is 8°C, and the average temperature of the warmest month is 25°C. The average BOD_5 is 200 mg/L with 70 percent being soluble. The reaction coefficient k is 0.23 d^{-1} at 20°C, and the value of ϕ is 1.06. Prepare a preliminary design for a facultative pond treatment system for the community to remove 90 percent of the soluble BOD.

SOLUTION

1. Compute the rate constants adjusted for temperature.

$$\text{Summer: } k_{25} = 0.23(1.06)^{25-20} = 0.31\ \text{d}^{-1}$$

$$\text{Winter: } k_{8} = 0.23(1.06)^{8-20} = 0.11\ \text{d}^{-1}$$

2. From Fig. 5-22, determine $k\theta$ when $S/S_0 = 0.10$ and the dispersion factor is 0.5.

$$k\theta = 4.0$$

$$\text{Summer: } \theta = \frac{4.0}{0.31} = 12.9 \text{ d}$$

$$\text{Winter: } \quad = \frac{4.0}{0.11} = 36.4 \text{ d}$$

Use longest time, $\theta = 36.4$ d

3. Compute volume of ponds.

$$V = \theta Q = 36.4 \text{ d} \times 3 \times 10^3 \text{ m}^3/\text{d} = 109{,}200 \text{ m}^3$$

Use three ponds (as shown in the accompanying sketch), each 36,400 m³, $\theta = 12$ d.

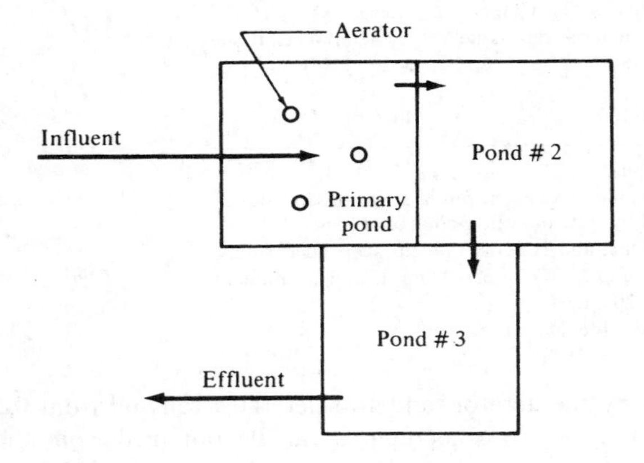

4. Use depth of 1.5 m for ponds.

$$\text{Area} = \frac{36{,}400}{1.5} = 24{,}267 \text{ m}^2 = 2.4 \text{ ha}$$

(Note: Add 1 m depth for sludge storage in primary pond.)

5. Assuming photosynthesis will not be sufficient to meet oxygen requirements in the primary pond throughout the year, size aeration equipment.

For primary pond summer conditions:

$$\theta = \frac{V}{Q} = \frac{36{,}400 \text{ m}^3}{5000 \text{ m}^2/\text{d}} = 7.3 \text{ d}$$

$$k\theta = 0.31 \times 7.3 = 2.3$$

From Fig. 5-22 with $d = 0.5$, $S/S_0 = 0.18$

BOD removed = 0.82 × 200 = 164 mg/L

Oxygen supplied = 2 × 0.164 kg/m³ × 5000 m³/d

= 1640 kg/d

Assume aerators transfer 1 kg O_2/kW · h

$$1640 \text{ kg } O_2/\text{d} \times \frac{1 \text{ d}}{24 \text{ h}} \times \frac{\text{kW} \cdot \text{h}}{1 \text{ kg } O_2} = 68.3 \text{ kW}$$

Use three aerators at 23 kW each.

5-13 ATTACHED-CULTURE SYSTEMS

Attached-culture systems employ reactors in which wastewater is contacted with microbial films attached to surfaces. Surface area for biofilm growth is increased by placing a porous medium in the reactor. When randomly packed solid medium is used, the reactor is called a *trickling filter*. The advent of modular synthetic media of high porosity and low weight enables a vertical arrangement of medium several meters high, leading to the term *bio-tower*. More recently, the use of rotating disks partially submerged in wastewater has led to the *rotating biological contactor* (*RBC*) process. Although other attached-culture systems, including submerged filters (anaerobic) and fluidized beds, may have application under certain conditions, the discussion here will be limited to trickling filters, bio-towers, and RBCs.

In the trickling filter and bio-tower, the medium is stationary and the wastewater is passed over the biofilm in intermittent doses. In the RBC, the medium moves the biofilm alternately through water and air. Because both systems maintain aerobic conditions at the biofilm surface, both are classified as aerobic processes.

In addition to the biological reactor, an attached-culture system usually includes both primary and secondary clarification. The primary clarifier may be omitted in bio-towers and RBC installations where plugging of the void spaces can be avoided by grinding the solids in the wastewater to sufficiently small sizes prior to application onto the medium.

System Biology

The biological metabolism of wastewater organics in attached-culture systems is remarkably similar to that in suspended-culture systems, the dissimilarities in reactor characteristics notwithstanding. The biological organisms that attach themselves to the solid surfaces of the medium come from essentially the same groups as those in activated-sludge systems. Most are heterotrophic organisms, with facultative bacteria being predominant. Fungi and protozoa are also abundant, and algae are present near the surface where light is available. Animals such as rotifers, sludge worms, insect larvae, snails, etc. may also be found. Nitrifying organisms are found in significant numbers only when the carbon content of the wastewater is low.

The organisms attach themselves to the medium and grow into dense films of a viscous, jellylike nature. Wastewater passes over this film in thin sheets with dissolved organics passing into the biofilm due to concentration gradients within the film. Suspended particles and colloids may be retained on the "sticky" surfaces where they are decomposed into soluble products. Oxygen from the wastewater and from air in the void spaces of the medium provide oxygen for aerobic reactions at the biofilm surface. Waste products from the metabolic processes diffuse outward and are carried away by the water or air currents moving through the voids of the medium. These processes are diagramed in Fig. 5-23.

Growth of the biofilm is restricted to one direction—outward from the solid surface. As the film grows thicker, concentration gradients of both oxygen and food develop. Eventually, both anaerobic and endogeneous metabolism occur at the biofilm-medium surface interface. The attachment mechanism is weakened, and the shearing action of the wastewater flowing across the film pulls it from its mooring and washes it away. This process, known as *sloughing*, is a function of both the hydraulic and organic loading rate. Biofilm is quickly reestablished in places cleared by sloughing.

The rate of food removal in attached-growth systems depends on many factors. These include wastewater flow rate, organic loading rate, rates of diffusivity of food and oxygen into the biofilm, and temperature. The depth of penetration of both oxygen and food is increased at higher loading rates. Oxygen diffusivity is usually the limiting factor. Aerobic zones of the biofilm are usually limited to a depth of 0.1 to 0.2 mm [5-10], with the remaining thickness being anaerobic.

The number of variables affecting the growth of biomass, and subsequently the rate of substrate utilization, makes mathematical modeling of attached-growth systems difficult. Biofilm growth, sloughing, and regrowth, and its aerobic-anaerobic nature, prevent application of equilibrium equations similar to those

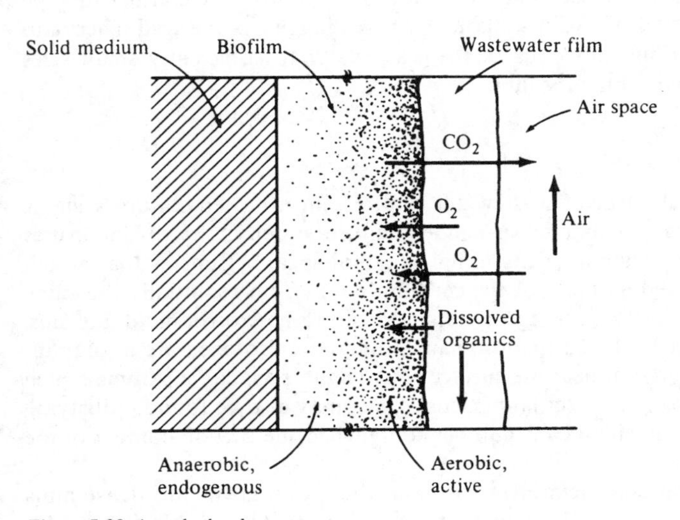

Figure 5-23 Attached-culture processes.

for suspended-culture systems [Eqs. (5-5) and (5-7)]. Design equations for attached-growth systems have been derived largely on an empirical basis.

Trickling Filters

The name *trickling filters* is applied to a reactor in which randomly packed solid forms provide surface area for biofilm growth. The system must contain equipment for distributing the wastewater over the medium and for removing the effluent. The term *filter* for this process is misleading, since few of the physical processes associated with filtration through granular media function in trickling filters. Instead, sorption and subsequent biological oxidation are the primary means of food removal.

Important characteristics of the medium include specific surface area and porosity. The *specific surface area* refers to the amount of surface area of the media that is available for biofilm growth. The *porosity* is a measure of the void space available for passage of the wastewater and air and for ventilation of product gases.

In most cases the medium in trickling filters is composed of crushed stone or slag. These materials provide hard, durable, chemically resistant surfaces for biofilm growth. Sizes ranging from 50 to 100 mm (2 to 4 in) provide specific surface areas of 50 to 65 m^2/m^3 (15 to 18 ft^2/ft^3), with porosities of 40 to 50 percent. Plastic media of various shapes may be used instead of the stone or slag, with sizable advantages in specific surface area and porosity. Areas up to 200 m^2/m^3 (57 ft^2/ft^3) and porosities of 95 percent are available with loose-bulk packing material. The use of modular media made from wooden slats or plastic sheets is also possible.

The application of wastewater onto the medium is accomplished by a rotating distribution system as shown in Fig. 5-24. Under a hydraulic head of about 1.0 m, jet action through the nozzles is sufficient to power the rotor. This arrangement results in intermittent dosing, with opportunity for air circulation through the pores between dosing. Dispersion of the wastewater is accomplished in the top few centimeters of randomly packed medium, resulting in uniform hydraulic loading throughout the remaining depth. Electrical motors may be necessary to drive the rotor where variable flows or insufficient head would result in uneven application.

The underdrain system is designed to carry away the treated wastewater and the sloughed biomass. Proprietary modules such as those shown in Fig. 5-25 are often used. These are designed to flow partially full to facilitate the circulation of air through the medium.

Several operational modes are available for trickling filters. Standard-rate filters have low hydraulic loading and do not include provision for recycling. High-rate filters maintain high hydraulic loading by recirculating portions of the effluent. Filters placed in series increase the effective depth, thus increasing the efficiency. A great number of possibilities exists for different flow regimes.

Figure 5-24 Rotating arm distributing wastewater over activated biofilter (*courtesy of Neptune Microfloc, Inc.*).

Many factors affect the operation of trickling filters, the most important being organic loading, hydraulic flow rates, and temperature of the water and ambient air. A high organic loading rate results in a rapid growth of biomass. Excessive growth may result in plugging of pores and subsequent flooding of portions of the medium. Increasing the hydraulic loading rate increases sloughing and helps to keep the bed open. Ranges of hydraulic and organic loading rates for trickling filters are shown in Table 5-8. These loading rates limit the depth of conventional trickling filters to about 2 m because of head loss through the randomly packed medium.

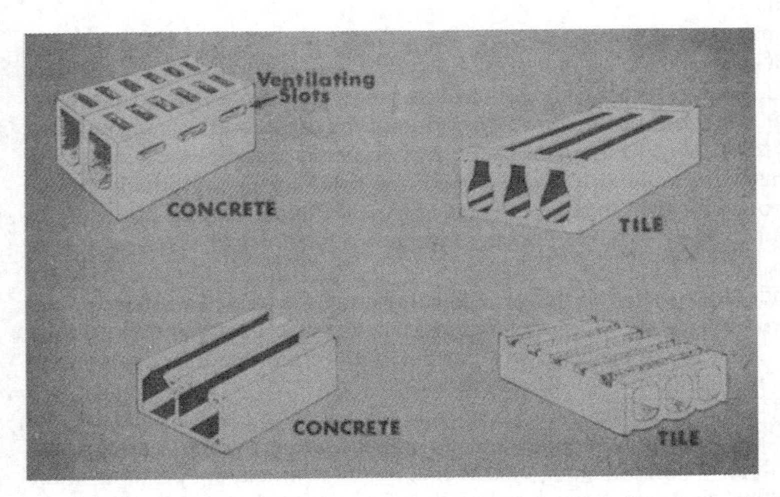

Figure 5-25 Typical blocks used in trickling filter underdrain system (*courtesy of Water Pollution Control Federation*).

Table 5-8 Typical design criteria for trickling filters

Item	Low-rate filter	Intermediate-rate filter	High-rate filter
Hydraulic loading, $m^3/m^2 \cdot d$	1–4	4–10	10–40
Organic loading, $kg/m^3 \cdot d$	0.08–0.32	0.24–0.48	0.32–1.0
Depth, m	1.5–3.0	1.25–2.5	1.0–2.0
Recirculation ratio	0	0–1	1–3; 2–1
Filter media	Rock, slag, etc.	Rock, slag, etc.	Rock, slag, synthetic materials
Power requirements, $kW/10^3\ m^3$	2–4	2–8	6–10
Filter flies	Many	Intermediate	Few, larvae are washed away
Sloughing	Intermittent	Intermittent	Continuous
Dosing intervals	Not more than 5 min (generally intermittent)	15 to 60 s (continuous)	Not more than 15 s (continuous)
Effluent	Usually fully nitrified	Partially nitrified	Nitrified at low loadings

Source: From Metcalf & Eddy, Inc. [5-36]

The biomass-water-air interfaces make trickling filters extremely sensitive to temperature variations. Effluent quality is thus likely to show drastic seasonal changes, due primarily to changes in ambient air temperature. Relative temperatures of the wastewater and the air also determine the direction of air flow through the medium. Cool water absorbs heat from the air, and the cooled air falls toward the bottom of the filter in a cocurrent fashion with the water. Conversely, warm water heats the air, causing it to rise through the underdrain and up through the medium. At temperature differentials of less than about 3 to 4°C, relatively little air movement results, and stagnant conditions prevent good ventilation. Extreme cold may result in icing and destruction of the biofilm.

Historically, trickling filters have played an important role in wastewater treatment. Their simplicity and low operating cost have made them an attractive option for small communities in warmer climates. However, modern effluent standards that demand high-quality effluent on a consistent basis make the use of the classical trickling filter questionable. Although multistage, high-rate filters can be designed to meet most secondary effluent standards, recent adaptations of the basic process, described in the following sections, have proven more economical in the construction of new facilities.

Bio-Towers

Bio-towers are essentially deep trickling filters. Lightweight, modular media formed by welding corrugated and flat polyvinyl chloride sheets together in alternating patterns provide structural rigidity for vertical stacking without the excessive weight that would result from stone or slag media. Additionally, the porosity and regular shapes provided by this medium overcome the head loss problem encountered in randomly packed reactors. Modules of this medium similar to that shown in Fig. 5-26*b* may be stacked to heights of up to 12 m to provide a large volume in a relatively small containment structure. Wooden lathes in alternating patterns, as shown in Fig. 5-26*c*, are sometimes used instead of a plastic medium. The pertinent characteristics of these media are given in Table 5-9.

Application of wastewater may be by a rotating distributor similar to that used in a trickling filter if the surface configuration of the bio-tower is round. Most often, application nozzles are stationary, with water being sprayed over the medium from a pipe grid as shown in Fig. 5-27. Underdrain systems are similar to those for trickling filters but must be designed for higher flow rates.

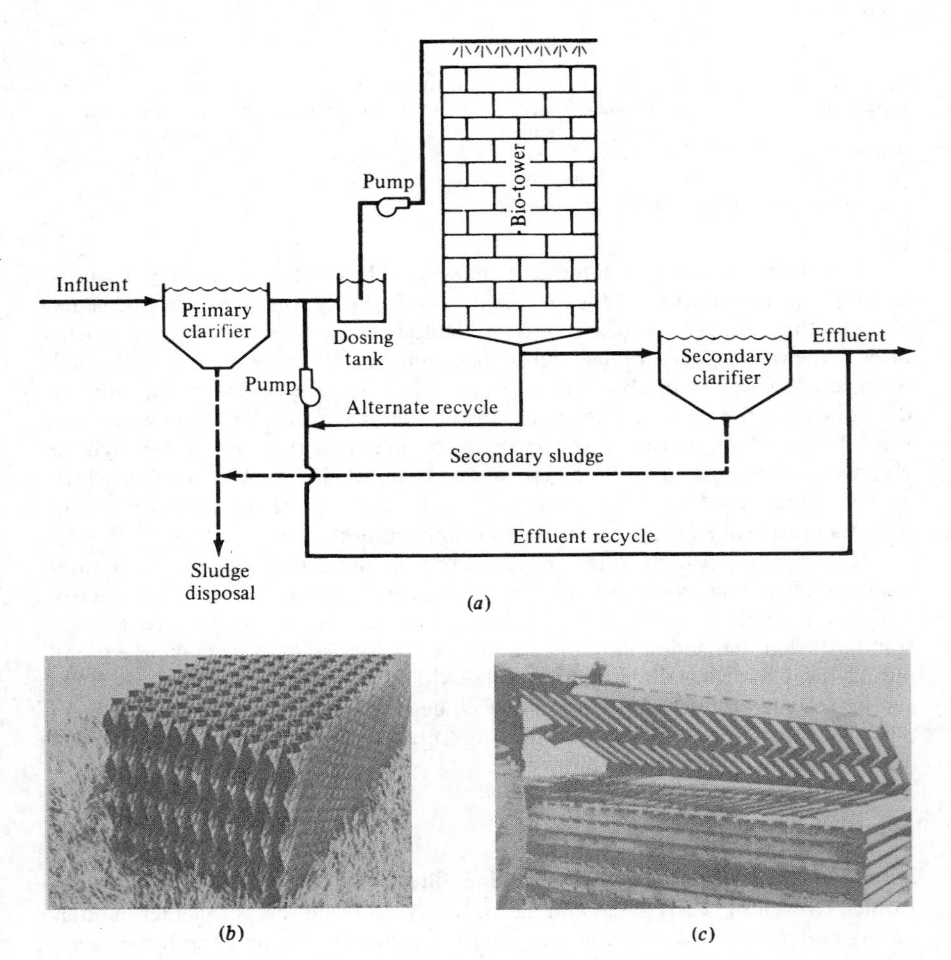

Figure 5-26 Bio-tower system: (*a*) diagrammatic sketch; (*b*) module of plastic medium (*courtesy of the Munters Corp.*); (*c*) wood biomedia modules (*courtesy of Neptune Microfloc, Inc.*).

Table 5-9 Properties of trickling filter media

Medium	Nominal size, mm	Mass/unit volume, kg/m³	Specific surface area, m²/m³	Void space, percent
River rock				
Small	25–65	1250–1450	55–70	40–50
Large	100–120	800–1000	40–50	50–60
Blast-furnace slag				
Small	50–80	900–1200	55–70	40–50
Large	75–125	800–1000	45–60	50–60
Plastic				
Conventional	600 × 600 × 1200*	30–100	80–100	94–97
High-specific surface	600 × 600 × 1200*	30–100	100–200	94–97
Redwood	1200 × 1200 × 500*	150–175	40–50	70–80

* Size of module of medium.
Source: From Metcalf & Eddy, Inc. [5-36]

Figure 5-27 Variflo fixed distribution nozzles over bio-tower media (*courtesy of Neptune Microfloc, Inc.*).

Bio-towers are operated in a fashion similar to high-rate trickling filters. The dispersion characteristics of the plastic modules are less effective than with random packing, and the hydraulic flow rate must be maintained at a high level to ensure that all surfaces are wetted throughout the entire depth. Direct recirculation of 1 to 3 times the inflow is commonly practiced. The diluted substrate results in endogenous respiration throughout most of the depth of the tower. Carbonaceous BOD is generally satisfied in the upper reaches of the medium. If the carbon content of the wastewater falls below about 20 mg/L, nitrifying bacteria become competitive and ammonia is converted to nitrate. A well-operated bio-tower should be able to produce a nitrified effluent.

Bio-towers have several advantages over classical trickling filters. The porosity and nature of the packing allow greater loading rates and virtually eliminate plugging problems. Increased ventilation minimizes odor problems under most operating conditions. The compact nature of the reactor allows for economical housing for operation in severe climates. Disadvantages include a relatively high pumping cost necessitated by the large recycle requirement and the head loss through the deep bed.

Design of bio-towers is usually based on formulas developed for trickling filters, with allowances being made for medium characteristics. The most commonly used formula was proposed by Eckenfelder [5-20] and is of the form

$$\frac{S_e}{S_0} = e^{-kD/Q^n} \tag{5-25}$$

where S_e = effluent substrate concentration, BOD_5, mg/L
S_0 = influent substrate concentration, BOD_5, mg/L
D = depth of the medium, m
Q = hydraulic loading rate, $m^3/m^2 \cdot min$
k = treatability constant relating to the wastewater and the medium characteristics, min^{-1}
n = coefficient relating to the medium characteristics

The values of the treatability constant k range from 0.01 to 0.1. Average values for municipal waste on modular plastic media are around 0.06 at 20°C. [5-23] Correction for other temperatures can be made by adjusting the treatability factor as follows [5-19]:

$$k_T = k_{20°C}(1.035)^{T-20} \tag{5-26}$$

Treatability factors should be determined from pilot-plant analysis of wastewater and selected medium. The coefficient n for modular plastic media can be taken as 0.5 without significant error. [5-6]

The above formula does not account for recirculation of wastewater. Because bio-towers almost universally employ recirculation, Eq. (5-25) must be modified as follows:

$$\frac{S_e}{S_a} = \frac{e^{-kD/Q^n}}{(1 + R) - Re^{-kD/Q^n}} \tag{5-27}$$

where S_a is the BOD_5 of the mixture of raw and recycled mixture applied to the medium

$$S_a = \frac{S_0 + RS_e}{1 + R} \tag{5-28}$$

and R is ratio of the recycled flow to the influent flow.

The design of bio-towers is illustrated by the following example.

Example 5-5: Designing a bio-tower A bio-tower composed of a modular plastic medium is to be used as the secondary-treatment component in a municipal wastewater treatment plant. Flow from the primary clarifier is 20,000 m^3/d with a BOD of 150 mg/L. Pilot-plant analysis has established a treatability constant of 0.055 min^{-1} for the system at 20°C, and the *n* factor can be taken as 0.5. Two towers are to be used, each with a square surface and separated by a common wall. The medium is to have a depth of 6.5 m, and the recirculation ratio is to be 2 to 1 during average flow periods. Determine the dimensions of the units required to produce an effluent with a soluble BOD_5 of 10 mg/L. Minimum temperature is expected to be 25°C.

SOLUTION

1. The influent concentration of BOD_5 is determined from Eq. (5-28).

$$S_a = \frac{150 + 2 \times 10}{1 + 2} = 56.7 \text{ mg/L}$$

2. The treatability constant must be adjusted for temperature [Eq. (5-26)].

$$\begin{aligned} k_{25} &= k_{20}(1.035)^{25-20} \\ &= 0.055(1.035)^5 \\ &= 0.065 \text{ min}^{-1} \end{aligned}$$

3. The loading rate is found by solving Eq. (5-27) for Q.

$$\frac{10}{56.7} = \frac{e^{-0.065 \times 6.5/Q^{0.5}}}{(1 + R) - Re^{0.065 \times 6.5/Q^{0.5}}}$$

$$\frac{10}{56.7}(1 + 2) = e^{-0.42/Q^{0.5}} + \frac{10}{56.7}(2)e^{-0.42/Q^{0.5}}$$

$$0.53 = 1.35e^{-0.42/Q^{0.5}}$$

$$0.39 = e^{-0.42/Q^{0.5}}$$

$$0.94 = 0.42/Q^{0.5}$$

$$Q^{0.5} = 0.45$$

$$Q = 0.20 \text{ m}^3/\text{m}^2 \cdot \text{min}$$

4. The surface area of each unit is determined as follows:

$$20{,}000 \text{ m}^3/\text{d} \times \frac{1 \text{ d}}{1440 \text{ min}} = 13.9 \text{ m}^3/\text{min}$$

$$\frac{13.9 \text{ m}^3/\text{min}}{2 \times 0.2 \text{ m}^3/\text{m}^2 \cdot \text{min}} = 34.8 \text{ m}^2$$

Each unit is square, so dimensions are

$$L = W = (34.8 \text{ m}^2)^{1/2} = 5.89 \text{ m, say } 6 \text{ m}$$

Each unit is 6.0 m × 6.0 m × 6.5 m deep. The system is shown schematically in the accompanying sketch.

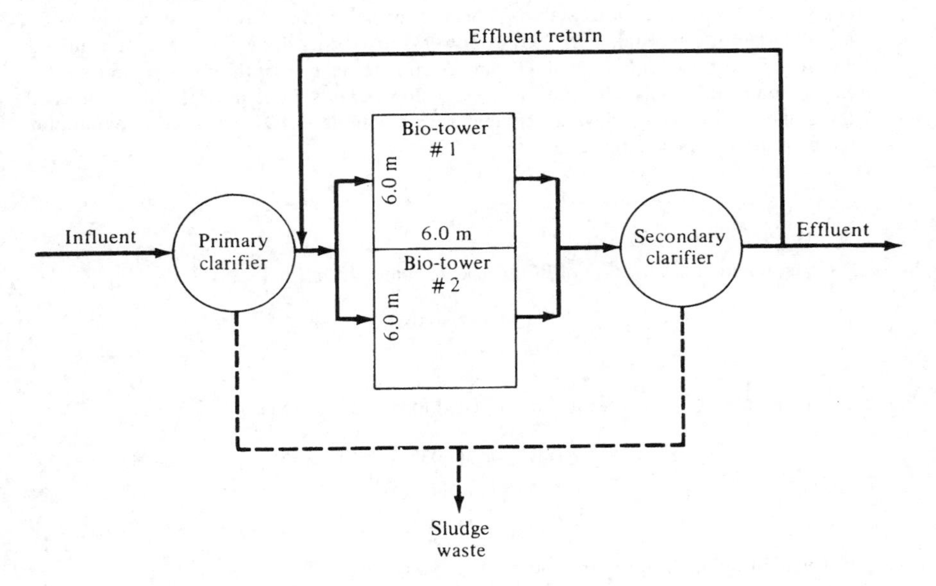

Rotating Biological Contactors

The rotating biological contactor (RBC) reactor is a unique adaptation of the attached-growth process. Media in the form of large, flat disks mounted on a common shaft are rotated through specially contoured tanks in which wastewater flows on a continuous basis. The system is shown in Fig. 5-28.

The medium consists of plastic sheets ranging from 2 to 4 m in diameter and up to 10 mm thick. Thinner materials can be used by sandwiching a corrugated sheet between two flat disks and welding them together as a unit. Spacing between flat disks is approximately 30 to 40 mm. The disks are mounted through the center on a steel shaft in widths up to 8 m. Each shaftful of medium, along with its tanks and rotating device, becomes a reactor module. Several modules may be arranged in parallel and/or in series to meet the flow and treatment requirements.

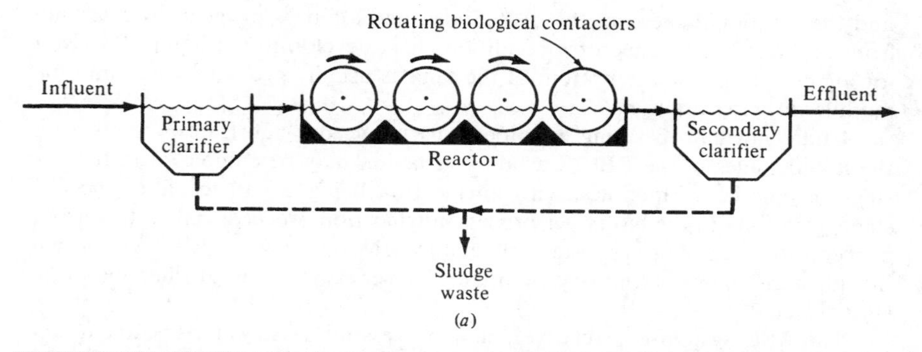

(*a*)

(*b*)

Figure 5-28 Rotating biological contactor system: (*a*) diagram of the rotating biological contactor system; (*b*) multiple installation (note covers on units in background) (*photo courtesy of Walker Processes Corp.*).

The disks are submerged in the wastewater to about 40 percent of their diameter and are rotated by power supplied to the shaft. Approximately 95 percent of the surface area is thus alternately immersed in the wastewater and then exposed to the atmosphere above the liquid. Rotational speed of the unit ranges from 1 to 2 r/min and must be sufficient to provide the hydraulic shear necessary for sloughing and to maintain enough turbulence to keep the solids in suspension as the wastewater passes through the tank.

Microorganisms growing on the medium surface remove food from the wastewater and oxygen from the air to sustain their metabolic processes. Growth and sloughing of the biofilm occur on a continuous basis as described earlier. Thickness of the biofilm may reach 2 to 4 mm, depending on the wastewater strength and the rotational speed of the disk. Since the biofilm is oxygenated externally from the wastewater, anaerobic conditions may develop in the liquid. Provision for air injection near the bottom of the tank is usually provided when multiple modules in series are used.

Under normal operating conditions, carbonaceous substrate is removed in the initial stages of the RBC. Carbon conversion may be completed in the first stage of a series of modules, with nitrification being completed after the fifth stage. [5-51] As in the bio-tower process, nitrification proceeds only after carbon concentrations have been substantially reduced. Most designs of RBC systems will include a minimum of four or five modules in series to obtain nitrification of the wastewater.

The RBC system is a relatively new process for wastewater treatment, and experience with full-scale applications is limited. The process appears to be well suited to the treatment of municipal wastewater, however. One module of 3.7 m in diameter by 7.6 m long contains approximately 10,000 m^2 of surface area for biofilm growth. This large amount of biomass permits short contact time, maintains a stable system under variable loading, and should produce an effluent meeting secondary-treatment standards. Recirculating effluent through the reactor is not necessary. The sloughed biomass is relatively dense and settles well in the secondary clarifier. Other advantages include low power requirement and simple operating procedures. A 40-kW motor is sufficient to turn the 3.7- by 7.6-m unit previously described. Powering the system by compressed air is even more economical and has the added benefit of aerating the wastewater.

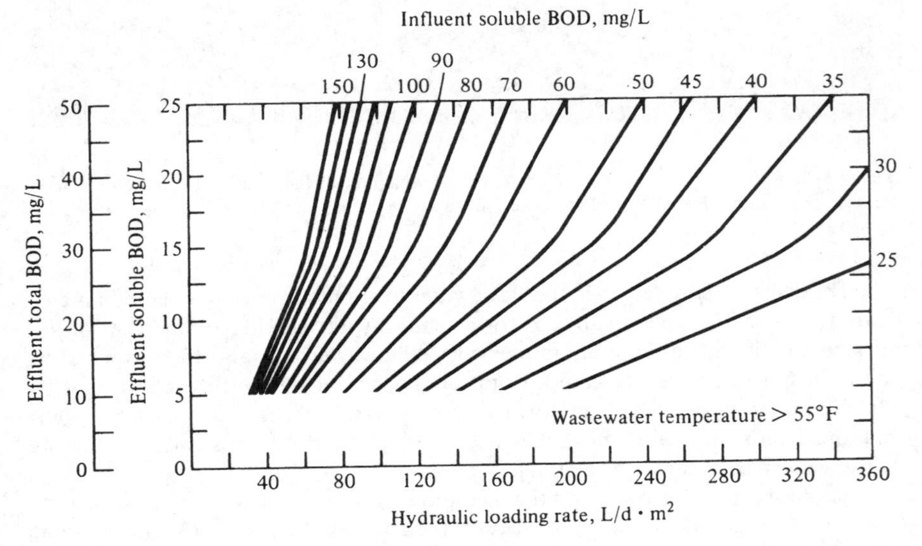

Figure 5-29 Efficiency and loading rate relationship for Bio-Surf medium treating municipal wastewater (*courtesy of Autotrol Corp.*).

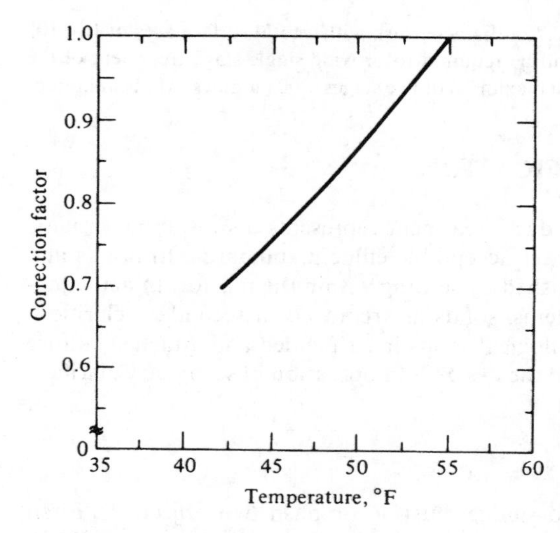

Figure 5-30 Temperature correction for loading curves in Fig. 5-29. Multiply loading rate by correction factors (*courtesy of Autotrol Corp.*).

Disadvantages of the system include a lack of documented operating experience, high capital cost, and a sensitivity to temperature. Covers must be provided to protect the media from damage by the elements and from excessive algal growths. Adequate housing also helps to minimize temperature problems in colder climates.

Design of an RBC unit is based on hydraulic loading rates. Graphs showing relationships between loading rates and efficiency similar to the graph shown in Fig. 5-29 can be obtained from manufacturers for specific media and various wastewater strengths. Required surface area is then translated into the number and size of the modules necessary. Temperature corrections can then be made using Fig. 5-30. The design of an RBC system is illustrated in Example 5-6. It should be emphasized that final design of an RBC system should be based on loading rates obtained from pilot-plant modeling as opposed to generalized figures such as those shown in Figs. 5-29 and 5-30.

Example 5-6: Designing a rotating biological contactor Determine the surface area required for an RBC system to treat the wastewater described in Example 5-5.

SOLUTION

1. Enter Fig. 5.29 with:
   Influent BOD = 150 mg/L
   Effluent soluble BOD = 10 mg/L
   The hydraulic loading rate is found to be 0.05 $m^3/m^2 \cdot d$.
2. Disk area is

$$A_d = \frac{20{,}000 \text{ m}^3/\text{d}}{0.05 \text{ m}^3/\text{m}^2 \cdot \text{d}} = 4 \times 10^5 \text{ m}^2$$

3. Assuming a 7.6-m shaft for a 3.7-m-diameter disk with a total surface area of 1×10^4 m^2, 40 modules in parallel will be required to provide single-stage treatment of the wastewater. For nitrification, a maximum of five stages (200 modules) will be required.

5-14 SECONDARY CLARIFICATION

The biomass generated by secondary treatment represents a substantial organic load and must be removed to meet acceptable effluent standards. In ponds and lagoons, this removal is accomplished by settling within the reactor. In activated-sludge and attached-culture systems, solids are removed in secondary clarifiers. Because the characteristics of biological solids in suspended and attached culture systems are significantly different, the design and operation of secondary clarifiers in these systems are also different.

Activated-Sludge Clarifiers

Secondary clarifiers for activated sludge must accomplish two objectives. First, they must produce an effluent sufficiently clarified to meet discharge standards. Secondly, they must concentrate the biological solids to minimize the quantity of sludge that must be handled. Because both functions are critical to successful operation, secondary clarifiers must be designed as an integral part of an activated-sludge system.

The biological solids in activated sludge are flocculent in nature and, at concentrations less than about 1000 mg/L, settle as a type-2 suspension. Most biological reactors, however, operate at concentrations in excess of 1000 mg/L, and thickening in the secondary clarifier results in even greater concentrations. A *concentrated suspension* was defined in Sec. 4-4 as a suspension in which particles are close enough together so that their velocity fields overlap with those of neighboring particles and a significant upward displacement of water occurs as particles settle. In concentrated suspensions, these and other factors act to prevent independent settling. Groups of particles settle at the same rate, regardless of size differences of the individual particles. The collective velocity of particles depends on several variables, the most obvious of which is the concentration of the suspension, the velocity being inversely proportional to the concentration.

In secondary clarifiers, the solid concentration must be increased from the concentration of the reactor X to the concentration of the clarifier underflow X_U. Settling velocities change correspondingly, resulting in zones with different settling characteristics. This phenomenon, known as *zone settling*, can be illustrated by a simple batch analysis in a column, as described below.

Batch analysis If a column is filled with a concentrated suspension and allowed to settle quiescently, the contents will soon divide into zones as shown in Fig. 5-31. In zone B, the initial concentration C_0 is preserved and settles at a uniform velocity characteristic of that concentration. The resulting clarified zone, zone A, is lengthened at this same velocity.

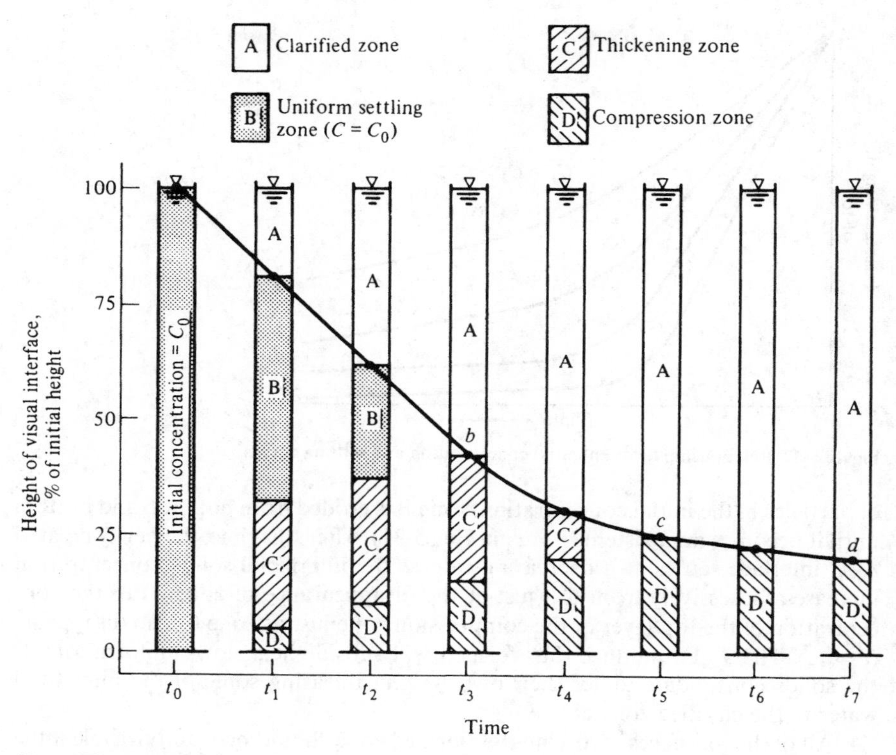

Figure 5-31 Zone settling.

Below the uniform velocity zone, two other zones develop. As the particles at the bottom come to rest on the floor of the cylinder, the particles immediately above fall on top of them, forming a zone in which particles are mechanically supported from below. This zone, labeled zone D in Fig. 5-31, is called the *compression zone*, and particles in this zone have only a slight velocity resulting from consolidation.

The area between zone D and zone B contains a concentration gradient ranging from slightly greater than C_0 just below zone B to slightly less than the concentration at the top of the compression zone. Collective velocities of particles in zone C, appropriately called the *thickening zone*, decrease in proportion to this concentration gradient.

As time progresses, the interfaces between the zones move relative to each other. Referring again to Fig. 5-31, the C–D interface moves upward as particles from zone C drop into zone D. As long as the concentration gradient in zone C remains unchanged, the width of this zone must also remain constant, and so the B–C interface is displaced upward at the same velocity as the C–D interface. Because the A–B interface moves downward at the uniform settling velocity

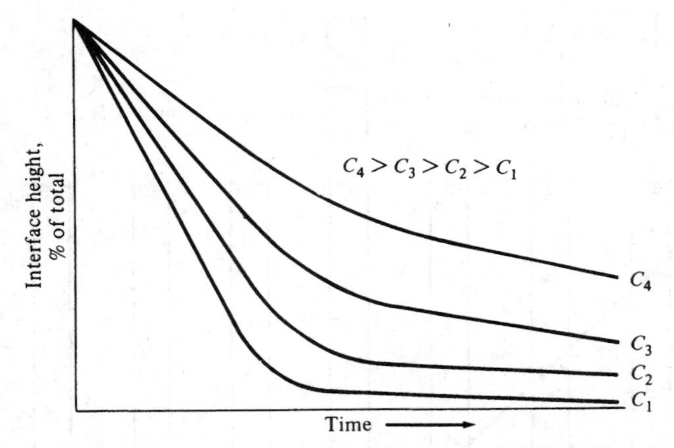

Figure 5-32 Relationship between initial concentration and settling curves.

of particles in the initial concentration, zone B is eroded from both top and bottom until it becomes nonexistent ($t = t_3$ in Fig. 5-31). After this time, the newly created A–C interface settles at a decreasing rate as the interfacial solids concentration increases successively from C_0 (just at the disappearance of zone B) to the concentration of the top layer of the compression zone just as zone C also disappears ($t = t_5$). The A–D interface thus formed will subside at a slow, uniform rate as the solids consolidate under their own weight, releasing some of the interstitial water to the clarified zone above.

All of the interfaces involving the clarified zone should be readily visible if the container used in the analysis is transparent. Other interfaces (B–C, C–D) will not be visible because concentration changes at these points are slight. The settling characteristics of activated sludge can be graphed by recording the visible interfacial height at succeeding time intervals. A plot of the interfacial height as a function of time, similar to that superimposed on Fig. 5-31, can then be drawn. The effect of varying the initial concentration of the activated sludge is illustrated by the family of curves shown in Fig. 5-32.

Continuous-flow analysis The zone settling principles just described for batch analysis are also applicable, within limits, to continuous-flow secondary clarifiers. An "idealized" secondary clarifier is shown in Fig. 5-33, with the appropriate zones labeled. If steady-state conditions are imposed with respect to flow rate and suspended-solids concentration for both the influent and the underflow, all of the zone will be maintained at static levels. Because the A–B interface is stationary, water in the clarified zone rises toward the overflow at a rate equal to the collective settling velocity of the C_0 concentration, thus satisfying the clarification function of the secondary clarifier.

The thickening function is accomplished via the concentration gradient in the thickening and compression zones and is more difficult to determine. The thickening function can be found by using the solids flux method first proposed by Coe

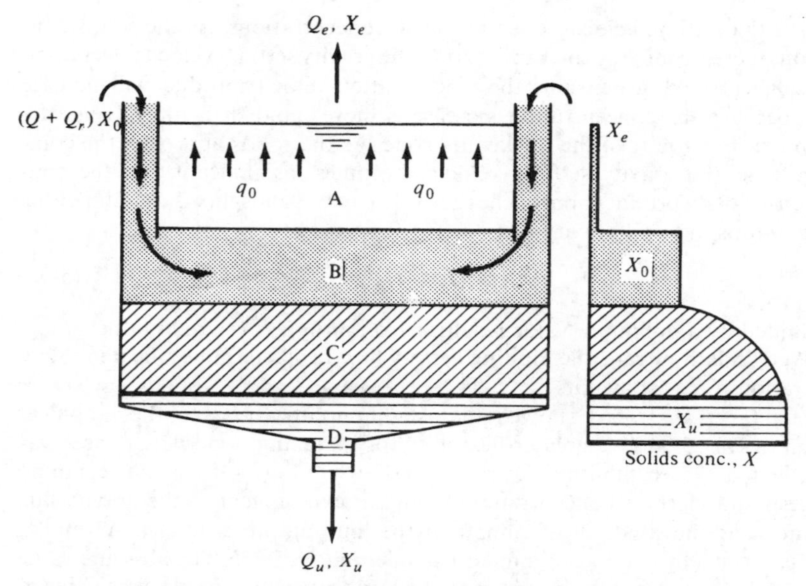

Figure 5-33 Zone settling in secondary clarifier. (*Adapted from Vesilind* [5-55].)

and Clevenger [5-12] and later modified by Yoshioka et al. [5-56], Dick and Ewing [5-16], Dick [5-15], and Dick and Young [5-17]. *Solids flux* is defined as the mass of solids per unit time passing through a unit area perpendicular to the direction of flow. In secondary clarifiers, it is the product of the solids concentration (mass/volume) times the velocity (length/time). The preferred units are kilograms per cubic meter (kg/m^3) times meter per hour (m/hr), or kilograms per square meter per hour ($kg/m^2 \cdot h$).

The downward velocity of solids in a secondary clarifier has two components: (1) the transport velocity due to the withdrawal of sludge, and (2) the gravity settling of the solids relative to the water. The transport velocity is a function of the underflow rate and the area of the tank.

$$v_u = Q_u/A \tag{5-29}$$

and the resulting solids flux for a clarifier operating at a given underflow rate is a linear function of the solids concentration.

$$G_u = v_u X_i = (Q_u/A)X_i \tag{5-30}$$

where G_u is the solids flux at the particular depth where the solids concentration is X_i. This relationship is shown graphically in Fig. 5-34.

The solids flux due to gravity settling is defined by

$$G_g = v_g X_i \tag{5-31}$$

where v_g is the settling velocity of solids at X_i concentrations. As the solids concentrations increase into the thickening zone, the gravity settling velocity decreases. In most concentrated suspensions, the concentration-velocity product will increase initially, because the concentration increase is more rapid than the velocity decrease in the upper part of the thickening zone. As the solids approach the compression zone, the gravity settling velocity becomes insignificant and the concentration-velocity product approaches zero. The total solids flux is the sum of the underflow transport and gravity flux

$$G_t = G_u + G_g \tag{5-32}$$

and is limited by a minimum value resulting from progressive gravity thickening. For a given underflow rate, the limiting gravity flux also determines the underflow concentration X_u, as shown in Fig. 5-34.

Yoshioka et al. [5-56] showed that slight modifications to the graphical approach of Fig. 5-34 give greater flexibility for matching underflow concentrations to their associated limiting flux rates. As shown in Fig. 5-35*a*, a line beginning at the desired underflow concentration X_u and drawn tangent to the gravity flux curve intersects the solids flux ordinate at the limiting flux rate. The Yoshioka method is verified by comparing similar triangles in Fig. 5-35*b*. The absolute value of the slope of the tangent line is the underflow velocity, while the abscissa value at the point of tangency is the limiting gravity flux concentration. The ordinate value corresponding to the point of tangency is the gravity solids flux, while the intercept, $G_L - G_g$, is the flux due to the underflow transport. The relationship between underflow velocity, limiting solids concentration, and limiting flux rate is readily demonstrated by this technique (Fig. 5-35*c*).

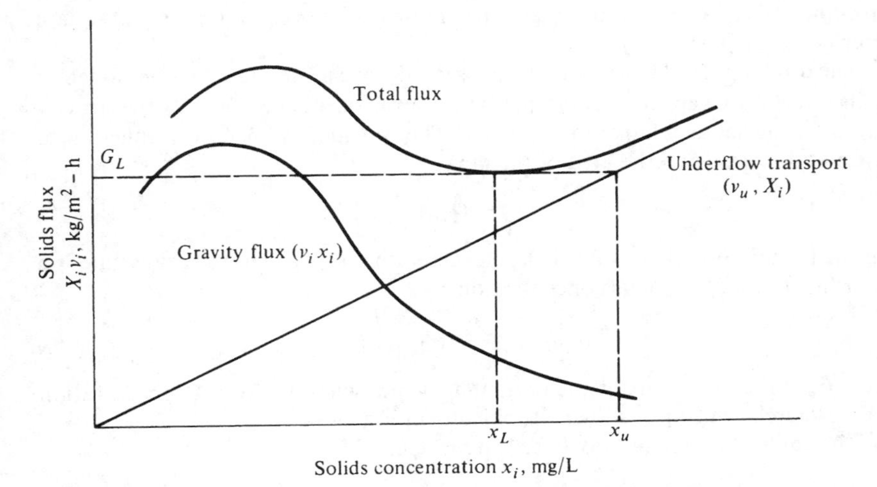

Figure 5-34 Solids flux as a function of solids concentration and underflow velocity.

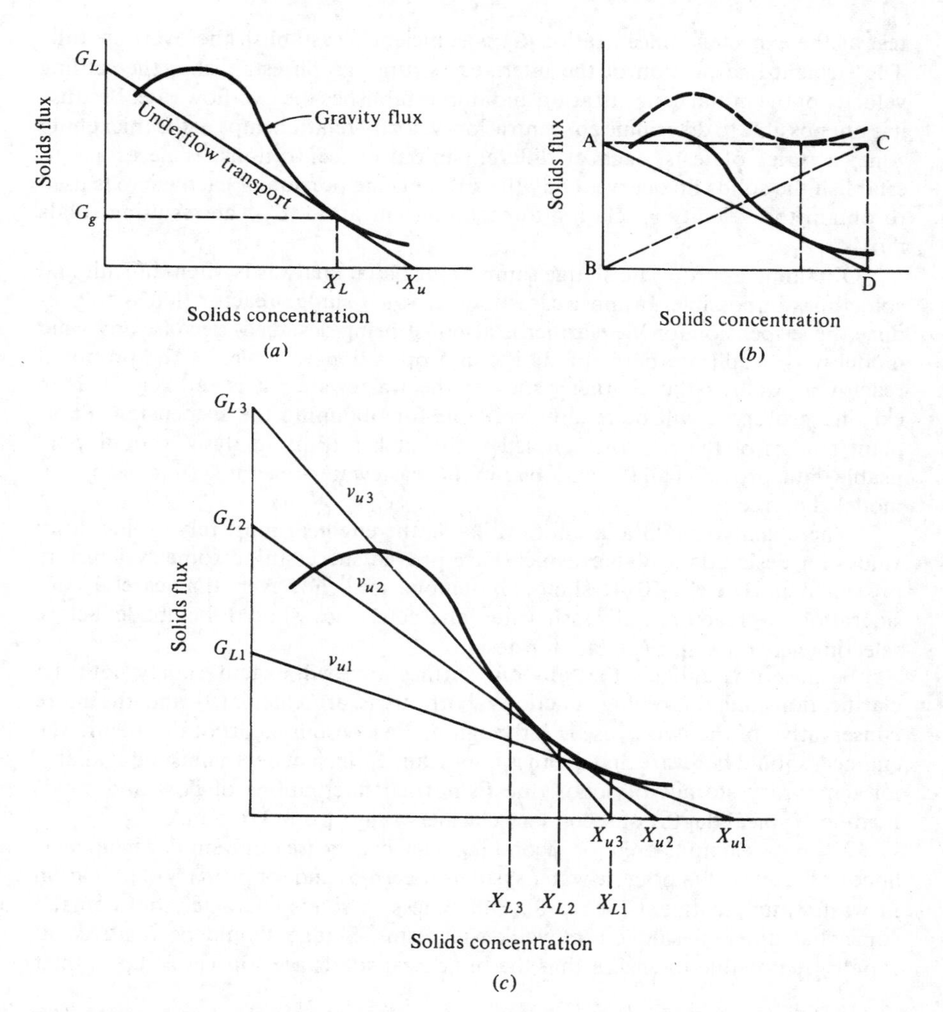

Figure 5-35 Yoshioka's graphical method for determining solids flux. (*a*) Yoshioka's modification; (*b*) verification of Yoshioka's modification. Note similarity of triangles ABD and ACD; (*c*) effects of underflow velocity on solids concentration.

Secondary clarifier design Secondary clarifiers must be designed for effluent clarification and solids thickening, both of which relate directly to the surface area. To determine the required surface area, an underflow concentration is selected and the overflow rate and limiting solids flux established or assumed for the particular activated sludge under consideration. Batch analysis similar to that previously described can be used to provide overflow rates and thickening characteristics, provided appropriate samples of activated sludge are available. A single test at the expected concentration C_0 is sufficient to establish the overflow rate. The straight-line portion of the interface vs. time graph establishes the settling velocity of the initial concentration and thus establishes the overflow rate. Because it is not possible to determine concentration-velocity relationships in the thickening zone, a series of tests, each at different initial concentrations, is necessary to establish the solids flux curve. Only the straight-line portion of each curve is used to obtain the velocity v_i relating to each concentration X_i. The resulting solids flux is $v_i X_i$.

Obtaining appropriate sludge samples for batch analysis is often difficult and sometimes impossible. In most cases the activated-sludge reactor that is to produce the suspension for the clarifier is also just being designed. Because any valid model must duplicate both the design and operation variables of the proposed reactor as well as the characteristics of the wastewater, it is unlikely that an existing prototype will be readily available for obtaining the suspension. Pilot-plant studies of the reactor, coupled with batch settling analysis, should yield usable data, provided all the variables in the wastewater-reactor system have been modeled correctly.

Where analytical data are not available, the engineer must rely on literature values for design data. Values which have proved successful in some systems are presented in Table 5-10. It should be emphasized, however, that careful consideration of reactor and wastewater characteristics should be made before selecting general empirical data for design.

Because it is unlikely that any one surface area will exactly satisfy both the clarification and thickening functions, both areas are calculated and the more conservative of the two is used. Although neither design incorporates depth, the engineer should be aware that depth is important. Sufficient depth must be available for temporary storage of solids due to normal fluctuations of flow and solids loading. Typical depths of secondary clarifiers range from 3 to 5 m.

The physical units used for secondary clarification are quite similar in appearance to those used in potable water systems (Sec. 4-5) and for primary clarification in wastewater treatment (Sec. 5-8). Differences in solids characteristics demand somewhat different sludge-removal mechanisms. Sludge should be removed as rapidly as possible to ensure that the biological solids are still viable upon their

Table 5-10 Design data for clarifiers for activated-sludge systems

	Overflow rate, $m^3/m^2 \cdot d$		Loading, $kg/m^2 \cdot h$		
Type of treatment	Average	Peak	Average	Peak	Depth, m
Settling following air-activated sludge (excluding extended aeration)	16–32	40–48	3.0–6.0	9.0	3.5–5
Settling following extended aeration	8–16	24–32	1.0–5.0	7.0	3.5–5

Source: Adapted from Metcalf & Eddy, Inc. [5-36]

Figure 5-36 Secondary clarifier with rapid-sludge-return system (*courtesy of FMC Corporation, Material Handling Systems Division*).

return to the aeration unit. A rapid sludge return also prevents anaerobic conditions from developing, with subsequent sludge flotation due to the release of gases. The sludge-return system must be capable of handling a wide range of flow. Underflow rates may exceed 100 percent of the wastewater flow under upset conditions, while normal underflow rates range from 20 to 40 percent of the wastewater flow.

A typical circular-tank secondary clarifier with rapid-sludge-return equipment is shown in Fig. 5-36. The sludge enters the "V" sections of the scraper as it rotates and is lifted vertically through the sludge-return pipes to a common conduit through which it is removed from the tank. Sludge is thus removed from the entire floor of the tank at each revolution of the scraper.

Early practice has tended toward the use of circular tanks, although the advent of rapid-sludge-removal mechanisms for rectangular tanks has resulted in an increase in their use. Physical parameters associated with the design of secondary clarifiers are given in Table 5-10. The design of secondary clarifiers is illustrated in Example 5-7.

Example 5-7: Designing a secondary clarifier for activated sludge A column analysis was run to determine the settling characteristics of an activated-sludge suspension. The results of the analysis are shown in the table below.

Conc MLSS, mg/L	1400	2200	3000	3700	4500	5200	6500	8200
Velocity, m/h	3.0	1.85	1.21	0.76	0.45	0.28	0.13	0.089

The influent concentration of MLSS is 3000 mg/L, and the flow rate is 8000 m^3/d. Determine the size of the clarifier that will thicken the solids to 10,000 mg/L.

SOLUTION

1. Calculate the solids flux from the above data:

$$G = \text{MLSS(kg/m}^3) \times \text{velocity (m/h)}$$

Conc mg/L	1400	2200	3000	3700	4500	5200	6500	8200
G kg/m$^2 \cdot$ h	4.20	4.07	3.63	2.8	2.03	1.46	0.9	0.73

2. Plot solids flux vs. MLSS concentration as shown in the accompanying figure. Draw a line from the desired underflow concentration, 10,000 mg/L, tangent to the curve and intersecting the ordinate. The value of G at the intersection, 2.4 kg/m$^2 \cdot$ h, is the limiting flux rate and governs the thickening function.

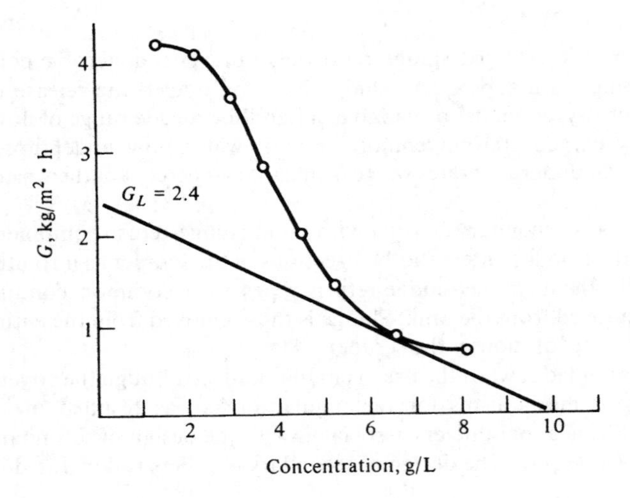

3. Determine total solids loading to the clarifier:

$$8000 \text{ m}^3\text{/d} \times \frac{\text{d}}{24 \text{ h}} \times \frac{3.0 \text{ kg}}{\text{m}^3} = 1000 \text{ kg/h}$$

4. Determine the surface area of the clarifier.

$$\frac{1000 \text{ kg/h}}{2.4 \text{ kg/h} \cdot \text{min}^2} = 416.7 \text{ m}^2$$

Assuming a circular shape

$$\text{Dia} = \left(\frac{4}{\pi} 416.7\right)^{1/2} = 23 \text{ m}$$

5. Check clarification function:

$$8000 \text{ m}^3/\text{d} \times \frac{\text{d}}{24 \text{ h}} = 333 \text{ m}^3/\text{h}$$

At 3000 mg/L, the settling velocity of the interface is 1.21 m/h.

$$\frac{333 \text{ m}^3/\text{h}}{1.21 \text{ m/h}} = 275 \text{ m}^2$$

Because 275 m^2 < 416.7 m^2, the thickening function governs the design.

Attached-Culture Systems Clarifier

Design of secondary clarifiers for attached-culture systems is similar to that for primary clarifiers. The clarification function is the important parameter because sludge thickening is not a factor. In fact, settling characteristics of the sloughed biofilm, or *humus* as it is often called, approach those of discrete particles. Overflow rates from 25 to 33 $m^3/m^2 \cdot d$ are commonly used, with a maximum of 50 $m^3/m^2 \cdot d$. [5-50] No sludge is recycled to the reactor, so the underflow is negligible compared to the overflow. Solids are often pumped to the primary clarifier where they are concentrated along with the raw wastewater solids for ultimate disposal.

The total quantity of solids generated by attached-culture systems is generally less than that generated by suspended-culture processes because of the endogenous nature of the biomass near the media. Solids production can be expected to range from 0.2 to 0.5 kg/kg BOD_5 removed from the liquid. Well-settled sludges range from 10 to 20 percent solids. [5-50] Liquid recirculation through high-rate trickling filters and bio-towers may increase the required size of the secondary clarifier substantially. This added volume may be avoided with modular plastic media by direct recirculation from the effluent of the reactor prior to the secondary clarifier.

5-15 DISINFECTION OF EFFLUENTS

The disinfection of wastewater is usually required where portions of the effluent may come in contact with humans. The processes available for disinfecting wastewater effluents are essentially the same as those described in Chap. 4 for potable water. The presence of much greater concentrations of suspended and dissolved material in the wastewater may result in interferences not found in potable water. Chemical oxidants are generally considered the most effective disinfectants, with required dosages being much higher than those used for cleaner water. Chlorine is the most common disinfectant in use, even though it may combine with certain constituents in the wastewater to produce haloform compounds.

Table 5-11 Chlorine dosages for various wastewaters

Wastewater type	Chlorine dosage (mg/L) to yield 0.2 mg/L free residual after 15-min contact time
Raw:	
Fresh to stale	6–12
Septic	12–25
Settled:	
Fresh to stale	5–10
Septic	12–40
Effluent chemical precipitation	3–6
Trickling filter	
Normal	3–5
Poor	5–10
Activated sludge	
Normal	2–4
Poor	3–8
Intermittent sand filter:	
Normal	1–3
Poor	3–5

Source: From Eckenfelder. [5-19]

Chlorination of wastewater effluents is accomplished in much the same manner as is the chlorination of potable water. Larger dosages are required since ammonium and a variety of other substances in wastewater exert a chlorine demand that must be met before a free residual is obtained. The exact amount of chlorine necessary also depends upon water temperature, contact time, and degree of kill necessary. Ranges of chlorine addition necessary to provide the needed free chlorine in various wastewaters are given in Table 5-11. Contact times of about 30 min at average flow, with a minimum of 15 min for peak flow, are common.

The use of chlorine for disinfection of wastewater effluents has come under close scrutiny due to the haloform-formation problem. It is quite likely that other disinfectants will be required or that disinfection practices will be limited to special cases in the future.

Sludge Treatment and Disposal

Wastewater treatment objectives are accomplished by concentrating impurities into solid form and then separating these solids from the bulk liquid. This concentration of solids, referred to as *sludge*, contains many objectionable materials and must be disposed of properly. Sludge disposal facilities usually represent 40 to 60 percent of the construction cost of wastewater-treatment plants, account for as much as 50 percent of the operating cost, and are the cause of a disproportionate share of operating difficulties.

5-16 SLUDGE CHARACTERISTICS

The quantity and nature of sludge depends on the characteristics of the wastewater and on the nature and efficiencies of the treatment processes. Primary settling removes the settleable fraction of the raw wastewater solids, usually 40 to 60 percent of the influent solids. The quantity of these solids, on a dry mass basis, can be determined by the following equation. [5-25]

$$M_p = \xi \times SS \times Q \tag{5-33}$$

where M_p = mass of primary solids, kg/d
ξ = efficiency of primary clarifier
SS = total suspended solids in effluent, kg/m^3
Q = flow rate, m^3/d

Primary sludge contains inorganic solids as well as the coarser fraction of the organic colloids. It contains a sizable fraction of the influent BOD, will become anaerobic within a few hours, and must be isolated to prevent nuisance problems.

Solids escaping primary settling are either solubilized or become entrained in the biomass during secondary treatment. Additional solids are generated by conversion of dissolved organics into cellular material. Secondary sludge is thus composed primarily of biological solids, the quantity of which can be estimated by the equation

$$M_s = Y' \times \text{BOD}_5 \times Q \tag{5-34}$$

where M_s = mass of secondary solids, kg/d
Y' = biomass conversion factor: fraction of food (BOD_5) incorporated into biomass, kg/kg
BOD_5 = BOD_5 removed by secondary treatment, kg/m^3
Q = flow rate, m^3/d

The value of Y' is a function of both the biomass conversion factor [Y in Eq. (5-5)] and the phase of the growth curve (Fig. 5-14) at which the particular system operates. More simply, it may be related to the food-to-biomass ratio as shown in Fig. 5-37. [5-25]

The consistency of wastewater sludges varies with the source. Primary sludge is more granular in nature than secondary sludge and is generally more concentrated. Consistency of secondary sludge is dependent on treatment processes and is more variable. Solids from attached-growth reactors are particulate in nature and consolidate better than the light, flooculent solids from suspended-culture systems. It is sometimes advantageous to mix primary and secondary sludge to facilitate further processing. The solids content of various sludges and sludge mixtures is given in Table 5-12 on a mass-per-volume basis, with each percent solids corresponding to 10,000 mg/L.

The organic content of both primary and secondary sludge is about 70 percent. Since the specific gravity of these organics is only slightly greater than 1, the unit

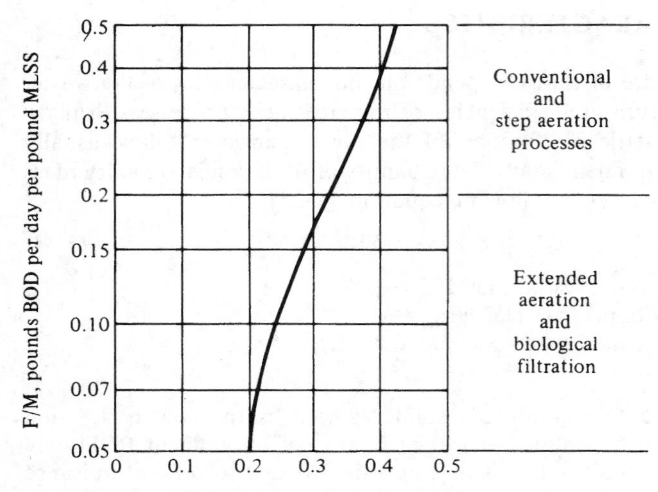

Figure 5-37 Generalized diagram of excess sludge production (Y' in Eq. 5-34) as a function of food-to-biomass ratio. Actual quantities would vary from plant to plant. (*From Hammer* [5-25].)

mass of sludge containing less than about 10 percent solids can be assumed to be equal to that of water without introducing significant error. The volume of wet sludge can therefore be approximated by the following equation:

$$V = M/1000 \cdot S \tag{5-35}$$

where V = volume of sludge produced, m^3/d
M = mass of dry solids, kg/d
S = solids content expressed as a decimal fraction
1000 = density of water, kg/m^3

Table 5-12 Typical solids content of sludges

Type of sludge	Sludge concentration, %: Unthickened	Sludge concentration, %: Thickened	Solids loading for gravity thickeners, $kg/m^2 \cdot d$
Separate			
Primary sludge	2.5–5.5	8–10	100–150
Trickling-filter sludge	4–7	7–9	40–50
Activated sludge	0.5–1.2	2.5–3.3	20–40
Pure-oxygen sludge	0.8–3.0	2.5–9	25–50
Combined			
Primary and trickling-filter sludge	3–6	7–9	60–100
Primary and modified-aeration sludge	3–4	8.3–11.6	60–100
Primary and air-activated sludge	2.6–4.8	4.6–9.0	40–80

Source: From Metcalf & Eddy, Inc. [5-36]

For a given solids-production rate, the volume of sludge varies inversely with the solids content as shown in Eq. (5-35). Within the concentration range of wastewater sludges, increasing the solids content by only a minimum percentage results in drastic reductions in the sludge volume. Because the size, and therefore cost, of sludge-disposal facilities is a function of the volume of sludge to be handled, considerable savings can be attained by volume reduction.

5-17 SLUDGE THICKENING

Several techniques are available for volume reduction. Mechanical methods such as vacuum filtration and centrifugation may be used where the sludge is subsequently to be handled in a semisolid state. These methods are commonly used preceding sludge incineration. Where further biological treatment is intended, volume reduction by gravity thickening and/or flotation is common practice. In both cases, the sludge remains in a liquid state.

Gravity thickeners are very similar in design and operation to the secondary clarifiers used in suspended-growth systems. The thickening function is the major design parameter, and tanks are generally deeper than secondary clarifiers to provide greater thickening capacity. A typical gravity thickener is shown in Fig. 5-38. The vertical "pickets" on the scraper cause a horizontal agitation which helps to release water trapped in the flocculent structure of the sludge, and are commonly used when suspended-culture system sludges are to be thickened.

A well-designed, well-operated gravity thickener should be able to, at least, double the solids content of the sludge, thereby eliminating half the volume. Solids contents for thickened sludges, along with commonly used loading rates for gravity thickeners, are included as part of Table 5-12. It should be noted that the design of gravity thickeners should be based on the results of pilot-plant analysis wherever possible, since successful loading rates are highly dependent on the nature of the sludge.

Figure 5-38 Typical gravity thickener. (*Courtesy of FMC Corporation, Material Handling Systems Division.*)

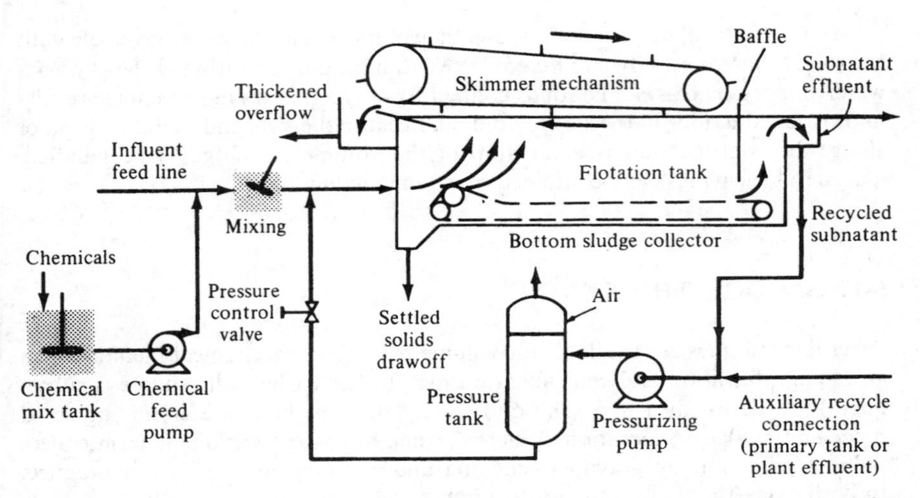

Figure 5-39 Typical dissolved-air flotation system for thickening activated sludge. (*From Metcalf & Eddy, Inc.* [5-36].)

As can be seen in Table 5-12, waste activated sludge does not thicken well in gravity thickeners and loading rates are significantly lower than for other sludge. Also, the effectiveness of gravity thickeners for primary sludge is diminished considerably by mixing with activated sludge. The light, flocculent nature of activated sludge lends itself quite well to thickening by dissolved air flotation, however, and the use of the process has been increasing in recent years.

In dissolved air flotation, a small quantity of water, usually secondary effluent, is subjected to aeration under a pressure of about 400 kPa (58 lb/in^2). This supersaturated liquid is then released near the bottom of a tank through which the sludge is passed at atmospheric pressure. The air is released in the form of very small bubbles that attach themselves to, or become entrapped in, the sludge solids, floating the solids to the surface. The thickened sludge is skimmed off at the top of the tank while the liquid is removed near the bottom and is returned to the aerator. A diagram of the system is shown in Fig. 5-39.

The capital and operating costs of sludge thickeners are justified when sludge digestion is practiced. The extent of volume reduction by sludge thickening is illustrated by the following examples.

Example 5-8: Sludge volume and solids content relationship Suppose the 1-L graduated cylinder in the figure below contains a sludge of 1 percent solids. From Eq. (5-35) the dry mass of the solids is

$$\begin{aligned} M &= 1000V \cdot S \\ &= 1000 \text{ kg/m}^3 \times 0.001 \text{ m}^3 \times 0.01 \\ &= 0.01 \text{ kg} \end{aligned}$$

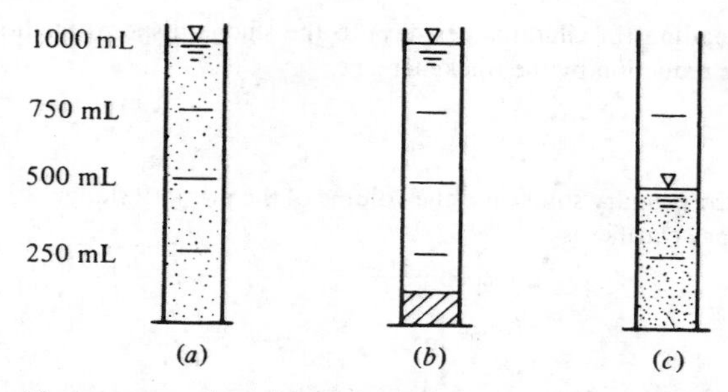

Now suppose that all of the solids are allowed to settle (*b* in figure) and that the liquid is decanted (*c* in figure) until the total volume is 500 mL. The new solids fraction is

$$
\begin{aligned}
S &= M/1000V \\
&= 0.01 \text{ kg}/1000 \text{ kg/m}^3 \times 0.0005 \text{ m}^3 \\
&= 0.02
\end{aligned}
$$

Thus, increasing the solids content by a factor of 2 (in this case only 1 percent) decreases the total volume by a factor of 2.

Example 5-9: Reducing the volume by sludge thickening A wastewater-treatment plant consists of primary treatment units followed by an activated-sludge secondary system. The primary and secondary sludges are mixed, thickened in a gravity thickener, and sent to further treatment. A schematic of the system is shown below.

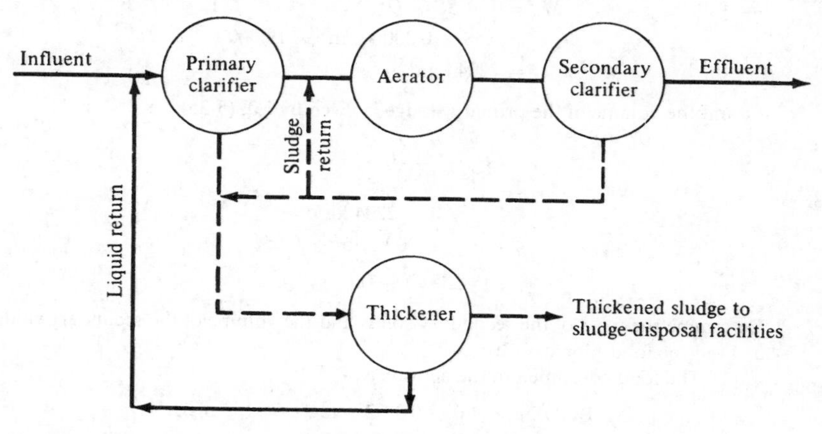

Wastewater, treatment plant, and sludge characteristics are as follows:

Wastewater		Treatment plant		Sludge	
Influent SS	200 mg/L	Primary clarifier diameter	25 m	Primary	5.0% solids
Influent BOD	225 mg/L	Aerator volume	2900 m^3	Secondary	0.75% solids
Effluent BOD	20 mg/L	MLSS in aerator	3500 mg/L	Thickened	4.0% solids
Flow	19,000 m^3/d				

Determine (*a*) the solids loading (in kilgrams per day) to the sludge disposal facilities and (*b*) the percent volume reduction by the thickener.

SOLUTION

1. Determine the mass of the primary solids and the volume of the primary sludge.
   a. The area of the primary clarifier is

$$A = \pi d^2/4$$

$$= \pi \times \frac{25\ \text{m}^2}{4}$$

$$= 491\ \text{m}^2$$

   b. The overflow rate is

$$\frac{19{,}000\ \text{m}^3/\text{d}}{491\ \text{m}^2} = 38.7\ \text{m/d}$$

   c. From Fig. 5-13 the efficiency of the clarifier is

$$SS = 58\%$$

$$\text{BOD} = 32\%$$

   d. The mass of primary solids removed is found by Eq. (5-33)

$$M_p = \xi \times SS \times Q$$
$$= 0.58 \times 0.200\ \text{kg/m}^3 \times 19{,}000\ \text{m}^3/\text{d}$$
$$= 2204\ \text{kg/d}$$

   and the volume of the primary sludge is given by Eq. (5-35).

$$V_p = \frac{M_p}{1000 \cdot S}$$

$$= \frac{2204\ \text{kg/d}}{1000\ \text{kg/m}^3 \times 0.05}$$

$$= 44.1\ \text{m}^3/\text{d}$$

2. Determine the mass of the secondary solids and the volume of the secondary sludge.
   a. Find the food-biomass ratio:
      (1) The food consumed in the aerator is:

$$\text{BOD in} = (1.0 - 0.32)225\ \text{mg/L} = 153\ \text{mg/L}$$

$$\text{BOD out} = \text{effluent BOD} = 20\ \text{mg/L}$$

$$\text{BOD consumed in the aerator} = 133\ \text{mg/L}$$

$$0.133\ \text{kg/m}^3 \times 19{,}000\ \text{m}^3/\text{d} = 2527\ \text{kg/d}$$

      (2) The biomass in the reactor is $3.5\ \text{kg/m}^3 \times 2900\ \text{m}^3 = 10{,}150\ \text{kg}$
      (3) The food-biomass ratio is

$$\frac{2527\ \text{kg/d}}{10{,}150\ \text{kg}} = 0.25\ \text{d}^{-1}$$

b. From Fig. 5-37, the biomass conversion factor is 0.35.

c. The mass of the secondary solids is found by Eq. (5-34).

$$\begin{aligned} M_s &= Y' \times \mathrm{BOD}_5 \times Q \\ &= 0.35 \times 0.133 \text{ kg/m}^3 \times 19{,}000 \text{ m}^3\text{/d} \\ &= 884 \text{ kg/d} \end{aligned}$$

d. The volume of the secondary sludge is

$$\begin{aligned} V_s &= \frac{M_s}{1000 \times S} \\ &= \frac{884 \text{ kg/d}}{1000 \text{ kg/m}^3 \times 0.0075} \\ &= 118 \text{ m}^3\text{/d} \end{aligned}$$

3. Determine the total mass of solids and the total volume of sludge to the thickener.

   a. $M_T = M_p + M_s = 2204 + 884 = 3088$ kg/d

   b. $V_T = V_p + V_s = 44.1 + 118 = 162.1$ m^3/d

4. Determine the total mass of solids and the total volume of sludge discharged from the thickener to the sludge disposal facilities.

   a. Assuming negligible solids in the thicker supernatant, the total mass of solids in the thickened sludge is 3088 kg/d.

   b. The total volume of the thickened sludge is

$$\begin{aligned} V_{\text{thick}} &= \frac{3088 \text{ kg/d}}{1000 \text{ kg/m}^3 \times 0.04} \\ &= 77.2 \text{ m}^3\text{/d} \end{aligned}$$

5. Determine the percent of volume reduction achieved by the thickener

$$\frac{162.1 - 77.2}{162.1} \times 100 = 52\%$$

5-18 SLUDGE DIGESTION

Concentrated wastewater sludges represent a considerable hazard to the environment and must be rendered inert prior to disposal. The most common means of stabilizing is by biological degradation. Because this process is intended to convert solids to noncellular end products, the term *digestion* is commonly applied to this process. Sludge digestion serves both to reduce the volume of the thickened sludge still further and to render the remaining solids inert and relatively pathogen-free. These goals can be accomplished by either anaerobic or aerobic digestion.

Anaerobic Digestion

Anaerobic digestion is by far the most common process for dealing with wastewater sludges containing primary sludge. Primary sludge contains large amounts of readily available organics that would induce a rapid growth of biomass if treated aerobically. Anaerobic decomposition produces considerably less biomass than aerobic processes. The principal function of anaerobic digestion, therefore, is to convert as much of the sludge as possible to end products such as liquids and gases, while producing as little residual biomass as possible.

Wastewater sludge contains a wide variety of organisms, and thus requires a wide variety of organisms for its decomposition. The literature relating to anaerobic sludge digestion often divides the organisms into broad groups, the acid formers and the methane formers. The *acid formers* consist of facultative and anaerobic bacteria and include organisms that solubilize the organic solids through hydrolysis. The soluble products are then fermented to acids and alcohols of low molecular weight. The *methane formers* consist of strict anaerobic bacteria that convert the acids and alcohols, along with hydrogen and carbon dioxide, to methane. Specific products in the metabolic process are shown in Fig. 5-40.

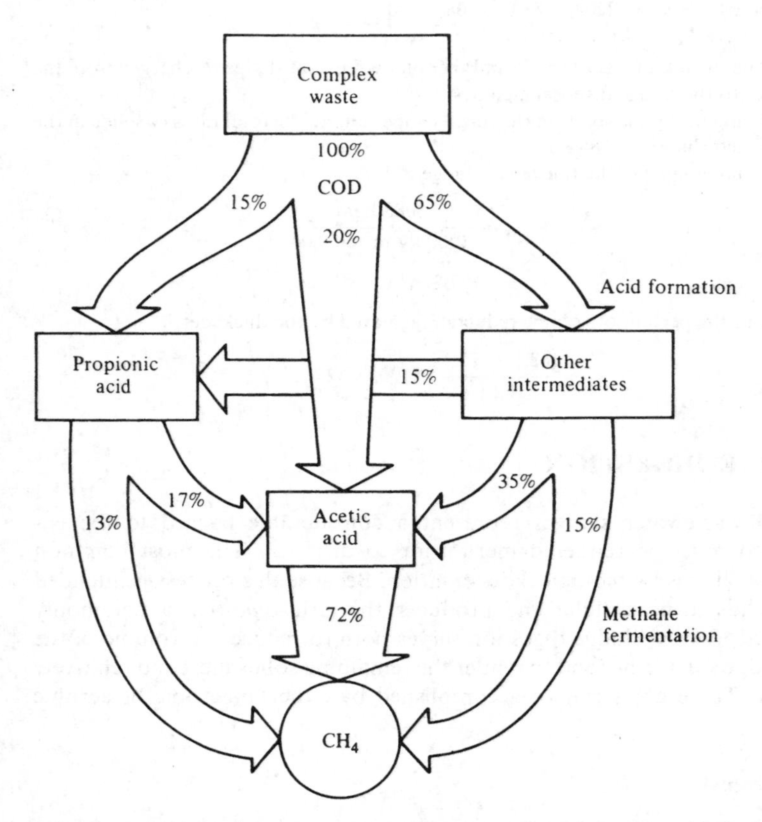

Figure 5-40 Pathways and products of anaerobic digestion of wastewater sludge. (*From McCarty* [*5-31*].)

Typically, about 50 to 60 percent of the organics are metabolized, with less than 10 percent being converted to biomass.

Reactors for anaerobic digesters consist of closed tanks with airtight covers. The completely mixed, continuous-flow model without solids recycle is usually assumed, although the first two conditions will seldom be met exactly. Although most larger installations utilize high-rate digestion, treatment plants processing less than 4000 m^3/day of wastewater often use standard-rate digestion for economic reasons or simplicity of operation.

A typical, standard-rate anaerobic digester consisting of a single-stage operation is shown in Fig. 5-41. The conical bottom facilitates sludge withdrawal while the "floating" cover accommodates volume changes due to sludge additions and withdrawals. The sludge separates in the reactor as shown, although some mixing occurs in the zone of active digestion and in the supernatant because of the withdrawal and return of heated sludge. Sludge is fed into the digester on an intermittent basis and the supernatant is withdrawn and returned to the secondary treatment unit. The digested sludge accumulates in the bottom, its

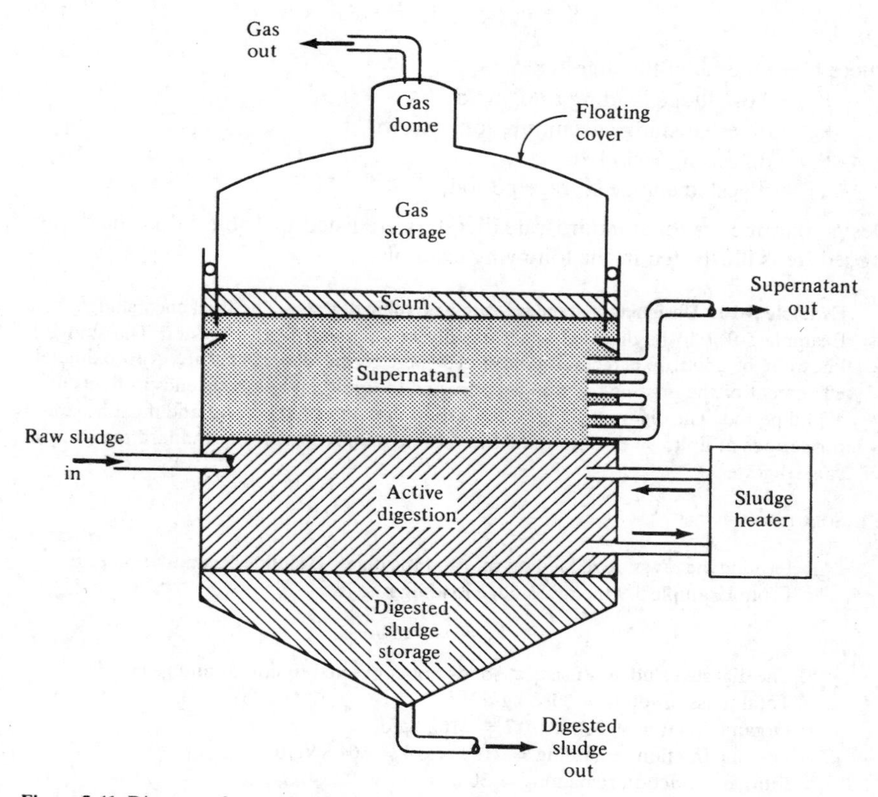

Figure 5-41 Diagram of a standard-rate anaerobic digester.

Table 5-13 Design parameters for anaerobic digesters

Parameter	Standard-rate	High-rate
Solids retention time, d	30–90	10–20
Volatile solids loading, $kg/m^3/d$	0.5–1.6	1.6–6.4
Digested solids concentration, %	4–6	4–6
Volatile solids reduction, %	35–50	45–55
Gas production (m^3/kg VSS added)	0.5–0.55	0.6–0.65
Methane content, %	65	65

removal often being determined by subsequent sludge disposal facilities rather than by operational needs of the digester.

The standard-rate digester volume is determined by loading rates, digestion period, solids reduction, and sludge storage. These are related by the following equations.

$$V = \frac{V_1 + V_2}{2} t_1 + V_2 t_2 \tag{5-36}$$

where V = volume of the digester, m^3
V_1 = raw sludge loading rate, m^3/d
V_2 = digested sludge accumulation rate, m^3/d
t_1 = digestion period, d
t_2 = digested sludge storage period, d

Design parameters for standard-rate digesters are listed in Table 5-13. The design procedure is illustrated in the following example.

Example 5-10: Designing a standard-rate anaerobic digester The thickened sludge from Example 5-9 is to be digested anaerobically in a standard-rate digester. The sludge is known to be about 70 percent organic and 30 percent inorganic in nature. Approximately 60 percent of the organic fraction is converted to liquid and gaseous end products after a 30-d period. The digested sludge has a solids content of 5.0 percent and must be stored for periods of up to 90 d. Determine the volume requirement for a standard rate, single-stage digester.

SOLUTION

1. Determine the raw sludge loading rate and the digested sludge accumulation rate.
   a. From Example 5-9 the raw sludge loading rate is

$$V_1 = 77.2 \text{ m}^3/\text{d}$$

   b. The digested sludge consists of solids not converted to liquids and gases.
   Total mass of solids = 3088 kg/d
   Organic fraction = 3088 × 0.7 = 2162 kg/d
   Organic fraction remaining = 2162 × 0.4 = 864.8 kg/d
   Inorganic fraction remaining = 3088 × 0.3 = 926.4 kg/d
   Total mass remaining = 864.8 + 926.4 = 1791.2 kg/d

Digested sludge accumulation rate:

$$V_2 = \frac{1791.2 \text{ kg/d}}{1000 \text{ kg/m}^3 \times 0.05} = 35.8 \text{ m}^3\text{/d}$$

2. Determine the digester volume from Eq. (5-36).

$$V = \frac{V_1 + V_2}{2} t_1 + V_2 t_2$$

$$= \frac{(77.2 + 35.8)}{2} \text{ m}^3\text{/d} \times 30 \text{ d} + 35.8 \text{ m}^3\text{/d} \times 90 \text{ d}$$

$$= 4917 \text{ m}^4$$

High-rate digesters are more efficient and often require less volume than single-stage digesters. The contents are mechanically mixed to ensure better contact between the organics and the microorganisms and the unit is heated to increase the metabolic rate of the microorganisms, thus speeding up the digestion process. Optimum temperature is around 35°C (95°F).

Because no dewatering occurs in the high-rate system, the volume of sludge is essentially unchanged, although the solids content is reduced. Dewatering of the sludge is necessary and may be accomplished by any of the mechanical dewatering operations described in Sec. 5-19. An alternative dewatering system is a second-stage digester similar to that used in standard-rate operations. A high-rate two-stage system is shown in Fig. 5-42. Little gas is generated in the second stage, but the influent is supersaturated with gases that are released in the second-stage reactor. Consequently, the second-stage reactor is usually covered and is equipped for gas recovery. The second-stage reactor is not heated. Design of volume requirements for high-rate, two-stage digesters is illustrated in the following example.

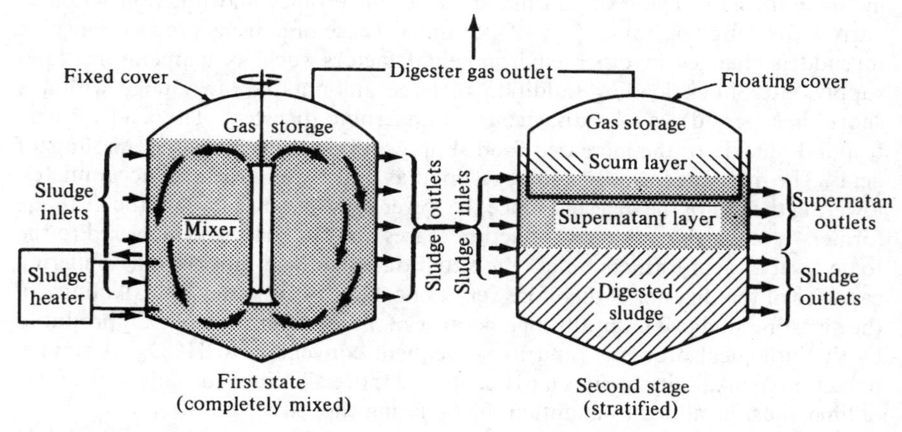

Figure 5-42 Diagram of high-rate, two-stage anaerobic sludge digester. (*From Linsley and Franzini* [*5-30*].)

Example 5-11: Designing a high-rate, two-stage anaerobic digester A high-rate, two-stage digester is to be designed for the sludge described in Example 5-10. A digestion time of 10 d in the first stage results in the destruction of approximately 60 percent of the organics. Dewatering in the second stage occurs within 3 d with the stored sludge having a solids content of about 5.0 percent solids. Determine the volume of the first- and second-stage digesters and compare the total volume to that of the single-stage digester in Example 5-10.

SOLUTION

1. The volume of the first-stage digester is

$$\begin{aligned} V &= V_1 t_1 \\ &= 77.2\ \text{m}^3/\text{d} \times 10\ \text{d} \\ &= 772\ \text{m}^3 \end{aligned}$$

2. The volume of the second-stage digester is

$$\begin{aligned} V &= \frac{V_1 + V_2}{2} t_1 + V_2 t_2 \\ &= \frac{77.2 + 35.8}{2}\ \text{m}^3/\text{d} \times 3\ \text{d} + 35.8\ \text{m}^3/\text{d} \times 90\ \text{d} \\ &= 3392\ \text{m}^3 \end{aligned}$$

3. Compare total volume to single-stage volume.

$$\begin{aligned} \text{Volume of single-stage (from Example 5-10)} &= 4917\ \text{m}^3 \\ \text{Total volume of two-stage} = 772 + 3392 &= 4164\ \text{m}^3 \\ \hline \text{Difference} &= 753\ \text{m}^3 \end{aligned}$$

Operation of anaerobic digesters is complicated by the delicate nature of the methane formers. These organisms are strict anaerobics and function within a narrow pH range of from 6.5 to 7.5 pH units. These organisms are also sensitive to sudden changes in other environmental factors such as temperature, food supply, etc. Shock loading (addition of large amounts of raw sludge within a short time period) can be disastrous to anaerobic digesters. The acid formers respond quickly to the increased food supply and produce increased amounts of acids. The methane formers cannot respond as quickly and the acid accumulates, lowering the pH of the digester. Once the lower pH-tolerance level of the methane formers is reached, methane production ceases and the pH can be lowered to the toxic level of acid formers unless the situation is rectified quickly. The buffering capacity of the digester is therefore very important. Fortunately, the alkalinity of the digesting sludge is naturally high because of the solubilization of CO_2 produced by the biological processes, and its subsequent conversion to HCO_3. A sudden reduction in alkalinity heralds a pH drop, and more alkalinity, usually in the form of lime, must be added to maintain the buffering capacity.

Product gases from anaerobic digestion typically contain 65 to 70 percent methane, 25 to 30 percent CO_2, and trace amounts of other gases. Approximately

Table 5-14 Composition of typical supernatant from anaerobic digesters

	Primary plants, mg/L	Trickling filters, mg/L	Activated-sludge plants, mg/L
Suspended solids	200–1000	500–5000	5000–15,000
BOD_5	500–3000	500–5000	1000–10,000
COD	1000–5000	2000–10,000	3000–30,000
Ammonia as NH_3	300–400	400–600	500–1000
Total phosphorus as P	50–200	100–300	300–1000

Source: From Benefield and Randall. [5-6]

1 m^3 of gas is produced per kilogram of solids digested. The heat content of the methane is approximately 36,000 kJ/m^3 (970 BTU/ft^3). The digester gas is usually combusted to provide space heating in the treatment plant buildings, to heat water for laboratory use, and to heat the digester if a two-stage system is used. These uses often consume less than one-half of the methane. The remainder could be used to drive an electrical generator and the resulting power used within the plant. Banerji and O'Connor [5-5] report that a significant portion of the energy necessary to operate a wastewater-treatment plant can be derived from the methane produced by anaerobic digesters. The conversion process requires expensive equipment and is a costly operation and maintenance item, however. Most plants simply flare the excess methane.

The supernatant withdrawn from the digester contains large amounts of solubilized organics and solids, as shown in Table 5-14. This material must be circulated back through the plant for further treatment. Solids withdrawn from the bottom of the digester should be relatively inert. Procedures for disposing of this material are discussed in a following section.

Aerobic Digestion

Sludge can also be stabilized by aerobic digestion. Generally restricted to biological sludges in the absence of primary sludge, this process is essentially a continuation of the aeration process, with the volume being reduced by thickening in the secondary clarifier and sludge thickener. The most common application of aerobic digestion involves stabilizing sludge wasted from extended aeration systems. Since an external food source is not supplied, aerobic digestion is an endogenous respiration process in which the organisms are forced to metabolize their own protoplasm. The result is a mineralized sludge in which any remaining organics are principally cell walls and other cell fragments not readily biodegradable.

Aerobic digestion is not as sensitive to environmental factors as is its anaerobic counterpart and is not as subject to upsets. Unlike the anaerobic process, aerobic digestion is energy-consumptive. The digested sludge is relatively inert but dewaters poorly. It is often necessary to dispose of the entire volume of sludge in a rather dilute state.

Table 5-15 Typical design parameters for aerobic digestion

Parameter	Value
Retention time, θ_c	
Activated sludge only	15–20 d
Activated sludge plus primary	20–25 d
Air required (diffused air)	
Activated sludge only	20–35L/min · m³
Activated sludge plus primary	55–65L/min · m³
Power required (surface air)	0.02–0.03 kW/m³
Solids loading	1.6–3.2 kg VSS/m³ · d

Source: From Steele and McGhee. [5-50]

Design criteria for aerobic digestion are given in Table 5-15. The design approach is essentially the same as for activated-sludge reactors.

5-19 SLUDGE DISPOSAL

Several options are available for the ultimate disposal of wastewater sludges. These include incineration, placement in a sanitary landfill, and incorporation into soils as a fertilizer or soil conditioner.

Raw (undigested) sludges can be incinerated, provided the water content is sufficiently reduced. Supplemental fuel is necessary to initiate and maintain combustion and municipal solid waste may be used for this purpose. Raw or digested sludge can also be disposed of in sanitary landfills, provided appropriate measures are taken to contain leachate and to isolate the sludge from the environment. These subjects are covered more fully in a later chapter on solid-waste disposal.

Land application of wastewater sludges has been practiced for many years, modern applications being limited to digested sludge. The nutrient value of the sludge is beneficial to vegetation, and its granular nature may serve as a soil conditioner. Its application has been limited to ground used for forage crops for nonhuman consumption, although the possibility of its use on ground used to grow edible produce is still being investigated. Metal toxicity in plants and water pollution from excess nitrates appear to be the limiting factors in land application of sludges. Sludges may be applied in a liquid state by spraying, ridge and furrow, or by direct injection beneath the soil. Injection under grasslands is illustrated in Fig. 5-43. Dewatered sludge may be spread on the land and cultivated into the soil by conventional agricultural equipment.

With the exception of irrigation practices, sludge disposal is greatly facilitated by volume reduction through dewatering. Dewatering may be accomplished by

Figure 5-43 Injection of wastewater sludge beneath grasslands. Note minimal disturbance of sod (*courtesy of Rickel Manufacturing Company*).

mechanical means such as centrifugation, vacuum filtration, filter pressing, or by air drying. Solids content achievable by various dewatering techniques is shown in Table 5-16.

Air drying of digested sludges is possible in climates with significant evaporation potential. Since a well-digested sludge is essentially inert, it can be handled and stored in the open air without creating nuisance conditions. Air-drying facilities include drying beds similar to those shown in Fig. 5-44. The sand and underdrain system may be omitted in dry climates where evaporation from the surface is sufficient to dispose of the liquid. Dried sludge is removed in cake form by solids-handling equipment.

Another popular form of dewatering of digested sludge is the sludge pond. Not to be confused with oxidation ponds used in secondary treatment processes,

Table 5-16 Solids content of dewatered sludge

Method	Approximate solids content, %
Vacuum filtration	20–30
Centrifuge	20–25
Filter press	30–45
Drying beds	40
Ponds	30

Figure 5-44 Typical open-air sludge drying bed. (*Photo courtesy of R. L. Sanks.*)

sludge ponds function as settling basins with long retention times. The solids consolidate in the bottom while the supernatant is periodically removed from the top and recycled for retreatment. When the solids have accumulated to a preselected depth, the pond is taken out of service and allowed to dry out. The dried sludge is then removed for final disposal.

Advanced Wastewater Treatment

The quality of effluent provided by secondary treatment may not always be sufficient to meet discharge requirements. This is often the case when large quantities of effluent are discharged into small streams or when delicate ecosystems are encountered. In these instances, additional treatment to polish the effluent from secondary systems will be required, or an alternative method of wastewater disposal must be found.

Additional treatment, usually referred to as *tertiary treatment*, often involves the removal of nitrogen and phosphorus compounds, plant nutrients associated with eutrophication. Further treatment may be required to remove additional suspended solids, dissolved inorganic salts, and refractory organics. Combinations of the above processes can be used to restore wastewater to potable quality, although at considerable expense. Referred to as *reclamation*, this complete treatment of wastewater can seldom be justified except in water-scarce areas where some form of reuse is mandated.

The term *advanced treatment* is frequently used to encompass any or all of the above treatment techniques, and this term would seem to imply that advanced treatment follows conventional secondary treatment. This is not always the case, as some unit operations or unit processes in secondary or even primary treatment may be replaced by advanced-treatment systems. Advanced-treatment processes and operations are described in the following section of this chapter. Because treatment systems are selected to meet discharge or reuse criteria with respect to specific parameters, the discussion is arranged according to treatment objectives.

5-20 NUTRIENT REMOVAL

The role of excess nutrients in entrophication was discussed in Chap. 3. Although the quantities of nutrients contributed by wastewater discharges may be less than those contributed by agricultural runoff and other sources, the point-source nature of wastewater discharges makes them more amenable to control techniques. Thus, wastewater-treatment plants that discharge to water bodies that are delicately balanced with respect to nutrient loads may have nutrient limitations imposed on their effluents. The nutrients most often of interest are nitrogen and phosphorous compounds. Processes for removing these nutrients from wastewater are discussed in the following paragraphs.

Nitrogen Removal

In domestic wastewater, nitrogen compounds result from the biological decomposition of proteins and from urea discharged in body waste. This nitrogen may be bound in complex organic molecules and is referred to simply as *organic nitrogen* rather than by specific compound. Organic nitrogen may be biologically converted to free ammonia (NH_3^0) or to the ammonium ion (NH_4^+) by one of several different metabolic pathways. These two species, together termed *ammonia nitrogen*, exist in equilibrium according to the following relationship:

$$NH_4^+ \rightleftharpoons NH_3^0 + H^+ \tag{5-37}$$

Ammonia nitrogen, the most reduced nitrogen compound found in wastewater, will be biologically oxidized to nitrate as follows if molecular oxygen is present.

$$NH_4^+ + \tfrac{3}{2}O_2 \longrightarrow NO_2^- + 2H^+ + H_2O \tag{5-38}$$

$$NO_2^- + \tfrac{1}{2}O \longrightarrow NO_3^- \tag{5-39}$$

These reactions result in the utilization of about 4.6 mg of O_2 per each mg of NH_4^+-N oxidized, with about 7.1 mg of alkalinity needed to neutralize the acid (H^+) produced.

In raw wastewater, the predominant forms of nitrogen are organic nitrogen and ammonia. Biological treatment may result in conversion to nitrate, provided the processess are aerobic and provided the treatment periods are long enough. Contact times in most secondary treatment systems, though sufficient to complete the conversion from organic nitrogen to ammonia, may not be sufficient for significant nitrification. Because of oxygen demand exerted by ammonia and because of other environmental factors, removal of ammonia may be required. The most common processes for removing ammonia from wastewater are (1) stripping with air and (2) biological nitrification-denitrification.

Air stripping Air stripping operations consist of converting ammonium to the gaseous phase and then dispersing the liquid in air, thus allowing transfer of the ammonia from the wastewater to the air according to the principles outlined in Sec. 3-4. The gaseous phase NH_3^0 and the aqueous phase NH_4^+ exist together in equilibrium as indicated in Eq. (5-37). The relative abundance of the phases depends upon both the pH and the temperature of the wastewater. As seen in Fig. 5-45, the pH must be in excess of 11 for complete conversion to NH_3^0. Since this is well above the normal pH for wastewater, pH adjustment is necessary prior to air stripping. For economic reasons, lime is the most common means of raising the pH. An unavoidable consequence of lime addition is the softening of the wastewater. Enough lime must be added to precipitate the alkalinity and to add the

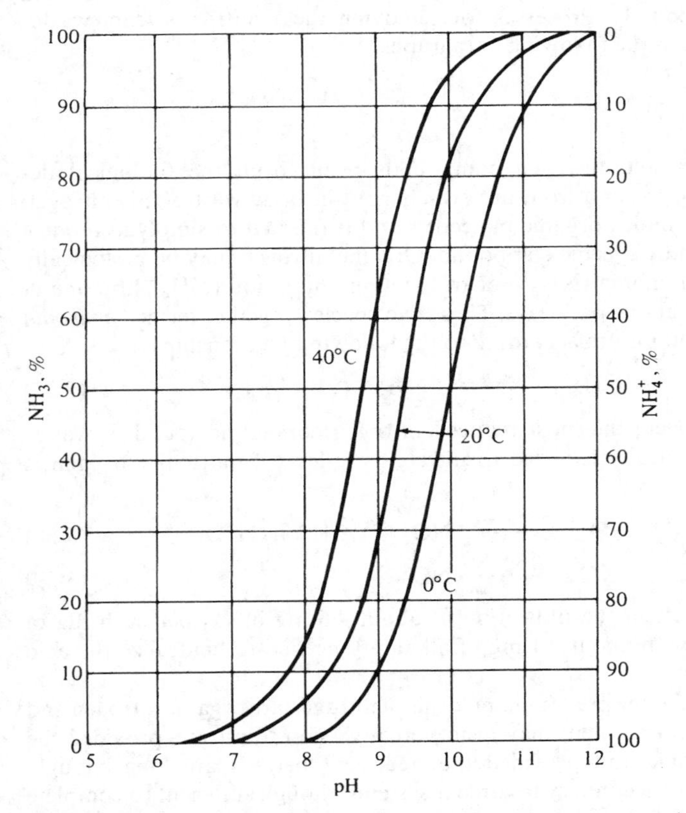

Figure 5-45 Effects of pH and temperature on distribution of ammonia and ammonium ion in water. (*From EPA* [5-43].)

excess OH^- ions for pH adjustment. An amount equivalent to the alkalinity plus 1.5 mequiv/L is usually sufficient to bring the pH to approximately 11.5.

Once the conversion to ammonia has been completed, stripping, or degasification, can proceed. The most efficient reactor has been found to be a countercurrent spray tower similar to the one shown in Fig. 5-46. Large quantities of air are required, and a fan must be included to draw air through the tower. Packing is usually provided to minimize film resistance to gas transfer by continuously forming, splashing, and reforming drops.

Design parameters for ammonia-stripping reactors include air-to-liquid ratios, tower depth, and loading rates. Common design practice is to use air-to-wastewater ratios ranging from about 2000 to 6000 m^3 of air per cubic meter of wastewater, with more air being required at lower temperatures. Tower depths are seldom less than 7.5 m, and hydraulic loading rates vary from about 40 to 46 L/min per square meter of tower. [5-14]

Air stripping is one of the most economical means of nitrogen removal, particularly if lime precipitation of phosphate is also required, because chemical conditioning can be concurrent. There are serious limitations to the process, however. As air temperature approaches freezing, a drastic reduction in efficiency is observed, and preheating of the air is not practical because of the large volume required. Furthermore, towers cannot operate in subfreezing weather because of icing. In cold climates, alternative methods of nitrogen removal must be provided during winter.

Other problems associated with ammonia stripping include noise and air pollution and scaling of the packing media. Noise and odor problems caused by the roar of the fans and the dispersion of ammonia gas can be minimized by locating the facility away from the populated area. Precipitation of calcium carbonate scale

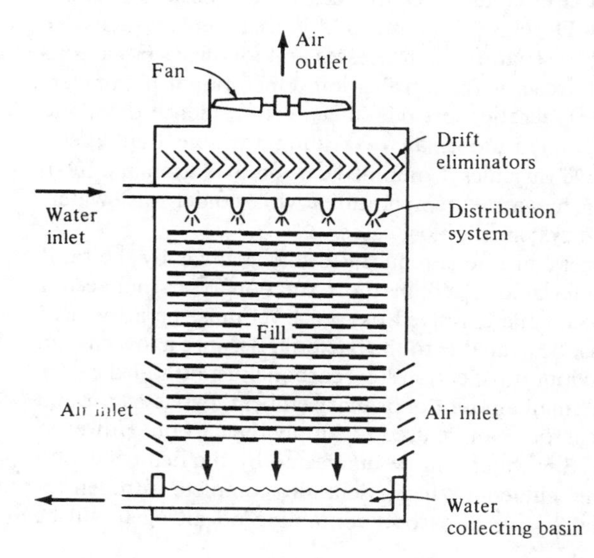

Figure 5-46 Diagram of countercurrent tower for ammonia stripping. (*From EPA* [*5-43*].)

on the packing media as a result of wastewater softening can be minimized by the use of smooth-surface polyvinyl chloride (PVC) pipe as packing material, though occasional cleaning of the packing media is still required.

Nitrification-denitrification Ammonia nitrogen can be converted to gaseous nitrogen, N_2, by biological processes. In this form, nitrogen is essentially inert and does not react with the wastewater itself or with other constituents of the wastewaters. Since N_2 is the principal constituent of air, treated wastewater is likely to be already saturated with molecular nitrogen and the additional N_2 is simply released to the atmosphere.

Biological conversion of ammonia to nitrogen gas is not a direct process but consists of two separate steps. The ammonia must first be oxidized to nitrate and then reduced to molecular nitrogen. These reactions require different environmental conditions and must be carried out in separate reactors.

The organisms responsible for nitrification are the autotrophic bacteria, nitrosomonas and nitrobacter. Equations (5-38) and (5-39) represent catabolic reactions that supply energy. Anabolic reactions use carbon dioxide and/or bicarbonate as a carbon source and may be represented by the following equation. [5-32]

$$4CO_2 + HCO_3^- + NH_4^+ + H_2O \longrightarrow C_5H_7O_2N + 5O_2 \tag{5-40}$$

Although some ammonia is converted to biomass by this reaction, the catabolic reactions are the principal ammonia conversion processes.

Nitrification can be accomplished in both suspended-culture and attached-culture reactors. Under favorable circumstances, nitrification can be accomplished along with carbonaceous BOD removal in secondary treatment systems. In other cases it is more efficient to separate the processes and follow carbonaceous BOD removal with a separate reactor for nitrification. Operational parameters of importance include pH, DO, aeration periods, mean cell-residence time, and carbon-to-nitrogen ratios. Temperature is an overriding variable that affects optimum ranges of all the above variables. Combined carbon oxidation and nitrification operations are possible at warmer temperatures, while colder wastewaters will require separate treatment systems in most cases.

In the denitrification process, nitrate is reduced to nitrogen gas by the same facultative, heterotrophic bacteria involved in the oxidation of carbonaceous material. For reduction to occur, the dissolved oxygen level must be at or near zero, and a carbon supply must be available to the bacteria. Because a low carbon content is required for the previous nitrification step, carbon must be added before denitrification can proceed. A small amount of primary effluent, bypassed around secondary and nitrification reactors, can be used to supply the carbon. However, the unnitrified compounds in this water will be unaffected by the denitrification process and will appear in the effluent. When essentially complete nitrogen removal is required, an external source of carbon containing no nitrogen will be required. The most commonly used external carbon source is methanol, CH_3OH. When methanol is added, the denitrification reaction is

$$NO_3^- + \tfrac{5}{6}CH_3OH \longrightarrow \tfrac{1}{2}N_2 + \tfrac{5}{6}CO_2 + \tfrac{7}{6}H_2O + OH^- \quad (5\text{-}41)$$

Theoretically, each milligram per liter of nitrate should require 1.9 mg/L of methanol. Under treatment plant conditions, however, about 3.0 mg/L of methanol is required for each milligram per liter of nitrate, making this process an expensive one. The interested reader is referred to Metcalf & Eddy, Inc. [5-36] and EPA literature [5-43] for design criteria.

Phosphorus Removal

Phosphorus is a ubiquitous constituent of municipal wastewater, averaging around 10 mg/L in most cases. The principal forms are organically bound phosphorus, polyphosphates, and orthophosphates. Organically bound phosphorus originates from body and food waste and, upon biological decomposition of these solids, is released as orthophosphates. Polyphosphates are used extensively in synthetic detergents and often contribute up to one-half the phosphorus in wastewater. Polyphosphates can be hydrolyzed to orthophosphates. Thus the principal form of phosphorus in wastewater is assumed to be orthophosphates, although the other two forms may coexist.

Orthophosphates consist of the negative radicals PO_4^{3-}, HPO_4^{2-}, and $H_2PO_4^-$ and may form chemical combinations with cations or positive radicals. In most cases the compounds are quite soluble, and phosphate removal in conventional primary treatment is negligible. Because phosphorus is a component of microbial cells, some phosphate may be removed in the biomass in secondary treatment processes. However, microorganisms need relatively little phosphorus as compared with carbon and nitrogen, and less than 3 mg/L of phosphorus is usually removed in conventional secondary treatment. When effluent requirement necessitates greater removal efficiencies, additional treatment must be provided.

The principal means of phosphorus removal is chemical precipitation. At slightly acidic pH, orthophosphates combine with trivalent aluminum or iron cations to form a precipitate.

$$Al^{3+} + (H_nPO_4)^{(3-n)-} \longrightarrow AlPO_4\downarrow + nH^+ \quad (5\text{-}42)$$

$$Fe^{3+} + (H_nPO_4)^{(3-n)-} \longrightarrow FePO_4\downarrow + nH^+ \quad (5\text{-}43)$$

Because domestic wastewater usually contains only trace amounts of iron and aluminum, the addition of these materials is necessary. Salts of these metals, such as those discussed in Sec. 4-6, can be added for this purpose.

At higher pH values, calcium forms an insoluble complex with phosphate. The addition of lime can provide both the calcium and the pH adjustment necessary.

$$5Ca(OH)_2 + 3(H_nPO)_4)^{(3-n)-} \longrightarrow Ca_5(OH)(PO_4)_3\downarrow + nH_2O + (9-n)OH^- \quad (5\text{-}44)$$

This reaction requires a pH of at least 9.0 for significant phosphorus removal. Higher pH values generally increase removal efficiencies. However, recarbonation may be necessary to lower the pH after the precipitation process has removed the phosphorus.

Chemical requirements for phosphate precipitation exceed the stoichiometric requirements indicated in Eqs. (5-42) through (5-44). Aluminum and iron salts react with alkalinity in the wastewater to produce metallic hydroxide flocs [$Al(OH)_3$ and $Fe(OH)_3$] and may increase the required dosages by up to a factor of 3. Fortunately, this increase is not totally wasted, as the metallic hydroxides assist in the flocculation and removal of the metallic-phosphate precipitate, along with other suspended and colloidal solids in the wastewater, and are thus useful in the treatment process. At high pH values calcium reacts completely with wastewater alkalinity to form calcium carbonate. Lime additions equivalent to the alkalinity plus that required for phosphate precipitation and pH adjustment are required.

Phosphorus removal can be incorporated into primary or secondary treatment or may be added as a tertiary process. Selection of the point of application depends on efficiency requirements, wastewater characteristics, and the type of secondary treatment employed. The advantages and disadvantages of each system are summarized in Table 5-17. Where effluent phosphorus concentrations of up to 1.0 mg/L are acceptable, the use of iron or aluminum salts in the secondary

Table 5-17 Comparison of point of application for phosphorus removal systems

Primary	Secondary	Tertiary
	Advantages	
Applicable to all plants	Lowest capital	Lowest phosphorus in effluent
Increased BOD and suspended solids removal	Lower chemical dosage than primary	Most efficient metal use
Lowest degree of metal leakage	Improved stability of activated sludge	Lime recovery possible
	Polymer not required	Separation of organic and inorganic sludge
	Disadvantages	
Least efficient utilization of metal	Careful pH control to get phosphorus < 1 mg/L	Highest capital cost
Polymer required for flocculation	Overdose of metal may cause low pH toxicity	Highest metal leakage
Sludge more difficult to dewater than primary sludge	Cannot use lime because of excessive pH	

Source: Adapted from Kugelman. [5-28]

system is often the process of choice, while high pH precipitation by lime in a tertiary unit is required to obtain very low levels of effluent phosphorus. Where nitrogen removal by ammonia stripping is also practiced, tertiary lime precipitation at a pH of 11.5 serves in both processes.

5-21 SOLIDS REMOVAL

Removal of suspended solids, and sometimes dissolved solids, may be necessary in advanced wastewater-treatment systems. The solids removal processes employed in advanced wastewater treatment are essentially the same as those used in the treatment of potable water, although application is made more difficult by the overall poorer quality of the wastewater.

Suspended Solids Removal

As an advanced treatment process, suspended-solids removal implies the removal of particles and flocs too small or too lightweight to be removed in gravity settling operations. These solids may be carried over from the secondary clarifier or from tertiary systems in which solids were precipitated.

Several methods are available for removing residual suspended solids from wastewater. Removal by centrifugation, air flotation, mechanical microscreening, and granular-media filtration have all been used successfully. In current practice, granular-media filtration is the most commonly used process. Basically, the same principles that apply to filtration of particles from potable water apply to the removal of residual solids in wastewater. These principles were discussed in Sec. 4-8, and that discussion will not be repeated here. Differences in operational modes for application of these principles to wastewater filtration vs. potable water filtration may range from slight to drastic, however, and the most commonly used wastewater filtration techniques are discussed below.

Sand filters have been used to polish effluents from septic tanks, Imhoff tanks, and other anaerobic treatment units for decades. Because they are alternately dosed and allowed to dry, the term *intermittent sand filters* has been applied to this type of unit. The process is essentially the slow sand filter described in Sec. 4-8. More recently, this type of filter has been applied to the effluent from oxidation ponds with considerable success. Effluent concentrations of less than 10 mg/L of BOD and suspended solids have been reported at filtering rates of 0.37 to 0.56 $m^3/m^2 \cdot d$. Filter runs in excess of 1 month are possible. [5-26]

Use of intermittent sand filters in tandem with conventional secondary treatment has not been very successful. [5-14] The nature of the solids from these processes results in rapid plugging at the sand surface, necessitating frequent cleaning and thus high maintenance costs. The use of intermittent filters for tertiary treatment is usually restricted to plants with small flows.

Granular-media filtration is usually the process of choice in larger secondary systems. Dual or multimedia beds prevent surface plugging problems and allow for longer filter runs. Loading rates depend on both the concentration and nature of solids in the wastewater. Filtering rates ranging from 12 to 30 m^3/m^2 day have been used with filter runs of up to 1 d. More detailed information on the design of high-rate filters for advanced wastewater systems can be found in Culp et al. [5-14] and Metcalf & Eddy. [5-36]

Other recent innovations in filtration practices hold promise for advanced wastewater treatment. *Moving bed filters* have been developed which are continuously cleaned, and the rate of cleaning can be adjusted to match the solids loading rate. Another modification called the *pulsed-bed filter*, uses compressed air to periodically break up the surface mat deposited on a thin bed of fine filter media. Only after a thick suspension of solids has accumulated on the bed, requiring frequent pulsing, is the filter backwashed.

Both the moving bed and the pulsed-bed filters have the capability of filtering raw wastewater. A much higher percentage of solids can be removed by filtration than can be removed in primary settling. The filter effluent, containing lower levels of mostly dissolved organics, responds very well to conventional secondary treatment. The filtered solids can be thickened and treated by anaerobic digestion, with a resultant increase in overall methane production, a possible source of energy for use within the plant.

Dissolved Solids Removal

Both secondary treatment (Secs. 5-9 to 5-12) and nutrient removal (Sec. 5-19) decrease the dissolved-organic-solids content of wastewater. Neither process, however, completely removes all dissolved organic constituents, and neither process removes significant amounts of inorganic dissolved solids. Further treatment will be required where substantial reductions in the total dissolved solids of wastewater must be made.

Ion exchange, microporous membrane filtration, adsorption, and chemical oxidation can be used to decrease the dissolved solids content of water. These processes, described in Chap. 4 (Sec. 4-10), were developed to prepare potable water from a poor-quality raw water. Their use can be adopted to advanced wastewater treatment if a high level of pretreatment is provided. The removal of suspended solids is necessary prior to any of the processes described in Sec. 4-10. Removal of the dissolved organic material (by activated carbon adsorption) is necessary prior to microporous membrane filtration to prevent the larger organic molecules from plugging the micropores.

Advanced wastewater treatment for dissolved solids removal is complicated and expensive. Treatment of municipal wastewater by these processes can be justified only when reuse of the wastewater is anticipated.

Wastewater Disposal and Reuse

An insignificant volume of the influent wastewater accompanies sludges and other materials disposed of during wastewater-treatment processes. The bulk of the wastewater remains to be disposed of after the treatment processes have been completed. Ultimate receptors of treated wastewaters include surface water and groundwater bodies, land surfaces, and, in some instances, the atmosphere. Recognition of the value of wastewater as a water resource has resulted in an increase in the reuse of treated effluents, particularly in water-scarce regions. Portions of the reused wastewater may appear as effluent for disposal after reuse.

Disposal sites or reuse facilities must be found within a reasonable distance of the wastewater-treatment plant because of the cost of transporting the effluent over long distances. Because of the possibility that wastewater may contain a few viable pathogens even after extensive treatment, both disposal and reuse must be accomplished with due caution.

5-22 WASTEWATER DISPOSAL

The most common method of wastewater disposal is by dilution in surface waters. The response of receiving streams to wastewater discharges was discussed in Chap. 3, and the effects are related to the dilution factor and to the quality of the effluent. In most cases, secondary treatment is sufficient to prevent problems. However, where adequate dilution is not available, or where discharge is to a delicate ecosystem, advanced wastewater treatment may be required. Advanced treatment prior to disposal in surface waters most often involves the removal of nutrients. In a few instances, advanced treatment may be necessary to remove colloidal solids.

In climates where evaporation from water surfaces exceeds precipitation, it may be possible to dispose of wastewater by discharge to the atmosphere in vapor form. Evaporation systems are essentially oxidation ponds, with surface areas being designed for total influent evaporation. Except for arid areas where the net evaporation is significant, large surface areas are required, thus limiting evaporation systems to small flows in rural settings.

For cities in coastal areas, ocean disposal offers an economically attractive form of disposal. The effluent is transported out to sea by pipelines along the ocean floor and discharged at multiple points through a manifold. The length of the pipeline will depend primarily on ocean currents and the quantity of waste involved. Although raw wastewater has been disposed of in this fashion without causing appreciable problems, it is desirable to eliminate floating debris, oils and greases, and recognizable objects from the wastewater prior to disposal. It is essential to remove large objects which could plug the pipeline or the manifold orifices.

Land application of wastewater may be considered a disposal technique, a form of wastewater reuse, or both. The most common forms of land application are irrigation and rapid infiltration. Wastewater may be used to supply both the water and nutrient needs of plants. Use for this purpose may be prompted by economics relating to either the agricultural aspects or to the wastewater disposal aspects. In either case, direct discharge to surface streams is avoided. Rapid infiltration results in the discharge of the wastewaters to groundwater bodies rather than to surface waters. In addition to wastewater disposal, objectives may include groundwater recharge as described in a later section of this chapter.

Irrigation

Wastewater may be applied to land surfaces to provide both water and nutrients to enhance plant growth. Although some of the effluent may be lost to evaporation or to percolation beyond the reach of plant roots, most of the water is incorporated into plant tissue or is transpired to the atmosphere. Wastewater effluents have been used successfully in both argiculture and silviculture and have been used to maintain vegetation in parks, on golf courses, and along freeways and airport run-

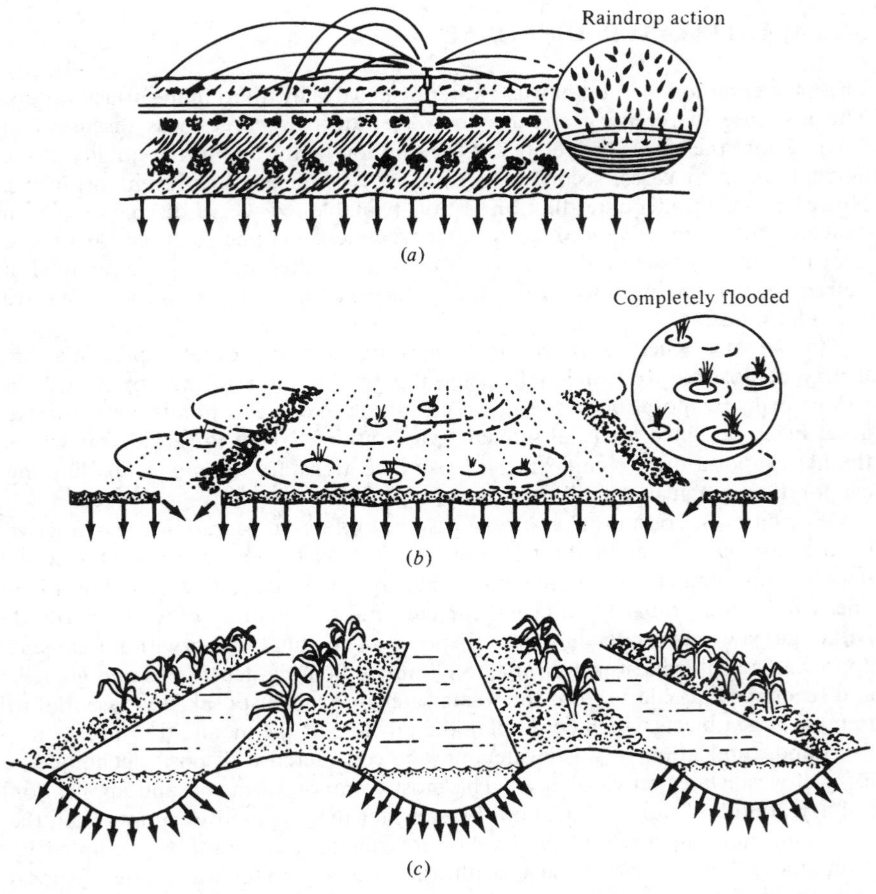

Figure 5-47 Irrigation techniques using municipal wastewater: (*a*) sprinkling; (*b*) flooding; (*c*) ridge-and-furrow techniques. (*From Pounds and Crites* [*5-41*].)

ways. Land application has become commonplace in semiarid areas where irrigation is necessary to sustain desirable vegetation.

Land application of wastewater can be by sprinkling, flooding, or ridge-and-furrow techniques, as shown in Fig. 5-47. Sprinkle irrigation is the most common method, with application rates varying from 2.5 to 10 cm per week, depending on climate, soil characteristics, and the water and nutrient requirement of the plants.

The degree of pretreatment prior to land application varies with the nature of the crop to be grown. Crops grown for animal consumption or for seed production can generally accept lower-quality effluent than crops grown for human consumption. In most cases, secondary treatment is required. Wastewater should not be used to irrigate vegetables that are eaten raw.

Wastewater irrigation systems may be owned by the mincipality, or contractual arrangements can be made with local farmers for effluent utilization. In either case, it is often possible to recover part of the cost of pretreatment of the wastewater from the cash crop or from sale of the effluent. When wastewater is reused for landscape irrigation of public property, savings of potable water supplies may be a significant advantage.

There are, however, several disadvantages to the use of wastewater effluent for irrigation purposes. The seasonal nature of irrigation water needs may result in large storage requirements. If the system is to be operated by the municipality, land and equipment must be purchased at considerable expense. Where large, high-pressure sprinklers are used, aerosols can be formed which may transport viral pathogens. Large-scale irrigation systems must be located away from heavily populated areas, and the cost of conveyance systems to the site is often significant. Distribution systems for irrigation of parks, greenbelts, and other publicly owned areas can be expensive if such areas are widely dispersed and if the transport system must be constructed through developed areas.

Rapid Infiltration

The rapid infiltration process involves spreading wastewater in shallow, unlined earthen basins and allowing the liquid to pass through the porous bottom and percolate toward the groundwater, as shown in Fig. 5-48. Wastewater is applied at the maximum rate at which the soil can carry it away. Intermittent "resting" periods must be provided in which the soil is allowed to dry and reestablish aerobic conditions. Application cycles of 10 to 20 d with 1- to 2-week resting periods are common. The bottom surface may be raked or disked prior to each application cycle to disperse solids and prevent an impermeable layer from forming.

Many of the rapid infiltration systems in current use were designed primarily to dispose of unwanted wastewater. More recently the process has been used as a means of aquifer recharge or as an advanced wastewater treatment, with the percolate being collected for reuse. Collection may be by horizontal flow to surface streams, or by wells or drain tiles installed for this purpose. The soil acts essentially as a filter for tertiary treatment.

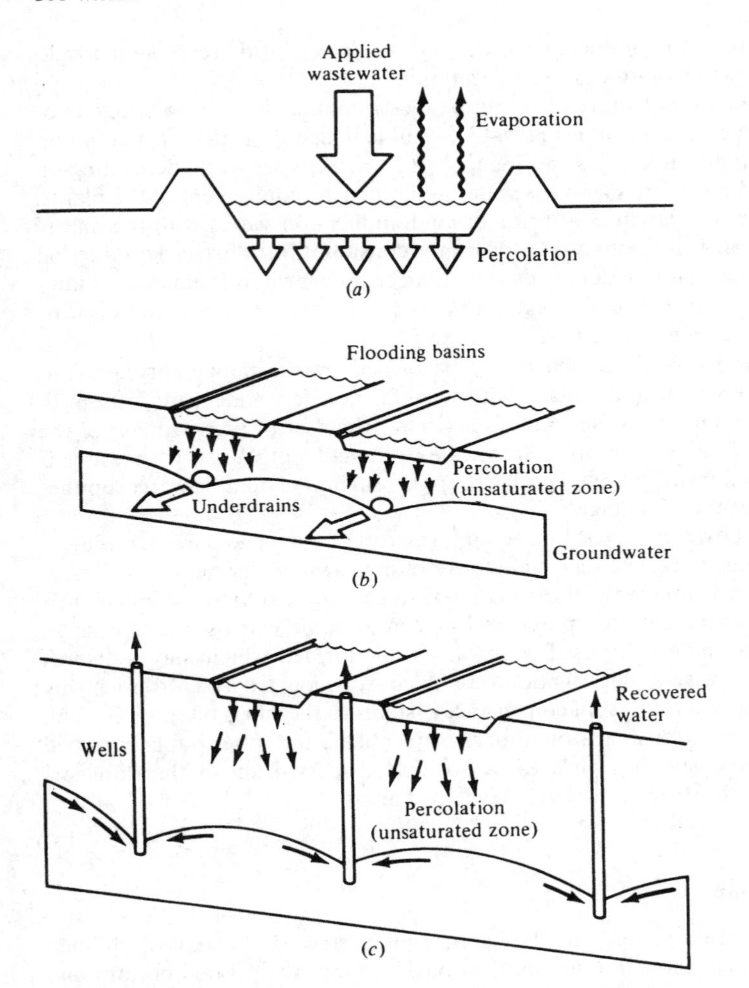

Figure 5-48 Rapid infiltration of wastewater. (*a*) Percolation to groundwater; (*b*) recovery by underdrain tiles; (*c*) recovery by wells. (*From U.S. EPA* [5-42].)

5-23 WASTEWATER REUSE

Reuse of treated wastewater may be dictated by any of several circumstances. In water-scarce areas, wastewater may constitute a major portion of the available resource. Where delicate ecosystems necessitate stringent effluent requirements, reuse of the wastewater may help to offset the cost of advanced wastewater treatment, or a reuse that will accept a lower level of treatment may obviate the need for the expense of tertiary treatment prior to discharge. Wastewater has been reused for several purposes, which include creation or enhancement of recreational facilities, industrial water supplies, groundwater recharge, and direct reuse in potable supplies.

Recreational Facilities

Water-quality requirements for recreational uses are quite stringent, and some form of advanced wastewater-treatment techniques will almost invariably be required prior to wastewater reuse for this purpose. Indeed, where body-contact activities such as swimming and water skiing are included, the quality of the water resource must approach that of drinking water with respect to most parameters. Recreational water should be aesthetically pleasing and essentially free of toxicants and pathogenic organisms. Recreational waters composed chiefly or entirely of wastewater effluents are possible, provided a sufficient degree of treatment is provided.

Two examples of wastewater reuse in recreational facilities often cited in the literature are the Santee project and the Indian Creek Reservoir, both in California. Both facilities provide a high quality of recreational water, but by different treatment processes.

At the Santee facility secondary effluent is first polished in a tertiary oxidation pond and then pumped into a canyon and allowed to flow horizontally through approximately 1000 m of sand and gravel material before being recovered. The recovered water is then routed through a series of three lakes surrounded by a public park. Fishing and boating are allowed on the first two lakes. The third lake is chlorinated and used as a swimming facility. The reclaimed water is of sufficient quality to meet California standards for body-contact recreation. [5-14]

Indian Creek Reservoir receives treatment effluent from the South Tahoe Public Utilities District advanced wastewater-treatment plant. The first full-scale advanced wastewater-treatment plant to be built in the United States, the Tahoe facility includes nutrient removal, filtration, and activated carbon adsorption. The reservoir contains about 27×10^6 m^3 of water, essentially all treated effluent, and provides a variety of water-based activities, including swimming and water skiing. The impoundment also supports excellent trout fishing. [5-13]

Surplus waters from both the Santee and Tahoe facilities are used for irrigation purposes. Inspired by the success of these two projects, other municipalities are planning recreational use as one step in the reuse of wastewater. The city of Denver has an ambitious plan for wastewater recycling, a portion of which includes recreational facilities. [5-24] The Fairfax County Water Authority has included an intermediate reservoir between its advanced wastewater-treatment plant and the Occoquan Reservoir, which forms a part of the Washington, D.C., water supply. Recreational activities are included as a beneficial use. [5-13]

Advanced treatment of wastewater solely for the purpose of creating a recreational resource could seldom be justified on an economic basis. However, when advanced wastewater treatment is required for other reasons, intermediate use of reclaimed water for recreation can prove to be a viable scheme and may improve public acceptance of wastewater reuse in general.

Industrial Water Supply

In terms of total volume, industrial water use outranks all other water-use categories in the United States. Additionally, industrial water requirements are growing more rapidly than are municipal or agricultural requirements. An increase in the use of wastewater effluents for industrial water supplies parallels this growth.

The quality of water required for various industrial processes varies greatly. Cooling water generally has the lowest quality constraints, while boiler water has the highest. The degree of treatment given wastewaters will obviously be dictated by the intended industrial use. Cooling processes, which constitute the largest water requirement in most industries, may be able to use secondary effluent directly, although additional solids removal is desirable and additional treatment with biocides may be necessary to prevent biofouling of surfaces. Advanced wastewater treatment may be provided by the wastewater authority prior to delivery to the industry, or industry may receive secondary effluent and provide treatment processes designed to meet their particular needs.

A wide variety of industries make use of municipal effluents, the most common being the power-generating industry and petrochemical plants. In Concord, California, an industrial complex consisting of Phillips Petroleum, Shell Oil, Stauffer Chemical, Monsanto Chemical, and Pacific Gas and Electric receives effluent from the Central Contra Costa Sanitary District. The advanced wastewater-treatment plant provides about 64,000 m^3/d of high-quality effluent to the industries. In Odessa, Texas, a petrochemical industry receives secondary effluent from the city's wastewater plant and provides additional treatment as necessary. After use in the industry, the wastewater is reused for secondary recovery operations in the oil fields. [5-3]

Groundwater Recharge

Wastewater can become a part of groundwater as an inadvertent consequence of land application for irrigation or from rapid infiltration systems designed for wastewater disposal. As discussed in this section, however, groundwater recharge will be considered a planned activity with well-defined objectives. These objectives may include stabilizing the groundwater table, creating hydrostatic barriers to prevent saltwater intrusion into freshwater aquifers, and storing water for future use.

In areas where groundwater is used extensively for agricultural, industrial, and municipal purposes, water may be withdrawn from an aquifer more rapidly than it can be replenished by natural means. In addition to the depletion of the resource, the drop in the water table may result in subsidence of the area as the pores in the drained part of the aquifer collapse. Should this occur, the storage and hydraulic conductivity of the acquifer may be altered. This process can be slowed, stopped, or even reversed by recharge with reclaimed wastewater. [5-4]

In coastal areas, salt water from the ocean may wedge underneath the freshwater aquifer because of its greater density. Drawdown from wells exacerbates the problem and can result in saltwater contamination at the well. Injection of wastewater between the pumping well and the source of the salt water may serve to create a hydrostatic barrier that will push the salt water backward. This process is shown in Fig. 5-49.

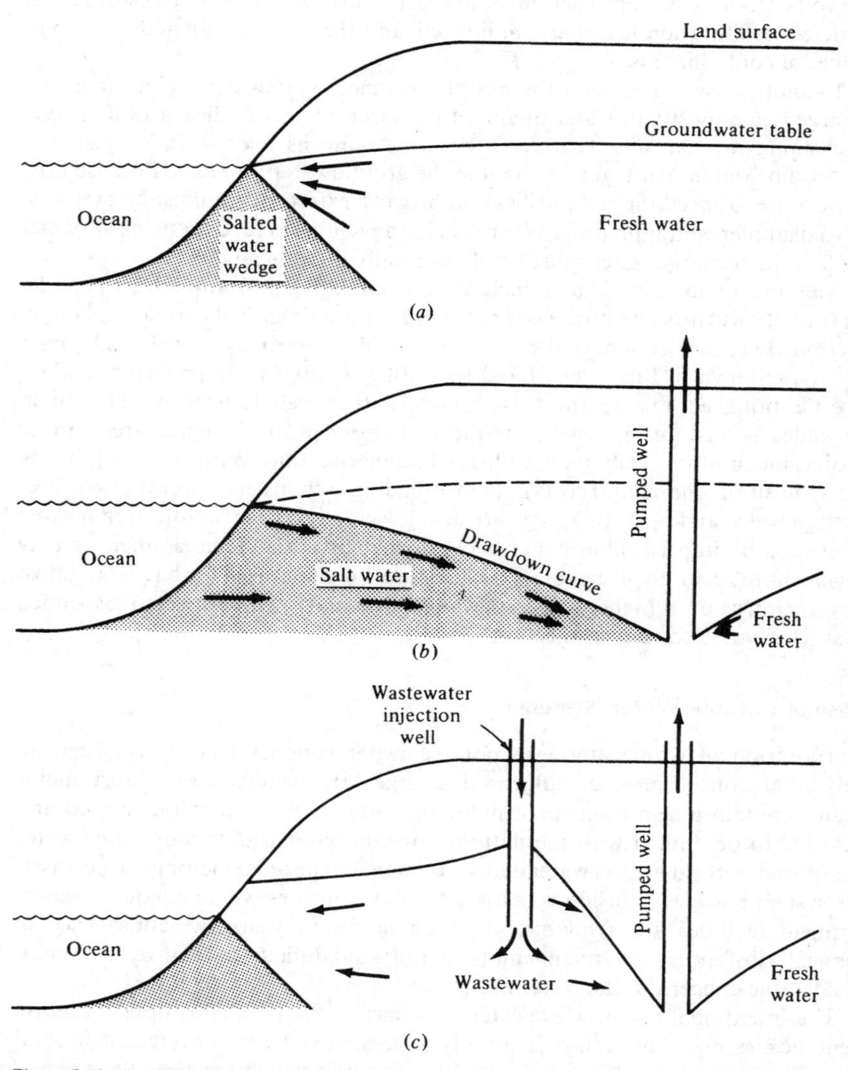

Figure 5-49 Use of treated wastewater to block saltwater intrusion. (*a*) Natural conditions; (*b*) saltwater intrusion from overpumping; (*c*) wastewater barrier.

Storage of wastewater in the aquifer is incidental to both of the above processes, but groundwater recharge systems may also be designed with water storage as their primary function. This storage may function much the same as storage in surface reservoirs, the water table falling during periods of high pumping and rising during periods of low withdrawal. When aquifer characteristics are favorable to storage, this method has several advantages over surface storage reservoirs. Extensive construction is avoided, surface use is not disturbed or restricted, evaporation losses are minimized, and the water is isolated from most sources of contaminants.

Liabilities associated with the use of reclaimed wastewater for groundwater recharge relate mostly to water quality. Like surface water bodies, acquifers have self-cleaning mechanisms. However, these mechanisms may work very slowly, and certain contaminants may remain in the groundwater for years. Because little can be done to speed the self-purification process, extreme care must be exercised to avoid aquifer contamination. Where part of an aquifer is used for drinking-water supplies, the recharge water must be of essentially potable quality.

Methods of aquifer recharge include land spreading and subsequent percolation (essentially the same process as rapid infiltration described earlier) and direct injection. Direct injection is the reverse of widthdrawal by a well and pump system, as shown in Fig. 5-49*c*. Land spreading is usually the preferred method since additional aerobic treatment is provided in the aerated soil above the aquifer. Suspended solids are removed near the surface where the plugged area can be restored much more easily than a plugged aquifer section. With the exception of land acquisition, the capital costs of landspreading systems are lower than those of injection wells, and operating costs are also lower. [5-45] Direct injection may be necessiated by impermeable strata between the surface and the aquifer, or may provide more accurate placement if the reclaimed water is used for barriers against saltwater intrusion. A higher quality of water, particularly with respect to suspended solids, is required for direct injection.

Reuse in Potable Water Systems

Incorporation of wastewater into potable water supplies has always been an inadvertent consequence of effluent discharge into watercourses. Most major streams contain a significant percentage of water that was previously used and discarded to be diluted with the natural flow and later withdrawn as raw water for a second or third use. As water demands increase, the reuse factor also increases. This system has been considered satisfactory in countries where adequate water-treatment facilities are available, although in recent years the appearance of chemical substances that are difficult to identify and difficult to remove has caused considerable concern to the water industry.

The intentional use of wastewater as a part of the potable supply is a more recent occurrence. This reuse is usually necessitated by a shortage of natural water. Reuse may be direct or indirect. *Direct reuse* is usually referred to as *closed loop* or *pipe-to-pipe recycling*, which indicates that the treated effluent from the wastewater-treatment system is piped directly to the influent of the water-treatment plant. *Indirect reuse* involves storage of treated effluent in natural or artificial water bodies for a period of time prior to withdrawal and incorporation into the water supply. Indirect reuse is the more acceptable practice at the present time.

Direct reuse of wastewater has been practiced at Windhoek, Southwest Africa, since 1969. After secondary treatment, wastewater is stored in maturation (holding) ponds and then treated as shown in Fig. 5-50. This system is operated at high-use periods of the year and during drought conditions and has constituted as much as 50 percent of the potable supply. [5-54]

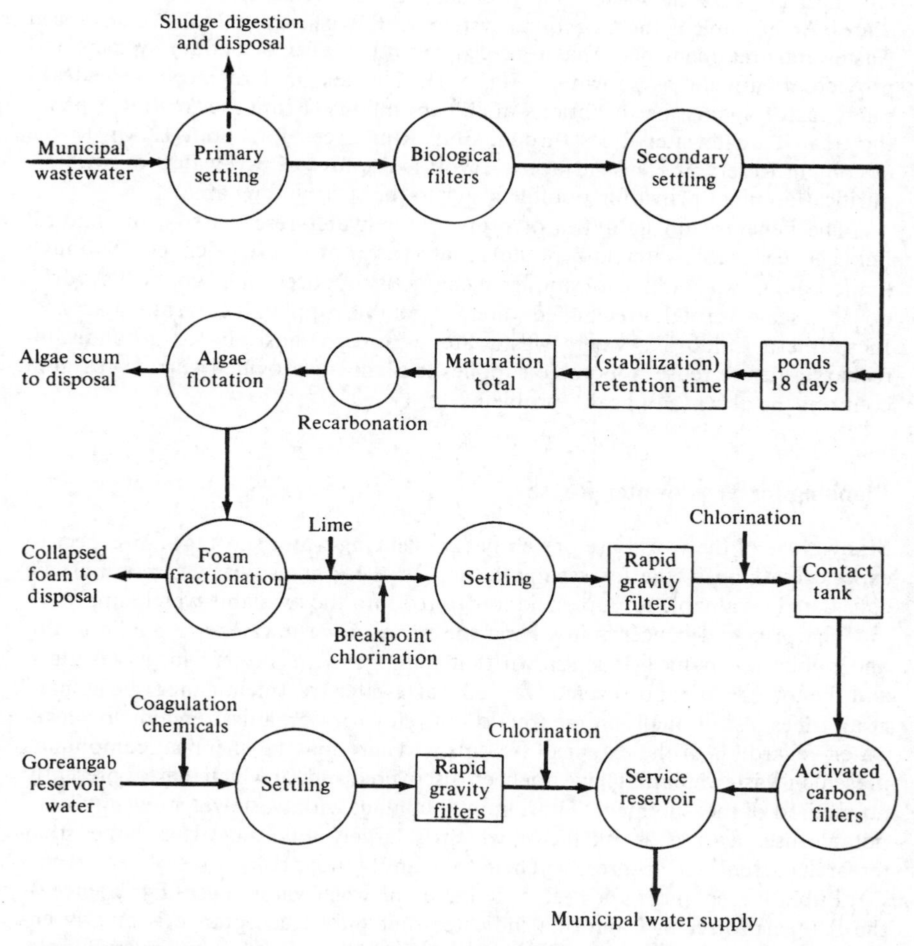

Figure 5-50 Schematic of wastewater-treatment plant incorporating direct reuse. (*From Clayton and Pybus* [*5-11*].)

Indirect reuse separates the wastewater-treatment plant from the water-purification plant by a carefully controlled natural link. The most common approach is by storage in surface reservoirs or in aquifers for varying periods of time. In surface reservoirs, the wastewater is subjected to sunlight, aeration, biological action, and other processes that reduce the chance of transmission of pathogens. [5-13] Dilution by runoff water may or may not be desirable, depending on its quality. Water stored in aquifers is subjected to filtration through the soil material, biological action, and adsorption and ion-exchange processes. Wastewater stored in aquifers is less likely to become recontaminated than is surface water.

Indirect reuse of wastewater is practiced at several places in the United States. An example is the Occoquan system near Washington, D.C. An advanced wastewater-treatment plant has replaced several smaller secondary systems and provides treatment as shown in Fig. 5-51. The terminal reservoir provides a safety factor against perturbations in effluent quality. From the treatment plant, the treated wastewater flows through Bull Run Creek for about 12 km to the Occoquan Reservoir. This surface reservoir is a source of raw water for a water-purification plant providing potable water to the surrounding area.

The benefits and liabilities of using groundwater reservoirs as the natural link between wastewater and potable water are not so well defined. Although limited use of wastewater for aquifer recharge is practiced throughout the world, no large-scale use of this reclaimed water for potable supplies is currently practiced. In southern California, where wastewater is used extensively for groundwater recharge, future reuse in potable supplies is planned, provided current research confirms the absence of health problems. [5-4]

Planning for Wastewater Reuse

Many areas of the world are presently experiencing water shortages or expect to experience them in the foreseeable future. In these areas, wastewaters must be considered a valuable resource and integrated into the available water supply.

The principal concerns involving the reuse of wastewater are public health and public acceptance. It is known that pathogens are present in wastewaters, and the total removal by even advanced wastewater treatment cannot be assured at all times. Additionally, some fraction of refractory organics remains in wastewater, regardless of the extent of treatment. There may be chemical compounds present in wastewater that have not been discovered and for which there is presently no method of measurement. Thus, human contact with wastewater, even in non-potable uses, carries a risk factor which is largely unknown. It is hoped that research currently in progress will help to quantify those risks.

Public acceptance is a necessary factor in wastewater reuse. Experience at the Santee project in California indicates that public acceptance is greatly enhanced by informing and involving the public at all stages of planning and implementation of wastewater reuse. Following this lead, the City of Denver has

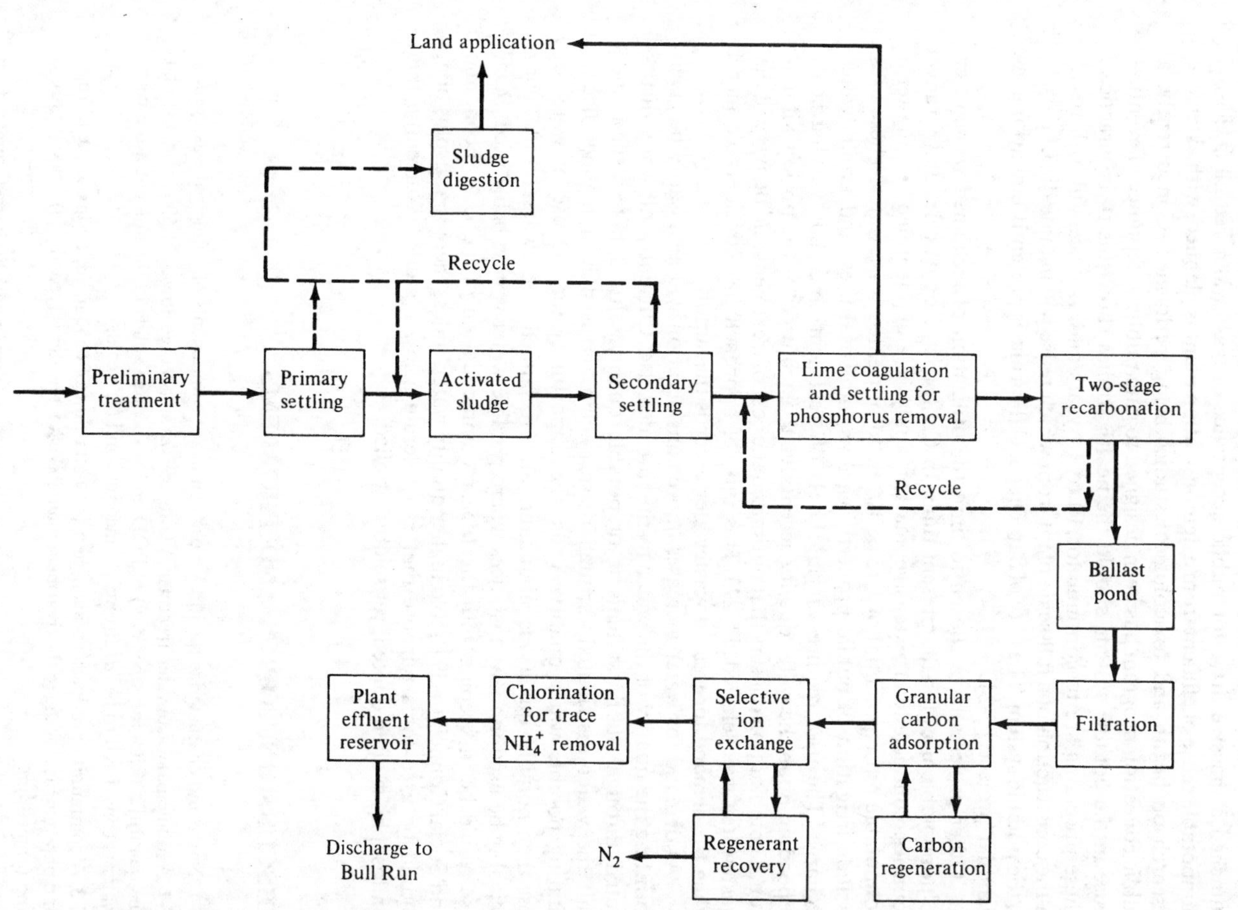

Figure 5-51 Schematic of Occoquan wastewater treatment incorporating indirect reuse. (*From U.S. EPA* [5-43].)

launched a massive drive for public acceptance of wastewater recycling. [5-24] Nonpotable reuse of effluent from a demonstration plant is planned, with extensive research on health and toxicological studies being performed. Concurrently, a public education program has been designed to gain public acceptance of eventual reuse in the potable system, should the health studies show this to be practical. These programs are to continue for 10 to 15 years and, if successful, will result in the construction of a full-scale plant from which reuse will include direct recycle to the potable system. The City of San Diego, California, is presently embarking on a similar project. [5-2]

Other projects have chosen to limit the utilization of reclaimed wastewater to nonpotable uses for the present time. In California, Los Angeles and Orange counties conducted an extensive market analysis and identified a long-term nonpotable reuse potential of 1.15×10^6 m^3/d of treated wastewater. These uses include irrigation of public property, various industrial uses, and groundwater recharge. The uses require varying levels of treatment, and thus varying costs depending upon the water quality acceptable to each user. At the project's initiation, a combination of users that optimized the cost of treatment and delivery of the wastewater was selected. [5-27] This type of approach has considerable merit when the demand for treated wastewater exceeds the supply.

As demand for water increases, more consideration will necessarily be given to fitting the quality of the water to the intended use. Currently, all water distributed through public systems is of potable quality, although less than one-half the water distributed through these systems is used in a manner necessitating potable water quality. Thus, the opportunity for the use of water of less-than-potable quality is abundant, and reclaimed wastewater could conceivably be used in many instances where potable water is now being used. Such use would be in keeping with the 1958 recommendation of the U.N. Economic and Social Council [5-39]: "No higher quality water, unless there is a surplus of it, should be used for a purpose that can tolerate a lower grade." Currently, this philosophy is often quoted, yet seldom applied.

DISCUSSION TOPICS AND PROBLEMS

5-1 Name and characterize the three most significant components of municipal wastewater.

5-2 A municipal wastewater treatment plant receives an average flow of 11,500 m^3/d. Estimate the quantities (kilograms per day) of BOD and suspended solids in the influent if the wastewater is considered to be (*a*) strong, (*b*) medium, and (*c*) weak.

5-3 A community produces an average wastewater flow of 6550 m^3/d. Estimate the nitrogen and phosphorus loading to the treatment plant if the wastewater is typically (*a*) weak, (*b*) strong, and (*c*) medium.

5-4 A municipal wastewater-treatment plant receives a seasonal discharge from a fruit-processing plant. Influent flows and strengths of the wastewater when the industry is both on- and off-line are shown below. Determine the contribution of each constituent by the industry.

	Industry on-line	Industry off-line
Flow, m^3/d	18,750	13,275
BOD_5, mg/L	300	215
SS, mg/L	420	240
Ammonia, mg/L	64	15
Chloride, mg/L	29	41
Alkalinity, mg/L	57	125

5-5 What is an effluent-limited stream? A water-quality-limited stream?

5-6 Differentiate between unit operations and unit processes.

5-7 Define and describe the components of (*a*) primary treatment, (*b*) secondary treatment, and (*c*) tertiary treatment.

5-8 What are the common engineered methods of removing solids from wastewater? Describe and define each of these methods.

5-9 What are the major types and sources of grit in municipal wastewaters? Describe treatment methods used to remove grit.

5-10 A channel-type grit chamber has a flow-through velocity of 0.29 m/s, a depth of 0.8 m, and a length of 10 m. For inorganic particles with specific gravity of 2.5, determine the largest-diameter particle that can be removed with 100 percent efficiency.

5-11 A channel-type grit chamber is to be installed in a wastewater-treatment plant processing 8550 m^3/d. The flow-through velocity is to be controlled at 0.33 m/s by a downstream proportioning weir. Determine the channel dimensions for a depth to width ratio of 1 : 1.5.

5-12 Determine the appropriate dimensions for an aerated grit chamber processing 23,500 m^3/d of municipal wastewater. Also calculate the total air flow.

5-13 What are the most common devices used for measuring flows in a wastewater-treatment plant?

5-14 Describe unit operations used in primary sedimentation.

5-15 A municipal wastewater-treatment plant processes an average flow of 14,000 m^3/d. The peak flow is 1.75 times the average. The wastewater contains 190 mg/L BOD_5 and 210 mg/L suspended solids at average flow and 225 mg/L BOD_5 and 365 mg/L suspended solids at peak flow. Determine the following for a primary clarifier with a 20-m diameter.

(*a*) Surface overflow rate and the approximate removal efficiency for BOD_5 and suspended solids at average flow

(*b*) Surface overflow rate and the approximate removal efficiency for BOD_5 and suspended solids at peak flow

(*c*) Mass of solids (kilograms per day) that is removed as sludge for average and peak flow conditions.

5-16 A wastewater-treatment plant must process an average flow of 24,500 m^3/d, with peak flows of up to 40,000 m^3/d. Design criteria for surface overflow rates have been set by the state regulatory agency at a maximum of 40 m/d for average conditions and 100 m/d for maximum conditions. Determine the dimensions of the primary clarifier if it is a

(*a*) Circular basin

(*b*) Long-rectangular basin ($L = 3\ W$)

(*c*) Square cross-flow tank

If the influent suspended solids is 200 mg/L at average flow and 230 mg/L at peak flow, determine the mass of solids (kilograms per day) removed by the primary clarifier.

5-17 A large wastewater-treatment plant processes 200,000 m^3/d of municipal wastewater. The design overflow rate is 50 m/d, and four units in parallel are to be constructed. Concrete tanks are to be used and the cost of pouring circular sidewalls is 1.2 times the cost of pouring straight sidewalls. Determine the percent savings in construction costs in each instance if (*a*) long-rectangular tanks ($L = 4W$) or (*b*) square cross-flow tanks are used (with common walls) instead of circular tanks.

5-18 Determine the weir-loading rates in Prob. 5-16 if a simple weir is used at the periphery of the circular tank, at the end of the long-rectangular tank, and along one side of the square tank.

5-19 Define: (*a*) biomass, (*b*) lag phase, (*c*) log-growth phase, (*d*) stationary phase, (*e*) endogenous phase, (*f*) suspended cultures, (*g*) attached cultures, and (*h*) flocs.

5-20 Name, define, and describe the most common method of quantifying biomass.

5-21 What external factors may affect the rate of biomass production and food utilization?

5-22 Explain the basic concept of the activated-sludge process and indicate the advantages and disadvantages of the two major kinds of activated-sludge reactors.

5-23 A tapered aeration system similar to that shown in Fig. 5.17*b* is used to treat 12,500 m^3/d of municipal wastewater. The wastewater has received primary treatment and has a BOD_5 of 140 mg/L and a suspended solids of 125 mg/L. The system is to be operated in the following way.

Soluble BOD_5 in effluent $\leq$ 5 mg/L
Average solids concentration in the reactor = 2000 mg/L
Mean cell-retention time = 10 d

The biological constants have been determined by pilot-plant analysis and are:

$$Y = 0.55 \frac{\text{kg biomass}}{\text{kg BOD utilized}}$$

$$k_o = 0.05\ \text{d}^{-1}$$

(*a*) Determine the length of the reactor if it is 5 m wide and 5 m deep.

(*b*) Assume an effluent suspended-solids concentration of 30 mg/L; the BOD_5 of the solids is 0.65 mg BOD/1.0 mg *SS*. Determine the total BOD in the effluent.

5-24 Determine the average biomass concentration in a conventional activated-sludge reactor similar to that shown in Fig. 5.16*b* under the following conditions.

Flow = 18,300 m^3/d
Influent BOD = 160 mg/L
Effluent BOD = 5 mg/L

$$\text{Cell yield coefficient } Y = \frac{\text{kg biomass}}{\text{kg BOD utilized}}$$

Endogeneous decay coefficient = 0.04 d^{-1}
Tank volume = 6100 m^3
Mean cell-residence time = 9 d

5-25 Determine the volumetric loading rate of the system described in Prob. 5-23.

5-26 Determine the food-mass ratio of the system described in Prob. 5-24.

5-27 A wastewater flow having the characteristics of that in Prob. 5-24 is to be treated in a completely mixed activated-sludge system similar to that of Fig. 5-16*a*. The reactor is to operate at a concentration of 3000 mg/L MLSS, and the secondary clarifier is designed to thicken the sludge to 12,000 mg/L. For a mean cell-residence time of 8 d, determine

(*a*) The volume of the reactor
(*b*) The mass of the solids and the wet volume of sludge wasted each day
(*c*) The sludge recycle ratio

5-28 A completely mixed activated-sludge plant is to treat 10,000 m^3/d of industrial wastewater. The wastewater has a BOD_5 of 1200 mg/L that must be reduced to 200 mg/L prior to discharge to a municipal sewer. Pilot-plant analysis indicates that a mean cell-residence time of 5d maintaining MLSS concentration of 5000 mg/L produces the desired results. The value for Y is determined to be 0.7 kg/kg and the value of k_d is found to be 0.03 d^{-1}. Determine

(*a*) The volume of the reactor
(*b*) The mass and volume of solids wasted each day
(*c*) The sludge recirculation ratio

5-29 The activated-sludge system shown in the sketch below is operating at equilibrium. Determine the volume of sludge that must be wasted each day if wastage is accomplished from (*a*) Point A and (*b*) Point B.

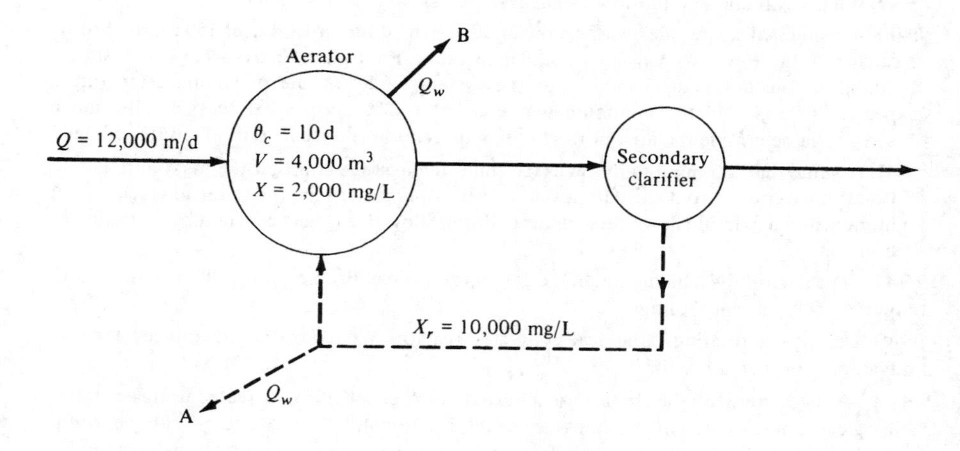

5-30 Why are aeration devices a vital part of biological reactors? Name and describe the two major aeration techniques, indicating the kinds of biological reactors in which they are most often used.

5-31 What other suspended-culture biological systems are available for treating wastewater besides the activated-sludge processes?

5-32 An oxidation-ditch activated-sludge system receives 7500 m^3/d of municipal wastewater. The BOD in the wastewater is 210 mg/L and no primary treatment is provided. The oxidation ditch is 3 m deep, 7 m wide, and 400 m long. The reactor is operated at 1800 mg/L MLSS and the biological constants are Y = 0.5 kg/kg and k_d = 0.06 d^{-1}. Determine the mean cell-residence time for 90 percent BOD removal.

5-33 A wastewater flow of 5000 m^3/d is treated in a facultative oxidation pond that is 2.0 m deep with a surface area of 20 ha. The wastewater has a soluble BOD_5 of 150 mg/L and a reaction rate coefficient of 0.30 d^{-1}. Determine the soluble BOD of the effluent. (Assume a completely mixed reactor without solids recycle).

5-34 Rework Prob. 5-33 with the 20-ha surface area being equally divided between three ponds.

5-35 A wastewater flow of 3550 m^3/d is to be treated in a facultative pond system. The reaction rate coefficient at the average operating temperature is 0.35 d^{-1}. The pond is expected to operate at a dispersion factor of 0.5. Determine the surface area required for 85 percent removal of soluble BOD for a pond depth of 2 m with (*a*) a single-cell pond and (*b*) a four-cell system.

5-36 Wastewater from a poultry-processing plant averages 1000 m^3/d with a soluble BOD_5 of 1000 mg/L. This waste is to be treated in a three-cell facultative pond system in which the 20°C reaction rate constant k has been found to be 0.5 d^{-1}. The coldest monthly average temperature is expected to be 10°C. For a dispersion coefficient of 0.25, determine the surface area required to meet an effluent standard of 50 mg/L soluble BOD.

5-37 Rework Prob. 5-35*b* with surface aerators being placed in the primary pond. Determine the power requirement for the aerators if the oxygen transfer rate is 0.9 kg O_2/kW · h.

5-38 Rework Prob. 5-36 with surface aerators being placed in the primary pond. Determine the power requirement for the aerators if the oxygen transfer rate is 1.05 kg O_2/kW · h.

5-39 What advantages do bio-towers have over classical trickling filters?

5-40 A municipal wastewater with a flow of 17,550 m^3/d and a BOD_5 of 150 mg/L is to be treated in a bio-tower with plastic modular medium. Pilot-plant analysis has established a treatability constant of 0.05 min^{-1} for the system at 20°C. The maximum temperature expected is 23°C, and the minimum temperature is 13°C. For a 2:1 recycle ratio and a 7.0-m depth, determine the area of the tower required to produce a 20 mg/L BOD_5 effluent.

5-41 Assume that the minimum flow is 0.6 times the average and that the maximum flow is 2 times the average. From the data given in Prob. 5-40, determine the removal efficiency for minimum and maximum flow rate with the hydraulic flow rate Q held constant by adjusting the recycle ratio Q_r.

5-42 Repeat Prob. 5-41, but maintain the 2:1 recycle ratio of Prob. 5-40 and allow the hydraulic flow rate Q to vary accordingly.

5-43 Describe a rotating biological contactor reactor. What are the advantages and disadvantages of such a reactor?

5-44 A wastewater with the characteristics given in Prob. 5-40 is to be treated using a rotating biological contactor system. Assume that the information in Figs. 5-29 and 5-30 applies to the selected medium. The medium is manufactured in 8-m shaft lengths, with each shaft containing 1.2×10^4 m^2 of surface area. Determine the number of modules for complete nitrification of the wastewater.

5-45 A wastewater with the characteristics of that given in Prob. 5-23 is to be treated by an RBC system. Assume a minimum temperature of 10°C and the RBC characteristics of Prob. 5.44. Determine the number of required units for 90% BOD removal.

5-46 What are the two objectives that should be met by secondary carifiers for activated sludge?

5-47 Using the information in Table 5-10, determine the size of a secondary clarifier to follow a conventional activated-sludge reactor. The influent flow is 13,250 m^3/d and the recycle ratio Q_r/Q is 0.5. The solids concentration in the effluent from the reactor is 2500 mg/L.

5-48 A secondary clarifier is to be designed to remove the biomass from a completely mixed activated-sludge reactor. Average flow conditions (influent plus recycle) is 11,200 m^3/d, the minimum flow is one-half this amount, and the peak flow is 2.1 times the average. Using Table 5-10, design a secondary clarifier to meet these conditions.

5-49 A settling analysis is run on the contents of an activated-sludge reactor with the following results:

Concentration, mg/L	1200	2200	3800	6100	8200	11,000
Velocity, m/h	5.8	3.2	1.6	0.6	0.4	0.09

Given the following information, determine the concentration of solids and the flow rate of the secondary clarifier underflow.

(*a*) Flow from the reactor is 9500 m^3/d with a solids content of 3000 mg/L.
(*b*) The secondary clarifier has a diameter of 17 m.

5-50 A settling analysis is run on sludge from an extended aeration activated-sludge reactor with the following results:

Concentration, mg/L	1000	2000	3000	4000	5000	6000
Settling velocity, m/h	2.8	1.4	0.4	0.2	0.1	0.06

Under equilibrium conditions, flow to the secondary clarifier is 4200 m^3/d with solids content of 2000 mg/L. For a preselected solids flux rate of 2.5 $kg/m^2 \cdot h$, determine the required diameter of the clarifier.

5-51 A secondary clarifier processes a total flow of 10,000 m^3/d from a conventional activated-sludge reactor. The concentration of solids in the flow from the reactor is 2600 mg/L. The results of a settling analysis on the sludge is given below.

Concentration, mg/L	1490	2600	3940	5425	6930	9100	12,000
Settling velocity, m/h	5.50	3.23	1.95	1.01	0.55	0.26	0.14

For equilibrium conditions and a solid flux rate of 6 $kg/m^2 \cdot h$, determine the underflow rate, the underflow solids concentration, and the overflow rate.

5-52 When is disinfection of wastewater effluents required? Why has the wisdom of using chlorine for disinfection of wastewater come under question?

5-53 What is the organic content of primary and secondary sludge?

5-54 Name and describe the most common methods available for volume reduction of sludge.

5-55 A wastewater-treatment plant consists of primary treatment followed by an activated-sludge secondary system. Sludges from the primary clarifier and waste-activated sludge from the underflow are mixed and thickened in a gravity thickener. The primary sludge contains 1250 kg of dry solids per day with a 4 percent solids content. The waste-activated sludge contains 525 kg of dry solids per day and has a solids content of 1.2 percent. After thickening, the mixture has a solids content of 3.0 percent.

Calculate (*a*) the volume of sludge that must be processed after thickening and (*b*) the percent volume reduction in the thickener.

5-56 A high-rate aeration system produces 1140 m^3/d of waste sludge. The sludge is wasted directly from the aerator and has a solids content of 3300 mg/L. This sludge is thickened by a dissolved air flotation unit to 3.0 percent solids. Determine the volume of the thickened sludge.

5-57 What is sludge digestion? What are the two basic types of sludge digestion units?

5-58 A wastewater-treatment plant consists of primary treatment followed by a completely mixed activated-sludge secondary system. The primary and secondary sludges are mixed, thickened, and treated by anaerobic digestion. The system is shown schematically in the accompanying figure. The wastewater, treatment plant, and sludge characteristics of interest are given below.

Wastewater	Treatment plant	Sludges
Influent SS = 240 mg/L	Dia. of primary clarifier = 25 m	Primary sludge = 3.8% solids
Influent BOD = 210 mg/L	Aeration basin volume = 3600 m^3	Waste secondary sludge = 0.95% solids
Effluent BOD = 10 mg/L	MLSS in aeration = 2800 mg/L	Thickened sludge = 2.6% solids
Flow = 14,350 m^3/d		

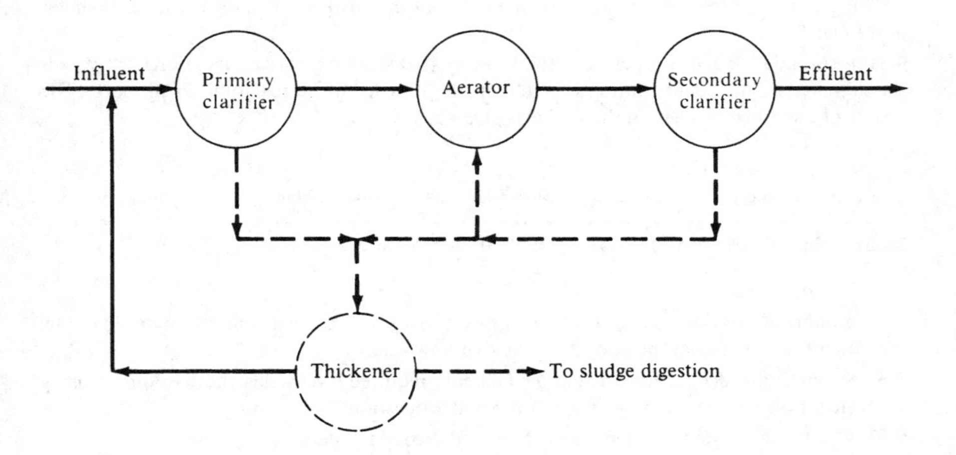

Determine (*a*) the solids loading to the digesters (kg/d and m^3/d), (*b*) the percent volume reduction in the thickener, and (*c*) the volume of supernatant returned from the thickener to the primary clarifier.

5-59 The thickened sludge in Prob. 5-56 is processed in a standard-rate anaerobic digester. The digestion period is 30 d and the sludge must be stored for 3 mo between final disposal events. Organic content of the sludge is 75 percent and 55 percent of the organics is converted to gaseous or liquid end products. The solids content of the digested sludge is 6 percent.

Determine (*a*) the required reactor volume, (*b*) the volatile solids loading rate ($kg/m^3 \cdot d$), and (*c*) the gas production (m^3/day).

5-60 Rework Prob. 5-58 for a high-rate two-stage system employing a mixed, heated first stage with a digestion period of 10 d and a second stage with a thickening period of 4 d.

5-61 A wastewater-treatment plant consists of primary treatment plus secondary treatment in a bio-tower. The underflow from the secondary clarifier is returned to the influent of the primary clarifier where it assists in removing the raw solids by adsorption and settling. The wastewater flow is 22,500 m^3/d with 250 mg/L of suspended solids. The secondary underflow contains 1180 kg/d of biological solids. Virtually all of the secondary solids plus 60 percent of the raw solids is removed. The underflow from the primary clarifier has a solids content of 5 percent. These solids are to be further treated in a high-rate anaerobic digester. The first stage is heated and mixed and requires 15 d for complete digestion. The sludge is dewatered to 35 percent solids by a filter press before final disposal. Assuming an organic content of 70 percent and digestion of 50 percent of the organics, determine the volume of the reactor and the volume of the dewatered solids.

5-62 Name and describe the most common methods of sludge disposal.

5-63 Name and describe the most common methods for removal of nutrients during tertiary treatment of municipal wastewater.

5-64 Draw a flow diagram in schematic form of a wastewater-treatment plant that includes primary (including preliminary) treatment, secondary treatment by conventional activated sludge, and nitrogen removal by air stripping. Sludge treatment is by two-stage anaerobic digestion. Identify each unit in the system and briefly state its purpose. State the destination of all materials leaving the plant. Identify all points of chemical addition and name the chemical.

5-65 Repeat Prob. 5-64 for the following treatment system:

(*a*) Primary treatment (including preliminary)
(*b*) Secondary treatment by rotating biological contactors
(*c*) Nitrogen removal by nitrification-denitrification

5-66 Name and describe the primary methods of removing suspended and dissolved solids during tertiary treatment of municipal wastewater.

5-67 Repeat Prob. 5-64 for the following treatment system

(*a*) Primary (including preliminary)
(*b*) Completely mixed activated-sludge secondary
(*c*) Nitrogen and phosphorus removal
(*d*) Advanced solids removal by granular-medium filtration
(*e*) Refractory organic removal by activated carbon

5-68 Name and describe the common methods of disposing of wastewater effluent from treatment plants.

5-69 A municipal wastewater is to be treated and discharged into a stream that empties into a pristine mountain lake that is used for recreational purposes. Draw a schematic flow diagram of a treatment plant to prepare the wastewater for discharge.

5-70 Treated municipal wastewater is to be injected into an aquifer to form a hydrostatic barrier against saltwater intrusion. Withdrawal of the injected water for domestic use is not anticipated. Draw a schematic diagram of a treatment plant to prepare the wastewater for this purpose.

5-71 Discuss the advantages and disadvantages of wastewater-treatment facilities designed to turn waste discharges into potable water.

REFERENCES

5-1 American Society of Civil Engineers and Water Pollution Control Federation: *Wastewater Treatment Plant Design*, ASCE, New York, 1977.

5-2 American Water Works Association: "Recycle and San Diego," *Municipal Wastewater Reuse News*, AWWA Research Foundation, Denver, August 1980.

5-3 American Water Works Association: "Reuse of Municipal Wastewater in Industry," *Municipal Wastewater Reuse News*, AWWA Research Foundation, Denver, November 1980.

5-4 Asano, T., and K. L. Wassermann: "Groundwater Recharge Operations in California," *J AWWA*, **72**(7):380 (July 1980).

5-5 Banerji, S. K., and J. T. O'Connor: "Designing More Energy Efficient Wastewater Treatment Plants," *Civ Eng*, **47**(7):76 (September 1977).

5-6 Benefield, L. D., and C. W. Randall: *Biological Process Design for Wastewater Treatment*, Prentice-Hall, Englewood Cliffs, N.J., 1980.

5-7 Besselievre, E. B.: *The Treatment of Industrial Wastes*, McGraw-Hill, New York, 1969.

5-8 Bouwer, Herman: "Renovating Municipal Wastewater by High-Rate Infiltration for Groundwater Recharge," *J AWWA*, **66**(3):159 (March 1974).

5-9 ———, R. C. Rice, J. C. Lance, and R. G. Gilbert: "Rapid Infiltration Research at Flushing Meadows Project, Arizona," *J WPCF*, **52**(10):2457 (October 1980).

5-10 Clark, J. W., Warren Viessman, Jr., and M. J. Hammer: *Water Supply and Pollution Control*, 3d ed., Harper & Row, New York, 1977.

5-11 Clayton, A. J., and P. J. Pybus: "Windhoek Reclaiming Sewage for Drinking Water," *Civ Eng*, **42**:103 (September 1972).

5-12 Coe, H. S., and G. H. Clevenger: "Determining Thickener Unit Areas," *Trans AIME*, **55**(3):356 (1916).

5-13 Culp, G. L., R. L. Culp, and C. L. Hamann: "Water Resource Preservation by Planned Recycling of Treated Wastewater," *J AWWA*, **65**(10):641 (October 1973).

5-14 ———, Wesner, G. M., and G. L. Culp: *Handbook of Advanced Wastewater Treatment*, Van Nostrand Reinhold, New York, 1978.

5-15 Dick, R. I.: "Role of Activated Sludge Final Settling Tanks," *J San Eng Div, ASCE*, **96**:423 (1970).

5-16 ——— and B. B. Ewing: "Evaluation of Activated Sludge Thickening Theories," *J San Eng Div, ASCE*, **93**(SA4):9 (1967).

5-17 ——— and K. W. Young: "Analysis of Thickening Performance of Final Settling Tanks," *Proc 27th Ind. Waste Conference*, Purdue University, 1972, p. 33.

5-18 Eckenfelder, W. W., Jr.: *Industrial Water Pollution Control*, McGraw-Hill, New York, 1966.

5-19 ———: *Principles of Water Quality Management*, CBI Publishing, Boston, 1980.

5-20 ———: "Trickling Filter Design and Performance," *J. San Eng Div, ASCE*, **87**(SA6):87 (1961).

5-21 Finer, S. E.: *The Life and Times of Edwin Chadwick*, Methuen, London, 1952.

5-22 Gaudy, A. F., and E. T. Gaudy: *Microbiology for Environmental Scientists and Engineers*, McGraw-Hill, New York, 1980.

5-23 Germain, J. E.: "Economical Treatment of Domestic Waste by Plastic-Media Trickling Filters," *J WPCF*, **38**(2):192 (1966).

5-24 Hadeed, S. J.: "Potable Water from Wastewater—Denver's Program," *J WPCF*, **49**(8):1757 (August 1977).

5-25 Hammer, M. J.: *Water and Wastewater Technology*, Wiley, New York, 1975.

5-26 Harris, S. E., J. H. Reynolds, D. W. Hill, D. S. Filip, and E. J. Middlebrooks: "Intermittent Sand Filtration for Upgrading Waste Stabilization Pond Effluents," *J WPCF*, **49**(1):83 (January 1977).

5-27 Horne, F. W., R. L. Anderton, and F. A. Grant: "Water Reuse: Projecting Markets and Costs," *J AWWA*, **73**(2):66 (February 1981).

5-28 Kugelman, I. J.: "Status of Advanced Waste Treatment," in H. W. Gehm and J. I. Bregman (eds.), *Handbook of Water Resources and Pollution Control*, Van Nostrand, New York, 1976.

5-29 Lawrence, A. W., and McCarty, P. L.: "Unified Basis for Biological Treatment Design and Operation," *J San Eng Div, ASCE,* **96**(SA3): 757 1970.
5-30 Linsley, R. K., and Franzini, J. B.: *Water Resources Engineering*, 3d ed., McGraw-Hill, New York, 1979.
5-31 McCarty, P. L.: "Anaerobic Waste Treatment Fundamentals," *Public Works*, 95: 107 (September 1964).
5-32 ———: "Biological Processes for Nitrogen Removal: Theory and Applications," *Proc Twelfth Sanitary Engineering Conference*, University of Illinois, Urbana, 1970.
5-33 McKinney, R. E.: *Microbiology for Sanitary Engineers*, McGraw-Hill, New York, 1962.
5-34 Mara, D. D.: *Sewage Treatment in Hot Climates*, Wiley, New York, 1976.
5-35 Marais, G. V. R.: "Faecal Bacterial Kinetics in Stabilization Ponds," *J Env Eng Div, ASCE*, **100**: 119 (1974).
5-36 Metcalf & Eddy, Inc.: *Wastewater Engineering: Treatment, Disposal, Reuse*, 2d ed., McGraw-Hill, New York, 1979.
5-37 Monod, J.: "The Growth of Bacterial Cultures," *Ann Rev Microbiology*, vol. 3, 1949.
5-38 Nemero, N. L., *Liquid Wastes of Industry: Theories, Practices, and Treatment*, Addison-Wesley, Reading, Mass., 1971.
5-39 Okun, D. A.: "Planning for Water Reuse," *J AWWA*, **65**(10): 617 (October 1973).
5-40 Parker, H. W.: *Wastewater Systems Engineering*, Prentice-Hall, Englewood Cliffs, N.J., 1975.
5-41 Pounds, C. E., and R. W. Crites: *Wastewater Treatment and Reuse by Land Application*, U.S. EPA, Cincinnati, Ohio, 1973.
5-42 *Process Design Manual for Land Treatment of Municipal Wastewater*, U.S. Environmental Protection Agency, Technology Transfer, October 1977.
5-43 *Process Design Manual for Nitrogen Control*, U.S. Environmental Protection Agency, Technology Transfer, October 1977.
5-44 *Process Design Manual for Suspended Solids Removal*, U.S. Environmental Protection Agency, Technology Transfer, January 1975.
5-45 Roberts, P. V.: "Water Reuse for Recharge: An Overview," *J AWWA*, **72**(7): 375 (July 1980).
5-46 Sawyer, C. N., and McCarty, P. L.: *Chemistry for Environmental Engineering*, McGraw-Hill, New York, 1978.
5-47 Schroeder, E. D.; *Water and Wastewater Treatment*, McGraw-Hill, New York, 1977.
5-48 Simon, A. L.: *Practical Hydraulics*, 2d ed., Wiley, New York, 1976.
5-49 Steel, E. W.: *Water Supply and Sewerage*, McGraw-Hill, New York, 1960.
5-50 ——— and J. J. McGhee: *Water Supply and Sewerage*, 5th ed., McGraw-Hill, New York, 1979.
5-51 Stover, E. L., and D. F. Kincannon: "One-Step Nitrification and Carbon Removal," *Water and Sewage Works*, **66**, June 1975.
5-52 Sunderstron, D. W., and H. E. Klei: *Wastewater Treatment*, Prentice-Hall, Englewood Cliffs, N.J., 1979.
5-53 Thirumurthi, D.: "Design Principles of Waste Stabilization Ponds," *J San Eng Div, ASCE*, **95**: 311 (1969).
5-54 van Vuuren, L. R. J., A. J. Clayton, and D. C. van der Post: "Current Status of Water Reclamation at Windhoek," *J WPCF*, **52**(4): 661 (April 1980).
5-55 Vesilind, P. A.: *Treatment and Disposal of Wastewater Sludges*, 2d ed., Ann Arbor Science, Woburn, Mass., 1979.
5-56 Yoshioka, N., et al.: "Continuous Thickening of Homogeneous Flocculated Slurries" (English abstract), *Chem Eng*, **21**, Tokyo, 1957.
5-57 Wong-Chong, G. M., and R. C. Loehr: "The Kinetics of Microbial Nitrification," *Water Research*, **9**: 1099 (1975).

CHAPTER

SIX

ENVIRONMENTAL ENGINEERING HYDRAULICS DESIGN

Issues related to water and wastewater quality and treatment have been examined in Chaps 2 through 5. It is the purpose of this chapter to introduce the reader to the physical facilities needed to meet water-supply and wastewater-management objectives. To do this the chapter is organized into four major sections dealing with water distribution systems, wastewater collection systems, pumps and pumping stations, and treatment plant hydraulics.

Water Distribution Systems

To deliver water to individual consumers with appropriate quality, quantity, and pressure in a community setting requires an extensive system of pipes, storage reservoirs, pumps, and related appurtenances. The term *distribution system* is used to describe collectively the facilities used to supply water from its source to the point of usage.

6-1 METHODS OF DISTRIBUTING WATER

Depending on the topographic relationship between the source of supply and the consumer, water can be transported by canals, flumes, tunnels, and pipelines. Gravity, pumping, or a combination of both may be used to supply water to the consumers (see Fig. 6-1) with adequate pressure.

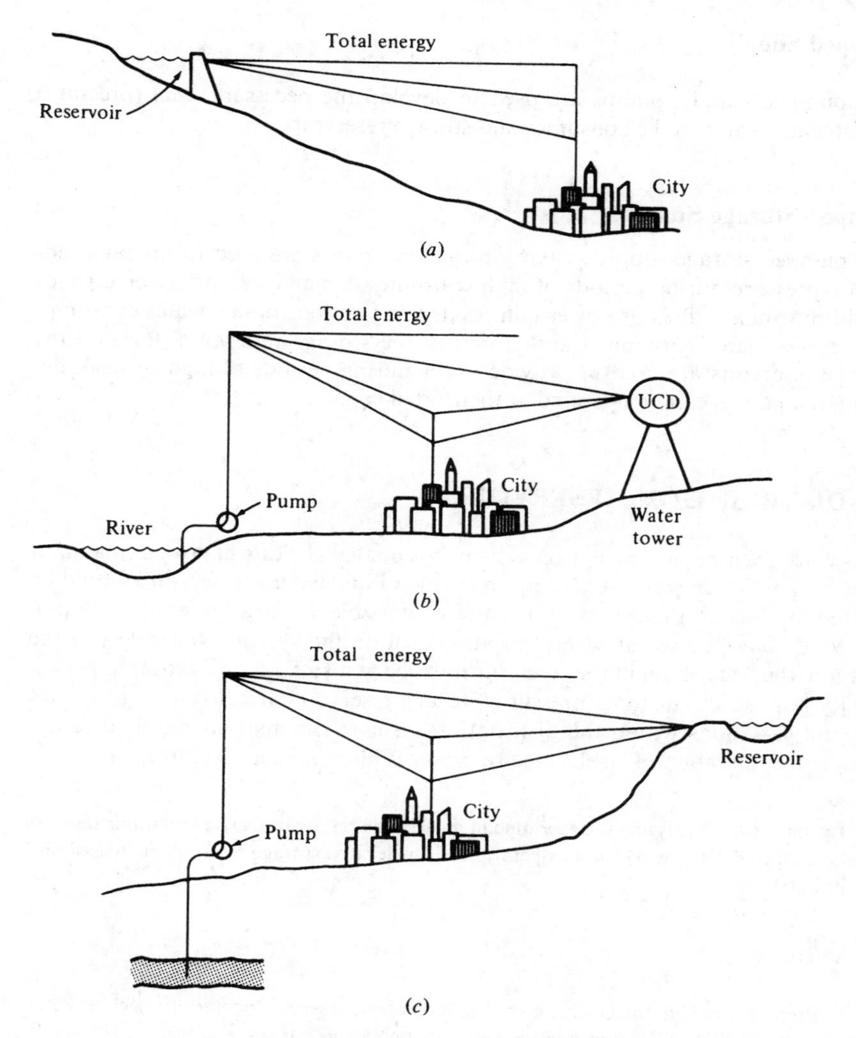

Figure 6-1 Typical distribution systems: (*a*) gravity, (*b*) pumped, and (*c*) combined.

Gravity Supply

Where the source of supply is at a sufficient elevation above the consumer so that the desired pressure can be maintained, a gravity supply can be used. In gravity systems, it is often possible to supply water to one or more storage reservoirs within the system. Where a gravity supply can be used, it has proven to be quite economical.

Pumped Supply

In a pumped supply, pumps are used to develop the necessary *head* (pressure) to distribute water to the consumer and storage reservoirs.

Pumped-Storage Supply

In a pumped-storage supply system, storage reservoirs are used to maintain adequate pressure during periods of high consumer demand and under emergency conditions such as fires or power failures. During periods of low water consumption, excess water is pumped and stored in the storage reservoirs. Because the storage reservoirs are used to provide water during periods of high or peak demand, the pumps can be operated at their rated capacity.

6-2 DISTRIBUTION RESERVOIRS

Reservoirs are used in distribution systems to equalize the rate of flow, to maintain pressure, and for emergencies. To optimize their intended use, reservoirs should be located as close to the center of demand as possible. In large cities, distribution reservoirs may be used at several locations within the system. Regardless of the location, the water level in the reservoir must be at a sufficient elevation to permit gravity flow at an adequate pressure. Storage reservoirs are also used to reduce pressure variations within the distribution system. The analysis required to determine the operation of an elevated reservoir is illustrated in Example 6-1.

Example 6-1: Analyzing the operation of an elevated reservoir Derive equations that can be used to define the hydraulic operation of an elevated storage reservoir such as shown in Fig. 6-1*b*.

SOLUTION

1. Prepare a definition sketch for the analysis of the reservoir operation. Such a sketch is shown below. The terms in the sketch are defined as follows:

Q_1 = pump discharge, m^3/s

$Q_2(Q_3)$ = discharge to (from) reservoir, m^3/s

Q_D = municipal discharge (demand), m^3/s

E_2, E_3 = energy at load center under various conditions of operation (includes pressure and velocity head), m

h_{f_1}, h_{f_2}, etc. = head loss due to friction, m

z_1, z_2, z_3 = elevations above a reference datum, m

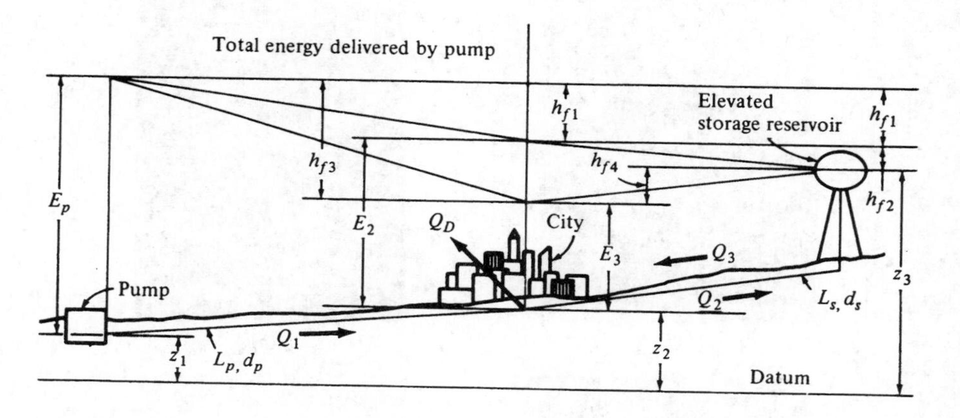

2. Define the three conditions of flow that can exist.
   a. When the municipal demand is low, the discharge from the pump will supply the municipal demand; the excess pump discharge will be diverted to the storage reservoir.
   b. When the municipal demand is high, discharge from both the pump and the storage reservoir will be used to meet the demand.
   c. At some point of operation, the pump discharge will just equal the municipal demand and there will be no flow from the elevated storage reservoir.
3. Write equations that can be used to solve the three flow conditions defined in step 2.
   a. Low demand:

$$Q_1 - Q_D = Q_2$$

$$z_1 + E_p = z_3 + h_{f1} + h_{f2}$$

   b. High demand:

$$Q_1 + Q_3 = Q_D$$

$$z_1 + E_p = z_2 + E_3 + h_{f3}$$

$$z_3 = z_2 + E_3 + h_{f4}$$

   c. No flow from storage reservoir:

$$Q_1 = Q_D$$

$$z_1 + E_p = z_2 + E_3 + h_{f1}$$

$$z_2 + E_3 = z_3$$

COMMENT To solve the equations developed in step 3 for a high-demand situation, a trial value of E_3 is assumed and the computed values of discharge are compared to the demand. The computation is repeated until the equation of continuity ($Q_1 + Q_3 = Q_D$) is satisfied.

In the approach described above, it was assumed that z_3 remains constant. In actual practice, z_3 will vary with time. To solve the problem with a varying value of z_3 it is necessary to develop a relationship between the storage volume and the water surface elevation.

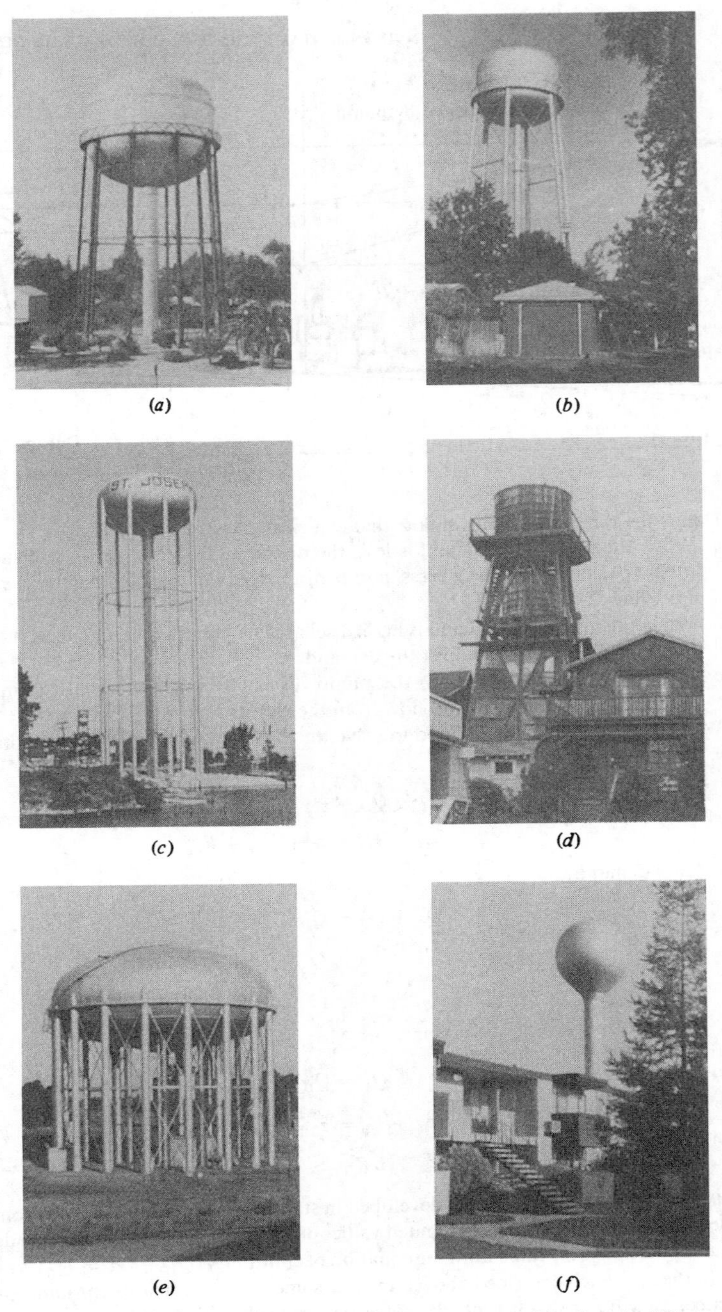

(a) (b) (c) (d) (e) (f)

Figure 6-2 Typical shapes of elevated water-storage tanks. (Photo of St. Joseph, MI storage tank, *courtesy Consoer Townsend & Associates, Inc.*)

Types of Reservoirs

Depending on the topography and local environmental conditions, storage reservoirs may be located above, on, or below the ground surface. Underground reservoirs are usually constructed of reinforced concrete. Small ground-level reservoirs are usually earth-lined with gunite, asphalt, or some synthetic membrane. Large surface reservoirs are concrete-lined. Most large surface reservoirs are covered to prevent contamination by birds, animals, and humans. Open distribution reservoirs should be fenced to keep out trespassers.

To obtain the necessary head within the distribution system, water towers and elevated reservoirs are often used. Water towers, located at ground level, can be constructed of prestressed concrete or steel. Elevated water-storage reservoirs are usually constructed of steel. Common shapes for elevated storage tanks are illustrated in Fig. 6-2.

Capacity of Storage Reservoirs

The capacity of storage reservoirs can be determined analytically or graphically. In either case a mass balance is the basis of the analysis. Both methods of analysis are illustrated in Example 6-2.

Example 6-2: Determining reservoir storage capacity Determine the capacity of a storage reservoir required to maintain a constant water supply (draft) of 2×10^6 m^3/m given the following monthly mean-runoff values:

Month	Runoff Q_R 10^6 m^3	Month	Runoff Q_R 10^6 m^3
1	9.0	10	0.4
2	10.8	11	0.5
3	4.2	12	0.9
4	2.8	13	1.1
5	1.2	14	2.0
6	1.1	15	5.5
7	0.9	16	10.5
8	0.5	17	3.5
9	0.6	18	2.5

SOLUTION

1. Set up a table for the computations for the graphical and numerical solution of the problem. The required computations are shown in the accompanying table. The entries in the columns are as follows:
   a. The month and the corresponding runoff are entered in columns 1 and 2, respectively.
   b. The cumulative runoff is computed and entered in column 3.
   c. The water supply draft is entered in column 4.

d. The deficit (runoff − water supply draft) is computed and entered in column 5. A minus sign means that the water supply draft exceeds the runoff and a deficit exists.

e. The cumulative deficit is computed and entered in column 6. The numbers in parentheses represent the cumulative surplus. The maximum cumulative deficit represents the required reservoir capacity.

2. Prepare a graphical analysis of the problem. The required graphical solution is shown in the figure below. Key points in the construction of the graphical solution are as follows:

a. First, the cumulative runoff data from column 3 in the table are used to plot the runoff curve.

b. Next, a line is drawn from the origin at a slope equal to the monthly water supply draft.

c. To determine the required capacity of the storage reservoir a line is drawn parallel to the water supply draft line, but starting at the point of tangency at the beginning of the dry period. The maximum distance between the draft line from the point of tangency and cumulative runoff represents the required capacity of the storage reservoir. As shown in the figure, the capacity value is 10.8×10^6 m^3, which is the same as the value given in the table.

COMMENT The graphical method for reservoir sizing illustrated in the figure was developed by W. Ripple sometime before 1883 when he published the method. [6-12]

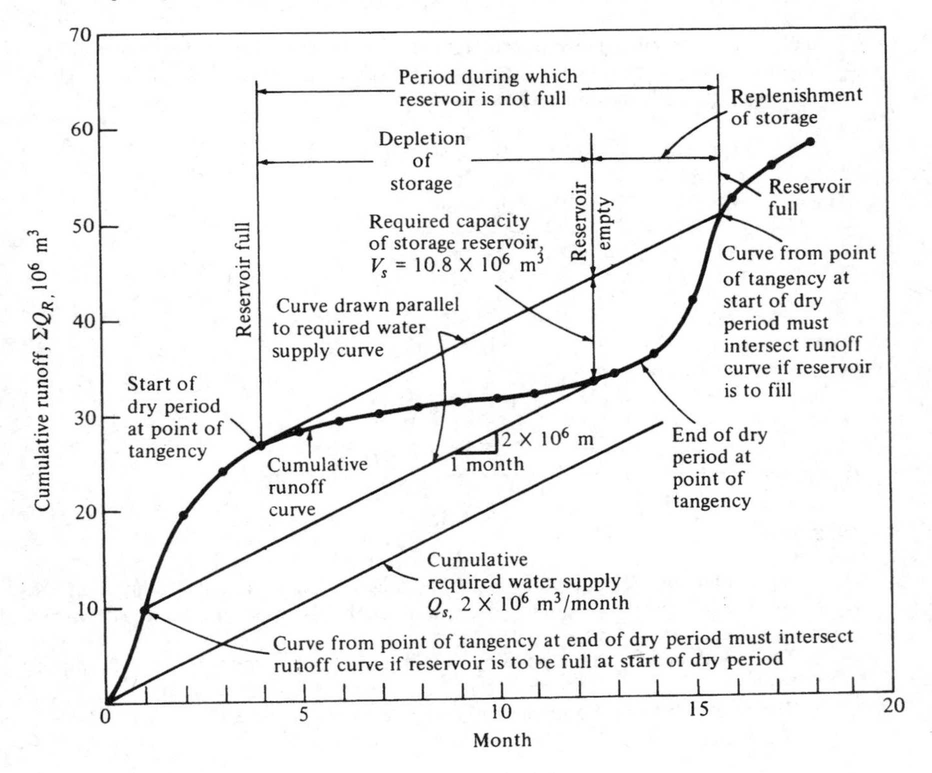

Computation of required storage

Month	Runoff, Q_R 10^6 m^3	Cumulative runoff	Water supply, Q_s 10^6 m^3	Deficit $(Q_R - Q_s)$ 10^6 m^3	Cumulative deficit, $\Sigma(Q_R - Q_s)$ 10^6 m^3
1	9.0	9.0	2.0	7.0	0.0(7.0)
2	10.8	19.8	2.0	8.8	0.0(15.8)
3	4.2	24.0	2.0	2.2	0.0(18.0)
4	2.8	26.8	2.0	0.8	0.0(18.8)*
5	1.2	28.0	2.0	−0.8	−0.8
6	1.1	29.1	2.0	−0.9	−1.7
7	0.9	30.0	2.0	−1.1	−2.8
8	0.5	30.5	2.0	−1.5	−4.3
9	0.6	31.1	2.0	−1.4	−5.7
10	0.4	31.5	2.0	−1.6	−7.3
11	0.5	32.0	2.0	−1.5	−8.8
12	0.9	32.9	2.0	−1.1	−9.9
13	1.1	34.0	2.0	−0.9	−10.8†
14	2.0	36.0	2.0	0.0	−10.8
15	5.5	41.5	2.0	3.5	−7.3
16	10.5	52.0	2.0	8.5	0.0(1.2)‡
17	3.5	55.5	2.0	1.5	0.0(2.7)
18	2.5	58.0	2.0	0.5	0.0(3.2)

* Reservoir is full at beginning of dry period.

† Maximum deficit at or near the end of the dry period. The cumulative maximum deficit represents the required reservoir storage capacity.

‡ Reservoir is refilled during the 16th month.

6-3 DISTRIBUTION SYSTEMS

The series of interconnected pipes used to supply water to the consumer is known as a *distribution network*. Several network configurations have been used. Each of these is described below.

Branching System

The branching type of water distribution network is shown in Fig. 6-3*a*. As shown, the structure of such a system is similar to a tree. The trunk line is the main source of water supply. Service mains are connected to the trunk line, and submains are connected to the service mains. In turn, building connections used to provide service to individual residences and buildings are connected to the submains.

Although such a system is simple to design and build, it is not favored in modern waterworks practice for the following reasons: (1) bacterial growths and sedimentation may occur in the branch ends due to stagnation; (2) it is difficult to maintain a chlorine residual at the dead ends of the pipe; (3) when repairs must

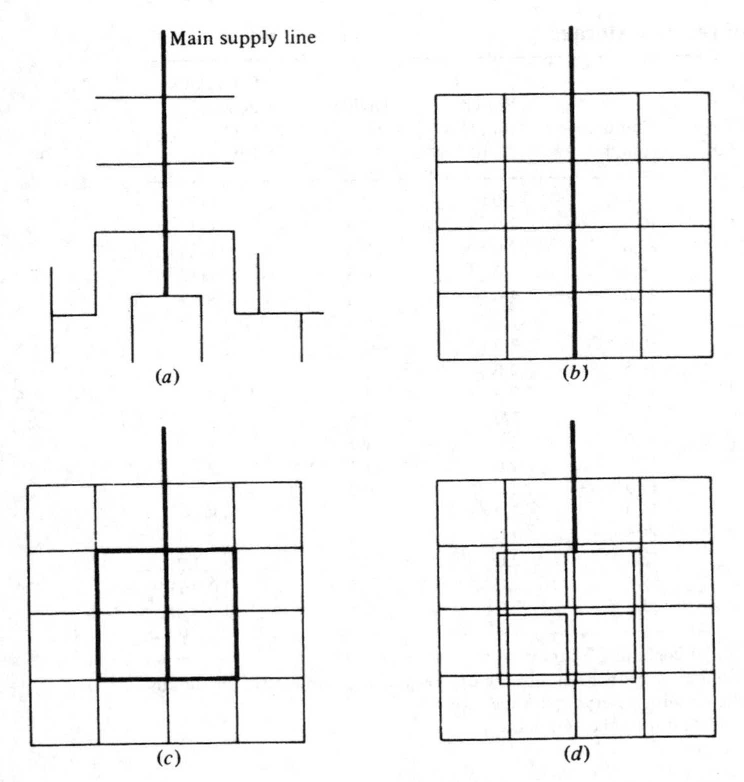

Figure 6-3 Water distribution systems: (*a*) branching with dead ends; (*b*) grid pattern; (*c*) grid pattern with loops; (*d*) grid pattern with dual mains.

be made to an individual line, service connections beyond the point of repair will be without water until the repairs are made; and (4) the pressure at the end of the line may become undesirably low as additional extensions are made. The latter problem is common in many less-developed countries.

Grid System

The distinguishing feature of the grid system is that all of the pipes are interconnected and there are no dead ends (see Fig. 6-3*b*). In such a system, water can reach a given point of withdrawal from several directions. The grid system overcomes all of the difficulties of the branching system discussed previously. One disadvantage is that the determination of the pipe sizes is somewhat more complicated.

Several variations of the grid system are also in use. Two of the most common are the grid pattern with loops (see Fig. 6-3*c*) and the grid pattern with dual mains (see Fig. 6-3*d*). In the former, additional loops are added to improve the distribution of water. Loops are usually added to serve business districts and other high-risk areas. The loops may be constructed with separate pipes or by enlarging some of the pipes in the existing grid. The main advantage of the dual-main system (Fig. 6-3*d*) is that breaks in mains do not limit the usefulness of fire hydrants.

To help protect against freezing, pipes are usually placed on the north and east sides of streets in the northern hemisphere. In the southern hemisphere water pipes are normally placed on the south and east sides of the streets. In all cases, pipes should be buried below the frost line.

6-4 DISTRIBUTION SYSTEM COMPONENTS

The principal components of the distribution network are pipes, valves, fire hydrants, and service (building) connections. Storage reservoirs and pumps are considered separately. Typically, the requirements for sizes and placement are specified by local code. Representative values and data are reported in Table 6-1.

Pipes

A variety of materials has been used for the pipes in water distribution networks. The most common materials are steel, cast iron, and reinforced concrete. The type

Table 6-1 Representative data of distribution system components

Item	Value
Pipes	
Smallest pipes in grid	150 mm (6 in)
Smallest branching pipes (dead ends)	200 mm (8 in)
Largest spacing of 150-mm (6-in) grid [200-mm (8-in) pipe used beyond this value]	180 m (600 ft)
Smallest pipes in high-value district	200 mm (8 in)
Smallest pipes on principal streets in central district	300 mm (12 in)
Largest spacing of supply mains or feeders	600 m (2000 ft)
Valves	
In single- and dual-main systems	Three at crosses, two at tees
Largest spacing on long branches	250 m (800 ft)
Largest spacing in high-value district	150 (500 ft)
Fire hydrants	
Areas protected by hydrants	See Table 6-2
Largest spacing when fire flow exceeds 300 L/s (5000 gal/min)	60 m (200 ft)
Largest spacing when fire flow is as low as 60 L/s (1000 gal/min)	90 m (300 ft)

of pipe material is important as it will affect the ancillary equipment needed for its installation and maintenance.

Valves

The types of valves used most commonly are gate valves, air-relief valves, and check valves. As noted in Table 6-1 and shown in Fig. 6-4 three gate valves are used at all crosses and two gate valves are used at all tees. The principal function of

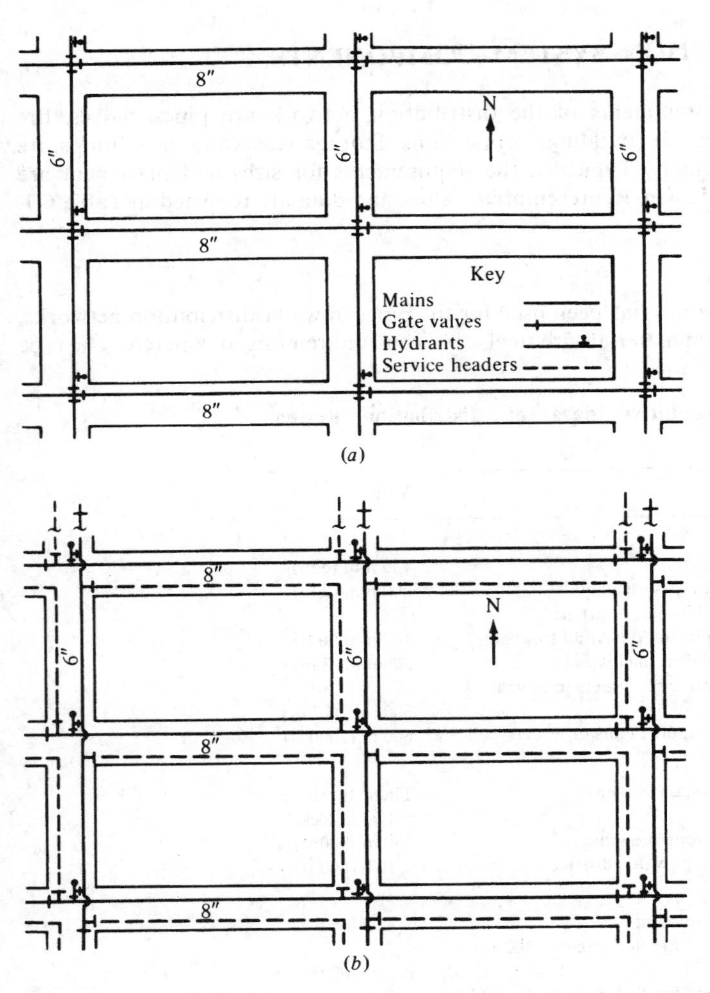

Figure 6-4 Section of grid distribution system: (*a*) single-main system and (*b*) dual-main system. (*From Fair et al.* [6-2].)

these valves is to isolate subsections of the system for repairs and maintenance. Drain valves should be provided at low points in the system. To remove air from pipelines, air-relief or release valves are placed at high spots in the pipeline. Check valves are used to limit the flow of water to a single direction.

Fire Hydrants

Fire hydrants are placed on mains to provide locations where fire hoses and pumper trucks can be connected to the water source needed for fire fighting. Three types of hydrants are used: flush, wall, and post. As the name implies, flush hydrants are placed in a chamber that is even with the surface of the street or sidewalk. Wall hydrants project from the walls of buildings and are used extensively in commercial districts. Post hydrants extend from the main to about 1 m above the street or sidewalk.

Post hydrants are usually placed on a concrete block to eliminate settling and are braced to resist the lateral forces of the flowing water. Typically, hydrants are provided with one or more 60-mm ($2\frac{1}{2}$-in) hose outlets and a 100-mm (4-in) pumper-truck connection. In cold climates, the operating valve is located below ground level so that the barrel contains no water except when in use. A drain valve opens automatically when the hydrant valve is closed, to permit the escape of water after use and to avoid damage by freezing. In warm climates, the hydrant barrel may contain water at all times, and an individual valve is provided for each outlet.

Service (Building) Connections

The service (building) connection is that portion of a water supply system that lies between the water supply main in the street and the take offs for the various plumbing fixtures at the point of usage.

6-5 CAPACITY AND PRESSURE REQUIREMENTS

The capacity of a water distribution system must be sufficient to meet the requirements for fire fighting in conjunction with domestic, commercial, and industrial demands and for other system uses and losses. It is important to note that capacity and pressure must be considered simultaneously. For example, water must rise to the upper stories of low-rise buildings in sufficient quantity and pressure, especially when fire fighting is considered. It should be noted that a pumped supply is used in most modern high-rise buildings. Also, the capacity and pressure available at hydrants must be sufficient for fire-fighting purposes.

Capacity

With the exception of that needed for fire fighting, the capacity of the distribution system must be sufficient to meet the peak demand based on domestic, commercial, industrial, and other miscellaneous uses and system losses. The ultimate peak demand would be the combination of the peak fire-fighting and peak conventional-consumer demand. In practice, however, most distribution systems are not sized to provide the ultimate peak demand. The reasons for this are (1) the probability that the peak fire and consumer demand will occur simultaneously is low and (2) most distribution systems are sized for the future so excess capacity is available.

In general, most distribution systems are sized to meet the fire demand and a consumer demand of 150 to 200 L/capita · d (40 to 50 gal/capita · d) in excess of the yearly average value. Alternatively, the consumer demand may be taken as the maximum daily demand (150 percent of the average daily demand). In the United States, the general fire-fighting requirements are based on the recommendations of the Insurance Services Office, New York. The required fire flow is estimated using the following equation.

$$F = 320C\sqrt{A} \tag{6-1}$$

where F = required fire flow, m^3/d
C = coefficient related to type of construction
A = the total floor area (including all stories, but excluding basements) in the building under consideration, m^2. For fire-resistive buildings the six largest successive floor areas are used if the vertical openings are unprotected; if the vertical openings are protected properly, only the three largest successive floor areas are considered.

Values for the coefficient C are 1.5 for wood frame construction, 1.0 for ordinary construction, 0.8 for noncombustible construction, and 0.6 for fire resistive construction. Interpolation between these values is used for construction that does not fall into one of the four categories. The computed value is then adjusted up or down for (1) occupancy, (2) sprinkler protection, and (3) exposure. The maximum fire flow determined using Eq. 6-1 shall not exceed 43,600 m^3/d for wood frame construction and for ordinary and heavy timber construction, and 32,700 for noncombustible construction and for fire resistive construction for any one location. The required fire flow rate must be available in addition to the coincident maximum daily flow rate. The duration during which the required fire flow should be available varies from 2 to 10 h as summarized in Table 6-2. Because a city will be penalized in its fire insurance rates if the needed flows cannot be met for the specified durations, most cities provide storage reservoirs to meet fire demands.

Pressure

For typical residential rates of demand, a static pressure of 275 kPa, gage (40 lb/in^2, gage) is considered to be normal. The minimum recommended pressure is about 140 kPa, gage (20 lb/in^2, gage). In business districts, pressure values in the range of 350 to 550 kPa, gage (50 to 80 lb/in^2, gage) are common. For high-rise buildings (greater than three stories) water is pumped to storage tanks located on intermediate floors, on the roof, or in towers.

Table 6-2 Duration of required fire flows based on flow*

Fire flow		Duration,
gal/min	m^3/d	h
2,000 or less	10,900	2
3,000	16,400	3
4,000	21,800	4
5,000	27,300	5
6,000	32,700	6
7,000	38,200	7
8,000	43,600	8
9,000	49,100	9
10,000 and greater	54,500	10

* Adapted from *Guide for Determination of Required Fire Flow*, 2d ed., Insurance Service Office, New York, 1974.

With the advent of the modern fire-fighting equipment, the pressure that must be maintained at a fire hydrant rarely needs to be greater than 350 to 400 kPa, gage (50 to 60 lb/in^2, gage). The exception is in small towns where full-time fire departments with new equipment cannot be afforded. When pumper trucks are used, the pressure at the fire hydrant should not be allowed to drop below about 70 kPa, gage (10 lb/in^2, gage). This low pressure should be maintained to prevent untreated water from entering the water distribution system by seepage or pipe failure caused by vacuum collapse.

6-6 DESIGN OF DISTRIBUTION SYSTEMS

The design of a water distribution system for a new area can be outlined as follows. (The analysis of existing systems is considered in the following section.)

1. Obtain a detailed map of the area to be served on which topographic contours (or controlling elevations) and the locations of present and future streets and lots are identified.
2. Based on the topography, select possible locations for distribution reservoirs. If the area to be served is large, it may be divided into several subareas to be served with separate distribution systems.
3. Estimate the average and peak water use for the area or each subarea, allowing for fire fighting and future growth.
4. Estimate pipe sizes on the basis of water demand and local code requirements.
5. Lay out a skeleton system of supply mains leading from the distribution reservoir or other source of supply.
6. Analyze, using one of the several methods discussed in the following section, the flows and pressures in the supply network for fire flows. A separate analysis

should be performed for each subarea. Also several configurations should be examined for each area under various conditions of withdrawal.

7. Adjust pipe sizes to reduce pressure irregularities in the basic grid.
8. Add distribution mains to the grid system. Distribution mains that serve fire hydrants should be at least 150 mm (6 in) in diameter in residential areas and 200 or 250 mm (8 or 10 in) in diameter in commercial and high-risk industrial areas.
9. Reanalyze the hydraulic capacity of the system.
10. Add street mains for domestic service. These mains usually vary in size from 50 to 100 mm (1 to 4 in) in diameter.
11. Locate the necessary valves and fire hydrants.
12. Prepare final design drawings and quantity takeoffs.

6-7 HYDRAULIC ANALYSIS OF DISTRIBUTION SYSTEMS

The purpose of a hydraulic analysis of a distribution system is to assess flows (including direction) and the associated pressure distribution that develops within the system under various conditions of withdrawal. Several methods are available. These include (1) sectioning, (2) the circle method, (3) relaxation, (4) pipe equivalence, (5) digital computer analysis, and (6) electrical analogy. The characteristics of each of these methods are summarized in Table 6-3. The method of sections and the use of digital computer analysis are considered further below.

Table 6-3 Methods of analysis for water distribution systems

Method	Description
Method of sections	Water-distribution-system grid is cut with a series of sections, and the capacity of the cut pipes is compared to the downstream demand.
Circle method	The pipes in a distribution system tributary to a central fire hydrant or group of hydrants are cut with a circle, and the capacity of the pipes to meet the fire demand is assessed.
Relaxation	A trial-and-error procedure in which systematic corrections are applied to (1) an initial set of assumed flows or (2) an initial set of assumed heads until the flow network is balanced hydraulically.
Pipe equivalence	The pipes in a complex distribution system are replaced with a single pipe of equivalent capacity.
Digital computer analysis	Algorithms are written to solve Eqs. 6-2, 6-3, and 6-4 simultaneously throughout the network. The algorithms are solved using modern high-speed digital computers. Numerous commerical programs are available for the solution of water-distribution flow problems.
Electrical analogy	The distribution system is modeled with electrically equivalent components. For example, nonlinear resistors are used to simulate pipe friction. If the current inputs and withdrawals are proportional to the water flow, then the head losses will be proportional to measured voltage drops.

Method of Sections

The method of sections was developed by Allen Hazen [6-2] as a quick method for checking the correctness of network pipe sizes. A similar procedure was proposed by Pardoe. [6-10] Although the method is approximate, it is extremely useful in analyzing pipe networks if its limitations are appreciated. The principal steps involved in the application of this method are as follows.

1. Cut the network with a series of lines selected with due regard to varying pipe sizes and district characteristics. The lines need not be straight or regularly spaced. Typically the first series of lines will cut across the network at right angles to the direction of flow. Additional cut lines may be oriented in other important directions. For more than one source of supply, a curved cut line should be used to intercept the flow from each source of supply (see Fig. 6-5).
2. Estimate the amount of water that must be supplied to the areas beyond each cut line (i.e., downstream). The water demand is composed of the fire demand and the normal coincident draft due to domestic, commercial, industrial and other uses. In most networks, the coincident draft will decrease from section to section. The fire demand will remain high until high-demand or high-valve areas are left behind.
3. Estimate the capacity of the distribution network at each cut line or section. This can be done as follows:
   a. Count and tabulate the number of pipes of each size that were cut. Only those pipes that provide water in the direction of flow should be counted.

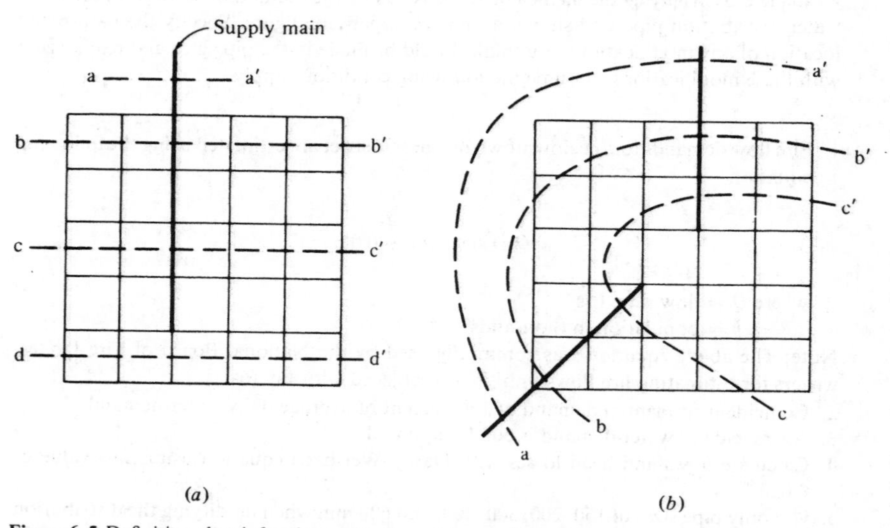

Figure 6-5 Definition sketch for the application of the method of sections.

b. Determine the average available hydraulic gradient. This will depend on system pressures and allowable flow velocities. For example, if a flat grid is 10,000 m wide in the direction of flow, if the pressure available at the transmission pipe connection is 415 kPa, gage (60 lb/in^2, gage), and if the minimum allowable pressure is 140 kPa, gage (20 lb/in^2, gage) the average hydraulic gradient is 0.0028 [(415 − 140)/10,000]. Hydraulic gradients and velocities between 0.001 m/m and 0.003 m/m and 0.6 to 1.25 m/s (2 to 4 ft/s), respectively, are common.

4. For the calculated hydraulic gradient, determine the capacity of the existing cut pipes and total capacity.
5. Determine the difference between the required and the existing capacity.
6. If the existing capacity is inadequate, select pipe sizes and paths that will offset any deficiencies at the required hydraulic gradient. The capacity of the system can be increased by replacing small pipes with larger pipes or adding pipes to the grid. Experience with the system sometimes helps in selecting the pipe sizes, but such experience is not necessary. If excess capacity is found, pipe sizes may be reduced using the same procedures.
7. Determine the size of the equivalent pipe for the reinforced system and estimate the flow velocity. High velocities should be limited to avoid water-hammer problems by reducing the hydraulic gradient.
8. Check the pressure requirements against the reinforced system.

Application of the method of sections is illustrated in Example 6-3.

Example 6-3: Applying the method of sections Using the method of sections analyze the water-distribution-pipe grid shown in the accompanying figure. Specify the nature and location of any modifications you think should be made in the pipe grid and reanalyze it with these modifications. Assume the following conditions apply.

1. Fire flow demands for the downtown business district are estimated using the following equation

$$Q = 65\sqrt{p}(1 - 0.01\sqrt{p})$$

where Q = flow rate, L/s
p = population in thousands

Note: The above equation was commonly used by the National Board of Fire Underwriters for estimating fire flows until it was replaced with Eq. (6-1).

2. Coincident residential demand of 150 percent of average daily water demand.
3. Average daily water demand is 500 L/capita · d.
4. Calculate flows and head losses with Darcy–Weisbach equation using an f value of 0.020.
5. Use only pipe sizes of 150, 200, 300, 400, and 600 mm when modifying the distribution system grid.

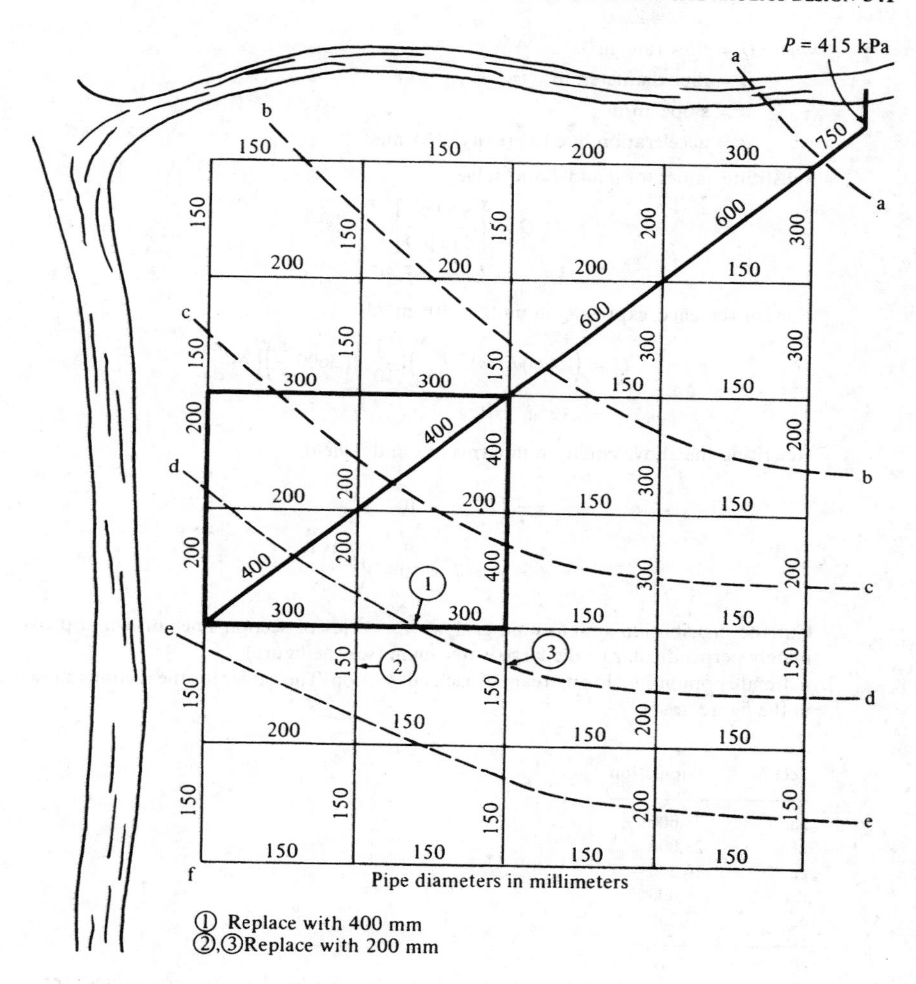

SOLUTION

1. Modify the Darcy–Weisbach equation to obtain slope and diameter form.

$$h = f\,\frac{L}{d}\,\frac{V^2}{2g}$$

$$\frac{h}{L} = s = \frac{f}{d}\,\frac{V^2}{2g}$$

$$s = \frac{f}{2g}\,\frac{Q^2}{A^2 d}$$

$$Q = \left(\frac{g\pi^2}{8f}\right)^{1/2} d^{5/2} s^{1/2}$$

where Q = flow rate, m^3/s
d = pipe diameter, m
s = slope, m/m
g = acceleration due to gravity, 9.81 m/s^2

Substitute values for g and f and solve.

$$Q = \left[\frac{9.81\pi^2}{8\,(0.02)}\right]^{1/2} d^{5/2}s^{1/2}$$

$$= 24.60d^{5/2}s^{1/2}$$

For convenience, express Q in units of 10^3 m^3/d.

$$Q = \left(24.60d^{5/2}s^{1/2}\,\frac{m^3}{s}\right)\left(\frac{1}{10^3}\right)\left(3600\,\frac{s}{L}\right)\left(24\,\frac{h}{d}\right)$$

$$= 2125d^{5/2}s^{1/2}$$

Rewriting the above equation in terms of s and d yields

$$s = (2.215 \times 10^{-7})\frac{Q^2}{d^5}$$

$$d = (4.667 \times 10^{-2})\left(\frac{Q^2}{s}\right)^{1/5}$$

2. Cut the distribution-system pipe grid with a series of section lines drawn approximately perpendicular to the large water main (see the figure).
3. Estimate population downstream of each cut section. The values for the sections shown in the figure are:

Section	Population
aa	28,000
bb	23,000
cc	16,500
dd	9,000
ee	3,750

4. Estimate water demand downstream of each cut section. The required values are:

		Demand, 10^3 m^3/d		
Section	Population	Coincident	Fire	Total
aa	28,000	21.00	28.14	49.14
bb	23,000	17.25	28.14	45.39
cc	16,500	12.38	28.14	40.52
dd	9,000	7.88	28.14	34.89
ee	3,750	2.81	5.45*	8.27

* Based on residential fire demand of 63 L/s (1000 gal/min).

a. Coincident demand $= \dfrac{(1.5)(500 \text{ L/capita} \cdot \text{day})(16{,}500 \text{ people})}{10^3 \text{ L/m}^3}$

$= 12.38 \times 10^3 \text{ m}^3/\text{d}$

b. The fire demand for the downtown business district is based on a population of 28,000.

5. Determine the available hydraulic gradient across the distribution system.

$$s = \frac{\dfrac{P_1}{\gamma} - \dfrac{P_2}{\gamma}}{L}$$

where P_1 = pressure in supply main at head end of distribution system
$= 415 \times 10^3$ Pa (given)

P_2 = minimum pressure required at farthest end of distribution system
$= 140 \times 10^3$ Pa (20 lb/in^2)

L = length of main supply pipe across system
= 9025 m

$$s = \frac{\dfrac{415 - 140 \text{ N/m}^2}{9810 \text{ N/m}^3} \times 10^3}{9025 \text{ m}} = 0.0031 \text{ m/m}$$

6. Count and tabulate the number of pipes of each size cut by each section.

For example, for section bb: 1 – 0.6-m diameter
1 – 0.3-m
2 – 0.2-m
4 – 0.15-m

7. Calculate the capacity of each pipe cut by section bb using the Darcy–Weisbach equation and the available hydraulic gradient. For example, for the 0.6-m-diameter pipe:

$$Q = 2125 d^{5/2} s^{1/2} \qquad \text{(see step 1)}$$
$$= 2125(0.6)^{5/2}(0.0031)^{1/2} = 32.99 \times 10^3 \text{ m}^3/\text{d}$$

The capacities of the pipes cut by section bb are:

1 – 0.6	32.99×10^3 m^{3}d
1 – 0.3	5.83
2 – 0.2	4.23
4 – 0.15	4.12
	47.17×10^3 m^{3}d

If the sum is greater than the demand there is sufficient capacity across this section. For section bb the supply is greater than the demand (45.39×10^3 m^3/d).

8. Calculate the diameter of a single equivalent pipe using the diameter form of the Darcy–Weisbach equation (see step 1).

$$d = (4.667 \times 10^{-2})\left(\frac{47.17^2}{0.0031}\right)^{1/5} = 0.692 \text{ m}$$

9. Calculate the actual hydraulic gradient when the capacity at the section line equals the demand using the slope form of the Darcy–Weisbach equations (see step 1).

$$s = (2.215 \times 10^{-7})\frac{45.39^2}{0.692^5} = 0.00288$$

10. Using the actual hydraulic gradient, recalculate the capacity at the section line. It should equal the demand. For example, at section bb:

1 – 0.6	31.80 × 10^3 m^3/d
1 – 0.3	5.62
2 – 0.2	4.08
4 – 0.15	3.89
	45.48 × 10^3 m^3/d

11. Check for excess velocities in the pipes using the actual pipe capacities from step 10.

$$V = \frac{Q}{A} = \frac{\dfrac{10^3 \text{ m}^3}{d}(10^3)\dfrac{\text{h}}{3600 \text{ s}}\dfrac{\text{d}}{24 \text{ h}}}{\dfrac{Ld^2}{4}}$$

$$= (1.474 \times 10^{-2})\frac{Q}{d^2}$$

where Q = flow rate, 10^3 m^3/d

d = pipe diameter, m

V = fluid velocity, m/s

For example, for the 0.6-m-diameter pipe

$$V = (1.474 \times 10^{-2})\frac{31.80}{0.6^2} = 1.30 \text{ m/s}$$

If the calculated velocities are too high, modifications to the distribution-system grid may be necessary.

12. Complete the necessary computations for the remaining sections. The required computations for the distribution system shown in the figure are summarized in the accompanying table. Based on the calculations in the table, sections dd and ee have insufficient capacity. Although many modifications to the distribution system grid are possible, three that will correct the insufficient capacity problems are shown in the figure. Sections dd and ee were recalculated and the new results are shown in the table as sections dd and ee (revised).

Summary computations for the application of the method of sections

Section	Demand, 10^3 m^3/d	Pipes		Capacity at $s = 0.0031$, 10^3 m^3/d	Equivalent diameter, m	Slope, s when capacity = demand	Capacity check, 10^3 m^3/d	Velocity when capacity = demand, m/s
		No.	Dia, m					
aa	44.14	1	0.75	57.64	0.750	0.00225	49.10	1.29
bb	45.39	1	0.6	32.99	0.692	0.00228	31.80	1.30
		1	0.3	5.83			5.62	0.92
		2	0.2	4.23			4.08	0.75
		4	0.15	4.12			3.98	0.65
				47.17			45.48	
cc	40.52	2	0.4	23.95	0.667	0.00275	22.55	1.04
		2	0.3	11.66			10.98	0.90
		3	0.2	6.35			5.98	0.73
		1	0.15	1.03			0.97	0.64
				42.99			40.48	
dd	34.89	1	0.4	11.97	Insufficient capacity			
		1	0.3	5.83				
		4	0.2	8.47				
		2	0.15	2.06				
				28.33				
ee	8.27	1	0.2	2.12	Insufficient capacity			
		5	0.15	5.16				
				7.28				
dd (revised)	34.89	2	0.4	23.95	0.618	0.00299	23.52	1.08
		5	0.2	10.58			10.39	0.77
		1	0.15	1.03			1.01	0.66
				35.56			34.92	
ee (revised)	8.27	3	0.2	6.35	0.364	0.00237	5.55	0.68
		3	0.15	3.09			2.70	0.59
				9.44			8.25	

Digital Computer Analysis

Most distribution networks are now analyzed using digital computer programs. In writing a computer program to solve network flow problems, the following equations must be satisfied simultaneously throughout the network.

At each junction:

$$\sum Q_{\text{inflow}} = \sum Q_{\text{outflow}} \tag{6-2}$$

For each complete circuit:

$$\Sigma H = 0 \tag{6-3}$$

For each pipe:

$$H = kQ^n \tag{6-4}$$

In the more sophisticated of the network computer programs, Eqs. 6-2, 6-3, and 6-4 are solved simultaneously using one of several matrix inversion techniques. Several solution techniques are presented and analyzed in Jeppson. [6-4]

Perhaps the greatest advantage offered by the use of computers is that many more solutions can be developed at a reasonable cost to assess the response of the system to varying inputs. Also, real-time analysis can be used to study the effects of varying pump operation plans. At the present time almost all consulting firms and most industries have in-house computer programs or have access to such programs offered by several of the national computing services. The key issue is not in writing such a program, but in understanding what problems should be solved.

6-8 CROSS-CONNECTIONS IN DISTRIBUTION SYSTEMS

A cross-connection occurs when the drinking-water supply is connected to some source of pollution. For example, if a community has a dual water distribution system, one for fire fighting and the other for domestic consumption, the two may be interconnected so that domestic water may be used to supplement the other system in case of fire. Such an arrangement is dangerous, for contaminated water from the fire-fighting supply may get into the drinking-water system even though the two systems are normally separated by closed valves. The preferred method for interconnecting dual systems is the air break, although double check valves are sometimes used.

Cross-connections may occur in private residences, apartment houses, and commercial buildings, especially with old-style plumbing fixtures. If the water inlet of a plumbing fixture is below the overflow drain or rim, a reduced pressure in the water system may cause back siphonage. Other sources of cross-connections around a household include bathtubs, fish ponds, swimming pools with underrim inlets, and lawn sprinklers that become submerged when used.

6-9 CONSTRUCTION OF WATER DISTRIBUTION SYSTEMS

The basic requirements of pipes for water distribution systems are adequate strength and maximum corrosion resistance. Although cast-iron, cement-lined steel, plastic, and asbestos cement compete in the small sizes, steel and reinforced concrete are more competitive in the larger sizes. In cold climates, pipes should be far enough below ground to prevent freezing in winter. For even the coldest parts of the United States, a depth of 1.5 m (5 ft) is generally more than adequate. In warm climates, the pipes need be buried only sufficiently deep to avoid damage from traffic loads. Service connections to cast-iron or asbestos-cement pipe are made by tapping the distribution main with a special tapping machine. A corporation cock is then installed with a flexible gooseneck pipe leading to the service pipe. The gooseneck prevents damage if there is unequal settlement between the main and the service pipe. Service pipes leading from the main to the consumer are usually of copper tubing or galvanized steel. For single-family dwellings, 20- to 30-mm ($\frac{3}{4}$- to $1\frac{1}{4}$-in) pipe is common, but larger service sizes may be needed for apartment houses or business establishments.

Filling A New System

When a new pipe is first filled, all hydrants and valves are opened so that air can escape freely. Filling is done slowly and may require several days for large systems. Excessive pressures can develop if the air is not properly taken out of the system. When a steady, uninterrupted stream issues from a hydrant, it is closed. The procedure is continued until all valves and hydrants are closed and the system is full of water.

Leakage

The amount of leakage from distribution systems will vary with the care exercised in construction and the age and condition of the system. Leakage values from 5 to 25 L/mm of pipe diam · km · d (50 to 250 gal/in of pipe diam · mi · d) are common for new systems. The test is made by closing off a length of pipe between valves and all service connections to the pipe. Water is introduced through a special inlet, and normal working pressure is maintained for at least 12 h while leakage is measured. In an operating system the total loss is estimated from the difference between measured input to the system and metered deliveries to the customers. There are several possible methods of locating a specific leak. Patented leak detectors use audiophones to pick up the sound of escaping water or the disturbance in an electrical field caused by saturated ground near the leak. Similar devices may be used to locate the pipe itself if the exact location is unknown. If pressure gages are installed along a given length of pipe from which there are no takeoffs, a change in slope of the hydraulic gradient can be taken as an indication of a leak. In some instances the escaping water itself or unusually lush vegetation may be used to spot the location of a leak. The location of all pipes, valves, and appurtenances should be entered on maps. This information is essential in case repairs are ever required.

Disinfection of New Systems

While pipe is being handled and placed, there are many opportunities for pollution. Hence, it is necessary to disinfect a new system or an existing system after repairs or additions. Disinfection is usually accomplished by introducing chlorine, calcium hypochlorite, or chlorinated lime in amounts sufficient to give an immediate chlorine residue of 50 mg/L. The chemical is introduced slowly and permitted to remain in the system for at least 12 and preferably 24 h before it is flushed out. The flushing may be accomplished by opening several fire hydrants.

Maintenance of Distribution Systems

The hydraulic efficiency of pipes will diminish with time because of tuberculation, encrustation, and sediment deposits. Flushing will dislodge some of the foreign matter, but to clean a pipe effectively a scraper must be run through it. The scraper may be forced through by water pressure or pulled through with a cable. Cleaning, even though costly, may pay off with increased hydraulic efficiency and increased pressures throughout the system. The effects of cleaning may last only a short time, and in many cases pipes are lined with cement mortar after cleaning to obtain more permanent results.

6-10 PUMPING REQUIRED FOR WATER SUPPLY SYSTEMS

In some cases, gravity can be used as the driving force to bring water from its source to the consumer. In most cases, however, some form of pumping will be required. Pumps are required to deliver water from wells and where necessary to lift water to distribution reservoirs and elevated tanks. Often booster pumps must be installed on the mains to increase the pressure. Pumps and pump stations are considered in greater detail later in this chapter.

Wastewater Collection

Once used for its intended purposes, the water supply of a community is considered to be wastewater. The individual pipes used to collect and transport wastewater are called *sewers*, and the network of sewers used to collect wastewater from a community is known as a *collection system*.

The purpose of this section is to define the types of collection systems that are used, the appurtenances used in conjunction with sewers, the flow in sewers, the design of sewers, the materials of construction, and the construction and maintenance of sewers.

6-11 TYPES OF COLLECTION SYSTEMS

The three general types of collection systems commonly used in the United States are sanitary, stormwater, and combined. The characteristics of each of these types of sewers are discussed below.

Sanitary Sewers

Often identified as separate sewers, sanitary sewers were developed to remove domestic wastes from residential areas. Originally, the flow in sanitary sewers was by gravity. More recently, both pressure and vacuum sewers have been used to serve areas where gravity sewers would be difficult and costly to install and maintain.

Stormwater Sewers

Sewers intended solely for the collection of stormwater are known as stormwater sewers. Usually larger than sanitary sewers, separate stormwater sewers are constructed to eliminate pollution problems associated with the discharge of untreated wastewater from combined sewers into watercourses and receiving waters. More recently, the treatment of stormwater has developed into a separate and specialized field. For this reason the design of stormwater sewers is not considered in this section. Detailed information on stormwater sewers may be found in Refs. [6-5, 6-7, 6-14, and 6-16].

Combined Sewers

Domestic wastewater and stormwater are collected together in combined sewers. Although the use of combined sewers persists in many of the older cities in the United States, they are seldom constructed today. They are still used extensively in many parts of the world, however.

6-12 TYPES OF SEWERS

The types and sizes of sewers used will vary with size of the collection system and the location of the wastewater-treatment facilities. The principal types of sewers found in most collection systems are described by function in Table 6-4 and illustrated graphically in Fig. 6-6.

Table 6-4 Types of sewers in a typical collection system

Type of sewer	Purpose
Building	Building sewers, sometimes called *building connections*, connect to the building plumbing and are used to convey wastewater from the buildings to lateral or branch sewers, or any other sewer except another building sewer. Building sewers normally begin outside the building foundation. The distance from the foundation wall to where the sewer begins depends on the local building regulations.
Lateral or branch	Lateral sewers form the first element of a wastewater collection system and are usually in streets or special easements. They are used to collect wastewater from one or more building sewers and convey it to a main sewer.
Main	Main sewers are used to convey wastewater from one or more lateral sewers to trunk sewers or to intercepting sewers.
Trunk	Trunk sewers are large sewers that are used to convey wastewater from main sewers to treatment or other disposal facilities or to large intercepting sewers.
Intercepting	Intercepting sewers are larger sewers that are used to intercept a number of main or trunk sewers and convey the wastewater to treatment or other disposal facilities.

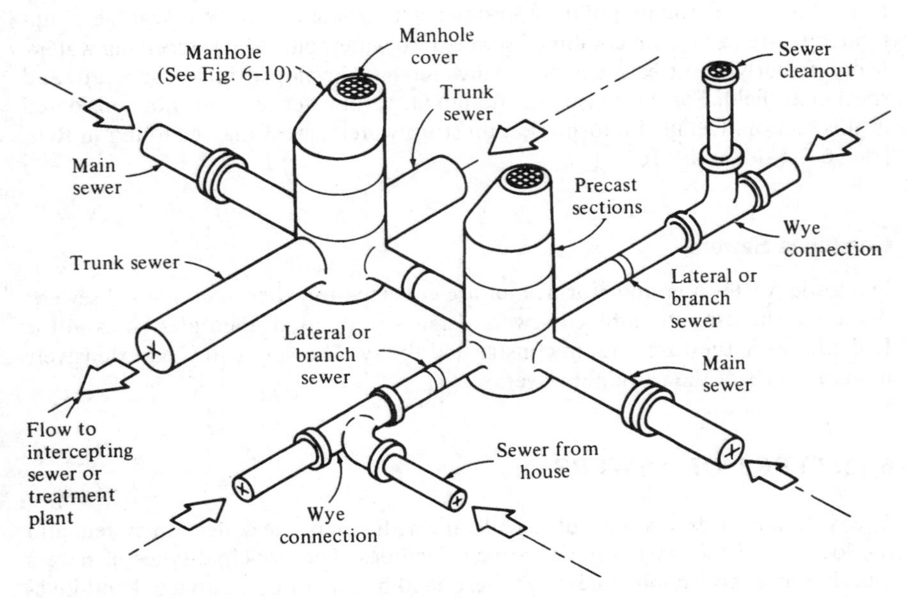

Figure 6-6 Definition sketch for types of sewers used in collection systems.

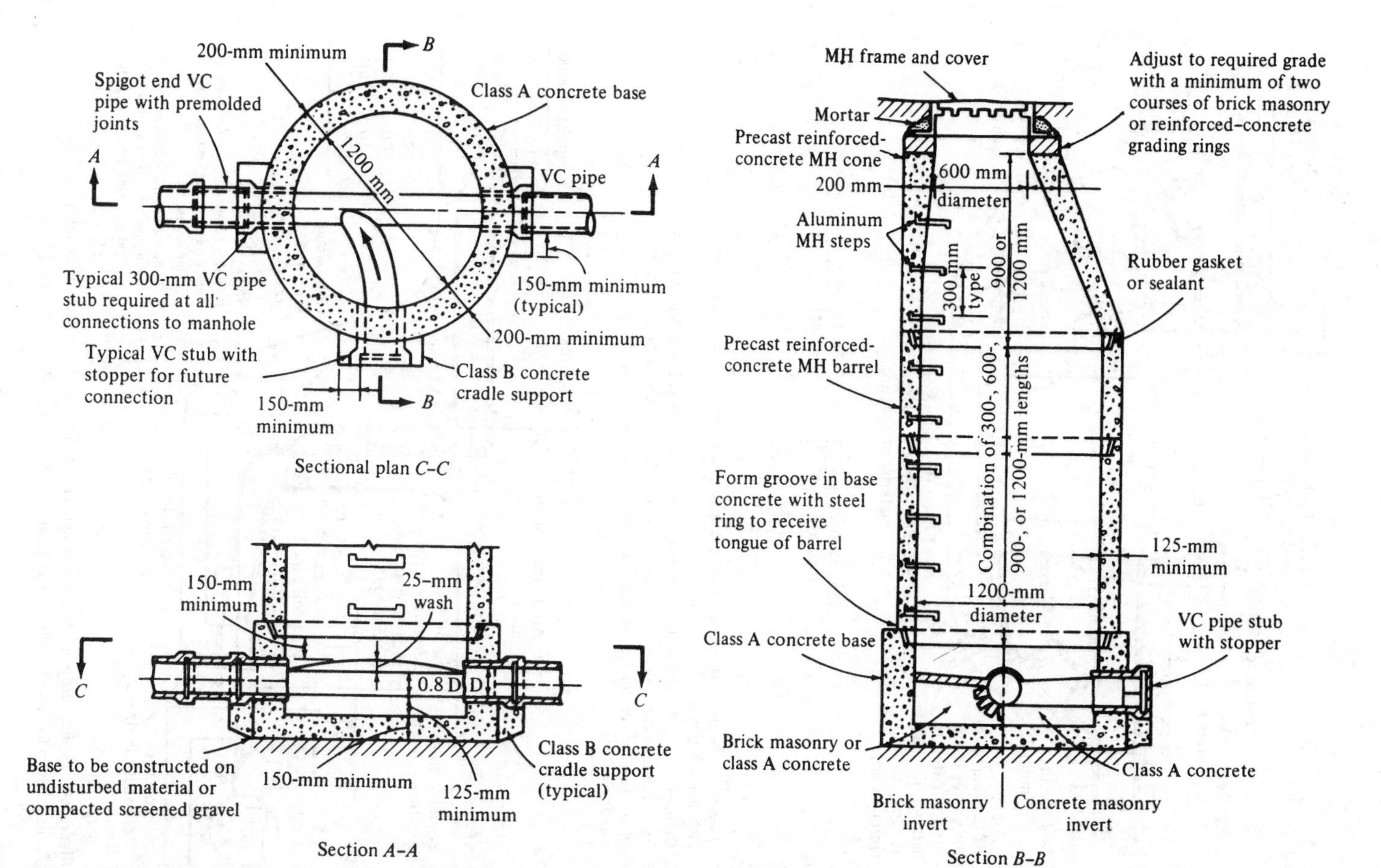

Figure 6-7 Typical manhole used for reinforced-concrete sewer pipe. (*From Metcalf & Eddy, Inc.* [6-8].)

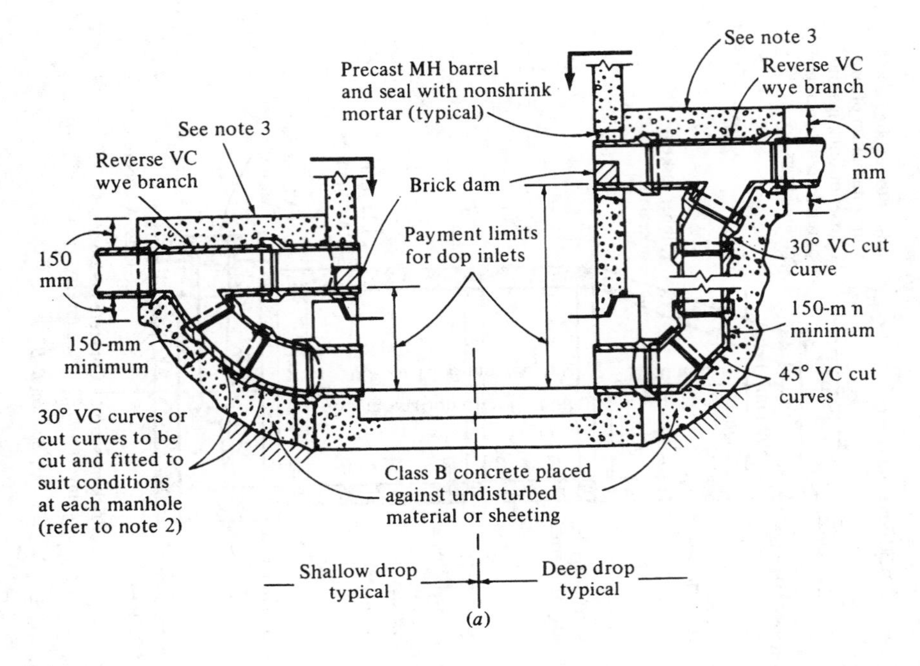

Notes:

1. Drop pipe to be same diameter as sewer discharging into manhole for up to and including including 300-mm size.
2. Deeper drop may be constructed with straight pipe between wye branch and curve.
3. Extend encasement to first joint beyond excavation for drop connection.
4. Dimensions and construction of drop manhole to be similar to typical manhole except as shown.

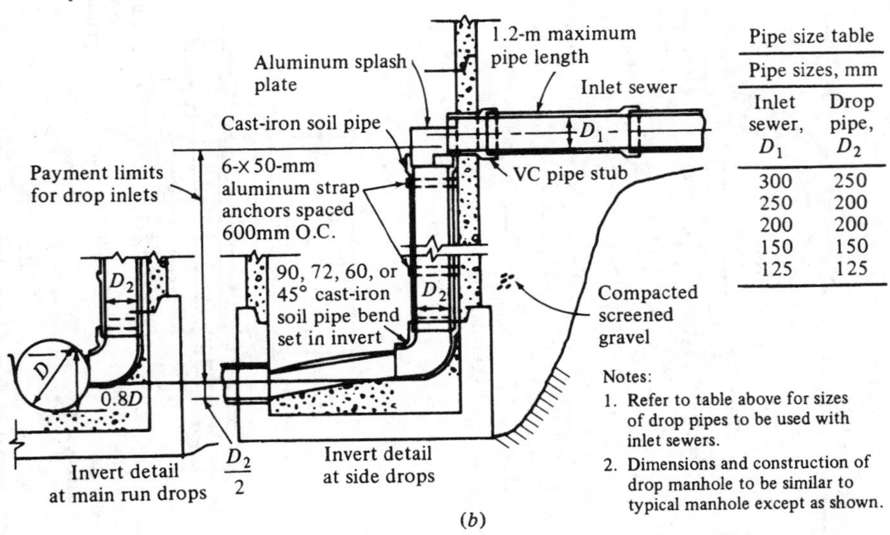

Pipe size table

Pipe sizes, mm	
Inlet sewer, D_1	Drop pipe, D_2
300	250
250	200
200	200
150	150
125	125

Figure 6-8 Typical drop inlets for vitrified clay pipe used in collection systems: (*a*) outside drop, (*b*) inside drop for sewer 600 mm and smaller. (*From Metcalf & Eddy, Inc.* [6-8].)

6-13 COLLECTION SYSTEM APPURTENANCES

The principal appurtenances of sanitary sewers are manholes, drop inlets to manholes, building connections, and junction chambers. Depending on local topography, special structures may be required.

Manholes

Manholes are used to interconnect two or more sewers (see Fig. 6-6) and to provide entry for sewer cleaning. For sewers that are 1200 mm (48 in) and smaller, manholes should be located at changes in size, slope, or direction. In larger sewers these changes can be made without using a manhole. A typical manhole for reinforced-concrete pipe is shown in Fig. 6-7.

Drop Inlets to Manholes

Where the difference in elevation between the incoming and outgoing sewer exceeds 0.5 m (1.5 ft), flow from incoming sewer can be dropped to the elevation of the outgoing sewer with a drop inlet such as shown in Fig. 6-8.

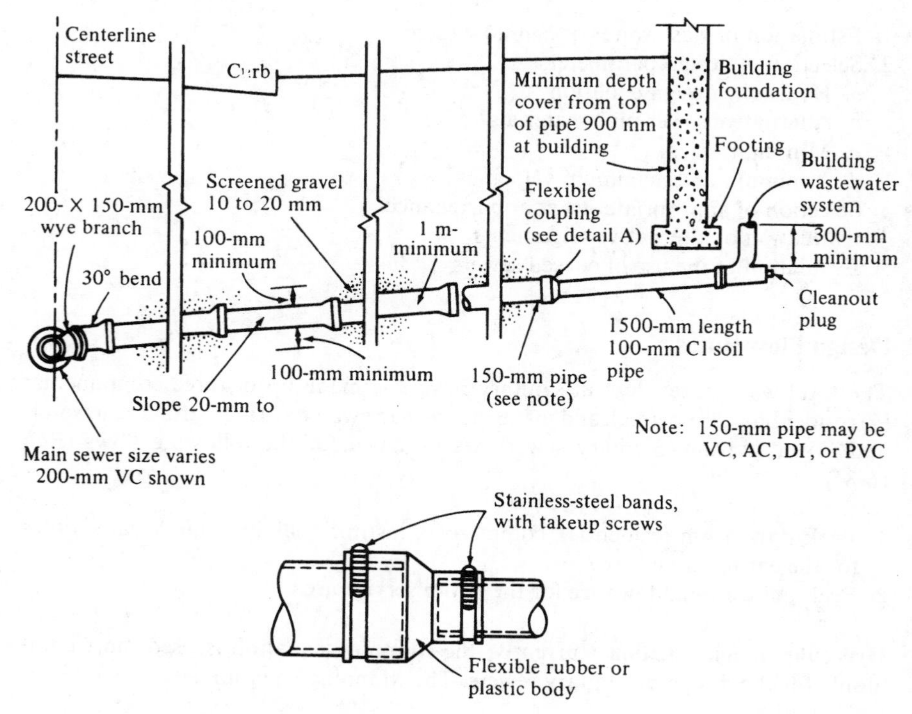

Figure 6-9 Typical building connection. (*From Metcalf & Eddy, Inc.* [6-8].)

Building Connections

The sewers leading from individual houses or buildings to the municipal collection system are known as building connections. A typical house connection is shown in Fig. 6-9.

Junction Chambers

As the diameter of intersecting sewers (e.g., main and trunk sewers) continues to increase as wastewater from more of the service area is collected, precast manholes can no longer be used. When this situation occurs, special junction chambers are constructed to connect the intersecting sewers.

6-14 BASIC CONSIDERATIONS IN THE DESIGN OF SEWERS

In planning and designing sanitary sewers the following factors must be considered separately for each installation:

1. Estimation of wastewater design flow rates
2. Selection of design parameters
   a. Hydraulic design equation
   b. Alternative sewer pipe materials
   c. Minimum sizes
   d. Minimum and maximum velocities
3. Selection of appropriate sewer appurtenances
4. Evaluation of alternative alignments
5. Evaluation of the use of curved sewers

Design Flow Rates

The total wastewater flow in sanitary sewers is made up of three components: (1) residential, commercial, and institutional wastewater, (2) industrial wastewater, and (3) infiltration. Sanitary sewers are designed for the following flows (Ref. [6-8]):

1. Peak flows from residential, commercial, institutional, and industrial sources for the entire service area
2. Peak infiltration allowance for the entire service area

Hydraulic design equation Currently, the Manning equation is used most commonly for the design of sanitary sewers. The Manning equation is

$$V = \frac{1}{n} R^{2/3} S^{1/2} \tag{6-5}$$

where V = velocity, m/s
n = friction factor
R = hydraulic radius

$$= \frac{\text{cross-sectional area of flow, m}^2}{\text{wetted perimeter, m}}$$

S = slope of energy grade line, m/m

The recommended n value for the design of new and existing well-constructed sewers is 0.013. An n value of 0.015 is recommended for the analysis of older sewers. The graphs presented in Figs. 6-10 and 6-11 have been prepared to simplify the use of the Manning equation in the design of sewers. Also because many sewers do not flow full, the relationship between hydraulic elements for flow at full depth and at other depths in circular sewers is illustrated in Fig. 6-12. Developed using the Manning equation, Fig. 6-12 is used to obtain the values of V, Q, A, R, and n at a given depth ratio based on the corresponding values of V_f, Q_f, A_f, R_f, and n_f when the pipe is flowing full.

Sewer Pipe Materials and Sizes

The principal materials used in the manufacture of sewer pipe are asbestos cement, ductile iron, reinforced concrete, prestressed concrete, polyvinyl chloride, and vitrified clay. Information on the sizes of pipes made with these materials is presented in Table 6-5.

Minimum sewer sizes are usually specified in local building codes. The smallest sewer used should be larger than the building sewer connections so that objects passed through the building sewer will not clog the municipal sewer. Building sewer connections vary in size from 100 to 150 mm (4 to 6 in). The minimum size recommended for gravity sewers is 200 mm (8 in), although 150-mm (6-in) connections have been used in some communities.

Minimum and Maximum Velocities

When the velocity of flow in a sewer is low, there is a tendency for the solids present in wastewater to settle out. Because the deposited solids may accumulate and ultimately block the flow, sufficient velocity should be developed on a regular basis to flush out any deposited solids. Based on past experience, current practice is to design sanitary sewers with appropriate slopes to maintain a minimum flow velocity of 0.6 m/s (2.0 ft/s) when the sewer is flowing full or half full. To prevent the deposition of sand and gravel a velocity of 0.75 m/s (2.5 ft/s) is recommended. To avoid damaging sewers it is recommended that the maximum flow velocities be limited to values equal to or less than 3.0 m/s (10 ft/s).

Minimum Slopes

Minimum slopes are often used to avoid extensive excavation where the slope of the ground surface is flat. In general, minimum slopes based on Manning's equation

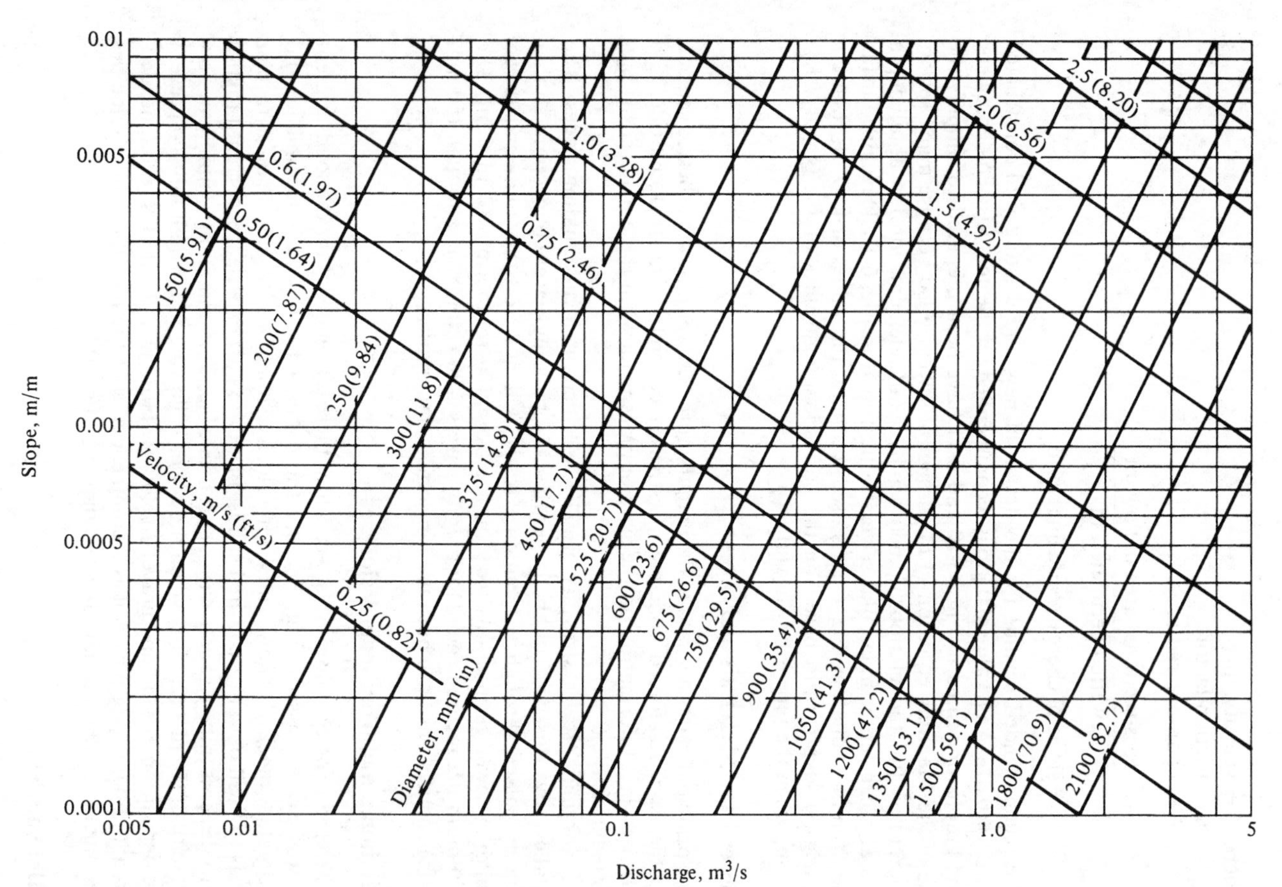

Figure 6-10 Nomograph for solution of Manning's equation for $n = 0.013$.

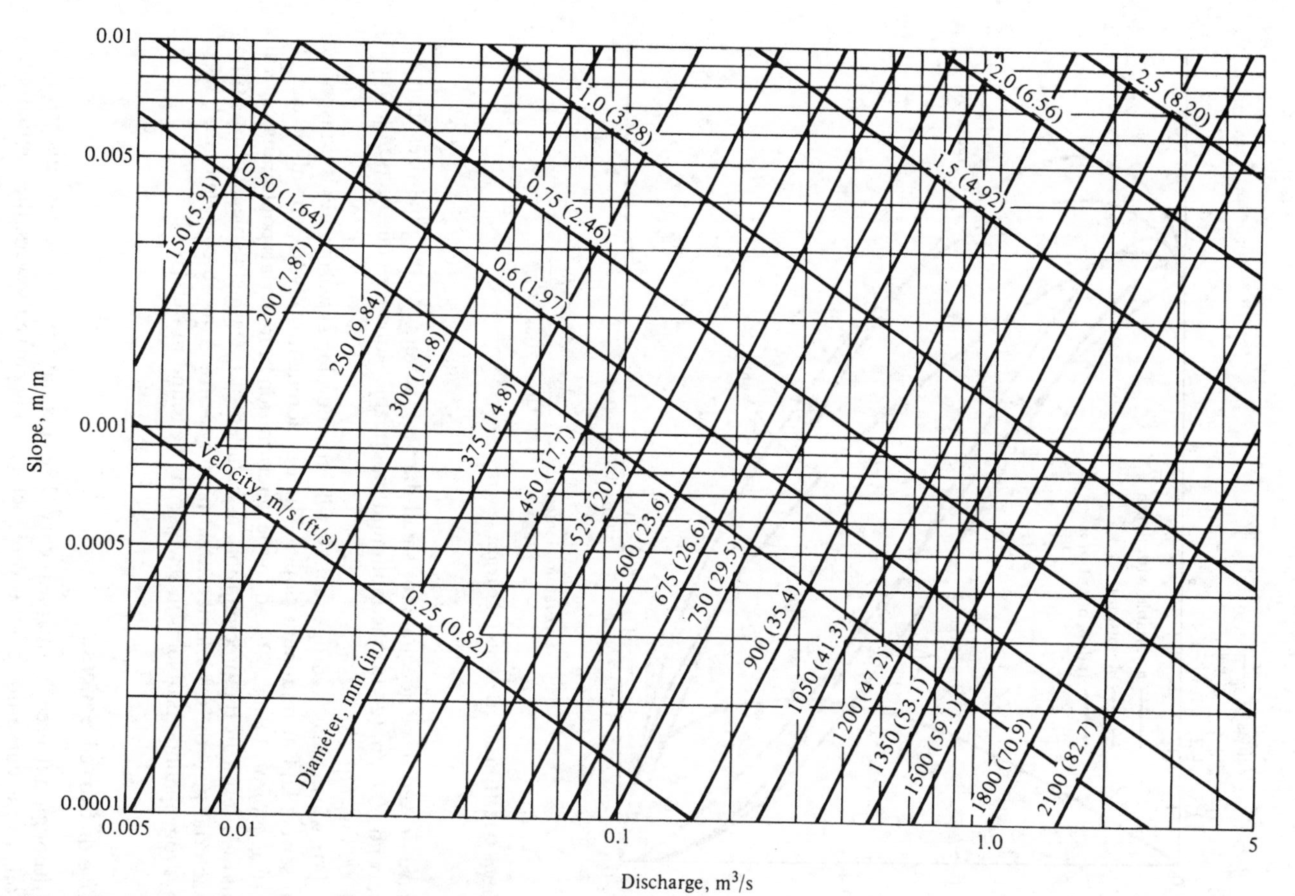

Figure 6-11 Nomograph for solution of Manning's equation for $n = 0.015$.

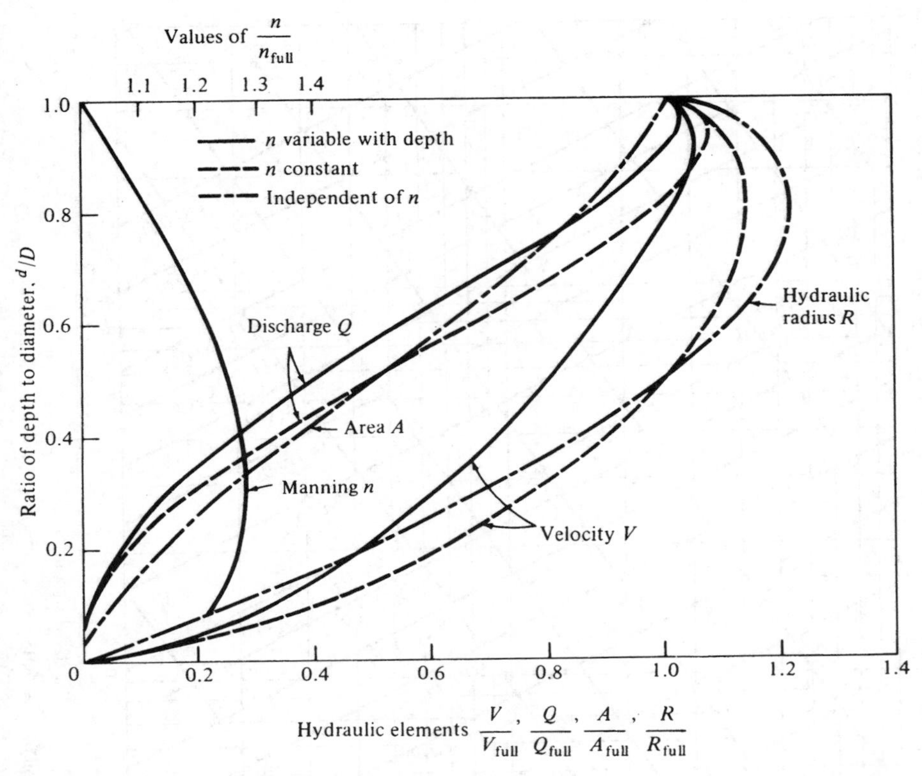

Figure 6-12 Hydraulic elements for circular sewers.

have proved to be adequate for small-diameter sewers. As the pipe sizes increase beyond 600 mm (24 in) the minimum practicable slope for construction is about 0.0008 m/m.

In warm areas, hydrogen sulfide will often develop as wastewater is transported in sewers laid at minimum slopes. The development of hydrogen sulfide can (1) cause odor problems, (2) lead to the deterioration of materials containing cement, and (3) result in the precipitation, as sulfides, of trace metals needed for proper bacterial growth in biological treatment systems. A more complete analysis of hydrogen-sulfide problems in sewers may be found in Ref. [6-8].

Use of Curved Sewers

Although not used in the past, curved sewers have proved to be satisfactory as long as the curvature is not severe. Before using curved sewers, the compatibility of the cleaning equipment to be used for sewer maintenance must be assessed.

Table 6-5 Available size ranges and descriptions of commonly used pipe for gravity-flow sewers

Type of pipe	Available size range, mm (in)	Description
Asbestos cement (AC)	100–900 (4–36)	Weighs less than other commonly rigid pipes. May be susceptible to acid corrosion and hydrogen sulfide attack, but if properly cured with steam at high pressure (autoclave process), may be used even in environments with moderately aggressive waters or soils with high-sulfate content.
Ductile iron (DI)	100–1350 (4–54)	Often used for river crossings and where the pipe must support unusually high loads, where an unusually leakproof sewer is required, or where unusual root problems are likely to develop. Ductile-iron pipes are susceptible to acid corrosion and hydrogen sulfide attack, and therefore should not be used where the groundwater is brackish, unless suitable protective measures are taken.
Reinforced concrete (RC)	300–3600 (12–144)	Readily available in most localities. Susceptible to corrosion of interior if the atmosphere over waste-water contains hydrogen sulfide, or from outside if buried in an acid of high-sulfate environment.
Prestressed concrete (PC)	400–3600 (16–144)	Especially suited to long transmission mains without building connections and where precautions against leakage are required. Susceptibility to corrosion (the same as reinforced concrete).
Polyvinyl chloride (PVC)	100–375 (4–15)	A plastic pipe used for sewers as an alternative to asbestos-cement and vitrified-clay pipe. Lightweight but strong. Highly resistant to corrosion.
Vitrified clay (VC)	100–900 (4–36)	For many years the most widely used pipe for gravity sewers; still widely used in small and medium sizes. Resistant to corrosion by both acids and alkalies. Not susceptible to damage from hydrogen sulfide, but is brittle and susceptible to breakage.

* *Source*: From Metcalf & Eddy, Inc. [6-8]

Another reason for not installing curved sewers is that the use of laser-type surveying equipment for maintaining grade during construction is not feasible.

Sewer Ventilation

Ventilation in sewers is needed to avoid (1) the danger of asphyxiation of sewer maintenance employees, (2) the buildup of odorous gases, and (3) the development of explosive mixtures of sewer gases, principally methane and oxygen. Design considerations for the ventilation of sewers are discussed in detail in Ref. [6-8].

6-15 DESIGN OF SANITARY SEWERS

The design of sanitary sewers involves fieldwork, the preparation of maps and profiles, and detailed design computations. Each of these topics is considered briefly below. The detailed design of sanitary sewers is illustrated in Example 6-4.

Fieldwork

To design sanitary sewers properly, accurate and detailed maps should be available for the areas to be served. The location of streets, alleys, highways, railroads, public buildings and parks, streams, drainage ditches, and other features that may influence the design of the sewers should be identified. Accurate elevations are needed throughout the area to be served by the proposed sewer. Profiles are needed for all existing or proposed streets, alleys, and potential rights-of-way where sewers may be placed. In addition, detailed information must be available on the location of surface and subsurface utilities such as water and gas mains, electrical conduits, drain lines, and other underground structures.

In addition to the above information, soils data should also be available. Soils borings should be made to a depth of at least 1.5 m (5 ft) below the bottom of the sewer trench.

Preparation of Maps and Profiles

While the fieldwork is going on, work on the preparation of maps and profiles should proceed simultaneously. Thus, if any information is found to be missing it can be collected before the fieldwork is completed. Maps on a scale of 25 m to 10 mm (200 ft to 1 in) are acceptable for most purposes. Where additional detail is needed a scale of about 5 m or less to 10 mm (40 ft or less to 1 in) is often used. In preparing design profiles, street centerline elevations are shown at least every 15 m (50 ft) and at all locations where the surface slope changes abruptly.

Design Computations for Sanitary Sewers

The detailed design of sanitary sewers involves the selection of appropriate pipe sizes and slopes to transport the quantity of wastewater expected from the surrounding and upstream areas the next pipe in series, subject to the appropriate design constraints. The design procedure for sanitary sewers is illustrated in Example 6-4.

Example 6-4: Designing a gravity-flow sanitary sewer Design a gravity-flow trunk sanitary sewer for the area shown in the accompanying figure (*a*). The trunk sewer is to be laid along Peach Avenue starting at 4th Street and ending at 11th Street. Assume that the following design criteria have been developed based on an analysis of local conditions and codes.

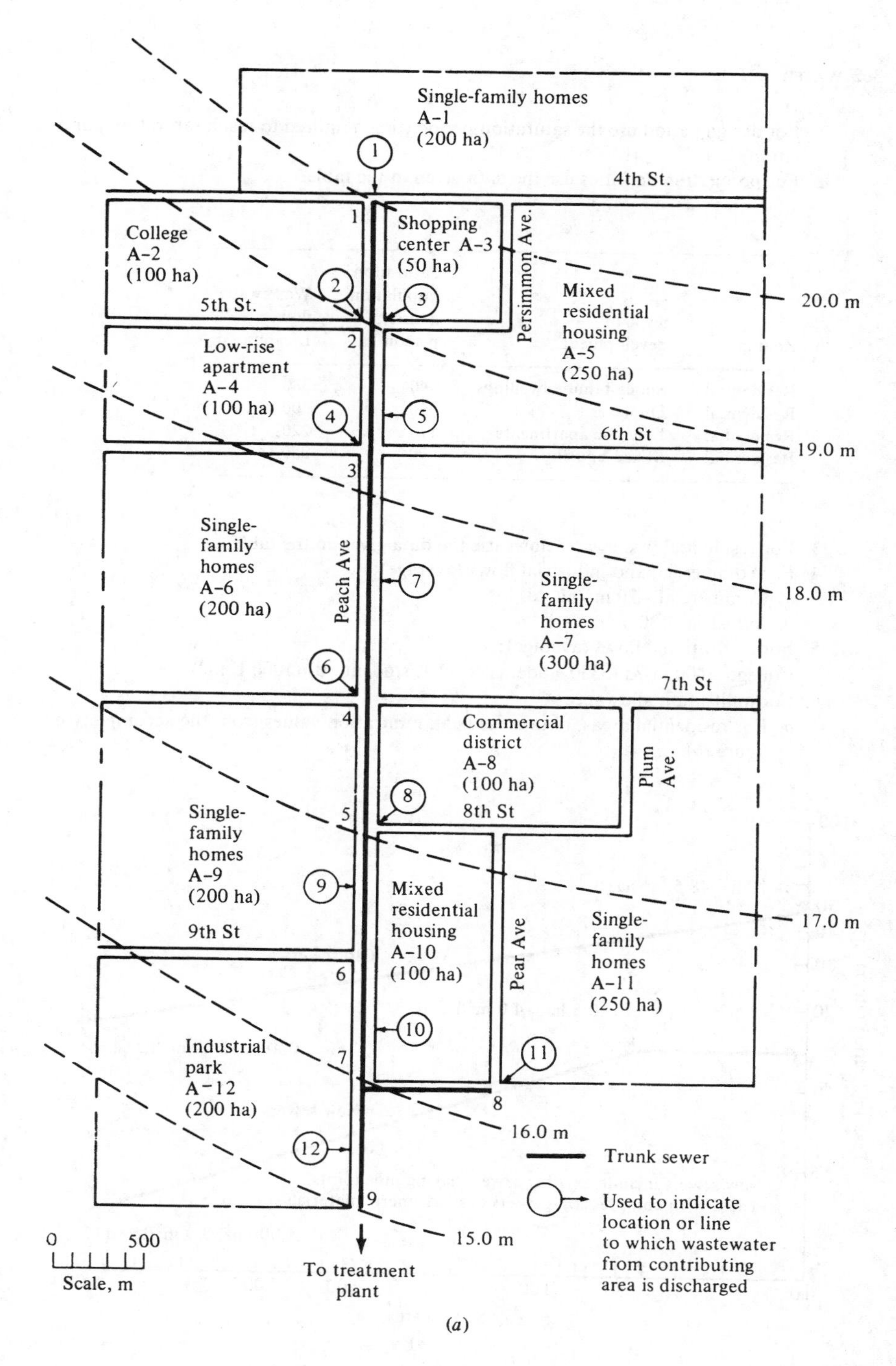
Single-family homes
A-1
(200 ha)
1
4th St.
College
A-2
(100 ha)
1
Shopping
center A-3
(50 ha)
Persimmon Ave.
2
3
Mixed
residential
housing
A-5
(250 ha)
20.0 m
5th St.
Low-rise
apartment
A-4
(100 ha)
2
4
5
6th St
19.0 m
3
Single-
family
homes
A-6
(200 ha)
Peach Ave
7
Single-
family
homes
A-7
(300 ha)
18.0 m
6
7th St
4
Commercial
district
A-8
(100 ha)
8th St
Plum
Ave.
8
5
Single-
family
homes
A-9
(200 ha)
9
9th St
Mixed
residential
housing
A-10
(100 ha)
Pear Ave
Single-
family
homes
A-11
(250 ha)
17.0 m
6
10
11
7
Industrial
park
A-12
(200 ha)
8
16.0 m
12
Trunk sewer
9
Used to indicate
location or line
to which wastewater
from contributing
area is discharged
15.0 m
0
500
Scale, m
To treatment
plant

(*a*)

1. For design period use the saturation period (time required to reach saturation population).
2. For population densities use the data given in the table.

Zoning	Type of development	Saturation population density, person/ha	Wastewater flow, L/capita · d
Residential	Single-family dwellings	40	380
Residential	Duplexes	60	300
Residential	Low-rise apartments	120	220
Residential	Mixed housing	70	250

3. For residential wastewater flows use the data given in the table.
4. For commercial and industrial flows (average):
   a. Commercial—20 m^3/ha · d.
   b. Industrial—30 m^3/ha · d.
5. For institutional flows (average):
   College—400 m^3/d (5330 students × 75 L/student · d)/(1000 L/m^3)
6. For infiltration allowance:
   a. For residential areas, obtain the peak infiltration values from the accompanying figure (*b*).

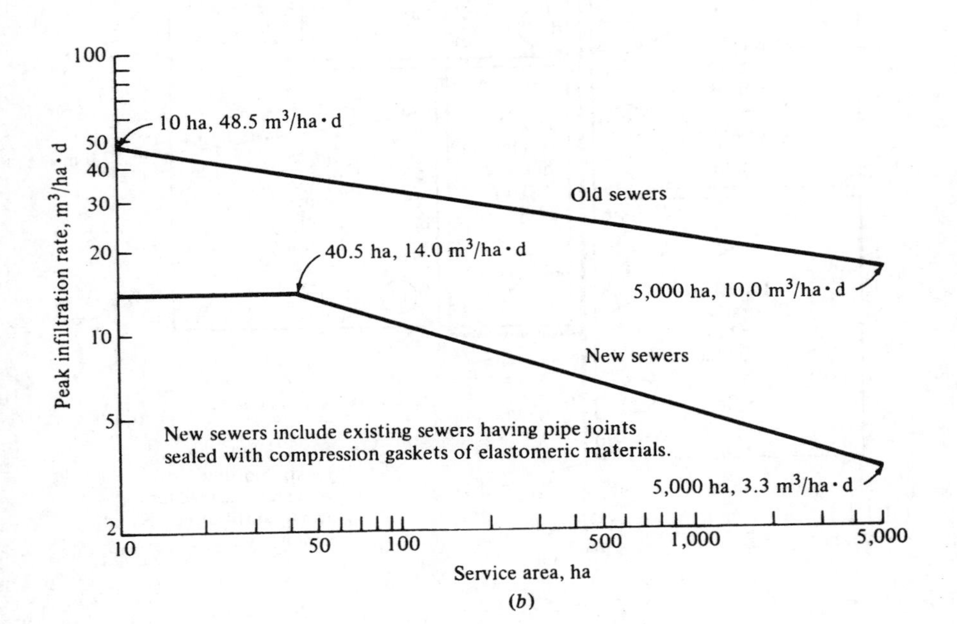

(*b*)

   b. For commercial, industrial, and institutional areas, also obtain the peak infiltration values from the figure (*b*). However, to take into account that the total length of sewers in these areas will generally be less than that in residential areas, use only 50 percent of the actual area to compute the infiltration allowance.
7. For inflow allowance assume that the steady-flow inflow is accounted for in the infiltration allowance.
8. Peaking factors:
   a. Residential—use the curve given in the accompanying figure (*c*).
   b. Commercial—1.8
   c. Industrial—2.1
   d. Institutional (school)—4.0

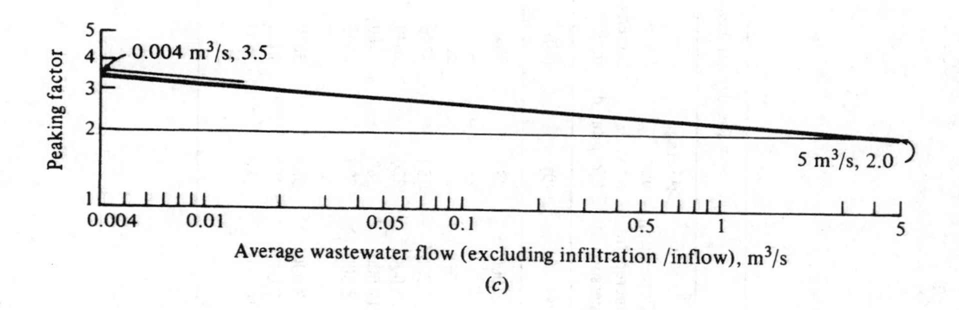

(*c*)

9. For the hydraulic design equation use the Manning equation with an n value of 0.013. To simplify the computations, use Fig. 6-10.
10. Minimum pipe size: The local building code specifies 200 mm (8 in) as the smallest pipe permissible for this situation.
11. Minimum velocity: To prevent the deposition of solids at low wastewater flows, use a minimum velocity of 0.75 m/s (2.5 ft/s) during the peak flow conditions.
12. Minimum cover (minimum depth of cover over the top of the sewer): As established by the local community building code, the minimum depth of cover is 2.0 m.

SOLUTION

1. Lay out the trunk sewer. Draw a line to represent the proposed sewer [see figure (*a*)].
2. Locate and number the manholes. Locate manholes at (1) changes in direction, (2) changes in slope, (3) pipe junctions, (4) upper ends of sewers, and (5) intervals from 90 to 120 m or less. Identify each manhole with a number [see figure (*a*)]. For the purpose of this example only the manholes at the major junctions have been numbered. In an actual design, intermediate manholes would be located and numbered.
3. Prepare a sewer design computation table. Based on the experience of numerous engineers, it has been found that the best approach for carrying out sewer computations is to use a computation table. The necessary computations for the sanitary sewer shown in figure (*a*) are presented in the accompanying table. Although the table is, for the most part, self-explanatory, the following comments are presented to clarify its development.

Sewer computation table

Location					Residential flows							
Line	From	To	Length of sewer, m	Subarea*	Area, ha	Population density, persons/ha	Population increment, persons	Average unit flow, L/capita · d	Flow increment, m^3/d	Cumulative average flow, m^3/d	Peaking factor	Cumulative peak flow, m^3/d (11 × 12)
(1)	(2)	(3)	(4)	(5)	(6)	(7)	(8)	(9)	(10)	(11)	(12)	(13)
1	1	2	707	A-1	200	40	8,000	380	3040	3,040	2.9	8,816
2	2	3	707	A-2	—	—	—	—	—	3,040	2.9	8,816
				A-3	—	—	—	—	—	3,040	2.9	8,816
				A-5	250	70	17,500	250	4375	7,415	2.7	20,021
3	3	4	1414	A-4	100	120	12,000	220	2640	10,055	2.6	26,143
				A-7	300	40	12,000	380	4560	14,615	2.6	37,999
4	4	5	707	A-6	200	40	8,000	380	3040	17,655	2.5	44,138
5	5	6	707	A-8	—	—	—	—	—	17,655	2.5	44,138
				A-9	200	40	8,000	380	3040	20,695	2.5	51,738
6	6	7	707	A-10	100	70	7,000	250	1750	22,445	2.5	56,113
7	8	7	707	A-11	250	40	10,000	380	3800	3,800†	2.9	11,020
8	7	9	707	707	A-12					26,245	2.5	65,613

Sewer computation table (*Continued*)

Location					Commercial flows					Industrial flows				
Line	From	To	Length of sewer, m	Subarea*	Area, ha	Average unit flow, $m^3/ha \cdot d$	Cumulative average flow, m^3/d	Peaking factor	Cumulative peak flow, m^3/d (16 × 17)	Area, ha	Average unit flow, $m^3/ha \cdot d$	Cumulative average flow, m^3/d	Peaking factor	Cumulative peak flow, m^3/d (21 × 22)
(1)	(2)	(3)	(4)	(5)	(14)	(15)	(16)	(17)	(18)	(19)	(20)	(21)	(22)	(23)
1	1	2	707	A-1	—	—	—	—	—	—	—	—	—	—
2	2	3	707	A-2	—	—	—	—	—	—	—	—	—	—
				A-3	50	20	1000	1.8	1800	—	—	—	—	—
				A-5	—	—	1000	1.0	1800	—	—	—	—	—
3	3	4	1414	A-4	—	—	1000	1.8	1800	—	—	—	—	—
				A-7	—	—	1000	1.8	1800	—	—	—	—	—
4	4	5	707	A-6	—	—	1000	1.8	1800	—	—	—	—	—
5	5	6	707	A-8	100	20	3000	1.8	5400	—	—	—	—	—
				A-9	—	—	3000	1.8	5400	—	—	—	—	—
6	6	7	707	A-10	—	—	3000	1.8	5400	—	—	—	—	—
7	8	7	707	A-11	—	—	—	—	—	—	—	—	—	—
8	7	9	707	A-12	—	—	3000	1.8	5400	200	30	6000	2.1	12,600

Sewer computation table *(Continued)*

Location					Institutional flows			Cumulative subtotals		Infiltration			
Line	From	To	Length of sewer, m	Subarea*	Cumulative average flow, m³/d	Peaking factor	Cumulative peak flow, m³/d (24 × 25)	Cumulative average flow, m³/d (11 + 16 + 21 + 24)	Cumulative peak flow, m³/d (13 + 18 + 23 + 26)	Area, ha	Cumulative area, ha	Peak unit infiltration allowance m³/ha · d	Cumulative infiltration allowance, m³/d (30 × 31)
(1)	(2)	(3)	(4)	(5)	(24)	(25)	(26)	(27)	(28)	(29)	(30)	(31)	(32)
1	1	2	707	A-1	—	—	—	3,040	8,816	200‡	200	8.0	1600
2	2	3	707	A-2	400	4.0	1600	3,440	10,416	50†	250	7.5	1875
				A-3	400	4.0	1600	4,440	12,216	25‡	275	7.5	2063
				A-5	400	4.0	1600	8,815	23,421	250	525	7.0	3675
3	3	4	1414	A-4	400	4.0	1600	11,455	29,543	100	625	6.5	4063
				A-7	400	4.0	1600	16,015	41,399	300	925	5.5	5088
4	4	5	707	A-6	400	4.0	1600	19,055	47,538	200	1125	5.0	5625
5	5	6	707	A-8	400	4.0	1600	21,055	51,138	50‡	1175	5.0	5875
				A-9	400	4.0	1600	24,095	58,738	200	1375	4.9	6738
6	6	7	707	A-10	400	4.0	1600	25,845	63,113	100	1475†	4.8	7080
7	8	7	707	A-11	—	—	—	3,800	11,020	250	250†	8.0	2000
8	7	9	707	A-12	400	4.0	1600	35,645	85,213	100‡	1825	4.0	7300

Sewer computation table (*Continued*)

Location					Design flows		Sewer design				Sewer layout			
											Ground surface elevation		Sewer pipe invert elevation	
Line	From	To	Length of sewer, m	Subarea*	Cumulative peak flow, m^3/d (28 + 32)	Cumulative peak flow,§ m^3/s	Sewer diameter, mm	Slope, m/m	Capacity when full, m^3/s	Velocity when full, m/s	At upper manhole	At lower manhole	Upper end	Lower end
(1)	(2)	(3)	(4)	(5)	(33)	(34)	(35)	(36)	(37)	(38)	(39)	(40)	(41)	(42)
1	1	2	707	A-1	10,416	0.121	450	0.0018	0.121	0.75	20.00	19.00	17.50	16.23
2	2	3	707	A-2	12,291	0.142	—	—	—	—	—	—	—	—
				A-3	14,279	0.165	—	—	—	—	—	—	—	—
				A-5	27,096	0.314	750	0.0009	0.330	0.75	19.00	18.33	15.93	15.29
3	3	4	1414	A-4	33,606	0.389	—	—	—	—	—	—	—	—
				A-7	46,487	0.538	900	0.0009	0.540	0.85	18.33	17.40	15.14	13.86
4	4	5	707	A-6	53,163	0.615	1050	0.0008¶	0.770	0.87	17.40	17.00	13.71	13.14
5	5	6	707	A-8	57,013	0.660	—	—	—	—	—	—	—	—
				A-9	65,476	0.758	1050	0.0008¶	0.770	0.87	17.00	16.50	13.14	12.58
6	6	7	707	A-10	70,193	0.812	1050	0.0009	0.820	0.95	16.50	16.00	12.58	11.94
7	8	7	707	A-11	13,020	0.151	525	0.0014	0.165	0.75	16.20	16.00	12.46	13.46
8	7	9	707	A-12	92,513	1.071	1200	0.0008¶	1.100	0.98	16.00	15.00	11.79	11.22

* See figure (*a*).
† Line 7 receives flow from subarea A-11 only.
‡ 50 percent of area (see assumption 6*b*).
§ $m^3/s = (m^3/d)/(86{,}400\ s/d)$.
¶ The minimum practical slope for construction is about 0.0008 m/m.

a. The entries in columns 1 through 5 are used to identify the sewer lines under consideration and to summarize the basic physical data from figure (*a*).

b. The entries in columns 6 through 13 are used to obtain the cumulative peak domestic flow (column 13). The area (column 6) is obtained from figure (*a*). The population density (column 7) and the unit flow data (column 9) were given. Peaking factors, obtained from figure (*c*), are entered in column 12.

c. The commercial area, the corresponding unit flow, and the cumulative average flows are entered in columns 14, 15, and 16, respectively. The given peaking factor for the commercial area is entered in column 17, and the computed cumulative peak commercial flows are entered in column 18.

d. The entries in columns 19 through 23 for the industrial flows are the same as described for the commercial flows (columns 14 through 18).

e. The institutional flows are entered in columns 24 through 26.

f. The cumulative average and peak flows are summarized in columns 27 and 28, respectively.

g. The infiltration allowance (columns 29 through 32) is determined using the curve for new sewers in figure (*b*).

h. The total cumulative peak design flow (column 33) is obtained by summing columns 28 and 32.

i. Sewer design information is summarized in columns 35 through 38. The required pipe sizes are estimated using Manning's equation with an n value of 0.013 (see Fig. 6-10). The capacity of the selected pipe and the velocity when full are tabulated in columns 37 and 38. In all cases the velocity should exceed 0.75 m/s (2.5 ft/s).

j. The necessary layout data for the sewer (columns 39 through 42) are obtained as follows: The ground surface elevations at the manhole locations entered in columns 39 and 40 are obtained by interpolation with the elevation data given in figure (*a*). The sewer invert elevations shown in columns 41 and 42 are obtained by trial and error with a sewer profile work sheet. The first step in preparing a work sheet is to plot the ground-surface elevations given in columns 39 and 40, working backwards from a convenient point. After the ground-surface profile is drawn, the next step is to begin sketching the invert and crown (inside bottom and inside top of the pipe, respectively) of each sewer section as the necessary elevation data are developed.

The method for establishing the invert elevations will be illustrated by analyzing selected sewer lines starting with line 1, which connects manholes 1 and 2. The first step is to locate the invert of the upper end of the pipe at such an elevation that the minimum cover requirement is satisfied, taking into account both the inside diameter of the pipe and its wall thickness. The upper invert elevation of the 450-mm pipe is set initially at elevation 17.5 m:

$$\begin{array}{ccccccc} \text{ground surface} & - & \text{depth of cover} & - & \text{pipe wall thickness} & - & \text{pipe diameter} \\ 20.00\text{ m} & - & 2.00\text{ m} & - & 0.05\text{ m} & - & 0.45\text{ m} \end{array}$$

The pipe thickness will vary with the type of sewer. For this example, 0.05 m will be used for all pipe sizes. The lower elevation is computed by subtracting the fall as follows:

$$\begin{array}{l} \text{Lower} \\ \text{invert} \\ \text{elevation} \end{array} = \begin{array}{l} \text{Upper} \\ \text{invert} \\ \text{elevation} \end{array} - \begin{array}{l} \text{Slope} \\ \text{of} \\ \text{sewer} \end{array} \times \begin{array}{l} \text{Length} \\ \text{of} \\ \text{sewer} \end{array}$$

For line 1:

$$\text{Lower invert elevation} = 17.50 \text{ m} - (0.0018 \text{ m/m})(707 \text{ m}) = 16.23 \text{ m}$$

If the depth of cover (remember to allow for the pipe wall thickness above the crown) for any section has become too shallow, repeat the process with a lower initial invert elevation or a steeper slope for that section.

When a manhole is located at a sewer junction, the outlet sewer elevation is fixed by the lowest inlet sewer. If the pipe size increases, the crowns of the two pipes must be matched at the manhole. This is done to avoid the backing up of wastewater into the smaller pipe. An example of this situation is the increase in size from 450 to 750 mm at manhole 2. For this case, the calculations are as follows:

Lower invert elevation of the 450-mm sewer is 16.23 m.
Upper invert elevation for the 750-mm-sewer (line 2) is 15.93 m (16.23 m + 0.45 m − 0.75 m).
Lower invert elevation for the 750-mm sewer is 15.29 m [15.93 m − (0.0009 m/m) × (707 m)].
These procedures are repeated until the elevations for the entire sewer are established.

COMMENT A computation table, such as the one shown in this example, not only saves time but also is useful for summarizing both the data and the computed results in an orderly sequence for subsequent use. The specific columns in a given computation table depend on the factors that must be considered in arriving at the peak design flows. Most sanitary and civil engineering consulting firms have developed tabulation forms of their own for sewer design computations. Although the forms may differ in specific details and in the order of presentation from this table, the same information is usually presented. Some engineering firms have developed computer programs for sewer design.

6-16 PREPARATION OF CONTRACT DRAWINGS AND SPECIFICATIONS

Once the sewer design computations have been completed, alternative alignments have been examined, and the most cost- and energy-effective alignment has been selected, the next step is to prepare contract drawings and specifications. Detailed contract drawings, including plans and profiles for each sewer, and specifications must be prepared before bids can be obtained to build the project. A typical plan and profile of a segment of a sewer line is shown in Fig. 6-13. The importance of preparing accurate and detailed drawings and specifications cannot be overemphasized. Perhaps the most compelling reason is that such careful preliminary work will likely ensure a successful project with a minimum number of change orders and without a lawsuit.

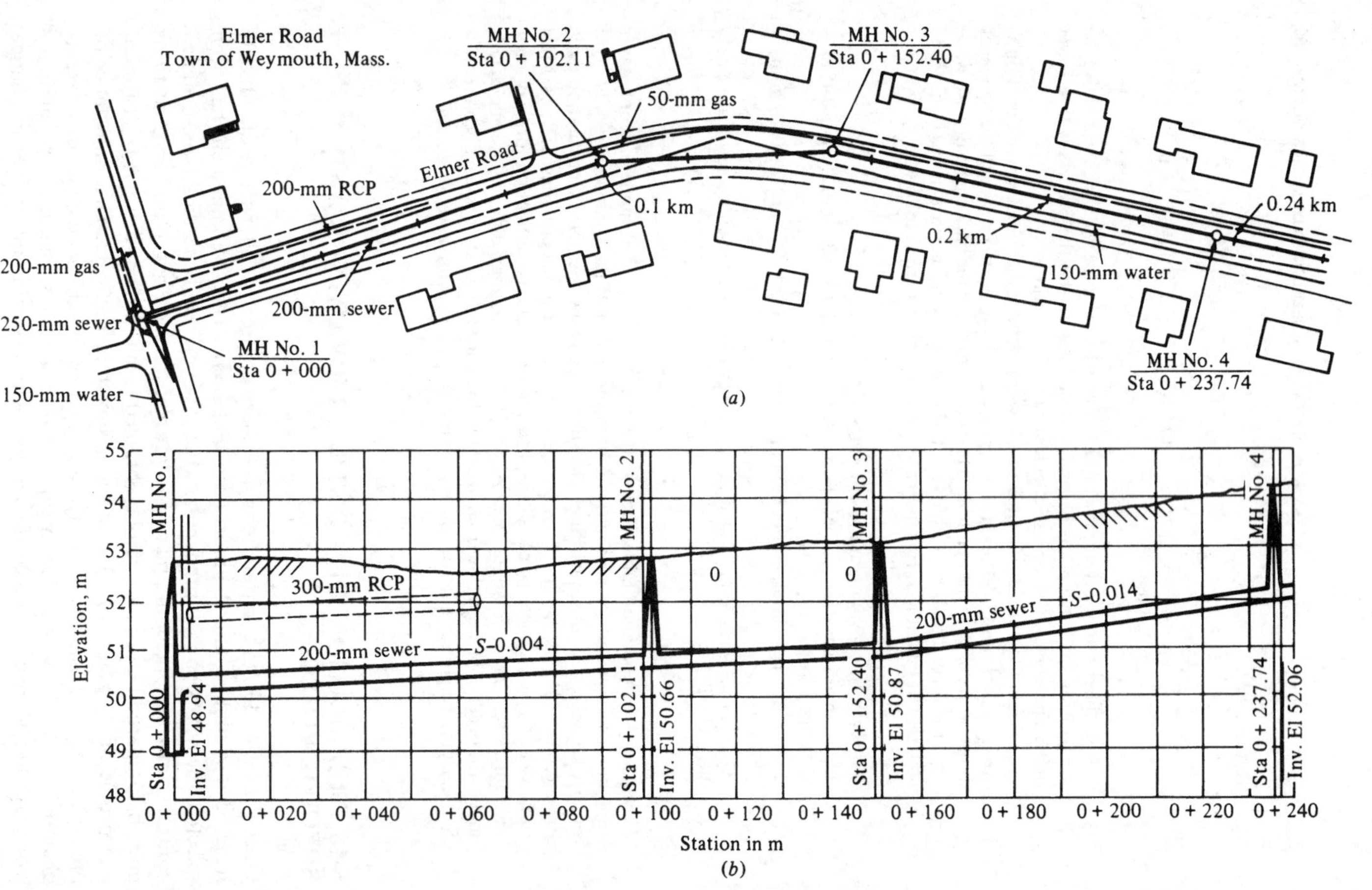

Figure 6-13 Typical sewer construction drawing: (*a*) plan and (*b*) profile. (*From Metcalf & Eddy, Inc.* [6-8].)

6-17 CONSTRUCTION OF SEWERS

There are many ways in which the actual construction of a sewer system may be accomplished, depending on the soil conditions encountered and the construction equipment available for the job. The important thing is that the finished sewer perform the function for which it is intended with a minimum of maintenance cost. To attain this end, three conditions should be met: (1) the pipe should be handled carefully, properly bedded, and backfilled in such a manner that there is a minimum of breakage during and after construction; (2) joints should be made with sufficient care to eliminate excessive infiltration; and (3) the line and slope of the sewer should be free of irregularities that might favor the accumulation of solid wastes with resultant clogging of the pipe.

6-18 MAINTENANCE OF SEWERS

Sewer maintenance involves keeping sewers clear of obstructions and in proper working order. Most stoppages in sewers are caused by the accumulation of settled material, tree roots, accumulation of grease, or collapse of the sewer pipe. Roots usually enter through the pipe joints. Good design and proper joint construction is the best preventive measure. Most cities have ordinances requiring the use of grease traps on service connections where wastewater may contain large amounts of grease. Collapse of the pipe is unlikely if adequate cover is provided and reasonable care is exercised to avoid breakage during and after construction. [6-9]

Where flushing is inadequate to remove an obstruction, sewers are cleaned with special tools attached to cables or jointed rods and pushed or pulled through the sewer from a manhole or other point of entry. The type of tool depends on the cause of the obstruction. Cutting tools are used to remove roots, scoops or scrapers are used to remove grit and sludge, and brushes are effective in removing grease. The use of a little copper sulfate in a sewer is often effective in killing roots without damaging the tree.

Occasionally, explosions may occur in sewers. The most common sources of explosive gases are inflammable and volatile liquids in the wastewater or leakage of domestic gas from an adjacent main. Gases given off by the decomposition of wastes are rarely the cause of explosions. However, many sewer-maintenance workers have been asphyxiated in gas-filled sewers. In no case should a worker be permitted to enter a sewer until proper tests for the presence of dangerous gases have been made. Whenever a worker enters a sewer, there should be a second person at the surface who can give emergency aid if required. [6-5, 6-8]

6-19 DESIGN OF STORMWATER SEWERS

The design procedure for stormwater sewers is essentially the same as that used for the design of sanitary sewers. The major difference is that the quantity of stormwater to be removed from a service area is determined on the basis of a hydrological analysis. Details on the analysis and design of stormwater sewers may be found in Refs. [6-5, 6-7, 6-14, and 6-16].

Water and Wastewater Pumping

Some form of pumping is used in most water supply and wastewater collection systems. As noted earlier, in water supply systems pumps are required to deliver water from wells, to lift water to distribution reservoirs and elevated tanks, and to increase pressure in distribution systems. In wastewater systems, pumps are used to avoid deep excavations, to convey wastewater over hills and other terrain where gravity sewers cannot be used, and at treatment plants to provide sufficient head for plant operation. The movement of water and wastewater from one location to another is the most common application of pumps in both types of systems.

Because pumping is so important in the operation of water supply and collection systems, it is the purpose of this section (1) to examine the types of pumps and pump drivers that are commonly used, (2) to review pump application terminology, (3) to review pump characteristics and their applications, (4) to review the analysis of pump systems and the selection of pumps, and (5) to review the design of water and wastewater pump stations.

6-20 PUMPS

The types of pumps used most commonly in the water and wastewater systems are described in this section. In general, pumps may be classified according to their (1) principle of operation, (2) field of application (i.e., liquids handled), (3) operational duty (i.e., head and capacity), (4) type of construction, and (5) method of drive. With respect to the principle of operation, pumps may be classified as *kinetic-energy pumps* or *positive-displacement pumps*. The term *turbo machine* is also used to describe kinetic-energy pumps. The principal types of pumps included under these two classifications are shown in Fig. 6-14.

Kinetic-Energy Pumps

The principal subclassification of kinetic-energy pumps is *centrifugal*, which, in turn, is divided further into three groups:

1. Radial-flow pumps
2. Mixed-flow pumps
3. Axial-flow pumps

The above classifications are derived from the manner in which the fluid is displaced as it moves through the pump. Thus, the fluid is displaced radially in a radial-flow pump, axially in an axial-flow pump, and both radially and axially in a mixed-flow pump.

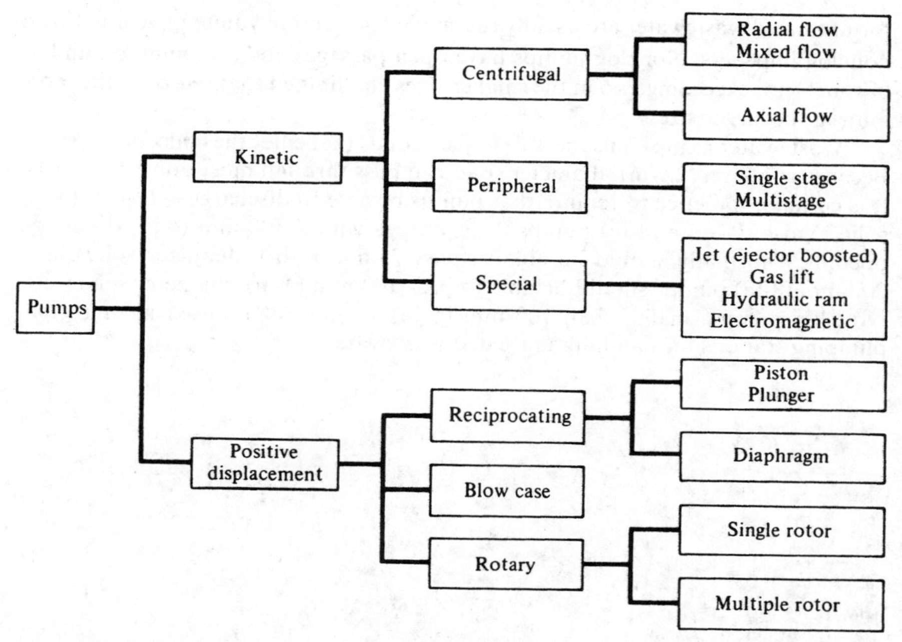

Figure 6-14 Principal types of pumps. (*Adapted from Hydraulic Institute Standards* [6-3].)

The principal components of kinetic-energy pumps are:

1. The rotating element called the *impeller* that imparts energy to the liquid being pumped.
2. The shaft on which the impeller is mounted.
3. The pump casing that includes the inlet and outlet passages for leading the liquid being pumped into and out of the pump, and the recuperating section which receives the liquid discharged from the impeller and directs it to the outlet passage. The function of the recuperating section is to convert a portion of the kinetic energy of the fluid into pressure energy. Typically this is accomplished by means of a volute or a set of diffusion vanes. In a volute casing, the size of the channel surrounding the impeller increases gradually to the size of the pump discharge nozzle, and most of the conversion of velocity to pressure occurs in the conical discharge nozzle. In a diffusion casing, the impeller discharges into a channel provided with guide vanes. The conversion of velocity to pressure occurs within the vane passages.
4. The frame which supports the pump casing.

Because rags and trash in wastewater (even though screened) would quickly clog the small passages in typical clearwater radial-flow pumps, the pumps used for untreated wastewater are usually the single-end suction volute type, fitted with nonclog impellers. Nonclog pumps have open passages and a minimum number of vanes (not exceeding two in the smaller sizes and limited to three, or at the most four, in the larger sizes).

Wastewater pumps must be able to pass solids that enter the collection system. Because a 70-mm (2.5-in) -diameter solid can pass through most domestic toilets, it is common practice to require that pumps be able to discharge a 75-mm (3-in) solid. Most 100-mm (4-in) pumps—i.e., pumps with a 100-mm (4-in) discharge opening—normally should be able to pass 75-mm (3-in) -diameter solids, and 200-mm (8-in) pumps should be able to pass 100-mm (4-in) -diameter solids, etc. Nonclog pumps smaller than 100 mm (4 in) should not be used in municipal pumping stations for handling untreated wastewater.

Figure 6-15 Typical screw pump used to pump wastewater.

Positive-Displacement Pumps

Positive-displacement pumps are usually divided into two major categories: reciprocating (piston or diaphragm) pumps and rotary pumps. Pneumatic ejectors and the Archimedean screw pump are also included under this category.

Piston-type reciprocating pumps utilize a reciprocating piston or plunger in a cylinder to draw a fluid in on the suction side and to discharge it under pressure on the discharge side. In a diaphragm pump, the reciprocating element is a flexible diaphragm. In both of these pumps check valves are used to control the pump suction and discharge.

In rotary positive-displacement pumps, the essential working element is a rotor that may have the form of an impeller, vane, lobe, or any other suitable configuration. The principal types of rotary positive-displacement pumps are (1) eccentric rotor screw (progressive cavity), (2) gear, (3) lobe, (4) peristaltic, (5) piston, (6) screw, (7) vane, and (8) flexible vane.

Pneumatic ejectors are often used for raising wastewater from building sumps. The ejector consists of an airtight tank into which wastewater flows by gravity and out of which the wastewater is forced automatically whenever sufficient wastewater has accumulated to raise a float and open the compressed air-inlet valve.

The screw pump is based on the Archimedean screw principle in which a revolving shaft fitted with one, two, or three helical blades rotates in an inclined trough and pushes the wastewater up the trough (see Fig. 6-15). Screw pumps are commonly used in wastewater-treatment plants to pump untreated wastewater and return waste activated sludge.

6-21 PUMP DRIVE UNITS

The most commonly used drives for pumps are direct-connected electric motors (see Fig. 6-16). Constant-speed electric motors coupled to variable-speed devices are also used extensively. Internal-combustion engines and turbines are often installed to ensure that the pumps can operate during electric-power outages or where wastewater gas or other gas is available for fuel.

Electric Motors—Direct Connected

Electric direct-connected motors may be constant-, multi- or variable-speed. Each is described below.

Constant-speed pumps may be driven by squirrel-cage induction motors, wound-rotor induction motors, or synchronous motors. Squirrel-cage induction motors and synchronous motors operate at a constant speed, but wound-rotor induction motors can operate at different speeds by varying the resistance of the rotor or secondary circuit. Squirrel-cage motors will normally be selected for constant-speed pumps because of their simplicity, reliability, and economy.

Figure 6-16 Examples of electric motors used to drive pumps.

Multiple-speed operation can be obtained with squirrel-cage or wound-rotor motors. For squirrel-cage motors, the choice of speeds is restricted to two or more of the speeds listed in Table 6-6. If the lower speed of a two-speed motor is one-half the higher speed, a single- or two-winding motor can be used. If the lower speed is not one-half, a two-speed motor with two windings is required. When operating a pump at two constant speeds, the advantage of the squirrel cage (or synchronous motor) is that the motor operates at maximum efficiency at both speeds. However, the wound-rotor motor operates at maximum efficiency only at full speed.

If the operating conditions vary in pumping stations, *variable-speed* operation of the pumps may be desirable. Variable-speed (stepless) motor operation has been possible for many years using liquid resistors with wound-rotor motor controls. With the development of solid electronic controls a number of methods of achieving variable-speed control for both squirrel-cage and wound-rotor motors are now available. They include (1) variable-frequency drive, (2) variable voltage, (3) wound-rotor motor, solid-state control, and (4) wound-rotor, regenerative secondary control.

Electric Constant-Speed Motors Coupled to Variable-Speed Devices

Worldwide, the most common way to obtain variable-speed pump operation is to use a constant-speed electric motor coupled to a variable-speed device. Variable-speed devices, inserted between the motor and the pump, may be classified as mechanical, magnetic, and fluid.

Table 6-6 Approximate operating speeds of constant-speed motors on 60-cycle alternating current

	Motor speed, r/min	
Poles	Synchronous	Induction
2	—	3500–3550
4	1800	1750–1770
6	1200	1150–1170
8	900	870–905
10	720	690–705
12	600	585
14	514	500
16	450	435
18	400	390
20	350	350
22	327	318
24	300	290
26	277	268
28	257	249

The most common mechanical variable-speed devices include cone drives, gear drives, and belt drives. A magnetic coupling (also known as an eddy current clutch) is installed between the motor and pump. A magnetic coupling consists of a constant-speed member (drum) and a rotor. The drum is driven by the constant-speed electric motor. The rotor is used to drive the pump. Fluid couplings placed between the motor and pump are used to obtain variable-speed operation with pumps. The most common fluid couplings may be classified as hydrostatic, hydrokinetic, and hydroviscous.

Internal-Combustion Engines and Gas Turbines

In large pumping stations, internal-combustion engines are used as a source of standby power for driving the pumps and the critical electrical controls if the power fails. Internal-combustion engines usually drive generators so that power is available not only for the pump but also for the auxilliary equipment and the control system. The power generated by these engines can also be used in any of the available pumps instead of being connected to a single pump. Diesel engines or spark-ignited engines fueled with either natural or propane gas are commonly used for this service. Gasoline engines are installed occasionally, but are not common because of the problems with fuel storage.

At treatment plants where sludge gas is available, either dual-fuel diesel engines or spark-ignited gas engines can be used. Dual-fuel diesel engines are operated with a mixture of diesel oil and gas. Spark-ignited engines can be operated with sludge gas. These engines would normally be supplied with dual carburetors and a separate source of alternate fuel, such as natural or propane gas, to provide power when the sludge gas is not available.

Gas turbines have been used as high-speed drives for pumps, especially in large-capacity mobile pumping units. In larger sizes, gas turbines are competitive with steam turbines.

Fluid-Driven Pumps

Fluid-driven pumps are also becoming more common throughout the world. The most common fluid drives are powered with compressed gas, pressurized water or oil, and steam.

6-22 PUMP APPLICATION TERMINOLOGY AND USAGE

The purpose of this section is to present the basic terminology used to define pump performance and to consider its usage in the solution of pump problems. Terminology to be considered in this discussion includes (1) capacity, (2) head, (3) pump efficiency, and (4) power input to the pump. [6-8]

Capacity

The capacity (flow rate) of a pump is the volume of fluid pumped per unit of time, which usually is measured in cubic meters per second (m^3/s), liters per second (L/s), gallons per minute (gal/min), million gallons per day (Mgal/d), or cubic feet per second (ft^3/s).

Head

The term *head* refers to the elevation of a free surface of water above or below a reference datum. Terms applied specifically to the analysis of pumps and pump systems are illustrated graphically in Figs. 6-17 and 6-18 and are defined briefly below.

1. *Static suction head* (h_s) is the difference in elevation between the suction liquid level and the centerline of the pump impeller. If the suction liquid level is below the centerline of the pump impeller, it is a *static suction lift*.
2. *Static discharge head* (h_d) is the difference in evaluation between the discharge liquid level and the centerline of the pump impeller.
3. *Static head* (H_{stat}) is the difference in elevation between the static discharge and static suction liquid levels ($h_d - h_s$).
4. *Manometric suction head* (H_{ms}) is the suction gage reading (expressed in meters) measured at the suction nozzle of the pump referenced to the centerline of the pump impeller.

5. *Manometric discharge head* (H_{md}) is the discharge gage reading (expressed in meters) measured at the discharge nozzle of pump referenced to the centerline of the pump impeller.
6. *Manometric* (H_m) is the increase of pressure head (expressed in meters) generated by the pump ($H_{ms} + H_{md}$).
7. *Friction head* (h_{fs}, h_{fd}) is the head of water that must be supplied to overcome the frictional loss caused by the flow of fluid through the pipe system. The frictional head loss in the suction (h_{fs}) and discharge (h_{fd}) piping system may be computed with the Hazen–Williams or Darcy–Weisbach equations.
8. *Velocity head* is the kinetic energy contained in the liquid being pumped at any point in the system and is given by

$$\text{Velocity head} = \frac{V^2}{2g} \tag{6-6}$$

where V = velocity of fluid, m/s (ft/s)
g = acceleration due to gravity, 9.81 m/s^2 (32.2 ft/s^2)

9. *Minor head loss* is the term applied to the head of water that must be supplied to overcome the loss of head through fittings and valves. Minor losses in the

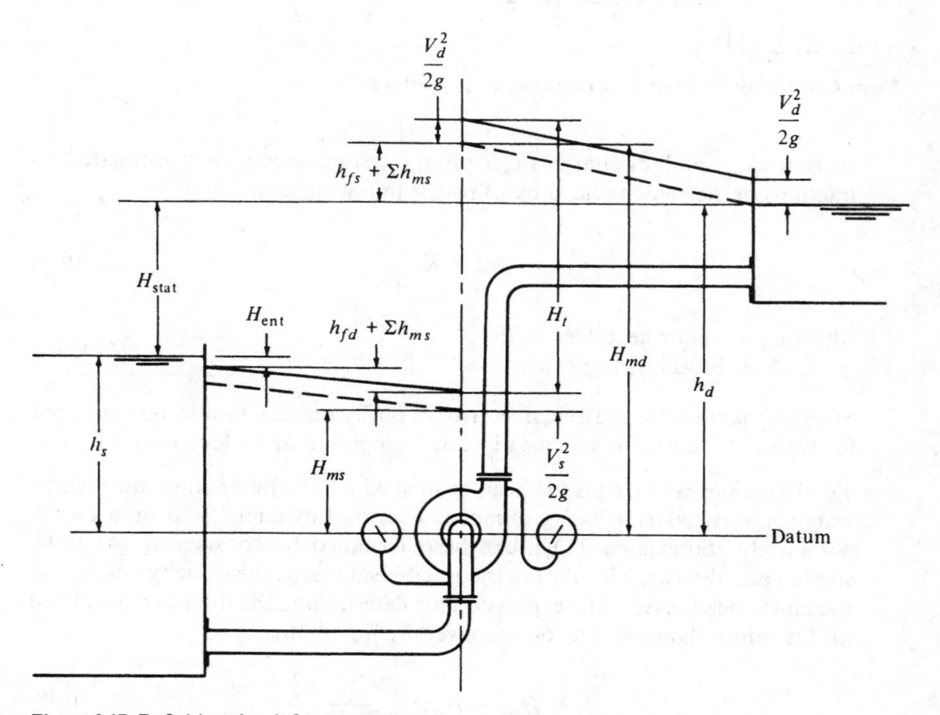

Figure 6-17 Definition sketch for a pump installation with a suction head.

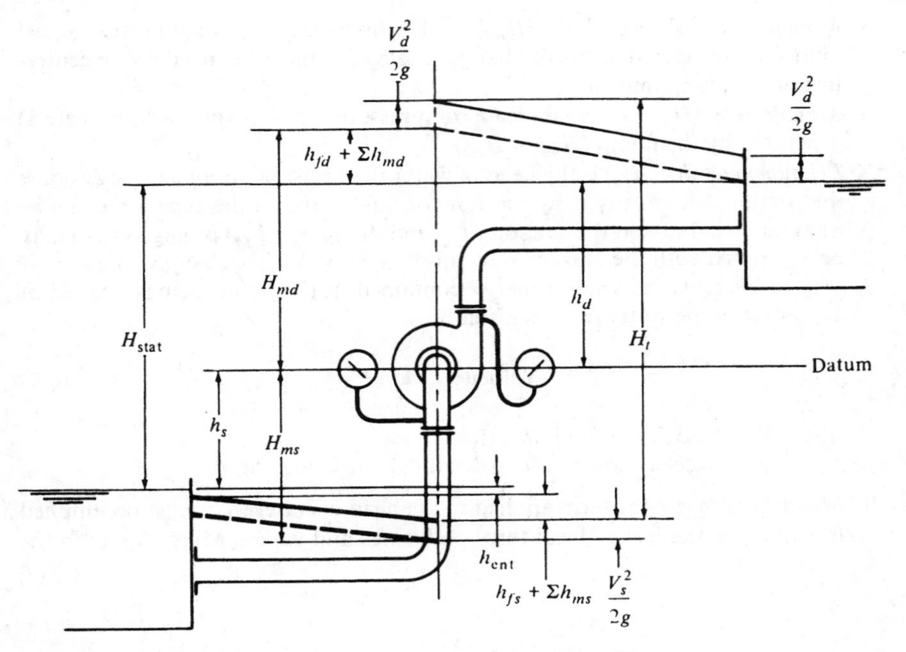

Figure 6-18 Definition sketch for a pump installation with a suction lift.

suction (h_{ms}) and discharge (h_{md}) piping system are usually estimated as fractions of the velocity head by using the following expression:

$$h_m = K\frac{V^2}{2g} \tag{6-7}$$

where h_m = minor head loss, m (ft)
K = head loss coefficient

Standard textbooks and reference works on hydraulics should be consulted for typical K values for various pipeline fittings and appurtenances.

10. *Total dynamic head* (H_t) is the head against which the pump must work when water or wastewater is being pumped. The total dynamic head on a pump, commonly abbreviated TDH, can be determined by considering the static suction and discharge heads, the frictional head losses, the velocity heads, and the minor head losses. The expression for determining the total dynamic head for the pump shown in Fig. 6-18 is given by Eq. (6-10).

$$H_t = H_{md} - H_{ms} + \frac{V_d^2}{2g} - \frac{V_s^2}{2g} \tag{6-8}$$

where

$$H_{md} = h_d + h_{fd} + \Sigma h_{md} \tag{6-9}$$

$$H_{ms} = h_s - h_{ent} - h_{fs} - \Sigma h_{ms} - \frac{V_s^2}{2g} \tag{6-10}$$

where H_t = total dynamic head, m (ft)
$H_{md}(H_{ms})$ = manometric discharge (suction) head measured at discharge (suction) nozzle of pump referenced to the centerline of the pump impeller, m (ft)
$V_d(V_s)$ = velocity in discharge (suction) nozzle, m/s (ft/s)
g = acceleration due to gravity, 9.81 m/s^2 (32.2 ft/s^2)
h_d (h_s) = static discharge (suction) head, m (ft)
h_{ent} = suction entrance loss, m (ft)
h_{fd} (h_{fs}) = frictional head loss in discharge (suction) piping, m (ft)
h_{md} (h_{ms}) = minor fitting and valve losses in discharge (suction) piping system, m (ft)

As noted previously, the reference datum for writing Eq. (6-10) is taken as the elevation of the centerline of the pump impeller. In accordance with the standards of the Hydraulic Institute [6-3], distances (heads) above datum are considered positive; distances below datum are considered negative.

In terms of the static head, Eq. (6-10) can be written as

$$H_t = H_{stat} + h_{ent} + h_{fs} + \Sigma h_{ms} + h_{fd} + \Sigma h_{md} + \frac{V_d^2}{2g} \tag{6-11}$$

where H_t = total dynamic head, m (ft)
H_{stat} = total static head, m (ft)
$= h_d - h_s$

The energy (Bernoulli's) equation can also be applied to determine the total dynamic head on the pump. The energy equation written between the suction and discharge nozzle of the pump is

$$H_t = \frac{P_d}{\gamma} + \frac{V_d^2}{2g} + z_d - \left(\frac{P_s}{\gamma} + \frac{V_s^2}{2g} + z_s\right) \tag{6-12}$$

where H_t = total dynamic head, m (ft)
P_d (P_s) = discharge (suction) gage pressure, kN/m^2 (lb$_f$/ft^2)
γ = specific weight of water, N/m^3 (lb$_f$/ft^3)
V_z (V_s) = velocity in discharge (suction) nozzle, m/s (ft/s)
g = acceleration due to gravity, 9.81 m/s^2 (32.2 ft/s^2)
z_d (z_s) = elevation of discharge (suction) gage above datum, m (ft)

Pump Efficiency

Pump performance is measured in terms of the capacity that a pump can discharge against a given head and at a given efficiency. The pump capacity is a function of the design. Information on the design is furnished by the pump manufacturer in a series of curves for a given pump. Pump efficiency E_p, the ratio of the useful power output to the power input, is given by

$$E_p = \frac{\text{power delivered to fluid}}{\text{power input to pump}} = \frac{P_0}{P_1} \tag{6-13}$$

Pump efficiencies usually range from 60 to 85 percent. The energy losses in a pump may be classified as volumetric, mechanical, and hydraulic. *Volumetric losses* occur because the small clearances necessary between the pump casing and the rotating element can leak. *Mechanical losses* are caused by mechanical friction in the stuffing boxes and bearings, by internal disk friction, and by fluid shear. Frictional and eddy losses within the flow passages account for the *hydraulic losses*.

Power Input

In practice, power input to the pump is computed using the following equation:

$$P_I = \frac{\text{power delivered to fluid}}{E_p} = \frac{\gamma Q H_t}{E_p} \tag{6-14}$$

where P_I = power input to pump, kW (kN · m/s)
γ = specific weight of liquid, kN/m^3
Q = capacity, m^3/s
H_t = total dynamic head [see Eq. (6-11)], m
E_p = pump efficiency

When the flow rate is given in gallons per minute, million gallons per day, or cubic feet per second and the head is given in feet, then the power input to the pump can be computed using Eqs. (6-15), (6-16), and (6-17), respectively.

$$P_I = \frac{(62.4\ \text{lb/ft}^3)(Q\ \text{gal/min})(H_t\ \text{ft})}{(60\ \text{s/min})(7.48\ \text{gal/ft}^3)(550\ \text{ft}\cdot\text{lb/s}\cdot\text{hp})E_p}$$

$$= \frac{(Q\ \text{gal/min})(H_t\ \text{ft})}{3960\ E_p} \tag{6-15}$$

$$P_I = \frac{(62.4\ \text{lb/ft}^3)(Q\ \text{Mgal/d})(H_t\ \text{ft})(10^6\ \text{gal/Mgal})}{86{,}400\ \text{s/d}\ (7.48\ \text{gal/ft}^3)(550\ \text{ft}\cdot\text{lb/s}\cdot\text{hp})E_p}$$

$$= \frac{(Q\ \text{Mgal/d})(H_t\ \text{ft})}{5.696\ E_p} \tag{6-16}$$

$$P_I = \frac{(62.4\ \text{lb/ft}^3)(Q\ \text{ft}^3/\text{s})(H_t\ \text{ft})}{(550\ \text{ft}\cdot\text{lb/s}\cdot\text{hp})E_p}$$

$$= \frac{(Q\ \text{ft}^3/\text{s})(H_t\ \text{ft})}{8.814\ E_p} \tag{6-17}$$

Application of the terminology and equations used to define pump performance is illustrated in Example 6-5.

Example 6-5: Finding energy requirements for pumping A water pump is discharging at a rate of 0.25 m^3/s. The diameters of the discharge and suction nozzles are 300 and 350 mm, respectively. The reading on the discharge gage located 0.25 m above the centerline of the impeller is 150 kN/m^2; the reading on the suction gage located at the centerline of the impeller is 20 kN/m^2. Determine (1) the total dynamic head, (2) the power input required by the pump, and (3) the power input to the motor. Assume the efficiency of the pump and motor are 65 and 90 percent, respectively.

SOLUTION

1. Determine the head on the pump using the energy equation [Eq. (6-12)]. The reference datum is the centerline of the pump impeller.

   a. The values for the individual terms in the energy equations are as follows:

$$\frac{P_d}{\gamma} = \frac{150{,}000 \text{ N/m}^2}{9810 \text{ N/m}^2} = 15.29 \text{ m}$$

$$V_d = \frac{Q_d}{A_d} = \frac{0.25 \text{ m}^3/\text{s}}{(\pi/4)(0.3 \text{ m})^2} = 3.54 \text{ m/s}$$

$$\frac{V_d^2}{2g} = \frac{(3.54 \text{ m/s})^2}{2(9.81 \text{ m/s}^2)} = 0.64 \text{ m}$$

$$z_d = 0.25 \text{ m}$$

$$\frac{P_s}{\gamma} = \frac{20{,}000 \text{ N/m}^2}{9810 \text{ N/m}^2} = 2.04 \text{ m}$$

$$V_s = \frac{Q_s}{A_s} = \frac{0.25 \text{ m}^3/\text{s}}{(\pi/4)(0.35 \ m)^2} = 2.60 \ m/s$$

$$\frac{V_s^2}{2g} = \frac{(2.60 \text{ m/s})^2}{2(9.81 \text{ m/s}^2)} = 0.34 \text{ m}$$

$$z_s = 0$$

   b. The total dynamic head is obtained by substituting the above values in Eq. (6-12).

$$H = \frac{P_d}{\gamma} = \frac{V_d^2}{2g} + z_d + \left(\frac{P_s}{\gamma} + \frac{V_s^2}{2g} + z_s\right)$$

$$= 15.29 \text{ m} + 0.64 \text{ m} + 0.25 \text{ m} - (2.04 \text{ m} + 0.34 \text{ m} + 0)$$

$$= 13.8 \text{ m}$$

2. Using Eq. (6-14) determine the power input required by the pump.

$$P_1 = \frac{\gamma Q H}{E_p}$$

$$= \frac{(9.810\ \text{kN/m}^3)(0.25\ \text{m}^3/\text{s})(13.8\ \text{m})}{0.65}$$

$$= 52.1\ \text{kW}$$

3. Determine the power input to the motor.

$$P_m = \frac{P_1}{E_m}$$

$$= \frac{52.1\ \text{kW}}{0.90}$$

$$= 57.9\ \text{kW}$$

6-23 PUMP OPERATING CHARACTERISTICS AND CURVES

The operating characteristics of pumps depend on their size, speed, and design. Pumps of similar size and design are produced by many manufacturers, but they vary somewhat because of the design modifications made by each manufacturer. Important basic relationships that can be used to characterize and analyze pump performance under varying conditions include the affinity laws, type numbers, and net positive suction head. To aid in the selection of an appropriate pump for a given service, pump manufacturers provide characteristic curves for their pumps.

Affinity Laws

For the same pump operating at different speeds the diameter does not change, and the following relationships can be derived for centrifugal pumps.

$$\frac{Q_1}{Q_2} = \frac{n_1}{n_2} \tag{6-18}$$

$$\frac{H_1}{H_2} = \frac{n_1^2}{n_2^2} \tag{6-19}$$

$$\frac{P_1}{P_2} = \frac{n_1^3}{n_2^3} \tag{6-20}$$

These relationships, known collectively as the *affinity laws*, are used to determine the effect of changes in speed on the capacity, head, and power of a pump.

The effect of changes in speed on the pump characteristic curves is obtained by plotting new curves with the use of the affinity laws. The new operating point, the intersection of the pump and system head-capacity curves, will be given by the intersection of the new pump head-capacity curve with the system head-capacity curve, and not by application of the affinity laws to the original operating point only.

Changes in Impeller Diameters

To cover a wide range of flows economically with a minimum number of pump sizes and impeller designs, manufacturers customarily offer a range of impeller diameters for each size casing (see Pump Characteristic Curves below). In general, these impellers have identical inlets and only the outside diameter is changed, usually by machining down the diameter. The following relationships for determining the effect of changes in the diameter of the impeller hold approximately, but with less accuracy than the affinity laws.

$$\frac{Q_1}{Q_2} = \frac{D_1}{D_2} \qquad (6\text{-}21)$$

$$\frac{H_1}{H_2} = \frac{D_1^2}{D_2^2} \qquad (6\text{-}22)$$

$$\frac{P_1}{P_2} = \frac{D_1^3}{D_2^3} \qquad (6\text{-}23)$$

In some cases, two or more impeller designs may be available, each in a range of sizes, for the same casing. Because these impellers are not geometrically similar, the affinity laws do not hold.

Type Number (Specific Speed)

For a geometrically similar series of pumps operating under similar conditions, the following relationship obtained is defined as the type number (specific speed).

$$n_s = \frac{nQ^{1/2}}{H^{3/4}} \quad \text{or} \quad \frac{nQ^{1/2}}{(gH)^{3/4}} \qquad (6\text{-}24)$$

where n_s = type number
n = speed, r/min
Q = capacity, m^3/s (gal/min)
H = head, m (ft)

Although the second form of the type-number equation is correct when applied with a consistent set of units, the first form is used in the United States. [6-1]

For any pump operating at any given speed, Q and H are taken at the point of maximum efficiency. When using Eq. (6-24) for pumps having double-suction impellers, one-half of the discharge is used, unless otherwise noted. For multistage pumps, the head is the head per stage. The variations in maximum efficiency to be expected with variations in size (capacity) and design (type number) are shown in Fig. 6-19. The progressive changes in impeller shape as the type number increases are shown along the bottom of Fig. 6-19.

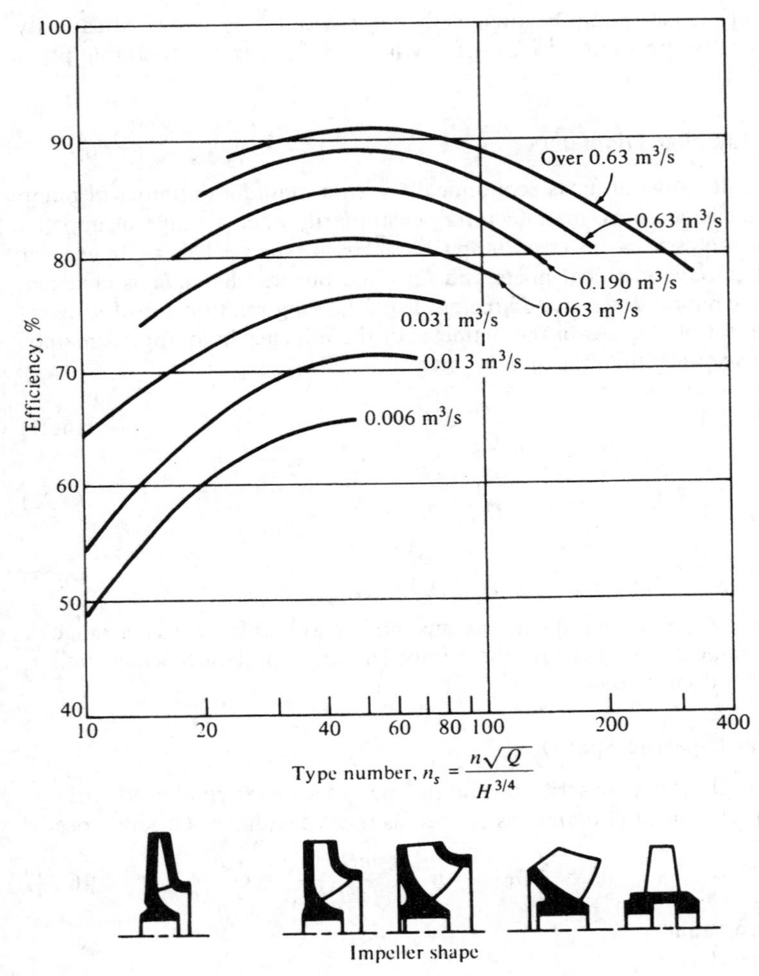

Figure 6-19 Pump efficiency versus type number and pump capacity.

Pump design characteristics, cavitation parameters, and abnormal operation under transient conditions can be correlated satisfactorily with the type number. Further consideration of the type-number equation reveals the following:

1. If larger units of the same type are selected for about the same head, the operating speed must be reduced.
2. If units of higher specific speed are selected for the same head and capacity, they will operate at a higher speed; hence the complete unit, including the driver, should be less expensive. However, long-term operation and maintenance costs will generally be higher.

Net Positive Suction Head (NPSH)

When pumps operate at high speeds and at a capacity greater than the best efficiency point (bep), pump cavitation is a potential danger: cavitation reduces pump capacity and efficiency and can damage the pump. It occurs in pumps when the absolute pressure of the inlet drops below the vapor pressure of the fluid being pumped. Under this condition, vapor bubbles form at the inlet, and when the vapor bubbles are carried into a zone of higher pressure, they collapse abruptly and the surrounding fluid rushes to fill the void with such force that a hammering action occurs. The high localized stresses that result from the hammering action can pit the pump impeller.

When determining if cavitation will be a problem, the NPSH available ($NPSH_A$) at the eye of the impeller is determined. The available $NPSH_A$ is then compared to the NPSH required by the pump ($NPSH_R$) to prevent cavitation. The $NPSH_A$ is the total energy available at the inlet flange of the pump, above the vapor pressure of the water, expressed in feet (meters). In effect, the $NPSH_A$ is the head available to push liquid into the pump to replace liquid discharge by the pump. The $NPSH_A$ is found by adding the term ($P_{atm}/\gamma - P_{vapor}/\gamma$) to the total energy head available at the suction side of the pump.

$$NPSH_A = \pm h_s - h_{ent} - h_{fs} - \sum h_{ms} + \frac{P_{atm}}{\gamma} - \frac{P_{vapor}}{\gamma} \tag{6-25}$$

where P_{atm} = atmospheric pressure, $kN/m^2(lb_f/ft^2)$
P_{vapor} = vapor pressure of water, $kN/m^2(lb_f/ft^2)$
γ = specific weight of water, $kN/m^3(lb_f/ft^3)$

The NPSH required by the pump is determined by tests of geometrically similar pumps operated at constant speed and rated capacity but with varying suction heads. The onset of cavitation is indicated by a drop in efficiency as the head is reduced. The application of Eq. (6-25) is illustrated in Example 6-6.

Example 6-6: Determining net positive suction head Determine the available net positive suction head ($NPSH_A$) for the pump installation shown in Fig. 6-17. Assume the following data are applicable

$$h_s = 2.0 \text{ m}$$
$$h_{ent} = 0.10 \text{ m}$$
$$h_{fs} = 0.25 \text{ m}$$
$$\sum h_{ms} = 0.15 \text{ m}$$
$$\text{Temp} = 20°\text{C}$$

SOLUTION

1. Determine the vapor pressure at 20°C.

$$P_{vapor} = 2.34 \text{ kN/m}^2 \text{ (see Appendix C)}$$

2. Substitute known quantities in Eq. (6-25) and solve for NPSH_A.

$$\text{NPSH}_A = h_s - h_{\text{ent}} - h_{fs} - \sum h_{ms} + \frac{P_{\text{atm}}}{\gamma} - \frac{P_{\text{vapor}}}{\gamma}$$

$$= -2.0\text{ m} - 0.1\text{ m} - 0.25\text{ m} - 0.15\text{ m} + \frac{101.3\text{ kN/m}^2}{9.789\text{ kN/m}^3} - \frac{2.34\text{ kN/m}^2}{9.787\text{ kN/m}^3}$$

$$= -2.0\text{ m} - 0.1\text{ m} - 0.25\text{ m} - 0.15\text{ m} + 10.35\text{ m} - 0.24\text{ m}$$

$$= 7.61\text{ m}$$

COMMENT The computed value of NPSH_A is compared to the value required for the pump (NPSH_R) to determine if the pump can be operated safely without cavitation.

Pump Characteristic Curves

Pump manufacturers provide information on the performance of their pumps in the form of characteristic curves, commonly called *pump curves*. In most pump

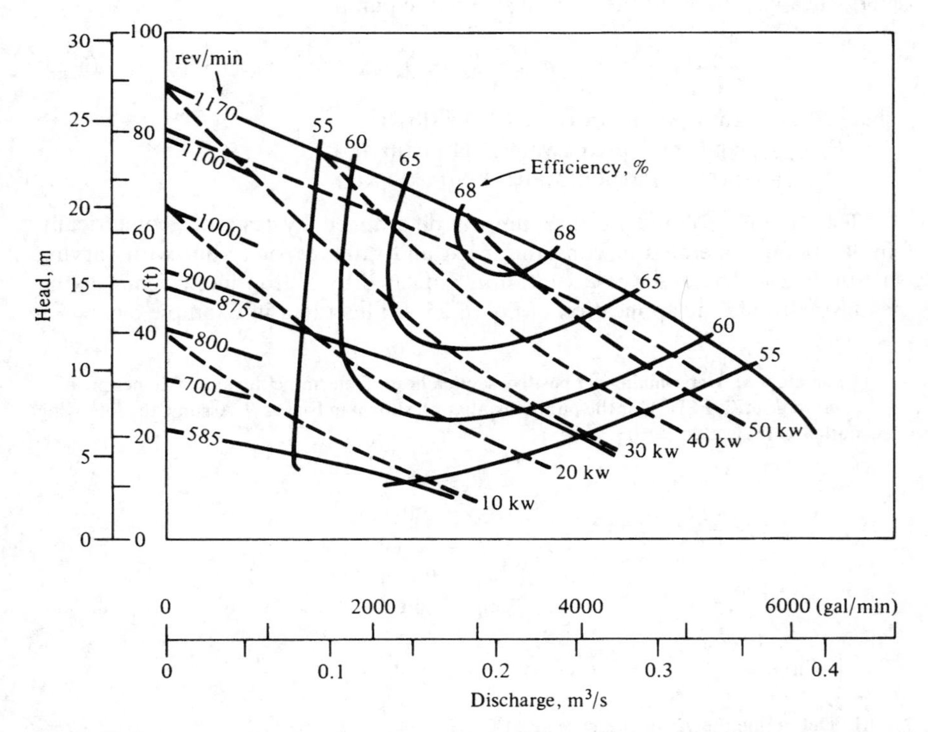

Figure 6-20 Typical pump characteristic curves for a 375-mm-diameter impeller variable-speed pump. (*Courtesy of Smith and Loveless.*)

curves, the total dynamic head H_t in meters (feet), the efficiency E in percent, and the power input P in kilowatts (horsepower) are plotted as ordinates against the capacity (flow rate) Q in cubic meters per second (gallons per minute or million gallons per day) as the abscissa (see Fig. 6-20). The general shape of these curves varies with the type number. Characteristic curves for typical radial-flow, mixed-flow volute, mixed-flow propeller, and axial-flow centrifugal pumps are shown in Fig. 6-21. The variables have been plotted as a percentage of their values at the best efficiency point (bep).

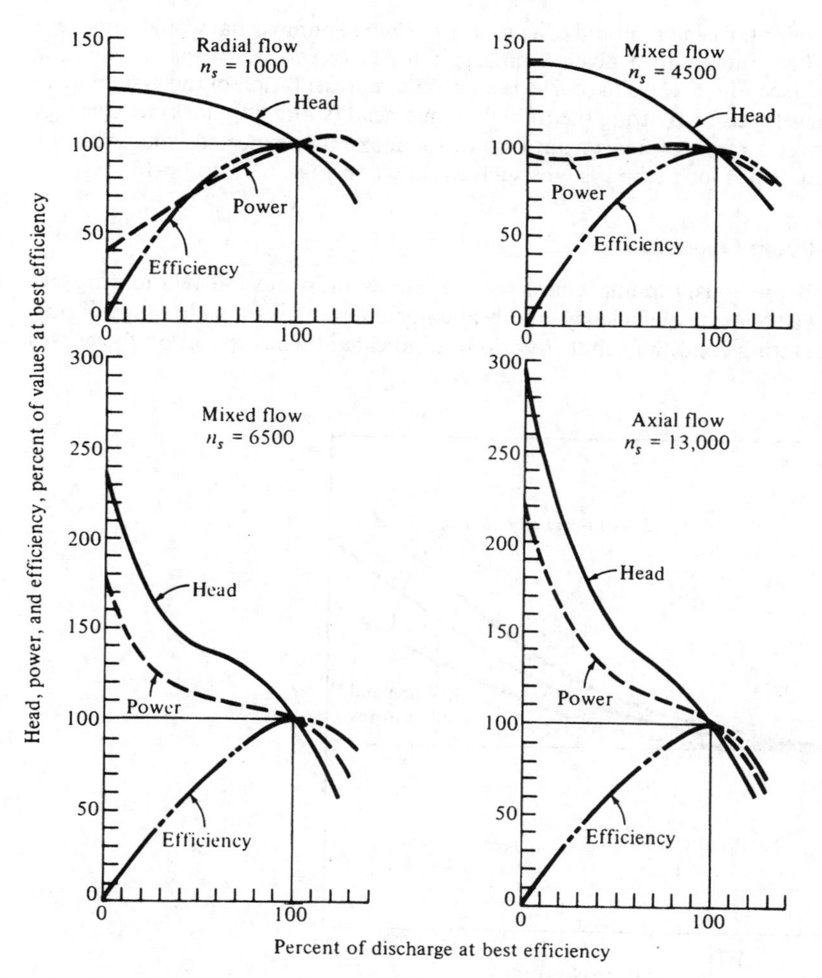

Figure 6-21 Typical characteristic curves for centrifugal pumps: (*a*) radial-flow; (*b*) mixed-flow volute; (*c*) mixed-flow propeller; (*d*) axial-flow.

6-24 ANALYSIS OF PUMP SYSTEMS

System analysis for a pumping installation is conducted to select the most suitable pumping units and to define their operating points. System analysis involves calculating system head-capacity curves for the pumping system and the use of these curves in conjunction with the head-capacity curves of available pumps. Both single-pump and multiple-pump systems are considered.

System Head-Capacity Curve

To determine the head required of a pump, or group of pumps, that would discharge various flow rates into a given piping system, a system head-capacity curve is prepared (see Fig. 6-22). This curve is a graphic representation of the system head and is developed by plotting the total dynamic head (static lift plus kinetic energy losses) over a range of flows from zero to the maximum expected value with the use of Eq. (6-11) for pump systems such as shown in Figs. 6-17 and 6-18.

Single-Pump Operation

As noted previously, pump characteristic curves illustrate the relationship between head, capacity, efficiency, and brake horsepower over a wide range of possible operating conditions, but they do not indicate at which point on the curves

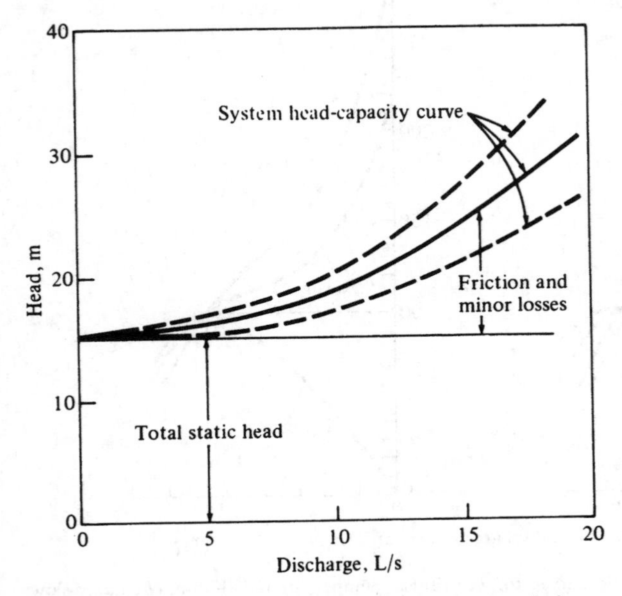

Figure 6-22 Typical head-capacity curve for a pump installation.

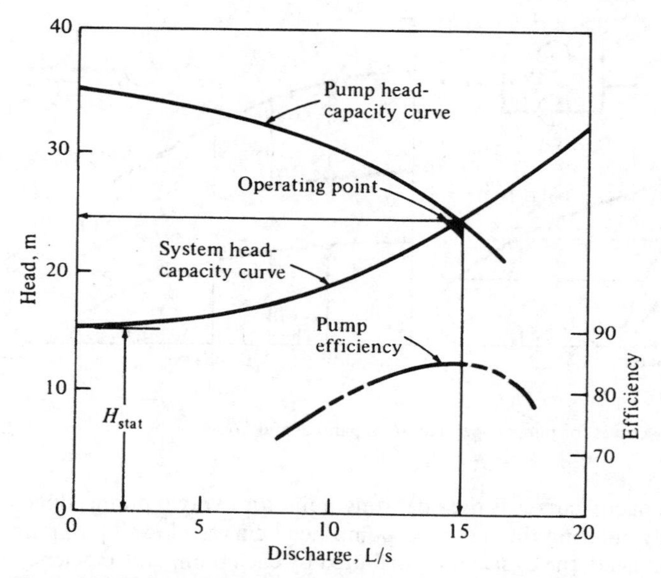

Figure 6-23 Definition sketch for determination of pump operating point.

the pump will operate. The operating point is found by plotting the pump head-capacity curve on the system head-capacity curve (see Fig. 6-23). The intersection of the pump head-capacity curve and the system head-capacity curve represents the head and capacity that the pump will produce if operated in the given piping system. This point is also known as the *pump operating point*.

Multiple-Pump Operation

In most pump stations, two or more pumps usually operate in parallel. Situations will also be encountered where pumps operate in series. In pumping stations where two or more pumps may operate either individually or in parallel and discharge into the same header and force main, the following method for determining the pump operating point is recommended:

1. The friction losses in the suction and discharge piping of individual pumps are omitted from the system head-capacity curve.
2. Instead, these losses are subtracted from the head-capacity curves of the individual pumps to obtain modified pump head-capacity curves, which represent the head-capacity capability of the pump and its individual valves and piping combines.
3. When two or more pumps operate in parallel, the combined pump head-capacity curve is found by adding the capacities of the modified curves at the same head (see Fig. 6-24*a*). The point of intersection of the combined pump-head curve with

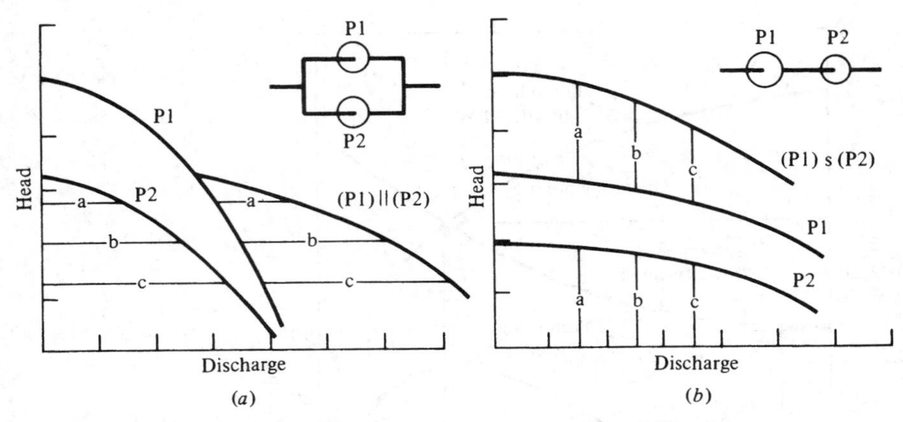

Figure 6-24 Head-capacity curves for pumps operated in (*a*) parallel and (*b*) series.

the system head-capacity curves is the operating point for the two pumps operating in parallel. By entering the modified pump-head curves of each pump at the operating-point head, the capacity contributed by each pump, the efficiency of each pump, and the brake horsepower required under these conditions can be determined. To find the total head at which each individual pump will operate, proceed vertically at constant capacity from the modified pump head-capacity curve to the actual head-capacity curve. The pump specifications or purchase order must be drawn so that the pump will produce this head. Each pump can operate at several points on the head-capacity curve, with the head increasing and the discharge decreasing as more pumps go into operation. An effort should be made to limit these operating points to a range of flows between 60 and 120 percent of the bep.

Often one or more booster pumps may be installed in the suction line or the force main leading from a pumping station to meet specific site conditions. Pumps installed in series with existing pumps are used to increase the head capacity of the pumping station. When two or more pumps operate in series, the combined head-capacity curve is found by adding the head of each pump at the same capacity. This procedure is illustrated in Figure 6-24*b*. When a booster pump is added to a force main fed by parallel pumps, the combined head-capacity curve is found by adding the head of the booster pump to the modified head of the parallel pumps at a given capacity. The analysis of a typical pump system is illustrated in Example 6-7.

Example 6-7: Analyzing a pump system Two centrifugal pumps are available for use in the pump system shown in the accompanying figure (*a*). Using the data given below, determine the system discharge when each pump is operated separately and when both pumps are operated in parallel.

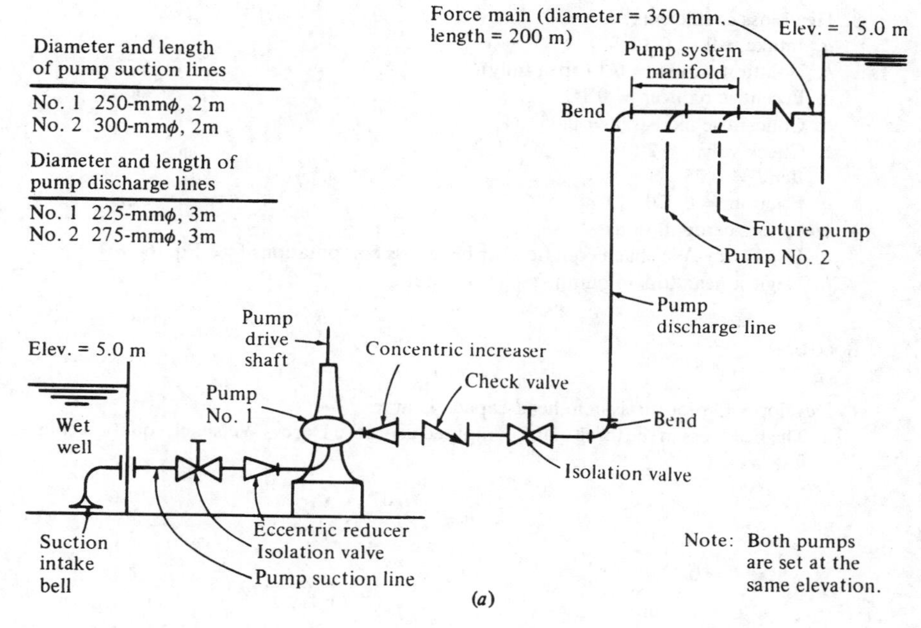

(*a*)

1. Pump no. 1:

   Nominal impeller size = 225 mm

   Operating speed = 1150 r/min

Q, m^3/s	H, m
0	30.0
0.1	28.0
0.2	18.5
0.25	8.0

2. Pump no. 2:

   Nominal impeller size = 275 mm

   Operating speed = 700 r/min

Q, m^3/s	H, m
0.0	40.0
0.1	39.2
0.2	35.0
0.3	26.0
0.4	10.0

3. Head loss coefficients:
   a. Intake bell = 0.3
   b. Isolation valves = 0.1 (open fully)
   c. Eccentric reducer = 0.1
   d. Concentric increaser = 0.05
   e. Check valve = 2.5
   f. Bend = 0.25
   g. Friction = 0.020
4. Head loss computations:
   a. Use Darcy–Weisbach equation for head loss computations [see Eq. (6-26)].
   b. Neglect head loss in pump system manifold.

SOLUTION

1. Develop and plot the system head-capacity curve.
   a. The head loss in the force main, computed using the Darcy–Weisbach equation, is as follows

$$h = f\frac{L}{D}\frac{V^2}{2g}$$

where $f = 0.020$

$L = 250$ m

$D = 0.35$ m

$$V = Q\Big/\frac{(3.14 \times (0.35)^2)}{4}$$

$g = 9.81$ m/s^2

   b. Prepare a head loss computation table.

Q, m^3/s	h, m	h_{exit},* m	h_{stat},† m	h_{total}
0.0	0.00	0.00	10.0	10.00
0.1	0.79	0.05	10.0	10.84
0.2	3.15	0.22	10.0	13.37
0.3	7.09	0.50	10.0	17.59
0.4	12.60	0.88	10.0	23.48
0.5	19.68	1.38	10.0	31.06

* $h_{exit} = \dfrac{V^2}{2g}$

† Static head = 10.0 (15.0 − 5.0)

   c. The system head-capacity curve is plotted in figure (*b*).

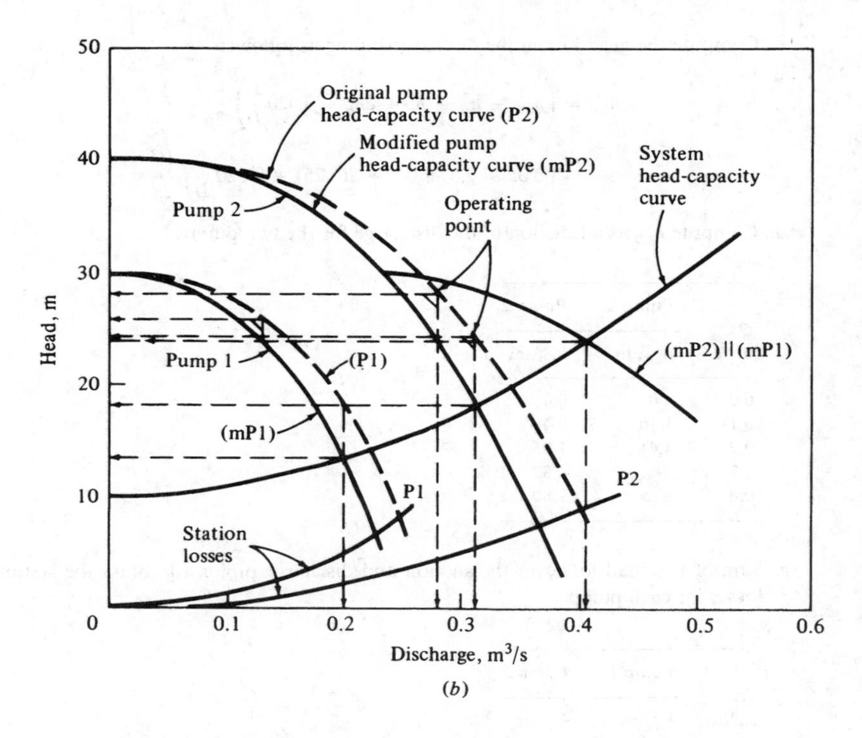

(*b*)

2. Plot the original pump head-capacity curves. The head-capacity curves for the given pumps are plotted on figure (*b*).
3. Determine the station losses for the pumps.
   a. Compute the head loss in the suction piping as follows:

$$h_{Ls} = h_{\text{ent}} + h_v + h_{\text{red}} + h_{fs}$$

$$= \left(0.3 + 0.1 + 0.1 + 0.020\frac{L}{D}\right)\frac{V^2}{2g}$$

   b. Compute h_{Ls} as a function of the discharge for the two pumps

Q, m³/s	Pump 1 h_{Ls}, m	Pump 2 h_L, m
0.0	0.0	0.0
0.1	0.14	0.06
0.2	0.56	0.26
0.3	—	0.58
0.4	—	1.03

c. Compute the head loss in the discharge piping as follows:

$$h_{LD} = \left(h_{\text{inc}} + h_{cv} + h_v + 2h_b + 0.020\frac{L}{D}\right)\frac{V^2}{2g}$$

$$= \left(0.05 + 2.5 + 0.1 + 2(0.25) + 0.020\frac{L}{D}\right)\frac{V^2}{2g}$$

d. Compute h_{LD} as a function of the discharge for the two pumps.

	Pump 1	Pump 2
Q, m^3/s	h_{LD}, m	h_{LD}, m
0.0	0.0	0.0
0.1	1.10	0.48
0.2	4.41	1.95
0.3	—	4.39
0.4	—	7.80

e. Sum of the head losses in the suction and discharge piping to obtain the station losses for each pump.

	Pump 1	Pump 2
Q, m^3/s	h_{st}, m	h_{st}, m
0.0	0.0	0.0
0.1	1.24	0.54
0.2	4.97	2.21
0.3	—	4.97
0.4	—	8.83

4. Plot the station losses and develop the modified pump curves.
   a. The station losses are plotted as shown in figure (*b*).
   b. The modified pump curves are obtained by subtracting the station losses from the original pump head-capacity curves. The modified pump curves are designated ($mP1$) and ($mP2$).
5. Determine the system discharges and corresponding heads.
   a. Referring to figure (*b*) the following values are obtained:

Pump(s)	Q, m^3/s	H, m
1	0.200	13.5
2	0.312	18.1
(2)∥(1)	0.405	24.0

b. Determine the pump discharges and operating heads at the given discharge values.

Pump(s)	Q, m^3/s	H, m
1	0.200	18.0
2	0.312	24.2
(1)*	0.133	26.0
(2)*	0.272	28.0

* When pumps (1) and (2) are operated in parallel.

COMMENT A wider range of discharges could be achieved if the smaller pump was converted to variable-speed operation.

6-25 PUMP STATIONS FOR WATER AND WASTEWATER

Pump stations for water and wastewater will vary in configuration depending on the service requirements. Because the design of pumping stations is beyond the scope of the present discussion, the reader is referred to Refs. [6-1, 6-8, 6-11, and 6-15] for a more complete discussion of water and wastewater pump stations.

Hydraulic Analysis of Water and Wastewater Treatment

The primary purpose of this section is to delineate the steps involved in the hydraulic analysis of water- and wastewater-treatment plants. However, before considering the subject of treatment plant hydraulics, it is important to consider all of the steps involved in the design of water- and wastewater-treatment plants.

6-26 TREATMENT PLANT DESIGN

Once the required effluent quality has been defined, the steps involved in treatment plant design typically include: (1) synthesis of alternative flow sheets, (2) bench tests and pilot-plant studies, (3) selection of design criteria, (4) sizing of physical facilities, (5) preparation of solids balances, (6) layout of the physical facilities, (7) preparation of hydraulic profiles, and (8) preparation of construction drawings, specifications, and cost estimates. Because of the importance of each of these steps, each is considered separately in the following discussion.

Synthesis of Alternative Treatment Process Flow Sheets

A flow sheet can be defined as the grouping together of unit operations and processes to achieve a specific treatment objective. Alternate flow sheets will be developed on the basis of the characteristics of the water and wastewater to be treated, the treatment objectives and, if available, the results of bench and pilot-scale tests. The best alternative flow sheets are selected after they have all been evaluated in terms of their performance, physical implementation, energy requirements, and cost. Typical examples of such flow sheets are shown in Figs. 4-1 and 4-2 and Figs. 5-2 and 5-3.

Bench Tests and Pilot-Plant Studies

The purpose of conducting bench tests and pilot-plant studies is (1) to establish the suitably of alternative unit operations and processes for treating a given water or wastewater and (2) to obtain the data and information necessary to design the selected operations and processes. Bench tests, as the name implies, are small-scale tests that can be conducted in the laboratory. Typically they are used to establish approximate chemical dosages and to obtain kinetic coefficients. Continuous pilot-plant studies are conducted to verify the results of bench tests.

Selection of Design Criteria

After one or more alternative flow sheets have been developed, the next step in design involves selection of design criteria. Design criteria are selected on the basis of theory, published data in the literature, the results of bench tests and pilot-scale studies, and the past experience of the designer.

Sizing of Unit Operations and Processes

Once design criteria have been selected, the next step is to size the required unit operations and processes so that the physical facilities required for their implementation can be determined. Depending on site constraints, it may be necessary to change from a circular to a rectangular basin, for example.

Solids Balances

After design criteria have been selected and the unit operations and processes sized, solids balances should be prepared for each selected process flow sheet. Ideally, solids balances should be prepared for the average and peak flow rates. The preparation of a solids balance involves the determination of the quantities of solids entering and leaving each unit operation or process. These data are especially important in the design (sizing) of the sludge-processing facilities.

Plant Layout

Using the information on the size of the facilities determined on the basis of the selected criteria, various plant layouts are developed within the constraints of the physical site. In laying out the various facilities, special attention should be given to minimizing pipe lengths, to grouping together related facilities, and to the need for future expansion.

Hydraulic Profiles

Once the treatment facilities and interconnecting piping have been sized preliminarily, hydraulic profiles should be developed for peak and average flow rates. The preparation of hydraulic profiles is considered in detail in the following section.

Construction Drawings and Specifications

The final step in the design process involves the preparation of construction drawings, specifications, and cost estimates. Because the clarity with which the construction drawings are presented will affect both the bid prices and final plant operation, the importance of this step cannot be overstressed. Construction specifications have been more or less standardized. The key issue is to make sure that specifications are complete so that costly change orders can be eliminated. Finally, the engineer's cost estimate is used as a guide in evaluating the bids submitted by the various contractors.

6-27 PREPARATION OF HYDRAULIC PROFILES

Hydraulic profiles are prepared for three reasons: (1) to ensure that the hydraulic gradient is adequate for flow through the treatment facilities, (2) to establish the head needed for pumps, where required, and (3) to ensure that plant facilities will not be flooded or backed up during periods of peak flow. Preparing hydraulic profiles involves careful consideration of the frictional and minor head losses that can occur in piping systems and of the head losses associated with control structures. These head losses are considered separately below. Application of the information on head losses in the preparation of hydraulic profiles is illustrated in the final part of this section.

Frictional Head Loss

The frictional head loss that occurs as water and wastewater flows through pipes can be computed with several equations. The recommended equation is the Darcy-Weisbach as given below.

$$h_f = f\frac{L}{D}\frac{V^2}{2g} \tag{6-26}$$

where h_f = head loss, m (ft)
f = coefficient of friction
L = length of pipe, m (ft)
D = diameter of pipe, m (ft)
V = mean velocity, m/s (ft/s)
g = acceleration due to gravity, 9.81 m/s^2 (32.2 ft/s^2)

The values of the friction factor are obtained from a Moody diagram. A representative value used for most friction computations is 0.020.

Minor Head Losses

As noted earlier in the section on pumps, minor head losses are produced when various control devices are inserted in piping systems. Valves are the most common control devices used in piping systems. Minor head losses also occur at pipe joints, pipe interconnections, pipe expansions and contractions, and pipe entrances and exits. For practical purposes minor head losses are usually estimated as a fraction of the velocity head in the downstream pipe section using Eq. (6-7).

$$h_m = K \frac{V^2}{2g} \tag{6-7}$$

Typical K values for various kinds of control devices and pipe configurations may be found in Refs. [6-6, 6-8, and 6-13] and in manufacturers' literature.

Head Losses from Control Structures

The most common control structures used in both water- and wastewater-treatment plants are weirs of one sort or another. The formulas used most commonly for rectangular and vee-notch weirs are give below.

For rectangular weirs, the Francis equation is used most commonly. The Francis equation is

$$Q = 1.84\,(L - 0.1\,nh)h^{3/2} \tag{6-27}$$

where Q = discharge, m^3/s (ft^3/s)
1.84 = numerical constant
L = length of crest of weir, m (ft)
n = number of end contractions
h = head on weir crest, m (ft)
3.33 = value of numerical constant for U.S. customary units

For 90° triangular weirs the general equation is:

$$Q = 0.55\,h^{5/2} \tag{6-28}$$

where Q = discharge, m^3/s (ft^3/s)
0.55 = numerical constant
h = head on weir crest, m (ft)
2.5 = value of numerical constant for U.S. customary units.

The application of Eqs. (6-26) and (6-27) is illustrated in Example 6-8. A more complete review of the equations used for the analysis of control structures may be found in Refs. [6-6, 6-8, 6-13, and 6-15].

Example 6-8: Preparing a hydraulic profile Prepare a hydraulic profile for peak flow conditions and set control elevations for the portion of a treatment plant shown in the accompanying figure. The following data and assumptions are applicable.

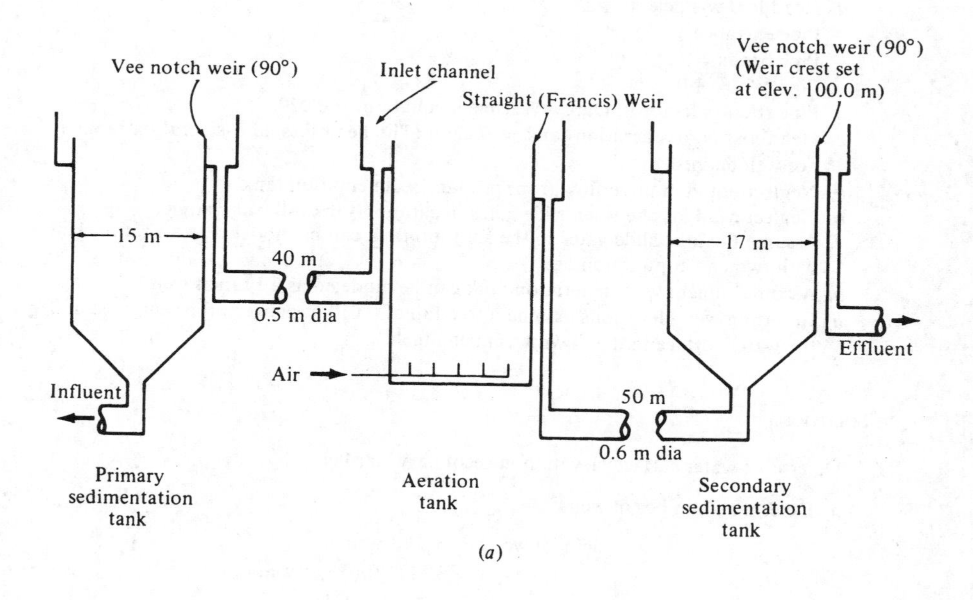

(*a*)

1. Flow rates:
   a. Average flow = 8000 m^3/d
   b. Peak flow = 16,000 m^3/d
   = 0.185 m^3/s
2. Primary sedimentation tank:
   a. Diameter at weir circle = 15 m
   b. Weir spacing = 0.3 m
   c. Weir type = 90° vee notch
   d. Weir depth = 0.1 m
   e. Return flows from sludge-processing facilities = $0.15Q$
3. Aeration tank:
   a. Inlet type = slide gates
   b. Number of gates = 6
   c. Width of slide gate = 0.20 m
   d. Return activated sludge discharged to influent channel at peak flow = $0.25Q$
   e. Length of aeration tank effluent weir = 15 m
   f. Weir type = straight sharp-crested

4. Secondary sedimentation tank:
   a. Weir crest elevation = 100 m
   b. Diameter at weir circle = 17 m
   c. Weir spacing = 0.3 m
   d. Weir type = 90° vee notch
   e. Weir depth = 0.1
   f. Underflow = $0.4Q$
5. Head loss computations:
   a. Head loss coefficients
      Pipe entrance = 0.5
      Pipe bends = 0.4
      Pipe exit = 1.0
   b. Pipe friction factor in Darcy–Weisbach equation = 0.020
   c. Head loss across aeration tank = 0.02 m (The head loss across aeration tanks is not well defined.)
   d. Neglect liquid in underflow from primary sedimentation tank.
   e. Neglect head loss between slide gates in aeration-tank influent channel.
   f. Assume the inlet slide gates to the aeration tank can be modeled as a Francis weir with two end contractions.
   g. Assume effluent weir in aeration tank can be modeled as a Francis weir.
   h. In setting weir elevations assume a free-fall of 0.010 m between the weir crest and the water surface in the downstream channel.

SOLUTION

1. Determine water surface elevation in secondary clarifier.

   a. Determine number of weirs.

$$\begin{aligned}\text{No. of weirs} &= \pi D/(\text{d/weir})\\ &= 3.14\,(17)/(0.3\text{ m/weir})\\ &= 177.9\text{, say }178\end{aligned}$$

   b. Determine flow per weir.

$$\begin{aligned}q/\text{weir} &= (16{,}000\text{ m}^3/\text{d})/178\\ &= 89.89\text{ m}^3/\text{d}\cdot\text{weir}\\ &= 0.00104\text{ m}^3/\text{s}\cdot\text{weir}\end{aligned}$$

   c. Determine head on vee-notch weirs.

$$q = 0.55h^{5/2}$$

$$\begin{aligned}h &= (Q/0.55)^{2/5}\\ &= (0.00104/0.55)^{2/5}\\ &= 0.081\text{ m}\end{aligned}$$

   d. Determine water surface elevation in secondary clarifier.

$$\begin{aligned}\text{Elev.} &= 100.0\text{ m} + 0.081\text{ m}\\ &= 100.081\text{ m}\end{aligned}$$

2. Determine water surface elevation in aeration-tank effluent channel.
   a. Summarize head losses and coefficient values.
      (1) Exit loss, $k_{ex} = 1.0$
      (2) Bend losses, 2 at $k_b = 0.4$
      (3) Friction loss in pipe, $f = 0.020$
      (4) Entrance loss, $k_{en} = 0.5$
   b. Determine velocity in pipe connecting the aeration tank to the secondary sedimentation tank.

$$\begin{aligned} V &= Q/A \\ &= 1.4(0.185\ \text{m}^3/\text{s})/3.14(0.3\ \text{m})^2 \\ &= 0.92\ \text{m/s} \end{aligned}$$

   c. Determine head loss in piping system connecting the aeration tank to the secondary sedimentation tank.

$$\begin{aligned} h &= \left(k_{ex} + 2k_b + f\frac{L}{D} + k_{en}\right)\frac{V^2}{2g} \\ &= \left(1 + 2(0.4) + 0.020\frac{50\ \text{m}}{0.6\ \text{m}} + 0.5\right)\frac{(0.92)^2}{2(9.81)} \\ &= 0.171\ \text{m} \end{aligned}$$

   d. Determine water surface elevation in aeration-tank effluent channel.

$$\begin{aligned} \text{Elev.} &= 100.081\ \text{m} + 0.171\ \text{m} \\ &= 100.252\ \text{m} \end{aligned}$$

3. Set the elevation of effluent discharge weir and determine water surface elevation in aeration tank near the effluent discharge weir.
   a. Set the elevation of the effluent weir in the aeration tank. As given in the problem statement, the free-fall distance between the weir crest and the water surface elevation in the effluent channel is 0.010 m. Thus

$$\text{Elev.} = 100.252\ \text{m} + 0.010\ \text{m} = 100.262\ \text{m}$$

   b. Determine the head on the effluent weir assuming two end contractions.

$$Q = 1.84(L - 0.1nh)h^{3/2}$$

$$1.4(0.185\ \text{m}^3/\text{s}) = 1.84[15\ \text{m} - 0.1(2)h]h^{3/2}$$

$$h = 0.044\ \text{m (by trial-and-error analysis)}$$

   c. Determine water surface elevation in aeration tank near effluent discharge weir.

$$\text{Elev.} = 100.262\ \text{m} + 0.044\ \text{m} = 100.306\ \text{m}$$

4. Set elevation of slide gates and determine water surface elevation in influent channel to aeration tank.
   a. Assume a head loss of 0.020 m across the aeration tank. Also assume a free fall of 0.010 m between the crest of the slide gate and the water surface in the aeration tank.

b. Set the elevation of the crest of the slide gate.

$$\text{Elev.} = 100.306 \text{ m} + 0.020 \text{ m} + 0.010 \text{ m} = 100.336 \text{ m}$$

c. Determine the head on the slide gates.

(1) The flow per slide gate $= 1.4(0.185 \text{ m}^3/\text{s})/6 = 0.043 \text{ m}^3/\text{s}$

(2) Determine head on slide gate assuming slide gate is a Francis weir with two end contractions.

$$Q = 1.84(L - 0.1nh)h^{3/2}$$

$$0.043 \text{ m}^3/\text{s} = 1.84(0.5 \text{ m} - 0.1(2)h)h^{3/2}$$

$$h = 0.139 \text{ m (by trial-and-error analysis)}$$

d. Determine water surface elevation in influent channel to aeration tank.

$$\text{Elev.} = 100.336 \text{ m} + 0.139 \text{ m} = 100.475 \text{ m}$$

5. Determine water surface elevation in primary-sedimentation-tank effluent channel.
   a. Summarize head losses and coefficient values. See step 2*a*.
   b. Determine velocity in pipe connecting the primary sedimentation tank to the aeration tank inlet channel.

$$\begin{aligned} V &= Q/A \\ &= 1.15(0.185 \text{ m}^3/\text{s})/3.14(0.25 \text{ m})^2 \\ &= 1.08 \text{ m/s} \end{aligned}$$

c. Determine head loss in piping system connecting the primary sedimentation tank to the aeration tank inlet channel.

$$\begin{aligned} h &= \left(k_{ex} + 2k_b + f\frac{L}{D} + k_{en}\right)\frac{V^2}{2g} \\ &= \left(1 + 2(0.4) + 0.020\frac{40}{0.5} + 0.5\right)\frac{1.0^2}{2(9.81)} \\ &= 0.232 \text{ m} \end{aligned}$$

d. Determine water surface elevation in primary sedimentation tank effluent channel.

$$\text{Elev.} = 100.475 \text{ m} + 0.232 \text{ m} = 100.707 \text{ m}$$

6. Set elevation of primary effluent weirs and determine water surface elevation in primary sedimentation tank.
   a. Set elevation of vee-notch weirs in primary sedimentation tank.

$$\text{Elev.} = 100.707 \text{ m} + 0.010 \text{ m} = 100.717 \text{ m}$$

b. Determine number of weirs.

$$\begin{aligned} \text{No. of weirs} &= \pi D/(\text{d/weir spacing}) \\ &= 3.14(15 \text{ m})/(0.3 \text{ m/weir}) \\ &= 157 \end{aligned}$$

c. Determine flow per weir.

$$
\begin{aligned}
q/\text{weir} &= 1.15(16{,}000\ \text{m}^3/\text{d})/157 \\
&= 117.2\ \text{m}^3/\text{d} \cdot \text{weir} \\
&= 0.00136\ \text{m}^3/\text{s} \cdot \text{weir}
\end{aligned}
$$

d. Determine head on vee-notch weirs.

$$q = 0.55h^{5/2}$$

$$
\begin{aligned}
h &= (Q/0.55)^{2/5} \\
&= (0.00136/0.55)^{2/5} \\
&= 0.091\ \text{m}
\end{aligned}
$$

e. Determine water surface elevation in primary sedimentation tank.

$$\text{Elev.} = 100.717\ \text{m} + 0.091\ \text{m} = 100.808\ \text{m}$$

7. Prepare a hydraulic profile showing the computed elevations. See the accompanying figure (*b*).

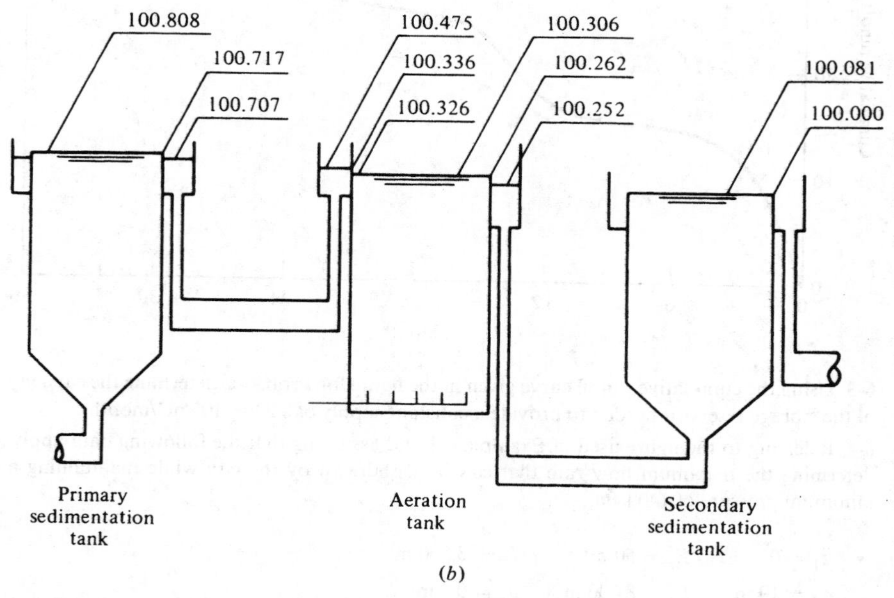

(*b*)

COMMENT In this example a distance of 0.01 m was used as a free-fall at each of the control weirs. Where the loss of head is critical, some designers will allow the weirs to become submerged at peak-flow conditions. Submerged inlets and outlets have also been used in many plants. A trade-off analysis should also be made to select the optimum pipe sizes to use to interconnect treatment units. The cost of a larger pipe size should be compared to the cost of energy needed to overcome the differential head loss associated with the smaller pipe size. In most situations, the maximum pipe size will be limited by the minimum velocity required to avoid the deposition of solids.

DISCUSSION TOPICS AND PROBLEMS

6-1 Determine the maximum monthly water supply that can be taken from a stream using the data from Example 6-2. What is the capacity of the required storage reservoir?

6-2 Determine the maximum monthly water supply that can be obtained from a stream with the cumulative runoff record shown in the accompanying figure. What is the capacity of the required storage reservoir?

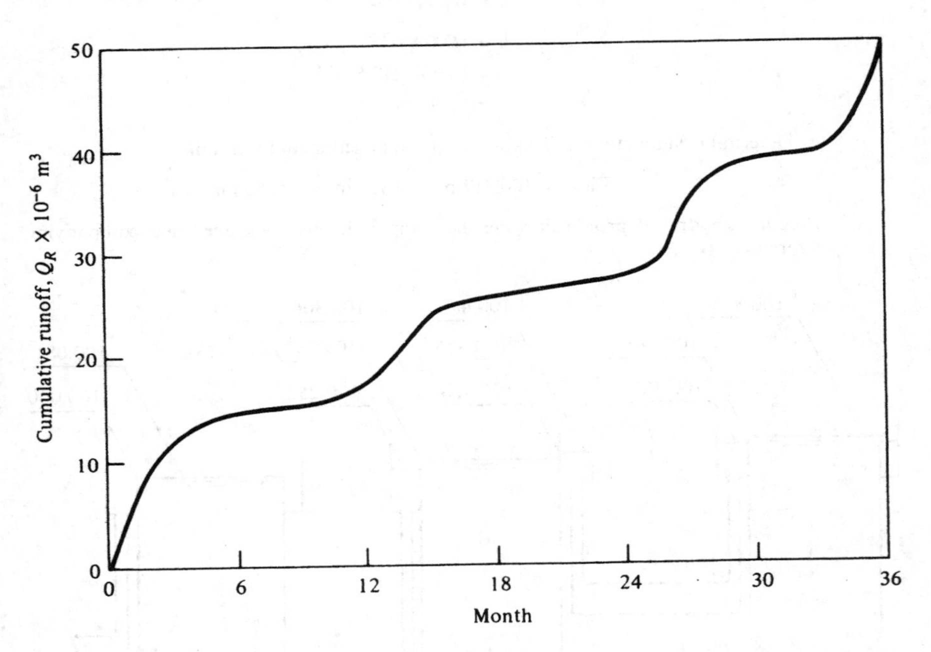

6-3 Using the cumulative runoff curve given in the figure for Prob. 6-2, determine the capacity of the storage reservoir needed to provide a constant supply of 0.33×10^6 m^3/month.

6-4 Referring to the figure used in Example 6-1 and assuming that the following data apply, determine the maximum flow rate that can be withdrawn by the city while maintaining a minimum pressure of 140 kPa.

$z_1 = 0$	$E_p = 50$ m	$L_s = 3{,}500$ m
$z_2 = 10$ m	$L_p = 8{,}000$ m	$d_p = 0.5$ m
$z_3 = 35$ m	$d_p = 0.5$ m	$f = 0.020\ (d_p, d_s)$

6-5 Referring to the same figure and assuming that the data given in Prob. 6-4 apply, determine E_3 when the city demand is equal to 0.3 m^3/s.

6-6 Estimate the required fire flow for a school of wood frame construction with a total floor area of 1600 m^2. Assume the fire flow must be increased by 10 percent due to unfavorable exposure.

6-7 Estimate the required fire flow for your classroom building.

6-8 Why is the equation used in Example 6-3 no longer favored for estimating fire flows in business districts?

6-9 What is the origin of the term "fire plug"?

6-10 Referring to the figure used in Example 6-3 and assuming that the data given in that example are applicable, what modifications must be made in the water distribution system to serve an additional population of 2500 persons located beyond the lower boundary of the existing system?

6-11 Referring to the figure and the population information given in Prob. 6-15, and to the data given in Example 6-3, estimate the size of the pipes for a water distribution system for the service area shown. Assume a water main will be placed in each street and that a loop will be placed around the service area. Assume the water supply main, to be placed in Peavy Avenue, will enter the service area from 65th Street.

6-12 A 2.0-m circular sewer is laid on a slope of 0.0008 m/m. If n is equal to 0.013 at all depths of flows, determine:

(*a*) Q and V when the sewer is flowing full

(*b*) Q and V when the sewer is flowing at a depth of 0.3 m

(*c*) Q and V when the flow in the sewer is at 0.6 of its capacity

(*d*) V and depth of flow when $Q = 1.0\ m^3/s$

6-13 Solve Prob. 6-12 assuming that n is variable and is equal to 0.015 when the pipe is flowing full.

6-14 A rectangular sewer 1.25 m wide and 15 m high has been laid on a slope of 0.0055.

(*a*) What is the maximum flow rate if the Manning's n value for the sewer is 0.013?

(*b*) What are the dimensions of an egg-shaped sewer laid at the same slope that has the same flow capacity?

6-15 Develop a preliminary design, including flow rates, pipe sizes, and pipe slopes for a trunk sewer to be laid in the development shown in the accompanying figure from 6th to 1st Streets along Peavy Avenue. Assume the following data apply.

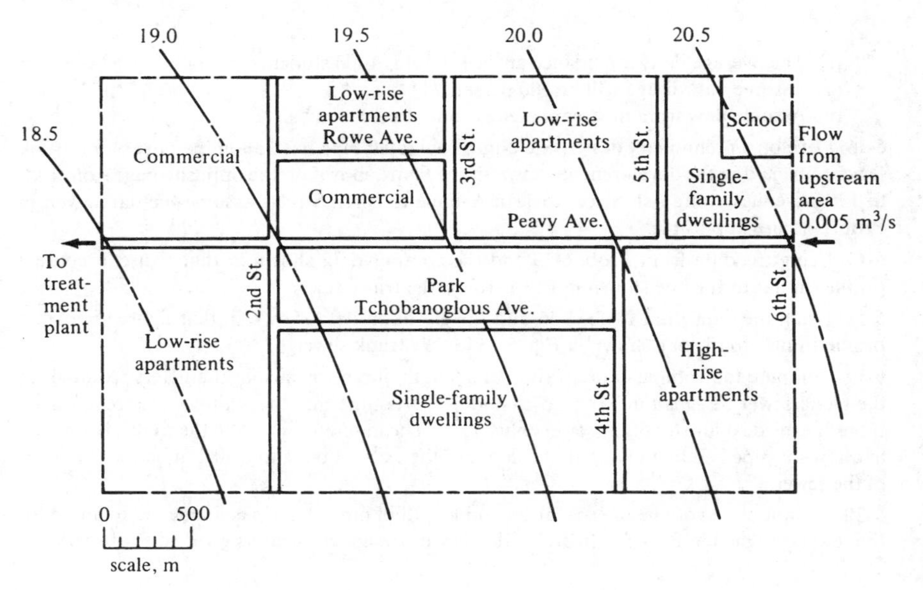

(*a*) Saturation population and flow data:

Zoning	Type of development	Saturation population density, person/ha	Wastewater flow,* L/capita · d
Residential	Single-family dwellings	30	380
Residential	Duplexes	50	310
Residential	Low-rise apartments	100	260
Residential	High-rise apartments	160	220

* Refer to figure (*c*) in Example 6-4 for peaking factors.

(*b*) Wastewater flow and peaking data for commercial and industrial areas and institutional facilities:

Area/ facility	Flow basis	Value	Peaking factor
Commercial	m^3/ha · d	32	1.5
Industrial	m^3/ha · d	60	2.2
School	L/student · d	80	3.5
Hospital	L/bed · d	60	3.5
Park	L/person · d	30	4.0

(*c*) The average daily attendance at the school is 1000 students.
(*d*) Assume park usage will be 500 persons/d.
(*e*) Assume flow from upstream area is equal to 0.005 m^3/s.

6-16 Develop a preliminary design, including flow rates, pipe sizes, and pipe slopes for a trunk sewer to be laid in the development shown in the figure shown on the opposite page from 10th to 13th Avenues along Ash Street to 11th Avenue to Birch Street. Assume the data given in Prob. 6-15 are applicable.

6-17 Using the data from Prob. 6-15 and the contour data shown in that figure, prepare a profile similar to the one shown in Fig. 6-16 for the trunk sewer.

6-18 Using the data from Prob. 6-16 and the elevation data shown in that figure, prepare a profile similar to the one shown in Fig. 6-13 for the trunk sewer.

6-19 Compute the volume of excavation and length of pipe of various diameters required for the trunk sewer designed in Probs. 6-15 and 6-17. Assume that the width of the trench is 1.4 times the inside diameter of the sewers plus 0.3 m. The minimum width of the trench is 1 m and to allow for pipe bedding material, the depth of the excavation is to be 0.2 m below the invert of the sewer.

6-20 Compute the volume of excavation and length of pipe of various diameters required for the sewer designed in Probs. 6-16 and 6-18. Use the design constraints given in Prob. 6-19.

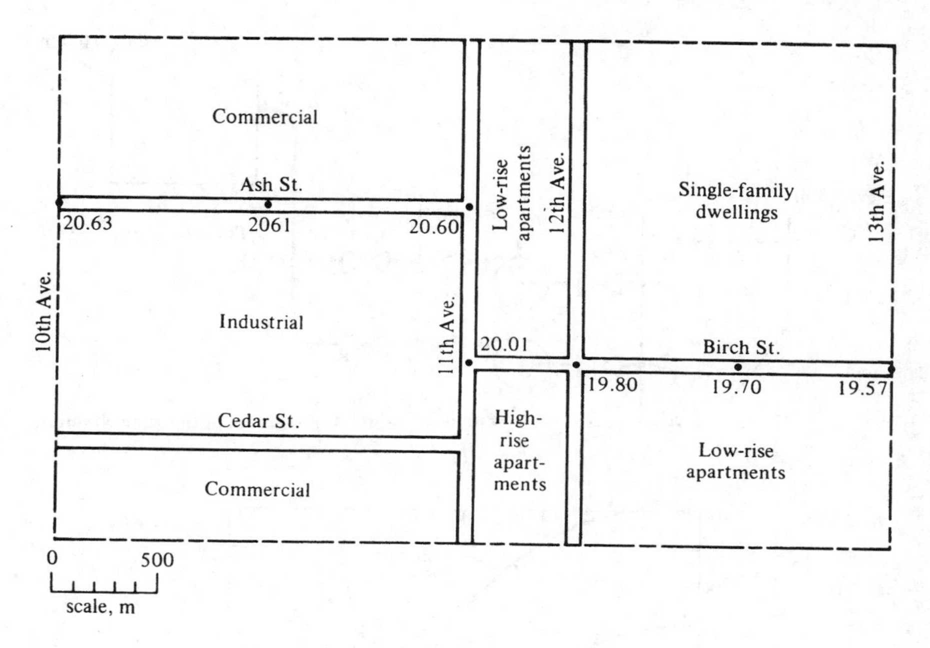

6-21 A wastewater pump has a 300-mm discharge and a 350-mm suction. The reading on the discharge gage located at the pump centerline is 140 kPa (kN/m^2). The reading on the suction gage located 0.75 m below the pump centerline is 20 kPa (kN/m^2). If the total head on the pump is 15 m, determine (1) the pump discharge and (2) the energy input to the motor, assuming a pump efficiency of 82 percent and a motor efficiency of 91 percent.

6-22 Solve Prob. 6-21, but assume that the total head is 10 m and the reading on the discharge gage is 100 kPa (kN/m^2).

6-23 A centrifugal pump with an impeller diameter of 0.25 m delivers 0.02 m^3/s against a head of 18 m at a power input of 4 kW when operating at 1170 r/min. If it is assumed that the efficiency remains the same, determine the (1) head, (2) discharge, and (3) power input for a geometrically similar pump with an impeller diameter of 0.30 m operating at 870 r/min.

6-24 A mixed-flow volute pump is to operate at a head of 5 m and discharge 0.17 m^3/s. It is to be driven by a direct-coupled squirrel-cage induction motor operating on 6-cycle (60-Hz) current. If the specific speed is not to exceed 100, what should be the operating speed? What efficiency could be expected, and how much power will be required?

6-25 If the diameter of the impeller in pump no. 2 in Example 6-7 were changed from 275 to 250 mm, what would be the maximum discharge that could be expected with two pumps operating in parallel?

6-26 Using the data given below in conjunction with the pumping system schematic shown in figure (*a*) on page 410:

(*a*) Determine the system pumping capacity when pumps 1 and 2 are operating in parallel. The pump performance curves for pumps 1 and 2 are given in figure (*b*). Ignore losses other than friction in developing the system curve. Minor losses should be considered in developing the modified pump curves.

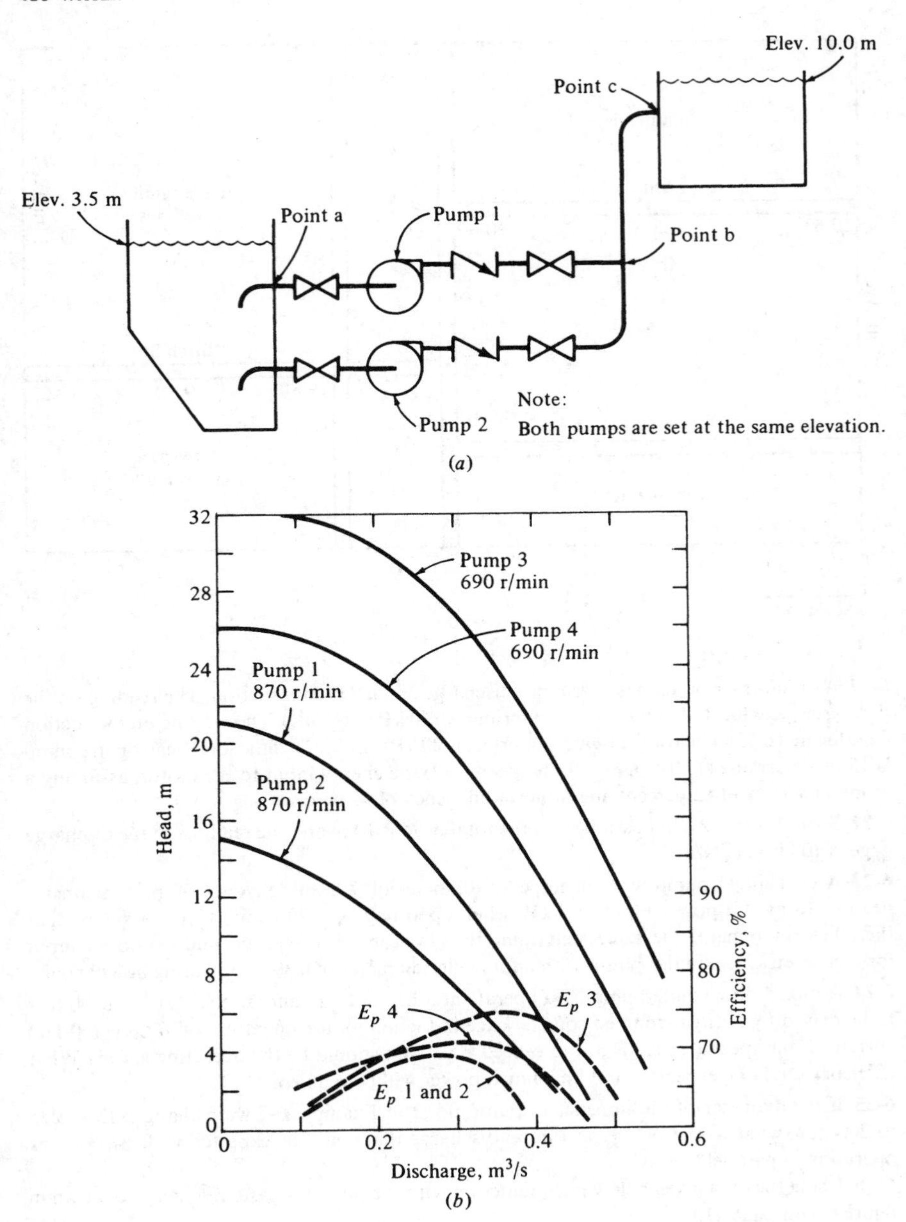

Elev. 10.0 m
Point c
Elev. 3.5 m
Point a
Pump 1
Point b
Pump 2
Note:
Both pumps are set at the same elevation.
(a)
Pump 3
690 r/min
Pump 4
690 r/min
Pump 1
870 r/min
Pump 2
870 r/min
E_p 4
E_p 3
E_p 1 and 2
Head, m
Efficiency, %
Discharge, m³/s
0
0.2
0.4
0.6
4
8
12
16
20
24
28
32
70
80
90
(b)

(*b*) What is the kilowatt requirement for each pump at the above operating point?

(*c*) At what reduced speed must pump 1 be operated alone to pump 0.20 m^3/s to the reservoir? What is the corresponding head?

Pipe	D, m	L, m
a–b	0.35	30 (same for both pumps)
b–c	0.50	800

Head loss computations:

$$k_{\text{ent}} = 0.3$$

$$k_{\text{valve}} = 0.5$$

$$k_{\text{check valve}} = 2.5$$

$$f = 0.020 \text{ (for all piping)}$$

6-27 Solve Prob. 6-26 using pump curves 3 and 4 given in figure (*b*) of Prob. 6-26.

6-28 Determine the available net positive suction head ($NPSH_A$) for the pumping system given in Example 6-6.

6-29 Solve Prob. 6-26 for the pumping system given in Prob. 6-24.

6-30 Develop the hydraulic profile for the peak-flow condition for the portion of the wastewater-treatment plant shown in the figure on page 412. Assume that 33 percent of the plant inflow is recycled from the secondary sedimentation tanks to the head end of the aeration tank. The pertinent data and information are shown as follows:

$$Q_{\text{avg}} = 10{,}000 \text{ m}^3/\text{d}$$

$$Q_{\text{peak}} = 18{,}000 \text{ m}^3/\text{d}$$

Primary sedimentation:

Number of tanks = 2

Diameter – 13.75 m

Secondary sedimentation:

Number of tanks = 2

Diameter = 15 m

Aeration-tank effluent weir:

Type-sharp-crested, straight weir

Weir length = 4 m

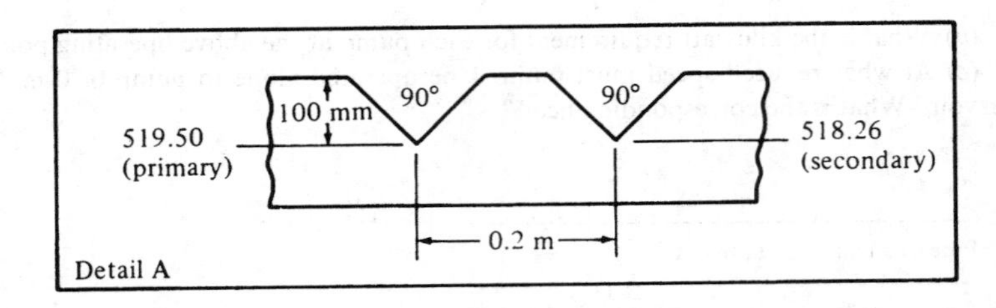

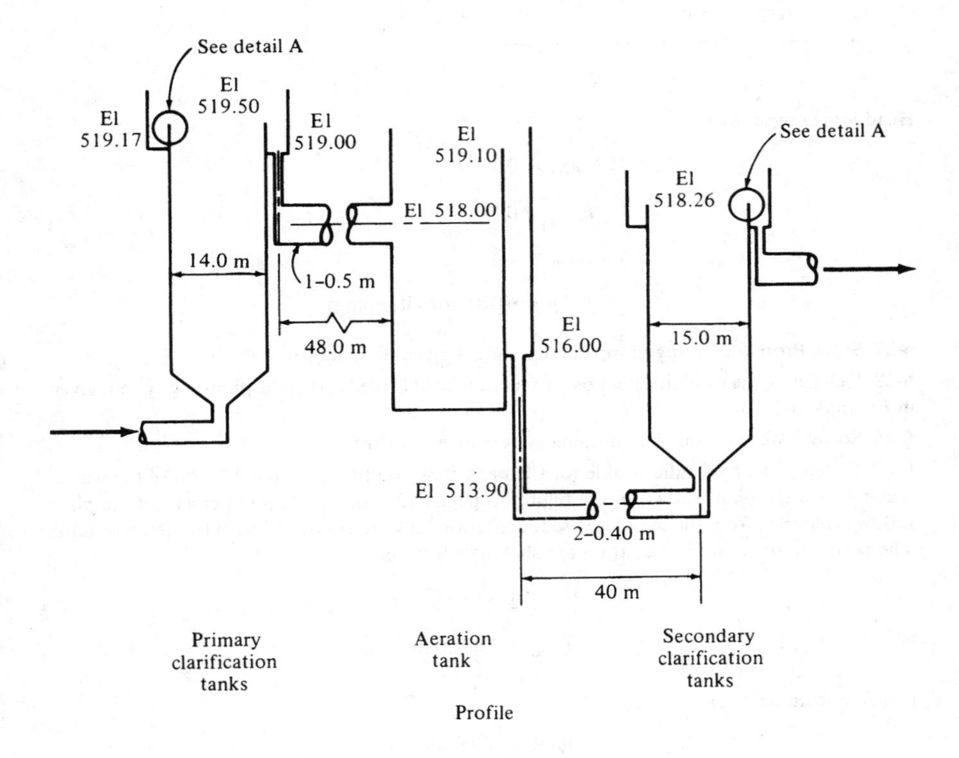

REFERENCES

6-1 Addison, H.: *Centrifugal and Other Rotodynamic Pumps*, 3d ed., Chapman and Hall, London, 1966.

6-2 Fair, G. M., J. C. Geyer, and D. A. Okun: *Water and Wastewater Engineering*, Volume 1: *Water Supply and Wastewater Removal*, Wiley, New York, 1968.

6-3 *Hydraulic Institute Standard For Centrifugal, Rotary, and Reciprocating Pumps*, 13th ed., Hydraulic Institute, Cleveland, Ohio, 1975.

6-4 Jeppson, R. W.: *Analysis of Flow in Pipe Networks*, Ann Arbor Science, Ann Arbor, Mich., 1976.

6-5 Joint Committee of the American Society of Civil Engineers and the Water Pollution Control Federation: *Design and Construction of Sanitary Sewers*, ASCE Manuals and Reports on Engineering Practice no. 37, New York, 1969.

6-6 King, H. W., and E. F. Brater: *Handbook of Hydraulics*, 5th ed., McGraw-Hill, New York, 1963.
6-7 Lager, J. A., and W. G. Smith: *Urban Stormwater Management and Technology: An Assessment*, EPA-670/2-74-040, Cincinnati, Ohio, December 1974.
6-8 Metcalf & Eddy, Inc.: *Wastewater Engineering: Collection and Pumping of Wastewater*, rev. by G. Tchobanoglous, McGraw-Hill, New York, 1981.
6-9 Metcalf & Eddy, Inc.: *Wastewater Engineering: Treatment, Disposal, Reuse*, 2d ed., McGraw-Hill, New York, 1979.
6-10 Pardoe, W. S.: "Computing Head Loss in Gridiron Distribution System," *Engineering News-Record*, **93**(13):516 (1924).
6-11 *Pumping Station Design for the Practicing Engineer: Volume I Fundamentals, Volume II Wastewater, Volume III Water*, Conference Proceedings, Department of Civil Engineering and Engineering Materials, Montana State University, Bozeman, 1981.
6-12 Ripple, W.: "The Capacity of Storage Reservoirs for Water Supply," *Proc Inst Civ Eng*, vol. 71, 1883.
6-13 Vennard, J. K., and R. L. Street: *Elementary Fluid Mechanics*, 5th ed., Wiley, New York, 1975.
6-14 Wanielista, M. P.: *Stormwater Management: Quantity and Quality*, Ann Arbor Science, Ann Arbor, Mich., 1978.
6-15 Water Pollution Control Federation: *Design of Wastewater and Stormwater Pumping Stations*, Manual of Practice no. FD-4, New York, 1981.
6-16 Whipple, W., et al.: *Stormwater Management In Urbanizing Areas*, Prentice-Hall, Englewood Cliffs, N.J., 1983.

PART TWO

AIR

CHAPTER

SEVEN

AIR QUALITY: DEFINITIONS, CHARACTERISTICS, AND PERSPECTIVES

Though most people tend to associate air pollution problems with the coming of the industrial revolution, such problems, in one form or another, have plagued the human race for centuries. The earliest pollutants noted in the atmosphere were probably of natural origin. Smoke, fumes, ash, and gases from volcanoes and forest fires; sand and dust from windstorms in arid regions; fog in humid, low-lying areas; and natural terpene hazes from pine trees in mountainous regions were part of our environment long before human-induced, or anthropogenic, problems came on the scene.

Indeed, several of the above-mentioned natural problems would surely have qualified as "air pollution" under this generally accepted definition of pollution of ambient, or outdoor, air:

> Air pollution is the presence in the outdoor atmosphere of one or more air contaminants (i.e., dust, fumes, gas, mist, odor, smoke, or vapor) in sufficient quantities, of such characteristics, and of such duration as to be or to threaten to be injurious to human, plant, or animal life or to property, or which reasonably interferes with the comfortable enjoyment of life or property. [7-6]

Except in such extreme cases as volcanic eruption, pollution from natural sources does not usually, by itself, pose problems severe enough to endanger life and property. Ultimately, human activities are to blame for pollution problems that threaten to make portions of the earth's atmosphere an inhospitable environment.

Air Pollution—Past, Present, and Future

7-1 HISTORICAL OVERVIEW

When Prometheus stole fire from the gods, he unwittingly stole a curse as well as a blessing, for fire generates smoke and smoke is one of the earliest anthropogenic air pollutants. When the smoke from isolated wood-burning fires of early cave dwellers became the smoke from coal-burning furnaces in heavily populated cities, the effects of pollution became severe enough to alarm some of the inhabitants of those cities. In A.D. 61 the philosopher Seneca described "the heavy air of Rome" and "the stink of the smoky chimneys thereof" [7-73], and in 1273 King Edward I was bothered enough by the smoke and fog mixture that brooded over London to prohibit the burning of "sea coal." [7-82]

Despite King Edward's warning, heavy deforestation made wood a scarce commodity, and Londoners increased rather than decreased their reliance on coal. By the time Queen Elizabeth I ascended the throne, the city's notorious pea-soup fogs had become smog, that foul condition that occurs when *smoke* and *fog* intermingle, though the word itself was probably not coined until several centuries later. Apparently the Queen had an allergy to coal smoke as well as an aversion to it, and she moved out of the city into the cleaner air of Nottingham. Toward the end of Elizabeth's reign, a law was passed prohibiting the burning of coal when Parliament was sitting. [7-37]

By 1661, over a century later, compliance was apparently still not obtained, for John Evelyn's pamphlet, *Fumifugium: or the Inconvenience of the Aer and Smoake of London Dissipated, together with some Remedies Humbly Proposed.* recommended removal of all smoke-producing plants from London and the planting of greenbelts around the city, two solutions to air pollution that are still being recommended today. [7-72] His words apparently fell on empty, if polluted, air, and even a 1772 reissue of his pamphlet failed to effect the needed improvements in air quality.

As shown in Table 7-1, air pollution problems surfaced intermittently in the late 1880s. There was some corresponding evidence of interest in overcoming air pollution, including enactment of smoke-control laws in Chicago and Cincinnati in 1881. [7-12] Yet enforcement of these and similar laws was difficult, and nothing was really done about cleaning up the air or preventing further pollution. By the turn of the century, members of England's Smoke Abatement League were calling in experts to testify to the ill effects of air pollution, and 23 of the 28 largest cities in the United States had passed smoke control laws. [7-12] Still, despite the efforts of these groups and others, nothing was done to prevent pollution of the atmosphere.

In 1930, an inversion trapped smog in Belgium's highly industrialized Meuse Valley. Sixty-three persons died, and several thousand others became ill. Some 18 years later, similar conditions led to one of the first major air-pollution disasters in the United States. Seventeen people died and 43 percent of the population of

Table 7-1 Reported disease morbidity and mortality occurring during air pollution episodes

Year and month	Location	Excess deaths reported	Reported illness
1873, Dec. 9–11	London, England		
1880, Jan. 26–29	London, England		
1892, Dec. 28–30	London, England		
1930, December	Meuse Valley, Belgium	63	6000
1948, October	Donora, Pennsylvania	17	6000
1948, Nov. 26–Dec. 1	London, England	700–800	
1952, Dec. 5–9	London, England	4000	
1953, November	New York, New York		
1956, Jan. 3–6	London, England	1000	
1957, Dec. 2–5	London, England	700–800	
1958	New York, New York		
1959, Jan. 26–31	London, England	200–250	
1962, Dec. 5–10	London, England	700	
1963, Jan. 7–22	London, England	700	
1963, Jan. 9–Feb. 12	New York, New York	200–400	
1966, Nov. 23–25	New York, New York		

Source: From *Air Quality Criteria*. [7-3]

Donora, Pennsylvania, became ill. Around this same time, Californians were beginning their efforts to legislate against smog, citing automobile emissions, rather than coal smoke, as the smoke portion of smog in their major cities. Still, the first conclusive paper on air pollution and automobile emissions was not published until 1955.

That was 3 years after the London smog disaster of 1952 made it impossible to ignore any longer the serious consequences of air pollution. On Thursday, December 4, of that year, a high-temperature air mass moved over southern England, creating a temperature inversion and causing a white fog to settle over London. As particulate and sulfur dioxide levels rose because of the extensive use of coal-fired heating and power production systems, the fog began to blacken. The high-pressure area stalled, and pollutant buildup worsened, since there were no air currents to disperse the smog. Zero visibility greeted early risers on Friday, and one observer recalled that "a white shirt collar became almost black within 20 minutes." [7-84]

The smog was intensely irritating to the human respiratory system, and most people soon developed red eyes, burning throats, and nagging coughs. Soon, reports of smog-related deaths began to come in. The elderly and people with chronic respiratory problems were dying, as were young and otherwise healthy persons whose work kept them out in the smog. Before the fog lifted on December 9, 4000 deaths attributed to air pollution had been recorded, enough to move Britons to pass the Clean Air Act in 1956.

One year earlier, the United States had passed the Air Pollution Control Act of 1955 (Public Law 84-159), but the act did little more than move *toward* effective legislation. Revised in 1960 and again in 1962, the act was supplanted by the Clean Air Act of 1963 (PL 88-206) which encouraged state, local, and regional programs for air pollution control but reserved the right of federal intervention, should pollution from one state endanger the health and welfare of citizens residing in another state. This act also began the development of air quality criteria upon which the air quality and emissions standards of the 1970s were based.

The Clean Air Act of 1970 (PL 91-604) gave the job of implementing regulations to the newly created Environmental Protection Agency (EPA). The act set primary and secondary ambient-air-quality standards. Primary standards, based on air quality criteria, allowed for an extra margin of safety to protect public health, while secondary standards, also based on air quality criteria, were established to protect public welfare—plants, animals, property, and materials. Further discussion of these standards, including a table that presents them (Table 7-19), will be found later in this chapter.

The Clean Air Act Amendments of 1977 (PL 95-95) further strengthened the existing laws and set the nation's course toward cleaning up our atmosphere. [7-81] These amendments were themselves reassessed and amended during the energy crunch of the late 1970s. Though it is likely that further changes will be made, it is probable that the move toward air pollution control has gathered enough momentum and public support to maintain a course which ensures a cleaner, healthier atmosphere for generations to come.

7-2 GLOBAL IMPLICATIONS OF AIR POLLUTION

If the earth is to remain inhabitable, the long-range effects of pollution of the atmosphere which blankets the planet can never again be neglected. The isolated examples discussed above give evidence of what can happen when any one community or area experiences heavy exposure to atmospheric contaminants for significant periods of time. What is less well known are the global implications of the various forms of pollution being introduced into the atmosphere.

Composition and Structure of the Atmosphere

A necessary first step toward understanding air pollution and its control is understanding the composition and structure of the atmosphere. The total mass of each gas in the atmosphere is given in Table 7-2. Varying amounts of most of these gases may be found in each of the four major layers of the atmosphere—the troposphere, the stratosphere, the mesosphere, and the thermosphere (see Fig. 7-1).

As indicated in Table 7-3 the air in the troposphere, the air which we breathe, consists, by volume, of about 78 percent nitrogen (N_2), 21 percent oxygen (O_2), 1 percent argon (Ar), and 0.03 percent carbon dioxide (CO_2). Also present are traces of other gases, most of which are inert.

Table 7-2 Mass of different gases in the world's atmosphere

Gas or vapor	Trillions of tonnes
Nitrogen (N_2)	3900
Oxygen (O_2)	1200
Argon (Ar)	67
Water vapor (H_2O)	14
Carbon dioxide (CO_2)	2.5
Neon (Ne)	0.065
Krypton (Kr)	0.017
Methane (CH_4)	0.004
Helium (He)	0.004
Ozone (O_3)	0.003
Zenon (Xe)	0.002
Dinitrogen oxide (N_2O)	0.002
Carbon monoxide (CO)	0.0006
Hydrogen (H_2)	0.0002
Ammonia (NH_3)	0.00002
Nitrogen dioxide (NO_2)	0.000013
Nitric oxide (NO)	0.000005
Sulfur dioxide (SO_2)	0.000002
Hydrogen sulfide (H_2S)	0.000001

Source: From Giddings. [7-21]

The layer of greatest interest in pollution control is the troposphere, since this is the layer in which most living things exist. One of the more recent changes in the trophosphere involves the phenomenon of *acid rain*. Acid rain, or *acid deposition*, results when gaseous emissions of sulfur oxides (SO_x) and nitrogen oxides (NO_x) interact with water vapor and sunlight and are chemically converted to strong acidic compounds such as sulfuric acid (H_2SO_4) and nitric acid (HNO_3). These compounds, along with other organic and inorganic chemicals, are deposited on the earth as aerosols and particulates (*dry deposition*) or are carried to the earth by raindrops, snowflakes, fog, or dew (*wet deposition*).

The effects of acid deposition vary according to the sensitivity of the ecosystems upon which the deposits fall. In some highly buffered areas acidic compounds could be deposited for years without causing any appreciable increase in soil or surface-water acidity, but the same dposition could cause sharp increases in acidity in poorly buffered areas.

Acid rain has caused considerable damage to buildings and monuments in highly industrialized areas, but damage is not limited to the immediate area of industrialization. Tall stacks disperse pollutants into the upper reaches of the troposphere where they may remain for days, often being carried long distances. For example, there is substantial evidence that prevailing winds may be carrying significant amounts of emissions of SO_x and NO_x from industrialized areas of the midwest into New England and Canada, where the emissions may return to

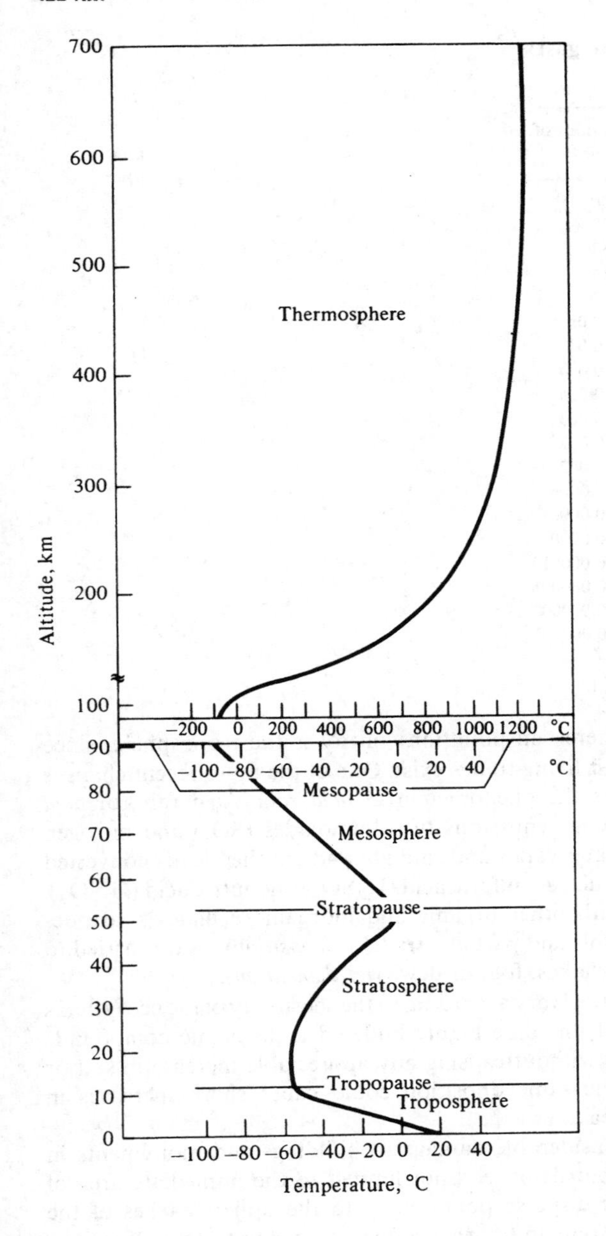

Figure 7-1 Temperature profile of the atmosphere. (*From Miller* [7-*42*].)

Table 7-3 Concentration of atmospheric gases in clean, dry air at ground level

Gas	Concentration, ppm by volume	Concentration, % by volume
Nitrogen (N_2)	280,000	78.09
Oxygen (O_2)	209,500	20.95
Argon (Ar)	9,300	0.93
Carbon dioxide (CO_2)	320	0.032
Neon	18	0.0018
Helium (He)	5.2	0.00052
Methane (CH_4)	1.5	0.00015
Krypton (Kr)	1.0	0.0001
Hydrogen (H_2)	0.5	0.00005
Dinitrogen oxide (N_2O)	0.2	0.00002
Carbon monoxide (CO)	0.1	0.00001
Zenon (Xe)	0.08	0.000008
Ozone (O_3)	0.02	0.000002
Ammonia (NH_3)	0.006	0.0000006
Nitrogen dioxide (NO_2)	0.001	0.0000001
Nitric oxide (NO)	0.0006	0.00000006
Sulfur dioxide (SO_2)	0.0002	0.00000002
Hydrogen sulfide (H_2S)	0.0002	0.00000002

Source: From Giddings. [7-21]

earth as acid rain. Pollutants that are generated in one country and deposited in another have become a matter of international concern and of international negotiation. [7-12]

Other changes in the atmosphere as a whole may not be quite so obvious. For example, the twentieth century has seen widespread use of radioactive materials, and concern over the long-range effects of release of these substances into the atmosphere has led to investigation of possible methods of safe disposal by deep burial in the earth or ocean. [7-17]

Recently, the National Academy of Science reported that the ozone (O_3) layer in the stratosphere is being depleted as ozone reacts with chlorine released from the fluorocarbons used as aerosol spray propellants. Since the O_3 in the atmosphere reduces the ultraviolet radiation that reaches the earth's surface, and since ultraviolet radiation at high levels can damage plants and animals, loss of O_3 represents a potentially serious problem. In light of this danger, some industrialized nations have banned the use of fluorocarbons.

The amount of tropospheric carbon dioxide (CO_2) is reported to be increasing at a rate of 1.8 mg/m^3 per year, a process that may not be reversible. Furthermore, this increase has been accompanied by an equivalent decrease in atmospheric oxygen (O_2). Currently, there are more than 700 billion tons of carbon in the form of CO_2 in the atmosphere. Each year this figure increases by 2.3 billion tons, the equivalent of a 3 percent increase every decade. [7-17]

Fossil fuel consumption and agricultural, forestry, and land-use practices of various types contribute to the CO_2 buildup. As Chap. 8 indicates, CO_2 strongly absorbs long-wave (infrared) terrestrial radiation, and continued CO_2 buildup could lead to a significant enough rise in earth's surface temperatures to melt the Arctic ice pack. If the warming trend can be confirmed and positively linked to CO_2 buildup, then global action such as reforestation may eventually have to be pursued to remove CO_2 from the atmosphere. [7-17]

Global Programs for Pollution Control

Because of the possible detrimental consequences of gradual global increase in air pollution, in the early 1970s both the World Health Organization (WHO) and the World Meteorological Organization (WMO) initiated air-monitoring programs. The WHO program is active in urban and industrial areas, so-called impact areas. Its main thrust is protection of the health of urban populations. Concentrating on background air pollution measurements, WMO predicts the potential for long-term global and regional changes in air pollution concentrations as these affect climate.

The WHO and WMO programs were given further support when, at the second session of its governing council, the United Nations Environmental Program (UNEP) authorized the implementation of the Global Environmental Monitoring System (GEMS). The principal goals of the GEMS program in the air pollution area are the expansion of human health warning systems, the assessment of global atmospheric pollution and its impac on climate, and the assessment of critical problems associated with agricultural and land-use practices. [7-38].

The GEMS is highly dependent upon the monitoring activities of the WHO project and the WMO network. The WHO project measures on the microscale (urban) and the WMO network measures on the macroscale (continental and global), and comparisons of measurements lend valuable insights. For example, typical monthly means for SO_2 range from 20 to 100 $\mu g/m^3$ on the local and urban scale, from 10 to 40 $\mu g/m^3$ on the continental scale, and are less than 5 $\mu g/m^3$ on the global scale. [7-15] Only through worldwide cooperation of those involved in air pollution research can such data be brought together, and only when pollutants are observed on local, regional, continental, and global scales can we hope to see the total ramifications of the air pollution problem, a vital first step toward solving that problem.

7-3 UNITS OF MEASUREMENT

For GEMs and other such groups, those in charge of air pollution control must often compare measurements of air-contaminant emissions and ambient-air-contaminant concentrations with the standards set forth in air pollution control regulations. In the past, this has been difficult because there was no standardization of units, but a move to report units in the metric system should make such comparisons much easier in the future.

For uniformity, the Environmental Protection Agency has recommended using the following units for particulates and gaseous pollutants. *Particulate fallout*, or *dustfall*, is to be expressed in milligrams per square centimeter per time interval (i.e., $mg/cm^2 \cdot mo$ or $mg/cm^2 \cdot y$). In *particulate counting*, notations will be given as number of particles per cubic meter of gas or million particles per cubic meter ($10^6/m^3$). Measurements for *suspended particulates* and *gaseous contaminants* are to be given on a mass per unit volume basis, such as micrograms per cubic meter ($\mu g/m^3$). Formerly concentrations of gaseous contaminants were usually given in parts per million (ppm), parts per hundred million (pphm), or parts per billion (ppb) by volume. Thus designations in micrograms per cubic meter may be followed by equivalent concentration on a parts-per-million (ppm) basis—e.g., 80 $\mu g/m^3$ (0.03 ppm) of sulfur dioxide. For gases, parts per million (ppm) can be converted to micrograms per cubic meter ($\mu g/m^3$) by use of the following formula:

$$\mu g/m^3 = \frac{\text{ppm} \times \text{g mol mass} \times 10^3}{\text{L/mol}} \tag{7-1}$$

The liters per mole designation Eq. (7-1) is necessarily influenced by the temperature and pressure of the gas. According to Avogadro's law, one mole of any one gas occupies the same volume as one mole of any other gas at the same temperature and pressure. At 273 K (0°C) and 1 atm pressure (760 mmHg), standard conditions for many chemical reactions, this volume is 22.4 L/mol. However, most regulations for air quality determinations are referenced at 25°C and 760 mmHg, while most source measurements are referenced at 21.1°C and 760 mmHg. To convert to liters per mole at these or other conditions, the following formula can be used:

$$\frac{V_1 P_1}{T_1} = \frac{V_2 P_2}{T_2} \tag{7-2}$$

where V_1, P_1, and T_1 relate to the above conditions of 22.4 L/mol at 273 K and 760 mmHg, and V_2, P_2, and T_2 relate to the actual conditions being considered. Use of Eq. (7-2) is illustrated in Example 7-1.

Example 7-1: Determining volume, temperature, and pressure relationships Determine the volume occupied by 2 mol of gas at 25°C and 820 mmHg.

SOLUTION

1. Using Eq. (7-2):

$$\frac{V_1 P_1}{T_1} = \frac{V_2 P_2}{T_2}$$

$$\frac{2 \text{ mol} \times 22.4 \text{ L/mol} \times 760 \text{ mmHg}}{273 \text{ K}} = \frac{V_2\, 820 \text{ mmHg}}{(273 + 25) \text{ K}}$$

2. Solving for V_2:

$$V_2 = \frac{2 \times 22.4 \times 760 \times 298}{273 \times 820}$$

$$= 45.32 \text{ L}$$

Once liters-per-mole calculations have been made, Eq. (7-1) can be used to convert parts per million to micrograms per cubic meter, as shown in Example 7-2.

Example 7-2: Converting from parts per million (ppm) to mass per volume ($\mu g/m^3$) A sample of air analyzed at 0°C and 1 atm pressure is reported to contain 9 ppm of CO. Determine the equivalent CO concentration in micrograms per cubic meter and milligrams per cubic meter.

SOLUTION

1. Using Eq. (7-1):

$$\mu g/m^3 = \frac{\text{ppm} \times \text{g mol mass} \times 10^3}{\text{L/mol}}$$

2. The gram molecular mass of CO is

$$12 + 16 = 28 \text{ g/mol}$$

3. At 0°C and 1 atm of pressure (760 mmHg) the volume of the gas is 22.4 L/mol.
4. Substituting into step 1:

$$\mu g/m^3 = \frac{9 \times 10^{-6} \times 28 \text{ g/mol} \times 10^3 \text{ L/m}^3 \times 10^6 \ \mu g/g}{22.4 \text{ L/mol}}$$

$$\text{CO} = 11{,}250 \ \mu g/m^3 = 11.25 \text{ mg/m}^3$$

The move toward uniformity in reporting units will reduce confusion in understanding and analyzing air pollution reports. It could also assist in pointing up discrepancies in monitoring reported by various investigators. [7-33]

7-4 SOURCES OF POLLUTANTS

All air contains natural contaminants such as pollen, fungi spores, salt spray, and smoke and dust particles from forest fires and volcanic eruptions. It contains also naturally occurring carbon monoxide (CO) from the breakdown of methane (CH_4); hydrocarbons in the form of terpenes from pine trees; and hydrogen sulfide (H_2S) and methane (CH_4) from the anaerobic decomposition of organic matter.

In contrast to the natural sources of air pollution there are contaminants of anthropogenic origin. The total anthropogenic emissions (on a weight basis) of the five most prevalent air pollutants for the years 1940 to 1980 are given in Table 7-4.

The use of fossil fuels for heating and cooling, for transportation, for industry, and for energy conversion, and the incineration of the various forms of industrial,

Table 7-4 National air pollutant emissions, by pollutant, 1940–1980*

Year	Total suspended particulates	Percent of 1974	Sulfur oxides	Percent of 1974	Nitrogen oxides	Percent of 1974	Hydro-carbons	Percent of 1974	Carbon monoxide	Percent of 1974
1940	21.9	181	17.4	64	6.5	32	13.9	58	74.7	73
1950	23.2	192	19.6	73	9.3	46	17.5	74	32.8	81
1960	20.2	167	19.2	71	12.7	63	21.6	91	90.8	89
1970	17.6	145	27.9	103	18.5	92	27.1	114	110.9	108
1971	16.4	136	26.5	98	19.0	95	25.4	111	110.5	108
1972	14.9	123	27.3	101	20.1	100	26.7	112	109.7	107
1973	13.9	115	28.4	105	20.4	101	26.2	110	107.4	105
1974	12.1	100	27.0	100	20.1	100	23.8	100	102.5	100
1975	10.1	83	25.6	95	19.6	98	22.8	96	98.1	96
1976	9.4	78	26.4	98	20.9	104	23.7	100	100.4	98
1977	8.5	70	26.4	98	21.3	106	23.8	100	97.8	95
1978	8.6	71	24.8	92	21.5	107	24.4	103	96.7	94
1979	8.5	70	25.3	94	21.5	107	23.4	98	92.6	90
1980	7.8	64	23.7	88	20.7	103	21.8	92	85.4	83
Change 1970–1980	−56%		−15%		+12%		−20%		−23%	

* In million tonnes per year.

Source: From *Twelfth Annual Report*. [7-12]

municipal, and private waste all contribute to the pollution of the atmosphere. So do the handling and processing operations of various and sundry industries. The sources of these pollutants are so numerous and varied that they have been categorized into four main groups—*mobile transportation* (i.e., motor vehicles, aircraft, railroads, ships, and the handling and/or evaporation of gasoline), *stationary combustion* (i.e., residential, commercial, and industrial power and heating, including steam-powered electric power plants), *industrial processes* (i.e., chemical, metallurgical, and pulp-paper industries and petroleum refineries), and *solid-waste disposal* (i.e., household and commercial refuse, coal refuse, and agricultural burning). [7-28]

The activities that accounted for most of the overall air contaminants in 1980 are shown in Table 7-5. It will be noted that while transportation was the single largest source of air pollution, fuel combustion in stationary sources (for power and heating) was the second major contributor. Power generation and heating accounted for about 80 percent of the oxides of sulfur and 51 percent of the oxides of nitrogen emitted to the ambient air, while industrial processes contributed 50 percent of the hydrocarbons. Not included in Table 7-5 are the photochemical oxidants because they are secondary pollutants formed in the atmosphere by the action of sunlight on hydrocarbons and oxides of nitrogen.

The total emissions (million tonnes per year) of the five major air pollutants for 1977 are illustrated in Fig. 7-2. Of these pollutants, transportation produced approximately 83 percent of the total carbon monoxide, 41 percent of the total hydrocarbons, 40 percent of the oxides of nitrogen, 9 percent of the particulates, and 3 percent of the sulfur oxide emitted. In 1980, transportation produced a slightly lower percentage of total carbon monoxide and hydrocarbons, a slightly higher percent of oxides of nitrogen, and a significantly higher percent of particu-

Table 7-5 Sources of air pollutants, 1980*

	Pollutants					
Source	CO	Particulates	SO_x	HC	NO_x	Total
Transportation	69.1	1.4	0.9	7.8	9.1	88.3
Fuel combustion from stationary sources (power, heating)	2.1	1.4	19.0	0.2	10.6	33.3
Industrial processes	5.8	3.7	3.8	10.8	0.7	24.8
Solid-waste disposal	2.2	0.4	0.0	0.6	0.1	3.3
Miscellaneous (forest fires, agricultural burning, etc.)	6.2	0.9	0.0	2.4	0.2	9.7
Total	85.4	7.8	23.7	21.8	20.7	159.4

* In million tonnes per year.
Source: From *Twelfth Annual Report*. [7-12]

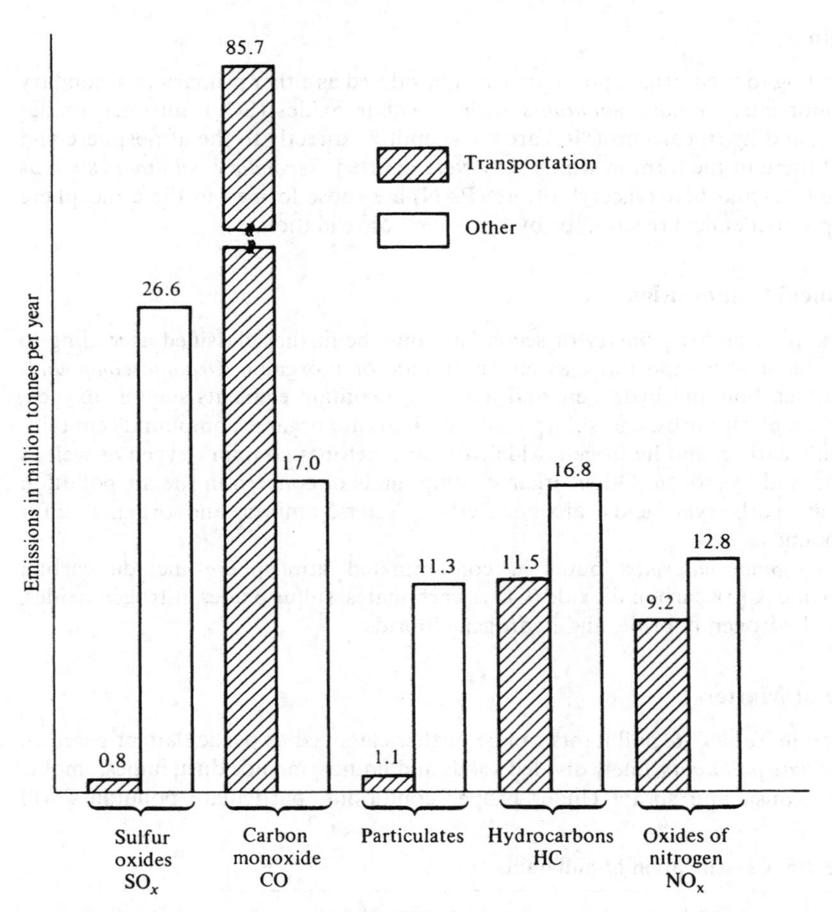

Figure 7-2 Pollution emissions in the United States—1977. (*From U.S. EPA* [7-76].)

lates. Transportation was responsible for 56 percent (by weight) of all the contaminants emitted to the atmosphere in 1977, as compared to 55 percent in 1980. [7-12]

Further discussion of the specific sources of some of these contaminants will be found under the appropriate parameter headings, and emission factors and emission inventories for various activities will be discussed in the final section of this chapter.

Classifications of Pollutants

All air pollutants may be classified according to origin, chemical composition, and state of matter. For clarity, these classifications will be used to structure the discussion of air pollution parameters.

Origin

According to their origin, pollutants are considered as either primary or secondary contaminants. *Primary pollutants* such as sulfur oxides (SO_x), nitrogen oxides (NO_x), and hydrocarbons (HC) are those emitted directly to the atmosphere and found there in the form in which they were emitted. *Secondary pollutants* such as ozone (O_3) and peroxyacetyl nitrate (PAN) are those formed in the atmosphere by a photochemical reaction or by hydrolysis or oxidation.

Chemical Composition

Pollutants, whether primary or secondary, may be further classified according to their chemical composition, as either organic or inorganic. *Organic compounds* contain carbon and hydrogen, and many also contain elements such as oxygen, nitrogen, phosphorus, and sulfur. Hydrocarbons are organic compounds containing only carbon and hydrogen. Aldehydes and ketones contain oxygen as well as carbon and hydrogen. Other organic compounds of concern in the air pollution field are carboxylic acids, alcohols, ethers, esters, amines, and organic sulfur compounds.

Inorganic materials found in contaminated atmosphere include carbon monoxide (CO), carbon dioxide (CO_2), carbonates, sulfur oxides, nitrogen oxides, ozone, hydrogen fluoride, and hydrogen chloride.

State of Matter

As seen in Table 7-6, pollutants can be further classified as particulate or gaseous. *Particulate pollutants*, finely divided solids and liquids, include dust, fumes, smoke, fly ash, mist, and spray. Under proper conditions, particulate pollutants will

Table 7-6 Classification of pollutants

Major classes	Subclasses	Typical members of subclasses
Particulates	Solid	Dust, smoke, fumes, fly ash
	Liquid	Mist, spray
Gases		
Organic	Hydrocarbons	Hexane, benzene, ethylene, methane, butane, butadiene
	Aldehydes and ketones	Formaldehyde, acetone
	Other organics	Chlorinated hydrocarbons, alcohols
Inorganic	Oxides of carbon	Carbon monoxide, carbon dioxide
	Oxides of sulfur	Sulfur dioxide, sulfur trioxide
	Oxides of nitrogen	Nitrogen dioxide, nitric oxide
	Other inorganics	Hydrogen sulfide, hydrogen fluoride, ammonia

Source: Adapted from Simon. [6-69]

settle out of the atmosphere. *Gaseous pollutants*, formless fluids that completely occupy the space into which they are released, behave much as air and do not settle out of the atmosphere. In addition to substances that are gases at normal temperature and pressure, gaseous pollutants include vapors of substances that are liquid or solid at normal temperatures and pressures. Among common gaseous pollutants are carbon oxides, sulfur oxides, nitrogen oxides, hydrocarbons, and oxidants.

7-5 PARTICULATES

As noted earlier, air quality parameters fall into two broad categories—particulate matter, which may be liquid or solid, and gaseous matter. Federal criteria documents identify particulates as any dispersed matter, solid or liquid, in which the individual aggregates are larger than a single small molecule (about 0.002 μm in diameter) but smaller than about 500 μm. [7-46]

Particulates may be classified and discussed according to their physical, chemical, or biological characteristics. *Physical characteristics* include size, mode of formation, settling properties, and optical qualities. *Chemical characteristics* include organic or inorganic composition, and *biological characteristics* relate to their classification as bacteria, viruses, spores, pollens, etc.

Physical Characteristics

Size Size is one of the most important physical properties of particulates. The size range for some common natural and anthropogenic air-borne particulates is shown in Fig. 7-3. Particle sizes are measured in micrometers (μm), with one micrometer being equal to 10^{-6} meters, 1/1000 of a millimeter, or 1/25,400 of an inch. Particles larger than 50 μm can be seen with the unaided eye, while those smaller than 0.005 μm may only be observed through an electron microscope.

Human hair ranges from 5 to 600 μm in diameter, while the particles of major interest in air pollution studies range from 0.01 to 100 μm in size. Particles smaller than 1 μm do not tend to settle out rapidly. Metallurgical fumes, cement dust, fly ash, carbon black, and sulfuric acid mist all fall into the 0.01- to 100-μm range.

Mode of formation Particles may be classified according to their mode of formation as dust, smoke, fumes, fly ash, mist, or spray. The first four are solid particles, while the last two are liquid.

Dust, small, solid particles created by the breakup of larger masses through processes such as crushing, grinding, or blasting, may come directly from the processing or handling of materials such as coal, cement, or grains. It may be a by-product of a mechanical process such as the sawing of wood, or made up of residue of a mechanical operation, such as sandblasting. Capable of temporary

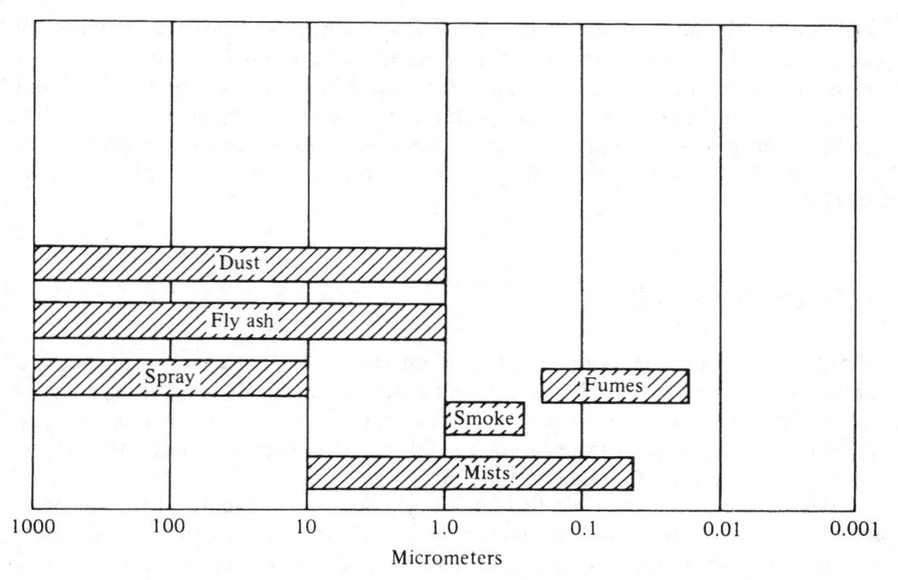

Figure 7-3 Sizes of particulates in micrometers. (*Adapted from Federal Register* [7-54].)

suspension in air or other gases, dusts do not diffuse. They settle under the influence of gravity. As indicated in Fig. 7-3, dusts may range in size from 1.0 to 10000 μm.

Smoke, fine, solid particles resulting from the incomplete combustion of organic particles such as coal, wood, or tobacco, consists mainly of carbon and other combustible materials. Smoke particles have diameters ranging from 0.5 to 1 μm.

Fumes are fine, solid particles (often metallic oxides such as zinc and lead oxides) formed by the condensation of vapors of solid materials. Fumes may be from sublimation, distillation, calcination, or molten metal processes, and they range in size from 0.03 to 0.3 μm. Fumes flocculate and coalesce, then settle out.

Fly ash consists of finely divided, noncombustible particles contained in flue gases arising from combustion of coal. Inherent in all coal, these mineral or metallic substances are released when the organic portion of coal is burned. Fly ash shares characteristics of all three of the other solid particulates discussed. Like dust, it has particles that range in size from 1.0 to 1000 μm; like smoke, it results from burning; and like fumes, it consists of inorganic metallic or mineral substances.

Mist consists of liquid particles or droplets formed by the condensation of a vapor, the dispersion of a liquid (as in foaming or splashing), or the enactment of a chemical reaction (such as the formation of sulfuric acid mist). Mists are usually less than 10 μm in diameter. If mist concentration is high enough to obscure visibility, the mist is called a fog.

Spray consists of liquid particles formed by the atomization of parent liquids, such as pesticides and herbicides. Spray particles range in size from 10 to 1000 μm.

Settling properties Settling characteristics are one of the most important properties of particulates, since settling is the major natural self-cleansing process for removal of particulates from the atmosphere. Particulates can generally be classified as suspended or settleable. *Suspended particulates* vary in size from less than 1 μm to approximately 20 μm. They remain suspended in the atmosphere for long periods of time. *Settleable particles*, or *dustfall*, are larger and heavier and settle out close to their sources. They are generally greater than 10 μm in size.

Stokes' law, as given in Chap. 4, is a mathematical equation used to calculate the terminal settling velocity of waterborne particulates. The equation takes into account the gravitational as well as the frictional forces, or drag forces, acting upon the particulates. [7-23] This equation can be used to calculate the terminal settling velocity of airborne particulates, though it is well to remember that Stokes' law is based on quiescent settling and does not consider the turbulent nature of the atmosphere. Furthermore, it only applies generally to particulate matter between 0.1 and 100 μm, and it can only be used when the Reynolds number is less than 1. (As noted in Chap. 4, the Reynolds number for a particle relates its inertial forces to viscous forces for the medium in which it is settling.)

In the following equation, the density and viscosity of air are substituted for the density and viscosity of water. For most applications, the density of air is negligible compared to the density of the particle and can be omitted from the equation.

$$v_t = \frac{g(\rho_p - \rho_a)\, d_p^2}{18\mu} \tag{7-3}$$

where v_t = terminal settling velocity, m/s
g = gravitational constant, m/s^2
ρ_p = density of the particle, kg/m^3
ρ_a = density of air, kg/m^3 (approximately 1.2 kg/m^3 for air at sea level)
d_p = diameter of particle, m
μ = viscosity of air, N · s/m^2 or kg/m · s (see Fig. C-2 in the appendix)

Example 7-3: Determining the settling velocities of particulates in air Determine the settling rate of a fog cloud composed of 1-μm particles in air at 27°C.

SOLUTION

1. From Fig. C-2 in the appendix, the viscosity of air at 27°C is:

$$1.85 \times 10^{-5}\ \text{N} \cdot \text{s/m}^2 = 1.85 \times 10^{-5}\ \text{kg/m} \cdot \text{s}$$

2. Using Eq. (7-3)

$$v_t = \frac{9.81\ \text{m/s}^2 \times 1 \times 10^3\ \text{kg/m}^3 \times (1 \times 10^{-6}\ \text{m})^2}{18\,(1.85 \times 10^{-5}\ \text{kg/m} \cdot \text{s})}$$

$$= 2.95 \times 10^{-5}\ \text{m/s}$$

3. Check Reynolds number:

$$N_{Re} = \frac{vdp}{\mu}$$

$$= \frac{2.95 \times 10^{-5}\ \text{m/s} \times 1 \times 10^{-6}\ \text{m} \times 1.2\ \text{kg/m}^3}{1.85 \times 10^{-5}\ \text{kg/m} \cdot \text{s}}$$

$$= 1.91 \times 10^{-6}$$

Stokes' law applies and the settling velocity is 2.95×10^{-5} m/s, which explains why fog appears to float in the atmosphere.

The surface properties of particulates, including adsorption, absorption, chemisorption, and adhesion, are particularly important factors in the settling process of particles of less than 1 μm. Settling of even smaller particles, those less than 0.1 μm in diameter, tends to be affected by a phenomenon known as *Brownian motion*. The random movement, or Brownian motion, of particles in the submicron range causes them to collide with the surrounding molecules, then to coagulate, flocculate, and eventually settle out.

Optical qualities Reduction in visibility is one of the most obvious effects of air pollution, and the scattering of light by particulate matter is primarily responsible for that reduction. Particles in the range of visible light (0.38 to 0.76 μm) are the most effective in visibility reduction.

However, since visibility is affected by particle shape and surface characteristics, as well as by the distribution of the particles by size, accurate calculation of transmission and scattering of light is a highly complex procedure. Although the following formula does not take all factors into consideration, but merely assumes that all particulate matter in the atmosphere is of uniform size and that scattering alone accounts for the light attenuation, it does show that, under highly specialized circumstances, a mathematical relationship can be derived between visibility and particulate matter.

$$V = \frac{5.2\ \rho r}{KM} \qquad (7\text{-}4)$$

where V = visibility, km
ρ = density of particle, kg/m^3
r = particle radius, μm
K = scattering area ratio, dimensionless (values can be found in the literature)
M = mass concentration of particle, μg/m^3

From 5 to 20 percent of the suspended particulates involved in light scattering consist of sulfuric acid mist and sulfates. SO_2 concentrations of 26 μg/m^3 (0.10 ppm) in air, with a comparable concentration of particulate matter and a relative humidity of 50 percent, will reduce visibility to 8 km (5 mi). [7-49] Decreased visibility of this severity interferes with safe operation of aircraft and automobiles, slowing airport operations because of the need to maintain greater distances between aircraft. At least 15 to 20 plane crashes each year are attributed to poor visibility caused by smoke, dust, and haze. [7-52]

Chemical Characteristics

There is great variation in the chemical composition of the particulates found in the atmosphere. Atmospheric particulates contain both organic and inorganic components. Some of the more common organics found in particulates include phenols, organic acids, and alcohols. Common inorganics found in particulates include nitrates, sulfates, and metals such as iron, lead, manganese, zinc, and vanadium.

Biological Characteristics

The biological particles in the atmosphere include protozoa, bacteria, viruses, fungi, spores, pollens, and algae. Microorganisms generally survive for only a short time in the atmosphere because of the lack of nutrients and ultraviolet radiation from the sun. However, certain bacteria and fungi form spores and can survive for long periods. Many spores and pollens are adapted for aerial dispersion and are found at elevations above 300 m (1000 ft). Some, especially the blue-green algae, have been found at altitudes up to 2000 m (6600 ft). [7-35]

Effects of Particulates

Effects on human health At high concentrations, suspended particulate matter poses health hazards to humans, particularly those susceptible to respiratory illness. As indicated in Table 7-7, the nature and extent of the ill effects that may

Table 7-7 Particulate matter—effects on health

Concentration, $\mu g/m^3$	Accompanied by	Time	Effect
750	715 $\mu g/m^3$ SO_2	24-h average	Considerable increase in illness
300	630 $\mu g/m^3$ SO_2	24-h average	Acute worsening of chronic bronchitis patients
200	250 $\mu g/m^3$ SO_2	24-h average	Increased absence of industrial workers
100–130	120 $\mu g/m^3$ SO_2	Annual mean	Children likely to experience increased incidence of respiratory disease
100	Sulfation rate above 30 $mg/cm^2/mo$	Annual geometric mean	Increased death rate for those over 50 likely
80–100	Sulfation rate above 30 $mg/cm^2/mo$	2-yr geometric mean	Increased death rate for those 50 to 69 yrs

Source: From *Air Quality Criteria*. [7-46]

be linked to suspended particulates depend upon the concentration of particulates, the presence of other atmospheric contaminants (notably sulfur oxides), and the length of exposure.

The human respiratory system is depicted in Fig. 7-4, and the ways in which that system defends itself against the invasion of foreign substances is indicated in Table 7-8. The success or failure of the respiratory defense systems depends, in part, upon the size of the particulates inhaled and the depth of their penetration into the respiratory tract. [7-30]

Approximately 40 percent of the particles between 1 and 2 μm in size are retained in the bronchioles and alveoli (see Fig. 7-5). Particles ranging in size from 0.25 to 1 μm show a decrease in retention, because many particles in this range are breathed in and out again. However, particles below 0.25 μm show another increase in retention because of Brownian motion, which results in impingement.

Lead Health problems associated with lead particulates have been of special interest in recent years. Nearly all human lead exposures originate from inhalation and ingestion of lead-containing particulates. Lead is emitted into the atmosphere as elemental lead (Pb), oxides of lead (PbO, PbO_2, Pb_xO_3), lead sulfates and lead sulfides ($PbSO_4$, PbS), alkyl lead [$Pb(CH_3)_4$, $Pb(C_2H_5)_4$], and lead halides. [7-58]

Clinical, epidemiological, and toxicological studies have demonstrated that exposure to lead adversely affects human health. [7-53] The three systems in the body most sensitive to lead are the blood-forming system, the nervous system, and

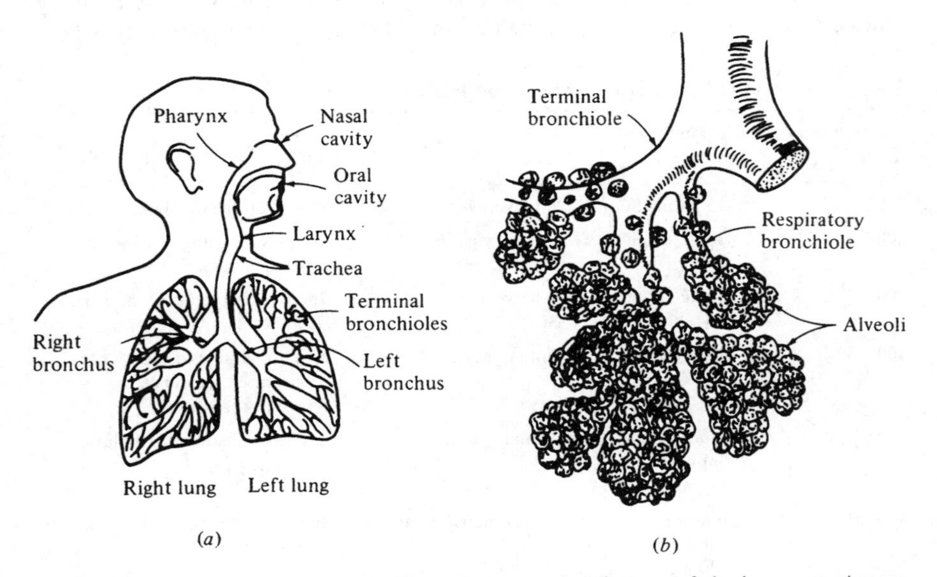

Figure 7-4 The respiratory system: (*a*) The major anatomical features of the human respiratory system. (*b*) The terminal bronchial and alveolar structure of the human lung.

Table 7-8 Particulate size and respiratory defense mechanism

Particle size	Description	Mechanism
Over 10 μm	Coarse dust, fly ash, (visible to the naked eye)	Hairs at the front of the nose remove all particles over 10 μm.
2 to 10 μm	Fumes, dust, smoke particles	Movement of cilia sweeps mucus upward, carrying particles from windpipe to mouth, where they can be swallowed.
Less than 2 μm	Aerosols, fumes	Lymphocytes and phagocytes in the lung attack some submicron particles.

Source: From U.S. EPA. [7-30]

the renal system. Reproductive, endocrine, hepatic, cardiovascular, immunologic, and gastrointestinal functions may also be affected by lead. In children, blood levels from 0.8 to 1.0 $\mu g/L$ can inhibit enzymatic systems. [7-58]

Lead poisoning can be either acute or chronic, with chronic lead poisoning occurring more frequently. The symptoms of acute lead poisoning include vomiting, colic, constipation or bloody diarrhea, with such central-nervous-system effects as insomnia, irritability, convulsions, and even death. Symptoms of chronic lead poisoning include headache, weakness, lassitude, constipation, and a blue line along the gums.

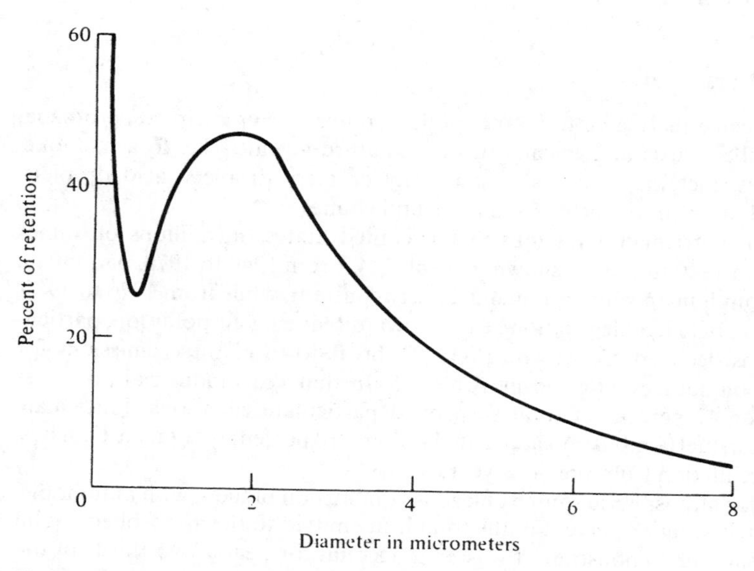

Figure 7-5 Retention of particulates in lungs. (*From Maxcy-Rosenau* [7-39].)

Aeroallergens Airborne substances causing allergies are called *aeroallergens*; pollens and spores are the cause of allergic reactions in sensitive persons. Ragweed pollen is one of the worst allergens. About 20 μm in diameter, ragweed pollen will usually be deposited within 90 m (100 yd) of the parent plant, and hayfever or asthma sufferers coming within that range may suffer severe allergic reactions. Other aeroallergens of biological origin include yeasts, molds, and animal fur, feathers, or hair. Finely powdered industrial materials can also cause allergic reactions in sensitive persons. [7-68]

Effects on plants and animals Little information is available on the detrimental effects of particulate matter on vegetation. Dry cement-kiln dust appears to cause little damage if deposited on a leaf surface, yet in the presence of moisture, such dust imparts damage and consequential growth inhibition to plant tissues. Of course, the dust coating on leaves reduces photosynthesis, and the increased plugging of stomata reduces plant growth. Animals who eat plants coated with particulates containing fluorides, arsenic, or lead may suffer some ill effects. [7-61]

Effects on materials Particulate matter can damage materials by soiling clothing and textiles, corroding metals (especially at relative humidities above 75 percent), eroding building surfaces, and discoloring and destroying painted surfaces. For example, at particulate-matter concentrations ranging from 130 to 180 $\mu m/m^3$ and in the presence of SO_2 and moisture, corrosion of steel and zinc panels is from three to four times greater than in areas close to background particulate levels (60 $\mu g/m^3$). [7-61]

Sources of Particulates

Particulates can either be natural—i.e., pollen, spores, bacteria, viruses, protozoa, fungi, plant fibers, rusts, and volcanic dust—or anthropogenic—i.e., fly ash, smoke, soot particles, metallic oxides and salts, oily or tarry droplets, acid droplets, silicates, and other inorganic dusts, and metallic fumes.

The total particulate emissions in the United States (in millions of tonnes per year) from 1940 to 1980 is shown in Table 7-4. From 1940 to 1970, particulate emissions from human sources remained fairly constant, while from 1970 to 1980, the early years of federal legislation geared toward reducing air pollution, particulate emissions declined 55 percent. [7-12] A breakdown of these emissions for 1980, based on sources, is given in Table 7-5. In that year, industrial processes accounted for 47 percent of total suspended particulate emissions. Particulate emissions from fuel combustion accounted for only 18 percent of total particulates, a dramatic reduction from previous years.

The major atmospheric source of lead is combustion of fuels, with automobile exhaust contributing 88 percent of the total lead emitted to the atmosphere. Solid waste disposal and combustion of waste oil account for nearly two-thirds of the lead from stationary sources. [7-58]

Detection and Analysis

Settleable particulates are usually measured by means of a dustfall bucket, while suspended particulate matter is often sampled by means of a high-volume sampler or a paper-tape sampler. Opacity and pollen tests are also used to evaluate the amount of suspended particulates in the air.

Settleable particulates A *dustfall bucket*, an open-mouthed container made of glass, polyethylene, or some other inert material, is used to collect settleable atmospheric particulates. After collection, the particles are weighed and the results reported in milligrams per square centimeter per month ($mg/cm^2/m$). Though dustfall measurement is inexpensive and easy to carry out, the dustfall bucket has limited value as an air-monitoring tool.

Suspended particulates The *high-volume sampler*, the workhorse of air monitoring, consists of a specially housed vacuum-sweeper motor with an attached filter holder (see Fig. 7-6). The filter holder consists of two parts: the cone-shaped stainless-steel filter support screen and an open rectangular face plate of cast iron with a sponge-rubber gasket. To sample, a flash-fired glass-fiber filter is placed between the filter-support screen and the gasket face plate. When the adapter is screwed onto the blower unit, a circular rubber gasket is used to make an air-tight seal.

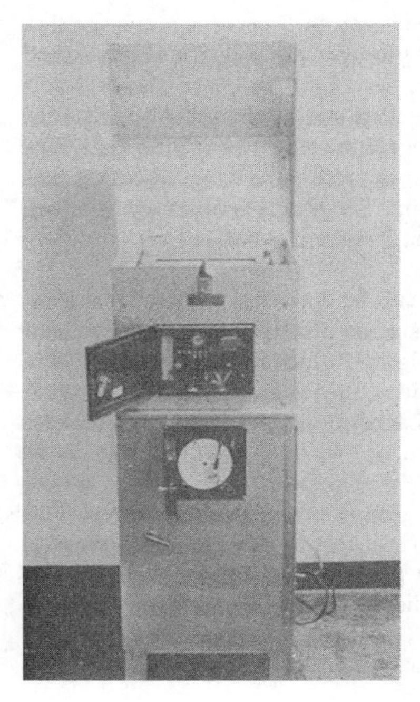

Figure 7-6 High-volume sampler.

The rate of air flow through the filter is measured with a "visi-float" flow meter calibrated in liters per minute (L/min). Samples are usually collected for a 24 h period, with flow meter readings being taken at the beginning (allowing a 5-min warm-up period) and at the end of each sampling period. The average flow rate is determined for each sampler from the calibration curve.

The total suspended particulates (TSP) can be determined wth a high-volume sampler, and the filter can be used to determine benzene-soluble organics, benzo(a)pyrene, some 20 metals, fluorides, radioactive materials, and biologically active nonmetals such as sulfate, nitrate, and ammonium. [7-24]

The *cascade impactor* used in conjunction with the high-volume sampler separates the particulates into various size ranges, thus permitting physiological evaluation of the health effects of an aerosol.

The *paper-tape sampler*, composed of a pump, a filtering mechanism, and a timing device, can be used to measure the soiling effects of particulate matter in the atmosphere. [7-43] Such a sampler draws air through a strip of filter paper at a specified rate (usually 7 L/min) and for a specified time (usually 2h, though it may be as short a time as 10 min or as long a time as $3\frac{1}{2}$ h).

Opacity Opacity is defined as the degree to which emissions reduce the transmission of light and obscure the view of an object in the background. As noted earlier, reduction in visibility is one of the most serious problems caused by airborne particulates. [7-41]

The opacity of emissions from stationary sources is determined by a qualified observer certified as able to assign opacity readings in 5-percent increments to 24 different black plumes and 25 different white plumes, with an error not to exceed 7.5-percent opacity in each category.

Facing away from the sun, the qualified observer stands at a sufficient distance to provide a clear view of the emission and records such pertinent information as time of day, estimated distance to the plume, approximate wind direction and speed, description of sky, and plume background. The plume is observed at 15-sec intervals, and opacity is recorded to the nearest 5 percent. [7-70]

Lead Lead concentrations in the atmosphere can be determined by air sampling with a high-volume sampler and analysis by atomic absorption. Nonurban lead concentrations in air are generally below 0.5 $\mu g/m^3$, while airborne lead levels in urban areas range from 1 to 2 $\mu g/m^3$. In large cities, the range is from 3 to 5 $\mu g/m^3$; near expressways, it may exceed 20 $\mu g/m^3$. Large industrial stationary sources may have airborne lead levels of over 300 $\mu g/m^3$. [7-4, 7-58]

Pollen A wide variety of airborne and spore samplers are available for pollens and spore sampling. The Durham sampler is a simple device that has extensive use in atmospheric allergen sampling. It is a gravity-slide device. A 1- by 3-in microscope slide is coated with a thin film of silicone stopcock grease (or Vaseline) and is then exposed, usually for a 24-h period. The slides are taken to a laboratory where they are stained before counting. The pollen grains are then counted through a microscope, and the number of pollen grains per square centimeter is determined. The following scale is often used to evaluate the results.

0 to 5 grains/cm^2 = slight
6 to 15 grains/cm^2 = moderate
16 to 25 grains/cm^2 = heavy
More than 25 grains/cm^2 = extremely heavy

Generally, a count of 7 or above is the point at which hayfever sufferers have symptoms. [7-27]

The most often utilized units for particulate sampling have been discussed. However, other types of units are available for use. These include centrifugal devices, impingers and impactors, scrubbers, electrostatic precipitators, and thermal precipitators.

Standards and Control

The national air quality standards are based on 1970 air-quality-criteria documents. National primary and secondary ambient-air-quality standards for particulate matter are shown in Table 7-9. However, some state agencies have additional particulate regulations as they relate to settleable particulates and the coefficient of haze. Four new volumes of preliminary criteria documents were published for particulate matter and sulfur oxides in April of 1980; revised standards were to be published in 1983.

Although the control of particulate matter can be undertaken either at the source or by dilution, the idea that "dilution is the solution to pollution" is no longer applicable and cannot be considered a viable control method. Control at the source, the only acceptable method, may depend upon principles of sedimentation, centrifugation, impaction, filtration, or electrical charge. Details of control procedures based upon these principles are described in Chap. 9.

Lead Ambient-air-quality standards for lead have been established to prevent the potential development of health problems. In 1978, the EPA established a 1.5 $\mu g/m^3$ lead standard, averaged over a calendar quarter. This was based on the safe lead

Table 7-9 National ambient-air-quality standards for particulate matter

Pollutant	Maximum 24-h concentration,* $\mu g/m^3$	Annual (geometric mean)
Primary	260	75
Secondary	150	60

* Not to be exceeded more than once a year.
Source: From *Twelfth Annual Report*. [7-12]

levels in the blood for a population of young children. Automobile exhausts, the greatest source of lead, can be reduced by eliminating the lead from gasoline. In addition to requiring the reduction of lead in gasoline from 450 to 130 mg/L (1.7 to 0.5 g/gal), the federal government encouraged the development of lead-trap devices which are capable of reducing overall particulate lead emissions from motor vehicles by 85 to 95 percent. [7-57] High-efficiency fine-particulate controls such as electrostatic precipitators, fabric filters, and wet scrubbers can be used to reduce lead emissions from stationary sources. [7-58]

7-6 HYDROCARBONS

Organic compounds containing only carbon and hydrogen are classified as hydrocarbons. Most of the major chemicals in gasoline and other petroleum products are hydrocarbons, which are divided into two major classifications—aliphatic and aromatic.

Aliphatic hydrocarbons The aliphatic hydrocarbon group contains alkanes, alkenes, and alkynes. The *alkanes*, saturated hydrocarbons (i.e., methane) are fairly inert and generally not active in atmospheric photochemical reactions.

The *alkenes*, often called *olefins*, are unsaturated and highly reactive in atmospheric photochemistry. The reactivity of alkenes such as ethylene makes them much more important in the study of air pollution than alkanes, because in the presence of sunlight they react with nitrogen dioxide at high concentrations to form secondary pollutants such as peroxyacetyl nitrate (PAN) and ozone (O_3). The third series of the aliphatics are the *alkynes*, which, though highly reactive, are relatively rare and thus not of major concern in air-pollution studies.

Ethylene, produced in automobile exhausts, is among the few hydrocarbons that can cause plant damage. Exposure of orchids to 342 $\mu g/m^3$ (0.3 ppm) for 1 h or to 57 $\mu g/m^3$ (0.05 ppm) for 6 h produces adverse effects. Abnormalities of tomato and pepper plants occur at ethylene exposure levels of 11 $\mu g/m^3$ (0.01 ppm) over several hours. Ethylene will generally only be found in the ambient air of large urban areas. [7-25] Experimental tests on humans and animals with aliphatic hydrocarbon concentrations of 326.5 mg/m^3 (500 ppm) produced no harmful effect. [7-44]

Aromatic hydrocarbons Aromatic hydrocarbons are biochemically and biologically active, and some are potentially carcinogenic. All aromatics are derived from or related to benzene. Though aromatics do not display the reactivity characteristic of unsaturated aliphatic hydrocarbons, the polynuclear group of aromatic hydrocarbons is of concern in any study of air pollution because a number of these compounds have been shown to be carcinogenic. Increases in lung cancer in urban areas have been blamed on the polynuclear hydrocarbons from automative exhaust emissions. Benzo(a)pyrene has been shown to be the most carcinogenic hydrocarbon for test animals. Benz(e)acephenanthrylene and benzo(j)fluroranthene follow, and benzo(3)pyrene, benz(a)anthracene, and chrysene are all weakly carcinogenic. [7-26] The threshold limit value for industrial exposure to benzene compounds has gone from 80 mg/m^3 to 30 mg/m^3 (25 ppm to 10 ppm). A further attempt to reduce it to 3 mg/m^3 (1 ppm) failed.

Sources of Hydrocarbons

Hydrocarbons present in the atmosphere are from both natural and anthropogenic sources. Natural background levels of methane in the atmosphere range from 784 to 980 μg/m^3 (1.2 to 1.5 ppm) on a worldwide basis. According to the Continuous Air Monitoring Project (CAMP), annual averages of monthly maximum 1-h average hydrocarbons (as carbon) range from 3.918 to 8.326 μg/m^3 (8 to 17 ppm) in many major cities.

Most natural hydrocarbons are from biological sources, though small amounts of these hydrocarbons come from geothermal areas, coal fields, natural gas from petroleum fields, and natural fires. [7-44] The more complex, naturally produced hydrocarbons found in the atmosphere, such as volatile terpenes and isoprene, are produced by plants and trees. The terpene molecules combine to form aerosols that produce the "blue haze" over forested areas. Concentrations of natural hydrocarbons range from 1.3 to 6.5 μg/m^3 (0.002 to 0.01 ppm). The combined estimated natural emissions of these hydrocarbons is 7.4×10^8 tons per year, an amount much greater than the hydrocarbon tonnage produced by humans and discharged into the atmosphere. [7-44]

The anthropogenic sources and estimated quantities of hydrocarbons emitted into the atmosphere are shown in Table 7-10. As indicated in the table, industrial sources (notably refineries) have become the major anthropogenic source of hydrocarbons. Until recently, transportation, including incomplete combustion from car engines, along with evaporative emissions from fuel tanks, crankcases, and carburetors, contributed the largest percentage of hydrocarbons. Hydrocarbon

Table 7-10 Sources and quantities of hydrocarbon emission*

Source	1968	1970	1975	1977	1980
Transportation	16.6	16.8	10.4	11.5	7.8
Fuel combustion in stationary sources (power, heating)	0.7	0.5	1.3	1.5	0.2
Industrial processes	4.6	4.8	2.7	10.1	10.8
Solid-waste disposal and miscellaneous†	10.1	7.9	12.6	5.2	3.0
Total	32.0	30.0	27.0	28.3	21.8

* In million tonnes per year.

† Primarily forest fires, agricultural burning, and coal waste fires.

Source: From *First Annual Report* [7-8], EPA [7-79], and *Twelfth Annual Report* [7-12].

reduction in the transportation area is a result of automotive emission-control devices. Afterburners with catalysts burn the emitted hydrocarbons and carbon monoxide and release CO_2 and H_2O. Hydrocarbon emissions from solid-waste disposal and such miscellaneous activities as forest fires, agricultural burning, and coal waste fires have also shown a significant decrease over the past few years.

Detection and Analysis

It is necessary to separate methane and nonmethane components in the detection and analysis for hydrocarbons. Gas chromatography in combination with flame ionization is the most acceptable procedure capable of doing this. While gas chromatography provides the necessary sensitivity and specificity for hydrocarbons, it is still an expensive, sophisticated procedure and at present is limited to research or short-term air quality studies. [7-60]

Standards and Control

In the 1970s, hydrocarbon standards were set as a means of achieving photochemical oxidant standards. In one urban area, air quality data indicated that in the morning hours (6 to 9 A.M.), concentrations of 196 $\mu g/m^3$ (0.3 ppm) nonmethane hydrocarbons produced a maximum hourly average oxidant concentration of 65 $\mu g/m^3$ (0.1 ppm). [7-42] Based on this figure, the EPA set the nonmethane hydrocarbon air quality standard for that metropolitan area at 160 $\mu g/m^3$ (0.24 ppm) maximum 3-h (6–9 A.M.) average, not to be exceeded more than once a year. The nonmethane standard was used, as it has been established that methane does not participate in the photochemical reactions. However, in 1982, EPA rescinded hydrocarbon standards because the ozone formation process was deemed too complex to assume that attainment of any single hydrocarbon standard would ensure attainment of the ozone standard. [7-12]

Available control technology for hydrocarbons from stationary sources can be divided into five general classifications: incineration, adsorption, absorption, condensation, and substitution of other materials. [7-50]

Incineration processes for control of hydrocarbons are classified as either afterburners or catalytic afterburners. Afterburners can be used to burn dilute organic vapors by adding fuel to the gas stream. But, if the gases have sufficient heat value, the vapors can be burned or flared directly. The catalytic afterburner uses a solid active surface upon which combustion takes place. This can be done at much reduced temperatures and has the advantage of lower fuel disposal costs.

Carbon adsorption of hydrocarbon is a proven method of reducing emissions with the possible economic advantage of recovering the material. The activated carbon can be adversely affected by particulate matter in the gas stream.

Absorption of the hydrocarbons in a scrubbing solution is usually carried out in bubble-plate columns, jet scrubbers, packed towers, spray towers, and venturi scrubbers. [7-50]

In condensation, the temperature of the gaseous hydrocarbons is lowered so that they turn into a liquid, which can then be recovered for reuse. Condensation is generally used only for high-level hydrocarbon emissions.

The substitution of less reactive hydrocarbons has been utilized in California to reduce photochemical oxidant formation. For instance, trichlorethylene, which is less reactive than olefins or aldehydes, can be substituted where appropriate.

7-7 CARBON MONOXIDE

Colorless, tasteless, and odorless, carbon monoxide gas is chemically inert under normal conditions and has an estimated atmospheric mean life of about $2\frac{1}{2}$ months. [7-40] The total emission of CO on a mass basis in 1977 accounted for slightly over half (53 percent) of all the anthropogenic air pollutants.

Carbon monoxide at present ambient levels has little if any effect on property, vegetation, or materials. At higher concentrations, it can seriously affect human aerobic metabolism, owing to its high affinity for hemoglobin, the component of the blood responsible for the transport of oxygen. Carbon monoxide reacts with the hemoglobin (Hb) of blood to give carboxyhemoglobin (COHb), thus reducing the capability of the blood to carry oxygen. Since the affinity of hemoglobin for carbon monoxide is more than 200 times as great as its affinity for oxygen, CO can seriously impair the transport of O_2, even when present at low concentrations. The health effects observed in persons exposed to CO are indicated in Table 7-11. As COHb levels increase, effects become more and more severe.

As indicated in Fig. 7-7, the absorption of CO by the body increases with CO concentration, exposure duration, and the activity being performed. Carbon monoxide concentrations are especially high in congested urban areas where traffic is heavy and slow-moving. For example, automobile traffic alone in New York City produces an average of 3773 tonnes of CO daily, and Los Angeles

Table 7-11 Health effects of COHb at various levels in the blood

COHb level, %	Demonstrated effects
Less than 1.0	No apparent effect
1.0 to 2.0	Some evidence of effect on behavioral performance
2.0 to 5.0	Central nervous system effects; impairment of time interval discrimination, visual acuity, brightness discrimination, and certain other psychomotor functions
Greater than 5.0	Cardiac and pulmonary functional changes
10.0 to 80.0	Headaches, fatigue, drowsiness, coma, respiratory failure, death

Source: From Wolf. [7-83]

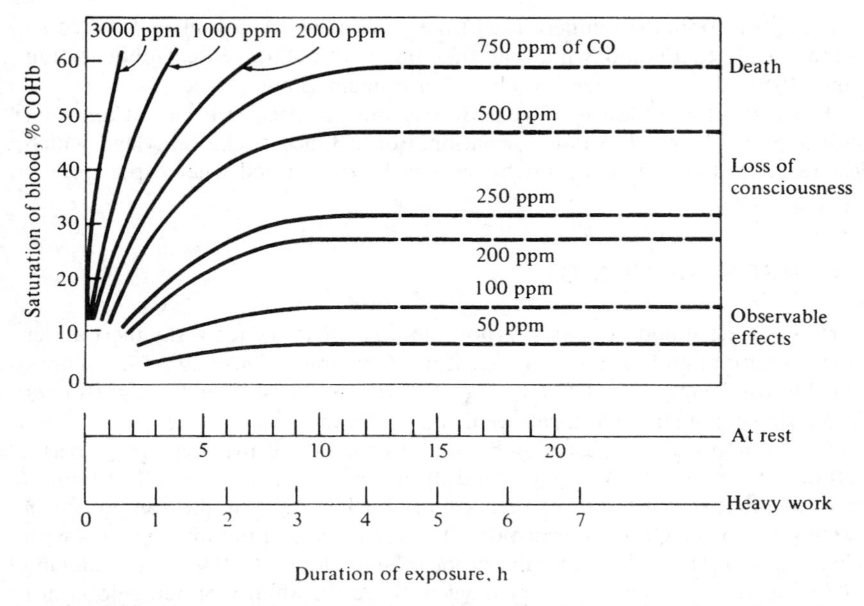

Figure 7-7 COHb levels in blood: correlation to atmospheric CO, duration of exposure, and type of physical activity. (*From Wolf* [*7-83*].)

traffic produces about 9090 tonnes/d. [7-68] Smaller cities also have CO problems from traffic. Carbon monoxide levels in downtown Louisville, Kentucky, during peak traffic periods range from 3.5 to 8.7 mg/m^3 (4 to 10 ppm). One mathematical model indicates that the CO concentration could reach 48 mg/m^3 (55 ppm) for 1 h at some street-crossing sites. [7-20] A person trapped in traffic at such a location for an hour would show a COHb blood level close to 2.3 percent. This exposure, according to Table 7-11, would affect the central nervous system, impairing a person's time-interval discrimination and brightness discrimination.

Sources of Carbon Monoxide

Carbon monoxide sources are both natural and anthropogenic. Argonne National Laboratory reports $3\frac{1}{2}$ billion tonnes of CO are produced in nature yearly by the oxidation of methane gas from decaying vegetation. Still another source is human metabolism. The exhalations of a resting person contain approximately 1 ppm CO. [7-56] Applied to the entire nation, this would total about 2.9 tonnes of CO produced each day (1058 tonnes per year).

Yet this production is still a great deal less than the estimated 69.1 million tonnes produced in 1980 by transportation sources—primarily gasoline-powered

Table 7-12 Sources and quantities of carbon monoxide

Source	Emissions, 10^6 tonnes/y				
	1968	1970	1975	1977	1980
Transportation	102.5	100.6	77.2	85.7	69.1
Fuel combustion in stationary sources (power and heating)	1.8	0.7	1.5	1.2	2.1
Industrial processes	7.7	10.3	7.7	8.3	5.8
Solid-waste disposal and miscellaneous	24.8	23.3	9.8	7.5	8.4
Total	136.8	134.8	96.2	102.7	85.4

Source: From EPA [7-80] and *Twelfth Annual Report* [7-12].

internal combustion engines. The quantities of CO emission from the four major groups are shown in Table 7-12. On a weight basis, the total estimated emission of CO from transportation in 1968, 1970, 1975, 1977, and 1980 was about 78.3 percent of the total CO emitted by all sources combined. In 1980, the next largest source of anthropogenic carbon monoxide was solid-waste disposal and miscellaneous causes, which included forest fires, structural fires, coal refuse, and agricultural burning.

On a mass basis, the emissions of carbon monoxide from anthropogenic sources have dropped from 137 million tonnes in 1968 to 85 million tonnes in 1980. This reduction has taken place mainly in the automotive area, owing to the initiation in 1968 of pollution control devices. [7-80] [7-78] Even at present levels of emission, were it not for the natural processes of removal, the CO content of the atmosphere would be increasing at the rate of about 0.5 ppm yearly.

Detection and Analysis

Numerous detection methods are available for both intermittent (grab) and continuous sampling of CO. These methods use gravimetric, chemical, electrochemical, and colorimetric processes. The reference method specified by EPA for monitoring ambient-air-quality standards for CO is nondispersive infrared (NDIR), and the approved equivalent methods are gas chromatography, catalytic conversion, and flame ionization detection. [7-54]

Standards and Control

The National Air Pollution Control Administration in 1970, in accordance with the Clean Air Act, issued "Air Quality Criteria for Carbon Monoxide," criteria which described the effects of CO on humans and the environment at various contaminant concentrations and exposure times. Based on these criteria and providing for a margin of safety, the air quality standards for CO were established as 10 mg/m^3 (9 ppm) as the maximum 8-h average and 40 mg/m^3 (35 ppm) as the maximum 1-h average. Neither average is to be exceeded more than once a year. These figures represent both primary and secondary standards for CO. [7-7]

Emergency situations may arise in which CO levels reach potentially deadly levels within a short period of time. Episode regulations have been developed to counteract any such events. First-stage control action begins if CO concentrations reach 34 mg/m^3 (30 ppm) as an 8-h average. If CO levels reach 46 mg/m^3 (40 ppm) on an 8-h-average, stringent control action is carried out—including shutting down industrial plants and closing traffic routes.

Adsorption, absorption, condensation, and combustion are four basic technical control methods used for CO, and use of these methods can control almost all carbon monoxide emissions. Control of CO at the source is far more desirable than control by dilution in the ambient air. Exhaust emission standards for

Table 7-13 Exhaust emission standards for autos and light trucks, 1968–1985

	Standards for new autos g/mi			Percent reduction of standards from precontrol vehicles		
Year	Hydrocarbons	Carbon monoxide	Nitrogen oxides	Hydrocarbons	Carbon monoxide	Nitrogen oxides
1968–1971	4.1	34.0	NA	50	62	NA
1972–1974	3.0	28.0	3.1	63	69	9
1975–1979	1.5	15.0	3.1	82	83	9
1980–1985	0.41	7.0	2.0	96	92	41

Light trucks

	Standards for new light trucks g/mi			Percent reduction of standards from precontrol vehicles		
Year	Hydrocarbons	Carbon monoxide	Nitrogen oxides	Hydrocarbons	Carbon monoxide	Nitrogen oxides
1968–1971	4.1	34.0	NA	50	62	NA
1972–1974	3.0	28.0	3.1	63	69	9
1975–1978	2.0	20.0	3.1	76	78	9
1979–1983	1.7	18.0	2.3	79	80	32
1984–1985	0.8	10.0	2.3	90	89	32

NA—Not applicable.

All standards are based on the current test procedure. Standards for hydrocarbons and carbon monoxide were not established until 1968; for nitrogen dioxides, not until 1972.

Source: From *Twelfth Annual Report*. [7-12]

carbon monoxide, as well as for hydrocarbons and nitrogen oxides (discussed later), are given in Table 7-13.

7-8 OXIDES OF SULFUR

The oxides of sulfur (SO_x) are probably the most widespread and the most intensely studied of all anthropogenic air pollutants. They include six different gaseous compounds: sulfur monoxide (SO), sulfur dioxide (SO_2), sulfur trioxide (SO_3), sulfur tetroxide (SO_4), sulfur sesquioxide (S_2O_3), and sulfur heptoxide (S_2O_7). [7-48] Sulfur dioxide (SO_2) and sulfur trioxide (SO_3) are the two oxides of sulfur of most interest in the study of air pollution.

Sulfur dioxide is a colorless, nonflammable, and nonexplosive gas with a suffocating odor. It has a taste threshold of 784 $\mu g/m^3$ (0.3 ppm) and an odor threshold of 1306 $\mu g/m^3$ (0.5 ppm). Sulfur dioxide is highly soluble in water (11.3 g/100 mL at 20°C), has a molecular weight of 64.06, and is about twice as heavy as air. [7-49] It is estimated that SO_2 remains airborne an average of 2 to 4 d, during which time it may be transported as far as 1000 km (621 m). Thus, the problem of SO_2 pollution can become an international one.

Relatively stable in the atmosphere, SO_2 acts either as a reducing or an oxidizing agent. [7-49] Reacting photochemically or catalytically with other components in the atmosphere, SO_2 can produce SO_3, H_2SO_4 droplets, and salts of sulfuric acid. As indicated by Eq. (7-5), SO_2 can react with water to form sulfurous acid, a weak acid which can react directly with organic dyes. This property is used to colorimetrically detect SO_2 in the atmosphere.

$$SO_2 + H_2O \rightleftharpoons H_2SO_3 \quad \text{(sulfurous acid)} \tag{7-5}$$

As shown in Eq. (7-6), SO_3 can react with water to form sulfuric acid droplets.

$$SO_3 + H_2O \rightleftharpoons H_2SO_4 \quad \text{(sulfuric acid)} \tag{7-6}$$

Effects of Sulfur Oxides

Effects on Human Health Sulfuric acid (H_2SO_4), sulfur dioxide (SO_2), and sulfate salts tend to irritate the mucous membranes of the respiratory tract and foster the development of chronic respiratory diseases, particularly bronchitis and pulmonary emphysema. Data gathered from several different sources, shown in Table 7-14, reveals some disparity in the information available as to the effects of SO_2 on human health. Generally the data are for healthy adults under experimental conditions in the laboratory, conditions not comparable to the ambient air where the synergistic effects of other pollutants would probably cause different reactions.

Table 7-14 Effects of SO_2 on humans

Concentration, ppm	Exposure time	Effects
0–0.6		No detectable response
0.15–0.25	1-4 d	Cardiorespiratory response
1.0–2.0	3–10 min	Cardiorespiratory response in healthy subjects
1.0–5.0		Detectable responses, tightness in chest
5.0	1 h	Choking and increased lung resistance to air flow
10.0	1 h	Severe distress, some nosebleeding
Greater than 20		Digestive tract affected, also eye irritation
400–500		Dangerous for short periods of time

Source: From *Air Conservation* [7-1] and *Air Quality Criteria* [7-49].

The effects of SO_2 concentrations in ambient air as related to the times of exposure are summarized in Fig. 7-8. The shaded area represents conditions resulting in excess deaths. The grid area represents conditions when significant detrimental health effects are reported. The speckled area represents conditions where detrimental health effects are suspected. The unshaded area represents exposures which will not likely result in significant problems. Chicago's annual concentrations of 470 $\mu g/m^3$ (0.18 ppm) SO_2 would, according to EPA data, result in excess deaths.

In a dusty atmosphere, SO_2 is particularly harmful because both sulfur dioxide and sulfuric acid molecules paralyze the hairlike cilia which line the respiratory tract. Without the regular sweeping action of the cilia, particulates are able to penetrate to the lungs and settle there. These particulates usually carry with them concentrated amounts of SO_2, thus bringing this irritant into direct, prolonged contact with delicate lung tissues. The SO_2-particulate combination has been cited as cause of death in several air pollution tragedies.

In 1930 during the Meuse Valley episode, the estimated maximum SO_2 concentration reached 20.9 mg/m^3 (8.0 ppm). At Donora, Pennsylvania, in 1948, the calculated SO_2 concentration was 1.3 mg/m^3 (0.5 ppm) to 5.2 mg/m^3 (2.0 ppm) over three to four days. During the 1952 London episode, the SO_2 concentrations averaged 1.5 mg/m^3 (0.57 ppm) for five days, with the maximum daily average being 3.4 mg/m^3 (1.3 ppm). In the New York incident of 1953, the SO_2 ranged from 0.5 to 2.2 mg/m^3 (0.2 to 0.86 ppm) for three consecutive days, and in the 1963 incident concentrations reached 1.20 mg/m^3 (0.46 ppm) for 15 days, with the recorded maximum being 3.9 mg/m^3 (1.50 ppm) for four hours. [7-48]

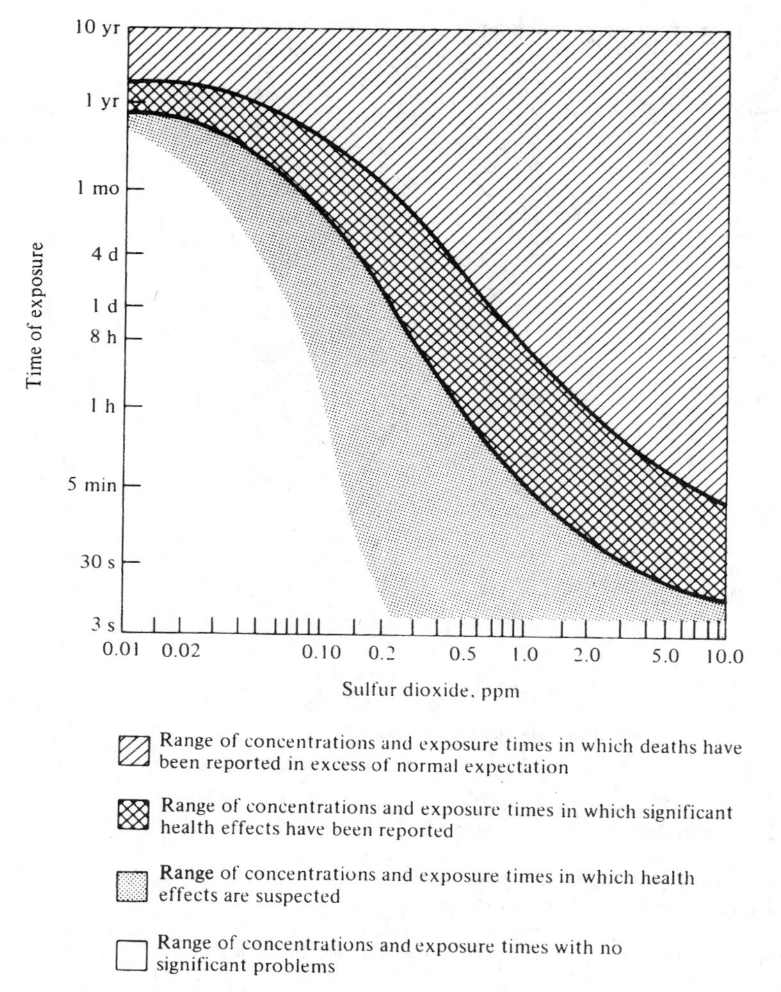

Figure 7-8 Effects of sulfur oxide pollution on human health. (*From Air Quality Criteria* . . . [*7-48*].)

Effects on plants Much work has been done on the responses of plants to SO_2. The relationship between SO_2 concentrations (ppm) in the ambient air, exposure time, and effects on vegetation are shown in Fig. 7-9. The shaded area represents exposures that cause damage, while the unshaded area represents exposure times of undetermined significance.

Injury to vegetation can be classified as acute or chronic. The SO_2 concentration in acute exposure is high for a short period, resulting in damage characterized by clearly marked dead tissue between the veins or on the margins of the leaves.

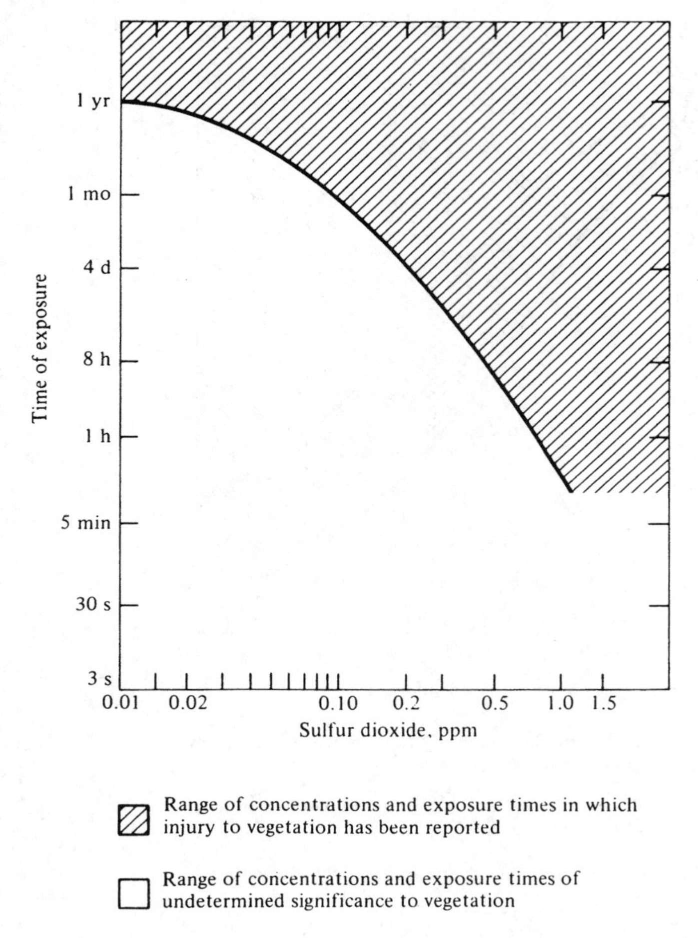

Figure 7-9 Effects of sulfur oxide pollution on vegetation. (*From Air Quality Criteria* . . . [7-49].)

Chronic injury comes from exposure to low concentrations for long periods of time, which causes brownish-red or bleached white areas on the blade of the leaf. The plant injury threshold for SO_2 is in the neighborhood of 0.8 mg/m^3 (0.3 ppm) to 1.0 mg/m^3 (0.4 ppm) exposure for 8 h.

Plants are particularly sensitive to SO_2 during periods of intense light, high relative humidity, adequate moisture, and moderate temperature. Plants are also generally more sensitive to SO_2 during the growing season, regardless of climatic conditions.

Plants vary widely in their susceptibility to SO_2. Alfalfa, cotton, and soybeans are crops particularly sensitive to SO_2, along with such vegetables as beans, lettuce, and spinach and such trees as apple, pine, and mulberry. Potatoes, onions, corn, and maples are more resistant to SO_2 damage.

Effects on materials Sulfuric acid aerosols will readily attack building materials, especially those containing carbonates such as marble, limestone, roofing slate, and mortar. The carbonates are replaced by sulfates which are water soluble, as the following equation indicates.

$$\underset{\text{limestone}}{CaCO_3} + H_2SO_4 \longrightarrow CaSO_4 + CO_2 + H_2O \qquad (7\text{-}7)$$

The calcium sulfate, or gypsum ($CaSO_4$), formed in this process is washed away, leaving a pitted, discolored surface. Many of our historical monuments, public buildings, cathedrals, and sculptures have deteriorated from exposure to the by-products of sulfur oxides. Cleopatra's Needle in Central Park, New York City, has deteriorated more since it was moved to New York in the nineteenth century than in all of the 2000 years or so that it stood in Egypt. [7-16]

Sulfuric-acid mists can also damage cotton, linen, rayon, and nylon. A few years ago when women in Toronto, Jacksonville, and Chicago began to notice their nylon hose were disintegrating, the cause was investigated and eventually laid to the presence of a fine sulfuric-acid mist in the air. Leather also weakens and disintegrates in the presence of excess amounts of by-products of SO_2. Paper absorbs SO_2, the SO_2 is oxidized to H_2SO_4, and the paper yellows and becomes brittle. This is one reason why London, Rome, and other large, industrialized cities store historic documents in carefully controlled environments.

Excess exposure to SO_2 accelerates corrosion rates for many metals such as iron, steel, zinc, copper, and nickel, especially at relative humidities over 70 percent. Steel panels exposed to an annual mean concentration of 78 $\mu g/m^3$ (0.03 ppm) accompanied by high particulate levels will lose more than 10 percent by mass in 1 year. [7-49] The accelerated corrosion is particularly noticeable in the fall and winter when more fuel is burned. Corrosion rates are from 1.5 to 5 times greater in polluted urban areas than in clean air areas.

Sources of Sulfur Oxides

The burning of solid and fossil fuel contributes more than 80 percent of anthropogenic SO_2 emissions. As indicated in Table 7-15, fuel combustion in stationary sources (primarily electric utilities) and industrial processes are the principal contributors of sulfur oxides from human sources.

Transportation contributes little to the anthropogenic SO_x in the atmosphere, because the sulfur content of gasoline is low (about 0.03 percent by mass). [7-71] Present concern about automotive catalytic converters oxidizing SO_2 to SO_3 is of small consequence when compared to the potential dangers of carbon monoxide and hydrocarbon emissions. However, the SO_3 can react with moisture in the air to produce H_2SO_4 mist.

Table 7-15 Sources and quantities of oxides of sulfur

	Emission, 10^6 tonnes/y				
Source	1968	1970	1975	1977	1980
Transportation	0.7	0.9	1.1	0.8	0.9
Fuel combustion in stationary sources (power and heating)	20.6	23.5	22.0	22.4	19.0
Industrial processes	6.1	5.3	2.9	4.2	3.8
Solid waste disposal and miscellaneous	0.6	0.3	—	—	—
Totals	28.0	30.0	26.0	27.4	23.7

Source: From *Third Annual Report* [7-9], EPA [7-76], and *Twelfth Annual Report* [7-12].

Detection and Analysis

Continuous and intermittent sampling are commonly used for SO_2 detection. The colorimetric, conductometric, and coulometric methods are the most commonly used direct methods, with the colorimetric and conductometric methods being used mainly in the United States for continuous monitoring. [7-49] Other direct monitoring techniques for SO_2 are spectroscopic, utilizing infrared or ultraviolet absorption, as well as flame photometric methods. An indirect method used to approximate SO_2 concentrations involves using Huey sulfation plates.

Background levels for SO_2 in clean, nonurban air range from 0 to 52 $\mu g/m^3$ (0 to 0.02 ppm). However, in heavy-industry areas, maximum 5-min concentrations of 8.7 mg/m^3 (3.33 ppm) have been noted. The cities with the highest values are generally located in the northeastern part of the United States, with highest reported levels of SO_2 in Chicago. Concentrations of SO_2 in the major cities follow seasonal trends, with increased burning of coal for power and heating causing average winter concentrations approximately twice the average summer concentrations.

The emissions of sulfur oxides on a mass basis in the United States increased from 20 million tonnes in 1940 to 30 million tonnes in 1970. [7-10] However, from 1970 to 1980, sulfur oxide emissions from all sources decreased 21 percent. This improvement is particularly significant in view of the fact that the amount of fossil fuels burned increased by about 10 percent over that same period. [7-12]

Standards and Control

National ambient air quality standards have been set for SO_2, the most common form of sulfur oxides. Sulfur dioxide can react with oxygen and moisture in the air to produce sulfates. Preliminary epidemiological studies indicate adverse health effects for sulfates at levels as low as 6 to 10 $\mu g/m^3$ for a 24-h average and 10 to 15 $\mu g/m^3$ for an annual average. [7-64]

The broad-based methods for control of sulfur oxide emissions include burning fuel with less sulfur, removing sulfur from fuel, converting coal by liquefaction or gasification, substitution of another energy source, cleaning up the combustion products, or dispersion by tall stacks. [7-51] These methods will be discussed further in Chap. 9.

7-9 OXIDES OF NITROGEN

Oxides of nitrogen (NO_x) include six known gaseous compounds: nitric oxide (NO), nitrogen dioxide (NO_2), nitrous oxide (N_2O), nitrogen sesquioxide (N_2O_3), nitrogen tetroxide (N_2O_4), and nitrogen pentoxide (N_2O_5). The two oxides of nitrogen of primary concern in air pollution are nitric oxide (NO) and nitrogen dioxide (NO_2), the only two oxides of nitrogen that are emitted in significant quantities to the atmosphere. Heavier than air, nitrogen dioxide (NO_2) is readily soluble in water, forming nitric acid and either nitrous acid or nitric oxide, as indicated in the following equations.

$$2NO_2 + H_2O \longrightarrow HNO_3 + HNO_2 \quad \text{(nitrous acid)} \tag{7-8}$$

$$3NO_2 + H_2O \longrightarrow 2HNO_3 + NO \quad \text{(nitric oxide)} \tag{7-9}$$

Both nitric and nitrous acid will fall out in the rain or combine with ammonia (NH_3) in the atmosphere to form ammonium nitrate (NH_4NO_3). In this instance, the NO_2 will produce a plant nutrient. A good absorber of energy in the ultraviolet range, NO_2 consequently plays a major role in the production of secondary air contaminants such as ozone (O_3).

Nitric oxide (NO) is emitted to the atmosphere in much larger quantities than NO_2. It is formed in high-temperature combustion processes when atmospheric oxygen and nitrogen combine according to the following reaction:

$$N_2 + O_2 \rightleftharpoons NO \tag{7-10}$$

Effects of Nitrogen Oxides

Effects on human health Nitric oxide (NO) is a relatively inert gas and only moderately toxic. Although NO, like CO, can combine with hemoglobin to reduce the oxygen-carrying capacity of the blood, NO concentrations are generally less than 1.22 mg/m^3 (1 ppm) in the ambient air and are thus not considered health hazards. However, NO is readily oxidized to NO_2, which does have biological significance. [7-45]

$$NO + \tfrac{1}{2}O_2 \rightleftharpoons NO_2 \tag{7-11}$$

NO_2 irritates the alveoli of the lungs; the response of the human respiratory system to short-term exposure to nitrogen dioxide is seen in Table 7-16.

Table 7-16 Summary of human responses to short-term NO_2 exposures

Effect	NO_2 concentration		Time to effect
	mg/m³	ppm	
Odor threshold	0.23	0.12	Immediate
Threshold for	0.14	0.075	Not reported
dark adaptation	0.50	0.26	Not reported
Increased airway	1.3–3.8	0.7–2.0	20 min*
resistance	3.0–3.8	1.6–2.0	15 min
	2.8	1.5	45 min†
	3.8	2.0	45 min‡
	5.6	3.0	45 min§
	7.5–9.4	4.0–5.0	40 min¶
	9.4	5.0	15 min
	11.3–75.2	6.0–40.0	5 min
Decreased pulmonary diffusing capacity	7.5–9.4	4.0–5.0	15 min

* Exposure lasted 10 min, effect on flow resistance was observed 10 min after termination of exposure.

† Effect was produced at this concentration in normal subjects with chronic respiratory disease.

‡ Effect occurred in subjects at rest with chronic respiratory disease.

§ Effect occurred in normal subjects at rest.

¶ Exposure lasted 10 min; maximal effect on flow resistance was observed 30 min later.

Short-term animal studies show reduced resistance to respiratory infection at exposures of 6.6 mg/m³ (3.5 ppm) for 2 h. [7-55] Experimental exposures of volunteers to 9.4 mg/m³ (5 ppm) NO_2, considerably above the 7.0 mg/m³ (3.75 ppm) peak recorded in Los Angeles, for 10 min produced a substantial but transient increase in the resistance of the lung's airways to air movement. [7-45] Concentrations from 47 to 141 mg/m³ (25 to 75 ppm) cause reversible pneumonia. [7-55] Nitrogen dioxide at high-level exposures of 285 mg/m³ (150 ppm) and above are fatal to humans.

A Chattanooga, Tennessee, study demonstrated an 18 percent relative excess of respiratory illness among family members living in an area in which mean 24-h NO_2 concentration measured over a 6-mo period was between 0.12 and 0.20 mg/m³ (0.062 and 0.109 ppm), in contrast to those living where typical background NO_2 levels were 0.0075 to 0.037 mg/m³ (0.004 to 0.020 ppm). The study also demonstrated an increased frequency of acute bronchitis among infants and schoolchildren when 24-h NO_2 concentrations over a 6-mo period were between 0.12 and 0.16 mg/m³ (0.063 and 0.083 ppm). [7-30, 7-45] The findings of the Chattanooga study would seem to indicate that any city whose mean 24-h concentrations of NO_2 are 0.11 mg/m³ (0.06 ppm) or greater for a 6-mo period could anticipate increased incidence of respiratory illness. However, there might well be various meteorological and other factors that would invalidate such a prediction.

Effects on Plants and Materials There is no evidence that NO is damaging to plants outside the laboratory, and while NO_2 and primary pollutants can cause some injury to vegetation, PAN and O_3, secondary pollutants produced during photochemical reactions involving NO_x, are far more likely to be damaging to plants. Exposure to high levels of NO_2 can cause fading of textile dyes, yellowing of white fabric, and oxidation of metals. The present primary and secondary standards for NO_2 are 100 $\mu g/m^3$ (0.5 ppm), well below the threshold level for detection of effects on vegetation. [7-55]

Secondary Pollutants and NO

Photochemically, the NO_x are one of two groups of chemical compounds which are the necessary ingredients for the production of photochemical smog. Stated in vastly oversimplified form,

$$\text{Hydrocarbons} + NO_x \xrightarrow{\text{Sunlight}} \text{Smog} \tag{7-12}$$

Actually, there are many complex reactions taking place, and the exact reactions which lead to smog are, as yet, unknown. It is known that NO_2 functions primarily as the light-energy absorber and that in the presence of smog there are elevated levels of the oxidants. The major process by which NO_2 is formed in the atmosphere is [7-55]

$$O_3 + NO \longrightarrow O_2 + NO_2 \tag{7-13}$$

Hydroperoxyl radicals may also react with NO to generate NO_2 and hydroxyl radicals.

$$\cdot HO_2 + NO \longrightarrow \cdot HO + NO_2 \tag{7-14}$$

And alkylperoxyl radicals can react to oxidize NO to generate alkyloxyl radicals and NO_2.

$$\cdot RO_2 + NO \longrightarrow \cdot RO + NO_2 \tag{7-15}$$

The effect is rapid cycling of NO_2, and no overall effect would result if it were not for a series of competing reactions involving the hydrocarbons. These reactions are shown in Fig. 7-10 and in Eq. (7-12). The hydrocarbon reactions cause the photolytic cycle to be unbalanced. The O (atomic oxygen) reacts with the hydrocarbons to produce a reactive intermediate species called alkylperoxyl radicals ($\cdot RO_2$). These free radicals react rapidly with NO to produce NO_2. This removes the NO from the cycle and, thus, the reaction that would remove O_3 from the system is eliminated, causing the O_3 concentration to increase in the atmosphere. The $\cdot RO_2$ can also react with O_2 and NO_2 to produce peroxyacyl nitrates (PAN).

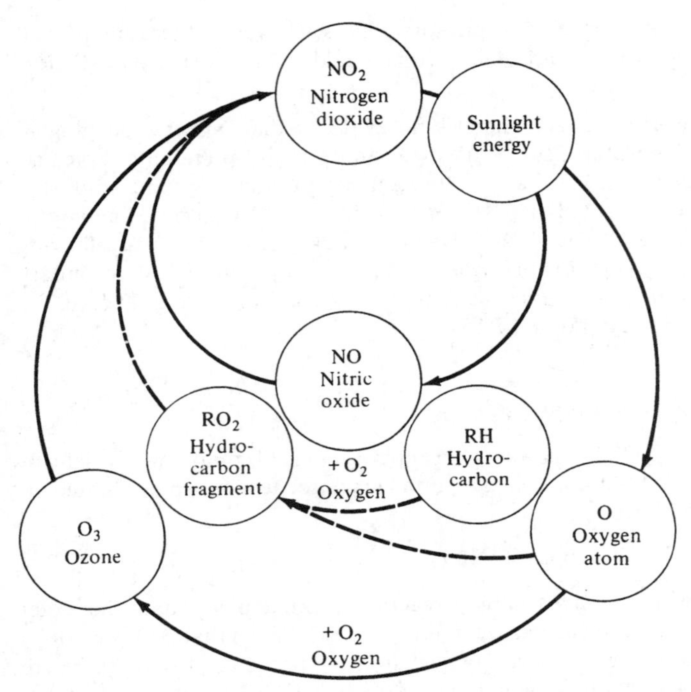

Figure 7-10 Interaction of atmospheric nitrogen dioxide photolytic cycle with hydrocarbons. (*From Air Quality Criteria* . . . [7-47].)

The end product of these photochemical reactions is photochemical smog consisting of air contaminants such as O_3, PAN, aldehydes, ketones, alkyl nitrates, and carbon monoxide. [7-45]

Sources of Nitrogen Oxides

Some oxides of nitrogen are produced naturally and others are anthropogenic in source. Small concentrations of the NO_x produced in the upper atmosphere by solar radiation reach the lower atmosphere through downward diffusion. Small amounts of NO_x are produced by lightning and forest fires. Bacterial decomposition of organic matter also releases NO_x into the atmosphere. In fact, the naturally occurring sources of NO_x produce approximately 10 times as much NO_x as do anthropogenic sources, which are concentrated in urban areas. Primary origins of human-induced NO_x are fuel combustion in stationary sources and in transportation. The anthropogenic sources and estimated quantities of NO_x emitted to the atmosphere are shown in Table 7-17.

The increased emission of NO_x from transportation (from 7.5 million to 10.1 million tonnes) in the first years is not only because of the increased number of

Table 7-17 Sources and quantities of oxides of nitrogen

	Emission, 10^6 tonnes/y				
Source	1968	1970	1975	1977	1980
Transportation	7.5	10.1	9.2	9.2	9.1
Fuel combustion in stationary sources (power and heating)	9.2	8.6	11.8	13.0	10.6
Industrial processes	0.2	0.2	—	0.7	0.7
Solid-waste disposal and miscellaneous	2.1	0.7	—	—	0.3
Total	19.0	19.6	21.0	23.1	20.7

Source: From *Third Annual Report* [7-9], EPA [7-76], and *Twelfth Annual Report* [7-12].

vehicles and miles traveled, but also because of the implementation of more stringent controls for carbon monoxide (CO) and hydrocarbons (HC), since early control devices generally resulted in significant increases in NO_x emissions. The increase in emissions from electric utilities rose because of increased demand for electricity. From 1970 to 1980, the nitrogen oxide emissions from all sources rose less than 10 percent.

Typical background levels of NO are about 3.7 to 5.6 $\mu g/m^3$ (2 to 3 ppb) and for NO_2 from 7.5 to 9.4 $\mu g/m^3$ (4 to 5 ppb). However, the NO_x concentrations in the urban areas can be up to 1000 times greater than those found in rural areas. [7-55] For example, record peak concentrations of total NO_x have been recorded in Los Angeles at 7.0 mg/m^3 (3.75 ppm) and in nearby Burbank at slightly over 3.8 mg/m^3 (2 ppm). New Orleans has registered maximum peak concentrations for NO_x of 1.2 mg/m^3 (0.63 ppm), and Phoenix has had maximum peak concentrations of 1.5 mg/m^3 (0.8 ppm). [7-65]

Both NO and NO_2 exhibit distinct diurnal variations depending upon solar radiation, meteorological phenomena, and traffic volume. The dramatic diurnal variation of NO, NO_2, and O_3 in Los Angeles in 1965 is shown in Fig. 7-11. Before daylight, NO and NO_2 remain at relatively low, stable concentrations. With increased morning activity, especially automobile use, concentrations of NO rise rapidly. Then, with increased solar activity, the following reactions take place: [7-55]

$$NO_2 + UV \rightleftharpoons NO + O \tag{7-16}$$

$$O + O_2 + M \rightleftharpoons O_3 + M \tag{7-17}$$

$$O_3 + NO \rightleftharpoons O_2 + NO_2 \tag{7-18}$$

The NO_2 concentrations climb and peak [Eq. (7-16)], then begin to decline as the photochemical oxidants start to accumulate. The photochemical oxidants

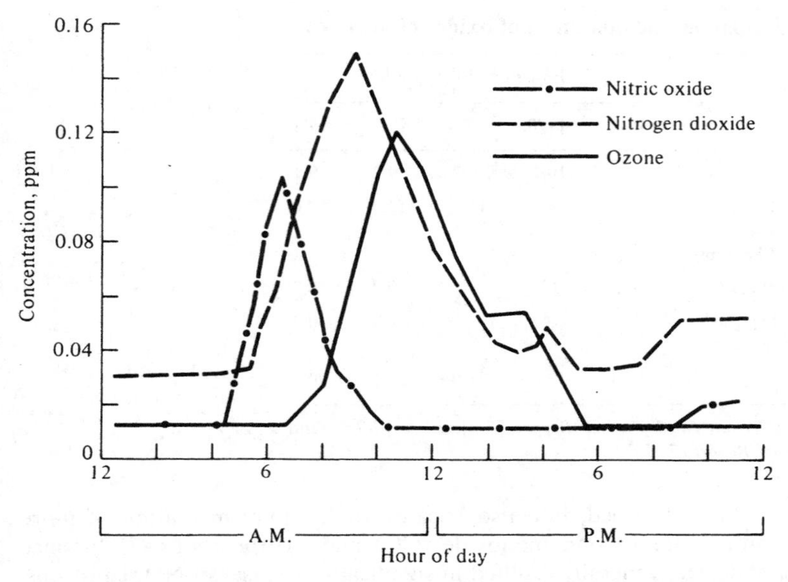

Figure 7-11 Diurnal variations of NO, NO_2, and O_3 concentrations in Los Angeles, July 19, 1965. (*From Air Quality Criteria* . . . [7-45].)

(mainly O_3) peak around noon [Eq. (7-17)]. Accumulated NO and O_3 react readily in the atmosphere [Eq. (7-18)] with an almost complete conversion of NO to NO_2. Though afternoon solar radiation is generally not intense enough to produce oxidants, the O_3 produced earlier will continue to react readily with the NO emitted by the late afternoon traffic, bringing about a slight rise in NO_2 at that time.

Maximum NO concentrations generally occur in the late fall and winter months, months characterized by maximum demand for heat energy, low wind speeds, and lessened solar radiation. Unlike NO, NO_2 does not tend to follow seasonal variations.

Detection and Analysis

Monitoring for NO requires that nitric oxide first be converted to NO_2 and its concentration determined indirectly. Two of the older methods of measuring for nitrogen oxides are the Griess–Ilosvay and the Jacobs–Hockheiser colorimetric methods. [7-45, 7-55] Many new methods for measuring NO and NO_2 have been developed, including chemiluminescense, ultraviolet spectrophotometry, electrical transducers, and dispersive infrared spectrometry. Other detection techniques under development use lasers, ion-selective electrodes, or pulsed UV fluorescence. [7-5]

Table 7-18 Episode regulations for NO_2

	Concentrations			
	1-h average		24-h average	
Level	$\mu g/m^3$	ppm	$\mu g/m^3$	ppm
Alert	1130	0.6	282	0.15
Warning	2260	1.2	565	0.3
Emergency	3000	1.6	750	0.4

Source: From *Kentucky Air Pollution Control Regulations*. [7-31]

Standards and Controls

In January of 1971, the federal Air Pollution Control Office set national primary and secondary ambient-air-quality standards for nitrogen dioxide of 100 $\mu g/m^3$ (0.05 ppm) as an arithmetic mean. Episode regulations have been developed for NO_2, and these regulations are presented in Table 7-18. These episode regulations are designed to prevent ambient-air pollutant concentrations from reaching levels which could cause significant harm to health.

In general, most control measures for NO_x emissions have been directed at modification of combustion conditions to decrease NO_x production and at utilization of various devices to remove NO_x from exhaust gas streams.

7-10 PHOTOCHEMICAL OXIDANTS

Oxidants or *total oxidants*, two terms used to describe levels of photochemical oxidants, generally indicate the net oxidizing ability of the ambient air. Ozone (O_3), the major photochemical oxidant, makes up approximately 90 percent of the oxidant pool. Other photochemical oxidants of concern in air pollution monitoring include nascent oxygen (O), excited molecular oxygen (O_2), peroxy-acetyl nitrate (PAN), peroxy-propinol nitrate (PPN), peroxy-butyl nitrate (PBN), nitrogen dioxide (NO_2), hydrogen peroxide (H_2O_2), and alkyl nitrates. [7-60]

Effects of Oxidants

Effects on human health Photochemical oxidants can cause coughing, shortness of breath, airway constriction, headache, chest tightness and soreness, impaired pulmonary function, altered red blood cells, pharyngitis, laryngitis, and eye, nose, and throat irritation. [7-6] Numerous clinical studies indicate that significant health effects are experienced in humans exposed to photochemical oxidant levels of 725 $\mu g/m^3$ (0.37 ppm).

Exposure of laboratory mice to high levels of ozone has resulted in damage to chromosomes. Because of experiments of this nature, ozone is considered to have effects similar to ionizing radiation. Chromosome breakage in human cell cultures was observed at exposures of 15,673 $\mu g/m^3$ (8.0 ppm) for 5 to 10 min; this would be equivalent to effects produced by 200 roentgens of radiation. [7-63]

Effects on plants The major components in the oxidant pool that cause damage to plants are O_3 and PAN. Both enter the plant leaf through the stomata and interfere with plant cell metabolism. Injury by O_3 is identified by stippling or flecking on the leaves, and sensitive species will show O_3 injury after an 8-h exposure of 59 $\mu g/m^3$ (0.03 ppm). Symptoms of injury from PAN exposure are bronzing, glazing, and silvering on the undersurface of the leaves. Sensitive species will be injured at PAN exposures of 20 $\mu g/m^3$ (0.01 ppm) for 5 h. [7-47]

Chronic oxidant exposure at levels from 94 to 282 $\mu g/m^3$ (0.05 to 0.15 ppm) will reduce crop yields for soybeans, corn, and radishes. Carnations, geraniums, and pinto beans show growth and flowering effects at chronic exposure levels of 94 to 282 $\mu g/m^3$ (0.05 to 0.15 ppm). The estimated agricultural losses for oxidant damage reach several hundred million dollars per year. [7-60]

Effects on materials Ozone is an extremely active compound and readily oxidizes paints, elastomers (such as rubber), textile fibres, and dyes. Technology is available to protect elastomers, fabrics, and dyes, but only at significantly high cost.

Sources of Oxidants

Ozone, the photochemical oxidant of major concern in air pollution, is produced in the upper atmosphere by solar radiation, and small concentrations of this gas diffuse downwards. Also, small concentrations are produced by lightning and forest fires. Background concentrations of ozone range from 39 to 78 $\mu g/m^3$ (0.02 to 0.04 ppm). [7-47] However, some researchers have found that background concentrations can reach the federal ambient-air-quality standard of 235 $\mu g/m^3$ (0.12 ppm). Anthropogenic precursors are usually involved where oxidant levels in rural areas have exceeded ambient standards. [7-60] Ozone concentrations often are higher in suburban and even rural areas than in urban areas because O_3 in urban areas is lost to reactions with NO.

The photochemical reactions of O_3 and the oxides of nitrogen were presented in the previous section. These photochemical systems can be reproduced in the laboratory, and the relationship of oxidant levels to hydrocarbon and oxides-of-nitrogen concentrations can be developed. However, because of the complex systems involved in atmospheric mixing, these laboratory relationships cannot be extrapolated to the ambient air. [7-47]

Detection and Analysis

The photochemical oxidant pool consists of both organic and inorganic gases capable of oxidizing specific reagents. The reagent used most frequently is a neutral-phosphate-buffered potassium iodide solution which is calibrated by bubbling through a known concentration of ozone. The oxidizing intensity of ambient air can be monitored by measuring the amount of iodine released from the potassium iodide solution.

$$O_3 + 2KI + H_2O \longrightarrow O_2 + 2KOH + I_2 \qquad (7\text{-}19)$$

In the past, the buffered KI method was recommended as the reference method for calibration. Ultraviolet photometry is currently recommended as the primary calibration method. [7-60]

Monitoring for oxidants where reducing agents such as SO_2 and H_2S are present will reduce the iodine released from the potassium iodide solution and thus reduce the oxidant readings. "Corrected" or "adjusted" oxidant concentrations indicate measurements which have been corrected for sulfur dioxide, nitrogen dioxide, and hydrogen sulfide. [7-60]

Other devices used for monitoring ozone and chemiluminescence and coulometric analysis. Long-path infrared spectroscopy and gas chromatography have been used to monitor PAN.

Standards and Control

In June of 1978, in an effort to reduce the ill effects of oxidants, EPA proposed to set the primary oxidant standard at 196 $\mu g/m^3$ (0.10 ppm); however, in early 1979, after reevaluation of medical and scientific evidence as well as public hearings, EPA changed the oxidant standard to an ozone (O_3) standard and set the new "highest one-hour per day" standard at 240 $\mu g/m^3$ (0.12 ppm).

Oxidant control strategy has traditionally been aimed at limiting hydrocarbon and nitrogen oxide emissions, the precursors of the oxidants. However, it is now known that even if no hydrocarbons or aldehydes are present in the atmosphere, significant concentrations of ozone can still be generated as long as CO and NO_x are present. [7-60] Currently, despite concerted efforts to control CO, HC, and NO_x emissions, quantities of these air contaminants sufficient to photochemically generate ozone are still present.

7-11 INDOOR AIR POLLUTION

Until recently, the principal actions taken to control air pollution have emphasized emissions to the atmosphere. However, since most people spend a large percentage of their time indoors, attention has recently been given to the health problems posed by air pollutants that originate from building materials, furnishings, equipment, and such human activities as cooking, cleaning, and smoking.

The National Academy of Sciences has identified several potential indoor air pollutants for which there is evidence of adverse health effects. Formaldehyde, tobacco smoke, radon and radon daughters, unvented combustion appliances, aeropathogens, pesticides, and asbestos are included in the NAS listings. Though there are many variables and unknowns associated with assessing the dangers these and other indoor air pollutants pose to human health, the National Academy of Sciences has concluded that the best approach to the alleviation of adverse health effects from these substances is by exposure reduction, with control strategies targeted at high-exposure groups.

Organizations such as the World Health Organization, the Air Pollution Control Association, and the New York Academy of Medicine have joined the Congressional Office of Technology Assessment, the General Accounting Office, and the National Commission on Air Quality in urging Congressional attention to the potential problem of indoor air pollution. [7-12]

Air-Quality Management Concepts

Air quality management is a term used to describe all the functions required to control the quality of the atmosphere. Among the essential elements of such a program are control regulations and a control strategy, legal authority to implement the control strategy, emission inventories, an atmospheric surveillance network, a data management system, agency staffing and funding, a system for analysis of complaints, and stack sampling operations. [7-29]

Ambient-Air-Quality and Emissions Standards

As noted earlier, national primary and secondary ambient-air-quality standards were set and incorporated into the Clean Air Act amendments of 1970. The primary standards, based on air quality criteria, allow an adequate margin of safety to protect public health, while the secondary standards, also based on air quality criteria, were established to protect public welfare, i.e., plants, animals, property, and materials. The primary and secondary air quality standards for the six contaminants included in the 1970 Clean Air Acts amendments, plus the lead standards published in the *Federal Register* are given in Table 7-19.

The 1970 national ambient-air-quality standards established the threshold levels below which no adverse effects are known to occur. To bring pollution levels below ambient-air-quality standards, national emission standards, based on availability of control technology, were designed. The 1977 amendments to the 1970 legislation essentially retained all original aspects of that legislation.

Air Quality Indexes

The Environmental Protection Agency, the Council on Environmental Quality, and several agencies of the Department of Commerce cooperated in developing a pollutant standards index (PSI) in order to integrate the several complex factors that make up "air quality." This index combines the ambient measures of the five major criteria pollutants into numbers ranging from 0 to 500. The pollutants rated are carbon monoxide, sulfur dioxide, total suspended particulates, photo-

Table 7-19 National ambient-air-quality standards

Pollutant	Primary*	Secondary*
Particulate matter		
Annual (geometric mean)	75	60
Maximum 24-h concentration†	260	150
Lead		
Averaged over 3 mon	1.5	Same as primary
Hydrocarbons		
Maximum 3-h (6–9 A.M.) concentration	160 (0.24 ppm)	Same as primary
Carbon monoxide		
Maximum 8-h concentration†	10 (9 ppm)	
Maximum 1-h concentration†	40 (35 ppm)	Same as primary
Sulfur oxides		
Annual (arithmetic mean)	80 (0.03 ppm)	
Maximum 24-h concentration†	365 (0.14 ppm)	
Maximum 3-h concentration†		1300 (0.5 ppm)
Nitrogen oxides		
Annual (arithmetic mean)	100 (0.05 ppm)	Same as primary
Photochemical oxidants		
Maximum 1-h concentration†	240 (0.12 ppm)	Same as primary

* All measurements are expressed in micrograms per cubic meter ($\mu g/m^3$) except those for carbon monoxide, which are expressed in milligrams per cubic meter (mg/m^3). Equivalent measurements in parts per million (ppm) are given for gaseous pollutants only.

† Not to be exceeded more than once a year.

Source: From *Twelfth Annual Report*. [7-12]

chemical oxidants or ozone, and nitrogen dioxide. If the concentration of any one of the five major and initial criteria pollutants rises to the level of its air quality standard at any monitoring station, the air quality in the area is deemed unhealthy for that particular day, even though concentrations of the other four criteria pollutants may be below the national standards.

Air quality is called "good" only when the ambient measures of all five criteria pollutants have an index value of 50 or less (i.e., a value less than half that allowed by the standards). A brief summary of the potential health effects related to different PSI levels is presented in Table 7-20, and the PSI for 24 standard metropolitan statistical areas (SMSAs) from 1973 through 1980 is given in Table 7-21. Although St. Louis appears on this table, that city's PSI is not considered reliable because the ozone data were in error, due to miscalibration of the monitoring equipment. The PSI of a given city can be affected by variables other than data inaccuracies (i.e., topography, tall buildings, and micrometeorology).

The number of days in which the PSI exceeded 100 and 200 is shown in Table 7-22 for 40 SMSAs. As indicated in this table, more than 30 percent of the 40

Table 7-20 Health effects associated with levels of pollutant standards index (PSI)

PSI value	Descriptor	Health effects	Warning
400 and above	Hazardous	Premature death of ill and elderly. Healthy people will experience adverse symptoms that affect their normal activity.	All persons should remain indoors, keeping windows and doors closed. All persons should minimize physical exertion and avoid traffic.
300–399	Hazardous	Premature onset of certain diseases in addition to significant aggravation of symptoms and decreased exercise tolerance in healthy persons.	Elderly and persons with existing diseases should stay indoors and avoid physical exertion. General population should avoid outdoor activity.
200–299	Very unhealthful	Significant aggravation of symptoms and decreased exercise tolerance in persons with heart or lung disease, with widespread symptoms in the healthy population.	Elderly and persons with existing heart or lung disease should stay indoors and reduce physical activity.
100–199	Unhealthful	Mild aggravation of symptoms in susceptible persons, with irritation symptoms in the healthy population.	Persons with existing heart or respiratory ailments should reduce physical exertion and outdoor activity.
50–99	Moderate		
0–49	Good		

Source: From *Twelfth Annual Report.* [7-12]

cities have fewer than 25 days of violation of standards and more than half have fewer than 50 days of violation of any one of the standards. Most of the cities with greatest numbers of days in violation are major population centers; the citizens of those cities in effect face health risks significantly higher than citizens of cities with lower numbers of days in violation.

Enforcement of Standards

Enforcement of the ambient-air-quality standards, the emission standards for new and existing stationary sources, and the emission standards for hazardous pollutants is the responsibility of state and local agencies. States can set more, but not less, rigorous performance or emission standards than those established by EPA. In addition to monitoring existing stationary sources, state air pollution control agencies must also review the plans for all new stationary sources.

State agencies also have the power to set emergency episode regulations designed to prevent ambient pollutant concentrations from reaching levels that could cause significant harm to human health. At the first level, the *alert level*, first-stage control action is to begin. At the *warning level*, limited restrictions are imposed

Table 7-21 Pollutant Standards Index (PSI) in 24 standard metropolitan statistical areas, 1973–1980*

	Days of year in PSI interval							
Standard Metropolitan Statistical Area[1]	1973	1974	1975	1976	1977	1978	1979	1980
Buffalo, New York								
PSI 0–99[2]	NA†	308	314	335	325	342	NA	348
PSI 100–199[3]	NA	47	40	23	37	20	NA	12
PSI 200–299[4]	NA	10	10	8	3	3	NA	6
PSI 300 or more[2]	NA	0	0	1	0	0	NA	0
Chicago, Illinois								
PSI 0–99	141	125	170	285	223	215	283	318
PSI 100–199	189	223	164	67	123	119	74	45
PSI 200–299	34	15	28	13	18	28	8	3
PSI 300 or more	1	2	3	1	1	3	0	0
Cincinnati, Ohio–Kentucky–Indiana								
PSI 0–99	334	355	345	325	302	335	336	341
PSI 100–199	30	10	18	40	59	29	26	25
PSI 200–299	1	0	2	1	4	1	3	0
PSI 300 or more	0	0	0	0	0	0	0	0
Denver, Colorado								
PSI 0–99	195	190	220	181	222	192	237	277
PSI 100–199	115	127	108	149	122	133	87	62
PSI 200–299	51	45	35	33	20	39	37	23
PSI 300 or more	4	3	2	0	1	1	4	4
Houston, Texas								
PSI 0–99	NA	330	317	302	315	271	247	265
PSI 100–199	NA	32	44	52	39	70	94	80
PSI 200–299	NA	3	4	12	11	24	23	21
PSI 300 or more	NA	0	0	0	0	0	1	0
Kansas City, Missouri–Kansas								
PSI 0–99	340	360	361	359	340	NA	NA	338
PSI 100–199	21	5	4	6	17	NA	NA	27
PSI 200–299	4	0	0	1	8	NA	NA	1
PSI 300 or more	0	0	0	0	0	NA	NA	0
Los Angeles, California								
PSI 0–99	88	72	93	98	112	140	117	145
PSI 100–199	125	129	134	126	136	121	126	109
PSI 200–299	147	163	137	142	117	101	112	110
PSI 300 or more	5	1	1	0	0	3	10	2

Table 7-21 (*continued*)

Standard Metropolitan Statistical Area[1]	Days of year in PSI interval							
	1973	1974	1975	1976	1977	1978	1979	1980
Louisville, Kentucky–Indiana								
PSI 0–99	280	244	180	206	262	271	307	307
PSI 100–199	71	94	154	146	90	86	56	56
PSI 200–299	14	27	31	14	13	8	2	3
PSI 300 or more	0	0	0	0	1	0	0	0
Memphis, Tennessee–Arkansas–Mississippi								
PSI 0–99	NA	339	346	342	343	328	NA	NA
PSI 100–199	NA	22	10	24	19	34	NA	NA
PSI 200–299	NA	4	7	0	2	3	NA	NA
PSI 300 or more	NA	0	2	0	1	0	NA	NA
Milwaukee, Wisconsin								
PSI 0–99	NA	336	325	334	332	331	332	350
PSI 100–199	NA	25	32	24	26	31	32	14
PSI 200–299	NA	4	8	8	7	3	1	2
PSI 300 or more	NA	0	0	0	0	0	0	0
New York, New York–New Jersey								
PSI 0–99	NA	NA	95	98	92	191	252	235
PSI 100–199	NA	NA	33	38	186	160	110	129
PSI 200–299	NA	NA	226	225	86	14	3	2
PSI 300 or more	NA	NA	11	4	1	0	0	0
Philadelphia, Pennsylvania–New Jersey								
PSI 0–99	NA	207	222	279	286	284	286	303
PSI 100–199	NA	132	124	80	69	71	75	59
PSI 200–299	NA	23	18	7	10	10	4	4
PSI 300 or more	NA	3	1	0	0	0	0	0
Portland, Oregon–Washington								
PSI 0–99	137	234	282	296	284	290	310	311
PSI 100–199	179	119	70	67	76	73	38	42
PSI 200–299	49	12	13	3	5	2	12	10
PSI 300 or more	0	0	0	0	0	0	5	3
San Bernardino–Riverside–Ontario, California								
PSI 0–99	173	133	152	182	183	220	158	195
PSI 100–199	117	118	99	86	74	77	102	78
PSI 200–299	71	107	112	88	108	67	102	92
PSI 300 or more	4	7	2	0	0	1	3	1

Table 7-21 (*continued*)

Standard Metropolitan Statistical Area	Days of year in PSI interval							
	1973	1974	1975	1976	1977	1978	1979	1980
Rochester, New York								
PSI 0–99	NA	362	363	358	361	359	359	363
PSI 100–199	NA	3	2	8	4	6	6	3
PSI 200–299	NA	0	0	0	0	0	0	0
PSI 300 or more	NA	0	0	0	0	0	0	0
Sacramento, California								
PSI 0–99	351	361	344	328	346	338	346	345
PSI 100–199	14	4	21	35	19	25	19	19
PSI 200–299	0	0	0	3	0	2	0	2
PSI 300 or more	0	0	0	0	0	0	0	0
St. Louis, Missouri–Illinois								
PSI 0–99	192	249	235	225	262	246	273	311
PSI 100–199	132	62	113	108	90	94	67	49
PSI 200–299	41	44	16	27	13	18	21	6
PSI 300 or more	0	10	1	6	0	7	4	0
Salt Lake City, Utah								
PSI 0–99	284	290	291	256	304	294	317	312
PSI 100–199	70	53	54	85	52	51	33	36
PSI 200–299	11	19	19	25	9	20	14	18
PSI 300 or more	0	3	1	0	0	0	1	0
San Diego, California								
PSI 0–99	329	337	317	292	320	327	259	NA
PSI 100–199	31	26	45	65	41	32	96	NA
PSI 200–299	5	2	3	9	4	6	10	NA
PSI 300 or above	0	0	0	0	0	0	0	NA
San Francisco, California								
PSI 0–99	335	334	336	321	341	343	342	358
PSI 100–199	26	29	27	44	24	21	23	8
PSI 200–299	3	2	2	1	0	1	0	0
PSI 300 or more	0	0	0	0	0	0	0	0
Seattle–Everett, Washington								
PSI 0–99	138	249	292	276	270	303	305	333
PSI 100–199	170	105	68	85	91	60	54	33
PSI 200–299	57	11	5	5	4	2	6	0
PSI 300 or more	0	0	0	0	0	0	0	0

Table 7-21 ***(continued)***

Standard Metropolitan Statistical Area[1]	Days of year in PSI interval							
	1973	1974	1975	1976	1977	1978	1979	1980
Syracuse, New York								
PSI 0–99	NA	362	360	359	358	353	361	362
PSI 100–199	NA	3	5	5	5	12	4	1
PSI 200–299	NA	0	0	2	2	0	0	3
PSI 300 or more	NA	0	0	0	0	0	0	0
Tampa–St. Petersburg, Florida								
PSI 0–99	NA	355	355	361	346	353	357	361
PSI 100–199	NA	8	9	5	19	10	6	5
PSI 200–299	NA	1	1	0	0	2	2	0
PSI 300 or more	NA	1	0	0	0	0	0	0
Washington, D.C.–Maryland–Virginia								
PSI 0–99	236	288	255	219	281	295	318	297
PSI 100–199	112	52	84	132	67	67	44	67
PSI 200–299	14	22	26	15	7	3	3	2
PSI 300 or more	3	3	0	0	0	0	0	0

Source: From *Twelfth Annual Report.* [7-12]

*PSI is a highly summarized health-related index based on direct measurements of five criteria pollutants: carbon monoxide, sulfur dioxide, total suspended particulates, photochemical oxidants or ozone, and nitrogen dioxide. The PSI for one day will rise above 100 when any one of the five criteria pollutants (at any one station in an SMSA) reaches a level judged to have adverse short-term effects on human health.

PSI is designed for the daily reporting of air quality to advise the public of potentially acute, but not chronic, health effects. PSI should not be used to rank cities. To properly rank the air pollution problems in different cities, one should rely not just on air quality data but should also include all data on population characteristics, daily population mobility, transportation patterns, industrial composition, emission inventories, meteorological factors, and the spatial representativeness of air monitoring sites.

The PSI analysis for 1973–1980 is based on standards applicable during 1980, not on standards applicable at the time of monitoring. The primary standard for ozone was relaxed in 1979 from 160 to 240 micrograms per cubic meter per hour.

† NA—Not available.

[1] A Standard Metropolitan Statistical Area is an area with an urban center of 50,000 persons or more, including the county containing that center and any neighboring counties that are closely associated with the central area by daily commuting ties. SMSAs contain not only urbanized areas, which occupy only 10% of the land, but also open space, forests, recreation areas, parks, and cropland.

The SMSAs shown here were chosen according to availability of trend data. For some of these SMSAs (for example, Buffalo, 1979), data are not included for years when carbon monoxide or ozone were monitored for less than 300 days. Other major SMSAs not shown here may have had many days of unhealthful air, but comparable data for 1973–1980 are not available in SAROAD.

[2] Note that the average PSI values do not represent an average SMSA, but an average of SMSAs.

[3] An index of 0–99 signifies good or moderate air quality.

[4] An index of 100–199 signifies unhealthful air quality.

[5] An index of 200–299 signifies very unhealthful air quality.

[6] An index of 300 or more signifies hazardous air quality.

Table 7-22 Ranking of 40 standard metropolitan statistical areas (SMSAs) using the Pollutant Standards Index (PSI), 1978–80

Severity level (days with PSI greater than 100)	SMSA	3-year average of number of days: Unhealthful Very unhealthful, and hazardous (PSI > 100)	Very unhealthful, and hazardous (PSI > 200)
More than 150 d	Los Angeles	231	113
	San Bernardino, Riverside, Ontario	174	89
100–150 d	New York	139	6
	Denver	130	36
	Pittsburgh	119	18
	Houston	104	23
50–99 d	Chicago	93	14
	St. Louis	89	19
	Philadelphia	74	6
	San Diego	72†	8†
	Louisville	70	4
	Phoenix	70†	6†
	Gary	68	33
	Portland	62	11
	Washington	62	3
	Jersey City	58*	0*
	Salt Lake City	58	18
	Seattle	52	3
	Birmingham	50*	8*
25–49 d	Cleveland	46	11
	Detroit	39	4
	Memphis	37*	3*
	Baltimore	36	2
	Indianapolis	34	2
	Cincinnati	28	1
	Milwaukee	28	2
	Kansas City	28*	1*
0–24 d	Sacramento	22	1
	Dallas	21	1
	Allentown	21	2
	Buffalo	20†	4†
	San Francisco	18	0
	Toledo	15	2
	Dayton	15	1
	Tampa	8	1
	Syracuse	7	1
	Norfolk	6†	0†
	Grand Rapids	6*	0*
	Rochester	5	0
	Akron	4	0

Source: From *Twelfth Annual Report*. [7-12]

* Based on 1 yr of data.

† Based on 2 yr of data.

on incinerators and vehicle operations. At the third level, the *emergency level*, stringent controls are imposed on open burning, incinerator operation, industrial plant operation, and automobile use. A comparison between the national primary ambient-air-quality standards and the emergency episode levels set for the state of Kentucky are given in Table 7-23.

State agencies must also monitor emissions for hazardous pollutants, those air contaminants which in the judgment of the administrator of EPA might cause or contribute to an increase in mortality or an increase in serious, irreversible, or incapacitating illness. States must adhere to the national emission standards set for five hazardous substances—asbestos, beryllium, mercury, vinyl chloride, and benzene. [5-75]

Basically, the primary responsibility for implementation of air-pollution-control standards is left at the state and local level. Still, if a state fails to develop and enforce a workable program for the prevention and control of air pollution, the Environmental Protection Agency can then act to assume the state's enforcement powers.

Emission Factors and Emission Inventories

State and local agencies responsible for implementation of air-pollution-control standards rely heavily upon emission factors and emission inventories. The *emission factor*, the statistical average of the mass of pollutants emitted from each source per unit quantity of material handled, processed, or burned, may be found in a

Table 7-23 Comparison of ambient-air-quality standards and emergency episode levels

Contaminant	National primary standards	Kentucky emergency episode levels
Particulate		
Maximum 24-h average concentration	260 $\mu g/m^3$	875 $\mu g/m^3$
SO_2		
Maximum 24-h average	365 $\mu g/m^3$ (0.5 ppm)	2100 $\mu g/m^3$ (0.8 ppm)
CO		
Maximum 8-h average	10 mg/m^3 (9 ppm)	46 mg/m^3 (40 ppm)
Ozone		
Maximum 1-h average	240 $\mu g/m^3$ (0.12 ppm)	980 $\mu g/m^3$ (0.5 ppm)
NO_2		
Maximum 1-h average	None	3000 $\mu g/m^3$ (1.6 ppm)
Maximum 24-h average	None	750 $\mu g/m^3$ (0.4 ppm)

Source: From *Kentucky Air Pollution Control Regulations* [7-32] and *Twelfth Annual Report* [7-12].

Table 7-24 Calculation of potential CO, HC, and NO_x emissions for light-duty vehicles, 1966–1978

Model	Fraction of vehicles in age group	No. of vehicles in age group	Annual mileage	Carbon monoxide (CO) emission rate		Hydrocarbon (HC) emission rate		Oxides of nitrogen (NO_x) emission rate	
				g/mi	g/d $\frac{3 \times 4 \times 5}{365}$	g/mi	g/d $\frac{3 \times 4 \times 7}{365}$	g/mi	g/d $\frac{3 \times 4 \times 9}{365}$
(1)	(2)	(3)	(4)	(5)	(6)	(7)	(8)		(9)
1978	0.08	1,906	15,900	20.3	1.685×10^6	1.3	1.079×10^5	1.6	1.328×10^5
1977	0.11	2,621	15,000	24.1	2.595×10^6	1.6	1.723×10^5	1.8	1.939×10^5
1976	0.11	2,621	14,000	28.2	2.835×10^6	1.9	1.910×10^5	2.7	2.714×10^5
1975	0.11	2,621	13,100	32.1	3.012×10^6	2.2	2.069×10^5	2.8	2.633×10^5
1974	0.10	2,382	12,200	68.7	5.470×10^6	5.7	4.538×10^5	3.0	2.388×10^5
1973	0.09	2,144	11,300	76.0	5.044×10^6	6.3	4.182×10^5	3.0	1.991×10^5
1972	0.09	2,144	10,300	82.8	5.010×10^6	6.9	4.175×10^5	4.4	2.662×10^5
1971	0.08	1,906	9,400	89.0	4.369×10^6	7.4	3.632×10^5	4.4	2.160×10^5
1970	0.07	1,668	8,500	94.7	3.678×10^6	7.9	3.068×10^5	4.4	1.709×10^5
1969	0.05	1,191	7,600	99.8	2.475×10^6	8.3	2.058×10^5	4.4	1.091×10^5
1968	0.04	953	6,700	104.3	1.824×10^6	8.7	1.522×10^5	4.4	0.770×10^5
1967	0.03	715	6,600	106.7	1.379×10^6	11.7	1.513×10^5	3.6	0.465×10^5
1966	0.04	953	6,200	108.7	1.760×10^6	12.1	1.959×10^5	3.6	0.583×10^5
		23,825			41.136×10^6 (41 tonnes/d)		33.428×10^5 (3.4 tonnes/d)		22.433×10^5 (2.3tonnes/d)

Source: From Davis [7-14] and EPA [7-78].

number of literature sources, with the most complete collection of factors being published by the Environmental Protection Agency in *Compilation of Air Pollutant Emission Factors* [7-77] and *Mobile Emission Factors.* [7-78] Auto emission factors for light-duty 1966–1978 vehicles are given in Table 7-24.

The purpose of an *emission inventory* is to locate the air pollution sources for a given area and to define the types and magnitude of pollution these sources are likely to produce, their projected emission of pollutants, and the frequency, duration, and relative contribution of pollutant emissions from each source. [5-18] Emission inventories are used (1) to plan developments in metropolitan areas, (2) to establish sampling programs and interpret the results of sampling activities, (3) to establish emission standards, (4) to provide basic input for simulation models, (5) to estimate air pollutant concentrations with various meteorological conditions, (6) to establish baseline levels of air pollutant concentrations and to relate these to future trends, (7) to indicate seasonal and geographical distribution of air pollutants in a study area, and (8) to assist in establishing priorities for a control program.

In carrying out an emission inventory, agencies classify the pollutants emitted into the community, classify the sources of those pollutants, determine the quality and quantity of the materials being handled, processed, or burned, determine the emission factors for those materials, and compute the rate at which each pollutant is emitted.

Traditionally, the five basic air pollutants generally included in an emission inventory included carbon monoxide, hydrocarbons, oxides of nitrogen, particulates, and oxides of sulfur. However, measurement of photochemical oxidants (i.e., ozone) has been substituted for hydrocarbon measurements in PSIs and in many emissions inventories. Such changes are indicative of the state of the art.

The sources of emissions include the four sources generally cited previously in this chapter—transportation, or mobile sources of combustion, stationary sources of combustion, industrial processes, and solid-waste disposal and miscellaneous activities. The quality and quantity of the materials being handled, processed, or burned in the four source groups may be determined through questionnaires, through direct contact with management, through chambers of commerce or research organizations, from periodicals and journals, from registration information available at state agencies, or by informed estimate. When the data have thus been gathered, the information may be used in conjunction with the emission factor to determine the pollution potential of each activity within a given community and to calculate the rate at which each pollutant is emitted. Example 7-4 gives calculations for an emission inventory for a city of 40,000 persons producing the specified amounts of material on a typical winter day.

Example 7-4: Deriving an emission inventory Utilizing the specified amounts of material on a typical winter day, calculate an emission inventory for a city of 40,000 citizens. The inventory should include particulates, sulfur oxides, carbon monoxide, hydrocarbons, and nitric oxides.

Sources of contaminants

Mobile sources	23,825 automobiles; 227,000 L (60,000 gal) of gas pumped.
Stationary sources	56,640 m^3 (2×10^6 ft^3) natural gas (sulfur content of 130 $g/10^6$ m^3) used in domestic and commercial heating; 36.3 tonnes of coal (4% sulfur, 10% ash) used in hand-fired commercial and domestic furnaces.
Industrial processes	Oil refinery (moving-bed catalytic cracking system) processed 40,000 bbl (6350 m^3) of petroleum products; iron and steel mill (ore-charged blast furnace with uncontrolled emissions) processed 90.8 tonnes.
Solid-waste disposal sources	Municipal incinerator with settling chamber and water spray burned 94.4 tonnes; 26.3 tonnes burned in open burning operations (landscape pruning).

SOLUTION

1. Mobile sources are calculated by determining probable automobile emissions for all five contaminants in the manner shown in Table 7-24. The contaminant calculations are then moved to an emission-factors summary sheet. Note that the summary sheet also contains a notation for hydrocarbons emitted during the pumping of the estimated 227,000 L of gas.
2. Using the formula

$$\text{Emission factor} \times \text{quantity used} = \text{emissions produced}$$

estimates are made for production of each of the five contaminants for all stationary sources, industrial processes, and solid-waste disposal sources. As an example, the emission factor for natural gas with a sulfur content of 130 $g/10^6$ m^3 is shown below. Each contaminant under consideration is listed. Because 56,640 m^3 of natural gas was used on the day being studied, each factor is multiplied by this quantity and the resultant emission is calculated.

Contaminant	Emissions factor $kg/10^3$ m^3	Quantity used 10^3 m^3	Emissions produced kg	tonnes
Particulates	240	6.35	1524	0.015
Sulfur oxides (as SO_2)	9.63	6.35	61	0.0006
Carbon monoxide (CO)	321	6.35	2038	0.02
Hydrocarbons (as CH_4)	128	6.35	813	0.008
Nitrogen oxides (as NO_2)	1284	6.35	8153	0.08

3. After similar approaches have been used to arrive at estimated emissions for other sources in the emission inventory, a summary of the inventory is prepared.

Summary of emission inventory (tonnes/d)

Source	Pollutants				Particulates or solids
	CO	HC	NO_x	SO_x	
Mobile					
Automobile	41.1	3.4	2.3	0.009	0.28
Crankcase and evaporative emissions		0.49			
Stationary					
Natural gas	0.02	0.008	0.08	0.0006	0.015
Coal	0.18	0.05	0.11	2.76	0.36
Industrial processes					
Oil refinery	69.0	1.58	0.09	1.09	0.31
Iron and steel mill	79.4				5.00
Solid-waste disposal					
Municipal incinerator	1.65	0.070	0.142	0.12	0.66
Open burning	0.54	0.18	0.018		0.15
Totals (rounded off)	191.9	5.8	2.7	4.0	6.8

In this example, carbon monoxide is the air contaminant with the highest emission rate, with industrial and mobile sources being the largest contributors of carbon monoxide. Particulates, hydrocarbons, oxides of sulfur, and oxides of nitrogen follow, in that order. Even through the oxides of sulfur measured by weight appear to be almost insignificant, they could actually present greater problems to the community than the other contaminants, because of their particular deleterious effects in combination with particulates and with other gaseous elements, as shown in Table 7-25. Thus, quantities do not present the total picture in terms of potential danger from pollutants. Still, calculations for the estimated emissions are a necessary step in developing an air-pollution-control program.

Surveillance and Control

Though emissions estimates are useful in developing an air-pollution-control program, they are based on assumptions and cannot take the place of an effective atmospheric surveillance system that garners actual data. Such a monitoring system is needed to determine when episode levels have been reached and to determine whether or not a control strategy is working. In such a system, the size of the sampling network will depend on the population and existing levels of pollution. Generally, the sampling locations should supply data that are representative for the entire air-quality region. Hence, specific sampling sites should be representative of the area, with some stations established at points of highest pollution levels to measure population exposure to specific source pollution and others placed away from such sources for nonurban or background readings.

Table 7-25 Summary of health effects of sulfur oxides when accompanied by other contaminants

Concentration μg/m³ (ppm)	Accompanied by	Time	Effect
1500 (0.52)	Elevated particulate matter	24-h average	Increased mortality
715 (0.25)	Smoke	24-h average	Increased mortality, sharp rise in illness rate for those over 54 yr
500 (0.19)	Low particulate matter	24-h average	Increased mortality rates
300–500 (0.11–0.19)	Low particulate matter	24-h average	Increased hospital admissions and absenteeism from work
600 (0.21)	Smoke	24-h average	Accentuation of chronic lung disease symptoms
105–265 (0.037–0.092)	Smoke	Annual mean	Increased frequency of respiratory symptoms and lung disease
120 (0.46)	Smoke	Annual mean	Increased frequency and severity of respiratory diseases
115 (0.04)	Smoke	Annual mean	Increased mortality

Source: From *Air Quality Criteria*. [7-49]

Surveillance systems can be an invaluable part of an agency program that includes a sound, comprehensive control strategy with reasonable control regulations. Such systems must take into account the meterological aspects of air pollution as discussed in Chap. 8, as well as the capabilities and expected effectiveness of the various control devices discussed in Chap. 9.

DISCUSSION TOPICS AND PROBLEMS

7-1 Name and describe the four major layers of the atmosphere.

7-2 What are the major components of the troposphere?

7-3 Name at least two international organizations initiating air-monitoring programs and discuss the importance of worldwide cooperation in solving air pollution problems.

7-4 In what units of measurement might you expect to see the following pollutants listed in an air-pollution survey report?

(*a*) Dustfall
(*b*) Suspended particulates
(*c*) SO_2

7-5 From which natural sources might the following pollutants arise: Hydrocarbons, CO, H_2S, CH_4, and particulate matter (dust, smoke, etc.)

7-6 Calculate the volume occupied by 4 mol of gas at 21.1°C and 760 mmHg.

7-7 Determine the volume of 3 mol of stack gas at 1400 mmHg and 1000°C.

7-8 Determine the volume of 6 mol of gas at 37°C and 700 mmHg.

7-9 Calculate the volume of 5 mol of methane at 31°C and 740 mmHg.

7-10 Gas from a thermal pool has an SO_2 content of 7 ppm at 760 mmHg and 50°C. Calculate the SO_2 concentration in micrograms per cubic meter and milligram per cubic meter.

7-11 The NO content of a sample of air measured at 22°C and 1 atm pressure is 14 ppm. Calculate the NO concentration in micrograms per cubic meter and milligrams per cubic meter.

7-12 The NO_2 content of a sample of stack gas measured at 950°C at 2 atm pressure was 9 ppm. Determine the NO_2 concentration in micrograms per cubic meter and milligrams per cubic meter.

7-13 The CO content of a sample of air measured at 20°C and 760 mmHg is 11 ppm. Calculate the CO concentration in micrograms per cubic meter and milligrams per cubic meter.

7-14 On a national basis, what are four principal sources of air pollution and what are the major types of pollutants produced by each source?

7-15 Which of the four main sources is the single largest source of air pollution?

7-16 What are primary pollutants? Secondary pollutants? Give examples.

7-17 Define and discuss the settling out potential of particulate and gaseous pollutants. Give examples of each.

7-18 Define dusts, smokes, mists, fumes, and vapors.

7-19 Determine the settling rate of fly ash composed of 10-μm particles in air at 21°C.

7-20 Calculate the settling rate of a dust cloud generated by a sandblasting operation if the average particle size is 17 μm in air at 27°C.

7-21 Determine the settling rate of smoke from a grass fire if the average particle size is 0.8 μm and the temperature is 12°C.

7-22 Calculate the settling rate of dust clouds in a desert duststorm if the average particle size is 200 μm and the temperature is 16°C.

7-23 In what size range do particles most effectively reduce vision?

7-24 Discuss the defense mechanisms of the human respiratory system against coarse dust, smoke particles, and fumes.

7-25 What is the major source of lead in our environment?

7-26 How are settleable particulates measured? Suspended particulates?

7-27 How is opacity measured?

7-28 A Hi-Vol sampler operated at 1.57 m/min. The sampling period was 24 h. The filter paper weighed 3.1690 g at the start of the run and 3.5882 g at the end of the sampling period. What is the concentration of the suspended particulate in micrograms per cubic meter?

7-29 Why are alkenes more important in the study of air pollution than alkanes?

7-30 What hazards are posed by the polynuclear group of aromatic hydrocarbons?

7-31 What are the primary natural and anthropogenic sources of the hydrocarbons found in the atmosphere? What is the major anthropogenic source?

7-32 In 1977, the total emission of CO on a weight basis accounted for slightly over half (53 percent) of all anthropogenic air pollutants. Discuss the short-term and long-range hazards posed by large increases in the CO content of the atmosphere.

7-33 List the oxides of sulfur and indicate which are of primary concern in air pollution.

7-34 What particular health hazards are posed by SO_2 in a dusty atmosphere? Discuss these hazards in relation to major air-pollution disasters of this century.

7-35 What is the mechanism by which H_2SO mist can cause destruction of limestone surfaces? Show the chemical reaction.

7-36 On a national basis, what are the principal sources of SO_x emissions to the atmosphere?

7-37 A power plant burns 20 tonnes of coal per hour, and the average sulfur content of the coal is 4.5 percent. What is the approximate emission of SO_2 in tonnes per day?

7-38 Which two of the gases that make up the oxides of nitrogen are of primary concern to air-pollution-control officials? Of these two gases, which one is emitted into the atmosphere in the largest quantities? Which one of these two gases is most toxic as far as humans are concerned?

7-39 What are two chemical compounds that are necessary ingredients for the production of photochemical smog?

7-40 What are the major natural and anthropogenic sources of nitrous oxides in the atmosphere?

7-41 Discuss the causes and relevance of diurnal variations in NO and NO_2.

7-42 The average car emits about 4 g of nitric oxide (NO) per mile. The average automobile in this country travels about 16,000 m/yr. Estimate the number of tonnes of NO emitted by each automobile per year. The number of automobiles in the United States is about 110 million. Calculate total NO emissions from automotive transportation in tonnes per year.

7-43 What is the major photochemical oxidant found in the atmosphere? What detrimental effects does this oxidant have on humans and animals? On plants? On materials?

7-44 What is a PSI? What are the six descriptors used in the PSI health-effects ratings? Which cities had a PSI greater than 300 in the year 1977? in 1980? Which city had the best air quality in 1979? Use information in Table 7-21.

7-45 Define emission inventory and discuss the methods used in such an inventory and the uses to which the inventory may be put.

7-46 What are the five basic air pollutants that have traditionally been included in an emission inventory? Which of these pollutants has been dropped from some inventories? Why?

7-47 The measurement of the dust size distribution in an industrial plant yielded the following data:

Particle size	% wt. frac.
0–1	8
1–2	10
2–3	12
3–4	15
4–5	19
5–6	14
6–7	13
7–8	9

What estimated fraction of these particles would be retained in the lungs if the particles are retained in accordance with Fig. 7-5? How many would be retained by a person doing light work and breathing at 20 L/min for an 8-h workday if the air contains 600 $\mu g/m^3$ of this suspended particulate?

7-48 Using Table 7-24 and assuming that all autos were driven 15,000 mi, estimate the total CO, HC, and NO_x emissions generated by 20 1968-model cars during the year 1979. Compare these figures to estimated emissions generated by 20 1978-model autos operated during the same year, each of which is also assumed to have been driven 15,000 mi. Based on these calculations, discusses the effectiveness of auto-emissions control devices over the decade 1968 to 1978.

REFERENCES

7-1 Air Conservation Commission: *Air Conservation*, Am Assoc for the Advancement of Science publ. no. 80, Washington, D.C., 1965.
7-2 *Air Pollution Control Regulations, Jefferson County, Kentucky*, Louisville, 1976.
7-3 *Air Quality Criteria*, staff report, Subcommittee on Air and Water Pollution, Committee on Public Works, U.S. Senate, 94–411, July 1968.
7-4 *Air Quality Criteria for Lead*, Office of Research and Development publ. no. EPA-600/8-77-017, Washington, D.C. 1977.
7-5 *Air Sampling Instruments for Evaluation of Atmospheric Contaminants*, American Conference of Governmental Industrial Hygienists, Cincinnati, Ohio.
7-6 Bishop, C. A.: "EJC Policy Statement on Air Pollution and Its Control," *Chem Eng Progr*, **53**(11):146 (1957).
7-7 *A Citizen's Guide to Clean Air*, Conservation Foundation, Washington, D.C., 1972.
7-8 Council on Environmental Quality: *First Annual Report of the Council on Environmental Quality*, Washington, D.C., 1970.
7-9 ———: *Third Annual Report of the Council on Environmental Quality*, Washington, D.C., 1972.
7-10 ———: *Fourth Annual Report of the Council on Environmental Quality*, Washington, D.C., 1973.
7-11 ———: *Eighth Annual Report of the Council on Environmental Quality*, Washington, D.C., 1977.
7-12 ———: *Twelfth Annual Report of the Council on Environmental Quality*, Washington, D.C., 1982.
7-13 Davis, Harold (Air Pollution Control District of Jefferson County, Kentucky): by personal communication, December 1978.
7-14 Davis, Mackenzie L.: *Air Resource Management Primer*, ASCE, New York, 1973.
7-15 deKoning, H. W., and A. Lohler: "Monitoring Global Air Pollution," *Env Sci Tech*, **12**(8):884 (1978).
7-16 *The Effects of Air Pollution*, Division of Air Pollution, Public Health Service, Washington, D.C., 1966.
7-17 Ember, Lois R.: "Global Environmental Problems: Today and Tomorrow," *Env Sci Tech*, **12**(8):874 (1978).
7-18 "The Emission Inventory," in *Community Air Pollution*, National Center for Air Pollution Control, Durham, N.C., 1964.
7-19 "EPA Proposed an Ambient Air Quality Standard for Lead," *Env Sci Tech*, **12**(3):245 (1978).
7-20 "The Evaluation and Application of An Urban Diffusion Model to Simulate Carbon Monoxide Concentrations in Downtown Louisville, Kentucky," Equitable Environmental Health, Woodbury, N.Y., 1976.
7-21 Giddings, J. Calvin: *Chemistry, Man, and Environmental Change*, Canfield, San Francisco, 1973.
7-22 Halliday, E. C.: "A Historical Review of Air Pollution," in *Air Pollution*, WHO, Geneva, 1961.
7-23 Henderson, J. J.: "Physical Properties of Aerosols," in *Atmospheric Sampling*, Nat Air Pol Cont Admin Res, Triangle Park, N.C., 1958.
7-24 *High Volume Air Sampler*, General Metal Works, Cleveland, Ohio, 1966.
7-25 Hindawi, T. J.: *Air Pollution Injury to Vegetation*, Nat Air Pol Admin pub. no. AP-71, Raleigh, N.C., 1970.
7-26 Hoffman, Dietrich, and Ernest L. Wynder: "Chemical Analysis and Carcinogenic Bioassays of Organic Particulate Pollutants," in Arthur C. Stern (ed.), *Air Pollution*, **11**:196 (1968).

7-27 Hunter, D. C., and H. C. Wohlers: *Air Pollution Experiments for Junior and Senior High School Classes*, 2d ed., Air Pol Cont Assoc, Pittsburgh, 1970.

7-28 Institute for Air Pollution Training: "Air Pollutants and Their Sources," in *Air Pollution Control Orientation Course (422-A)*, U.S. EPA, Atlanta, 1975.

7-29 ———: "Air Quality Management," in *Air Pollution Control Orientation Course (422-A)*, U.S. EPA, Atlanta, 1975.

7-30 ———: "Effects of Air Pollution," in *Air Pollution Control Orientation Course (422-A)*, U.S. EPA, Atlanta, 1975.

7-31 *Kentucky Air Pollution Control Regulations*, Bureau of Environmental Quality, Frankfort, 1975.

7-32 *Kentucky Air Pollution Control Regulations*, Bureau of Environmental Quality, Frankfort, 1979.

7-33 Lindstrom, C. A.: "Units of Measurement," in *Atmospheric Sampling 435*, Nat Air Pol Con Admin, Research Triangle Park, N.C., 1958.

7-34 Lloyd, W. G., and D. R. Rowe: "Catalytic Ozone Removal," unpublished, 1978.

7-35 McCrone, W. C., and J. G. Delly: *The Particle Atlas*, vol. II, 2d ed., Ann Arbor Science, Ann Arbor, Mich., 1973.

7-36 Magill, P. L., F. R. Holden, and C. Ackley: *Air Pollution Handbook*, McGraw-Hill, New York, 1956.

7-37 Mallette, Frederick S. (ed.): *Problems and Control of Air Pollution*, Reinhold, New York, 1955.

7-38 Martin, Brian, and Francesco Sella: "Earthwatching on a Macroscale," *Env Sci Tech*, **10**(3): 230 (1978).

7-39 Maxcy, K. F., and M. J. Rosenau: *Preventive Medicine and Public Health,* 9th ed., Appleton-Century-Crofts, New York, 1965.

7-40 *Medical and Biological Effects of Environmental Pollutants—Carbon Monoxide*, National Academy of Sciences, Washington, D.C., 1977.

7-41 "Method 9—Visual Determination of the Opacity of Emissions for Stationary Sources," *Federal Register*, vol. 39, no. 177, 1974.

7-42 Miller, A.: *Meterology*, Charlie Miller, Columbus, Ohio, 1966.

7-43 Moline, C. H.: "Sampling for Dustfall and Suspended Solids and Determination of Soiling Index," in *Community Air Pollution*, U.S. Dept. HEW Training Program, Durham, N.C., 1962.

7-44 National Air Pollution Control Administration, *Air Quality Criteria for Hydrocarbons*, publ. no. AP-64, Superintendent of Documents, U.S. Government Printing Office, Washington, D.C., 1970.

7-45 ———: *Air Quality Criteria for Nitrogen Oxides*, publ. no. AP-84, Superintendent of Documents, U.S. Government Printing Office, Washington, D.C., 1971.

7-46 ———: *Air Quality Criteria for Particulate Matter*, publ. no. FS 2.300 AP-49, Superintendent of Documents, U.S. Government Printing Office, Washington, D.C., 1970.

7-47 ———: *Air Quality Criteria for Photochemical Oxidants*, publ. no. AP-63, Superintendent of Documents, U.S. Government Printing Office, Washington, D.C., 1970.

7-48 ———: *Air Quality Criteria for Sulfur Oxides*, publ. no. 1619, Superintendent of Documents, U.S. Government Printing Office, Washington, D.C., 1967.

7-49 ———: *Air Quality Criteria for Sulfur Oxides*, publ. no. AP-50, Superintendent of Documents, U.S. Government Printing Office, Washington, D.C., 1970.

7-50 ———: *Control Techniques for Hydrocarbon and Organic Solvent Emissions from Stationary Sources*, publ. no. AP-68, Superintendent of Documents, U.S. Government Printing Office, Washington, D.C., 1970.

7-51 ———: *Control Techniques for Sulfur Oxide Air Pollutants*, Public Health Service publ. no. AP-52, 1969.

7-52 ———: *Danger in the Air: Sulfur Oxides and Particulates*, publ. no. 1, Washington, D.C., 1970.

7-53 "National Ambient Air Quality Standards for Lead," *Federal Register*, vol. 43, no. 194, 1978.

7-54 "National Primary and Secondary Ambient Air Quality Standards," *Federal Register*, vol. 36, no. 84, 1971.

7-55 *Nitrogen Oxides*, National Academy of Sciences, Washington, D.C., 1977.

7-56 *Occupational Exposure to Carbon Monoxide*, criteria document, National Institute for Occupational Safety and Health, 1972.

7-57 Office of Air Quality Planning and Standards: *Control Techniques for Lead Air Emissions*, vol. 1, Research Triangle Park, N.C., 1977.

7-58 ———: *Control Techniques for Lead Air Emissions*, vol. II, Research Triangle Park, N.C., 1977.

7-59 *The Oxides of Nitrogen in Air Pollution*, Bureau of Air Sanitation, Berkeley, Calif., 1966.

7-60 *Ozone and Other Photochemical Oxidants*, National Academy of Sciences, Washington, D.C., 1977.

7-61 Perkins, Henry C.: *Air Pollution*, McGraw-Hill, New York, 1974.

7-62 Peterson, Eugene K.: "Carbon Monoxide Affects Global Ecology," *Env Sci Tech*, **3**(11):1162 (1969).

7-63 *Photochemical Oxidants: An Air Pollution Health Effects Report*, American Lung Association, 1977.

7-64 *Position Paper on Regulation of Atmospheric Sulfates*, Office of Air Quality Planning and Standards, U.S. EPA, Research Triangle Park, N.C., 1975.

7-65 Rowe, D. R.: "Instrumental Methods for Monitoring Oxidants and Oxides of Nitrogen," report for Nat Air Pol Cont Admin, 1970.

7-66 ———: "Introduction to Air Pollution," script no. 1, Atmospheric Pollution for Science Teachers, National Science Foundation project no. EPP75-10694, 1975.

7-67 Sawyer, C. N., and P. L. McCarty: *Chemistry for Environmental Engineering*, 3d ed., McGraw-Hill, New York, 1978.

7-68 Shaheen, Esben I.: *Environmental Pollution: Awareness and Control*, Engineering Technology, Mahomet, Ill., 1974.

7-69 Simon, R. A.: "Classification and Definition of Air Pollutants," in *Community Air Pollution*, U.S. Dept. HEW Training Program, Durham, N.C., 1966.

7-70 "Standards of Performance for New Stationary Sources," *Federal Register*, vol. 39, no. 219, 1974.

7-71 Starkman, Ernest S.: "Emission Control and Fuel Economy," *Env Sci Tech*, **9**(9):824 (1975).

7-72 Stern, A. C. (ed.): *Air Pollution*, vol. 1, 2d ed., Academic Press, New York, 1968.

7-73 ———, Henry C. Wohlers, Richard W. Boubel, and William P. Lowry: *Fundamentals of Air Pollution*, Academic Press, New York, 1973.

7-74 Tebbens, B. D.: "Gaseous Pollutants in the Air," in Arthur C. Stern (ed.), *Air Pollution*, vol. I, Academic Press, New York, 1968.

7-75 U.S. Environmental Protection Agency: *Air Pollution and Your Health*, Office of Public Awareness OPA 54/8, Washington, D.C., 1979.

7-76 ———: *Cleaning the Air*, publ. no. 0-294-335, U.S. Government Printing Office, Washington, D.C., 1979.

7-77 ———: *Compilation of Air Pollution Emission Factors*, 3d ed., Office of Air Quality Planning and Standards publ. no. AP-42, Research Triangle Park, N.C., 1977.

7-78 ———: *Mobile Source Emission Factors*, Office of Transportation and Land Use Policy publ. no. EPA-400/9-78-005 and 006, Washington, D.C., 1978.

7-79 ———: *Trends in the Air Quality of the Nation*, publ. no. 0-221-430, U.S. Government Printing Office, Washington, D.C., 1977.

7-80 ———: *Trends in the Quality of the Nation's Air*, Office of Public Affairs publ. no. A-107, Washington, D.C., 1975.

7-81 Wark, Kenneth, and Cecil F. Warner: *Air Pollution: Its Origin and Control*, Dun-Dunnelley, New York, 1976.

7-82 Williamson, S. J. *Fundamentals of Air Pollution*, Addison-Wesley, Reading, Mass., 1973.

7-83 Wolf, Phillip C.: "Carbon Monoxide Measurement and Monitoring in Urban Air," *Env Sci Tech*, **5**(3):231 (1971).

7-84 Woolf, C.: Letter to the editor, *Arch Env Health*, **18**:715 (1969).

CHAPTER

EIGHT

METEOROLOGY AND NATURAL PURIFICATION PROCESSES

Pollution problems arise from the confluence of atmospheric contaminants, adverse meteorological conditions, and, at times, certain topographical conditions. The air pollution episodes mentioned in the previous chapter all involved meteorological conditions that restricted dispersion of contaminants, causing them to accumulate at harmful levels.

Because of the close relationship that exists between air pollution and certain atmospheric conditions, it is necessary for the environmental engineer to have a thorough understanding of meteorology. Even the most cursory exploration of the conditions which prevail in the greater Los Angeles basin, in the metropolitan Denver area, in Athens, or in any smog-troubled city will give a fair understanding of the cause-effect relationship between meteorological and topographical conditions and air pollution. That relationship will be explored in depth here, with special emphasis being given to the effect of meteorological elements on the dispersion of pollutants in the atmosphere and, conversely, to the influence of atmospheric contaminants on meteorological conditions.

Elemental Properties of the Atmosphere

The preceding chapter discussed the composition and structure of the earth's atmosphere. The source of all meteorological phenomena is a basic, but variable, ordering of the elemental properties of that atmosphere—heat, pressure, wind, and moisture. All weather, including pressure systems, wind speed and direction, humidity, temperature, and precipitation, ultimately results from variable relationships of heat, pressure, wind, and moisture.

8-1 SCALES OF MOTION

The interaction of these four elements may be observed on several different levels or scales. These *scales of motion* are related to mass movements of air which may be global, continental, regional, or local in scope. According to their geographic range of influence, the scales of motion may be designated as *macroscale*, *mesoscale*, or *microscale*. [8-30]

Macroscale

Atmospheric motion on the macroscale involves the planetary patterns of circulation, the grand sweep of air currents over hemispheres. These phenomena occur on scales of thousands of kilometers and are exemplified by the semipermanent high- and low-pressure areas over oceans and continents.

The sun's rays heat the earth near the equator to a greater extent than at the poles. If the rotation of the earth were discounted, the heated air at the equator would rise, and cool air from the poles would move in to take its place. This would set up two theoretical cells, involving only longitudinal motion [8-18], as shown in Fig. 8-1.

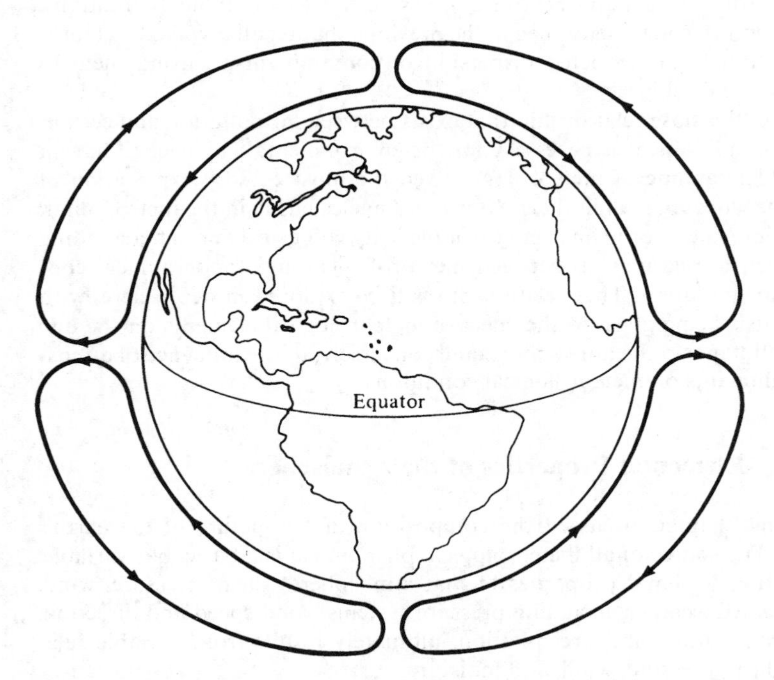

Figure 8-1 Halley's suggestion for general global circulation. (*From Linsley* [*8-18*].)

However, the west-to-east rotation of the earth must be taken into account, since it has a profound effect on air currents, deflecting the winds to the right in the northern hemisphere and to the left in the southern hemisphere. The effect of the earth's rotation on wind velocity and direction is called the *Coriolis force*, and this force has major significance in the formation of weather. [8-7]

Thus, air movement on the global scale is not simply in longitudinal directions, for the dual effect of heat differential between poles and equator and of the rotation of the earth along its axis establishes a more complicated pattern of air circulation. As indicated in Fig. 8-2, the general global circulation pattern (macroscale) is composed of three cells of air movement in each hemisphere. [8-18] It is under this dual influence of thermal convection and the Corolis force that high- and low-pressure areas, cold or warm fronts, hurricanes, and winter storms are formed.

One of the primary elements influencing air mass movement on this scale is the distribution of land and water masses over the surface of the earth. The great variance between conductive capacities of land and ocean masses accounts for the development of many of our weather systems. Over land masses, atmospheric temperature rises rapidly in the presence of solar radiation (day), then drops with

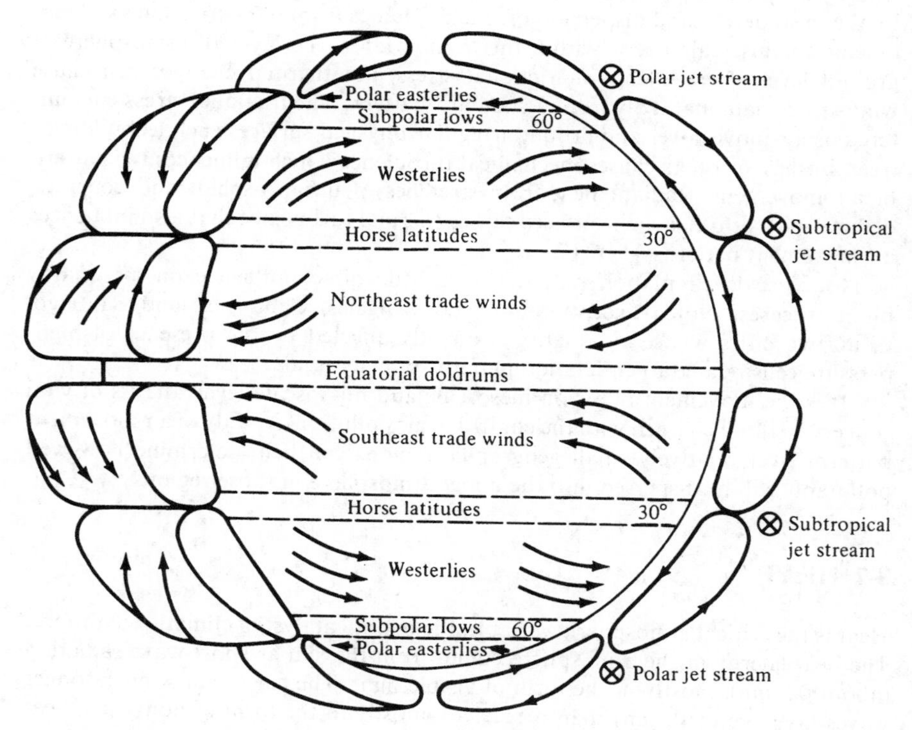

Figure 8-2 Schematic representation of general circulation. (*From Linsley* [*8-18*].)

equal rapidity in its absence (night), since landmasses quickly reradiate heat into the atmosphere. Conversely, air temperature over water rises and falls more slowly, since heat energy received by water penetrates to deeper layers than heat absorbed by land and is reradiated in lesser amounts.

Mesoscale

Secondary, or mesoscale, circulation patterns develop over regional geographic units, primarily because of the influence of regional or local topography. These phenomena occur on scales of hundreds of kilometers. Air movement on this scale is affected by the configuration of the earth's surface—the location of mountain ranges, of oceanic bodies, of forestation, and of urban development. Land or sea breezes, mountain or valley winds, migratory high- and low-pressure fronts, and urban heat islands are typical of the peculiarly local phenomena observable on this scale. [8-30]

Microscale

Microscale phenomena occur over areas of less than 10 k and can be exemplified by the meandering and dispersion of smoke plumes from industrial stacks. Phenomena on this scale occur within the *friction layer*, the layer of atmosphere at ground level where effects of frictional stress and thermal changes can cause winds to deviate markedly from a standard pattern. The frictional stress encountered as air moves over and around irregular physical surfaces such as buildings, trees, bushes, or rocks causes mechanical turbulence which influences the pattern of air movement. Radiant heat from stretches of urban asphalt and concrete, desert sands, or other such surfaces causes thermal turbulence that also influences air movement patterns. [8-13]

Macroscale circulation patterns have little direct influence on air quality in most cases. Noted exceptions are Los Angeles, California, and Santiago, Chili, two cities whose air quality is directly affected by the presence of high-pressure cells related to circulation patterns of macroscale.

It is the movement of air on mesoscale and microscale levels that is of vital concern to those charged with the control of air pollution. A study of air movement patterns over relatively small geographic regions can help determine how well pollutants will be dispersed into the upper atmosphere in those regions.

8-2 HEAT

Heat is the critical atmospheric variable, the major catalyst of climatic conditions. The heat energy in the atmosphere comes from the sun as short-wave radiation (about 0.5 μm), mostly in the form of visible light. The earth emits much longer waves (average of 10 μm) than it receives, mostly in the form of nonvisible heat radiation.

Some of the solar rays never reach the earth at all but are reflected back to space by individual particles in the air and by clouds. Solar rays may also be reflected back to space by the ground itself, with surface characteristics being a major factor in the rate of reflection. For example, desert sands, snow, and ice have a high rate of reflection, while forests and cultivated fields have a low rate.

Some of the sun's rays are scattered by intervening air molecules. It is this scattering of rays of different wave lengths that gives a clear sky its deep blue color. Scattering is more intense as the sun moves near the horizon, and it is this phenomenon that produces red sunrises and sunsets.

Some of the sun's rays are absorbed by ozone, water vapor, carbon dioxide, dust, and clouds of the lower atmosphere, but the earth's surface is the prime absorber of solar energy. Thus the troposphere is heated primarily from the ground, not from the sun. [8-29]

Tropospheric Heating

Four important ways in which heat transfer occurs in the troposphere are through the "greenhouse effect", the condensation-evaporation cycle, conduction, and convection.

Greenhouse effect As noted earlier, solar energy (light radiation) absorbed by the earth is converted to heat energy and emitted into space as long-wave (heat) radiation. Although water vapor and carbon dioxide are transparent to short-wave radiation, they are nearly opaque to long-wave radiation. Thus, much of the earth's reradiation is retained, raising the temperature of the atmosphere. This phenomenon is known as the *greenhouse effect*, taking its name from the principle of greenhouse construction, where glass operates in a fashion similar to carbon dioxide and water vapor, allowing solar rays to pass unhindered into the greenhouse, but blocking reverse radiation. [8-29]

Evaporation-condensation cycle Evaporation of water requires expenditure of energy, and the needed energy is absorbed from the atmosphere and stored in water vapor. Upon condensation, this heat energy is released. Because evaporation usually takes place on or near the earth's surface, while condensation normally occurs in the upper regions of the troposphere, the evaporation-condensation process tends to move heat from lower regions to higher regions.

On the macroscale, latent heat is transported from latitude belts where there is substantial precipitation. Below about 22° latitude, water vapor and latent heat are carried equatorward, while at higher latitudes they are carried poleward. [8-2]

Conduction Transfer of heat from earth to atmosphere is also accomplished through the process of *conduction*, heat transfer by direct physical contact of air and earth. As parcels of air move downward, they come into contact with the warmer ground and take heat from the earth into the atmosphere.

Convection *Convection*, a process initiated by the rising of warm air and the sinking of cold air, is a major force in transferring heat from earth to troposphere. As can be seen in Fig. 8-2, convection is a primary factor in movement of air masses on the macroscale.

Temperature Measurement

Maximum, minimum, and average temperatures are generally recorded at weather stations, and normal daily temperatures for a specific region can be calculated by averaging daily temperatures over a 10-, 20-, or 30-yr period. A temperature designation of particular interest to the environmental engineer is the *degree-days* of an area, since this figure is a measure of heating and fuel requirements and hence of air pollution potential from the burning of fossil fuels. The heating-degree-day of a region is calculated by subtracting the average daily temperatures for a year from a preselected "comfortable" temperature usually 18°C. The number of degrees by which the average daily temperature falls below this standard temperature yields the heating-degree-days for that region. [8-2]

Lapse Rates

In the troposphere, the temperature of the ambient air usually decreases with an increase in altitude. This rate of temperature change is called the *lapse rate*. This rate can be determined for a particular place at a particular time by sending up a balloon equipped with a thermometer. The balloon moves *through* the air, not with it, and the temperature gradient of ambient air, which the rising balloon measures, is called the *ambient lapse rate*, the *environmental lapse rate*, or the *prevailing lapse rate*. [8-2]

A specific parcel of air whose temperature is greater than that of the ambient air tends to rise until it reaches a level at which its own temperature and density equal that of the atmosphere that surrounds it. Thus, a parcel of artificially heated air (e.g., stack gas or automobile exhaust) rises, expands, becomes lighter, and cools. The rate at which the temperature decreases as the parcel gains altitude (the lapse rate) may be considerably different from the ambient lapse rate of the air through which the parcel moves. Thus, it is necessary to distinguish between the temperature decrease associated with the ambient lapse rate and the internal temperature decrease which occurs within a rising parcel of air or other gas.

The lapse rate for the rising parcel of air may be determined theoretically.* For calculation purposes, the cooling process within a rising parcel of air is assumed to be *adiabatic* (i.e., occurring without the addition or loss of heat). Under adiabatic conditions, a rising parcel of air behaves like a rising balloon, with the air in that distinct parcel expanding as it encounters air of lesser density until its own density

* See Petterssen [8-27], p. 106, for a discussion of the adiabatic lapse rate.

is equal to that of the atmosphere which surrounds it. This process is assumed to occur with no heat exchange between the rising parcel and the ambient air.

Expansion of air against its surroundings, like all other work, requires energy. As long as the rising parcel is very close to the earth's surface, it may receive some heat energy from the earth. As soon as it rises beyond that energy source, it must rely on its own store of heat for energy. An internal cycle is thus established; temperature within the air parcel decreases as heat energy is expended. [8-2] Since this process involves no transfer of heat from the rising parcel to the atmosphere which surrounds it, it is called *adiabatic cooling*.

Utilizing two basic concepts of physics, the ideal-gas law and the law of conservation of energy, it is possible to establish a mathematical ratio expressing temperature change against altitude gain under adiabatic conditions. [8-27] This rate of decrease is termed the *adiabatic lapse rate*. Dry air, expanding adiabatically, cools at 9.8°C per kilometer, the dry adiabatic lapse rate. [8-32] In wet, as in dry adiabatic process, a saturated parcel of air rises and cools adiabatically, but a second factor affects its temperature. Latent heat is released as water vapor condenses within the saturated parcel of rising air. Temperature changes of the parcel are then due to liberation of latent heat as well as to expansion of the air. Wet adiabatic lapse rate (6°C/km) is thus less than dry adiabatic lapse rate. [8-27] Since a rising parcel of effluent gases would seldom be completely saturated or completely dry, the adiabatic lapse rate generally falls somewhere between these two extremes.

Stability

Ambient and adiabatic lapse rates are a measure of atmospheric stability. Since the stability of the air reflects the susceptibility of rising air parcels to vertical motion, consideration of atmospheric stability or instability is essential in establishing the dispersion rate of pollutants.

The atmosphere is said to be *unstable* as long as a rising parcel of air remains warmer (or descending parcel remains cooler) than the surrounding air, since such a parcel will continue to accelerate in the direction of the displacement.

Conversely, when a rising parcel of air arrives at an altitude in a colder and denser state than the surrounding air, the resultant downward buoyancy force pushes the displaced parcel of air earthward and away from the direction of displacement. Under such conditions, the atmosphere is said to be *stable*. [8-2]

Stability is a function of vertical distribution of atmospheric temperature, and plotting the ambient lapse rate against the adiabatic lapse rate can give an indication of the stability of the atmosphere. Though dry, moist, or wet adiabatic lapse rates may be used in such a comparison, the dry adiabatic lapse rate (9.8°C/km) is used in Fig. 8-3 as the measure against which several possible ambient lapse rates are plotted. Thus, in Fig. 8-3, the boundary line between stability and instability is the dry adiabatic lapse rate.

When the ambient lapse rate exceeds the adiabatic lapse rate, the ambient lapse rate is said to be *superadiabatic*, and the atmosphere is highly unstable. When

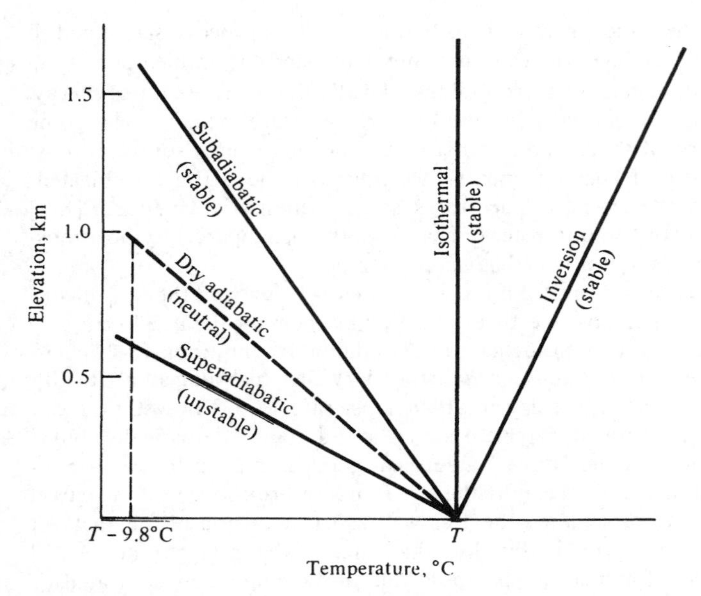

Figure 8-3 Relationship of the ambient lapse rates to the dry adiabatic rate.

the two lapse rates are exactly equal, the atmosphere is said to be *neutral.* When the ambient lapse rate is less than the dry adiabatic lapse rate, the ambient lapse rate is termed *subadiabatic* and the atmosphere is stable. If air temperature is constant throughout a layer of atmosphere, the ambient lapse rate is zero, the atmospheric layer is described as *isothermal,* and the atmosphere is stable. [8-2]

When temperature of the ambient air increases, rather than decreases, with altitude, the lapse rate is negative, or inverted, from the normal state. Negative lapse rate occurs under conditions commonly referred to as an *inversion,* a state in which warmer air blankets colder air. Thermal or temperature inversions represent a high degree of atmospheric stability. [8-2]

There are two types of inversions. The first is a *radiation inversion,* a phenomenon arising from the unequal cooling rates of the earth and the air above the earth. This type of inversion may extend a few hundred meters into the friction layer and is characteristically a nocturnal phenomenon that breaks up easily with the rays of the morning sun. A radiation inversion prompts the formation of fog and simultaneously traps gases and particulates, creating a concentration of pollutants.

The second type of inversion is the *subsidence inversion* that is usually associated with a high-pressure system. Such an inversion is caused by the characteristic sinking motion of air in a high-pressure cell. Air circulating around a stationary high descends slowly. As the air descends, it is compressed and heated, forming a blanket of warm air over the cooler air below and thus creating an inversion that prevents further vertical movement of air. This type of inversion may extend through the friction layer to heights of over 1500 m.

8-3 PRESSURE

In addition to heat, pressure is an important variable in meteorological phenomena. Because air has weight, the whole atmosphere presses down upon the earth beneath it. This pressure is commonly measured with a mercury barometer, an instrument which measures the weight, over a unit area, of a column of air extending to the top of the atmosphere. On the average, the atmosphere at latitude 45° and at a temperature of 0°C (32°F) is equivalent to a column of mercury 760 mm (29.9 in) high, and this, by international agreement, is called *one standard atmosphere*, or roughly one *bar*. Meteorologists usually express pressure in *millibars*, and one standard atmosphere is 1103 millibars. [8-29]

On weather maps such as the one shown in Fig. 8-4 pressure distribution throughout the atmosphere is represented by *isobars*, lines connecting points of equal atmospheric pressure. These lines delineate high- and low-pressure cells that influence the development of major weather systems.

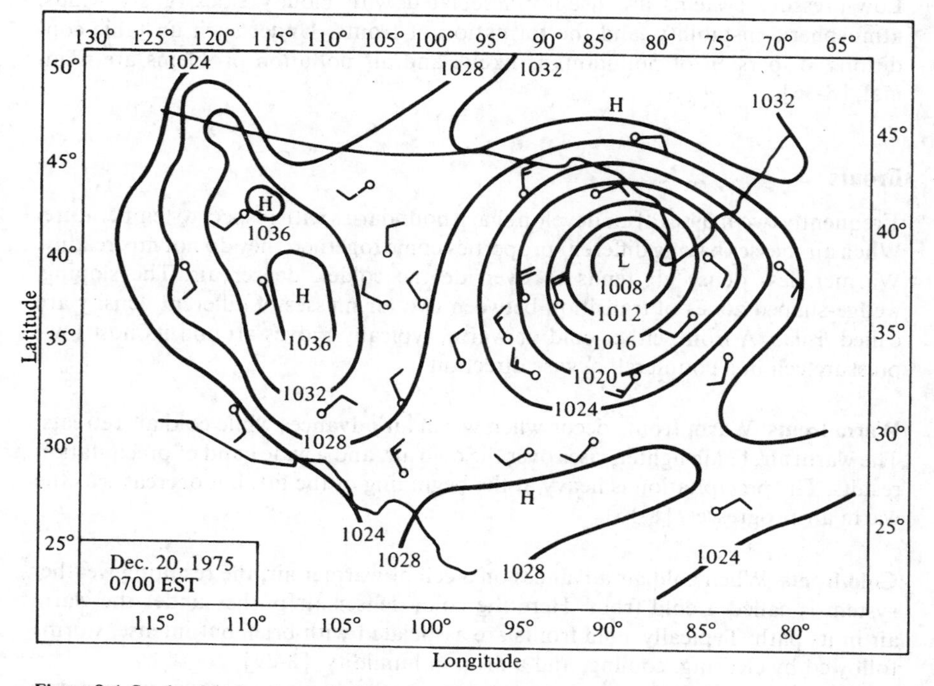

Figure 8-4 Sea-level isobars over the United States and observed wind velocities at selected stations. (*From Battan* [*8-2*].)

Pressure Systems

Pressure patterns over the earth are in constant flux as air pressure rises in some regions and falls in others. The location of continents, the differences in surface roughness and radiation, wind energy, and global circulation patterns all combine to force development of high- and low-pressure systems or cells. The circulation or movement of these high- and low-pressure systems is responsible for many weather changes. [8-36]

High-pressure systems High-pressure systems are related to clear skies, light winds, and atmospheric stability. In a typical high-pressure system in the northern hemisphere, the vertical motion of air is downward and the horizontal motion is clockwise. High-pressure systems reflect the relative uniformity of air masses. Under such stable conditions, temperature and humidity vary little over great areas, and any weather change is gradual. Under these stable conditions, dispersion is restricted and pollutants are likely to build to undesirable levels.

Low-pressure system In a typical low-pressure system in the northern hemisphere, horizontal air movement is counterclockwise and vertical movement is upward. Low-pressure systems are usually associated with cloudy skies, gusty winds, atmospheric instability, and the formation of fronts. Under such unstable conditions, dispersion of pollutants is likely, and air pollution problems are minimal. [8-36]

Fronts

Frequently two masses of air develop sharp boundaries with respect to temperature. When air masses having different properties come together, they do not mix readily. Warmer, less dense air tends to override the colder, denser air. The sloping, wedge-shaped zones of transition between two air masses of different density are called *fronts*. A front, either cold or warm, typically moves around its host low-pressure cell in a counterclockwise direction.

Warm fronts Warm fronts occur when warm air advances while cold air retreats. The warm air, being lighter, rises over the cold air, and a wide band of precipitation results. The precipitation is heavy at the beginning of the lift, but decreases as the warm air progresses. [8-16]

Cold fronts When cold air advances on a cell of warmer air, the resulting weather system is called a cold front. Here the cold, denser air pushes under the warm air in its path. Typically, cold fronts are associated with brief, but intense, storms followed by clearing, cooling, and a drop in humidity. [8-29]

When the transition zone between warm and cold air does not move one way or the other, that zone is called a *stationary front*. [8-2]

8-4 WIND

Wind is simply air in motion. On the macroscale, the movement originates in unequal distribution of atmospheric temperature and pressure over the earth's surface and is significantly influenced by the earth's rotation. The direction of wind flow is characteristically from high pressure to low, but the Coriolis force tends to deflect air currents out of these expected patterns.

On the mesoscale and microscale, topographical features critically influence wind flow. Surface variations have an obvious effect on the velocity and direction of air movement. Monsoons, sea and land breezes, mountain-valley winds, coastal fog, windward precipitation systems, urban heat islands—all are ready examples of the influence of regional and local topography on atmospheric conditions.

Wind channeling in river valleys is a phenomenon affecting many large urban areas. Updraft or up-mountain breezes, which predominate during warmer portions of the day, and nocturnal downdraft are weather phenomena peculiar to mountain valleys.

The variance of the conductive capacity of land and water accounts for another effect of topography on wind direction. Because land warms and cools more rapidly than neighboring bodies of water, the characteristic coastal winds fall into a pattern of daytime sea breezes and evening land breezes.

In the friction layer at the earth's surface, winds are often gusty and changeable, primarily due to locally generated mechanical or thermal turbulence. [8-15] Once free of the impediments of the friction layer, velocity of air movement generally increases, and winds aloft usually blow more steadily and more parallel to the isobars than do those in lower regions. An empirical formula relating wind speed to height in the friction layer, that zone of air beneath 700 to 1000 m (2,000 to 3000 feet), is

$$\frac{v}{v_0} = \frac{z^k}{z_0} \tag{8-1}$$

where v = wind speed at height z, m/s
v_0 = wind speed at anemometer level z_0, m/s
k = coefficient, approximately $\frac{1}{7}$

Wind speed is usually measured by an anemometer, an instrument typically consisting of three or four hemispherical cups arranged around a vertical axis. The faster the rate of rotation for the cups, the higher the speed of the wind. A three-cup anemometer and a wind vane for observation of wind direction are shown in Fig. 8-5.

When data for wind speed and direction are placed on a *wind rose*, they yield a graphic picture of the direction, frequency, and velocity of the winds in a particular location. For example, the wind rose shown in Fig. 8-6 indicates that the greatest percentage of winds in that month were from the southwest and that speeds of up to 8 m/s were recorded for winds from that direction. Relatively infrequent easterly winds were occasionally as high as 8 m/s. [8-38]

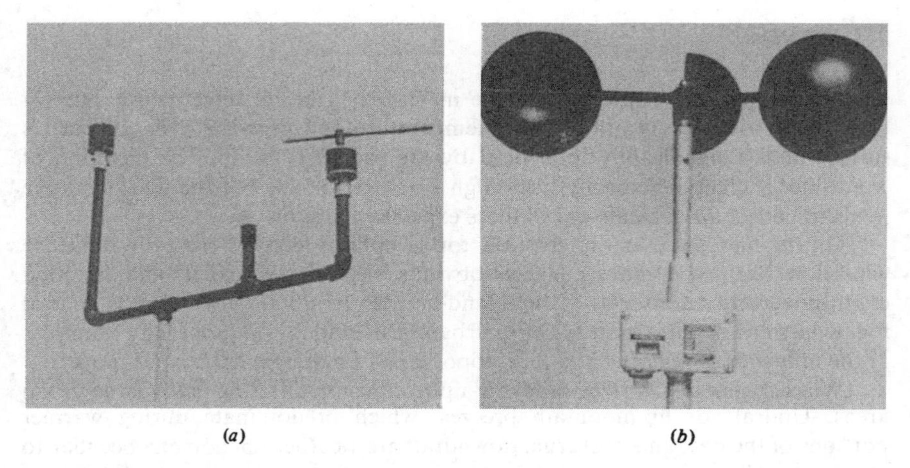

Figure 8-5 Wind measurement instruments: (*a*) three-cup anemometer and wind vane; (*b*) close-up of three-cup anemometer.

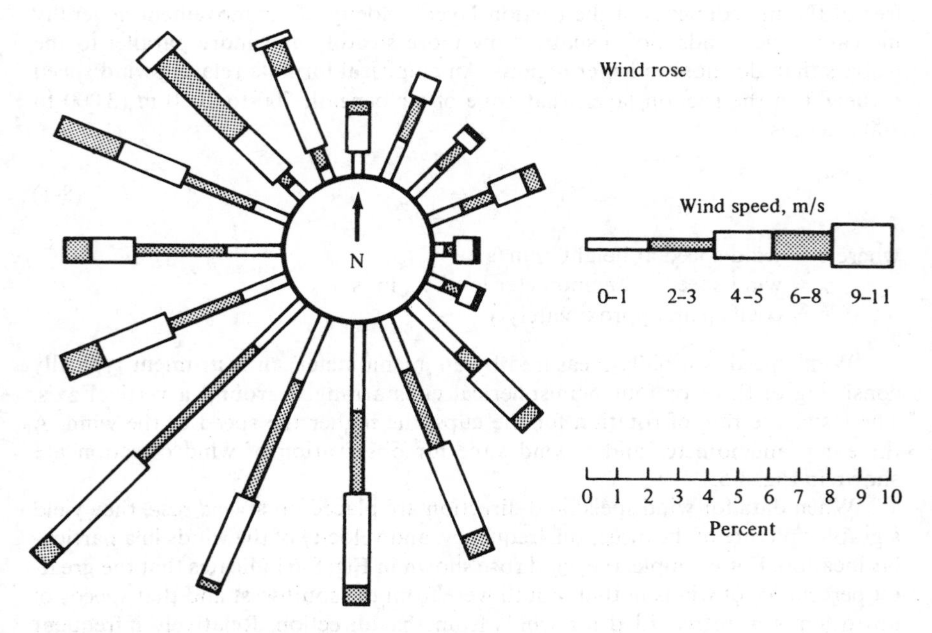

Figure 8-6 Wind rose. (*From Meteorological Aspects . . .* [8-22].)

8-5 MOISTURE

Evaporation to condensation to precipitation is a constantly repeating cycle in our environment, as discussed in Chap 2. Moisture is first transferred from the earth's surface into the atmosphere. This water vapor then condenses and forms clouds. The cycle completes itself as the condensed vapor is returned to the earth's surface in some form of precipitation—rain, hail, snow, or sleet.

Topography plays an important role in moisture distribution. Mountains tend to force the rise of moisture-laden air, resulting in heavier precipitation on the windward side of a range.

8-6 RELATIVE HUMIDITY

The amount of water vapor present in the atmosphere is measured in terms of humidity. The higher the temperature of the air, the more water vapor it can hold before it becomes saturated. At ground level, a temperature increase of 11.1°C roughly doubles the moisture capacity of the atmosphere. Saturated air at 16°C thus contains approximately twice as much water vapor as saturated air at 5°C.

Relative humidity is measured by an instrument called a *psychrometer*. The dry-bulb thermometer of a psychrometer indicates the temperature of the air, while the wet-bulb thermometer measures the amount of cooling that occurs as the moisture on the bulb evaporates. With the difference in the two readings (called the *wet-bulb depression*) and the dry-bulb temperature, one can obtain relative humidity readings from psychrometric tables. [8-2]

Influence of Meteorological Phenomena on Air Quality

The meteorological phenomena which have been briefly discussed to this point exert a critical influence on air quality. As noted earlier, under adverse atmospheric conditions, the presence of air contaminants gives rise to problems of air pollution. Understanding the relationship between atmospheric conditions and air pollution problems can enable the environmental engineer to minimize the adverse effects of the relationship.

As a river or stream is able to absorb a specific contaminant load without undesirable results, so the atmosphere can assimilate a certain amount of air contamination without ill effects. Dilution of air contaminants in the atmosphere is an important process in preventing undesirable levels of pollutants in the ambient air. Atmospheric dispersion of air contaminants is the result of ventilation, atmospheric turbulence, and molecular diffusion. However, gaseous and particulate air contaminants are primarily dispersed into the ambient air through wind action and atmospheric turbulence, much of it on the microscale level.

8-7 LAPSE RATES AND DISPERSION

By comparing the ambient lapse rate to the adiabatic lapse rate, it may be possible to predict what will happen to gases emitted from a stack. In the following examples, the *dry* adiabatic lapse rate is used, but prediction of plume patterns is more likely to be accurate if the moisture content of the stack gas is taken into account when ambient and adiabatic lapse rates are compared.

When the ambient lapse rate is superadiabatic (greater than the adiabatic), the turbulence of the air itself causes the atmosphere to serve as an effective vehicle of dispersion. As indicated in Fig. 8-7*a*, the resultant plume is designated a "looping" plume. In this highly unstable atmosphere, the stream of emitted pollutants undergoes rapid mixing, and any wind causes large eddies which may carry the entire plume down to the ground, causing high concentrations close to the stack before dispersion is complete. [8-26] In areas where conditions make looping plumes likely, higher stacks may be needed to prevent premature contact with the ground.

When the ambient lapse rate is equal to or very near the dry adiabatic lapse rate, the plume issuing from a single chimney or smokestack tends to rise directly into the atmosphere until it reaches air of density similar to that of the plume itself. This type emission, called a *neutral plume*, is seen in Fig. 8-7*b*. [8-11]

However, this neutral plume tends to "cone" (see Fig. 8-7*c*) when wind velocity is greater than 20 mi/h [8-9] and when cloud cover blocks solar radiation by day and terrestrial radiation by night.

When the ambient lapse rate is subadiabatic (less than the dry adiabatic), the atmosphere is slightly stable. Under such conditions, there is limited vertical mixing, and the probability of air pollution problems in the area is increased. The typical plume in such a situation is said to be "coning," since it assumes a conelike shape about the plume line, as shown in Fig. 8-7*c*. While the dispersion rate is faster for a looping plume, the distance at which a coning plume first reaches the ground is greater. [8-26]

When the lapse rate is negative, as in the presence of an inversion, the dispersion of stack gas is minimal, because of lack of turbulence. In the extremely stable air, a plume spreads horizontally, with little vertical mixing, and is said to be "fanning" (Fig. 8-7*d*), and in flat country such a plume may be visible for miles downwind of its source. [8-9] In areas where radiation inversions are common, construction of stacks high enough to allow for discharge of emissions above the inversion layer is recommended. This solution is not practical for subsidence inversions, since they usually extend to much greater heights.

Extenuating circumstances can often alleviate or aggravate the pollution possibilities accompanying negative lapse rate conditions. For instance, when the lapse rate is superadiabatic above the emission source and inversion conditions exist below the source, the plume is said to be "lofting." As shown in Fig. 8-7*e*, a lofting plume has minimal downward mixing, and the pollutants are dispersed downwind without any significant ground-level concentrations. As long as stack height remains above the inversion, lofting will continue, but lofting is usually a

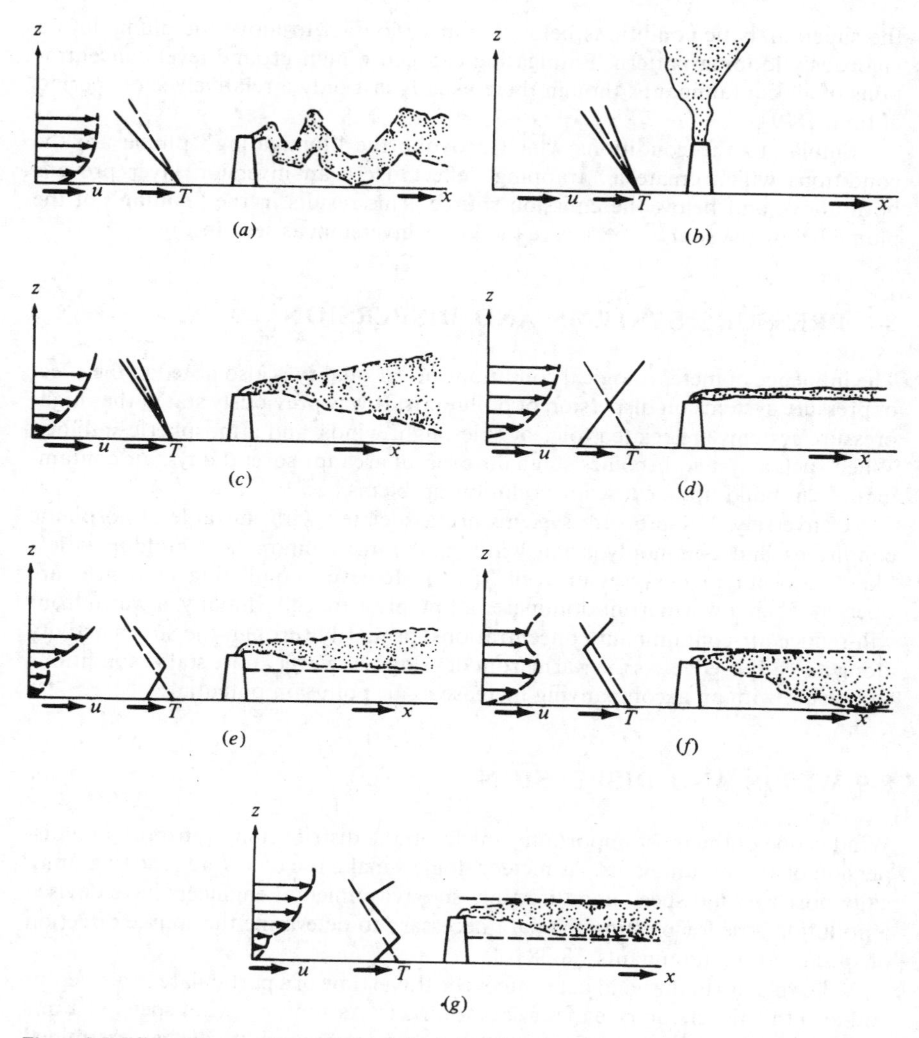

Figure 8-7 Effect of lapse rate on plume behavior (*a*) looping, (*b*) neutral, (*c*) coning, (*d*) fanning, (*e*) lofting, (*f*) fumigating, and (*g*) trapping.

transitional situation. If the inversion grows past the stack height, lofting will change to fanning. [8-26]

When an inversion layer occurs a short distance above the plume source and superadiabatic conditions prevail below the stack, the plume is said to be "fumigating" (Fig. 8-7*f*). Fumigating often begins when a fanning plume breaks up into a looping plume, as when morning sun breaks up a radiation inversion and the superadiabatic conditions below the inversion act to move the plume into a vigorously looping pattern. Fumigating can cause high ground-level concentrations of air contaminants, though these usually last only a relatively short period of time. [8-9]

Similar to the conditions which provoke the "fumigating" plume are the conditions which create a "trapping" effect. Here an inversion layer prevails both above and below the emission source. This results in the "coning" of the plume below the source and above the lower inversion, as seen in Fig. 8-7*g*.

8-8 PRESSURE SYSTEMS AND DISPERSION

The influence of meteorological conditions on air quality is also noted in the effect of pressure systems on dispersion of pollutants. It was previously stated that high-pressure systems are related to clear skies, light winds, and atmospheric stability. When such a system becomes stagnant over an area for several days, air contaminants can build up to cause air pollution problems.

Conversely, low-pressure systems are associated with unstable atmospheric conditions and commonly bring winds and rain; contaminant buildup is less likely to occur in low-pressure cells. [8-6] However, conflicting influences are operant when a warm front dominates a low-pressure cell. Initially, a warm front will reduce air-contaminant concentrations, primarily through the storm activity along its leading edge. As the warm front develops, however, more stable conditions will result, with an accompanying increase in air pollution potential.

8-9 WINDS AND DISPERSION

Wind is one of the most important vehicles in the distribution, transport, and dispersion of air contaminants. As meteorologists make use of a wind rose to graphically portray wind speed and direction, so environmental engineers have devised a pollution rose for plotting the data necessary to determine the source direction of specific air contaminants. [8-38]

The velocity of the wind determines the travel time of a particulate to a receptor and also the dispersion rate of air contaminants. Assuming a wind speed of 1 m/s and a source emitting 5 g of air contaminants per second, it can be determined that contaminant concentration in this plume is 5 g/m^3. If the wind velocity increases to 5 m/s, then the contaminant concentration from the same source is reduced to a single gram per cubic meter. Concentration of air contaminants in a plume is inversely proportional to wind velocity.

Frequently, topographic conditions will have a profound effect on winds and thus on air quality. This is seen in the wind channeling effect of a valley. Here, because of a particular geographic structure, air movement is predominantly up or down the valley, and dispersion of ambient contaminants outside the valley may be limited.

The differing conductive capacity of landmass and water mass gives rise to the alternating flow of sea breezes and land breezes, a pattern which can contribute to air pollution problems. The Los Angeles area frequently experiences this pattern of air movement, which carries the contaminants toward the ocean in the evening, only to return the polluted air to the urban basin when the direction of the wind shifts back toward land with the morning sun.

8-10 MOISTURE AND DISPERSION

Moisture content and form in the atmosphere can have a profound effect upon the air quality of a region. The presence and amount of water vapor in the atmosphere affects the amount of solar radiation received and reflected by earth. Water vapor serves to scatter or absorb radiation energy, and hence humidity has a major influence on air quality.

Precipitation serves as a cleansing agent for the atmosphere, removing particulates and soluble gases in a process called *washout*. Though the beneficial effects of washout are obvious, there are also some detrimental effects. When rainfall removes sulfur dioxide (SO_2) from the air, it may react with the water to form H_2SO_3 (sulfurous acid) or H_2SO_4 (sulfuric acid). The resultant "acid rain" increases the rate of corrosion where air contaminants are present. [8-17] In addition, the unnaturally low pH of such rains may change the pH of rivers and streams and thus influence the species of algae and other plant life which predominate in those bodies of water. [8-37]

8-11 MODELING

A knowledge of meteorological phenomena and an understanding of the variable factors that build weather systems can be used as a basis for forecasting air pollution potential and for devising air-pollution prevention and abatement programs. With information from an emission inventory and with atmospheric dispersion rates, it is possible to estimate air contaminant levels with some degree of reliability.

Maximum Mixing Depth (MMD)

Operations likely to produce significant amounts of air pollution should be limited to those areas in which atmospheric dispersion processes are most favorable. A determination of the maximum mixing depth of an ambient environment could help establish whether an area is a proper site for contaminant-causing human activities.

Maximum mixing depth (MMD) can be estimated by plotting maximum surface temperature and drawing a line parallel to the dry adiabatic lapse rate from the point of maximum surface temperature to the point at which the line

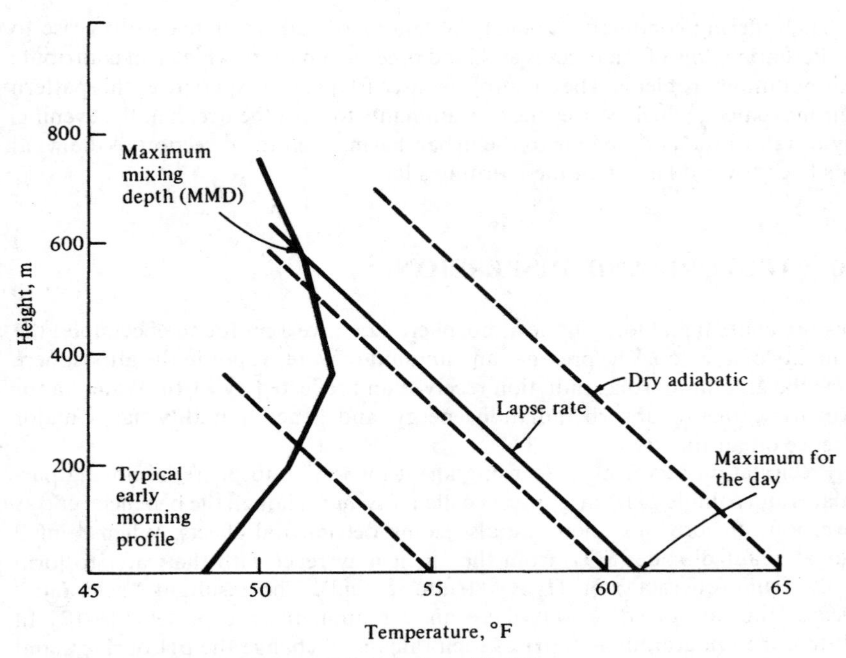

Figure 8-8 Maximum mixing depth. (*From Ledbetter* [8-15].)

intersects the ambient lapse rate for early morning. [8-15] In the instance plotted in Fig. 8-8, the MMD for the area was about 600 m.

Dispersion Models

Several empirical dispersion models have been developed. These models, or equations, are mathematical descriptions of the meteorological transport and dispersion of air contaminants in an area [8-5], and permit estimates of contaminant concentrations, either in the plume from an elevated or ground-level source. Among the most useful formula are those developed by Sutton, Bosanquet and Pearson, and Pasquill and Gifford. Most of the equations in use today are based on the following general equation which was suggested by Pasquill and modified by Gifford. [8-3, 8-4, 8-11, 8-13, and 8-25]

$$\frac{dC}{dt} = \frac{\partial x}{\partial x}\left(K_x \frac{\partial x}{\partial x}\right) + \frac{\partial}{\partial y}\left(K_y \frac{\partial x}{\partial y}\right) + \frac{\partial}{\partial z}\left(K_z \frac{\partial x}{\partial z}\right) \tag{8-2}$$

Equation (8-2) relates dispersion in the x (downwind) direction as a function of variables in all directions of a three-dimensional space. It assumes that the

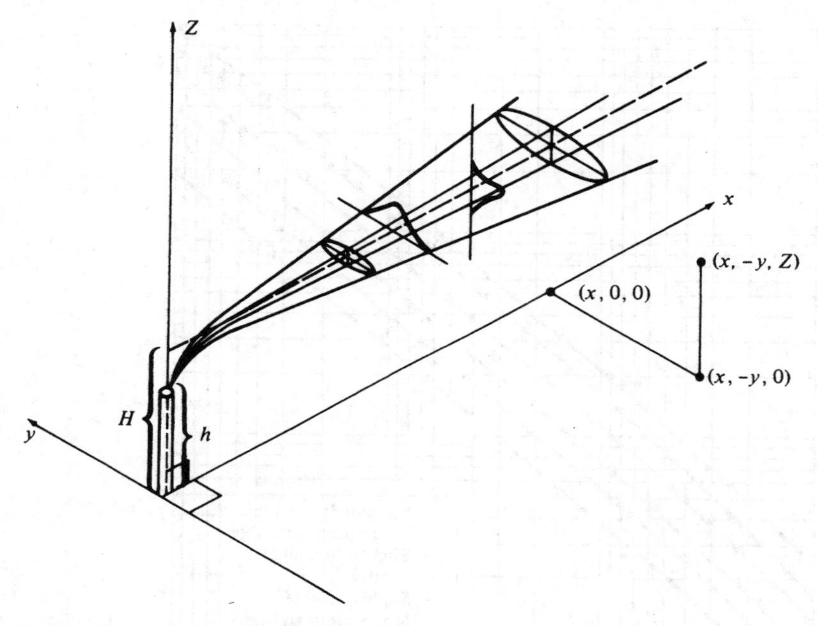

Figure 8-9 Coordinate system showing Gaussian distribution in the horizontal and vertical. (*From Turner* [8-34].)

plume has a Gaussian concentration distribution in both the z (vertical) and y (horizontal) directions, as shown in Fig. 8-9. [8-34]

The concentration (C) of a gas or aerosol ($<20\mu$) calculated at ground level for a distance downwind (x) is given by

$$C_{x,y} = \frac{Q}{\pi u \sigma_z \sigma_y} \exp\left[-\frac{1}{2}\left(\frac{H}{\sigma_z}\right)^2\right] \exp\left[-\frac{1}{2}\left(\frac{y}{\sigma_y}\right)^2\right] \tag{8-3}$$

where C = pollutant concentration, g/m³
Q = pollutant emission rate, g/s
π = pi, 3.14159
u = mean wind speed, m/s
σ_y = standard deviation of horizontal plume concentration, evaluated in terms of downwind distance x, m, (as shown in Fig. 8-10)
σ_z = standard deviation of vertical plume concentration evaluated in terms of downwind distance x, m, (as shown in Fig. 8-11)
exp = base of natural logs, 2.71828183
H = effective stack height, m
x = downwind distance along plume mean centerline from point source, m
y = crosswind distance from the centerline of the plume, m

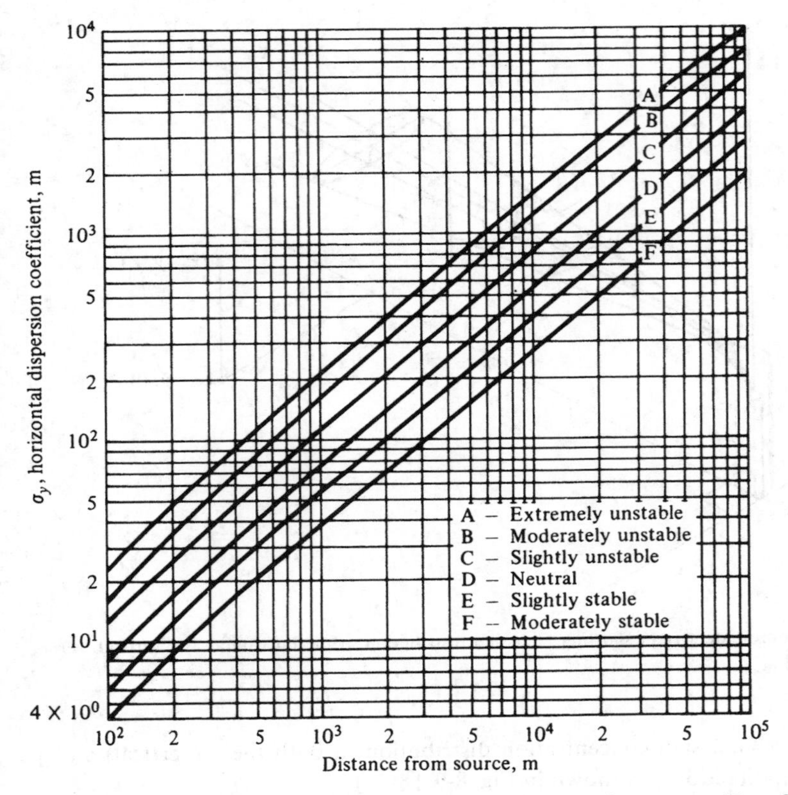

Figure 8-10 Lateral diffusion coefficient σ_y vs. downward distance from source. (*From Davis* [8-7].)

This equation may be simplified if concentrations along only the plume centerline are needed. In this case, $Y = 0$, and the equation is as follows:

$$C_{x,0} = \frac{Q}{\pi u \sigma_z \sigma_y} \exp\left[-\frac{1}{2}\left(\frac{H}{\sigma_z}\right)^2\right] \tag{8-4}$$

The equation may be further simplified if the effective stack height is zero, such as in a situation of ground-level burning.

$$C_{x,0} = \frac{Q}{\pi u \sigma_y \sigma_z} \tag{8-5}$$

Values for σ_y and σ_z are not only a function of downwind distances but are also a function of atmospheric stability. Values of σ_y and σ_z for various distances downwind (x), with various stability categories, are indicated in Figs. 8-10 and 8-11. Generalized categories are included in Table 8-1. [8-34]

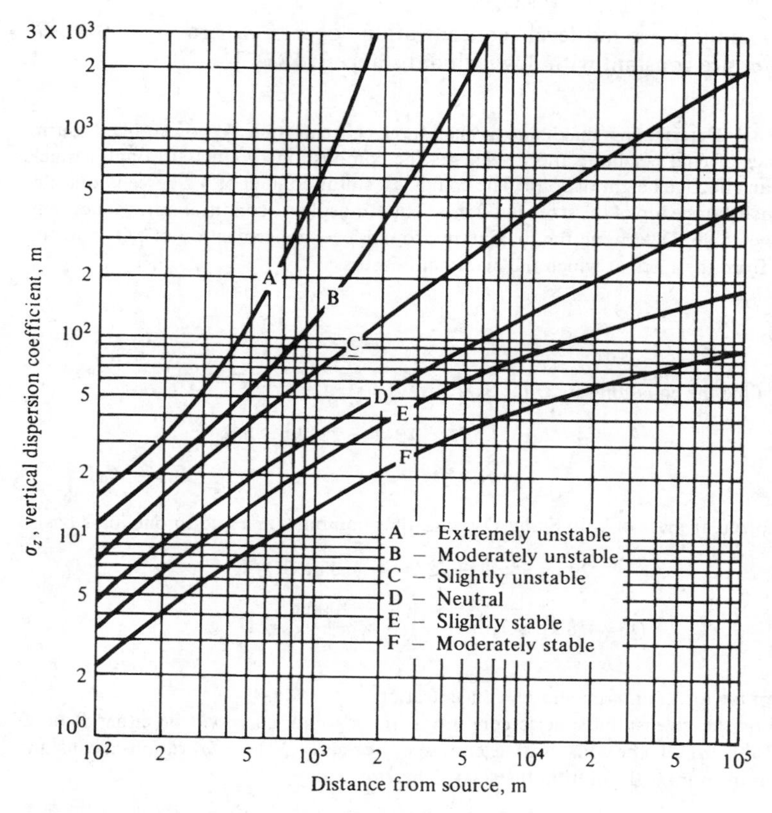

Figure 8-11 Vertical diffusion coefficient σ_z vs. downward distance from source. (*From Davis* [8-7].)

Table 8-1 Pasquill stability types

Surface wind speed, m/s	Day: Incoming solar radiation			Night	
	Strong	Moderate	Slight	Mostly overcast	Mostly clear
<2	A	A–B	B		
2	A–B	B	C	E	F
4	B	B–C	C	D	E
6	C	C–D	D	D	D
>6	C	D	D	D	D

A—Extremely unstable
B—Moderately unstable
C—Slightly unstable
D—Neutral
E—Slightly stable
F—Moderately stable

The maximum ground-level concentration occurs where $\sigma_z = 0.707H$, provided σ_z/σ_y are constant with downwind distance x. [8-22]

Example 8-1 Determining maximum ground-level concentration A power plant burns 5.45 tonnes of coal per hour and discharges the combustion products through a stack that has an effective height of 75 m. The coal has a sulfur content of 4.2 percent, and the wind velocity at the top of the stack is 6.0 m/s. The atmospheric conditions are moderately to slightly stable. Determine the maximum ground-level concentration of SO_2 and the distance from the stack at which the maximum occurs.

SOLUTION

1. Determine the emission rate Q for SO_2.

$$5.45 \times 10^3 \text{ kg/h} \times 0.042 = 229 \text{ kg S/h}$$

$$S + O_2 = SO_2$$

the molecular mass of both S and O_2 is 32, they combine on a one-to-one mass basis.

$$229 \text{ kg S} + 229 \text{ kg } O_2 = 458 \text{ kg } SO_2/\text{h}$$

$$Q = 458 \text{ kg } SO_2/\text{h} \times \frac{\text{h}}{3600 \text{ s}} \times \frac{1000 \text{ g}}{\text{kg}} = 127 \text{ g/s}$$

2. Determine location of maximum concentration.

For the given atmospheric conditions, the stability class will be either B or C (see Table 8-1). To be conservative, choose C. For class C, the σ_z/σ_y ratio is a constant for distances up to 1 km from the stack. Therefore,

$$\sigma_z = 0.707H = 0.707 \times 75 = 53 \text{ m}$$

From Fig. 8-11, σ_z reaches a value of 53 m at a distance of about 850 m from the stack with class C atmosphere. Thus,

$$x_{max} = 850 \text{ m}$$

3. Determine concentration at x_{max}.

From Fig. 8-10, $\sigma_y = 88$ m at $x = 850$ m.

$$C_{max} = \frac{127}{\pi \times 6 \times 53 \times 88} \exp - \frac{1}{2}\left(\frac{75}{53}\right)^2$$

$$= 5.31 \times 10^{-4} \text{ g/m}^3$$

$$= 531 \ \mu\text{g/m}^2$$

Example 8-2: Determining crosswind concentrations From the data in Example 8-1, determine the ground-level concentrations at a distance of 3.0 km downwind at the centerline of the plume and at a crosswind distance of 0.4 km on either side of the centerline.

SOLUTION

1. At 1.5 km;

$$\sigma_z = 170 \text{ m}$$

$$\sigma_y = 280 \text{ m}$$

2. The centerline concentration is

$$C_{(3,0)} = \frac{Q}{\pi u \sigma_z \sigma_y} \exp - \left[\frac{1}{2}\left(\frac{H}{\sigma_z}\right)^2\right]$$

$$= \frac{127}{\pi \times 6 \times 170 \times 280} \exp\left[-\frac{1}{2}\left(\frac{75}{170}\right)^2\right]$$

$$= 1.28 \times 10^4 \text{ g/m}^3$$

$$= 128\ \mu\text{g/m}^3$$

3. The concentration 0.4 km away from centerline is:

$$C_{(3,0.4)} = \frac{Q}{\pi u \sigma_z \sigma_y} \exp\left[-\frac{1}{2}\left(\frac{H}{\sigma_3}\right)^2\right] \exp\left[-\frac{1}{2}\left(\frac{Y}{\sigma_y}\right)^2\right]$$

$$= \frac{127}{\pi \times 6 \times 170 \times 280} \exp\left[-\frac{1}{2}\left(\frac{85}{170}\right)^2\right] \exp\left[-\frac{1}{2}\left(\frac{400}{280}\right)^2\right]$$

$$= 4.49 \times 10^{-5} \text{ g/m}^3$$

$$= 44.9\ \mu\text{g/m}^3$$

Stack Design

Meteorological data are necessary for the effective design of a stack. All factors must be weighed, including the fact that emissions from a tall stack designed to disperse contaminants into the upper atmosphere and away from the immediate area may result in fallout or washout far downwind. [8-23] Dispersion equations such as those in the foregoing section may be helpful. However, formulas alone are not enough, since many local variables must be considered if optimum stack design is to be achieved.

Location of nearby buildings may cause mechanical turbulence which may bring a plume to ground level, especially when the stack is downwind of the building and wind speeds are high. To avoid this problem, stacks should usually be designed 2 to $2\frac{1}{2}$ times the height of nearby structures. Since dispersion formulas commonly used are for flat, level terrain, irregular terrain must be taken into consideration when dispersion models are being adapted for use in specific cases. Heat islands and mechanical turbulence in metropolitan areas must also be considered. [8-9]

All of the above considerations concern single stacks designed for continuous emissions. When multiple stacks are preferred, still other factors must be taken into account. [8-32]. Different criteria may be used, too, for design of single stacks intended for short-term releases, for explosions, or for instantaneous release of nuclear fission products. [8-9, 8-31]

When the air contaminants are emitted from a stack, they rise above the stack before leveling out. The effective stack height H is not only the physical stack height h but includes the plume rise (Δh). [8-22]

$$H = h + \Delta h \tag{8-6}$$

See Fig. 8-12. Stack heights used in calculations such as those for Example 8-1 must be the effective stack height, and there are numerous equations for the calculation of plume height Δh. Holland's equation [8-22], often used for this determination, is given below.

$$\Delta h = \frac{v_s d}{u}\left[1.5 + \left(2.68 \times 10^{-3}\, p \frac{\Delta T d}{T_s}\right)\right] \tag{8-7}$$

where Δh = rise of plume above the stack, m
v_s = stack gas velocity, m/s
d = inside stack diameter, m
u = wind speed, m/s
p = atmospheric pressure, millibars
ΔT = stack gas temperature minus air temperature, K
T_s = stack gas temperature, K

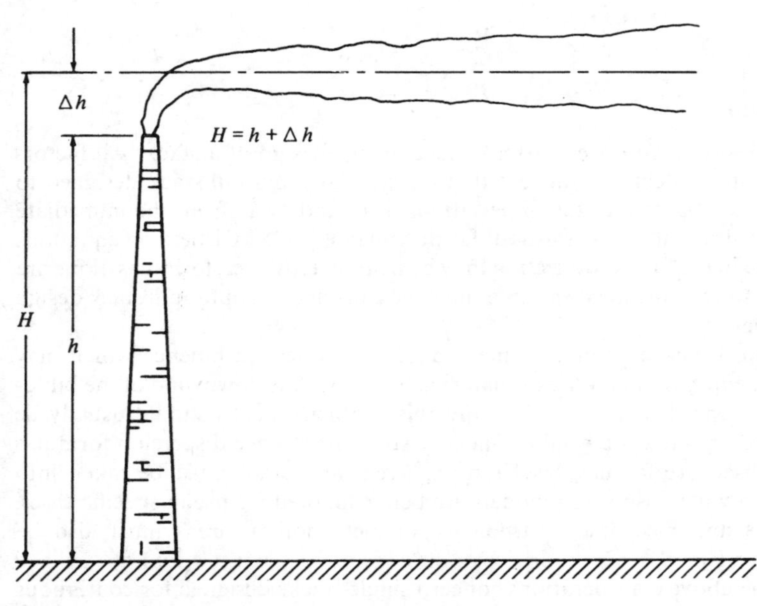

Figure 8-12 Effective stack height. (*From Smith* [*8-32*].)

The preceding calculations are suitable for neutral conditions. For unstable conditions, Δh should be increased by a factor of 1.1 to 1.2, and for stable conditions, Δh should be decreased by a factor of 0.8 to 0.9.

Davidson and Bryant [8-22] present another equation for plume rise.

$$\Delta h = d \left(\frac{v_s}{u}\right)^{1.4} \left(1 + \frac{\Delta T}{T_s}\right) \tag{8-8}$$

The terms in this equation have the same definition as those used in the Holland equation. Both equations frequently underestimate effective stack heights and provide a conservative estimate when used in the dispersion equation.

Example 8-3: Calculating effective stack height Determine the effective height of a stack given the following data.

(*a*) Physical stack is 203 m tall with 1.07 m inside diameter.
(*b*) Wind velocity is 3.56 m/s.
(*c*) Air temperature is 13°C.
(*d*) Barometric pressure is 1000 millibars.
(*e*) Stack gas velocity is 9.14 m/s.
(*f*) Stack gas temperature is 149°C.

SOLUTION

1. Convert temperatures to K.

$$T_a = 273 + 13 = 286$$

$$T_s = 273 + 149 = 422$$

2. Calculate ΔT.

$$\Delta T = 422 - 286 = 136 \text{ K}$$

3. Calculate ΔH using Eq. (8-7).

$$\Delta h = \frac{v_s d}{u}\left[1.5 + \left(2.68 \times 10^{-3} p \frac{\Delta T_d}{T_s}\right)\right]$$

$$= \frac{9.14 \times 1.07}{3.56}\left[1.5 + \left(2.68 \times 10^{-3} \frac{1000 \times 136 \times 1.07}{422}\right)\right]$$

$$= 6.6 \text{ m}$$

4. Calculate effective height.

$$H = h + \Delta h = 203 + 6.6 = 209.6 \text{ m}$$

Effects of Air Pollution on Meteorological Conditions

The interrelatedness of atmospheric conditions and air quality should be evident from the foregoing discussion of the effects that meteorological conditions can have upon the dispersion, concentration, or removal of atmospheric pollutants.

Since this is a bidirectional relationship, this chapter concludes with a discussion of situations in which polluted air modifies regional and global climatic conditions.

8-12 CHANGES ON THE MESOSCALE AND MICROSCALE

On the regional and local scale, reduced visibility, altered precipitation, and the "urban heat island" effect are among the changes in weather that have been attributed to air pollution.

Reduced Visibility

Reduced visibility is one of the first noticeable effects of air pollution on meteorological phenomena. In meteorological terms, visibility is the standard measure of the transparency of the atmosphere in the visible spectrum. [8-33] Reduced visibility presents safety hazards and is aesthetically displeasing.

Particulates in the size range of 0.38 to 0.76 μm and gas molecules (especially sulfur dioxide) are the major contaminants that contribute to reduced visibility. These pollutants absorb and scatter light. Scattering reduces visibility by decreasing the contrast between the object and the background sky. Scattering of light by small particles is also responsible for the reddish hue of sunsets. [8-30]

Altered Precipitation

Air contaminants either emitted to or formed in the atmosphere can result in increased precipitation. This phenomenon occurs because small particles act as nuclei, inducing the formation of raindrops, the same principle operant in cloud seeding. The demonstrable increase in precipitation is particularly noticeable over urban centers with high particulate emission. Furthermore, there is evidence that precipitation increases of at least 10 percent occurring as far as 30 to 35 km from an urban area can be ascribed to pollution. [8-33]

It has been shown that the occurrence of fog over cities is double the incidence of fog over undeveloped areas, and cloud formation provides 10 percent more cover in skies over cities than in those over countryside. [8-12] High concentrations of SO_2 are linked to increased fog over industrialized areas, and SO_2 and NO emissions have been linked to acid rain. (See Sec. 7-2)

Urban Heat Island

Air contaminants are responsible for marked reduction of solar radiation over cities. In some urban areas, from 15 to 20 percent less heat energy reaches the ground because of atmospheric pollution. [8-10, 8-22] However, countering this loss of energy is the heat-retention capability of urban facilities—the capabilities inherent in such building materials as asphalt, concrete, stone, and brick. Also counteracting this loss is the thermal increase noted in the atmosphere over cities

Table 8-2 Comparison of meteorological variables for urban and rural areas

Variable	Urban value relative to rural value, average
Daily temperature minima	2.5°C higher
Relative humidity, summer	8% lower
Total wind movement	25% less
Solar radiation	15% less
Total precipitation	10% greater
Cloud cover, all types	10% greater
Frequency of fog, winter	100% greater

Source: From Lowry and Brubel. [8-20]

during cold weather. This increase in atmospheric temperature is directly attributable to the increase of fuel burned, an increase which can be estimated from the heating-degree-days of the area in which the city is located. It is estimated that New York City generates and radiates through the combustion process 2.5 times the amount of energy received from solar radiation alone on a typical winter's day. [8-29] Since the increase in temperature brought about by the burning of fossil fuels outweighs the decrease in temperature brought about by the screening effect of particulate matter, cities experience higher average temperatures and less measurable snowfall than do nearby rural areas. [8-12, 8-19] Figures reflecting these factors are found in Table 8-2. The phenomenon arising from these factors is called the *urban heat island effect*.

8-13 CHANGES ON THE MACROSCALE

Since small changes in global temperature can affect the advance-and-retreat pattern of polar ice caps, with resultant alterations in the level of oceans, the potential effect of human activities on the temperature of the earth must be carefully assessed. [8-14, 8-24]

As noted in Chap. 7, there is speculation that increased levels of carbon dioxide from the burning of fossil fuels will intensify the greenhouse effect and raise the temperature of the earth. However, a slight *decrease* in temperature since 1945, despite increased use of fossil fuels, seems to bear out the theory that reduced solar energy due to scattering and absorption effects of increased amounts of particulate matter will, on the macroscale, more than offset the heat-trapping effects of carbon dioxide. [8-30] Indeed, there are those who feel that continued discharge of particulate matter into the atmosphere will lead to another ice age. [8-28]

In conclusion, the global effects of air pollution upon meteorological variables cannot be described in simplistic terms. At present, there is documented evidence of the impact of human activities upon the weather in and around metropolitan areas. Evidence of a cause-effect relationship between air pollution and global changes in weather is less conclusive, yet projections indicate that continued fouling of the atmosphere could result in meteorological changes on the macroscale. The uncertainty of the nature, extent, and consequences of these changes should be ample reason to reduce current, and prevent future, pollution of the atmosphere.

DISCUSSION TOPICS AND PROBLEMS

8-1 Choose a workable dictionary definition of meteorology and discuss its relevance in relation to air pollution.

8-2 Name the elemental properties of the atmosphere.

8-3 Discuss the scales of motion in relation to their influence on air quality.

8-4 What is the Coriolis force and what part does it play in weather patterns?

8-5 How is the heating-degree-day of a region calculated?

8-6 Explain the relationship between the adiabatic lapse rate of a rising plume of stack gas and the ambient lapse rate.

8-7 What is the dry adiabatic lapse rate? The wet adiabatic lapse rate? Explain why they differ.

8-8 Explain the relationship between ambient and adiabatic lapse rates and atmospheric stability. What is the boundary line between stability and instability in Fig. 8-3?

8-9 A rising parcel of dry air has a temperature of 15°C at sea level. Assuming a dry adiabatic lapse rate, determine the temperature at 1000 m.

8-10 A rising plume of stack gas has a temperature of 1000°C at 200 m. Assuming a dry adiabatic lapse rate, determine the temperature at 800 m.

8-11 A parcel of dry air rising over a grass fire has a temperature of 60°C at 10 m. Assuming a dry adiabatic lapse rate, determine the temperature at 200 m.

8-12 Name and define the two types of thermal inversions. Which type prompts the formation of fog?

8-13 What weather is usually associated with high-pressure systems? Low-pressure systems? Discuss the weather generated by such systems in relation to air pollution problems.

8-14 Define:

cyclone
anticyclone
relative humidity
wind rose
sea breeze
land breeze
washout
rainout
cold front
warm front
psychrometer
acid rain
superadiabatic rate

8-15 Sketch the following plume phenomena and discuss each sketch in relation to dry adiabatic lapse rate: (*a*) looping, (*b*) fanning, (*c*) trapping, (*d*) neutral, (*e*) lofting, (*f*) coning, and (*g*) fumigating.

8-16 How is maximum mixing depth (MMD) determined? Of what significance is MMD in air pollution control?

8-17 Under ordinary circumstances, what should the height of a stack be in relation to the height of nearby structures?

8-18 Determine σ_z at the point of maximum concentration downwind from a source with effective stack height of 60 m.

8-19 Determine σ_z at the point of maximum concentration downwind from a source with effective stack height of 80 m.

8-20 Determine σ_z at the point of maximum concentration downwind from a source with effective stack height of 120 m.

8-21 What is meant by effective stack height and what factors control this?

8-22 A coal-fired power plant burns 24,000 tonnes of coal per day. The coal has a sulfur content of 4.2 percent. The physical stack height is 200 m. The inside diameter of the stack is 8 m. The stack-gas exit velocity is 18.3 m/s; the gas has a temperature of 140°C. The ambient air temperature is 8°C. The atmospheric pressure is 1000 millibars, and the average wind speed is 4.5 m/s.

Compute the effective stack height. What is the maximum ground-level concentration of SO and how far is this from the plant? Does this plant by itself cause concentrations at the receptor in excess of the annual ambient-air-quality standard? Use moderately unstable conditions.

8-23 A coal-burning power plant burns 6.25 tonnes of coal per hour and discharges the combustion products through a stack that has an effective height of 80 m. The coal has a sulfur content of 4.7 percent, and the wind velocity at the top of the stack is 8.0 m/s. Atmospheric conditions are moderately to slightly stable. Determine the maximum ground-level concentration of SO_2 and the distance from the stack at which the maximum occurs.

8-24 Assuming a wind speed of 1 m/s and a source emitting 0.7 g of air contaminants per second, what is the contaminant concentration in the plume? What is the contaminant concentration in the plume assuming wind speed increases to 8 m/s and contaminant emission rate remains the same?

8-25 A power plant burns 7.30 tonnes of coal per hour and discharges the combustion products through a stack with an effective stack height of 75 m. The coal has a sulfur content of 4.1 percent, and the wind velocity at the top of the stack is 3.0 m/s. Atmospheric conditions are stable. Determine the maximum ground-level concentration of SO_2 and the distance from the stack at which the maximum occurs.

8-26 From the data in Prob. 8-22, determine the ground-level concentrations at a distance of 2.5 km downwind at the centerline of the plume and at a crosswind distance of 0.3 km on either side of the centerline.

8-27 From the data in Prob. 8-23, determine the ground-level concentrations at a distance of 3.25 km downwind at the centerline of the plume and at a crosswind distance of 0.5 km on either side of the centerline.

8-28 From the data in Prob. 8-24, determine the ground-level concentrations at a distance of 4.2 km downwind at the centerline of the plume and at a crosswind distance of 0.4 km on either side of the centerline.

8-29 Determine the effective height of a stack, given the following data:

(*a*) Physical stack is 180 m tall with a 0.95-m inside diameter.
(*b*) Wind velocity is 2.75 m/s.
(*c*) Air temperature is 20°C.
(*d*) Barometric pressure is 1000 millibars.
(*e*) Stack gas velocity is 11.12 m/s.
(*f*) Stack gas temperature is 160°C.

8-30 Determine the effective height of a stack, given the following data:
(*a*) Physical stack is 170 m tall with a 1.25-m inside diameter.
(*b*) Wind velocity is 5.17 m/s.
(*c*) Air temperature is 18°C.
(*d*) Barometric pressure is 1000 millibars.
(*e*) Stack gas velocity is 8.75 m/s.
(*f*) Stack gas temperature is 128°C.

8-31 Determine the effective height of a stack, given the following data:
(*a*) Physical stack is 230 m tall, with a 1.85-m inside diameter.
(*b*) Wind velocity is 6.5 m/s.
(*c*) Air temperature is 7°C.
(*d*) Barometric pressure is 1000 millibars.
(*e*) Stack gas velocity is 12.3 m/s.
(*f*) Stack gas temperature is 190°C.

8-32 What two air contaminants are of major concern in reduction of visibility?

8-33 Name, define, and discuss three meteorological changes on the regional and local scale that have been attributed to air pollution.

8-34 Discuss the suspected relationship between specific types of air pollution and meteorological changes on the macroscale.

REFERENCES

8-1 Anderson, D. M., et al.: *Pure Air for Pennsylvania*, Division of Air Pollution Control, Pennsylvania Dept. of Health, Harrisburg, 1961.
8-2 Battan, Louis J.: *Fundamentals of Meteorology*, Prentice-Hall, Englewood Cliffs, N.J., 1979.
8-3 Bosanquet, C. H., W. F. Carey, and E. M. Halton: "Dust Deposition from Chimney Stacks," *Proc Inst Mech Eng*, **162**: 355 (1950).
8-4 Briggs, G. A.: *Plume Rise*, A.E.C. Critical Review Series, #T10-25075, 1969.
8-5 Carson, J. E., and H. Moses: "The Validity of Several Plume Rise Formulas," *J Air Pol Cont Assoc*, **19**(11): 862 (1969).
8-6 *Community Air Pollution Control*, Nat Cen Air Pol Cont Training Program, Durham, N.C., 1972.
8-7 Davis, Mackenzie L.: *Air Resources Management Primer*, ASCE, New York, 1973.
8-8 *The Effects of Air Pollution*, Nat Cen Air Pol Con, Washington, D.C., 1967.
8-9 Faith, W. L., and Arthur A. Atkisson, Jr.: *Air Pollution*, 2d ed., Wiley, New York, 1972.
8-10 Georgii, H. W.: "The Effects of Air Pollution on Urban Climates," *Bull WHO* **40**: 624 (1969).
8-11 Gifford, F. A.: "Atmospheric Dispersion Calculations Using the Generalized Gaussian Plume Model," *Nuclear Safety*, **2**: 59 (December 1960).
8-12 "How the Cities Are Ruining Their Own Weather," *U.S. News & World Report*, September 17, 1973.
8-13 Institute for Air Pollution Training: *Meteorology in Air Pollution Control*, Air Pollution Control Orientation Course 442-A, U.S. EPA, Atlanta, Ga., 1969.
8-14 Landsberg, H. E.: "Man-Made Climatic Change," *Science*, **170**: 1265 (1970).
8-15 Ledbetter, J. O.: *Air Pollution—Part A Analysis*. Marcel Dekker, New York, 1972.
8-16 Lehr, P. E., R. W. Burnett, and H. S. Zim: *Weather*, Golden Press, New York, 1957.
8-17 Likens, Gene E.: "Acid Precipitation," *Chem Eng News*, special report, November 22, 1976, p. 29.
8-18 Linsley, R. K., Jr.: *Hydrology for Engineers*, McGraw-Hill, New York, 1958.
8-19 Lowry, W. P.: "The Climate of Cities," *Scientific American*, **217**(2): 15 (1967).
8-20 ——— and Boubel, R. W.: *Meteorological Concepts in Air Sanitation*, Oregon State University, Corvallis, 1967.

8-21 "Man's Impact on the Global Environment—Assessment and Recommendations for Action," M.I.T., Cambridge, 1970.

8-22 *Meteorological Aspects of Air Pollution*, Air Pollution Training Program, U.S. Dept. H.E.W. Division of Air Pollution, Cincinnati, Ohio, 1962.

8-23 Miller, S.: "The Building of Tall (and Not So Tall) Stacks," *J Air Pol Cont*, **9**(6):522 (1975).

8-24 Munn, R. E., and B. Bolin: "Global Air Pollution—Meteorological Aspects: A Survey," *Atmospheric Environment*, **5**:363 (1971).

8-25 Pasquill, F.: "Atmospheric Dispersion of Pollution," *Quarterly Journal Royal Meteorological Society*, **97**:369 (1971).

8-26 Perkins, Henry C.: *Air Pollution*, McGraw-Hill, New York, 1974.

8-27 Petterssen, Sverre.: *Introduction to Meteorology*, 2d ed., McGraw-Hill, New York, 1968.

8-28 Rasool, S. I., and S. H. Schneider: "Atmospheric Carbon Dioxide and Aerosols: Effects of Large Increases on Global Climate," *Science*, **173**: 138 (1971).

8-29 Riehl, Herbert: *Introduction to the Atmosphere*, McGraw-Hill, New York, 1972.

8-30 Seinfeld, John H.: *Air Pollution: Physical and Chemical Fundamentals*, McGraw-Hill, New York, 1975.

8-31 Slade, D. H. (ed.): *Meteorology and Atomic Energy*, U.S. Atomic Energy Commission, National Technical Information Service, U.S. Dept. of Commerce, Springfield, Va., 1968.

8-32 Smith, M. E. (ed.): *Recommended Guide for the Prediction of the Dispersion of Airborne Effluents*, ASME, New York, 1973.

8-33 Stern, Arthur C., Henry C. Wohlers, Richard W. Boubel, and William P. Lowry: *Fundamentals of Air Pollution*, Academic Press, New York, 1973.

8.34 Turner, D. B.: *Workbook of Atmospheric Dispersion Estimates*, Office of Air Programs, U.S. EPA, Research Triangle Park, North Carolina, 1970.

8-35 Wark, Kenneth, and Cecil F. Warner: *Air Pollution: Its Origin and Control*, Dun-Dunnelley, New York, 1976.

8-36 Williamson, S. J.: *Fundamentals of Air Pollution*, Addison-Wesley, Reading, Mass., 1973.

8-37 Vermeulen, A. J.: "Acid Precipitation in the Netherlands," *Env Sci Tech*, **12**(9):1017 (1978).

8-38 Vesilind, P. Aarne: *Environmental Pollution and Control*, Ann Arbor Science, Ann Arbor, Mich., 1978.

CHAPTER

NINE

ENGINEERED SYSTEMS FOR AIR POLLUTION CONTROL

The atmosphere, like a stream or a river, has natural, built-in self-cleansing processes without which the troposphere would quickly become unlivable for humans. Most of the air-pollution-control devices discussed in this chapter, for both stationary and mobile sources, make use of some of the principles involved in the natural atmospheric cleansing processes.

9-1 ATMOSPHERIC CLEANSING PROCESSES

Dispersion, gravitational settling, flocculation, absorption (involving washout and scavenging), rainout, and adsorption are some of the most significant natural removal mechanisms at work in the atmosphere. [9-35] Though not literally a removal mechanism, *dispersion* of pollutants by wind currents lessens the concentrations of pollutants in any one place. *Gravitational settling* is one of the most important natural mechanisms for removing particulates from the atmosphere, especially particles larger than 20 μm.

Gravitational settling also plays a key role in several of the other natural atmospheric cleansing processes. For example, through *flocculation*, particles smaller than 0.1 μm can be settled out. In this phenomenon, larger particles act as receptors for smaller ones. Two particles bump together to form a unit, and the process is repeated until a small floc particle is formed that will be large enough and heavy enough to settle out. (See Sec. 4-6.)

In the natural *absorption* process, particulates or gaseous pollutants are collected in rain or mist, then settle out with that moisture. This phenomenon, known as *washout* or *scavenging*, takes place below cloud level. The potential for scavenging gases and particulates depends on many factors, including the intensity of rainfall and the nature of the contaminants being scavenged. Under ordinary circumstances, only a fraction of the particles in the path of descent of a raindrop will be collected, with most small particles remaining in the air that flows around the falling drop. Recent research indicates that washout may be negligible for particles less than 1 μm in diameter. [9-30]

Gases may be dissolved without being chemically changed, or they may, in some instances, enter into chemical reactions with the rainwater. For example, SO_2 gas, which is simply dissolved into rain, falls with the droplets as SO_2. However, SO_2 may also react with rainwater to form H_2SO_3 (sulfurous acid) or H_2SO_4 (sulfuric acid) mists, mists known as "acid rains" and potentially far more harmful than the original SO_2.

Rainfall through uncontaminated air has a pH of 5.6 to 6.0. However, in the western hemisphere, pH values as low as 2 have been recorded for rainfall from a single precipitation event. [9-49] This low pH of rainfall can have far-reaching effects. As noted earlier, acid rain runoff can cause extensive erosion of some surfaces (notably limestone) and can change the pH in streams and rivers, thereby influencing the species of algae which predominate in those streams.

Rainout is another natural atmospheric cleansing process involving precipitation. Whereas washout occurs below cloud level when falling raindrops absorb pollutants, rainout occurs within clouds when submicron particulates serve as condensation nuclei around which drops of water may form. This phenomenon has resulted in increased rainfall and fog formation in urban areas.

Adsorption occurs primarily in the friction layer of the atmosphere, the layer closest to the earth's surface. In this phenomenon, gaseous, liquid, or solid contaminants are attracted (generally electrostatically) to a surface, where they are concentrated and retained. Natural surfaces such as soil, rocks, leaves, and blades of grass can adsorb and retain pollutants. Particles may be brought into contact with an adsorption surface by gravitational settling or by inertial impaction, a process by which particulates or gaseous pollutants are transported to surfaces by wind currents. Impaction is particularly effective for particles in the 10 to 15 μm range, and numerous small surfaces—such as blades of grass and leaves of trees—are more effective than larger surfaces in removing particulates in this size range. [9-30]

When the various and sundry natural atmospheric cleansing mechanisms are overwhelmed by gaseous and particulate emissions, the effects of air pollution become increasingly more evident. Clothing is soiled, particles are deposited on buildings and other surfaces, plants are damaged, visibility is reduced, and human respiratory problems are increased. To prevent these and other evidences of air pollution, it is necessary to establish control procedures or to install control devices. But even with the application of the best available technology, low-level emissions will still inevitably be made into the atmosphere, and these emissions must ultimately be removed by natural atmospheric cleansing mechanisms.

9-2 APPROACHES TO CONTAMINANT CONTROL

There are two broad approaches to the control of particulate and gaseous contaminants—dilution of the contaminants in the atmosphere and control of the contaminants at their source.

Dilution

As discussed in Chap. 8, dilution of contaminants in the atmosphere can be accomplished through the use of tall stacks. In 1955, before control regulations became stringent, there were only two stacks in excess of 150 m (500 ft) in the United States. By the end of 1975, with more and more emphasis being placed on air pollution control, there were 15 stacks in excess of 150 m (500 ft), and nearly 50 new stacks in this height range are being built each year. [9-25]

Tall stacks may penetrate the inversion layer and disperse the contaminants so that ground-level contaminants are greatly reduced. However, it must be remembered that what goes up must eventually come down, albeit often at considerable distance from its original source. For example, it is estimated that from 15 to 50 percent of the SO_2 found to be causing acid rain in Sweden originated in other countries, primarily Great Britain. [9-12] Ironically, tall stacks and other such dilution devices are really only a means of spreading air contaminants around our globe or diluting them to levels at which their harmful effects are less noticeable near their source. Seen in this light, dilution is, at best, a short-term control measure and is, at worst, a measure which tends to bring about highly undesirable long-range effects.

Control at Source

In terms of long-range control of air pollution, control of contaminants at their source is a more desirable and effective method than dilution. Control of pollutants at their source can be accomplished by several different means. The first, and obviously the most effective, is to prevent the contaminants from coming into being in the first place. In the case of contaminants associated with combustion processes, substituting an alternative power source (i.e., hydraulic, geothermal, or solar energy for fossil-fuel-derived energy) can prevent the creation of pollutants.

The remaining methods of at-source control can reduce contaminant emissions, but they cannot eliminate them entirely. For example, the traditional fossil fuel can be replaced by another fuel of lower air pollution potential, as when low-sulfur coal is used to reduce sulfur oxide emissions. Sometimes traditional, high-contaminant fuels can be altered in ways which reduce their pollution potential. Thus, coal or natural gas can be refined to desulfured, liquefied natural gas (LNG) or to liquid petroleum gas (LPG), products which produce relatively low emissions. [9-32]

Another method of control of pollutants at the source involves the proper use of existing equipment. Sometimes contaminant emissions can be greatly reduced by the proper operation and maintenance of equipment, especially equipment involved in combustion operations. For example, an automobile with a dirty air filter, a bad positive-crankcase-ventilation (PCV) system, an incorrect idle speed adjustment, an improper carburetor setting, fouled spark plugs, and broken spark plug wires will have higher pollutant emissions than one which is operating at top efficiency. In fact, regular, competent automotive inspection and maintenance can reduce hydrocarbons and carbon monoxide emissions from 20 to 50 percent. [9-2, 9-40]

Similarly, industry can decrease emissions from stationary sources by proper operation and maintenance of equipment. For example, excess fly ash emissions at power plants can be reduced by adjusting the boiler furnace air intake, and soot and carbon monoxide emissions from refinery flares can be reduced by increasing turbulence through injection of steam into the flame zone. [9-32] NO_x emissions from ore- and gas-fired boilers have been reduced by modification of combustion conditions. The introduction of low-excess-air firing, staged combustion, over-fire air, flue gas recirculation, and water injection can reduce NO_x emissions from 30 to 60 percent.

Changing the process being used is still another method of controlling emissions at their source. For example, replacing open-hearth furnaces with controlled basic oxygen furnaces or electric furnaces can reduce smoke, carbon monoxide, and fumes, while at the same time conserving energy. [9-32] A sawmill can eliminate particulate and gaseous contaminants by converting an open-pit or teepee slab incineration operation into a debarker-and-chipper system which converts most waste materials into usable form.

One of the most widely used methods of controlling emissions of air pollutants at their sources is to install control equipment designed according to some of the same basic principles by which natural removal mechanisms operate. These control devices, often needed in addition to one or more of the other approaches to source control, are designed to destroy, counteract, collect, or mask pollutants.

Since few such devices are effective in the control of both particulate and gaseous contaminants, control devices are usually designed to control either one or the other. In terms of volume of pollutants, the control of gaseous pollutants seems of primary importance. In 1978, the total estimated emissions of all air contaminants was 192.7 million tons; 93.5 percent of this total was attributable to gaseous emission and only 6.5 percent to particulate emissions. [9-8] However, in actual practice, much greater attention has been directed toward particulate control, perhaps because particulates are more easily seen. Therefore, control of particulates from stationary sources will be discussed first in this chapter. Control of gaseous contaminants from stationary sources will follow. Finally, an overview of control of particulate and gaseous contaminants from mobile sources will be presented.

Control Devices for Particulate Contaminants

Originating from a variety of sources, but primarily from industrial processes, airborne particulates exert a significant influence on atmospheric phenomena, plants, property, and human and animal health. In 1980, an estimated 7.8 million tonnes of suspended particulate matter was emitted to our atmosphere, with industry contributing 47 percent of that total. Progress is being made in control of particulates at their source, however, since a 56 percent overall reduction in particulates occurred between 1970 and 1980. [9-8]

Most of the control devices and physical principles involved in particulate control have been understood, developed, and used for some time. Control devices can be divided into five major groups: gravitational settling chambers, centrifugal collectors, wet collectors, electrostatic precipitators, and fabric filters. [9-3] Each of these devices is uniquely suited to specific applications, and the proper choice of method depends upon careful consideration of several factors. (See Table 9-1.) Such particle characteristics as size distribution, shape, density, stickiness, hygroscopicity, and electrical properties and such carrier gas properties as flow rate and particle concentration must be taken into consideration. Such operational factors as continuous or intermittent emission, overall efficiency desired, available space, ultimate waste-disposal methods required, and equipment limitations (i.e., pressure, temperature, and corrosion) must be balanced against the economic considerations of installation, operating, and maintenance costs. [9-46]

Table 9-1 Summary of particulate control devices

Device	Minimum particle size,* μm	Efficiency, % (mass basis)	Advantages	Disadvantages
Gravitational settler	> 50	< 50	Low pressure loss, simplicity of design and maintenance	Much space required; low collection efficiency
Centrifugal collector	5–25	50–90	Simplicity of design and maintenance Little floor space required Dry continuous disposal of collected dusts Low to moderate pressure loss Handles large particles Handles high dust loadings Temperature independent	Much head room required Low collection efficiency of small particles Sensitive to variable dust loadings and flow rates

Device	Minimum particle size,* μm	Efficiency, % (mass basis)	Advantages	Disadvantages
Wet collector			Simultaneous gas absorption and particle removal Ability to cool and clean high-temperature, moisture-laden gases Corrosive gases and mists can be recovered and neutralized Reduced dust explosion risk Efficiency can be varied	Corrosion, erosion problems Added cost of wastewater treatment and reclamation Low efficiency on submicron particles Contamination of effluent stream by liquid entrainment Freezing problems in cold weather Reduction in buoyancy and plume rise Water vapor contributes to visible plume under some atmospheric conditions
Spray tower	>10	<80		
Cyclonic	>2.5	<80		
Impingement	>2.5	<80		
Venturi	>0.5	<99		
Electrostatic precipitator	>1	95–99	99+ percent efficiency obtainable Very small particles can be collected Particles may be collected wet or dry Pressure drops and power requirements are small compared with other high-efficiency collectors Maintenance is nominal unless corrosive or adhesive materials are handled Few moving parts Can be operated at high temperatures (300 to 450°C)	Relatively high initial cost Precipitators are sensitive to variable dust loadings or flow rates Resistivity causes some material to be economically uncollectable Precautions are required to safeguard personnel from high voltage Collection efficiencies can deteriorate gradually and imperceptibly
Fabric filtration	<1	>99	Dry collection possible Decrease of performance is noticeable Collection of small particles possible High efficiencies possible	Sensitivity to filtering velocity High-temperature gases must be cooled to 100 to 450°C Affected by relative humidity (condensation) Susceptibility of fabric to chemical attack

* Collected at 90 percent efficiency.

Source: From Seinfeld. [9-41]

Table 9-2 Industrial process and control summary

Industry or process	Source of emissions	Particulate matter	Method of control
Iron and steel mills	Blast furnaces, steel-making furnaces, sintering machines	Iron oxide, dust, smoke	Cyclones, baghouses, electrostatic precipitators, wet collectors
Gray iron foundries	Cupolas, shake-out making	Iron oxide, smoke, oil dust, metal fumes	Scrubbers, dry centrifuge collectors
Nonferrous metallurgy	Smelters and furnaces	Smoke, metal fumes, oil, grease	Electrostatic precipitators, fabric filters
Petroleum refineries	Catalyst regenerators, sludge incinerators	Catalyst dust, ash from sludge	Cyclones, electrostatic precipitators, scrubbers, baghouses
Portland cement	Kilns, driers, material-handling systems	Alkali and process dusts	Fabric filters, electrostatic precipitators, mechanical collectors
Kraft paper mills	Recovery furnaces, line kilns, smelt tanks	Chemical dusts	Electrostatic precipitators, venturi scrubbers
Acid manufacture—phosphoric, sulfuric	Thermal processes, rock acidulating, grinding	Acid mist, dust	Electrostatic precipitators, mesh mist eliminators
Coke manufacture	Oven operation, quenching, materials handling	Coal and coke dust, coal tars	Meticulous design, operation and maintenance
Glass and fiberglass	Furnaces, forming and curing, handling	Acid mist, alkaline oxides, dust, aerosols	Fabric filters, afterburners

Source: From Jost. [9-16]

Table 9-2 indicates the type of particulates emitted by several different industries and the methods of control employed with those particulates. Most of these methods will be discussed in the remaining pages of this chapter.

9-3 GRAVITATIONAL SETTLING CHAMBERS

The gravitational settling of particles in a fluid was discussed in Sec. 4-5 and again in Sec. 7-5. The application of Stokes' law and similar equations to settling operations involving particles in air streams is essentially the same as that described in Sec. 4-5 for particles in water streams. Like settling basins in water and wastewater systems, settling chambers in air-pollution-control systems provide enlarged areas to minimize horizontal velocities and allow time for the vertical velocity to carry the particle to the floor.

The usual velocity through settling chambers is between 0.5 and 2.5 m/s (100 and 250 ft/min), although for best results the gas flow should be uniformly maintained at less than 0.3 m/s (60 ft/min). [9-3] Some settling chambers like the one in Fig. 9-1 are simply enlarged conduits, while others, such as the chamber in Fig. 9-2, have horizontal shelves and baffles. Spaced about 2.5 cm (1 in) apart, the horizontal shelves shorten the settling path of the particles and thus improve removal efficiency.

If we assume that Stokes' law applies we can derive a formula for calculating the minimum diameter of a particle collected at 100 percent theoretical efficiency in a chamber of length L. (In practice, efficiency is always less than 100 percent, since some reentrainment occurs.) [9-33]

$$\frac{v_t}{H} = \frac{v_h}{L} \tag{9-1}$$

where v_t = terminal settling velocity as defined in Eq. (7-3), m/s
H = height of settling chamber, m
v_h = horizontal flow-through velocity, m/s
L = length of settling chamber, m

Solving for v_t yields

$$v_t = \frac{v_h H}{L}$$

and substitution into Eq. (7-3)

$$\frac{v_h H}{L} = \frac{g(\rho_p - \rho_a)d_p^2}{18\mu}$$

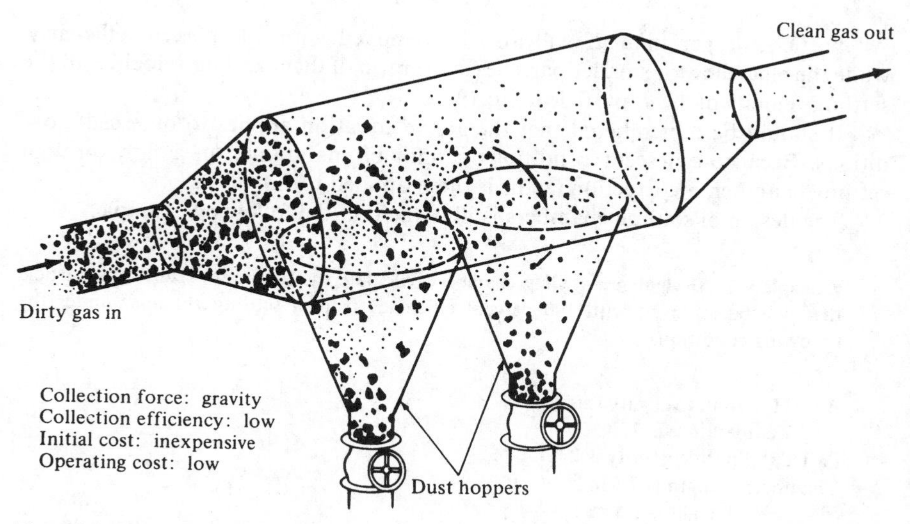

Figure 9-1 Gravitational settling chamber. (*From U.S. EPA* [9-15].)

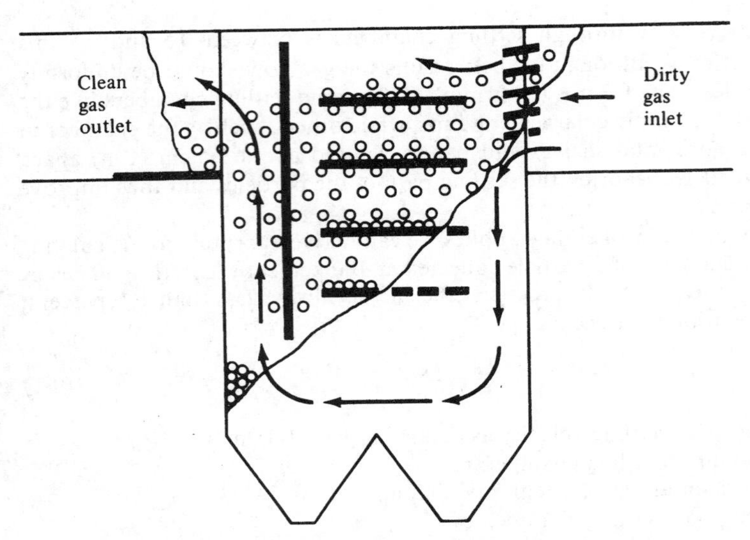

Figure 9-2 Baffled gravitational settling chamber. (*From U.S. EPA* [9-15].)

Because ρ_a is insignificant compared to ρ_p, it can be dropped from the equation. Solving for d_p gives an equation that predicts the largest-size particle that can be removed with 100 percent efficiency in a settling chamber of given dimensions.

$$d_p = \left(\frac{18\mu v_h H}{gL\rho_p}\right)^{1/2} \tag{9-2}$$

All particles larger than d_p will also be removed with 100 percent efficiency, while the efficiency for smaller particles is the ratio of their settling velocities to the settling velocity of the d_p particle in Eq. (9-2).

It should be remembered that the above equation applies to quiescent conditions. Because quiescent conditions cannot be maintained in a flow-through settling chamber, a correction factor is often applied to Eq. (9-2).

The design of settling chambers is illustrated in the following example.

Example 9-1: Designing a settling chamber Calculate the minimum size of the particle that will be removed with 100 percent efficiency from a settling chamber under the following conditions.

Air: Horizontal velocity is 0.3 m/s.
Temperature is 77°C.
Particle: Specific gravity is 2.0.
Chamber: Length is 7.5 m.
Height is 1.5 m.

SOLUTION

1. From Fig. C-2 in the appendix, the viscosity of air at 77°C is 2.1×10^{-5} kg/m · s.
2. Using Eq. (9-2) with a correction factor of 2.0:

$$d_p = \left(2 \frac{18\mu V_h H}{gL\rho_p}\right)^{1/2}$$

$$d_p = \left(\frac{2 \times 18 \times 2.1 \times 10^{-5} \text{ kg/m} \cdot \text{s} \times 0.3 \text{ m/s} \times 1.5 \text{ m}}{9.81 \text{ m/s}^2 \times 7.5 \text{ m} \times 2{,}000 \text{ kg/m}^3}\right)^{1/2}$$

$$d_p = 4.81 \times 10^{-5} \text{ m}$$

$$= 48.1\mu$$

Gravitational settling chambers are simple in design and operation, but they require a large space for installation and have relatively low efficiency, especially for removal of small particles. Although theoretically they should be able to remove particulates down to 5 or 10 μm, in actuality they are not practical for the removal of particles much less than 50 μm in size.

9-4 CENTRIFUGAL COLLECTORS

Centrifugal collectors employ a centrifugal force instead of gravity to separate particles from the gas stream. Because centrifugal forces can be generated that are several times greater than gravitational forces, particles can be removed in centrifugal collectors that are much smaller than those that can be removed in gravity settling chambers. Two general types of centrifugal collectors—cyclones and dynamic precipitators—are commonly used.

Cyclones

A cyclone collector such as the one shown in Fig. 9-3*a* consists of a cylindrical shell, conical base, dust hopper, and an inlet where the dust-laden gas enters tangentially. Under the influence of the centrifugal force generated by the spinning gas, the solid particles are thrown to the walls of the cyclone as the gas spirals upward at the inside of the cone. The particles slide down the walls of the cone and into the hopper.

The operating or separating efficiency of a cyclone depends on the magnitude of the centrifugal force exerted on the particles. The greater the centrifugal force, the greater the separating efficiency. The magnitude of the centrifugal force

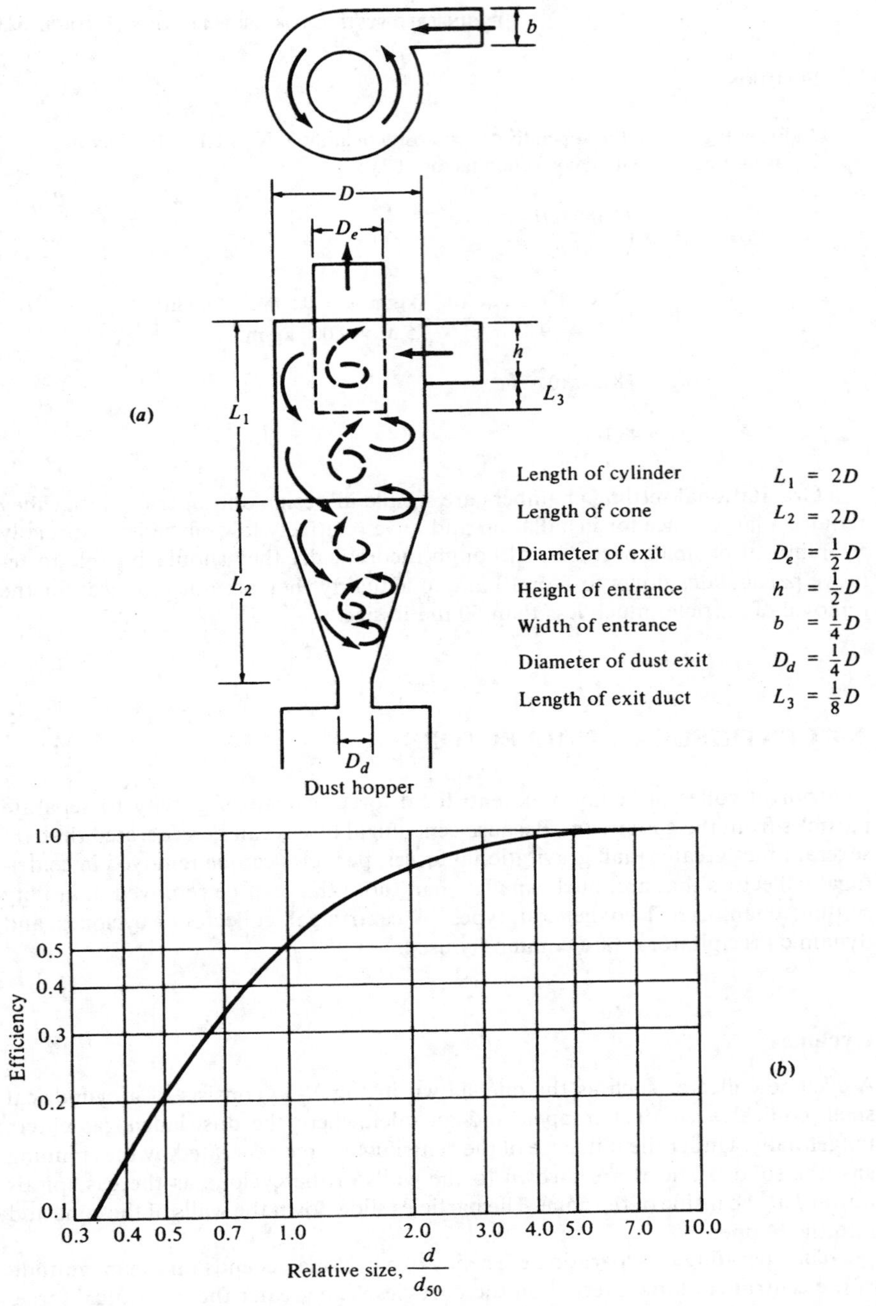

Figure 9-3 Standard-dimension cyclone collector. (*a*) Relative dimensions and operational mode. (*b*) Empirical efficiency for standard-dimension cyclone collector as a function of relative particle size. (*From Lapple* [*9-17*].)

generated depends on particle mass, gas velocity within the cyclone, and cyclone diameter, as shown in Eq. (9-3).

$$F_c = M_p \times \frac{v_i^2}{R} \tag{9-3}$$

where F_c = centrifugal force, N
M_p = particulate mass, kg
$\frac{v_i^2}{R}$ = centrifugal acceleration where v_i^2 equals particle velocity and R equals radius of the cyclone, m/s^2

From this equation it can be seen that the centrifugal force on the particles, and thus the collection efficiency of the cyclone collector, can be increased by decreasing R, the radius of the cyclone. [9-28] Large-diameter cyclones have good collection efficiencies for particles 40 to 50 μm in diameter. High-efficiency cyclones with diameters of 23 cm (9 in) or less have good efficiencies for particles from 15 to 20 μm. Multiple cyclones operating in parallel are necessary to treat large flows when small-diameter cyclones are used. The cleaning efficiency for units such as the one in Fig. 9-4 may be as high as 90 percent for particulates in the 5- to 10-μm range. The smaller radius of the cones not only increases the centrifugal force, but also reduces the distance the particles have to travel to reach the collection chamber. Small cyclones do have some disadvantages, such as problems with

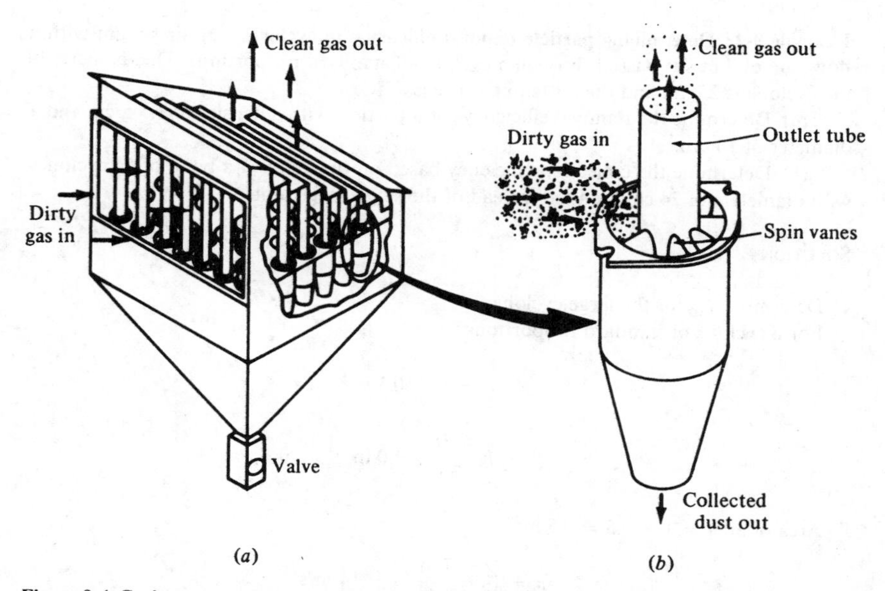

Figure 9-4 Cyclones arranged in parallel: (*a*) multiple cyclone and (*b*) collector element. (*From U.S. EPA* [9-15].)

equalizing gas flow to each cone, abrasion of tubes due to high velocity, and plugging of heavily loaded tubes. [9-28] Cyclones are usually built to standard relative dimensions, as shown in Fig. 9-3*a*.

Determination of collection efficiency with respect to particle size is not as straightforward for cyclones as for gravity settling chambers. Size-efficiency curves for cyclones are curvilinear rather than straight-line relationships. Although there are sizes beyond which all particles are collected with 100 percent efficiency, this size is approached asymptotically and may be quite large. For this reason, the reference particle is usually taken as the particle that will be removed with 50 percent efficiency on a weight basis. The size of this particle is determined by the following equation based on work by Lapple. [9-17]

$$d_{50} = \left(\frac{9\mu b}{2\pi N_e v_i \rho_p}\right)^{1/2} \tag{9-4}$$

where d_{50} = diameter of the particle that is collected with 50 percent efficiency, m
μ = gas viscosity, kg/m · s
b = width of cyclone inlet, m
N_e = number of effective turns within the cyclone
v_i = inlet gas velocity, m/s
ρ_p = density of the particulate matter, kg/m^3

The collection efficiency of particles both larger and smaller than d_{50} is determined by their ratio to d_{50}, as shown in Fig. 9-3*b*. These calculations are illustrated in the following example.

Example 9-2: Determining particle removal efficiency in cyclones An air stream with a flow rate of 7 m/s is passed through a cyclone of standard proportions. The diameter of the cyclone is 2.0 m, and the air temperature is 77°C.

(*a*) Determine the removal efficiency for a particle with a density of 1.5 g/cm and a diameter of 10 μm.

(*b*) Determine the collection efficiency based on the above if a bank of 64 cyclones with diameters of 24 cm are used instead of the single large unit.

SOLUTION

1. Determine d_{50} for the large cyclone.
For a cyclone of standard proportions

$$b = \frac{D}{4} = 0.5\ m$$

$$h = \frac{D}{2} = 1.0\ \text{m}$$

Area of inlet = 1 × 0.5 = 0.5 m^2

$$v_i = \frac{Q}{A} = \frac{7\ \text{m}^3/\text{s}}{0.5\ \text{m}^2} = 14\ \text{m/s}$$

At this inlet velocity, $N_e \simeq 5$ turns
At 77°C, $\mu = 2.1 \times 10^{-5}$ kg/m · s
Using Eq. (9-4).

$$d_{50} = \left(\frac{9 \times 2.07 \times 10^{-5}\ \text{kg/m}\cdot\text{s} \times 0.5\ \text{m}}{2\pi \times 5 \times 14\ \text{m/s}\cdot 1500\ \text{kg/m}^3}\right)^{1/2}$$

$$= 1.19 \times 10^{-5}\text{m}$$

$$= 12\mu\ \text{m}$$

$$\frac{d}{d_{50}} = \frac{10}{12} = 0.83$$

From Fig. 9-3*b* the efficiency for the 10-μm particle is found to be about 42 percent.

2. Determine the d_{50} for the small cyclone.

$$b = \frac{D}{4} = \frac{0.24}{4} = 0.06\ \text{m}$$

$$h = \frac{D}{2} = \frac{0.24}{2} = 0.12\ \text{m}$$

Area of inlet $= 0.06 \times 0.12 = 7.2 \times 10^{-3}\ \text{m}^2$
Area of all inlets $= 64 \times 7.2 \times 10^{-3}\ \text{m}^2$
$= 0.45\ \text{m}^2$

$$\text{Inlet velocity} = \frac{7\ \text{m}^3/\text{s}}{0.46\ \text{m}^2} = 15\ \text{m/s}$$

$$d_{50} = \left(\frac{9 \times 2.1 \times 10^{-5}\ \text{kg/m}\cdot\text{s} \times 0.06\ \text{m}}{2\pi \times 5 \times 15\ \text{m/s} \times 1500\ \text{kg/m}^3}\right)^{1/2}$$

$$= 4.0 \times 10^{-6}\ \text{m}$$

$$= 4.0\ \mu\text{m}$$

$$\frac{d}{d_{50}} = \frac{10}{4} = 2.5$$

From Fig. 9-3*b*, the efficiency for the 10-μm particle is about 88 percent.

Cyclone collectors are relatively inexpensive to construct and operate, and they can handle large volumes of gases at temperatures up to 980°C (1800°F). Pressure drops across these units are generally low and range from 2.5 to 20 cm (1 to 8 in) of water. Cyclones have been used successfully at feed and grain mills, cotton gins, cement plants, fertilizer plants, petroleum refineries, asphalt mixing plants, and other applications involving large quantities of gas containing relatively large particles.

Dynamic Precipitators

Dynamic precipitators are compact units that impart a centrifugal force to the particulate by the action of rotating vanes, a force about seven times that of a conventional cyclone of the same capacity. This type of unit, shown in Fig. 9-5,

Figure 9-5 Dynamic precipitator. (*Courtesy of American Air Filter, an Allis-Chalmers Company.*)

can serve as both an exhaust fan and a dust collector and is widely used in ceramics, food, pharmaceutical, and woodworking industries. It cannot handle wet, fibrous material, which can accumulate on the blades, and it requires a higher power input than does a centrifugal fan operating at the same volume and system resistance in a cyclone. [9-3]

9-5 WET COLLECTORS

Wet collectors, or scrubbers, remove particulate matter from gas streams by incorporating the particles into liquid droplets directly on contact. Either inertial impingement or interception during gravitational settling may be the contact mechanism.

According to the contact-power theory developed for scrubbers, collection efficiency for well-designed wet collectors of all types is a function of the energy consumed in the air-to-water contact process. The energy consumed is directly proportional to the pressure drop, and comparable performance can be expected from all well-designed wet collectors operating at or near the same pressure drop. [9-42] Average pressure drops from well-designed collectors of several different types are shown in Table 9-3. Applying the contact-power theory, more efficiency can be expected from a venturi-type wet collector with a 125-cm water gage (w.g.) pressure drop than from a spray chamber with a pressure drop of 1- to 4-cm w.g. In actual practice, most operations would not need the high efficiency and could not afford the high energy cost of a scrubber with a 125-cm w.g. pressured drop.

Table 9-3 Average pressure drops from well-designed collectors of several different types

Wet collector	Pressure drop, cm w.g.
Chamber	1–4
Centrifugal	5–15
Dynamic	15
Atomizing	
Orifice type	8–15
Venturi type	12–250

Source: From Shaheen. [9-42]

In addition to the pressure drop, the cleaning efficiency of wet collectors varies directly with the size of the particulates being collected. Generally, collectors operating at very low pressure loss will remove only medium- to coarse-size particles, while collectors operating at higher pressure losses (and therefore increased energy output) will be highly efficient at removing fine particles.

A graph of efficiency versus particle size for wet scrubbers operating at various pressure drops is given in Fig. 9-6. Consistent with the contact-power theory, efficiency in collection of particles in the smaller size range is substantially higher when more energy is expended in air-water contact. [9-42]

With virtually no particle reentrainment, wet collectors can provide efficient, relatively low-cost solutions to many air pollution problems, including the handling of hot, moist gases that many other systems cannot accommodate. However, they are not without drawbacks. High or fluctuating pressure drops can be a problem in some units, and maintenance costs can be high when corrosive materials are being collected. Further, wet collectors are not recommended for use where

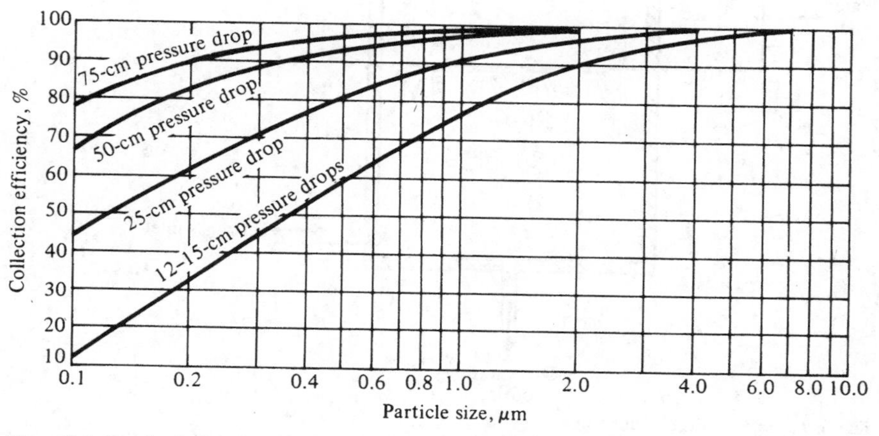

Figure 9-6 Graph of efficiency versus particle size for a wet scrubber. (*From Shaheen* [*9-42*].)

high plume rise is important, because wet plumes do not rise and disperse well. [9-33]

For all wet collectors, disposal of the wastewater in which the particulates have been collected poses problems. Often the wastewater will require some form of treatment (i.e., settling tank or pond, or centrifugal device) before being discharged into a receiving stream. In areas where water supplies are limited or water costs are unusually high, further treatment before recycling the water may be called for.

Three of the wet collectors most commonly used for control of particulate matter—the spray tower, the wet cyclone scrubber, and the venturi scrubber—are discussed in this section. Though wet collectors of other types (notably, packed tower scrubbers) may trap particles effectively, particles tend to clog such units too quickly to make them useful for particulate removal. Because the primary function of other types of wet collectors is to remove gaseous pollutants, they will be discussed in a later section of this chapter.

Spray Towers

Spray towers such as the one shown in Fig. 9-7 are low-cost scrubbers that can be used to remove both gaseous and particulate contaminants. The units cause very

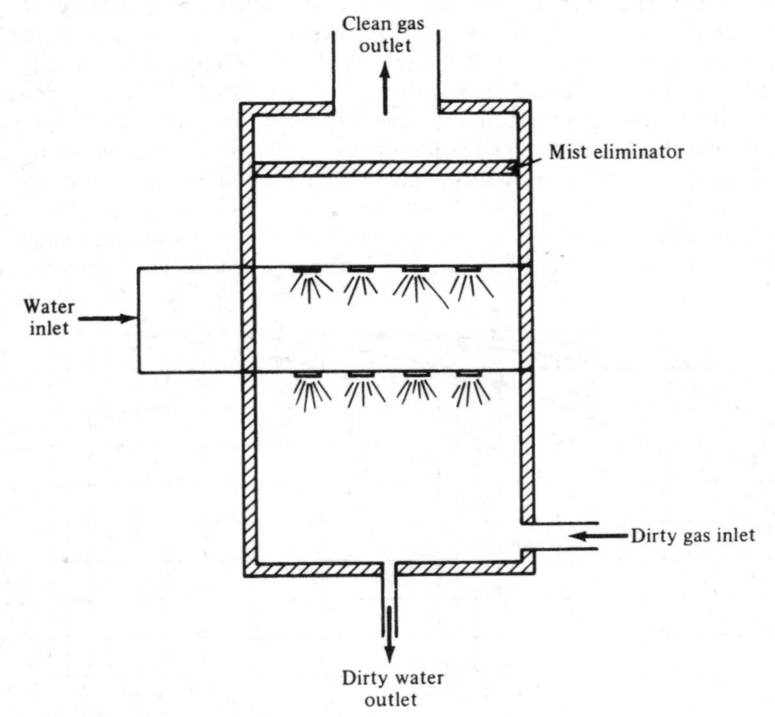

Figure 9-7 Spray tower scrubber.

little pressure loss and can handle large volumes of gases. As the gas flows upward, entrained particles collide with liquid droplets sprayed across the flow passage, and liquid droplets containing the particles settle by gravity to the bottom of the chamber. Spray towers are effective in removing particles in excess of 10 μm, and modifications can be made to improve efficiencies for smaller particles. [9-50]

Wet Cyclone Scrubbers

In a simple wet cyclone scrubber, high-pressure spray nozzles located in various places within the cyclone chamber generate a fine spray that intercepts the small particles entrained in the swirling gases. The particulate matter thrown to the wall by centrifugal force is then drained to the collection sump. One version of a cyclone scrubber is shown in Fig. 9-8. For droplets of 100 μm, efficiency approaches 100 percent, and 90 to 98 percent removal is achieved for droplets between 5 and 50 μm. [9-50] Particle removal depends on contact with the liquid droplets and is a

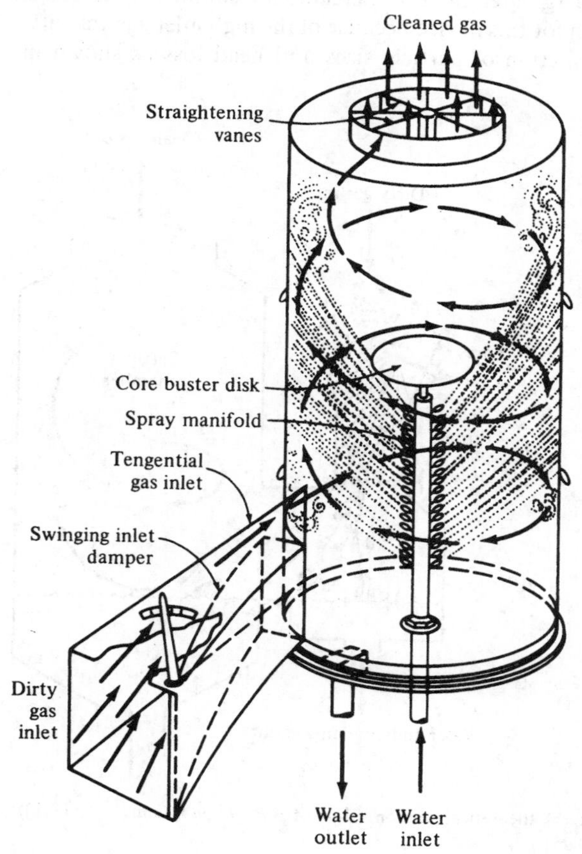

Figure 9-8 Wet cyclone scrubber. (*From Air Pollution Manual* . . . [9-*1*].)

function of the liquid flow rate and liquid droplet and particle sizes. Generally, efficiencies slightly higher than those obtained with the spray tower can be expected.

Venturi Scrubbers

Venturi scrubbers such as the one shown in Fig. 9-9 are most efficient for removing particulate matter in the size range of 0.5 to 5μm, that makes them especially effective for the removal of submicron particulates associated with smoke and fumes. At velocities from 60 to 180 m/s (200 to 600 ft/s), the contaminated gas passes through a duct that has a venturi-shaped throat section. A coarse water spray is injected into the throat, where it is atomized by the high gas velocities. The liquid droplets collide with the particles in the gas stream, and the water and particles fall down for later removal. Venturi scrubbers can efficiently remove gaseous as well as particulate contaminants (see Sec. 9-9). Although venturi scrubbers are highly efficient (greater than 90 percent) for submicron aerosols, power costs are relatively high for this device because of the high inlet gas velocity. Removal efficiencies are a function of particle sizes and head loss as shown in Fig. 9-6.

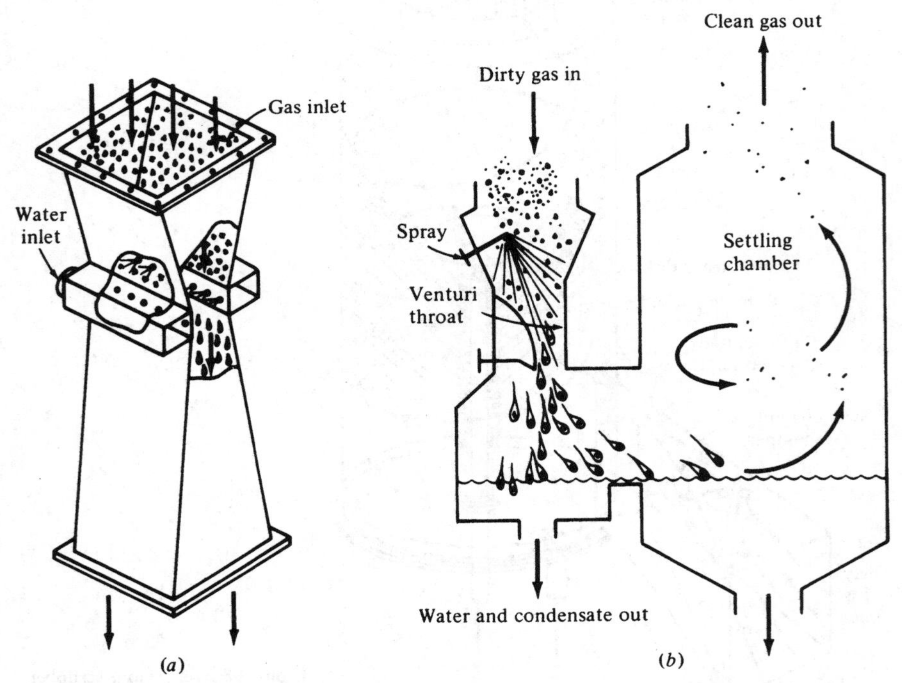

Figure 9-9 Venturi wet collector: (*a*) theoretical section (*from Air Pollution Manual* . . . [9-1]); (*b*) application (*from U.S. EPA* [9-15].)

9-6 FABRIC FILTERS (BAGHOUSE FILTERS)

In a fabric filter system, the particulate-laden gas stream passes through a woven or felted fabric that filters out the particulate matter and allows the gas to pass through. Small particles are initially retained on the fabric by direct interception, inertial impaction, diffusion, electrostatic attraction, and gravitational settling. After a dust mat has formed on the fabric, more efficient collection of submicron particles (99+ percent) is accomplished by sieving.

Filter bags, usually tubular or envelope-shaped, are capable of removing most particles as small as 0.5 μm and will remove substantial quantities of particles as small as 0.1 μm. Filter bags ranging from 1.8 to 9 m (6 to 30 ft) long, can be utilized in a baghouse filter arrangement similar to that shown in Fig. 9-10*a*. The upper ends are closed and the lower ends are attached to an inlet manifold A baghouse arrangement may be small enough to fit into an ordinary room or large enough to dwarf many industrial buildings. For example, one baghouse capable of removing 6 tons of fumes and dust per hour from an automotive foundry consists of 16 compartments housing 4000 bags, each of which is 19.4 cm (7.64 in) in diameter and 6.9 m ($22\frac{1}{2}$ ft) long. [9-50]

As particulates build up on the inside surfaces of the bags, the pressure drop increases. Before the pressure drop becomes too severe, the bags must be relieved of some of the particulate layer. Fabric filters can be cleaned intermittently, periodically, or continuously. Because intermittent cleaning means the unit must be shut down to prevent the discharge of raw gases directly into the atmosphere, that method is seldom used. In periodic cleaning portions of the filter device are shut down and cleaned for brief intervals, while the rest remains operational. In a continuous, automatic cleaning operation, cleaning of some parts of the filter occurs at all times. In any of these methods, a mechanical shaker, reverse-air flow (blow rings), or pulse jets may be used to remove the dust cake.

Continuously cleaned filters that use a traveling blow ring or reverse-air jets leave little filter cake in place at any time. Although this results in some loss of collection efficiency, the reduction in head loss allows faster filtering rates and decreases the size needed for the filter baghouse. [9-39]

There are some problems associated with the use of fabric filters. The possibility of explosion or fire exists if sparks are discharged in a baghouse area where organic dusts are being filtered. Space limitations may prohibit use of baghouses large enough to handle heavy loads. There is always a slight possibility of rupture or other adverse effects because of temperatures too high for the fabric medium or because of the moisture, acidity, ar alkalinity content of the particulate-laden gas stream. [9-10] Judicious fabric choice can minimize these problems; a summary of the limitations of some of the common filter fabrics used in industrial baghouses is given in Table 9-4.

The design of fabric filters is based on filtering rates, or air-to-cloth ratios. Filtering rates range from 0.5 to 5 m/min (m^3 air/min · m^2 cloth) depending on dust loading, fabric material, and method of cleaning. Sizing of the units is illustrated in the following example.

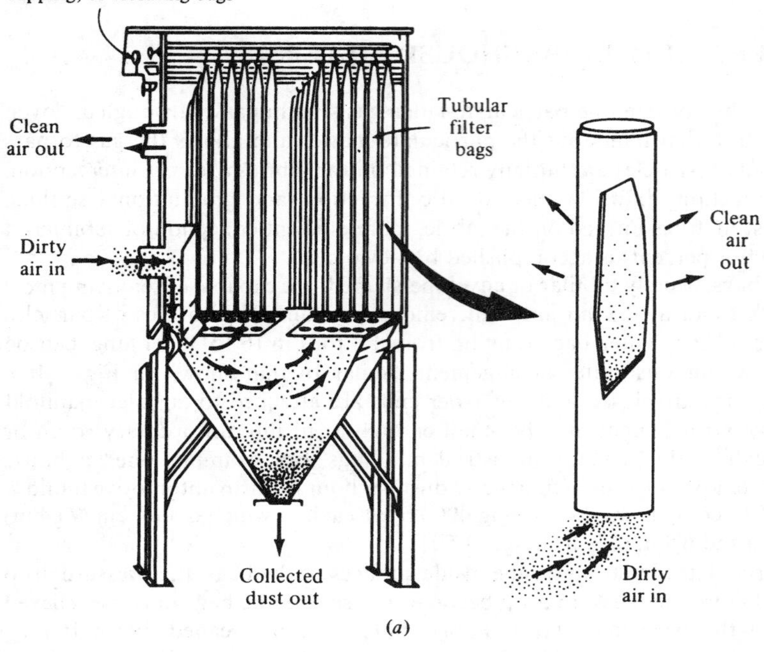

(*a*)

(*b*)

Figure 9-10 Baghouse filter. (*a*) Typical simple fabric filter baghouse design. (*From U.S. EPA* [*9-15*].) (*b*) Baghouse filters for removal of lead oxide particulates from Columbus, Georgia, plant of Chloride, Inc. (*Courtesy of American Air Filter, an Allis-Chalmers Company.*)

Table 9-4 Summary of data on the common filter media used in industrial baghouses

	Maximum temperature at baghouse inlet for continuous duty		Chemical resistance	
Fabric	Summary of published data, °C	Recommended maximum, °C	Acid	Alkali
Cotton	70–90	80	Poor	Fair
Dynel	65–80	80	Good	Good
Wool	80–110	105	Good	Poor
Nylon	90–140	105	Good	Poor
Orlon	90–175	135	Good	Fair
Dacron	120–175	135	Excellent	Good
Glass	260–370	290	Excellent	Excellent

Source: From Danielson. [9-10]

Example 9-3: Determining filter cloth area A fabric filter is to be constructed using bags that are 0.3 m in diameter and 6.0 m long. The baghouse is to receive 10 m^3/s of air, and the appropriate filtering velocity has been determined to be 2.0 m/min. Determine the number of bags required for a continuously cleaned operation.

SOLUTION

1. Determine the cloth area required.

$$10\ \text{m}^3/\text{s} \times \frac{60\ \text{s}}{\text{min}} = 600\ \text{m}^3/\text{min}$$

$$\frac{600\ \text{m}^3/\text{min}}{2\ \text{m/min}} = 300\ \text{m}^2$$

2. The area of one bag is:

$$\pi DH = \pi \times 0.3\ \text{m} \times 6.0\ \text{m} = 5.65\ \text{m}^2$$

3. The total number of bags is

$$\frac{300\ \text{m}^2}{5.65\ \text{m}^2} = 53.05, \quad \text{Use } 54$$

Fabric filters have many applications. They have a high collection efficiency over a broad range of particle sizes, extreme flexibility in design, the ability to handle large volumes of gases at relatively high speeds, reasonable operating pressure drops and power requirements, and the ability to handle a diversity of solid materials. [9-50] They are particularly useful in many high-volume operations such as cement kilns, foundries, steel furnaces, and grain-handling plants.

9-7 ELECTROSTATIC PRECIPITATORS (ESP)

Electrostatic precipitators can be classified as low-voltage two-stage units or high-voltage single-stage units. Low-voltage two-stage units operate at 6000 to 12,000 V and are employed mainly in conjunction with air-conditioning systems for hospitals and commercial installations. [9-28] They are used mainly to collect liquid particles and are not generally recommended for control of solid or sticky material.

Low-voltage precipitators have a separate ionizing zone located ahead of the collection plates, as seen in Fig. 9-11. The charged wires are located 2.5 to 5 cm (1 to 2 in) ahead of the parallel grounded plates. The corona discharge between the wires charges the particles suspended in the air flowing past them. The grounded collector plates are less than 2.5 cm (1 in) apart and alternately charged positive and negative. Liquid collects on the plate surfaces and drains by gravity to a collection chamber. Low-voltage precipitators have design capacities approaching 10 m^3/s (20,000 ft^3/min), with an air velocity of about 0.5 m/s (100 ft/min). [9-28]

High-voltage single-stage precipitators operate in the 30,000- to 100,000-V range and are used at large industrial plants such as coal-fired utility boilers. Four basic steps are required in the operation of a high-voltage single-stage electrostatic precipitator such as the one pictured in Fig. 9-12: (1) electrical charging of the particulates, (2) collection of charged particles on a grounded surface, (3) neutralization of the charge at the collector, and (4) removal of the particulate for disposal. [9-28, 9-33]

The electrical charge is imparted to the particulate by passing the particles through a high-voltage direct-current corona. The high-voltage field ionizes the

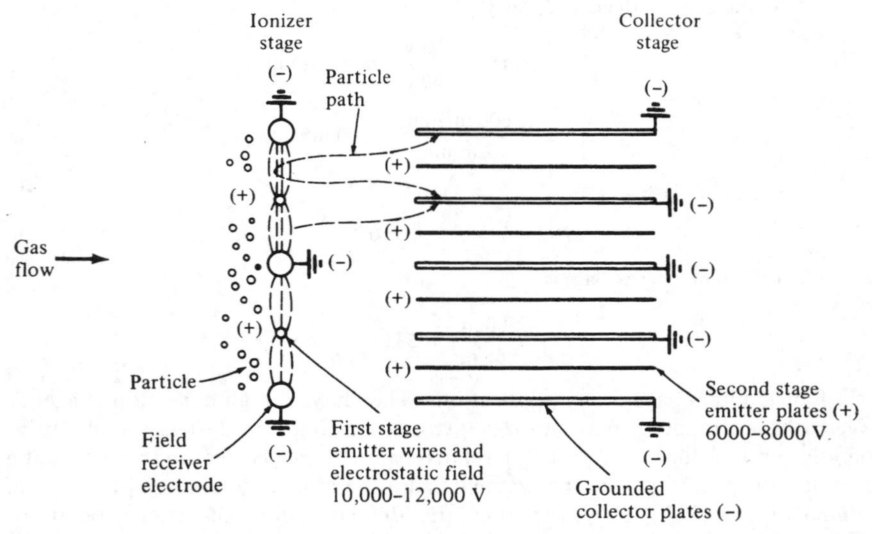

Figure 9-11 Low-voltage two-stage electrostatic precipitator. (*From Control Techniques . . .* [9-28].)

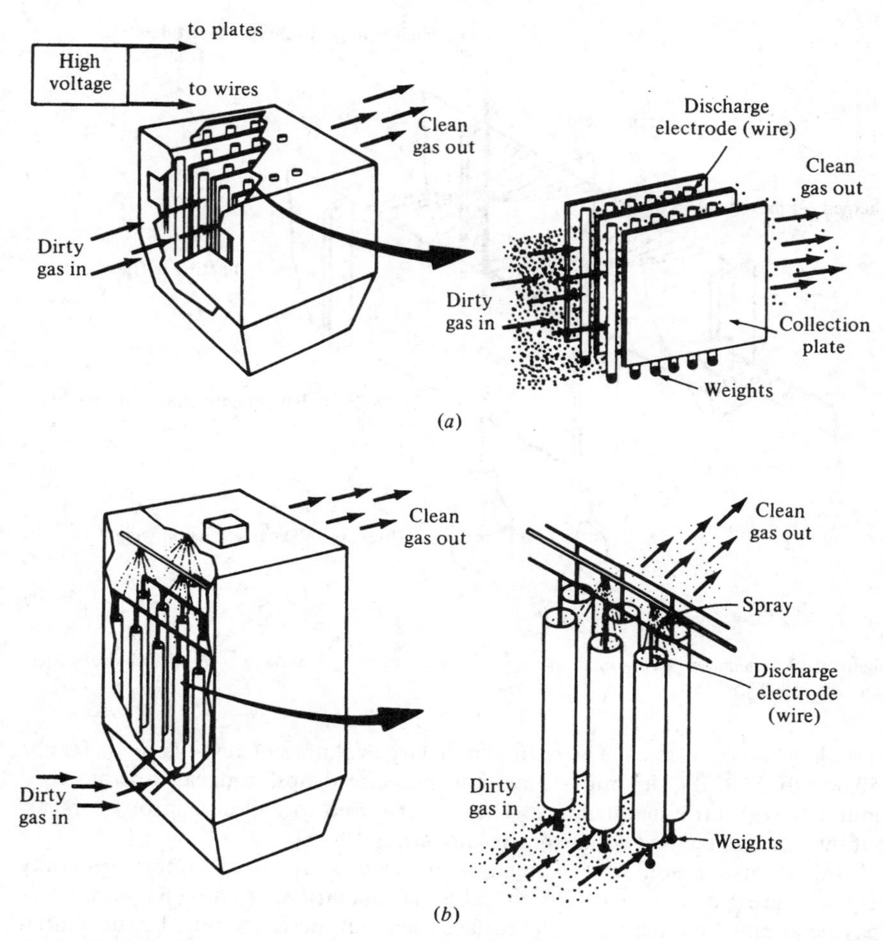

Figure 9-12 High-voltage electrostatic precipitators: (*a*) plate type and (*b*) tube type. (*From U.S. EPA* [9-*15*].)

gas molecules in the air stream, which in turn become attached to the particulate matter and give them a negative charge, as shown in Fig. 9-13. After being charged, the negative particles move toward the positive electrodes and are collected there. Their charge is neutralized at the moment of collection, and they can be removed from the collection surface by rapping, washing, or plain gravity. The energy used in separating particulates from a waste gas stream by means of an electrostatic precipitator is expended solely upon the particulate, not on the gas stream itself, as is the case in most other collection or control devices. [9-50]

Electrostatic precipitators have wide application. They are extremely efficient (99 percent or higher) for a wide range of particle sizes; even submicron-size

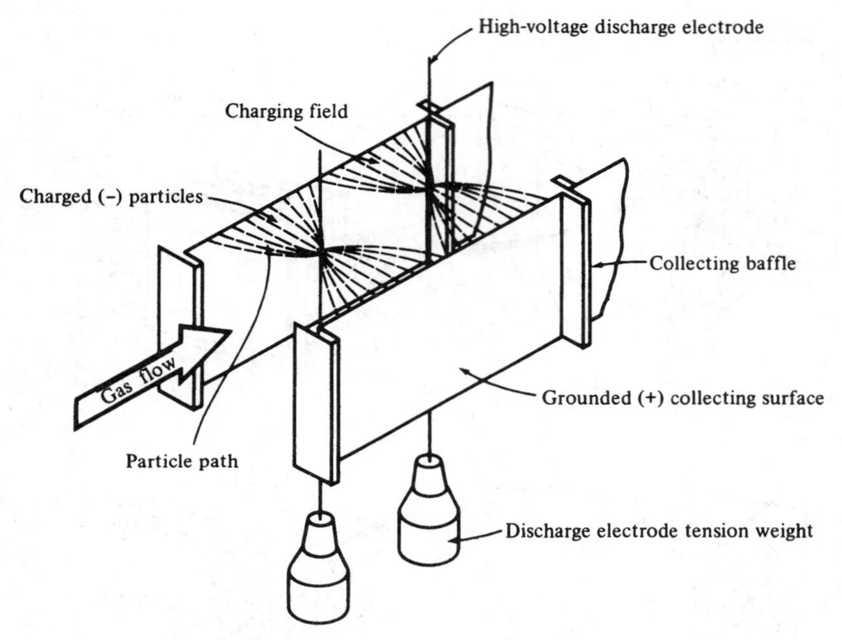

Figure 9-13 Schematic view of a plate-type electrostatic precipitator. (*From Control Techniques . . .* [*9-28*].)

particles can be collected. They can handle large volumes of gas—25 to 1000 m^3/s (50,000 to 2,000,000 ft^3/min)—have low pressure drops, and can operate continuously with little maintenance. They can be used to collect acid or tar mists, but they cannot be used with explosive materials. [9-28]

Initial installation cost is high, and electrostatic precipitators generally require a great deal of space for an industrial operation. Higher efficiency levels may be attained by increasing the surface area, but increases must be substantial in higher ranges. Electrostatic precipitation systems only operate at peak efficiency within a limited temperature range, and they may use excessive power if buildup of collected material causes "spark over." They may also become inefficient if buildup of collected material suppresses the corona discharge from the negative electrode.

The size-efficiency relationship for an electrostatic percipitator is a curvilinear function similar to that for a cyclone. The equation relating efficiency to particle size is a follows:

$$\xi = 1 - \exp - \left(\frac{Aw}{Q}\right) \tag{9-5}$$

where A = area of the collection plates, m^2
w = drift velocity of the charged particles, m/s
Q = flow rate of the gas stream, m^3/s

The drift velocity w is the velocity at which the particle approaches the collection plate. It is analogous to the terminal settling velocity in gravity settling, except the driving force is the electrical charge instead of gravity. The drift velocity can be expressed by

$$w = ad_p \tag{9-6}$$

where the parameter a has units of s^{-1} and is a function of the charging field, the carrier gas properties, and the ability of the particles to accept an electrical charge. The value of a is considered a constant for a given system but varies with system variables. The numerical value of w ranges from 0.04 to 0.2 m/s [9-50] and must be calculated or determined experimentally for any given system.

The sizing of electrostatic precipitators is illustrated in the following example.

Example 9-4: Determining plate-area requirements for an electrostatic precipitator An electrostatic precipitator is to be constructed to remove fly-ash particles from stack gases flowing at 10 m^3/s. Analysis of a similar system shows that the drift velocity can be taken as

$$w = 3.0 \times 10^5 d_p \text{ m/s}$$

Determine the plate area required to collect a 0.5 μm particle with

(*a*) 90 percent efficiency

(*b*) 99 percent efficiency

SOLUTION

1. Using Eq. (9-6)

$$w = 3.0 \times 10^5 \text{ s}^{-1} \times 5.0 \times 10^{-7} \text{ m}$$

$$= 0.15 \text{ m/s}$$

2. Using Eq. (9-5) for 90 percent efficiency

$$0.90 = 1 - \exp\left(-\frac{w}{Q} A\right)$$

$$= 1 - \exp\left(-\frac{0.15 \text{ m/s}}{10 \text{ m}^3/\text{s}} A\right)$$

$$1 - 0.9 = \exp(-0.015 \text{ 1/m}^2 A)$$

$$2.3 = 0.015 \text{ 1/m}^2 A$$

$$A = 153 \text{ m}^2$$

3. For 99 percent efficiency

$$1 - 0.99 = \exp(-0.015 \text{ 1/m}^2 \text{ A})$$

$$4.61 = 0.015 \text{ 1/m}^2 A$$

$$A = 307 \text{ m}^2$$

Because the efficiency equation is a logarithmic function, an increment to 99.9 percent efficiency would again double the required area.

Control Devices for Gaseous Contaminants

The principal gases of concern in air pollution control are the sulfur oxides (SO_x), carbon oxides (especially CO), nitrogen oxides (NO_x), organic and inorganic acid gases, and hydrocarbons (HC). Major treatment processes currently available for control of these and other gaseous emissions include adsorption, absorption, condensation, and combustion. [9-46]

It is not easy to decide which single or combined air pollution control technique to use for stationary sources. Gaseous pollutants can be controlled by a wide variety of devices, and choosing the most cost-effective, most efficient units requires careful attention to the particular operation for which the control devices are intended. Even after selecting what appears to be the best control technique, it is mandatory to monitor emissions carefully to ensure that predicted emission levels will meet emissions standards.

9-8 ADSORPTION

The pollution control process of gas adsorption involves passing a stream of effluent gas through a porous solid material (the *adsorbent*) contained in an adsorption bed. The surfaces of the porous solid material attract and hold the gas (the *adsorbate*) by either physical or chemical adsorption.

Physical adsorption, the condensation of gases and vapors on solids at temperatures above dew point, depends upon van der Waals force (an intermolecular attractive force). The amount of gas adsorbed relates to the ease of condensation of the gas—the higher the boiling point, the greater the amount adsorbed. Although physical adsorption is usually directly proportional to the amount of solid surface available, it is not limited to a single molecular layer, since a number of layers of molecules can build up on the surface. [9-4] Physical adsorption is accompanied by capillary condensation within the pores, which substantially increases the amount of gas that can be adsorbed. A small amount of heat is liberated during physical adsorption, and the process is relatively rapid and readily reversible. By lowering pressure or raising temperature, the adsorbed gas can be desorbed with no chemical change having occurred. an important consideration in air pollution control

In *chemical adsorption*, or *chemisorption*, the contaminant gas molecule forms a chemical bond with the adsorbent, and the gas is held strongly to the solid surface by valence forces. Much slower than physical adsorption because of the displacement of atoms that must occur in the molecules, chemisorption also liberates greater amounts of heat and requires more energy in the process. Indeed, at low temperatures, chemical adsorption may occur so slowly that it is hardly measurable.

Chemisorption results in the formation of a single layer of molecules on the solid surface, and the process is usually irreversible, because the chemical nature of the adsorbate will have been altered in the adsorption process. The amount of gas adsorbed in chemisorption processes depends upon pressure and temperature. [9-4]

Adsorbents

There are a great number of materials possessing adsorptive properties, and some of these and their uses are listed in Table 9-5. Preferential affinity for specific substances and surface-to-volume ratios are two key characteristics of solid adsorbents. Each adsorbent has a particular affinity for polar or nonpolar vapors. For example, alumina, bauxite, and silica gel have a higher affinity for water, a polar vapor, than for organic contaminants, hence their usefulness as drying agents. Conversely, activated charcoal preferentially adsorbs nonpolar organic compounds such as lower paraffin hydrocarbons.

For synthetic, silicate, or zeolite molecular sieves, selectivity or affinity is highly controlled because these crystalline structures are tailor-made to adsorb only certain molecules and only in specified amounts. In one instance, molecular sieves (specifically, crystalline metal aluminosilicates) proved capable of recovering 99 percent of SO_2 in a 182-tonne/d (200-ton/d) sulfuric acid plant, a feat which helped keep that plant's emission load to one-tenth the federal standard for new plants. In another instance, molecular sieves designed for use in a 50-tonne/d (55-ton/d) nitric oxide plant kept emissions to one-tenth the federal standard and enabled recovery rates that increased nitric acid production by 2.5 percent. However, use of molecular sieves to solve SO_2 problems for public utilities or NO_x problems from fossil-fueled steam-generating plants is not yet economically feasible. [9-27, 9-50]

While preferential affinity for specific substances is an important parameter of adsorbents, so, too, are surface-to-volume ratios. Large surface area per unit of weight is essential; the crucial surface area is that provided by the internal pores of the solid. Surface-to-volume ratios can be increased by activating some adsorbent. The most common adsorbent, activated carbon, is prepared by

Table 9-5 Types of adsorbents

Type adsorbent	Major uses
Activated carbon	Eliminating odors, purifying gases, recovering solvents
Alumina	Drying air, gases, and liquids
Bauxite	Treating petroleum fractions, drying gases and liquids
Bone char	Decolorizing sugar solutions
Decolorizing carbons	Decolorizing fats, oils, and waxes; deodorizing domestic water
Fuller's earth	Refining animal oils, lube oils, vegetable oils, fats, and waxes
Magnesia	Treating gasoline and solvents, removing metallic impurities from caustic solutions
Molecular sieves	Controling and recovering Hg, SO_2, and NO_x emissions
Silica gel	Drying and purifying gases
Strontium sulfate	Removing iron from caustic solutions

carbonizing wood, fruit pits, or coconut shells at very high temperatures and treating the substance with steam to burn away part of the carbon material and create a large internal pore structure that provides an exceptionally large internal surface area—an area from 10^5 to 10^6 m^2/kg (10^6 to 10^7 ft^2/lb). Activated alumina is aluminum oxide reactivated by heating to 175 to 325°C (350 to 600°F). [9-32, 9-50]

Adsorption Equipment

Adsorbers, the devices that contain the adsorbent solid through which the effluent gas must pass, can be designed with fixed, moving, or fluidized beds. The container for a simple *fixed bed* adsorption unit can be a vertical or horizontal cylindrical shell. The adsorbent, often activated carbon, is arranged on beds or trays in layers 1.3 cm (0.5 in) thick in thin-bed adsorbers and greater than 1.3 cm (0.5 in) thick in deep-bed adsorbers. If more than one bed is used, the beds can be arranged as shown in Fig. 9-14. [9-10]

An example of a *moving bed adsorber* is shown in Fig. 9-15. In this unit, the adsorption bed, activated carbon, is contained in a rotating drum. The filtered air, containing the gaseous contaminant, is moved by the fan into the rotating drum

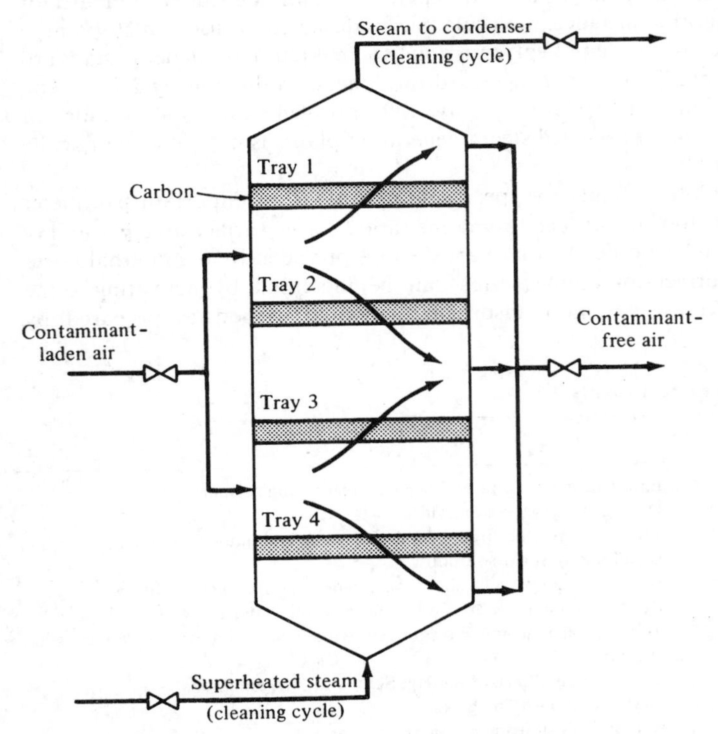

Figure 9-14 Multiple fixed-bed adsorber. (*From Danielson* [*9-10*].)

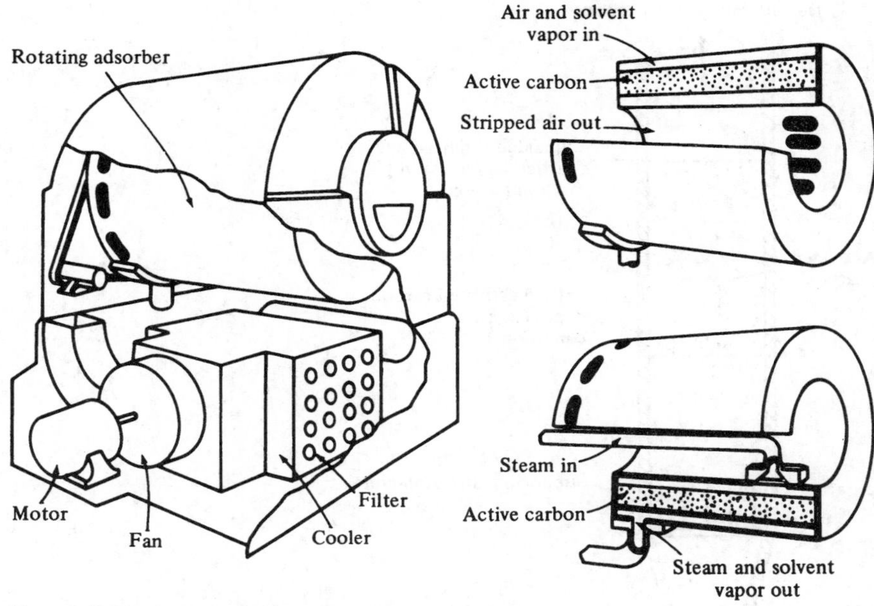

Figure 9-15 Moving-bed adsorber. (*From Danielson* [9-10].)

section. The vapor-laden air enters ports above the carbon bed, passes through the cylindrical activated carbon bed, enters the space in the inside of this drum, then leaves by ports at the ends of the drum. [9-10]

The *fluidized adsorber* shown in Fig. 9-16 contains a shallow, floating bed of adsorbent. The air flows upward, expanding the bed and suspending or fluidizing the adsorbent. The expanding and fluidizing of the adsorbent provides for intimate contact between the contaminated gas and the adsorbent and prevents channeling problems often associated with fixed beds. Once the gaseous contaminant has been adsorbed, the cleaned air stream passes through a dust collector before being discharged to the atmosphere.

The hydraulic principles involved in the fluidization of an adsorbent bed are essentially the same as those discussed in Sec. 4-8 for the fluidization of granular filter medium during backwashing. The different properties of the solid medium and the fluids must, of course, be taken into account.

Most adsorbtion units are highly efficient until a breakpoint occurs when the adsorbent becomes saturated with adsorbate. At this point, the concentration of pollutants in the exit gas stream begins to rise rapidly, and the adsorber must be regenerated or renewed. Adsorbers can be classified as *regenerative* or *nonregenerative*, depending upon whether the collected gas can be easily desorbed so that the adsorbent can be reused. Because nonregenerative adsorbent must be discarded after exhaustion and replaced with new material, the nonregenerative process is usually the more costly of the two. [9-26]

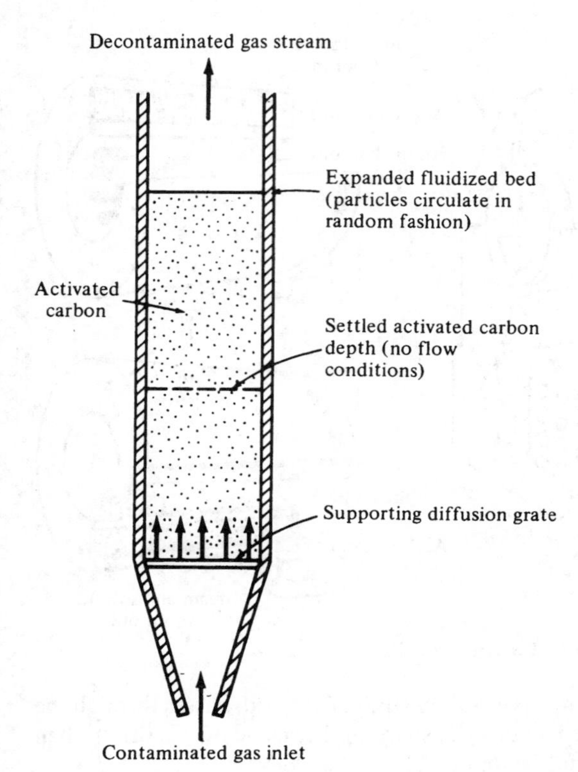

Figure 9-16 Fluidized adsorber bed configuration.

As noted earlier, physical adsorption leaves the adsorbent and adsorbate chemically unchanged and easily separated. In systems which rely on physical adsorption, regeneration of an adsorbent can be accomplished by use of superheated steam or circulating hot air. (If air with oxygen is used, temperatures must be kept below the carbon flash point.) The bed must be cooled before reuse. Burning, pressure reduction, and chemical treatment are other ways of removing or desorbing gases from adsorbents.

Adsorption equipment that allows for regeneration usually has more than one adsorption unit in operation so that the flow can be switched to an unsaturated bed while the saturated bed undergoes regeneration. The dual adsorption process shown in Fig. 9-17 provides for continuous operation.

Sometimes the gas is disposed of after it is removed from the effluent, but occasionally recovery of the gas itself, not just the adsorbent, may be economically worthwhile, as in the case of recovery of methyl chloroform from a movie-film processing plant and the recovery of ethyl alcohol vapors from a whiskey warehouse. [9-32]

Organic vapors that can be controlled by various adsorption processes include those discharged by the following industrial processes: dry cleaning, degreasing,

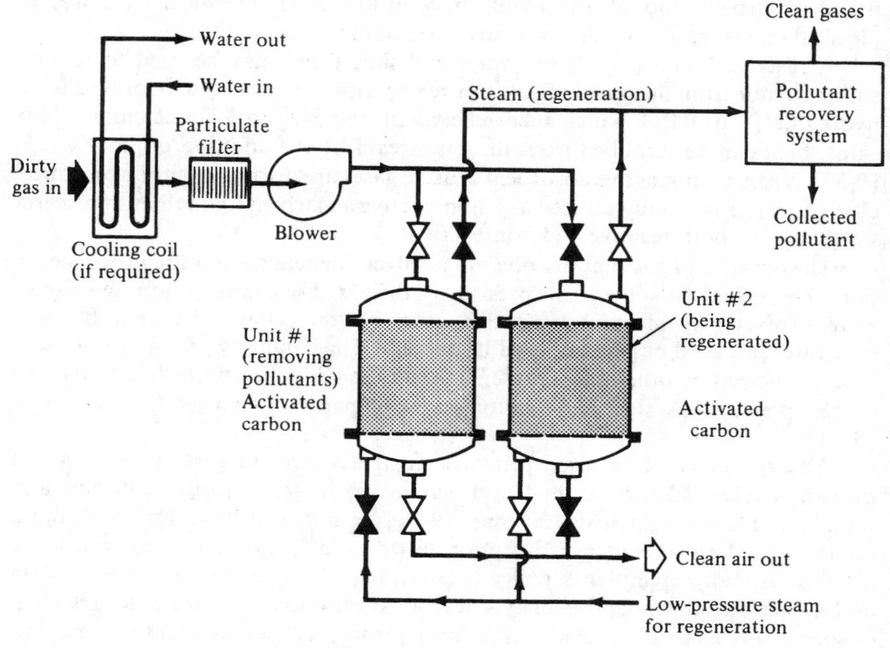

Figure 9-17 Diagrammatic sketch of two-unit fixed-bed adsorber. (*From U.S. EPA* [*9-15*].)

paint spraying, tank dipping, solvent extracting, and metal-foil coating. Emissions from plastics, chemical, pharmaceutical, rubber, linoleum, and transparent-wrap manufacturing processes may also be controlled by adsorption, as well as emissions from fabric impregnation processes. [9-26]

Recovery of valuable materials from some of these vapors is economically desirable, and adsorption allows for economically feasible recovery when concentrations of organic vapors are sufficiently high. [9-26] Sometimes, as in the Reinluft process, recovery of valuable by-products generated in the adsorption process is possible.

In the Reinluft process, flue gas containing SO_2 passes through a bed of activated carbon at temperatures of 93 to 149°C (200 to 300°F). The adsorbed SO_2 is oxidized to SO_3, and this substance then reacts with moisture in the gas stream to produce the valuable by-product H_2SO_4. [9-29]

9-9 ABSORPTION

Absorption, or scrubbing, involves bringing contaminated effluent gas (the *absorbate* or *solute*) into contact with a liquid absorbent (the *solvent*) so that one or more constituents of the effluent gas are removed, treated, or modified by the liquid absorbent. Liquid absorbents may utilize either chemical (*reactive*) or physical (*nonreactive*) change to remove pollutants.

A reactive liquid absorbent (water and limestone) may be used to remove sulfur dioxide from flue gases. The water reacts with the limestone to form calcium hydroxide [$Ca(OH)_2$], which then reacts with the SO_2 to form calcium sulfate salt, which can be scrubbed from the gas stream by the addition of more water. [9-32] When a nonreactive absorbent is used, gases are dissolved without chemical change. Water is usually utilized as a nonreactive absorbent, though in some cases water may be both reactive and nonreactive.

The amount of gas that dissolves in a solvent depends upon the properties of both the gas and the solvent, the pressure of the gas above the solution (see discussion of Henry's law in Sec. 3-4), the temperature of the system, and the turbulence, flow rate, and kind of packing used in the unit. The solubility of a gas is not materially affected by other gases present, as each gas dissolves according to its own partial pressure, as stated in Dalton's law of partial pressures (see Sec. 3-4). [9-4]

Absorption has been used primarily in the control of gases such as sulfur dioxide, oxides of nitrogen, hydrogen sulfide, hydrogen chloride, chlorine, ammonia, and some light hydrocarbons. [9-26] Removal of hydrocarbons from a contaminated air stream by absorption has been employed in many industries, notably, asphalt batch plants, coffee roasters, petroleum coker units, and varnish and resin cookers. Absorption of hydrocarbons is also used in industry for recovery of such products as acetic acid, chloroform, formic acid, amines, and ketones, but this recovery process does not usually remove enough of the contaminants to allow effluent gases to meet emission standards. [9-26] In such cases, adsorption and combustion processes may be used in combination with absorption processes.

Absorbent (Solvent)

As noted in Sec. 3-4, gas solubility differs from solvent to solvent. For instance, the solubility of nitrogen, oxygen, and carbon dioxide is two to ten times greater in ethanol, acetone, or benzene than in water. Solvents that are chemically similar to the solute generally provide good solubility, as do solvents of low viscosity. By choosing a solvent in which the absorbate is quite soluble, the rate of absorption can be increased and the amount of absorbent required can be decreased.

Other characteristics of the solvent should also be considered. Ideally, it should have a low freezing point and be low in toxicity, relatively nonvolatile, nonflammable, and chemically stable. Economical operation demands that it be relatively inexpensive, readily available, and noncorrosive, if possible, to reduce equipment repair and maintenance costs. [9-10]

Wet scrubbing units have been installed at a number of power plants to control SO_2 emissions, and removal efficiencies of 80 to 90 percent have been obtained. The main absorbents used in the various SO_2 absorption processes are aqueous solutions of the alkalies (sodium and ammonia) and the alkaline earths (calcium and magnesium). [9-43]

The leading alkali absorbents are sodium and ammonia. Solutions using sodium have the advantage over ammonia solutions, as they are not volatile and have no fume problems. An ammonia-solution system does have one important advantage in that ammonium sulfate as a by-product is more desirable than sodium sulfate. The use of ammonia solutions to absorb SO_2 is an old process, one widely used in the fertilizer industry long before SO_2 control was an air pollution concern. [9-43]

Alkaline earth compounds being used as solvents are magnesium oxide (MgO), calcium oxide (CaO), and calcium carbonate ($CaCO_3$). The magnesia processes involve scrubbing with a $Mg(OH)_2$ slurry, separation of the resulting $MgSO_3$, and calcination, yielding a stream of SO_2 and regenerated magnesia (MgO). The concentrated SO_2 can be used in H_2SO_4 production. In the plants that use this process to control SO_2 emissions and produce the valuable by-product H_2SO_4, the main problems encountered concern desposition of solids on surfaces in the scrubber. Currently, the bulk of absorption research and development is being applied to the exploration of possibilities related to the alkaline earth compounds. [9-43]

Absorption of NO_x from flue gases has been investigated in conjunction with control of sulfur oxides (SO_x). From 60 to 70 percent removal of NO_x along with simultaneous removal of 97 to 99 percent of the SO_x, has been reported. However, other studies using similar absorption solutions (limewater or limestone slurry) indicate only about 20 percent NO_x removal. [9-27]

Mass-Transfer Operations

The principles of mass transfer discussed in Sec. 3-4 apply to the absorption of contaminant gas from a carrier gas stream. Application of these principles is outlined below.

The most efficient mode of contacting the solute and solvent is the countercurrent flow system shown in Fig. 9-18*a*. In this mode, the mixture of gases containing the contaminant gas enters at the bottom of the tower and flows upward against the flow of the scrubbing liquid. The total gas flow is designated as G (mol/time) and, at the bottom of the tower, is composed of the carrier gas G'_B and the contaminant gas y_B associated with the carrier gas. At the top of the tower, the gas stream is composed of the carrier gas G'_A and any remaining contaminant gas y_A. Similarly, the total liquid flow is designated L (mol/time) and at the top of the tower is composed of the adsorbent L'_A plus any contaminant gas x_A that may be initially dissolved in the liquid. At the bottom of the tower, the liquid stream is composed of the liquid L'_B and the contaminant gas x_B dissolved in the liquid. If the carrier gas is insoluble in the liquid, then the carrier gas fraction and the liquid fraction will not change throughout the tower. That is

$$G'_B = G'_A = G' \tag{9-7}$$

and

$$L'_A = L'_B = L' \tag{9-8}$$

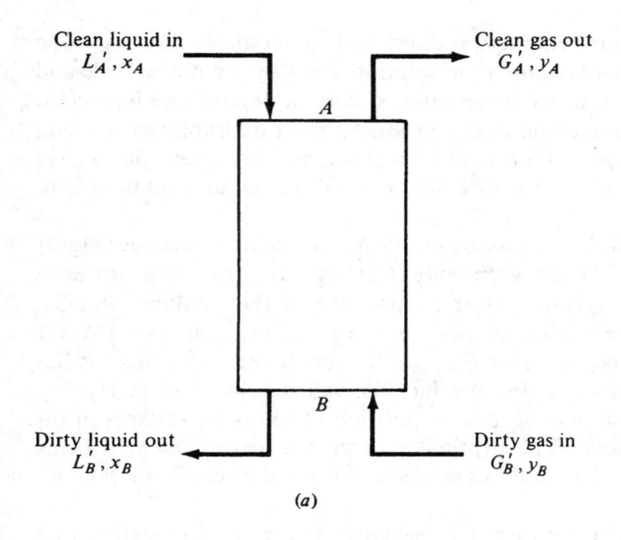

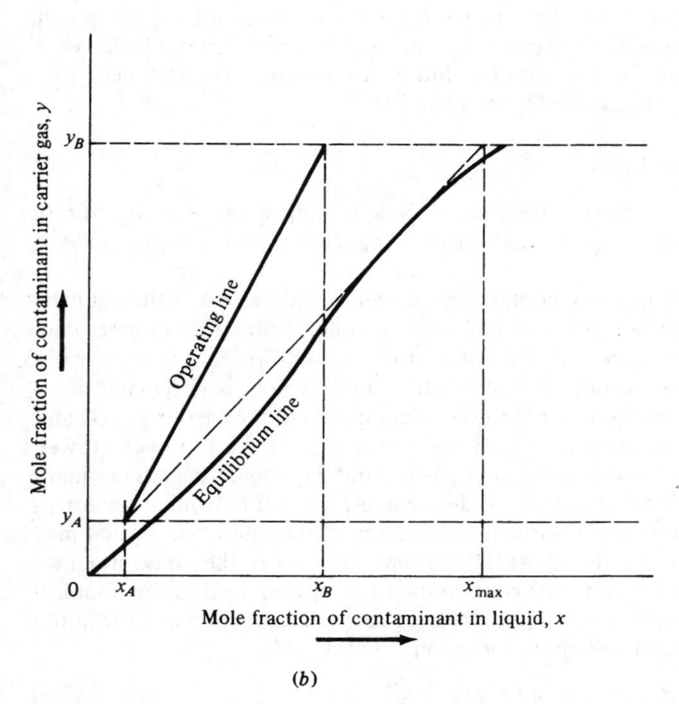

Figure 9-18 Absorption tower. (*a*) Definition sketch. (*b*) Equilibrium and operating lines.

As was discussed in Sec. 3-4, the equilibrium concentration of gas in a liquid is proportional to the partial pressure of the gas in contact with the liquid. Because the mole fraction of the contaminant gas in the carrier gas varies from bottom to top of the tower, the equilibrium concentration also varies. Equilibrium data for a particular gas in a given liquid under operating conditions of temperature and pressure can be obtained from published tables, or can be calculated using Henry's law. When the equilibrium data for all points of x and y between the bottom and top of the tower are plotted, an *equilibrium line* such as that shown in Fig. 9-18*b*, is obtained. For absorption to occur, the actual value of y must always be greater than the equilibrium value of y. Therefore, an *operating line* composed of corresponding x and y points must lie above the equilibrium line, as shown in Fig. 9-18*b*. The vertical distance between the operating and equilibrium lines represents the degree of saturation of the liquid with the contaminated gas.

If the mole fractions x and y are small, it can be shown that the operating line is a straight line with slope defined by

$$\frac{L'}{G'} = \frac{\dfrac{y_A}{1 - y_A} - \dfrac{y_B}{1 - y_B}}{\dfrac{x_a}{1 - x_a} - \dfrac{x_B}{1 - x_B}} \tag{9-9}$$

At the top of the tower, the operating line is fixed by the known x value and by the y value required to meet effluent standards. The slope of the operating line can be increased by increasing the liquid flow L', assuming a constant gas flow G'. The result is an increase in the difference between the actual and equilibrium value of the contaminant gas, resulting in an increase in the rate of absorption. On the other hand, a decrease in the rate of liquid flow results in a decrease in the slope of the operating line, and subsequently a decrease in the rate of absorption.

Operating lines with steep slopes and rapid absorption rates allow shorter towers to be used but require larger volumes of liquid. Conversely, operating lines with flatter slopes result in taller towers but more efficient use of the liquid. The minimum liquid flow is defined by the point at which the operating line crosses, or becomes tangent to, the equilibrium line. Although there is no theoretical maximum liquid flow, there are practical limits to the amount of liquid used. Obviously, a trade-off exists between tower height and liquid flow rate. A liquid flow rate of 1.5 times the minimum is often used in the design of scrubbers.

The following example illustrates the above principles with regards to liquid flow rates.

Example 9-5: Determining the solvent requirement in an absorption tower An absorption tower is to be used to remove SO_2 from the stack gas of a coal-fired furnace. The flow rate of the stack gas, measured at 1 atm pressure and 25°C, is 10 m^3/s, and the SO_2 content is 3.0 percent by volume. Removal of 90 percent of the SO_2 is required, and water, initially pure with respect to SO_2, is to be used as the liquid solvent. The equilibrium line for SO_2 and water can be estimated by $y = 30x$. Determine the flow rate of water that represents 150 percent of the minimum liquid requirement.

SOLUTION

1. Determine the x, y coordinates at the top and bottom of the tower.
   a. Top:

$$x = 0 \text{ (because the water is initially pure)}$$

$$y = 0.03 \times (1 - 0.90) = 0.003$$

   b. Bottom:

$$x = \text{unknown}$$

$$y = 0.03$$

2. These points are plotted on the diagram.

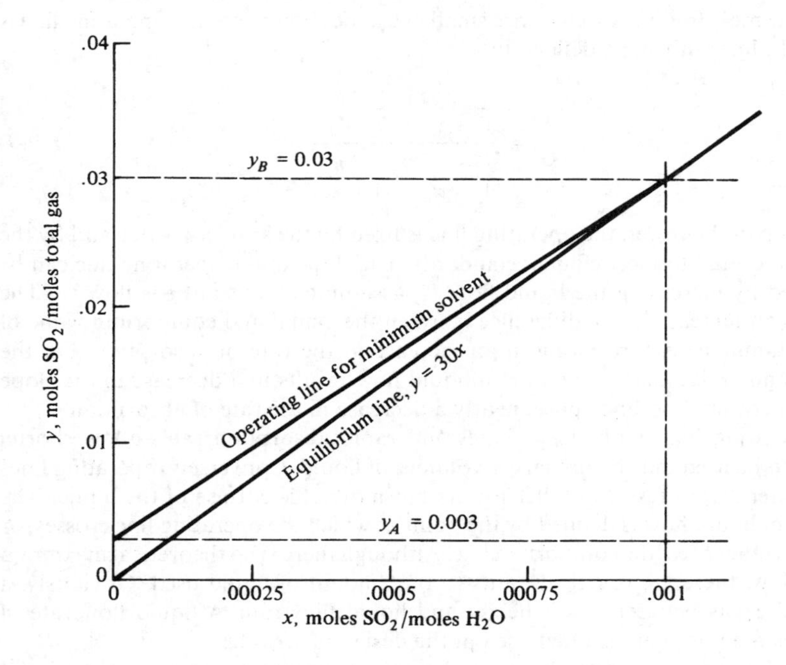

3. Plot the equilibrium line on diagram.

$$y = 30x \text{ is a straight line from the original at } y = 0.03$$

$$x = \frac{0.03}{30} = 0.001$$

4. The operating line representing the minimum water requirement is found by connecting the x, y coordinates at the top of the tower to the point $x = 0.001$, $y = 0.03$ at the bottom of the tower.

5. The approximate slope of the operating line can be determined from the diagram and can be found exactly by equation.

$$\frac{L'}{G'} = \frac{\dfrac{y_A}{1 - y_A} - \dfrac{y_B}{1 - y_B}}{\dfrac{x_A}{1 - x_A} - \dfrac{x_B}{1 - x_B}}$$

$$= \frac{\dfrac{0.003}{1 - 0.003} - \dfrac{0.03}{1 - 0.001}}{\dfrac{0}{1 - 0} - \dfrac{0.001}{1 - 0.001}}$$

$$= 27$$

6. Determine the mol/time flow rate of G'.

$$G' = 10 \text{ m}^3/\text{s} \times 0.97 = 9.7 \text{ m}^3/\text{s}$$

Eq. (7-2) is used to convert this volume to standard temperature and pressure.

$$\frac{V_1 P_1}{T_1} = \frac{V_2 P_2}{T_2}$$

$$V_1 = \frac{T_1 V_2 P_2}{T_2 P_1}$$

$$= \frac{273 \times 9.7}{(273 + 25)}$$

$$= 8.89 \text{ m}^3/\text{s}$$

$$G' = \frac{8.89 \text{ m}^3/\text{s}}{22.4 \text{ L/mol}} \times \frac{1000 \text{ L}}{\text{m}^3} = 397 \text{ mol/s}$$

7. Determine the minimum water flow rate.

$$L' = 27G = 27 \times 397 = 10{,}719 \text{ mol/s}$$

8. The actual flow rate should be

$$L'_A = 1.5 \times 10{,}719 = 16{,}079 \text{ mol/s}$$

One mole of water is 18 g.

$$18 \text{ g/mol} \times 16{,}079 \text{ mol/s} = 289{,}413 \text{ g/s}$$

or

$$0.289 \text{ m}^3/\text{s}$$

While the above example illustrates some of the principles of absorber design, there are many other factors that must be considered in actual practice. These include the physical characteristics of the absorbers and liquid injection systems, as well as the means of contacting the gas and the liquid. A detailed discussion of these factors is beyond the scope of this text but some general observations will be made.

A discussion of water and air contact systems was included in Sec. 4-3, and many of these principles apply to liquid and gas contact in absorbers. Spray towers are effective in situations where the gas film controls the rate of dissolution, because the gas film on the outside of the liquid drop is made thinner by shear. Generally, low turbulence in these units results in low head loss but poor liquid-gas contact, thus removal efficiencies are usually less than for other types of absorbers. Tray towers disperse the gas in the liquid and work best for liquid-film-control situations. Because thin water depths are used, tray towers may require many trays in order to effect the desired degree of treatment. Packed towers provide a combination of the principles in spray and tray towers. At low liquid loadings, the pores in the packing are open and the gas and liquid pass each other as in a spray tower. However, turbulence caused by flow around the medium increases gas-liquid contact and increases efficiency. At increased loading rates the pores begin to fill with liquid. This level of loading is referred to as the *load point*. If loading is increased, all of the pores become completely filled with liquid. This level of loading is called the *flood point* and results in an operation similar to a tray tower with a very large number of trays. Packed towers are usually operated between the load point and the flood point and are capable of removing gases that are gas-film-controlled, liquid-film-controlled, or mixed-film-controlled.

Absorption Units

Gas absorption units are designed to provide intimate contact between the gas and the liquid and to provide optimum diffusion of the gas into the solution. Several types of absorbers are used, including spray towers, plate or tray towers, packed towers, and venturi scrubbers. The selection of the best scrubber system involves consideration of the number of scrubbers to be used (a single unit that removes both particulate and gaseous contaminants or two separate units, one for particulates and another for gaseous matter). Some of the units discussed below are capable of handling both types of contaminants.

Spray towers Spray towers can handle fairly large volumes of gas with relatively little pressure drop and reasonably high efficiency of removal—as long as gaseous contaminant concentrations are fairly low. Spray towers are also effective for dual removal of particulate and gaseous contaminants, since they can handle gases with fairly high concentrations of particulates without plugging (see Sect. 9-5).

In general, the smaller the droplet size and the greater the turbulence, the more chance for absorption of the gas. Production of fine droplets requires the use of high-pressure spray nozzles that consume more energy than do low-pressure nozzles. Figure 9-19 shows an example of a spray tower in which the absorbing liquid, usually water, is sprayed through the contaminated gas and the absorbent-contaminant solution falls downward for removal while clean gas exits through an

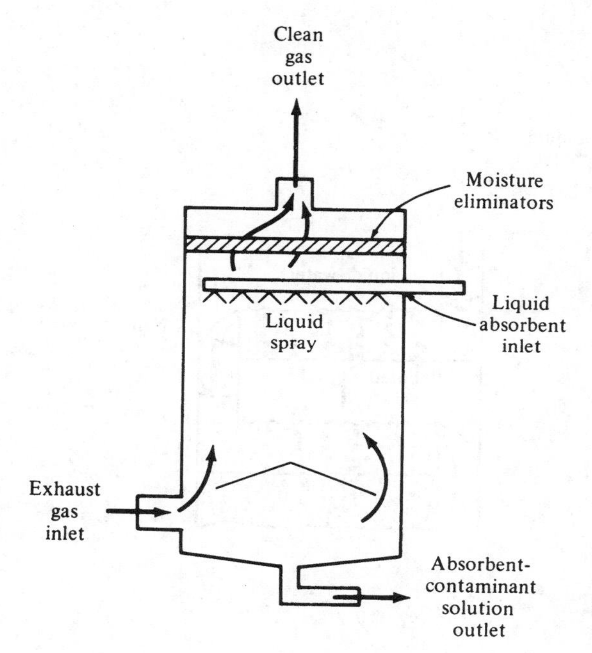

Figure 9-19 Spray tower. (*From Control Techniques . . .* [9-26].)

outlet valve in the top of the unit. [9-26] Moisture eliminators reduce the amount of moisture in the gases being released.

Since spray towers have much less gas-liquid interfacial area than most other types of absorbers, they are generally less effective in removal of gaseous contaminants. Nonetheless, they are effective enough for many operations and have the added advantage of being relatively inexpensive to install and operate.

Plate or tray towers Plate or tray towers contain horizontal trays or plates designed to provide large liquid-gas interfacial areas. In the perforated plate column shown in Fig. 9-20, the absorbent enters from the side of the column near the top and spills across the top sieve tray, the first in a series of horizontal perforated plates that are usually spaced 0.3 to 0.9 m (1 to 3 ft) apart and typically have a surface featuring 5-mm ($\frac{3}{16}$-in) holes on 13-mm ($\frac{1}{2}$-in)-square centers. The liquid flows across this tray, over a weir, and to a downpipe that directs the flow to the next tray down (see Fig. 9-20). This zigzag flow continues until the liquid reaches the bottom of the column.

The polluted air, introduced at one side of the bottom of the column, rises up through the openings in each tray, and it is the rising gas that prevents the liquid from draining through the openings rather than through the downpipe. Through repeated contact between air and liquid, gaseous contaminants (plus dust and other particulates) are removed, and the clean air emerges from the top of the column. [9-42]

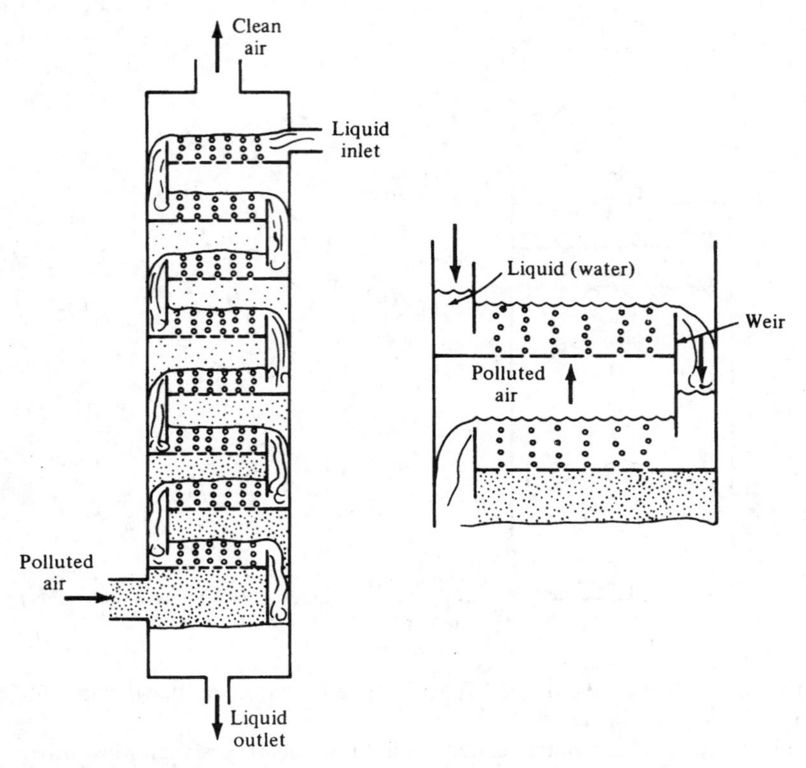

Figure 9-20 Sieve plate tower. (*From Shaheen* [*9-42*].)

In a bubble-cap tray column such as the one in Fig. 9-21, the contaminated gases rise upward until they strike the caps, at which point they are diverted downward and discharged as small bubbles from slots at the bottom of the caps. [9-26] As the gas continues upward, the water-gas contact continues in the fashion described until the clean gas emerges at the top. The contaminant-laden liquid flows to the bottom and is drawn off.

Packed towers In a packed tower such as the one in Fig. 9-22, packing is used to increase the contact time between vapor and liquid. The material chosen for packing has a large surface-to-volume ratio and a large void ratio that offers minimum resistance to gas flow. Lightweight and virtually unbreakable, Raschig rings, Berl saddles, Pall rings, Intalox saddles, and Tellerettes are suitable packing for such units (see Fig. 9-22*b*). [9-34]

In typical operations, the flow through a packed tower is countercurrent, with gas entering at the bottom of the tower and liquid entering at the top. Liquid

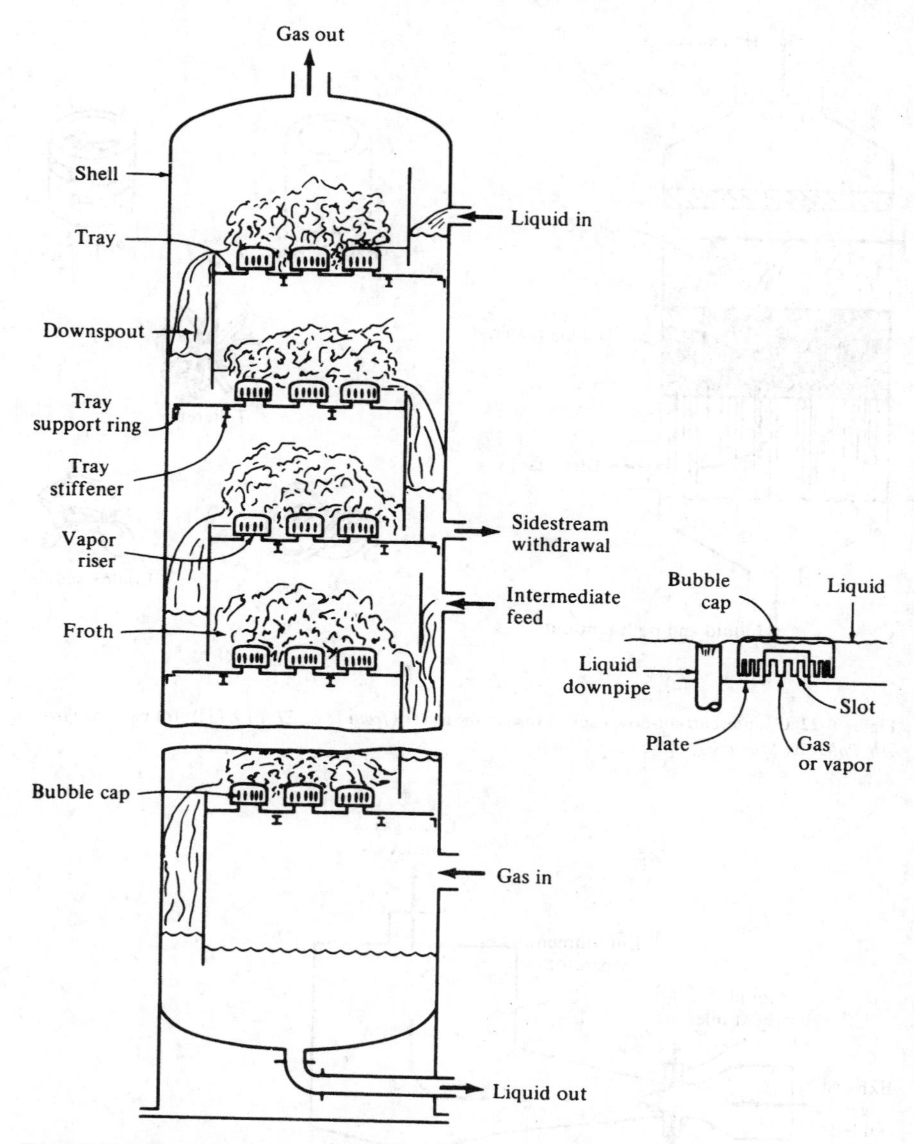

Figure 9-21 Schematic diagram of a bubble-cap tray tower. (*From Treybal* [9-47].)

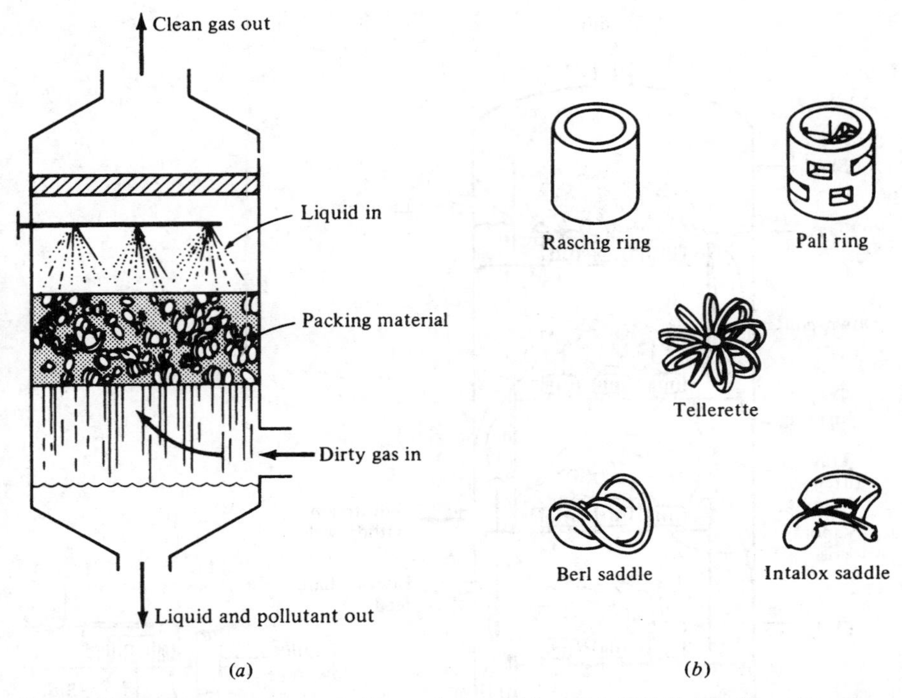

Figure 9-22 Countercurrent-flow packed tower: (*a*) tower (*from U.S. EPA* [*9-15*]); (*b*) packing (*from Air Pollution Manual* . . . [*9-1*]).

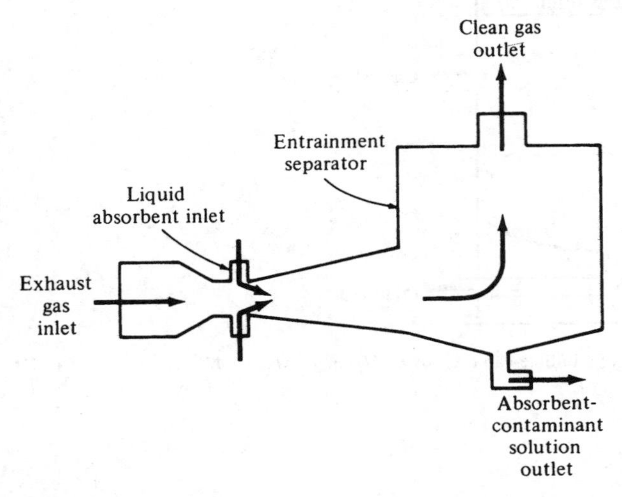

Figure 9-23 Venturi scrubber. (*From Control Techniques* . . . [*9-26*].)

flows over the surface of the packing in a thin film, affording continuous contact with the gases. [9-26]

Though highly efficient for removal of gaseous contaminants, packed towers become easily clogged when gases with high particulate loads are introduced. Therefore, they are often used after gases have already passed through a unit capable of removing most of the particulate matter entrained in the gases.

Venturi scrubbers Figure 9-23 is a schematic drawing for a venturi scrubber, a cocurrent unit in which the gas and the absorbing solution are brought into contact in or near the ventrui throat and move together into an entrainment separator. The gas-liquid mixture is then separated by the centrifugal force of the liquid droplets as the clean gas stream moves to the exit. Venturi scrubbers also remove particulate matter (see Sec. 9-5). [9-26]

9-10 CONDENSATION

A compound will condense at a given temperature if its partial pressure is increased until it is equal to or greater than its vapor pressure (the pressure exerted by a solid or liquid at equilibrium with its own vapor) at that temperature. If the temperature of a gaseous mixture is reduced to its saturation temperature, its vapor pressure equals its partial pressure and condensation will occur. While condensation may also be brought about by increasing pressure, this method is seldom used in air pollution control. [9-26]

There are two basic types of condensation equipment—surface and contact condensers. In a *surface condenser*, physical adsorption plays a key role, since contaminants are adsorbed onto a surface as the gaseous compound condenses. A surface condenser whose cooling medium is air or water is shown in Fig. 9-24;

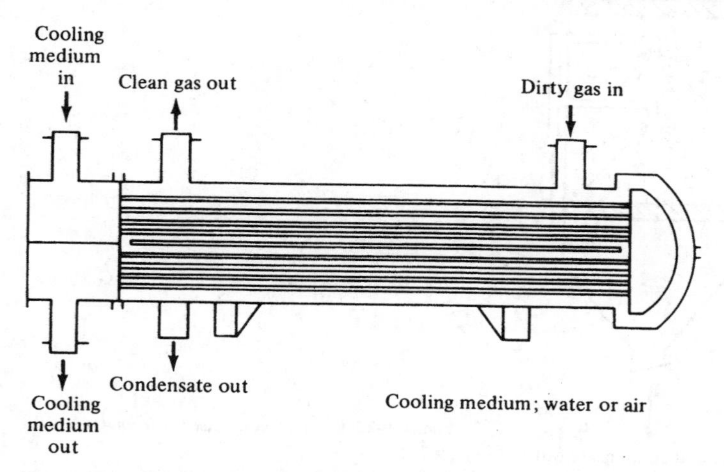

Figure 9-24 Surface condenser. (*From U.S. EPA* [9-15].)

the vapor to be condensed is separated from the cooling medium by a metal wall. As the cooling medium flows through the tubes, the vapor condenses on the surface of the tubes, the condensed vapor collects as a film of liquid, and the liquid drains off to storage. [9-15]

In a *contact condenser*, the vapor and cooling medium are brought into direct contact. A contact condenser in which a spray of cold water comes in direct contact with the contaminated air stream is shown in Fig. 9-25. The cooled vapor condenses, and the water and condensate mixture are removed, treated, and disposed of. Contact condensers are less expensive and more flexible than surface condensers, and they are more efficient in removing organic compounds. However, the fact that use of contact condensers can create a water pollution problem sometimes restricts their use. [9-15]

The specific application of the condensation process depends upon the amount and type of coolant that must be used, the waste liquid disposal problems that will result, and the amount of compound that is to be recovered. Generally considered as pretreatment devices for air pollution control, condensers are used in conjunction with afterburners, absorbers, or adsorption units. Representative applications of condensers in the air-pollution-control field are listed in Table 9-6. The widest application of condensers has been in the field of hydrocarbon emission control.

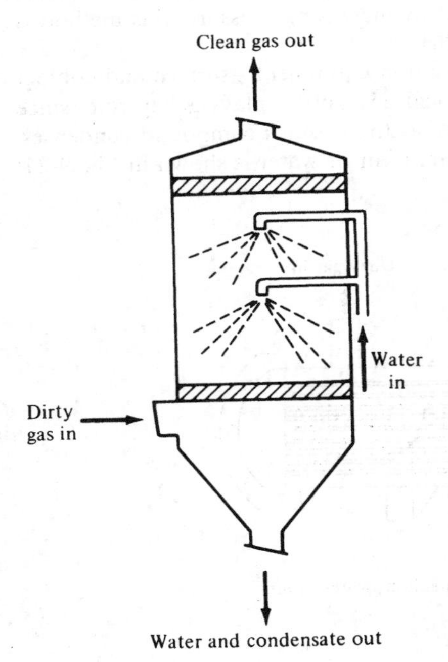

Figure 9-25 Contact condenser. (*From U.S. EPA* [9-*15*].)

Table 9-6 Representative applications of condensers in air pollution control

Petroleum refining	Petrochemical manufacturing	Basic chemical manufacturing	Miscellaneous industries
Gasoline accumulator vents	Polyethylene gas accumulator vents	Ammonia	Dry cleaning
Storage vessels	Styrene	Chlorine solution	Degreasers
Lube oil refining	Copper napthenates		Tar dipping
	Insecticides		
	Phthalic anhydride		
	Resin reactors		
	Solvent recovery		

Source: From *Control Techniques*. [9-26]

9-11 COMBUSTION

Though it is a major source of air pollution, combustion, or incineration, is also the basis for an important air-pollution-control process in which the objective is to convert the air contaminants (usually hydrocarbons or carbon monoxide) to innocuous carbon dioxide and water. [9-10] The combustion equipment used to control air pollution emissions is designed to push oxidation reactions as close as possible to completion, leaving a minimum of unburned compounds. For efficient combustion to occur, it is necessary to have the proper combination of four basic elements: oxygen, temperature, turbulence, and time.

During combustion, the supply of *oxygen* available will determine the end products obtained. Soot and carbon monoxide are by-products of combustion at low oxidation, while carbon dioxide is a by-product of combustion in the presence of sufficient oxygen. Although combustion can occur as soon as a substance reaches its kindling point, for air-pollution-control purposes, the *temperature* must be kept at ignition temperature, the point at which more heat is generated by the reaction than is lost to the surroundings. For carbon monoxide this may be from 610 to 657°C (1130 to 1215°F).

The *turbulence* needed to keep oxygen well mixed with the combustible substance may be provided by baffles or injection nozzles. Combustion chambers are designed to provide both turbulence and enough *time* for sufficient burning; often this is accomplished by increasing stack height. [9-32]

Depending upon the contaminant being oxidized, direct-flame combustion, thermal combustion (afterburners), or catalytic combustion methods can be used to control air pollution.

Direct-Flame Combustion

In direct-flame combustion, waste gases are burned directly in a combustor, with or without the addition of a supplementary fuel. In some cases, heat value and oxygen content of the waste gases are sufficient to allow them to burn on their own. In other cases, introducing air and or adding a small amount of supplemental fuel will bring the gaseous mixture to its combustion point.

Direct-flame combustion flares such as the one shown in Fig. 9-26 are frequently used in petrochemical plants and refineries. Flares are usually open-ended combustion units maintained outdoors at the end of a waste-gas stream at the top of a stack and equipped with pilots to ensure continuous burning. [9-15]

While flare burning is a relatively safe means of disposing of the large quantities of highly combustible waste gases, it is not an ideal disposal method in other respects. Some flares burn at sufficiently high temperatures and for sufficiently long periods to cause the formation of oxides of nitrogen (because of the nitrogen content of ambient air), thus creating a new air pollutant. Further, unless the fuel-air ratios and other factors are kept carefully controlled, the flares may produce

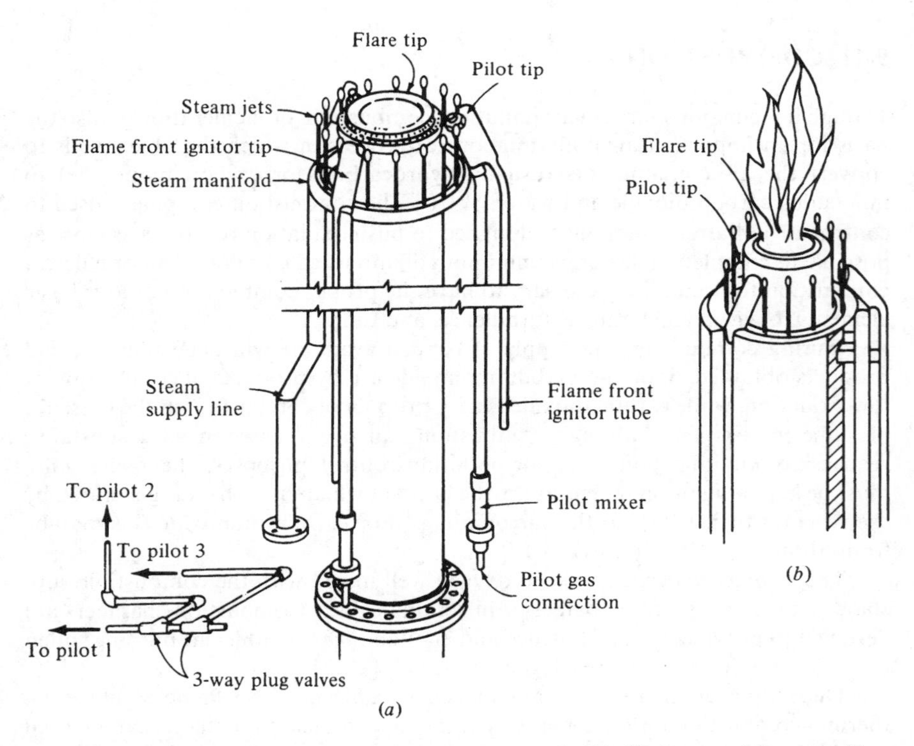

Figure 9-26 Steam injection type flare: (*a*) flare (*from Painter* [*9-32*]; (*b*) close-up of flare head (*from U.S. EPA* [*9-15*].)

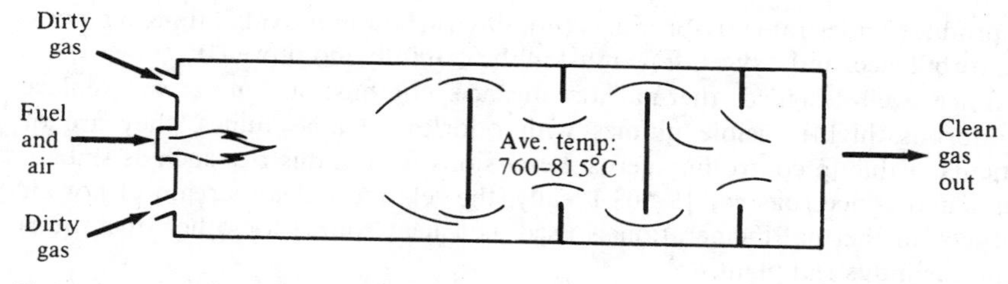

Figure 9-27 Thermal incinerator. (*From U.S. EPA* [9-15].)

visible smoke or soot. Finally, unless they are engineered as heat sources for waste-gas boilers or other operations, flares waste large amounts of heat energy, an important consideration in view of shrinking fossil-fuel resources. Direct-flame combustion processes are only economical when the waste gas itself contributes more than 50 percent of the total heating value required for incinerations. [9-50]

Thermal Combustion

In cases where the concentration of combustible gaseous pollutants is too low to make direct-flame incineration feasible, a thermal incinerator, or afterburner, such as the one in Fig. 9-27, may be the unit of choice. Generally, the waste gas is preheated, often by use of a heat-exchanger (a recuperator or regenerator such as the one shown in Fig. 9-28) utilizing heat produced by the thermal incinerator itself. [9-50] The preheated gas is directed into a combustion zone equipped with a burner supplied with supplemental fuel. The temperatures of operation depend upon the nature of the pollutants in the waste gas; common temperatures lie between 538 and 927°C (1000 and 1700°F), with operating temperatures up to 1093°C (2000°F) occasionally being utilized. [9-15] Thermal afterburners must be carefully designed to provide safe, efficient operation. Since incomplete burning

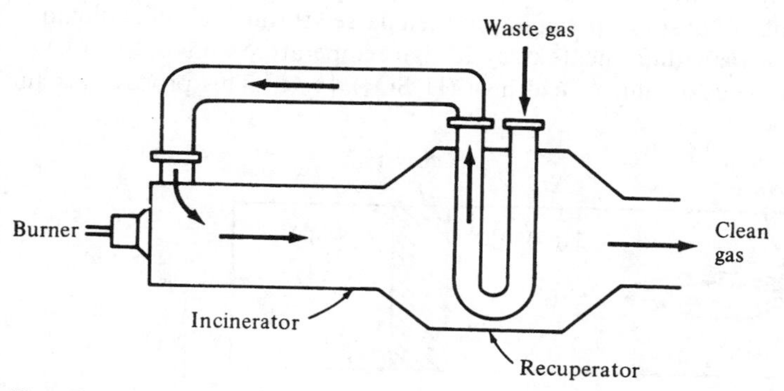

Figure 9-28 Thermal incinerator with recuperator.

can produce undesirable by-products (notably carbon monoxide), time, temperature, turbulence, and oxygen flow must all be carefully monitored.

Since well-designed thermal or furnace combustion units can replace malodorous, highly visible plumes with odorless steam plumes, they are of particular value in controlling aerosol emissions from industries such as smokehouses and coffee roasters. [9-50] Ideally, the relatively clean stream of hot air produced in thermal incineration is used as a heat source for other operations within the industrial plant.

Catalytic Combustion

Catalytic incineration is another method that can be used when combustible materials in the waste gas are too low to make direct-flame incineration feasible. A catalyst accelerates the rate of oxidation without itself undergoing a chemical change, thus reducing the residence, or dwell, time required for incineration. Thermal incineration may require residence times 20 to 50 times greater than catalytic incineration. [9-50]

A catalytic incinerator generally consists of a preheating section and a catalytic section, though cold catalytic systems are now available that operate at ambient temperatures, eliminating the need for a preheater. [9-22] A catalytic incinerator with a preheater is illustrated in Fig. 9-29. There is no direct flame in such a unit, though the catalyst surface glows. Often a fan is located in the afterburner housing to mix the gases and distribute them evenly over the catalyst.

Supplemental fuel usage for catalytic incinerators is generally lower than for thermal incinerators, thus reducing operating costs. While efficiency of well-designed catalytic combustion units can run between 95 and 98 percent, incinerator efficiency depends on many factors, including contaminant concentration, temperature of gas stream, oxygen concentration, contact time, and type of catalyst. Removal efficiences of greater than 95 percent are not uncommon.

Catalytic combustion processes have been used to control SO_2, NO_x, hydrocarbons, and carbon monoxide (CO). Monsanto Corporation has done the most significant work in the area of catalytic removal of SO_2. In their process, the gas is thoroughly cleaned of dust in a precipitator, then passed through a sulfur-dioxide-oxidation catalyst (vanadium pentoxide) at high temperature—454°C (850°F)—a process which yields a sulfuric acid mist (H_2SO_4). [9-18] This process has an

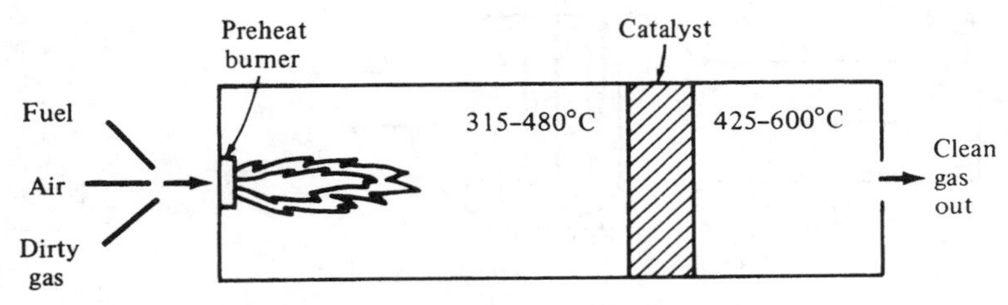

Figure 9-29 Catalytic incinerator. (*From U.S. EPA* [9-15].)

estimated capacity for removing 90 percent of the original SO_2 present in the flue gas. However, potential disadvantages include high capital cost, possible catalytic poisoning, and economic dependence upon marketing sulfuric acid produced in the process. [9-18].

The only large-scale application of catalytic removal of oxides of nitrogen is in the tail-gas treatment for nitric acid plants. The tail gas is mixed with CO and CH_4 before it is passed over the catalyst, and the end products are N_2, CO_2, and water. Platinum metals, found to be the most effective catalyst, can provide for over 90 percent reduction in NO_x emissions. [9-45]

Catalytic incineration of hydrocarbons can be done at low-contaminant concentrations. The major problems with these catalytic systems are their high cost of maintenance and catalytic poisoning. The hydrocarbon combustion catalyst consists of a basic material, such as activated alumina, impregnated with a metallic compound. Hydrocarbon catalytic systems can provide over 95 percent contaminant removal and are used in petroleum refining, chemical processing, and fabric-coating industries. [9-26]

Most catalytic systems that can convert CO to CO_2 operate at a minimum temperature of 200 to 220°C (400 to 600°F), for the oxidation of CO to CO_2 is thermodynamically favored by more than 60 K cal/mol. The problem has been in finding a way to reduce the activation energy barrier and permit the oxidation to proceed at lower temperatures. A Pd(II)/Cu(II) catalyst has been developed that has solved this problem and can oxidize CO to CO_2 at ambient temperature. [9-20, 9-21, 9-22]. Though this system has been known for some time, it has not found practical application, primarily because the overall reactions are quite slow at ambient temperatures. However, by the adjustment of the anion ratio, the catalytic activity can be sharply increased to allow for a more rapid conversion of CO to CO_2. [9-19]

Applications of this new catalyst for control of emissions from gas-fired furnaces and automotive exhausts have been investigated. In other research, this catalyst was incorporated into a catalytic cigarette filter that demonstrated a capability of removing up to 97 percent of the CO present in cigarette smoke. [9-38] In another study, a small air-intake catalytic filter reduced the CO inside the passenger compartment of automobiles to up to 46 percent of the ambient street-level concentrations. [9-37]

9-12 AUTOMOTIVE EMISSION CONTROL

On a weight basis, transportation accounted for 55 percent of all major air contaminants emitted to the atmosphere in 1980, including 81 percent of the total carbon monoxide, 36 percent of the volatile hydrocarbons, 44 percent of the oxides of nitrogen, 18 percent of the particulate matter, and 4 percent of the sulfur oxides. [9-8] These figures were high in 1960 when 60 million privately owned automobiles burned 58 billion gallons of gasoline in the United States. By 1980, with almost twice as many automobiles (one vehicle for every two persons in the country) burning over 100 billion gallons of gasoline, they were still high, despite the installation of emissions-control devices in many of the newer models. [9-5, 9-8] Inefficient operation, due to driving conditions and/or poor maintenance, accounts for some of the pollution problems. In general, numerous short, low-speed trips, accompanied by many startings and stoppings, make for inefficient automobile operation and therefore contribute to increased air pollution.

Over half the vehicle miles traveled in the United States each year are on urban streets, and 67 percent of all trips are job-related and less than 32 km (20 m) in length. The average automobile in large cities moves at speeds at which pollution potential is very great. [9-13] However, because driving conditions in most urban areas do not allow for increased speed, reduction in automotive emissions must be achieved by other means.

Ultimately, air pollution problems from mobile sources may be solved by indirect control measures, measures designed to prevent the generation of pollutants. The abolition of the internal combustion engine, the use of electric-powered vehicles, and the increased utilization of mass transit are some of the means of indirect control currently under discussion. While indirect measures may prove to be the long-range solution to air pollution problems caused by emissions from mobile sources, a full discussion of these methods of control is beyond the scope of this chapter.

Direct controls being utilized to solve air pollution problems arising from mobile sources include devices that reduce emissions from the crankcase, carburetor, fuel tank, and exhaust. Emissions from gasoline-powered vehicles come from these sources, with 20 percent coming from the crankcase, 15 percent from fuel tank and carburetor, and 65 percent from tail pipe. The crankcase, fuel tank, and carburetor are the major sources of hydrocarbon emissions, while almost all the carbon monoxide and oxides of nitrogen comes from the tail pipe. [9-33]

Positive Crankcase Ventilation (PCV) Systems

One of the first direct-control methods was initiated in 1961 in California for control of emissions from the crankcase. These emissions are mainly hydrocarbon blow-by, which goes past the piston rings into the crankcase. Soon after the California crankcase control mandates were issued, all American-made and some foreign-made cars were equipped with positive crankcase ventilation (PCV) systems. Basically, these systems close off the vent to the atmosphere and recycle the blow-by back into the engine intake, an operation regulated by the PCV valve. Incorporation of this simple, inexpensive device has reduced crankcase hydrocarbon emissions to negligible levels. [9-26, 9-33]

Adsorption Canisters

Diurnal changes in ambient temperatures result in expansion of the air-fuel mixture in a partially filled fuel tank, an expansion that expels gasoline vapor into the atmosphere. Operating losses from fuel tanks also occur during driving as the fuel is heated by the road surface. Further hydrocarbon emissions are caused by increased evaporation of the fuel when the car is sitting after use ("hot soak"). The direct control measure initiated in 1971 to control these hydrocarbon emissions was the installation of a canister filled with activated charcoal that adsorbs hydrocarbon emissions. The adsorbed vapors are then desorbed and fed back to the intake manifold during high-power operating conditions. [9-33]

Catalytic Converters

When the 1975 federal exhaust standards were released, catalytic-converter systems represented the only technology available that was capable of bringing tail-pipe emissions to the required levels. Generally, the catalysts used in the converters can be classified in three groups—oxidizing, reducing, and three-way.

Oxidizing catalytic systems accelerate the completion of the oxidation of carbon monoxide and hydrocarbons so that CO is converted to CO_2 and HC is converted to CO_2 and water vapor. In this system, metals such as platinum and palladium are used as catalysts. Since these catalysts can be poisoned by lead, sulfur, and phosphorus, only unleaded gasolines should be used in cars equipped with catalytic converters. [9-11]

Reducing catalytic conversion systems, generally using rhodium and ruthenium, accelerate the reduction of NO_x to N_2. Conversion to NH_3 is undesirable because the subsequent use of an oxidizing catalyst could recreate NO_x from the ammonia. Because the metals used as reducing catalysts are not available in the United States, securing them could pose problems over the next few decades. [9-11]

A *three-way catalytic converter system* using platinum and rhodium is under development. This system is capable of promoting reactions among the air contaminants at close to stoichiometric conditions. Here the HC and CO react with the oxygen in the NO_x to form H_2O, CO_2, N_2, and O_2. Such a system, used in conjunction with ignition, carburetor, and related controls, should make it possible to meet federal standards. [9-11]

DISCUSSION TOPICS AND PROBLEMS

9-1 Name five natural removal mechanisms at work in the atmosphere. Which is the most important particulate removal mechanism?

9-2 What are two broad approaches to control of air pollution emissions?

9-3 Name and discuss four methods of controlling emissions at the source.

9.4 What specific air-pollution-control devices are available for control of particulate emissions at their source? Indicate the size range of the particulate that each type of unit is capable of removing efficiently.

9-5 A settling chamber that is 12 m long, 2 m high, and 2 m wide processes 240 m^3/m of air at a temperature of 75°C. Determine the maximum size of the particle with specific gravity of 1.8 that is removed with a theoretical efficiency of 100 percent.

9-6 A series of horizontal trays is placed in the settling chamber described in Prob. 9-5. The trays are 10 m long, 20 cm apart, and reduce the cross-sectional area of the tank by 10 percent. Determine the smallest particle that can be removed with a theoretical efficiency of 100 percent.

9-7 Draw a size-efficiency curve for the system in Prob. 9-5.

9-8 Draw a size-efficiency curve for the system described in Prob. 9-6.

9-9 It is desired to construct a settling chamber to remove particles from an airstream of 120 m^3/min. The temperature of the air is 50°C, and the specific gravity of the particles is 2.5. The chamber is to be strapped to the ceiling of an industrial building, and the space is limited vertically to 2 m and horizontally to 1.5 m. Determine the length required to remove 100 percent of 50-μm particles.

9-10 A cyclone of standard dimensions with a diameter of 1.6 m processes 4.5 m^3/s of air with a temperature of 50°C. Determine d_{50} if the specific gravity of the particles is 1.2.

9-11 For the cyclone system described in Prob. 9-10, determine the collection efficiency for particles with diameters of (*a*) 5 μm and (*b*) 30 μm.

9-12 It is desired to design a cyclone that will remove a 15-μm particle with 50 percent efficiency from an airstream of 6.0 m^3/min. The temperature of the air is 75°C, and the specific gravity of the particles is 1.5. The cyclone is to have standard dimensions. Assume five turns.

9-13 Determine d_{50} for the gas-particle stream of Prob. 9-10 if a bank of 64 cyclones with diameters of 20 cm is used.

9-14 The size distribution of particles in an airstream is given as follows:

Size range	Average particle size	Percent by mass
0–10	5	18
10–20	15	32
20–30	25	24
30–50	40	16
>50	65	10

The airstream contains 11.2 g/m^3 of particles. Determine the overall efficiency if the d_{50} of the cyclone is 7.0 μm.

9-15 Discuss the advantages and disadvantages of wet collectors.

9-16 Under what indu. rial-plant operating conditions would use of a fabric filter be inappropriate?

9-17 Determine the filter cloth area to process a flow of 8.8 m^3/s of gas at a filtering velocity of 1.5 m/min.

9-18 The diameter of the bags used in the system described in Prob. 9-17 is 20 cm; the length is 5 m. How many bags will be required for continuous cleaning?

9-19 A filter baghouse must process 15 m^3/s of waste gas. The baghouse is to be divided into eight sections of equal cloth area so that one section can be shut down for cleaning and/or repairs while the others continue operating. Laboratory analysis indicates an air-to-cloth ratio of 9.0 $m^3/\text{min} \cdot m^2$ cloth will provide sufficient treatment. The bags are 0.25 m in diameter and 7.0 m long. Determine the number of bags and the physical arrangement to meet the above requirement.

9-20 Stack gas flows through an electrostatic precipitator at a rate of 12 m^3/s. The plate area is 250 m^2, and the drift velocity for the system has been found to be

$$w = 2.8 \times 10^5 d_p$$

Draw a size-efficiency curve for particles ranging in size from 0.1 to 10 μm.

9-21 An electrostatic precipitator must be designed to process 5 m^3/s of stack gas. The drift velocity of the fly ash particles has been determined to be

$$w = 1.5 \times 10^5 d_p$$

Determine the plate area required to remove particles with diameters of 0.7 μm with 95 percent efficiency.

9-22 If the flow rate of the system in Prob. 9-21 is suddenly increased to 7.5 m^3/s, determine the decrease in efficiency.

9-23 What are the principal gases of concern in air pollution control? What are the four primary types of treatment processes available for control of gases?

9-24 Define adsorption and differentiate between physical and chemical adsorption.

9-25 Name and describe three types of adsorbers.

9-26 Define absorption as it relates to air-pollution-control devices.

9-27 Name and describe four types of absorption units.

9-28 Equilibrium data for absorption of ammonia by water at 20°C is given as follows:

y	0.01	0.03	0.05	0.07	0.09
x	0.013	0.037	0.058	0.078	0.093

The gas flow (G' + ammonia) is 50 kg-mol/h at 20°C and is 10 percent by volume ammonia. Water that is initially pure with respect to ammonia is used as the solvent. Determine the water flow rate that is 1.3 times the minimum required to remove 96 percent of the ammonia.

9-29 Four hundred fifty kilograms per hour of a mixture of ethyl-methyl-ketone (butanone) in air (1.5 mol percent) is to be passed through a countercurrent absorber. The equilibrium equation for butanone in water is $y = 2.5x$.

Assume that the gas and water do not react chemically. What is the minimum flow of water-free butanone that must be used to absorb 95 percent of the butanone? (The molecular mass of the butanone is 72.)

9-30 Differentiate between surface condensers and contact condensers.

9-31 What are the four key factors affecting the efficiency of combustion as a pollution control device?

9 32 Name and describe three types of combustion units used in at-source control of air pollution.

9-33 Name and describe three control devices developed for control of automotive emissions.

REFERENCES

9-1 *Air Pollution Manual: Control Equipment, Part II*, American Industrial Hygiene Assoc., Detroit, 1968.

9-2 *Automotive Pollution—A Progress Report*: Atlantic-Richfield Company Products, Division of Research and Development, Harvey, Ill., 1971.

9-3 Brandt, A. D.: *Industrial Health Engineering*, Wiley, New York, 1947.

9-4 Brey, W. S.: *Principles of Physical Chemistry*, Appleton-Century-Crofts, New York, 1958.

9-5 Canter, L. W., and D. R. Rowe: "The Contribution of the Automobile to Air Pollution," unpubl., 1969.

9-6 Council on Environmental Quality: *Seventh Annual Report of the Council on Environmental Quality*, Washington, D.C., 1976.

9-7 ———: *Eighth Annual Report of the Council on Environmental Quality*, Washington, D.C., 1977.

9-8 ———: *Twelfth Annual Report of the Council on Environmental Quality*, Washington, D.C., 1982.

9-9 Crawford, Martin: *Air Pollution Control Theory*, McGraw-Hill, New York, 1976.

9-10 Danielson, J. A.: *Air Pollution Engineering Manual*, 2d ed., EPA Office of Air and Water Programs, Research Triangle Park, N.C., 1973.

9-11 Demmler, A. W.: "Automotive Catalysis," *Auto Eng*, **85**(3):29, 32 (1977).

9-12 Greenberg, D. S.: "Pollution Control: Sweden Sets Up an Ambitious New Program," *Science*, **166**:200 (Oct. 10, 1969).

9-13 Hackleman, E. C.: "Is An Electric Vehicle In Your Future?" *Env Sci Tech*, **11**(9):860 (1977).

9-14 Hanf, E. B., "A Guide to Scrubber Selection," *Env Sci Tech*, **4**(2):110 (1970).

9-15 Institute for Air Pollution Training, "Control Techniques for Gases and Particulates," EPA, Atlanta, 1971.

9-16 Jost, W., et al.: *Z. Phys. Chem. N.F.*, **45**:47 (1965).

9-17 Lapple, C. E.: "Processes Use Many Collection Types," *Chem Eng*, **58**(5):144 (May 1951).

9-18 Lawrence, W. F., and C. F. Cockrell: "Power Plant Emission Control," Coal Research Bureau, West Virginia University, Morgantown, 1971.

9-19 Lloyd, W. G., and D. R. Rowe: "Homogeneous Catalytic Oxidation of Carbon Monoxide," *Env Sci Tech*, **5**:11 (1971).

9-20 ——— and ———: "Palladium Compositions Suitable as Oxidation Catalysts," British Pat. 1,438,557, Larox Research Corp., 1976.

9-21 ——— and ———: "Palladium Compositions Suitable as Oxidation Catalysts, Canadian Pat. 1,011,531; 1,011,532, Larox Research Corp., 1977.

9-22 ——— and ———: "Palladium Compositions Suitable as Oxidation Catalysts," U.S. Pat. 3,849,336, 1974.

9-23 Ludwig, J. S.: "Progress in Control of Vehicle Emissions," unpubl., 1966.

9-24 Middleton, J. T., and W. Otto: "Air Pollution and Transportation," *Traffic Quarterly*, April 1968.

9-25 Miller, Stan, "The Building of Tall (and Not So Tall) Stacks," *Env Sci Tech*, **9**(6):522 (1975).

0-26 National Air Pollution Control Administration: *Control Techniques for Hydrocarbon and Organic Solvent Emissions for Stationary Sources*, document B, publ. AP-68, Washington, D.C., 1970.

9-27 ———: *Control Techniques for Nitrogen Oxide Emissions from Stationary Sources*, publ. AP-67, Washington, D.C., 1970.

9-28 ———: *Control Techniques for Particulate Pollutants*, publ. AP-51, Washington, D.C., 1969.

9-29 ———: *Control Techniques for Sulfur Oxide Air Pollutants*, publ. AP-52, Washington, D.C., 1969.

9-30 "Natural Cleansing Processes in the Atmosphere," in *Air Pollution Control Orientation Course* (422-A), Office of Air Programs, EPA, Research Triangle Park, N. C., 1971.

9-31 *Nitrogen Oxides*, National Academy of Sciences, Washington, D.C., 1977.

9-32 Painter, Dean E.: *Air Pollution Technology*, Reston, Reston, Va., 1974.

9-33 Perkins, H. C.: *Air Pollution*, McGraw-Hill, New York, 1974.

9-34 Rossano, August T., Jr.: *Air Pollution Control: Guidebook for Management*, McGraw-Hill, New York, 1969.

9-35 Rowe, D. R.: "Atmospheric Self-Cleansing Processes and Preparation of an Emission Inventory," script no. 2 in *Atmospheric Pollution for Science Teachers*, National Science Foundation proj. no. EPP75-10694, 1975.

9-36 ——— and Stephen Johnston: "Computer Speeds Adsorption Isotherm Analysis," *Water & Sewage Works J*, November 1978, p. 68.

9-37 ——— and William G. Lloyd, "Applications of Larox Catalysts to the Removal of Carbon Monoxide from Contaminated Air," unpubl., 1978.
9-38 ——— and ———: "Catalytic Cigarette Filter for Carbon Monoxide Reduction," *J Air Pol Cont Assoc*, **28**(3):253 (1978).
9-39 Sargent, G. D.: "Gas/Solid Separations," *Chem Eng*, deskbook issue, Feb. 15, 1971, p. 1971.
9-40 Schwartz, S. I.: "Reducing Air Pollution by Automobile Inspection and Maintenance: A Program Analysis," *J Air Pol Cont Assoc*, **23**(10):849 (1973).
9-41 Seinfeld, John: *Air Pollution*, McGraw-Hill, New York, 1975.
9-42 Shaheen, Esber I.: *Environmental Pollution: Awareness and Control*, Engineering Technology, Mahomet, Ill., 1974.
9-43 Slack, A. V.: "Removing SO_2 from Stack Gases," *Env Sci Tech*, **7**(2):110 (1973).
9-44 Starkman, E. S.: "Emission Control and Fuel Economy," *Env Sci Tech*, **9**(9):822 (1975).
9-45 Strauss, W.: *Air Pollution Control*, pt. 1, Wiley Interscience, New York, 1971.
9-46 Stumph, T. L.: "Control Equipment for Industrial Particulate Emission," *Community Air Pollution*, National Center for Air Pollution Control, Durham, N.C., 1966.
9-47 Treybal, R. E.: *Mass Transfer Operations*, McGraw-Hill, New York, 1955.
9-48 U.S. Environmental Protection Agency: *Cleaning the Air*, publ. no. 0-204-335, Washington, D.C., 1979.
9-49 Vermeulen, A. J.: "Acid Precipitation in the Netherlands," *Env Sci Tech*, **12**(9):1017 (1978).
9-50 Wark, Kenneth, and Cecil F. Warner: *Air Pollution: Its Origin and Control*, Dun-Dunnelley, New York, 1976.

PART THREE

SOLID WASTE

CHAPTER

TEN

SOLID WASTE: DEFINITIONS, CHARACTERISTICS, AND PERSPECTIVES

Solid wastes are all the wastes arising from human and animal activities that are normally solid and that are discarded as useless or unwanted. The term as used in this chapter is all-inclusive, and it encompasses the heterogeneous mass of throwaways from residences and commercial activities as well as the more homogeneous accumulations of a single industrial activity. To avoid confusion, the term *refuse*, often used interchangeably with the term *solid wastes*, is not used in this chapter.

The purpose of this chapter is threefold: (1) to identify the various types of solid wastes and their sources, (2) to examine the physical and chemical composition of wastes, and (3) to consider in general terms the elements involved in the management of these wastes. Engineering aspects of solid-waste management are considered in Chap. 11. The recovery of materials and energy from solid wastes is discussed in Chap. 12.

Types of Solid Wastes

The types and sources of solid wastes and the physical and chemical composition of solid wastes are considered in this section. The term *solid wastes* is all-inclusive and encompasses all sources, types of classifications, composition, and properties. As a basis for subsequent discussions, it will be helpful to define the various types of solid wastes that are generated. Three general categories are considered: (1) municipal wastes, (2) industrial wastes, and (3) hazardous wastes.

Table 10-1 Classification of materials comprising municipal solid waste

Component	Description
Food wastes	The animal, fruit, or vegetable residues (also called garbage) resulting from the handling, preparation, cooking, and eating of foods. Because food wastes are putrescible, they will decompose rapidly, especially in warm weather.
Rubbish	Combustible and noncombustible solid wastes, excluding food wastes or other putrescible materials. Typically, combustible rubbish consists of materials such as paper, cardboard, plastics, textiles, rubber, leather, wood, furniture, and garden trimmings. Noncombustible rubbish consists of items such as glass, crockery, tin cans, aluminum cans, ferrous and nonferrous metals, dirt, and construction wastes.
Ashes and residues	Materials remaining from the burning of wood, coal, coke, and other combustible wastes. Residues from power plants normally are not included in this category. Ashes and residues are normally composed of fine, powdery materials, cinders, clinkers, and small amounts of burned and partially burned materials.
Demolition and construction wastes	Wastes from razed buildings and other structures are classified as demolition wastes. Wastes from the construction, remodeling, and repairing of residential, commercial, and industrial buildings and similar structures are classified as construction wastes. These wastes may include dirt, stones, concrete, bricks, plaster, lumber, shingles, and plumbing, heating, and electrical parts.
Special wastes	Wastes such as street sweepings, roadside litter, catch-basin debris, dead animals, and abandoned vehicles are classified as special wastes.
Treatment-plant wastes	The solid and semisolid wastes from water, wastewater, and industrial-waste treatment facilities are included in this classification.

10-1 MUNICIPAL WASTES

It is important to note that the definitions of terms and the classifications used to describe the components of solid waste vary greatly in practice and in the literature. Consequently, the use of published data requires considerable care, judgment, and common sense. The definitions presented in Table 10-1 are intended to serve as a guide for municipal solid wastes.

10-2 INDUSTRIAL WASTES

Industrial wastes are those wastes arising from industrial activities and typically include rubbish, ashes, demolition and construction wastes, special wastes, and hazardous wastes.

10-3 HAZARDOUS WASTES

Wastes that pose a substantial danger immediately or over a period of time to human, plant, or animal life are classified as hazardous wastes. A waste is classified as hazardous if it exhibits any of the following characteristics: (1) ignitability, (2) corrosivity, (3) reactivity, or (4) toxicity. A detailed definition of these terms was first published in the *Federal Register* on May 19, 1980 (pp. 33, 121–122).

In the past, hazardous wastes were often grouped into the following categories: (1) radioactive substances, (2) chemicals, (3) biological wastes, (4) flammable wastes, and (5) explosives. The chemical category includes wastes that are corrosive, reactive, or toxic. The principal sources of hazardous biological wastes are hospitals and biological research facilities.

Sources of Solid Wastes

Knowledge of the sources and types of solid wastes, along with data on the composition and rates of generation, is basic to the engineering management of solid wastes.

10-4 MUNICIPAL WASTES

Sources and types of municipal solid wastes are reported in Table 10-2. In evaluating the sources of solid waste as reported in Table 10-2 it can be concluded that they are, for the most part, related to land use and zoning. The most difficult

Table 10-2 General sources of municipal solid wastes

Source	Typical facilities, activities, or locations where wastes are generated	Types of solid wastes
Residential	Single-family and multifamily dwellings, low-, medium-, and high-rise apartments, etc.	Food wastes, rubbish, ashes, special wastes
Commercial	Stores, restaurants, markets, office buildings, hotels, motels, print shops, auto repair shops, medical facilities and institutions, etc.	Food wastes, rubbish, ashes, demolition and construction wastes, special wastes, occasionally hazardous wastes
Open areas	Streets, alleys, parks, vacant lots, playgrounds, beaches, highways, recreational areas, etc.	Special wastes, rubbish
Treatment plant sites	Water, wastewater, and industrial treatment processes, etc.	Treatment-plant wastes, principally composed of residual sludges

source to deal with is open areas because in these locations the generation of wastes is a diffuse process.

10-5 HAZARDOUS WASTES

Hazardous wastes are generated in limited amounts throughout most industrial activities. In terms of generation, the concern is with the identification of the amounts and types of hazardous wastes developed at each source, with emphasis on those sources where significant waste quantities are generated. Unfortunately, very little information is available on the quantities of hazardous wastes generated in various industries.

The spreading of hazardous wastes by spillage must also be considered. The quantities of hazardous wastes that are involved in spillages usually are not known. After a spill, the wastes requiring collection and disposal are often significantly greater than the amount of spilled wastes, especially where an absorbing material, such as straw, is used to soak up liquid hazardous wastes or where the soil into which a hazardous liquid waste has percolated must be excavated. Both the straw and the liquid and the soil and the liquid are classified as hazardous wastes.

Properties of Solid Wastes

Information on the properties of solid wastes is important in evaluating alternative equipment needs, systems, and management programs and plans, especially with respect to the implementation of disposal and resource- and energy-recovery options.

10-6 PHYSICAL COMPOSITION

Information and data on the physical composition of solid wastes including (1) identification of the individual components that make up municipal solid wastes, (2) analysis of particle size, (3) moisture content, and (4) density of solid wastes are presented below. Sampling techniques used to obtain data on solid wastes are also discussed in this section.

Individual Components

Components that typically make up most municipal solid wastes and their relative distribution are reported in Table 10-3. Although any number of components could be selected, those listed in Table 10-3 have been selected because they are readily identifiable, are consistent with component categories reported in the literature, and are adequate for the characterization of solid wastes for most applications.

Table 10-3 Typical composition of municipal solid wastes

	Percent by mass			
Component	Range	Typical	Davis California*	Merida, Venezuela†
Food wastes	6–26	14	8.3	27.4
Paper	15–45	34	35.8	15.5
Cardboard	3–15	7	10.9	13.0
Plastics	2–8	5	6.9	4.6
Textiles	0–4	2	2.5	2.3
Rubber	0–2	0.5	2.5	0.4
Leather	0–2	0.5	0.7	1.3
Garden trimmings	0–20	12	10.8	5.8
Wood	1–4	2	1.9	3.6
Misc. organics	0–5	2	2.0	0.6
Glass	4–16	8	7.5	10.3
Tin cans	2–8	6	5.1	8.3
Nonferrous metals	0–1	1	1.6	0.1
Ferrous metals	1–4	2	2.2	1.2
Dirt, ashes, brick, etc.	0–10	4	1.3	5.6

* Based on measurements made during the month of October over a 5-year period (1978 through 1982).

† Based on measurements made during the month of July over a 3-year period (1978 through 1980).

Particle Size

The size of the component materials in solid wastes is of importance in the recovery of materials, especially with mechanical means such as trommel screens and magnetic separators. A general indication of the particle size distribution (by longest dimension and ability to pass a sieve) may be obtained from the data presented in Figs. 10-1 and 10-2.

Moisture Content

The moisture content of solid wastes usually is expressed as the mass of moisture per unit mass of wet or dry material. In the wet-mass method of measurement, the moisture in a sample is expressed as a percentage of the wet mass of the material; in the dry-mass method, it is expressed as a percentage of the dry mass of the material. In equation form, the wet-mass moisture content is expressed as follows:

$$\text{Moisture content } (\%) = \left(\frac{a - b}{a}\right) 100 \tag{10-1}$$

where a = initial mass of sample as delivered
b = mass of sample after drying

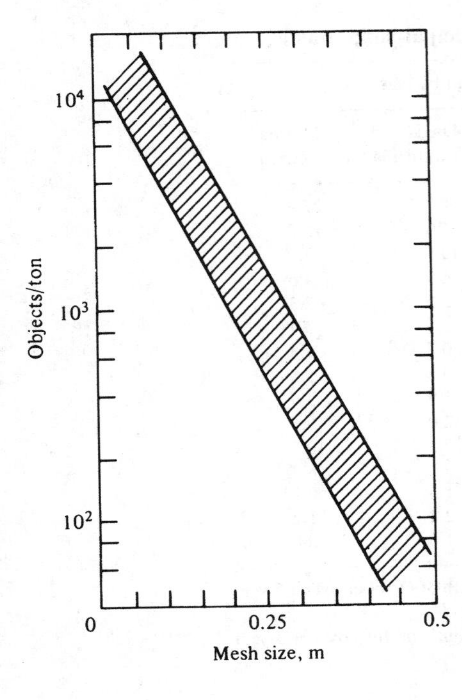

Figure 10-1 Typical sizes of individual components comprising solid wastes. (*Adapted from Winkler and Wilson* [*10-7*].)

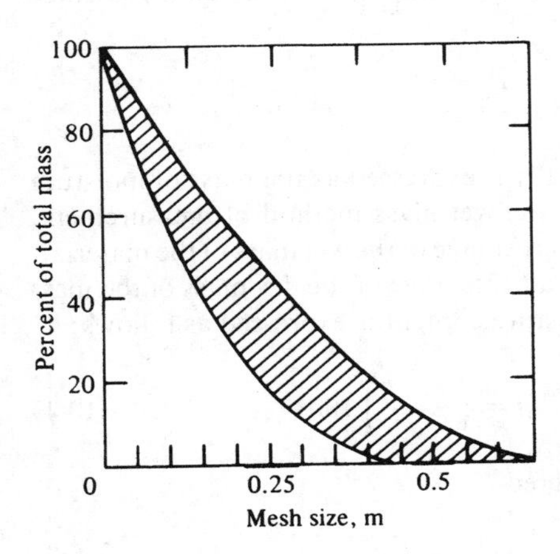

Figure 10-2 Number of individual components of a given size per tonne of municipal solid wastes. (*Adapted from Winkler and Wilson* [*10-7*].)

Table 10-4 Typical data on moisture content of municipal solid waste components

Component	Moisture, percent	
	Range	Typical
Food wastes	50–80	70
Paper	4–10	6
Cardboard	4–8	5
Plastics	1–4	2
Textiles	6–15	10
Rubber	1–4	2
Leather	8–12	10
Garden trimmings	30–80	60
Wood	15–40	20
Misc. organics	10–60	25
Glass	1–4	2
Tin cans	2–4	3
Nonferrous metals	2–4	2
Ferrous metals	2–6	3
Dirt, ashes, brick, etc.	6–12	8
Municipal solid wastes	15–40	20

Source: From Tchobanoglous et al. [10-5]

To obtain the dry mass, the solid-waste material is dried in an oven at 77°C (170°F) for 24 h. This temperature and time is used to dehydrate the material completely and to limit the vaporization of volatile materials.

Typical data on the moisture content for the solid-waste components are given in Table 10-4. For most industrial solid wastes, the moisture content will vary from 10 to 35 percent. The use of Eq. (10-1) is illustrated in Example 10-1.

Example 10-1: Estimating the moisture content of a solid-waste sample Estimate the moisture content of a solid-waste sample with the following composition:

Component	Percent by mass
Food wastes	15
Paper	45
Cardboard	10
Plastics	10
Garden trimmings	10
Wood	5
Tin cans	5

SOLUTION

1. Set up a computation table to determine the dry mass of the solid-waste sample using the data given in Table 10-4.

Component	Percent by mass	Moisture content, %	Dry mass,* kg
Food waste	15	70	4.5
Paper	45	6	42.3
Cardboard	10	5	9.5
Plastics	10	2	9.8
Garden trimmings	10	60	4.0
Wood	5	20	4.0
Tin cans	5	3	4.9
			79.0

* Based on 100-kg sample of waste.

2. Determine the moisture content using Eq. (10-1) and the data from step 1.

$$\text{Moisture content} = \left(\frac{100 - 79.0}{100}\right)100 = 21.0\%$$

COMMENT The composition of the solid-waste sample used in this example will be used in all of the examples in this chapter. By using the same composition throughout, the interrelationship of the various parameters can be established more clearly.

Density

Typical densities for various wastes as found in containers are reported by source in Table 10-5. Because the densities of solid wastes vary markedly with geographic location, season of the year, and length of time in storage, great care should be used in selecting typical values. Estimation of the density of a waste sample is illustrated in Example 10-2.

Example 10-2: Estimating the density of a solid-waste sample Estimate the "as-discarded" density of a solid-waste sample with the composition given in Example 10-1.

SOLUTION

1. Set up a computation table to determine the as-discarded volume of the solid waste sample using the data reported in Table 10-5.

Component	Percent by mass	Typical density, kg/m^3	Volume,* m^3
Food waste	15	290	0.52
Paper	45	85	5.29
Cardboard	10	50	2.00
Plastics	10	65	1.54
Garden trimmings	10	105	0.95
Wood	5	240	0.21
Tin cans	5	90	0.56
			11.07

* Based on a 1000-kg sample of waste.

2. Compute the density of a waste sample using the data from step 1.

$$\text{Density} = \frac{1000 \text{ kg}}{11.07 \text{ m}^3} = 90.33 \text{ kg/m}^3$$

Table 10-5 Typical densities for solid wastes components and mixtures

Item	Density, kg/m^3 Range	Density, kg/m^3 Typical
Component*		
Food wastes	120–480	290
Paper	30–130	85
Cardboard	30–80	50
Plastics	30–130	65
Textiles	30–100	65
Rubber	90–200	130
Leather	90–260	160
Garden trimmings	60–225	105
Wood	120–320	240
Misc. organics	90–360	240
Glass	160–480	195
Tin cans	45–160	90
Nonferrous metals	60–240	160
Ferrous metals	120–1200	320
Dirt, ashes, brick, etc.	320–960	480
Municipal solid wastes		
Uncompacted	90–180	130
Compacted (in compactor truck)	180–450	300
In landfill (compacted normally)	350–550	475
In landfill (well-compacted)	600–750	600

* Data for components is on an as-discarded basis.

Sampling Procedures

Perhaps the most difficult task facing anyone concerned with the design and operation of solid-waste management systems is to predict the composition of solid wastes that will be collected now and in the future. The problem is complicated because of the heterogeneous nature of waste materials and the fact that unpredictable externalities such as world oil prices can affect the long-term abundance of the individual waste components.

To assess the total mix of wastes components such as those listed in Table 10-1, the load-count and the mass-volume methods of analysis are recommended. These methods are considered in Sec. 11-3. The following technique is recommended where it is desired to assess the individual components within a given waste category (e.g., domestic wastes).

1. Unload a truckload of wastes in a controlled area away from other operations.
2. Quarter the wasteload.
3. Select one of the quarters and quarter that quarter.
4. Select one of the quartered quarters and separate all of the individual components of the waste into preselected components such as those listed in Table 10-3.
5. Place the separated components in a container of known volume and tare mass and measure the volume and mass of each component. The separated components should be compacted tightly to simulate the conditions in the storage containers from which they were collected.
6. Determine the percentage distribution of each component by mass and the as-discarded density (see Table 10-5). Typically, from 100 to 200 kg (200 to 400 lb) of waste should be sorted to obtain a representative sample. To obtain a more representative distribution of components, samples should be collected during each season of the year. Clearly, no matter how many samples are analyzed, common sense is needed in selecting the loads to be sorted, in analyzing the data, and in preparing projections.

10-7 CHEMICAL COMPOSITION

Information on the chemical composition of solid wastes is important in evaluating alternative processing and energy recovery options. If solid wastes are to be used as fuel, the four most important properties to be known are:

1. Proximate analysis
   a. Moisture (loss at 105°C for 1 h)
   b. Volatile matter (additional loss on ignition at 950°C)
   c. Ash (residue after burning)
   d. Fixed carbon (remainder)
2. Fusing point of ash

3. Ultimate analysis, percent of C (carbon), H (hydrogen), O (oxygen), N (nitrogen), S (sulfur), and ash
4. Heating value (energy value)

Typical proximate analysis data for the components in municipal solid wastes are presented in Table 10-6.

Energy Content

Typical data on the energy content and inert residue for solid wastes are reported in Table 10-7. Energy values may be converted to a dry basis by using Eq. (10-2).

$$\text{kJ/kg (dry basis)} = \text{kJ/kg (as discarded)} \frac{100}{100 - \%\ \text{moisture}} \tag{10-2}$$

The corresponding equation on an ash-free dry basis is:

$$\text{kJ/kg (ash-free dry basis)} = \text{kJ/kg (as discarded)} \frac{100}{100 - \%\ \text{ash} - \%\ \text{moisture}} \tag{10-3}$$

Application of the data in Table 10-7 and Eqs. (10-2) and (10-3) is illustrated in Example 10-3.

Table 10-6 Proximate and ultimate chemical analysis of municipal solid waste

	Value, percent*	
	Range	Typical
Proximate analysis		
Moisture	15–40	20
Volatile matter	40–60	53
Fixed carbon	5–12	7
Noncombustibles	15–30	20
Ultimate analysis (combustible components)		
Carbon	40–60	47.0
Hydrogen	4–8	6.0
Oxygen	30–50	40.0
Nitrogen	0.2–1.0	0.8
Sulfur	0.05–0.3	0.2
Ash	1–10	6.0
Heating value†		
Organic fraction, kJ/kg	12,000–16,000	14,000
Total, kJ/kg	8,000–12,000	10,500

* By mass.
† As-discarded basis.

Table 10-7 Typical data on inert residue and energy content of municipal solid wastes

Component	Inert residue,* percent		Energy,† kJ/kg	
	Range	Typical	Range	Typical
Food wastes	2–8	5	3,500–7,000	4,650
Paper	4–8	6	11,600–18,600	16,750
Cardboard	3–6	5	13,950–17,450	16,300
Plastics	6–20	10	27,900–37,200	32,600
Textiles	2–4	2.5	15,100–18,600	17,450
Rubber	8–20	10	20,900–27,900	23,250
Leather	8–20	10	15,100–19,800	17,450
Garden trimmings	2–6	4.5	2,300–18,600	6,500
Wood	0.6–2	1.5	17,450–19,800	18,600
Misc. organics	2–8	6	11,000–26,000	18,000
Glass	96–99†	98	100–250	150
Tin cans	96–99+	98	250–1,200	700
Nonferrous metals	90–99+	96		
Ferrous metals	94–99+	98	250–1,200	700
Dirt, ashes, brick, etc.	60–80	70	2,300–11,650	7,000
Municipal solid wastes			9,300–12,800	10,500

* After combustion.
† As-discarded basis.

Example 10-3: Estimating the energy content of a solid-waste sample Estimate the energy content of a solid-waste sample with the composition given in Example 10-1. What is the content on a dry basis and on an ash-free dry basis?

SOLUTION

1. Set up a computation table to determine the total as-discarded energy content of the solid-waste sample using the data in Table 10-7.

Component	Percent by mass	Energy,* kJ/kg	Total energy,† kJ
Food waste	15	4,650	69,750
Paper	45	16,750	753,750
Cardboard	10	16,300	163,000
Plastics	10	32,600	326,000
Garden trimmings	10	6,500	65,000
Wood	5	18,600	93,000
Tin cans	5	700	3,500
			1,474,000

* From Table 10-7, as-discarded basis.
† Based on 100-kg sample of waste.

2. Compute the unit energy content.

$$\text{Energy content} = \frac{1{,}474{,}000 \text{ kJ}}{100 \text{ kg}} = 14{,}740 \frac{\text{kJ}}{\text{kg}}$$

3. Determine the energy content on a dry basis.
   a. From Example 10-1, the moisture content of the waste is 21.0 percent.
   b. Using Eq. (10-2), the energy on a dry basis is

$$\text{kJ/kg (dry basis)} = 14{,}740 \frac{100}{100 - 21.0} = 18{,}658$$

4. Determine the energy content on an ash-free dry basis.
   a. Assume the ash content is equal to 5.0 percent.
   b. Using Eq. (10-3) the energy content on an ash-free dry basis is

$$\text{kJ/kg (ash-free dry basis)} = 14{,}740 \frac{100}{100 - 5 - 21} = 19{,}919$$

Chemical Content

Representative data on the ultimate analysis of typical municipal waste components are presented in Table 10-8. If energy values are not available, approximate values may be determined by using Eq. (10-4), known as the *modified Dulong formula*, and the data in Table 10-8.

$$\text{kJ/kg} = 337\text{C} + 1428\left(\text{H} - \frac{\text{O}}{8}\right) + 9\text{S} \qquad (10\text{-}4)$$

where C = carbon, percent
H = hydrogen, percent
O = oxygen, percent
S = sulfur, percent

Table 10-8 Typical data on ultimate analysis of the combustible components in municipal solid wastes

	Percent by mass (dry basis)					
Component	Carbon	Hydrogen	Oxygen	Nitrogen	Sulfur	Ash
Food wastes	48.0	6.4	37.6	2.6	0.4	5.0
Paper	43.5	6.0	44.0	0.3	0.2	6.0
Cardboard	44.0	5.9	44.6	0.3	0.2	5.0
Plastic	60.0	7.2	22.8	—	—	10.0
Textiles	55.0	6.6	31.2	4.6	0.15	2.5
Rubber	78.0	10.0	—	2.0	—	10.0
Leather	60.0	8.0	11.6	10.0	0.4	10.0
Garden trimmings	47.8	6.0	38.0	3.4	0.3	4.5
Wood	49.5	6.0	42.7	0.2	0.1	1.5
Misc. organics	48.5	6.5	37.5	2.2	0.3	5.0
Dirt, ashes, brick, etc.	26.3	3.0	2.0	0.5	0.2	68.0

Source: From Tchobanoglous et al. [10-5]

Use of the data in Table 10-8 and Eq. (10-4) is illustrated in Example 10-4.

Example 10-4: Estimating the overall chemical composition of a solid-waste sample Derive an approximate chemical formula for the organic portion of a solid-waste sample with the composition given in Example 10-1. Use the resulting chemical composition to estimate the energy content.

SOLUTION

1. Set up a computation table to determine the overall composition of the waste based on a 100-kg sample (see accompanying table).

Computation of the chemical composition of a waste sample

Component	Wet mass, kg	Dry mass, kg	Composition, kg					
			C	H	O	N	S	Ash
Food wastes	15	4.5	2.16	0.29	1.69	0.12	0.02	0.23
Paper	45	42.3	18.40	2.54	18.61	0.13	0.08	2.54
Cardboard	10	9.5	4.18	0.56	4.24	0.03	0.02	0.48
Plastics	10	9.8	5.88	0.71	2.23	—	—	0.98
Garden trimmings	10	4.0	1.91	0.24	1.52	0.14	0.01	0.18
Wood	5	4.0	1.98	0.24	1.71	0.01	—	0.06
Totals	95	74.1	34.51	4.58	30.00	0.43	0.13	4.47

2. Prepare a summary table of the above data.

Component	Mass, kg
Moisture	20.9*
Carbon	34.51
Hydrogen	4.58
Oxygen	30.00
Nitrogen	0.43
Sulfur	0.13
Ash	4.47

* (95.0 − 74.1).

3. Convert the moisture content reported in step 2 to hydrogen and oxygen.
   a. Hydrogen = $\frac{2}{18}20.9$ kg = 2.32 kg
   b. Oxygen = $\frac{16}{18}20.9$ kg = 18.58 kg

4. Prepare a revised summary table similar to the one prepared in step 2 using the data from step 3.

Component	Mass, kg	Percent by mass
Carbon	34.51	36.3
Hydrogen	6.90	7.3
Oxygen	48.58	51.1
Nitrogen	0.43	0.5
Sulfur	0.13	0.1
Ash	4.47	4.7
Total	95.02	100.0

5. Compute molar composition of the elements.

Element	Mass, kg	kg/mol	Moles
Carbon	34.51	12.01	2.873
Hydrogen	6.90	1.01	6.832
Oxygen	48.58	16.00	3.036
Nitrogen	0.43	14.01	0.031
Sulfur	0.13	32.06	0.004

6. Determine an approximate chemical formula with and without sulfur.
   a. Compute normalized mole ratios.

Element	Mol ratios	
	Sulfur = 1	Nitrogen = 1
Carbon	718.2	92.7
Hydrogen	1708.0	220.4
Oxygen	759.0	97.9
Nitrogen	7.8	1.0
Sulfur	1.0	0

   b. Chemical formula with sulfur

$$C_{718.2}H_{1708.0}O_{759}N_{7.8}S$$

   c. Chemical formula without sulfur:

$$C_{92.7}H_{220.4}O_{97.9}N$$

7. Estimate the energy content of the waste using Eq. (10-4) and the data from step 4.

$$\text{kJ/kg} = 337(36.3) + 1428\left(7.3 - \frac{51.1}{8}\right) + 95(0.1)$$

$$= 12{,}233 + 1{,}303 + 9.5 = 13{,}546$$

COMMENT Computations such as the above are especially important where the recovery of energy from solid wastes is being considered. The recovery of energy is considered in more detail in Chap. 12.

10-8 CHANGES IN COMPOSITION

To plan effectively for solid waste management, information and data on the expected future composition of the solid wastes are important. In addition to technological changes in areas such as food processing and packaging, changes in the world economy have also affected the composition of solid wastes. For example, prior to the energy crisis of 1974, the amount of ash in solid waste had all but disappeared. Yet today in many parts of the country there is an increase in the amount of ash present in solid waste. With tight economic conditions there is also an increase in the amount of waste oils in solid waste as more people begin to change their own automobile oil.

Solid-Waste Management: An Overview

Recognizing that our world is finite and that the continued pollution of our environment will, if uncontrolled, be difficult to rectify in the future, the subject of solid-waste management is both timely and important. The overall objective of solid-waste management is to minimize the adverse environmental effects caused by the indiscriminate disposal of solid wastes, especially of hazardous wastes. To assess the management possibilities it is important to consider (1) materials flow in society, (2) reduction in raw materials usage, (3) reduction in solid-waste quantities, (4) reuse of materials, (5) materials recovery, (6) energy recovery, and (7) day-to-day solid-waste management.

10-9 MATERIALS FLOW IN SOCIETY

An indication of how and where solid wastes are generated in a technological society is shown in the simplified materials-flow diagram presented in Fig. 10-3. Solid wastes (debris) are generated at the start of the process, beginning with the mining of raw material. Thereafter, solid wastes are generated at every step in the process as raw materials are converted to goods for consumption. It is apparent from Fig. 10-3 that one of the best ways to reduce the amount of solid wastes to be disposed is to reduce the consumption of raw materials and to increase the rate

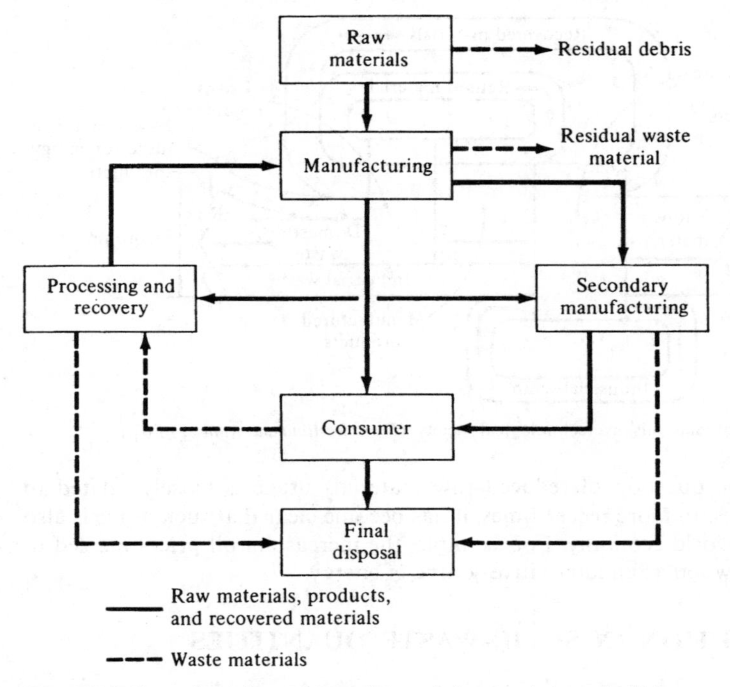

Figure 10-3 Generalized flow of materials and the generation of solid wastes in society. (*From Tchobanoglous et al.* [*10-5*].)

of recovery and reuse of waste materials. Although the concept is simple, effecting this change in a modern technological society has proved extremely difficult.

10-10 REDUCTION IN RAW MATERIALS USAGE

The general relationships shown in Fig. 10-3 can be quantified relatively, as shown in Fig. 10-4. To satisfy the principle of conservation of mass the input must equal the output. Clearly, if a reduction in the usage of raw materials is to occur either the input or output must be reduced. Raw materials usage can be reduced most effectively by reducing the quantity of municipal and industrial wastes. For example, to meet EPA mileage restrictions American cars are now (1984) on the average, 20 percent smaller than they were in the late 1950s and early 1960s. This reduction in size has also reduced the demand for steel by about 20 percent. The reduced demand for steel has in turn resulted in less mining for the iron ore used to make steel. While most people would agree that it is desirable to reduce the usage of raw materials, others would argue that as the usage of raw materials is decreased jobs in those industries also are decreased.

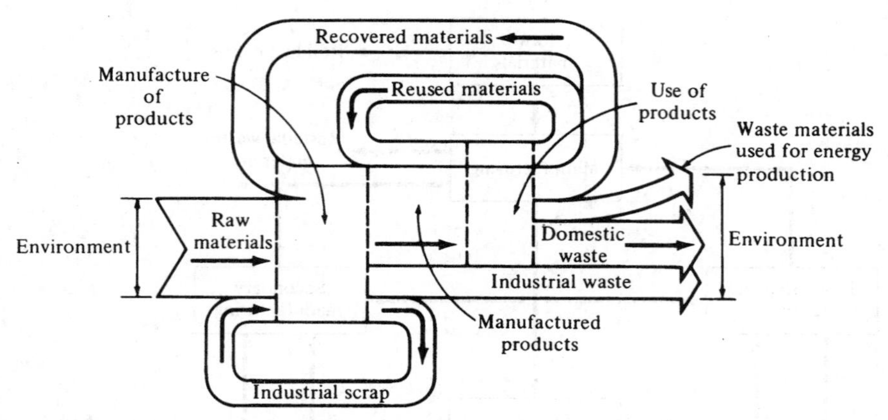

Figure 10-4 Flow of materials in a technological society. (*From Vesilind and Rimer* [*10-6*].)

Clearly, the question of reduced raw materials usage is closely related to national policies. In more recent times, it has become clear that such usage is also related to the world economy. For example, the increase in oil prices has led to more usage of wood as an alternative source of energy.

10-11 REDUCTION IN SOLID-WASTE QUANTITIES

Reductions in the quantities of waste can occur in several ways: (1) the amount of material used in the manufacture of a product can be reduced, (2) the useful life of a product can be increased, and (3) the amount of materials used for packaging and marketing of consumer goods can be reduced. For example, the quantity of automobile tires now disposed of on an annual basis could be cut almost in half if their useful life (or mileage) were doubled.

The many opportunities to reduce the quantities of municipal and industrial wastes aside, major reductions will occur with changes in our national priorities or as a result of the effects of external economic forces beyond our control. Although this view is somewhat pessimistic it adequately reflects the current situation with respect to the reduction of solid wastes. Further, this view is not meant to minimize the important efforts of individuals and concerned citizen groups.

10-12 REUSE OF SOLID-WASTE MATERIALS

Reuse (recycling) of waste materials now occurs most commonly in those situations where a product has utility in more than one application. For example, the paper bags used to bring home groceries are used to store household wastes prior to placing them in the containers used for their storage for collection. Soup and vegetable containers are used to store used cooking grease. Newspapers are used to start fires in fireplaces; they are also tightly rolled and used as logs for burning.

While all of the above uses are important, their impact on the generation of solid wastes is minimal. A much larger impact would occur if beverage containers were to be recycled. It has been estimated that about 60 billion beer and soft-drink containers are sold annually in the United States. Referring to Table 10-3 it can be seen that glass and nonferrous metals (mostly aluminum cans) consitute 9 percent of the total waste stream. Recycling of these containers would have a major impact on the quantity of wastes that must be disposed on an annual basis.

10-13 MATERIALS RECOVERY

A number of materials present in municipal and industrial solid wastes are suitable for recovery and reuse. Referring to the percentage distribution of the waste components reported in Table 10-3, it would appear that paper, cardboard, plastics, glass, nonferrous metals, and ferrous metals are the most likely candidates. With the exception of plastics, the above materials are those most commonly recovered from solid wastes. The estimated recovery of these materials is reported in Table 10-9. The opportunity for the further recovery of these materials is considered in Chaps. 11 and 12. Impediments to the recovery of energy and materials are examined in Ref. [10-40].

Table 10-9 Materials recovery in the United States in 1975 by category

Material category	Gross discards*	Material recycled	
		Quantity*	Percent
Paper	44.1	6.8	15.4
Glass	13.7	0.4	2.9
Metals	12.7	0.6	4.7
Ferrous	(11.3)	(0.5)	(4.4)†
Aluminum	(1.0)	(0.1)	(10.0)
Other nonferrous	(0.4)	(0.0)	(0.0)
Plastics	4.4	0.0	0.0
Rubber	2.8	0.2	7.1
Leather	0.7	0.0	0.0
Textiles	2.1	0.0	0.0
Wood	4.8	0.0	0.0
Other	0.1	0.0	0.0
Total nonfood product waste	85.4	8.0	9.3
Food waste	22.8	0.0	0.0
Yard waste	26.0	0.0	0.0
Miscellaneous inorganic wastes	1.9	0.0	0.0
Total	136.1	8.0	5.9

* Million tons per year (metric table unavailable).
† 4.4 = (0.5/11.3) × 100
Note: 1.0 ton = 0.907 tonnes.
Source: From U.S. EPA. [10-2]

10-14 ENERGY RECOVERY

Because about 70 percent of the components that comprise solid waste are organic, the potential for the recovery of energy is high. The energy contained in the organic matter must be converted to a form that can be used more easily. The recovery of heat by burning the organic material in solid waste is the option that is spoken of most frequently. In Chap. 12 several alternative energy-recovery technologies are examined.

10-15 DAY-TO-DAY SOLID-WASTE MANAGEMENT

While the issues that have been discussed previously are of great importance and provide a perspective on the waste problem in general, the fact remains that the day-to-day management of the municipal solid wastes is a complex and costly undertaking. Direct activities that must be considered and coordinated on a daily basis include waste generation rates, on-site storage, collection, transfer and transport, processing, and disposal. These activities are associated directly with the management of solid wastes. Indirect activities that are also an important part of a solid-waste management program include: financing; operations; equipment; personnel; cost accounting and budgeting; contract administration; ordinances and guidelines; and public communications. [10-1] The direct activities involved in the management of solid wastes are considered in Chap. 11. The indirect activities are beyond the scope of this text.

DISCUSSION TOPICS AND PROBLEMS

10-1 Obtain data on the percentage distribution of solid-waste components for your community or a nearby community. How do the values compare with the values in Table 10-3? Explain any differences.

10-2 Weigh and sort through a bag of solid waste at your residence and estimate the sizes of the various components. Determine the number of objects per tonne for several size categories (e.g., 51–100, 101–150, 151–200 mm, etc.). How do the values you obtained compare with the data in Fig. 10-1? Explain any major differences.

10-3 Estimate the moisture content for a waste sample (*a*) with the following composition:

	Percent by mass			
Component	(*a*)	(*b*)	(*c*)	(*d*)
Food waste	12	15	10	20
Paper	40	45	40	50
Cardboard	8	5	6	8
Plastic	4	5	6	6
Garden trimming	15	10	15	6
Wood	5	5	5	5
Inerts	16	15	18	5

10-4 Estimate the as-discarded density for a waste (*a*) with the composition given in Prob. 10-3.

10-5 Estimate the as-discarded energy content for a waste (*a*) with the composition given in Prob. 10-3.

10-6 Derive an approximate chemical formula for a waste (*a*) with the composition given in Prob. 10-3.

10-7 Assess what is being done on your campus to reduce the quantities of solid wastes collected for disposal. Have these efforts reduced the quantity of solid wastes generated?

10-8 Assess and discuss what is being done in your own or a nearby community to encourage the reuse of materials.

10-9 Has the recovery of materials been attempted in your community? What have been the results?

10-10 Discuss the merits of trying to pass mandatory recycling legislation.

REFERENCES

10-1 American Public Works Association: *Municipal Refuse Disposal*, 3d ed., Public Administration Service, Chicago, 1970.

10-2 Fourth Report to Congress: *Resource Recovery and Waste Reduction*, U.S. EPA-ASWMP, Washington, D.C., 1977.

10-3 Kaiser, E. R.: "Chemical Analysis of Refuse Compounds," *Proc. National Incinerator Conference*, ASME, New York, 1966.

10-4 Klingshirn, J. V., and O. W. Albrecht: "Impediments to Energy and Materials Recovery Facilities for Municipal Solid Wastes," project summary, EPA-600/S2-81-181, Cincinnati, Ohio, October 1981.

10-5 Tchobanoglous, G., H. Theisen, and R. Eliassen: *Solid Wastes: Engineering Principles and Management Issues*, McGraw-Hill, New York, 1977.

10-6 Vesilind, P. A., and A. E. Rimer: *Unit Operations in Resource Recovery Engineering*, Prentice-Hall, Englewood Cliffs, N.J. 1981.

10-7 Winkler, P. F., and D. G. Wilson: "Size Characteristics of Municipal Solid Wastes," *Compost Science*, **14**(5): (April 1973).

CHAPTER

ELEVEN

ENGINEERED SYSTEMS FOR SOLID-WASTE MANAGEMENT

The analysis of the activities associated with the management of solid waste from the point of waste generation to final disposal is the subject of this chapter. Following a brief overview of the activities involved in the management of solid waste, each activity is considered separately.

11-1 FUNCTIONAL ELEMENTS

The activities involved with the management of solid wastes from the point of generation to final disposal have been grouped into six functional elements: (1) waste generation; (2) on-site handling, storage, and processing; (3) collection; (4) transfer and transport; (5) processing and recovery; and (6) disposal. The functional elements are described in Table 11-1 and illustrated pictorially in Fig. 11-1. The interrelationship between the functional elements is shown in Fig. 11-2. By considering each functional element separately it is possible to identify the fundamental aspects and relationships involved in each element and to develop, where possible, quantifiable relationships for the purposes of making engineering comparisons, analyses, and evaluations.

Solid Waste Generation

Solid wastes, as noted previously, include all solid or semisolid material that is no longer considered of sufficient value to retain in a given setting. It should be noted that the wastes that are discharged may be of significant value in another setting. The types and sources and the physical and chemical composition of solid wastes have been considered previously in Chap. 10. Representative data on the quantities of solid wastes, the estimation of quantities, and the factors affecting the generation rates are considered below.

Table 11-1 Description of the functional elements of a solid waste management system

Functional element	Description
Waste generation	Those activities in which materials are identified as no longer being of value and are either thrown away or gathered together for disposal
On-site handling, storage, and processing	Those activities associated with the handling, storage, and processing of solid wastes at or near the point of generation
Collection	Those activities associated with the gathering of solid wastes and the hauling of wastes after collection to the location where the collection vehicle is emptied
Transfer and transport	Those activities associated with (1) the transfer of wastes from the smaller collection vehicle to the larger transport equipment and (2) the subsequent transport of the wastes, usually over long distance, to the disposal site
Processing and recovery	Those techniques, equipment, and facilities used both to improve the efficiency of the other functional elements and to recover usable materials, conversion products, or energy from solid wastes
Disposal	Those activities associated with ultimate disposal of solid wastes, including those wastes collected and transported directly to a landfill site, semisolid wastes (sludge) from wastewater treatment plants, incinerator residue, compost, or other substances from the various solid-waste processing plants that are of no further use

11-2 TYPICAL GENERATION RATES

Typical unit waste-generation rates for municipal and selected commercial and industrial sources are reported in Tables 11-2 and 11-3, respectively. Because waste-generation practices are changing so rapidly, the presentation of site-specific waste-generation data is meaningless. Where waste-generation data are not available, the data presented in Table 11-2 can be used for purposes of estimation with reasonable confidence.

Table 11-2 Typical per capita solid-waste generation rates in the United States

	Unit rate, kg/capita · d	
Source	Range	Typical
Municipal*	0.75–2.50	1.6
Industrial	0.4–1.60	0.9
Demolition	0.05–0.4	0.3
Other municipal	0.05–0.3	0.2
		3.0

* Includes residential and commercial.

Figure 11-1 Pictorial elements: (*a*) waste generation, (*b*) on-site storage, (*c*) collection, (*d*) transfer and transport, (*e*) on-site processing and (*f*) disposal.

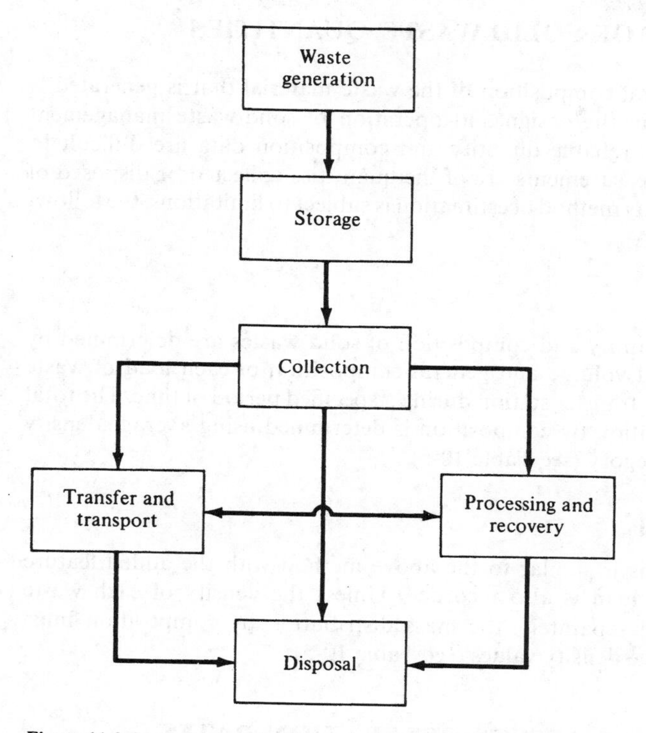

Figure 11-2 Interrelationship of functional elements comprising a solid-waste management system.

Table 11-3 Typical commercial and industrial unit waste-generation rates

Source	Unit	Range
Office buildings	kg/employee · d	0.5–1.1
Restaurants	kg/customer · d	0.2–0.8
Canned and frozen foods	tonnes/tonne of raw product	0.04–0.06
Printing and publishing	tonnes/tonne of raw paper	0.08–0.10
Automotive	tonnes/vehicle produced	0.6–0.8
Petroleum refining	tonnes/employee · d	0.04–0.05
Rubber	tonnes/tonne of raw rubber	0.01–0.3

11-3 ESTIMATION OF SOLID-WASTE QUANTITIES

The quantity and general composition of the waste material that is generated is of critical importance in the design and operation of solid waste management systems. Unfortunately, reliable quantity and composition data are difficult to obtain because most measurements are of the quantities collected or disposed of at a landfill. Although any method of estimation is subject to limitations, the following are recommended.

Load-Count Analysis

In this method, the quantity and composition of solid wastes are determined by recording the estimated volume and general composition of each load of waste delivered to a landfill or transfer station during a specified period of time. The total mass and mass distribution by composition is determined using average density data for each waste category (see Table 10-5).

Mass-Volume Analysis

This method of analysis is similar to the above method with the added feature that the mass of each load is also recorded. Unless the density of each waste category is determined separately, the mass distribution by composition must be derived using average density values (see Table 10-5).

11-4 FACTORS THAT AFFECT GENERATION RATES

Factors that influence the quantity of municipal wastes generated include (1) geographic location, (2) season of the year, (3) collection frequency (affects amount collected), (4) use of kitchen waste grinders, (5) characteristics of populace, (6) extent of salvaging and recycling, (7) public attitudes, and (8) legislation. The existence of salvage and recycling operations within a community definitely affects the quantities of wastes collected for disposal. Whether such operations affect the quantities generated is another question. Until more information is available, no definite statement can be made on this issue. Significant reductions in the quantities of solid wastes that are generated will occur when and if the public and consumer-oriented companies are willing to change—on their own volition—to conserve national resources and to reduce the economic burdens associated with the management of solid wastes.

On-Site Handling, Storage, and Processing

The handling, storage, and processing of solid wastes at the source before they are collected is the second of the six functional elements in a solid-waste management system.

11-5 ON-SITE HANDLING

On-site handling refers to the activities associated with the handling of solid wastes until they are placed in the containers used for their storage before collection. Depending on the type of collection service, handling may also be required to move loaded containers to the collection point and to return the empty containers to the point where they are stored between collections.

Domestic Solid Wastes

Typically, domestic wastes accumulated at several locations in and around low- and medium-rise residential dwellings are placed in larger storage containers to await removal by the waste-collection agency. Where curb collection is used, the resident is also responsible for placing the loaded larger storage container(s) at the curb and for returning the empty container(s) to their storage location next to or in the dwelling.

In high-rise apartments wastes are (1) picked up by building maintenance personnel or porters from each floor and taken to the basement service area, (2) taken to the basement service area by the tenants, or (3) bagged and placed by the tenants in specially designed chutes with openings located at each floor.

Commercial and Industrial Solid Wastes

In most office, commercial, and industrial buildings, solid wastes that accumulate in the offices or work locations usually are collected in relatively large containers mounted on rollers. Once filled, these containers are removed by means of the service elevator, if there is one, and emptied into (1) large storage containers, (2) compactors used in conjunction with the storage containers, (3) stationary compactors that can compress the material into bales or into specially designed containers, or (4) other processing equipment such as incinerators.

11-6 ON-SITE STORAGE

Factors that must be considered in the on-site storage of solid wastes include (1) the type of container to be used, (2) the container location, (3) public health and aesthetics, and (4) the collection methods to be used.

Containers

To a large extent, the types and capacities of the containers used depend on the characteristics of the solid wastes to be collected, the collection frequency, and the space available for the placement of containers. The types and capacities of containers now commonly used for on-site storage of solid wastes are summarized in Table 11-4.

Table 11-4 Data on the types and sizes of containers used for the on-site storage of solid wastes

Container type	Capacity		
	Unit	Range	Typical
Small capacity			
Plastic or metal (office type)	L	16–40	28
Plastic or galvanized metal	L	75–150	120
Barrel: plastic, aluminum, or fiber	L	75–250	120
Plastic containers with wheels (see Fig. 11-5)	L	300–380	340
Disposable paper bags standard leak-resistant and leak-proof	L	75–210	120
Disposable plastic bag	L	50–150	120
Medium capacity			
Side or top loading	m^3	0.75–9	3
Large capacity			
Open top, roll off (also called *debris boxes*)	m^3	9–38	27
Used with stationary compactor	m^3	15–30	23
Equipped with self-contained compaction mechanism	m^3	15–30	23
Trailer-mounted, open top	m^3	15–38	27
Enclosed, equipped with self-contained compaction mechanism	m^3	15–30	27

Because of increasing costs (including the cost of labor, workers' compensation, and fuel and equipment costs) there is a strong movement in the waste-collection field toward the use of large containers that can be emptied mechanically using a vehicle equipped with an articulated pickup mechanism (see Sec. 11-9). Where mechanical collection is to be used, the containers at the individual residences must be standardized to be compatible with the collection equipment.

Container Locations

In newer residential areas, containers for solid waste usually are placed by the side or rear of the house (see Fig. 11-1*b*). In older residential areas containers are located in alleys. In low-rise multifamily apartments large containers are often placed in specially designed and designated enclosures (see Fig. 11-3). In high-rise apartments storage containers are located in a basement or ground-floor service area.

The location of containers at existing commercial and industrial facilities depends on both the location of available space and service-access conditions. In newer facilities, specific service areas have been included for this purpose. Often, because the containers are not owned by the commercial or industrial activity, the locations and types of containers to be used for on-site storage must be worked out jointly between the industry and the public or private collection agency.

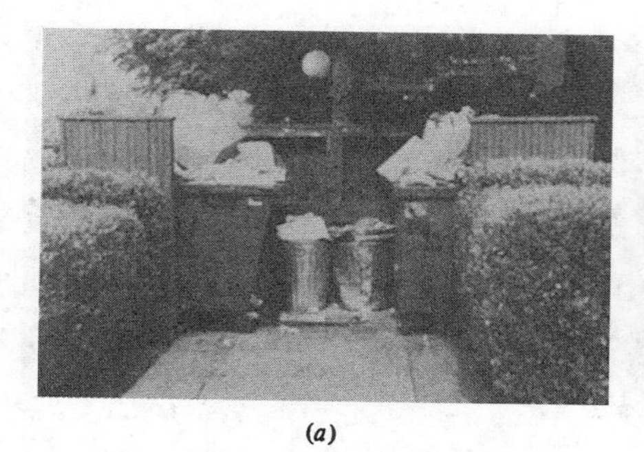

(a)

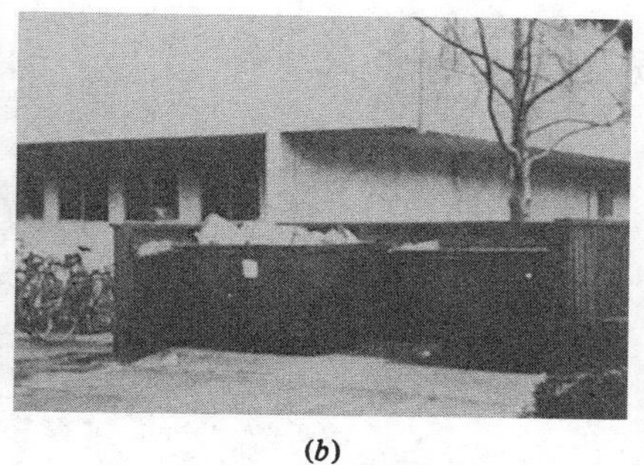

(b)

Figure 11-3 Typical storage locations for on-site containers for an apartment complex.

11-7 ON-SITE PROCESSING OF SOLID WASTES

On-site processing methods are used to recover usable materials from solid wastes, to reduce the volume, or to alter the physical form. The most common on-site processing operations include manual sorting, compaction, and incineration. These and other processing operations are considered later in this Chapter.

Collection of Solid Wastes

Information on collection, one of the most costly functional elements, is presented in four parts dealing with (1) the types of collection services, (2) the types of collection systems, (3) an analysis of collection systems, and (4) the general methodology involved in setting up collection routes.

11-8 COLLECTION SERVICES

The various types of collection services now used for municipal and commercial-industrial sources are described in this section.

Municipal Collection Services

Although a variety of collection services are used throughout the United States, the three most common are curb, alley, and backyard collection. Curb collection has gained popularity because labor costs for collection can be minimized (see Fig. 11-4). In the future, it appears that the use of large containers which can be emptied mechanically with an articulated container pickup mechanism (see Fig. 11-5) will be the most common method used for the collection of municipal wastes.

(*a*)

(*b*)

Figure 11-4 Collection of wastes placed at curb by homeowner. (*a*) Davis, California, (*b*) Venice, Italy.

(a)

(b)

Figure 11-5 (a) Collection of domestic wastes placed in large containers (340 L) with mechanical articulated pickup mechanism. (b) Close up of pickup mechanism. Containers are brought to the curb by the homeowner.

Commercial-Industrial Collection Services

The collection service provided to large apartment buildings, residential complexes, and commercial and industrial activities typically is centered around the use of large movable and stationary containers (see Fig. 11-6) and large stationary compactors (see Fig. 11-7). Compactors are of the type that can be used to compress material directly into large containers or to form bales that are then placed in large containers.

(a)

(b)

Figure 11-6 Large containers used for the collection of wastes from commercial establishments.

Figure 11-7 Stationary compactor used in conjunction with large container for the collection of wastes from commercial establishments.

11-9 TYPES OF COLLECTION SYSTEMS

Based on their mode of operation, collection systems are classified into two categories: hauled-container systems and stationary-container systems.

Hauled-Container Systems (HCS)

Collection systems in which the containers used for the storage of wastes are hauled to the processing, transfer, or disposal site, emptied, and returned to either their original location or some other location are defined as *hauled-container systems.* There are two main types of hauled-container systems: (1) tilt-frame container, and (2) trash-trailer. Typical data on the collection vehicles and containers used with these systems are reported in Table 11-5. The collector is responsible for driving the vehicle, loading full containers and unloading empty containers, and emptying the contents of the container at the disposal site. In some cases, for safety reasons, both a driver and helper are used.

Systems that use tilt-frame-loaded vehicles and large containers, often called *drop boxes*, (see Fig. 11-8) are ideally suited for the collection of all types of solid waste and rubbish from locations where the generation rate warrants the use of large containers. Open-top containers are used routinely at warehouses and construction sites. Large containers used in conjunction with stationary compactors

Table 11-5 Typical data on vehicles and containers used with various collection systems

Vehicle	Collection container type	Typical range of container capacities, m^3
Hauled-container systems		
Tilt-frame	Open top, also called *debris boxes*	8–40
	Used in conjunction with stationary compactor	10–30
	Equipped with self-contained compaction mechanism	15–30
Truck-tractor	Open-top trash-trailers	10–30
	Enclosed trailer-mounted containers equipped with self-contained compaction mechanism	15–30
Stationary-container systems		
Compactor, mechanically loaded	Open top and enclosed top and side-loading	0.6–8
Compactor, manually loaded	Small plastic or galvanized metal containers, disposable paper and plastic bags	75–200*†

* Liters.

† Loaded mass of container should not exceed 30 kg.

Figure 11-8 Typical tilt-frame collection vehicle loading large container.

are common at commercial and industrial services and at transfer stations. Because of the large volume that can be hauled, the use of tilt-frame hauled-container systems has become widespread, especially among private collectors servicing industrial accounts.

The application of trash-trailers is similar to that of tilt-frame container systems. Trash-trailers are better for the collection of especially heavy rubbish, such as sand, timber, and metal scrap, and often are used for the collection of demolition wastes at construction sites.

Stationary-Container Systems (SCS)

Collection systems in which the containers used for the storage of wastes remain at the point of waste generation, except when moved for collection are defined as *stationary-container systems.* Labor requirements for mechanically loaded stationary-container systems are essentially the same as for hauled-container systems. There are two main types of stationary-container systems: (1) those in which self-loading compactors are used and (2) those in which manually loaded vehicles are used.

Container size and utilization are not as critical in stationary-container systems using self-loading collection vehicles equipped with a compaction mechanism (see Fig. 11-9) as they are in the hauled-container system. Trips to the disposal site, transfer station, or processing station are made after the contents of a number of containers have been collected and compacted and the collection vehicle is full. Because a variety of container sizes and types are available, these systems may be used for the collection of all types of wastes.

The major application of manual transfer and loading methods is in the collection of residential wastes and litter. Manual methods are used for the collection of industrial wastes where pickup points are inaccessible to the collection vehicle.

(a)

(b)

Figure 11-9 Self-loading vehicle equipped with internal compactor.

11-10 DETERMINATION OF VEHICLE AND LABOR REQUIREMENTS

By separating the collection activities into unit operations, it is possible to develop design data and relationships that can be used to establish vehicle and labor requirements for the various collection systems.

Definition of Terms

The operational tasks for the hauled-container and stationary-container systems are shown schematically in Fig. 11-10. The activities involved in the collection of

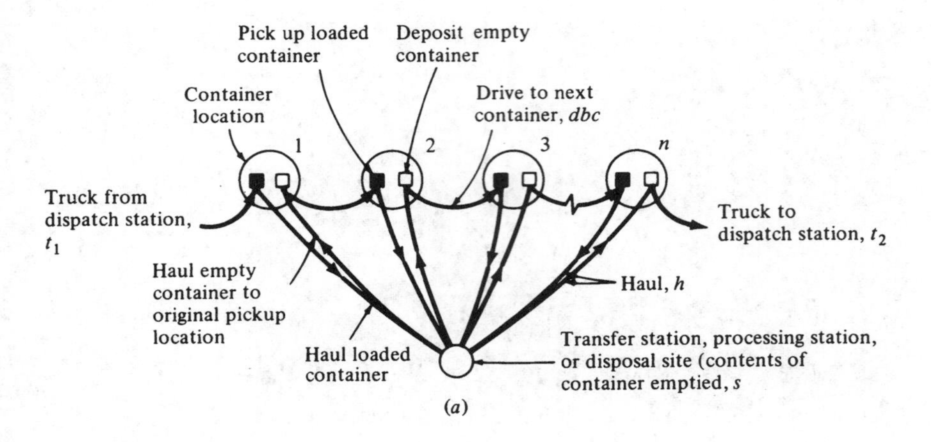

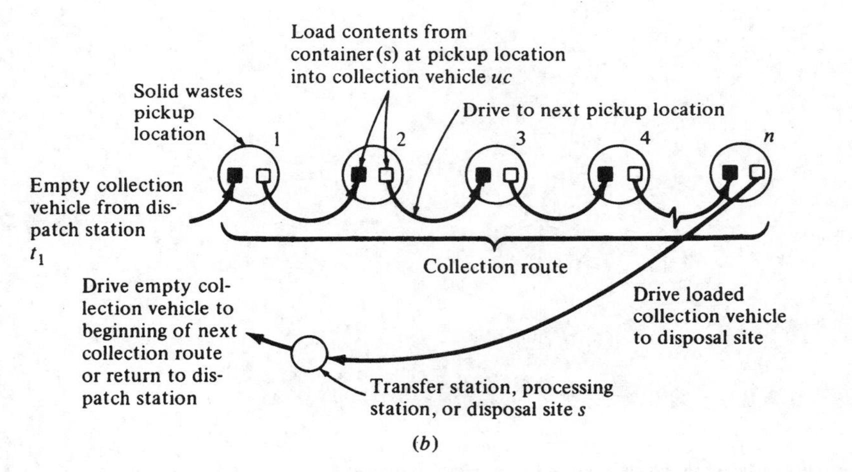

Figure 11-10 Definition sketch for waste collection systems: (*a*) hauled-container and (*b*) stationary-container.

solid wastes can be resolved into four unit operations: pickup, haul, at-site, and off-route. These unit operations are defined in Table 11-6.

Hauled-Container Systems

The time required per trip, which also corresponds to the time required per container, is equal to the sum of pick-up, at-site, and haul times and given by the following equation:

$$T_{hcs} = (P_{hcs} + s + a + bx) \tag{11-1}$$

Table 11-6 Definition of terms for the activities involved in the collection of solid wastes

Term	Definition
Pickup (P)	
Hauled-container system, P_{hcs}	The time spent picking up the loaded container, the time required to redeposit the container after its contents have been emptied, and the time spent driving to the next container.
Stationary-container system, P_{scs}	The time spent loading the collection vehicle, beginning with the stopping of the vehicle prior to loading the contents of the first container and ending when the contents of the last container to be emptied have been loaded.
Haul (h)	
Hauled-container system, h_{hcs}	The time required to reach the disposal site, starting after a container whose contents are to be emptied has been loaded on the truck, plus the time after leaving the disposal site until the truck arrives at the location where the empty container is to be redeposited. Time spent at the disposal site is not included.
Stationary-container system, h_{scs}	The time required to reach the disposal site, starting after the last container on the route has been emptied or the collection vehicle is filled, plus the time after leaving the disposal site until the truck arrives at the location of the first container to be emptied on the next collection route. Time spent at the disposal site is not included.
At-site (s)	The time spent at the disposal site, including the time spent waiting to unload as well as the time spent unloading.
Off-route (W)	All time spent on activities that are nonproductive from the point of view of the overall collection operation. Necessary off-route time includes (1) time spent checking in and out in the morning and at the end of the day, (2) time lost due to unavoidable congestion, and (3) time spent on equipment repairs and maintenance. Unnecessary off-route time includes time spent for lunch in excess of the stated lunch period and time spent on taking unauthorized coffee breaks, talking to friends, etc.

where T_{hcs} = time per trip for hauled-container system, h/trip
P_{hcs} = pickup time per trip for hauled-container system, h/trip
s = at-site time per trip, h/trip
a = empirical haul constant, h/trip
b = empirical haul constant, h/km
x = round-trip haul distance, km/trip

The pickup time per trip P_{hcs} is equal to:

$$P_{hcs} = pc + uc + dbc \tag{11-2}$$

where P_{hcs} = pickup time per trip, h/trip
pc = time required to pick up loaded container, h/trip
uc = time required to unload empty container, h/trip
dbc = average time spent driving between container locations, h/trip (determined locally)

Table 11-7 Typical values for haul constant coefficients a and b in Eqs. (11-1), (11-3), (11-4), and (11-8)

Speed limit, km/h	(mi/h)	a, h/trip	b, h/km	(h/mi)
88	(55)	0.016	0.011	(0.018)
72	(45)	0.022	0.014	(0.022)
56	(35)	0.034	0.018	(0.029)
40	(25)	0.050	0.025	(0.040)

The number of trips that can be made per vehicle per day with a hauled-container system, including a factor to account for off-route activities, is determined using Eq. (11-3):

$$N_d = [(1 - W)H - (t_1 + t_2)]/(P_{hcs} + s + a + bx) \tag{11-3}$$

where N_d = number of trips per day, trip/d
W = off-route factor, expressed as a fraction
H = length of workday, h/d
t_1 = time from garage to first container location, h
t_2 = time from last container location to garage, h

In deriving Eq. (11-3), it is assumed that off-route activities can occur at any time. Data that can be used in the solution of Eqs. (11-2) and (11-3) for various types of hauled-container systems are given in Tables 11-7 and 11-8. The off-route factor

Table 11-8 Typical data for computing equipment and labor requirements for hauled- and stationary-container collection systems

Collection: Vehicle	Collection: Loading method	Compaction ratio r	Pick up loaded container and deposit empty container, h/trip*	Empty contents of loaded container, h/container†	At-site time s, h/trip
Hauled-container systems					
Tilt-frame	Mechanical		0.40		0.127
Tilt-frame	Mechanical	2.0–4.0‡	0.40		0.133
Stationary-container systems					
Compactor	Mechanical	2.0–4.0		0.050	0.10
Compactor	Manual	2.0–4.0			0.10

* $pc + uc$ in Eq. (11-2).
† uc in Eq. (11-5).
‡ Containers used in conjunction with stationary compactor (see Fig. 11-7).

in Eq. 11-3 varies from 0.10 to 0.25; a factor of 0.15 is representative for most operations. Application of Eq. (11-3) is illustrated in Example 11-1.

Example 11-1: Analyzing a hauled-container collection system Solid waste from a new industrial park is to be collected in large containers (drop boxes), some of which will be used in conjunction with stationary compactors. Based on traffic studies at similar parks, it is estimated that the average time to drive from the garage to the first container (t_1) and from the last container (t_2) to the garage each day will be 15 and 20 min, respectively. If the average time required to drive between containers is 6 min and the one-way distance to the disposal site is 25 km (speed limit: 88 km/h), determine the number of containers that can be emptied per day, based on an 8-h workday.

SOLUTION

1. Determine the pickup time per trip using Eq. (11-2).

$$P_{hcs} = pc + uc + dbc$$

Use: $pc + uc = 0.4$ h/trip (see Table 11-8)

$dbc = 0.1$ h/trip (given)

$P_{hcs} = (0.4 + 0.1)$ h/trip

$= 0.5$ h/trip

2. Determine the time per trip using Eq. (11-1).

$$T_{hcs} = (P_{hcs} + s + a + bx)$$

Use: $P_{hcs} = 0.5$ h/trip (from step 1)

$s = 0.133$ (see Table 11-8)

$a = 0.016$ (see Table 11-7)

$b = 0.011$ (see Table 11-7)

$T_{hcs} = [0.5 + 0.133 + 0.016 + 0.011(50)]$ h/trip

$= 1.20$ h/trip

3. Determine the number of trips that can be made per day using Eq. (11-3).

$$N_d = [(1 - W)H - (t_1 + t_2)]/(P_{hcs} + s + a + bx)$$

Use: $W = 0.15$ (assumed)

$H = 8$ (given)

$t_1 = 0.25$ h (given)

$t_2 = 0.33$ h (given)

$N_d = [(1 - 0.15)8 - (0.25 + 0.33)]/1.20$ h/trip

$= (6.8 - 0.58)/1.20$

$= 5.18$ trips/d

N_d(actual) $= 5$ trips/d

4. Determine the actual length of the workday.

$$5 \text{ trips} = (1 - 0.15)H - 0.58/1.2$$

$$H = [(5 \times 1.2) + 0.58]/0.85$$

$$= 7.74 \text{ h (essentially 8 h)}$$

COMMENT If it is assumed that no off-route activities occur during times t_1 and t_2, then theoretically 5.26 trips/d could be made. Again, only 5 trips/d would be made in an actual operation. If, however, the number of trips per day that could be made were 5.8, for example, it may be cost-effective to pay the driver for the overtime and make 6 trips/d.

Stationary-Container Systems

For systems using mechanically self-loading compactors, the time per trip is:

$$T_{scs} = (P_{scs} + s + a + bx) \quad (11\text{-}4)$$

where T_{scs} = time per trip for stationary-container systems, h/trip
P_{scs} = pickup time per trip for stationary-container systems, h/trip

The pickup time for the stationary-container system is given by:

$$P_{scs} = C_t uc + (n_p - 1)(dbc) \quad (11\text{-}5)$$

where P_{scs} = pickup time per trip for stationary-container systems, h/trip
C_t = number of containers emptied per trip, container/trip
uc = average unloading time per container for stationary-container systems, h/container
n_p = number of container pickup locations per trip, locations/trip
dbc = average time spent driving between container locations, h/location (determined locally)

The term $n_p - 1$ accounts for the fact that the number of times the collection vehicle will have to be driven between container locations is equal to the number of containers less 1.

The number of containers that can be emptied per collection trip is related directly to the volume of the collection vehicle and the compaction ratio that can be achieved. This number is given by:

$$C_t = vr/cf \quad (11\text{-}6)$$

where C_t = number of containers emptied per trip, container/trip
v = volume of collection vehicle, m^3/trip
r = compaction ratio
c = container volume, m^3/container
f = weighted container utilization factor

The number of trips required per day is given by:

$$N_d = V_d/vr \quad (11\text{-}7)$$

where N_d = number of collection trips required per day, trips/d
V_d = daily waste generation rate, m^3/d

Where an integer number of trips are to be made each day, the proper combination of trips per day and the size of the vehicle can be determined by using Eq. (11-8) in conjunction with an economic analysis:

$$H = [(t_1 + t_2) + N_d(P_{scs} + s + a + bx)]/(1 - W) \qquad (11\text{-}8)$$

To determine the required truck volume, two or three different values for N_d are substituted in Eq. (11-8) and the available pickup times per trip are determined. Then, by trial and error, the truck volume required for each value of N_d is determined using Eqs. (11-5) and (11-6). From the available truck sizes, select the ones that most nearly correspond to the computed values. If the available truck sizes are smaller than the required values, compute the actual times per day that will be required using these sizes. The most cost-effective combination then can be selected. The application of the above equations is illustrated in Example 11-2.

Example 11-2: Analysing a stationary-container collection system Solid wastes from a commercial area are to be collected using a stationary-container collection system having 4-m^3 containers. Determine the appropriate truck capacity for the following conditions:

(*a*) Container size = 4 m^3
(*b*) Container utilization factor = 0.75
(*c*) Average number of containers at each location = 2
(*d*) Collection-vehicle compaction ratio = 2.5
(*e*) Container unloading time = 0.1 h/container
(*f*) Average drive time between container locations = 0.1 h
(*g*) One-way haul distance = 30 km
(*h*) Speed limit = 88 km/h (55 mi/h)
(*i*) Time from garage to first container location = 0.33 h
(*j*) Time from last container location to garage = 0.25 h
(*k*) Number of trips to disposal site per day = 2
(*l*) Length of workday = 8 h

SOLUTION

1. Using Eq. (11-8) determine the time available for each trip.

$$H = [(t_1 + t_2) + N_d(T_{scs})]/(1 - W)$$

where T_{scs} = time per trip
Use: $H = 8$ h

$t_1 = 0.33$ (given)
$t_2 = 0.25$ (given)
$N_d = 2.0$ (given)
$W = 0.15$ (assumed)
$T_{scs} = [(1 - 0.15)8 - (0.33 + 0.25)]/2$
$= 3.1$ h

2. Determine the pickup time per trip using Eq. (11-4).

$$T_{scs} = (P_{scs} + s + a + bx)$$

Use: $T_{scs} = 3.1$ hr

$s = 0.1$ h/trip (see Table 11-8)

$a = 0.016$ (see Table 11-7)

$b = 0.011$ (see Table 11-7)

$x = 60$ km (given)

$$P_{scs} = T_{scs} - (s + a + bx)$$
$$= 3.1 - [0.1 + 0.016 + 0.011(60)]$$
$$= 2.32 \text{ h/trip}$$

3. Using Eq. (11-5), determine the number of containers that can be emptied per trip.

$$P_{scs} = C_t uc + (n_p - 1)dbc$$

Use: $P_{scs} = 2.32$ h/trip

$uc = 0.1$ h/container (given)

$n_p = C_t/2$ (2 containers/location)

$dbc = 0.1$ h

$$C_t 0.1 + (0.5C_t - 1)0.1 = 2.32$$
$$0.15C_t = 2.42$$
$$C_t = 16.13, \text{ use } 16$$

4. Using Eq. (11-6), determine the required capacity of the collection truck.

$$C_t = vr/cf$$

Use: $C_t = 16$

$r = 2.5$ (given)

$c = 4\ m^3$ (given)

$f = 0.75$ (given)

$$v = \frac{16(4\ m^3)0.74}{2.5} = 19.2\ m^3$$

Use 20.0 m^3 or nearest larger standard size.

COMMENT The above analysis is exactly the same as would be used for the collection of domestic wastes using the collection system shown in Fig. 11-5.

Stationary-Container Systems (Manually Loaded)

The analysis and design of residential collection systems using manually loaded vehicles may be outlined as follows. If H hours are worked per day and the number of trips to be made per day is known or fixed, the time available for the pickup operation can be computed by using Eq. (11-8) because either all the factors are known or they can be assumed. Once the pickup time per trip is known, the number of pickup locations from which wastes can be collected per trip can be estimated as follows:

$$N_p = 60\ P_{scs} n/t_p \tag{11-9}$$

where N_p = number of pickup locations per trip, locations/trip
60 = conversion factor from hours to minutes, 60 min/h
P_{scs} = pickup time per trip, h/trip
n = number of collectors
t_p = pick up time per pickup location, collection · min/location

The pickup time for a two-person collection crew can be estimated using Eq. (11-10).

$$t_p = 0.72 + 0.18C_n + 0.014PRH \tag{11-10}$$

where t_p = average pickup time per pickup location, collection · min/location
C_n = average number of containers at each pickup location
PRH = rear-of-house pickup locations, percent

Equation (11-10) is typical of the types of equations derived from field observations for the pickup time per location. Usually, the first term in such equations represents the time spent driving between pickup locations. This value will, of course, depend on the characteristics of the residential area. Values for a one-person crew should be obtained from the field observations. Typically, the time per service is about 0.9 min/service where unlimited service is provided.

Once the number of pickup locations per trip is known, the proper size of collection vehicle can then be estimated as follows:

$$v = \frac{V_p N_p}{r} \tag{11-11}$$

where v = volume of collection vehicle, m^3/trip
V_p = volume of solid wastes collected per pickup location, m^3/location
N_p = number of pickup locations per trip, locations/trip
r = compaction ratio

In many housing areas, the collection frequency is twice per week. In terms of labor requirements, it has been found that the requirements for the second weekly collection are about 0.9 to 0.95 times those for the first weekly collection. In general, the labor requirements are not significantly different because container handling time is about the same for both full and partially full containers. Often this difference is neglected in computing the labor requirements.

11-11 COLLECTION ROUTES

Once the equipment and labor requirements have been determined, collection routes must be laid out so both the work force and equipment are used effectively. In general, the layout of collection routes is a trial-and-error process. There are no fixed rules that can be applied to all situations.

Some of the factors that should be taken into consideration when laying out routes are as follows: (1) existing company policies and regulations related to such items as the point of collection and frequency of collection must be identified, (2) existing system conditions such as crew size and vehicle types must be coordinated, (3) wastes generated at traffic-congested locations should be collected as early in the day as possible, (4) sources at which extremely large quantities of wastes are generated should be serviced during the first part of the day, and (5) scattered pickup points where small quantities of solid wastes are generated should, if possible, be serviced during one trip or on the same day, if they receive the same collection frequency.

Layout of Routes

The layout of collection routes is a four-step process. First, prepare location maps. On a relatively large-scale map of the area to be serviced, the following data should be plotted for each solid-waste pickup point: location, number of containers, collection frequency, and, if a stationary-container system with self-loading compactors is used, the estimated quantity of wastes to be collected at each pickup location. Second, prepare data summaries. Estimate the quantity of wastes to be collected from pickup locations serviced each day that the collection operation is to be conducted. Where a stationary-container system is used, the number of locations that will be serviced during each pickup cycle must also be determined. Third, lay out preliminary collection routes starting from the dispatch station or where the collection vehicles are parked. A route should be laid out that connects all the pickup locations to be serviced during each collection day. The route should be laid out so that the last location is nearest the disposal site. Fourth, develop balanced routes. After the preliminary collection routes have been laid out, the haul distance for each route should be determined. Next, determine the labor requirements per day and check against the available work times per day. In some cases it may be necessary to readjust the collection routes to balance the work load and the distance traveled. After the balanced routes have been established, they should be drawn on the master map. The layout of collection routes is illustrated in Example 11-3.

Example 11-3: Laying collection routes Lay out collection routes for the residential area shown in the accompanying figure. Assume the following data are applicable.

1. General
   a. Occupants per resident = 3.5
   b. Solid-waste generation rate = 1.6 kg/person · d
   c. Collection frequency
   d. Type of collection service = curb
   e. Collection crew size = one person
   f. Collection vehicle capacity = 20 m^3
   g. Compacted density of solid wastes in collection vehicle = 325 kg/m^3
2. Route constraints
   a. No U-turns in streets
   b. Collection from each side of street with stand-up right-hand-drive collection vehicle

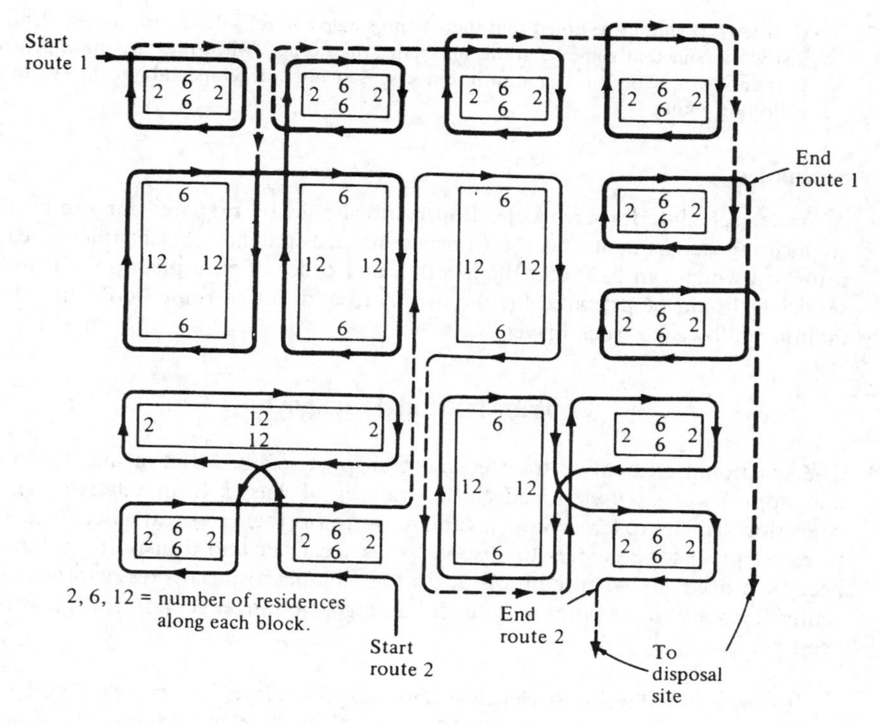

SOLUTION

1. Determine total number of residences from which wastes are to be collected.

$$\text{Residences} = 10(16) + 4(36) + 1(28) = 332$$

2. Determine the compacted volume of solid waste to be collected per week.

$$\begin{aligned}\text{Vol/wk} &= (332 \text{ residences} \times \\ &\quad 3.5 \text{ persons/residence} \times \\ &\quad 1.6 \text{ kg/person} \cdot \text{d} \times \\ &\quad 7 \text{ d} \cdot \text{wk})/325 \text{ kg/m}^3 \\ &= 40.0 \text{ m}^3\text{/wk}\end{aligned}$$

3. Determine the number of trips/wk.

$$\text{Trip/wk} = \frac{40.0 \text{ m}^3\text{/wk}}{20 \text{ m}^3\text{/trip}} = 2$$

4. Determine the average number of residences from which wastes are to be collected each day.

$$\text{Residences/trip} = \frac{332}{2} = 166$$

5. Lay out collection routes by trial and error using the route constraints cited above as a guide. The two routes are shown in the figure.

COMMENT It should be noted that there is no single correct solution to this problem. It just works out that some solutions are better than others when they are implemented. It is only with experience that an intuitive sense can be developed about the layout of collection routes.

Schedules

A master schedule for each collection route should be prepared for use by the engineering department and the transportation dispatcher. A schedule for each route, on which can be found the location and order of each pickup point to be serviced, should be prepared for the driver. In addition, a route book should be maintained by each truck driver.

Transfer and Transport

The functional element of transfer and transport refers to the means, facilities, and appurtenances used to effect the transfer of wastes from relatively small collection vehicles to larger vehicles and to transport them over extended distances to either processing centers or disposal sites. Transfer and transport operations become a necessity when haul distances to available disposal sites or processing centers increase to the point that direct hauling is no longer economically feasible. See Example 11-4.

Example 11-4: Economic comparison of transport alternatives Determine the break-even time for a stationary-container system and a separate transfer and transport system for transporting wastes collected from a metropolitan area to a landfill disposal site. Assume the following cost and system data are applicable.

1. Transportation costs:
   a. Stationary-container system using an 18-m^3 compactor = \$20/h
   b. Tractor-trailer transport unit with a capacity of 120 m^3 = \$25/h
2. Other costs:
   a. Transfer station operating cost, including amortization = \$0.40/m^3
   b. Extra cost for unloading facilities for tractor-trailer transport unit = \$0.05/m^3
3. Other data:
   a. Density of wastes in compactor = 325 kg/m^3
   b. Density of wastes in transport units = 150 kg/m^3

SOLUTION

1. Convert cost data to units of dollars/tonne · min.
   a. Stationary-container system:

$$\text{Operating cost} = (\$20.00/\text{h})/(60\ \text{min/h}) = \$0.33/\text{min}$$

$$\text{Tonnes/load} = \frac{18\ \text{m}^3 \times 325\ \text{kg/m}^3}{1000\ \text{kg/tonne}} = 5.85$$

$$\text{Operating cost} = (\$0.33/\text{min})/5.85\ \text{tonne} = \$0.0564/\text{tonne} \cdot \text{min}$$

b. Transfer-transport system:

$$\text{Operating cost} = (\$25.00/\text{h})/(60\ \text{min/h}) = \$0.42/\text{min}$$

$$\text{Tonnes/load} = \frac{120 \times 150}{1000} = 18$$

$$\text{Operating cost} = (\$0.42/\text{min})/18\ \text{tonnes} = \$0.0233/\text{tonne} \cdot \text{min}$$

c. Transfer station cost:

$$\text{Operating cost} = (\$0.40/\text{m}^3)/(0.150/\text{tonne}) = \$2.47/\text{tonne}$$

d. Unloading cost:

$$\text{Operating cost} = (\$0.05/\text{m}^3)/(0.150/\text{tonne}) = \$0.33/\text{tonne}$$

2. Prepare a plot of cost versus haul time in minutes and determine break-even time.
   a. Fixed cost for transfer and transport system:

$$\text{Cost/tonne} = \$2.67 + \$0.33 = \$3.00$$

   b. Variable costs at 100 min:
   (1) Stationary-container system

$$\text{Cost/ton} = (\$0.0564/\text{tonne} \cdot \text{min})100\ \text{min} = \$5.64$$

   (2) Transport system

$$\text{Cost/ton} = (\$0.0233/\text{tonne} \cdot \text{min})\ 100\ \text{min} = \$2.33$$

   c. The above data are plotted in the accompanying figure. As shown, the break-even time is equal to 83 min.

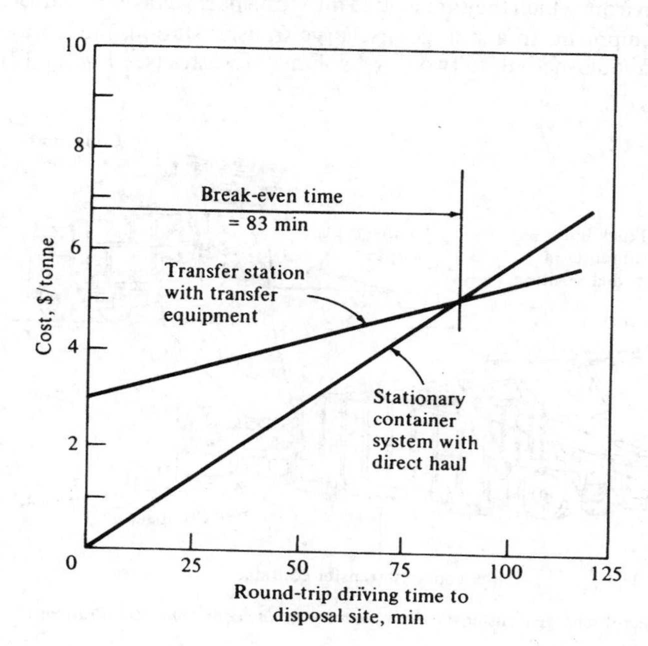

11-12 TRANSFER STATIONS

Important factors that must be considered in the design of transfer stations include: (1) type of transfer operation to be used, (2) capacity requirements, (3) equipment and accessory requirements, and (4) environmental requirements. Depending on the method used to load the transport vehicles, transfer stations may be classified into three types: (1) direct discharge, (2) storage discharge, and (3) combined direct and storage discharge.

Direct Discharge

In a direct-discharge transfer station, wastes from the collection vehicles usually are emptied directly into the vehicle to be used to transport them to a place of final disposition. To accomplish this, these transfer stations usually are constructed in a two-level arrangement. The unloading dock or platform from which wastes from collection vehicles are discharged into the transport trailers is elevated, or the transport trailers are located in a depressed ramp. Direct-discharge transfer stations employing stationary compactors are also popular (see Fig. 11-11).

Storage Discharge

In the storage-discharge transfer station, wastes are emptied either into a storage pit or onto a platform from which they are loaded into transport vehicles by various types of auxiliary equipment. In a storage-discharge transfer station, the storage volume varies from about one-half to two days' volume of wastes (see Fig. 11-12).

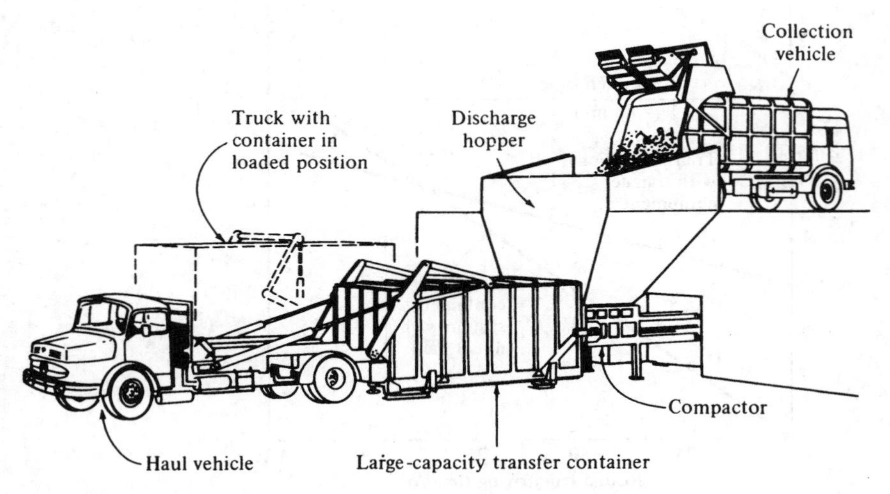

Figure 11-11 Typical direct-discharge transfer station. (*Courtesy Schindler Waggon AG, Prattein.*)

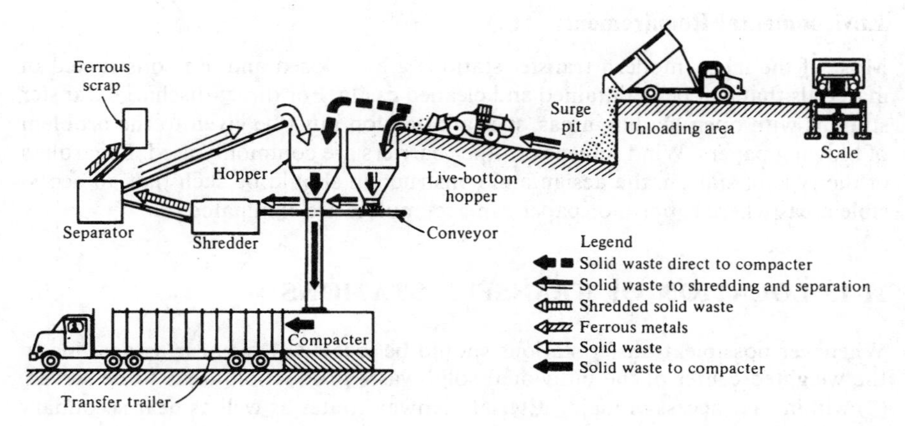

Figure 11-12 Typical storage-discharge transfer station. (*Courtesy of Municipality of Metropolitan Toronto, Department of Public Works.*)

Combined Direct and Storage Discharge

In some transfer stations, both direct-discharge and storage-discharge methods are used. Usually, these are multipurpose facilities designed to service a broader range of users than a single-purpose facility. In addition to serving a broader range of users, a multipurpose transfer station can also house a materials-salvage operation.

Capacity Requirements

The operational capacity of a transfer station must be such that the collection vehicles do not have to wait too long to unload. In most cases, it will not be cost-effective to design the station to handle the ultimate peak number of hourly loads. An economic trade-off analysis should be made between the annual cost for the time spent by the collection vehicles waiting to unload against the incremental annual cost of a larger transfer station and/or the use of more transport equipment. Because of the increased cost of transport equipment, a trade-off analysis must also be made between the capacity of the transfer station and the cost of the transport operation, including both equipment and labor components.

Equipment and Accessory Requirements

The types and amounts of equipment required vary with the capacity of the station and its function in the waste-management system. Specifically, scales should be provided at all medium and large transfer stations both to monitor the operation and to develop meaningful management and engineering data.

Environmental Requirements

Most of the large, modern transfer stations are enclosed and are constructed of materials that can be maintained and cleaned easily. For direct-discharge transfer stations with open loading areas, special attention must be given to the problem of blowing papers. Wind screens or other barriers are commonly used. Regardless of the type of station, the design and construction should be such that all accessible areas where rubbish or paper can accumulate are eliminated.

11-13 LOCATION OF TRANSFER STATIONS

Whenever possible, transfer stations should be located (1) as near as possible to the weighted center of the individual solid-waste production areas to be served, (2) within easy access of major arterial highway routes as well as near secondary or supplemental means of transportation, (3) where there will be a minimum of public and environmental objection to the transfer operations, and (4) where construction and operation will be most economical. Additionally, if the transfer-station site is to be used for processing operations involving materials recovery and/or energy production, the requirements for those operations must be considered.

11-14 TRANSFER MEANS AND METHODS

Motor vehicles, railroads, and ocean-going vessels are the principal means now used to transport solid wastes. Pneumatic and hydraulic systems have also been used. Still other systems have been suggested, but most have not been tested.

Motor Vehicle Transport

Motor vehicles used to transport solid wastes on highways should satisfy the following requirements: (1) the vehicles must transport wastes at minimum cost, (2) wastes must be covered during the haul operation, (3) vehicles must be designed for highway traffic, (4) vehicle capacity must be such that allowable weight limits are not exceeded, and (5) methods used for unloading must be simple and dependable. The maximum volume that can be hauled in highway transport vehicles depends on the regulations in force in the state in which they are operated.

In recent years, because of their simplicity and dependability, open-top trailers and semitrailers have found wide acceptance (see Table 11-9 and Fig. 11-13) for the transport of wastes. Some trailers are equipped with sumps to collect any liquids that accumulate from the solid wastes. The sumps are equipped with drains so that they can be emptied at the disposal site.

Methods used to unload the transport trailers may be classified according to whether they are self-emptying or require the aid of auxiliary equipment. Self-emptying transport trailers are equipped with mechanisms such as hydraulic

Table 11-9 Typical data on haul vehicles used at transfer stations

Type	Capacity per trailer,		Length of tractor and trailer units,
	m^3	Tonnes	m*
Tractor-trailer-trailer	54	11.4	19.8
	54	10.0	
Tractor-trailer	74	17.3	18.3
Tractor-compactor trailer	58	18	14.0

* Overall length will vary with the type of tractor (e.g., conventional or cab-over) and the turning radius of the trailers.

dump beds, powered diaphragms or moving floors that are part of the vehicle (see Fig. 11-14). Moving-floor trailers are an adaptation of equipment used in the construction industry. An advantage of the moving-floor trailer is the rapid turnaround time (typically 6 to 10 min) achieved at the disposal site without the need for auxiliary equipment. Unloading systems that require auxiliary equipment are usually of the "pull-off" type, in which the wastes are pulled out of the truck by either a movable bulkhead or wire-cable slings placed forward of the load. The disadvantage of requiring auxiliary equipment and work force to unload at the disposal site is relatively minor in view of the simplicity and reliability of these methods.

Another auxiliary unloading system that has proved very effective and efficient involves the use of movable, hydraulically operated tipping ramps located at the

Figure 11-13 Typical large vehicles used for the transport of solid wastes for disposal.

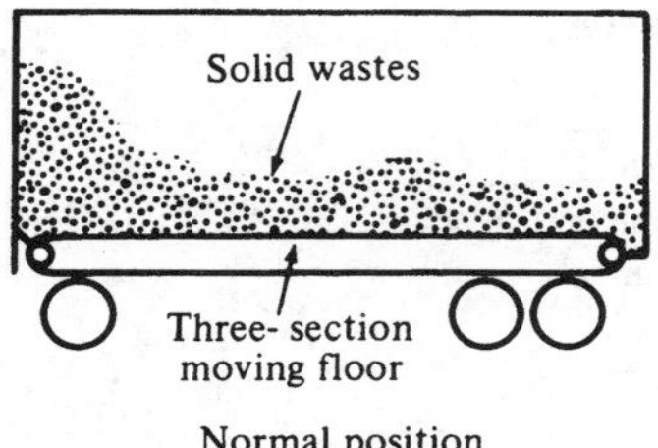

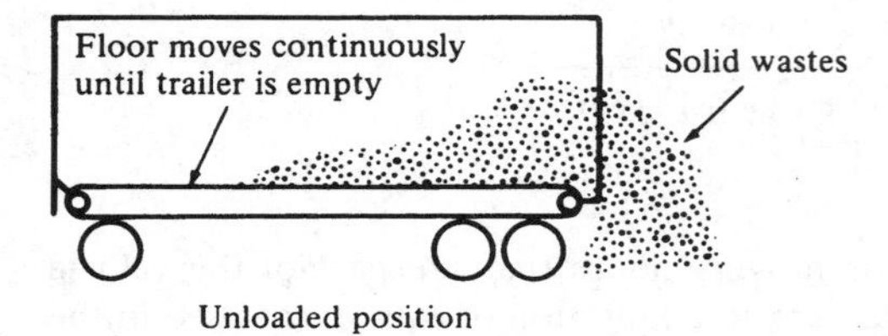

Figure 11-14 Typical self-emptying transport trailer.

disposal site (see Fig. 11-15). Operationally, the semitrailer of a tractor-trailer-trailer combination is backed up onto one of the tipping ramps; the tractor-trailer combination is backed up onto a second tipping ramp. The backs of the trailers are opened, and the units are then tilted upward until the wastes fall out by gravity. The time required for the entire unloading operation typically is about 5 min/trip.

Large-capacity containers and container trailers are used in conjunction with stationary compactors at transfer stations. In some cases, the compaction mechanism is an integral part of the container. When containers are equipped with a self-contained compaction mechanism, the movable bulkhead used to compress the wastes is also used to discharge the compacted wastes.

Railroad Transport

Although railroads were commonly used for the transport of solid wastes in the past, they are now used by only a few communities. However, renewed interest is again developing in the use of railroads for hauling solid wastes, especially to remote areas where highway travel is difficult and railroad lines now exist.

Water Transport

Barges, scows, and special boats have been used in the past to transport solid wastes to processing locations and to seaside and ocean disposal sites, but ocean disposal is no longer practiced by the United States. Although some self-propelled vessels (such as United States Navy garbage scows and other special boats) were once used, the most common practice was to use vessels towed by tugs or other special boats.

(*a*)

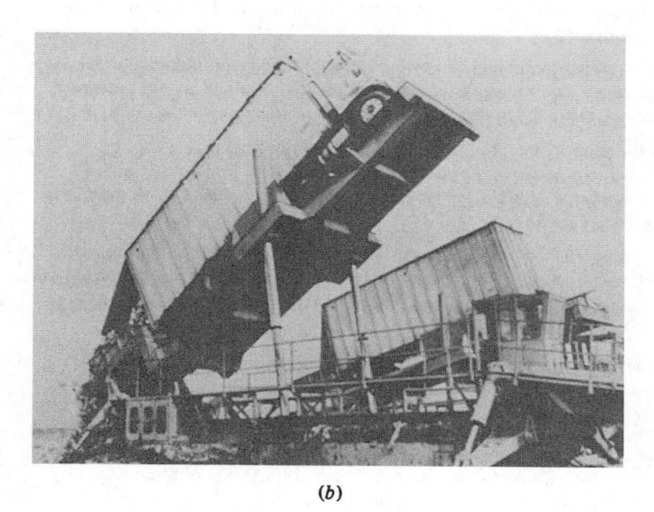

(*b*)

Figure 11-15 Hydraulically operated tipping platforms for unloading transport vehicles.

Pneumatic Transport

Both low-pressure air and vacuum conduit transport systems have been used to transport solid wastes. The most common application is the transport of wastes from high-density apartments or commercial activities to a central location for processing or for loading into transport vehicles. The largest pneumatic system now in use in the United States is at the Walt Disney World amusement park in Orlando, Florida.

Processing Techniques

Processing techniques are used in solid waste management systems to (1) improve the efficiency of solid-waste disposal systems, (2) to recover resources (usable materials), and (3) to prepare materials for the recovery of conversion products and energy. Processes used routinely to improve the efficiency of solid-waste systems and to recover materials manually are considered in this section. Me-

Table 11-10 Factors that should be considered in evaluating on-site processing equipment

Factor	Evaluation
Capabilities	What will the device or mechanism do? Will its use be an improvement over conventional practices?
Reliability	Will the equipment perform its designated functions with little attention beyond preventive maintenance? Has the effectiveness of the equipment been demonstrated in use over a reasonable period of time or merely predicted?
Service	Will servicing capabilities beyond those of the local building maintenance staff be required occasionally? Are properly trained service personnel available through the equipment manufacturer or the local distributor?
Safety of operation	Is the proposed equipment reasonably foolproof so that it may be operated by tenants or building personnel with limited mechanical knowledge or abilities? Does it have adequate safeguards to discourage careless use?
Ease of operation	Is the equipment easy to operate by a tenant or by building personnel? Unless functions and actual operations of equipment can be carried out easily, they may be ignored or "short-circuited" by paid personnel or by tenants.
Efficiency	Does the equipment perform efficiently and with a minimum of attention? Under most conditions, equipment that completes an operational cycle each time it is used should be selected.
Environmental effects	Does the equipment pollute or contaminate the environment? Where possible, equipment should reduce environmental pollution presently associated with conventional functions.
Health hazards	Does the device, mechanism, or equipment create or amplify health hazards?
Aesthetics	Does the equipment and its arrangement offend the senses? Every effort should be made to reduce or eliminate offending sights, odors, and noises.
Economics	What are the economics involved? Both first and annual costs must be considered. Future operation and maintenance costs must be assessed carefully. All factors being equal, equipment produced by well-established companies, having a proven history of satisfactory operation, should be given appropriate consideration.

Source: From Tchobanoglous et al. [11-8]

chanical systems used for the recovery of materials are considered in Chap. 12. Important processing techniques used routinely in municipal solid-waste systems include: compaction, thermal volume reduction (incineration), and manual separation of waste components. Factors that should be considered in evaluating on-site processing equipment are summarized in Table 11-10.

11-15 MECHANICAL VOLUME REDUCTION

Mechanical volume reduction is perhaps the most important factor in the development and operation of solid-waste management systems. Vehicles equipped with compaction mechanisms are used for the collection of most municipal solid wastes. To increase the useful life of landfills, wastes are compacted. Paper for recycling is baled for shipping to processing centers. When compacting a broad range of municipal solid wastes, it has been found that the final density (typically about 1100 kg/m^3) is essentially the same regardless of the starting density and applied pressure. This fact is important in evaluating the claims made by manufacturers of compacting equipment.

11-16 THERMAL VOLUME REDUCTION

The volume of municipal wastes can be reduced by more than 90 percent by incineration. In the past, incineration was quite common. However, with more restrictive air-pollution control requirements necessitating the use of expensive cleanup equipment only a limited number of municipal incinerators are currently in operation. More recently, increased haul distances to available landfill sites and increased fuel costs have brought about a renewed interest in incineration, and a number of new incinerator projects are now on the drawing boards. Incineration is considered further in Chap. 12.

11-17 MANUAL COMPONENT SEPARATION

The manual separation of solid waste components can be accomplished at the source where solid wastes are generated, at a transfer station, at a centralized processing station, or at the disposal site. Manual sorting at the source of generation is the most positive way to achieve the recovery and reuse of materials. The number and types of components salvaged or sorted (e.g., cardboard and high-quality paper, metals, and wood) depend on the location, the opportunities for recycling, and the resale market.

In Davis, California, residents, on a voluntary basis, manually separate newsprint, aluminum cans, and glass. The separated components are placed at the curb for collection with a special vehicle. The vehicle used for the collection of source-separated waste components is shown in Fig. 11-16. Waste paper is sold to an insulation manufacturer.

Figure 11-16 Collection of source-separated wastes with a special collection vehicle.

Ultimate Disposal

Disposal on or in the earth's mantle is, at present, the only viable method for the long-term handling of: (1) solid wastes that are collected and are of no further use, (2) the residual matter remaining after solid wastes have been processed, and (3) the residual matter remaining after the recovery of conversion products and/or energy has been accomplished. Landfilling is the method of disposal used most commonly for municipal wastes; landfarming and deep-well injection have been used for industrial wastes. Although incineration is often considered a disposal method, it is, in reality, a processing method (see Chap. 12).

11-18 LANDFILLING WITH SOLID WASTES

Landfilling involves the controlled disposal of solid wastes on or in the upper layer of the earth's mantle. Important aspects in the implementation of sanitary landfills include: (1) site selection, (2) landfilling methods and operations, (3) occurrence of gases and leachate in landfills, and (4) movement and control of landfill gases and leachate.

Site Selection

Factors that must be considered in evaluating potential solid-waste disposal sites are summarized in Table 11-11. Final selection of a disposal site usually is based on the results of a preliminary site survey, results of engineering design and cost studies, and an environmental impact assessment.

Table 11-11 Factors that must be considered in evaluating potential landfill sites

Factor	Remarks
Available land area	Site should have a useful life greater than 1 yr (minimum value).
Haul distance	Will have significant impact on operating costs.
Soil conditions and topography	Cover material must be available at or near the site.
Surface water hydrology	Impacts drainage requirements.
Geologic and hydrogeologic conditions	Probably most important factors in establishment of landfill site, especially with respect to site preparation.
Climatologic conditions	Provisions must be made for wet-weather operation.
Local environmental conditions	Noise, odor, dust, vector, and aesthetic factors control requirements.
Ultimate use of site	Affects long-term management for site.

Landfilling Methods and Operations

To use the available area at a landfill site effectively, a plan of operation for the placement of solid wastes must be prepared. Various operational methods have been developed, primarily on the basis of field experience. The principal methods used for landfilling dry areas may be classified as (1) area, (2) trench, and (3) depression.

The *area method* is used when the terrain is unsuitable for the excavation of trenches in which to place the solid wastes. The filling operation usually is started by building an earthen levee against which wastes are placed in thin layers and compacted (see Fig. 11-17). Each layer is compacted as the filling progresses until the thickness of the compacted wastes reaches a height varying from 2 to 3 m (6 to 10 ft). At that time, and at the end of each day's operation, a 150- to 300-mm (6- to 12-in) layer of cover material is placed over the completed fill. The cover material must be hauled in by truck or earth-moving equipment from adjacent land or from borrow-pit areas. In some newer landfill operations, the daily cover material is omitted. A completed lift, including the cover material, is called a *cell*. Successive lifts are placed on top of one another until the final grade called for in the ultimate development plan is reached. A final layer of cover material is used when the fill reaches the final design height.

The *trench method* of landfilling is ideally suited to areas where an adequate depth of cover material is available at the site and where the water table is well below the surface. To start the process (for a small landfill), a portion of the trench is dug with a bulldozer and the dirt is stockpiled to form an embankment behind the first trench. Wastes are then placed in the trench, spread into thin layers and compacted. The operation (depicted in Fig. 11-18) continues until the desired height is reached. Cover material is obtained by excavating an adjacent trench

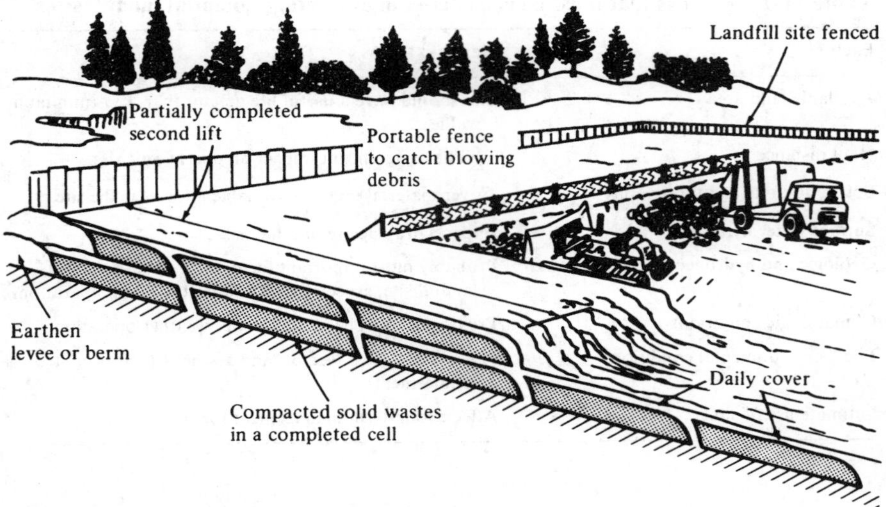

Figure 11-17 Area method of landfilling solid wastes.

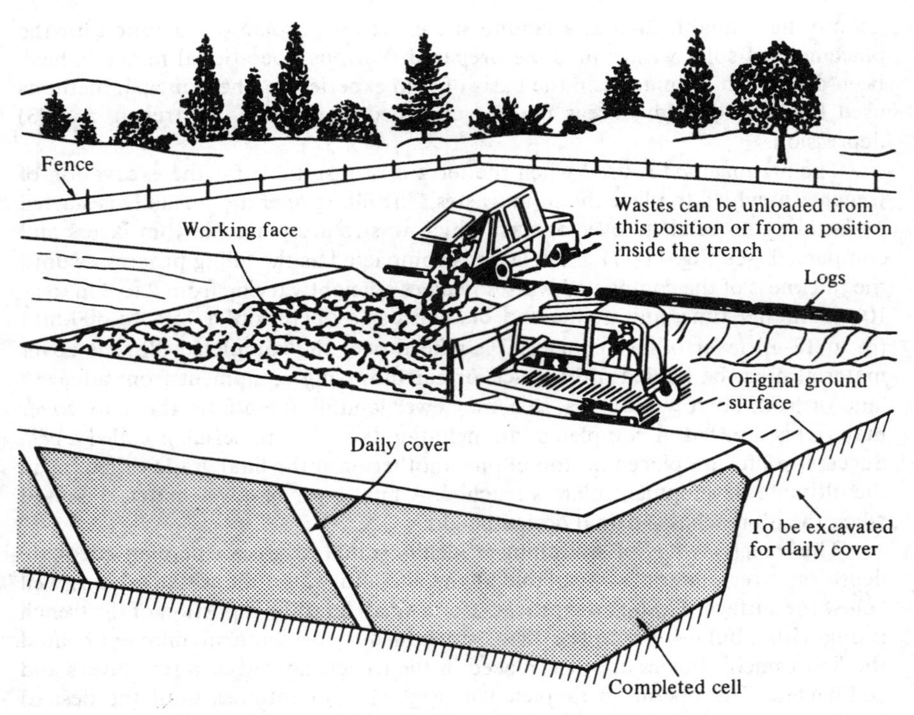

Figure 11-18 Trench method of landfilling solid wastes for small landfills.

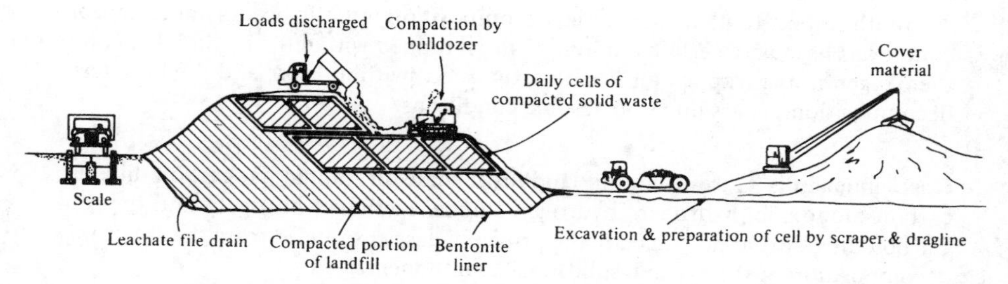

Figure 11-19 Pit method of landfilling solid wastes for large landfills. (*Courtesy of Municipality of Metropolitan Toronto, Department of Public Works.*)

or continuing the trench that is being filled. In large landfills, a dragline and one or more scrapers are used to excavate a deep rectangular pit (see Fig. 11-19).

At locations where natural or artificial depressions exist, it is often possible to use them effectively for landfilling operations. Canyons, ravines, dry borrow pits, and quarries have all been used for this purpose. The techniques to place and compact solid wastes in *depression landfills* vary with the geometry of the site, the characteristics of the cover material, the hydrology and geology of the site, and the access to the site. In a canyon site, filling starts at the head end of the canyon and ends at the mouth. This practice prevents the accumulation of water behind the landfill. Wastes usually are deposited on the canyon floor and from there are pushed up against the canyon face at a slope of about 2 to 1. In this way, a high degree of compaction can be achieved.

Because of the problems associated with contamination of local groundwaters, the development of odors, and structural stability, landfills in wet areas are seldom used. If wet areas such as swamps and marshes, tidal areas, and ponds, pits, or quarries must be used as landfill sites, special provisions must be made to contain or to eliminate the movement of leachate and gases from completed cells. Usually this is accomplished by first draining the site and then lining the bottom with a clay liner or other appropriate sealants. If a clay liner is used, it is important to continue operation of the drainage facility until the site is filled to avoid the creation of uplift pressures that could cause the liner to rupture from heaving.

Occurrence of Gases and Leachate in Landfills

The following biological, physical, and chemical events occur when solid wastes are placed in a sanitary landfill: (1) biological decay of organic materials, either aerobically of anaerobically, with the evolution of gases and liquids; (2) chemical oxidation of waste materials; (3) escape of gases from the fill; (4) movement of liquids caused by differential heads; (5) dissolving and leaching of organic and inorganic materials by water and leachate moving through the fill; (6) movement of dissolved material by concentration gradients and osmosis; and (7) uneven settlement caused by consolidation of material into voids.

With respect to item 1, bacterial decomposition initially occurs under aerobic conditions because a certain amount of air is trapped within the landfill. However, the oxygen in the trapped air is soon exhausted (within days), and the long-term decomposition occurs under anaerobic conditions.

Gases in landfills Gases found in landfills include air, ammonia, carbon dioxide, carbon monoxide, hydrogen, hydrogen sulfide, methane, nitrogen, and oxygen. Carbon dioxide and methane are the principal gases produced from the anaerobic decomposition of the organic solid-waste components.

The anaerobic conversion of organic compounds is thought to occur in three steps: the first involves the enzyme-mediated transformation (*liquefaction*) of higher-weight molecular compounds into compounds suitable for use as a source of energy and cell carbon; the second is associated with the bacterial conversion of the compounds resulting from the first step into identifiable lower-molecular-weight intermediate compounds; and the third step involves the bacterial conversion of the intermediate compounds into simpler end products, such as carbon dioxide (CO_2) and methane (CH_4). The overall anaerobic conversion of organic industrial wastes can be presented with the following equation:

$$C_aH_bO_cN_d \longrightarrow nC_wH_xO_yN_z + mCH_4 + sCO_2 + rH_2O + (d - nz)NH_3 \tag{11-12}$$

where $s = a - nw - m$

$r = c - ny - 2s$

The terms $C_aH_bO_cN_d$ and $C_wH_xO_yN_z$ are used to represent on a molar basis the composition of the material present at the start and the end of the process. If it is assumed that the organic wastes are stabilized completely, the corresponding expression is

$$C_aH_bO_cN_d + \left(\frac{4a - b - 2c + 3d}{4}\right)H_2O \longrightarrow \left(\frac{4a + b - 2c - 3d}{8}\right)CH_4 + \left(\frac{4a - b + 2c + 3d}{8}\right)CO_2 + dNH_3 \tag{11-13}$$

The rate of decomposition in unmanaged landfills, as measured by gas production, reaches a peak within the first 2 years and then slowly tapers off, continuing, in many cases, for periods up to 25 years or more. The total volume of the gases released during anaerobic decomposition can be estimated in a number of ways. If all the organic constituents in the wastes (with the exception of plastics, rubber, and leather) are represented with a generalized formula of the form $C_aH_bO_cN_d$, the total volume of gas can be estimated by using Eq. (11-13), with the assumption of complete conversion to carbon dioxide and methane.

Leachate in landfills Leachate may be defined as liquid that has percolated through solid waste and has extracted dissolved or suspended materials from it. In most landfills, the liquid portion of the leachate is composed of the liquid produced from the decomposition of the wastes and liquid that has entered the landfill from external sources, such as surface drainage, rainfall, groundwater, and water from underground springs. Representative data on the chemical characteristics of leachate are reported in Table 11-12. The leachate should be either contained within the landfill or removed for treatment.

Gas Movement

Under ideal conditions, the gases generated from a landfill should be either vented to the atmosphere or, in larger landfills, collected for the production of energy. In most cases, over 90 percent of the gas volume produced from the decomposition of solid wastes consists of methane and carbon dioxide. Although most of the methane escapes to the atmosphere, both methane and carbon dioxide have been found in concentrations of up to 40 percent at lateral distances of up to 120 m (400 ft) from the edges of landfills. If vented into the atmosphere in an uncontrolled

Table 11-12 Data on the composition of leachate from landfills

Constituent	Value, mg/L*	
	Range†	Typical
BOD_5 (5-day biochemical oxygen demand)	2000–30,000	10,000
TOC (total organic carbon)	1500–20,000	6,000
COD (chemical oxygen demand)	3000–45,000	18,000
Total suspended solids	200–1000	500
Organic nitrogen	10–600	200
Ammonia nitrogen	10–800	200
Nitrate	5–40	25
Total phosphorus	1–70	30
Orthophosphorus	1–50	20
Alkalinity as $CaCO_3$	1000–10,000	3,000
pH	5.3–8.5	6
Total hardness as $CaCO_3$	300–10,000	3,500
Calcium	200–3000	1,000
Magnesium	50–1500	250
Potassium	200–2000	300
Sodium	200–2000	500
Chloride	100–3000	500
Sulfate	100–1500	300
Total iron	50–600	60

* Except pH.

† Representative range of values. Higher maximum values have been reported in the literature for some of the constituents.

Source: From Tchobanoglous et al. [11-8]

manner, methane can accumulate (because its specific gravity is less than that of air) below buildings or in other enclosed spaces on, or close to, a sanitary landfill. With proper venting, methane should not pose a problem.

Because carbon dioxide is about 1.5 times as dense as air and 2.8 times as dense as methane, it tends to move toward the bottom of the landfill. As a result, the concentration of carbon dioxide in the lower portions of landfills may be high for years. Ultimately, because of its density, carbon dioxide will also move downward through the underlying formation until it reaches the groundwater. Because carbon dioxide is readily soluble in water, it usually lowers the pH, which in turn can increase the hardness and mineral content of the groundwater through the solubilization of calcium and magnesium carbonates.

Control of Gas Movement

The movement of gases in landfills can be controlled by constructing vents and barriers and by gas recovery.

Control of gas movement with vents and barriers The lateral movement of gases produced in a landfill can be controlled by installing vents made of materials that are more permeable than the surrounding soil. Typically, as shown in Fig. 11-20*a*, gas vents are constructed of gravel. The spacing of cell vents depends on the width of the waste cells but usually varies from 18 to 60 m (60 to 200 ft). The thickness of the gravel layer should be such that it will remain continuous even though there may be differential settling; 0.30 to 0.45 m (12 to 18 in) is recommended. Barrier or well vents (see Fig. 11-20*b*) also can be used to control the lateral movement of gases. Well vents (see Fig. 11-20*c*) are often used in conjunction with lateral-surface vents buried below grade in a gravel trench. Control of the downward movement of gases can be accomplished by installing perforated pipes in a gravel layer at the bottom of the landfill. If the gases cannot be vented laterally, it may be necessary to install gas wells and to vent the pumped gas to the atmosphere.

The movement of landfill gases through adjacent soil formations can be controlled by constructing barriers of materials that are more impermeable than the soil (Fig. 11-21). Some of the landfill sealants that are available for this use are identified in Table 11-13. Of these, the use of compacted clays is most common. The thickness will vary depending on the type of clay and the degree of control required; thicknesses ranging from 0.15 to 1.25 m (6 to 48 in) have been used.

Control of gas movement by recovery The movement of gases in landfills can also be controlled by installing gas recovery wells in complete landfills (see Fig. 11-22). Clay and other liners are used where landfill gas is to be recovered. In some gas-recovery systems, leachate is collected and recycled to the top of the landfill and reinjected through perforated lines located in drainage trenches. Typically, the rate of gas production is greater in leachate recirculation systems or where water is added.

Although gas-recovery systems have been installed in some large municipal landfills, the economics of such operations are, at present, not well defined. The

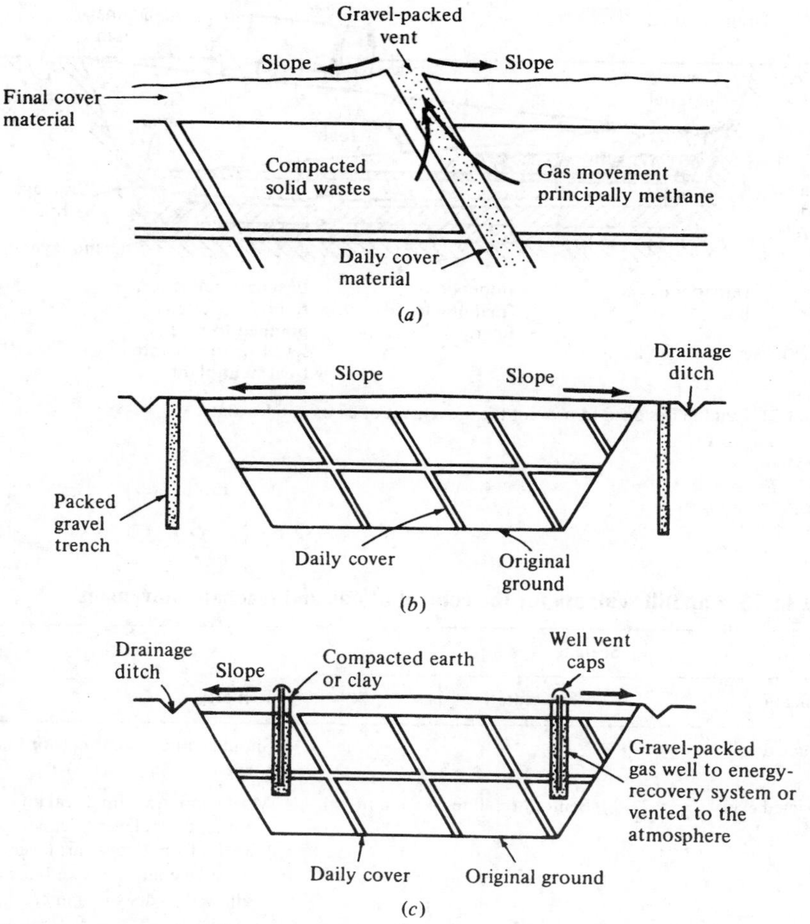

Figure 11-20 Typical methods for venting landfill gases: (*a*) cell; (*b*) barrier; (*c*) well.

cost of the gas cleanup and processing equipment may limit the recovery of landfill gases, especially from small landfills.

Leachate Movement

Under normal conditions, leachate is found in the bottom of landfills. From there, its movement is through the underlying strata, although some lateral movement may also occur, depending on the characteristics of the surrounding material. The rate of seepage of leachate from the bottom of a landfill can be estimated by Darcy's law by assuming that the material below the landfill to the top of the water table is saturated and that a small layer of leachate exists at the bottom of the fill. Under these conditions the leachate discharge rate per unit area is equal to the

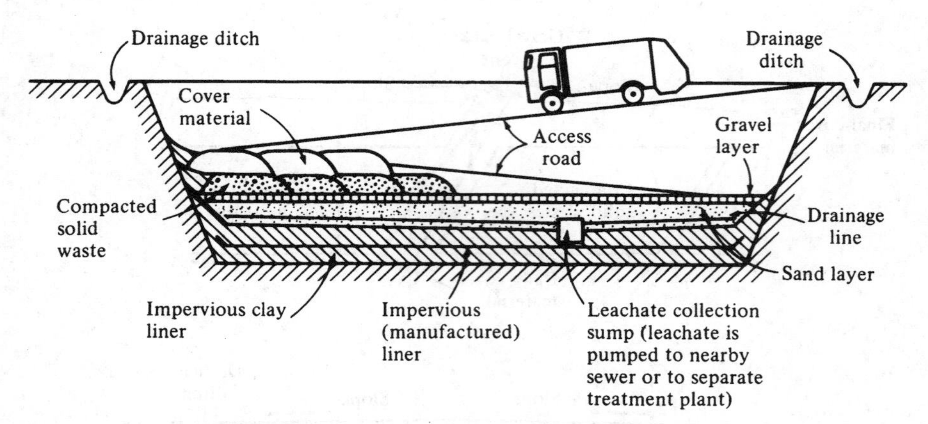

Figure 11-21 Landfill designed to prevent the movement of gases and leachate.

Table 11-13 Landfill sealants for the control of gas and leachate movement

Sealant		
Classification	Representative types	Remarks
Compacted soil		Should contain some clay or fine silt.
Compacted clay	Bentonites, illites, kaolinites	Most commonly used sealant for landfills; layer thickness varies from 6 to 48 in; layer must be continuous and not allowed to dry out and crack.
Inorganic chemicals	Sodium carbonate, silicate, or pyrophosphate	Use depends on local soil characteristics.
Synthetic chemicals	Polymers, rubber latex	Experimental, use not well established.
Synthetic membrane liners	Polyvinyl chloride, butyl rubber, hypalon, polyethylene, nylon-reinforced liners	Expensive, may be justified where gas is to be recovered.
Asphalt	Modified asphalt, asphalt-covered polypropylene fabric, asphalt concrete	Layer must be thick enough to maintain continuity under differential settling conditions.
Others	Gunite concrete, soil cement, plastic soil cement	

Source: From Tchobanoglous et al. [11-8]

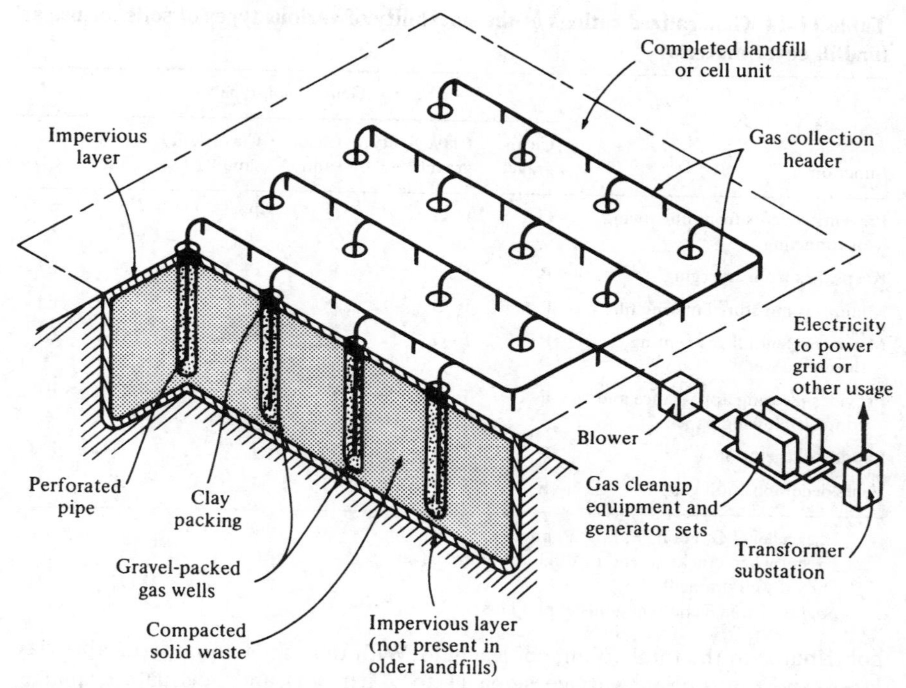

Figure 11-22 Well system used for the recovery of gases from landfills.

value of the coefficient of permeability K expressed in meters per day. The computed value represents the maximum amount of seepage that would be expected, and this value should be used for design purposes. Under normal conditions, the actual rate would be less than this value because the soil column below the landfill would not be saturated.

Control of leachate movement

As leachate percolates through the underlying strata, many of the chemical and biological constituents originally contained in it will be removed by the filtering and adsorptive action of the material composing the strata. In general, the extent of this action depends on the characteristics of the soil, especially the clay content. Because of the potential risk involved in allowing leachate to percolate to the groundwater, best practice calls for its elimination or containment. Ultimately, it may be necessary to collect and treat the leachate.

The use of clay has been the favored method of reducing or eliminating the percolation of leachate (see Fig. 11-21 and Table 11-13). Membrane liners have also been used, but they are expensive and require care so that they will not be damaged during the filling operations. Equally important in controlling the movement of leachate is the elimination of surface-water infiltration, which is the major

Table 11-14 Generalized ratings of the suitability of various types of soils for use as landfill cover material

Function	General soil type*					
	Clean gravel	Clayey-silty gravel	Clean sand	Clayey-silty sand	Silt	Clay
Prevents rodents from burrowing or tunneling	G	F–G	G	P	P	P
Keeps flies from emerging	P	F	P	G	G	E†
Minimizes moisture entering fill	P	F–G	P	G–E	G–E	E†
Minimizes landfill gas venting through cover	P	F–G	P	G–E	G–E	E†
Provides pleasing appearance and controls blowing paper	E	E	E	E	E	E
Supports vegetation	P	G	P–F	E	G–E	F–G
Vents decomposition gas‡	E	P	G	P	P	P

* E, excellent; G, good; F, fair; P, poor.
† Except when cracks extend through the entire cover.
‡ Only if well drained.
Source: From Tchobanoglous et al. [11-8]

contributor to the total volume of leachate. With the use of an impermeable clay layer, and appropriate surface slope (1 to 2 percent) and adequate drainage, surface infiltration can be controlled effectively. Generalized ratings for the suitability of various types of soil for use as a landfill cover are reported in Table 11-14.

Settlement and Structural Characteristics of Landfills

The settlement of landfills depends on the initial compaction, characteristics of wastes, degree of decomposition, and effects of consolidation when the leachate and gases are formed in the landfill. The height of the completed fill will also influence the initial compaction and degree of consolidation. The degree of consolidation can be modeled with a first-order equation.

11-19 DESIGN AND OPERATION OF LANDFILLS

Important design considerations in the design and operation of landfills include: (1) land requirements, (2) types of wastes that must be handled, (3) evaluation of seepage potential, (4) design of drainage and seepage control facilities, (5) development of a general operation plan, (6) design of solid-waste filling plan, and (7) determination of equipment requirements. The more important individual factors that must be considered in the design of a landfill are reported in Table 11-15. The last three items are considered further in the following discussion.

Table 11-15 Important factors that must be considered in the design and operation of solid-waste landfills

Factor	Remarks
Design	
Access	Paved all-weather access roads to landfill site; temporary roads to unloading areas.
Cell design and construction	Will vary depending on terrain, landfilling method, and whether gas is to be recovered.
Cover material	Maximize use of on-site earth materials; approximately 1 m^3 of cover material will be required for every 4 to 6 m^3 of solid wastes; mix with sealants to control surface infiltration. In some designs, intermediate cover is not used.
Drainage	Install drainage ditches to divert surface-water runoff; maintain 1 to 2 percent grade on finished fill to prevent ponding.
Equipment requirements	Vary with size of landfills.
Fire prevention	Water on-site; if nonpotable, outlets must be marked clearly; proper cell separation prevents continuous burn-through if combustion occurs.
Groundwater protection	Divert any underground springs; if required, install sealants for leachate control; install wells for gas and groundwater monitoring.
Land area	Area should be large enough to hold all wastes for a minimum of 1 yr but preferably 5 to 10 yr.
Landfilling method	Selection of method will vary with terrain and available cover.
Litter control	Use movable fences at unloading areas; crews should pick up litter at least once per month or as required.
Operation plan	With or without the codisposal of treatment plant sludges and the recovery of gas.
Spread and compaction	Spread and compact waste in 0.6-m (2-ft) layers.
Unloading area	Keep small, generally under 30 m (100 ft).
Operation	
Communications	Telephone for emergencies.
Days and hours of operation	Usual practice is 5 to 6d/wk and 8 to 10 h/d.
Employee facilities	Restrooms and drinking water should be provided.
Equipment maintenance	A covered shed should be provided for field maintenance of equipment.
Operational records	Tonnage, transactions, and billing if a disposal fee is charged.
Salvage	No scavenging; salvage should occur away from the unloading area; no salvage storage on-site.
Scales	Essential for record keeping.

Source: From Tchobanoglous et al. [11-8]

Landfill Operation Plan

The layout of the site and the development of a workable operating schedule are the main features of a landfill operation plan. In planning the layout of a landfill site, the location of the following must be determined: (1) access roads, (2) equipment shelters, (3) scales, if used, (4) storage sites for special wastes, (5) topsoil stockpile sites, (6) the landfill areas, and (7) plantings. A typical landfill operation plan is shown in Fig. 11-23.

Solid-Waste Filling Plan

The specific method of filling will depend on the characteristics of the site, such as the amount of available cover material, the topography, and the local hydrology and geology. To assess future development plans, it will be necessary to prepare a detailed plan for the layout of the individual solid-waste cells (see Fig. 11-24). On the basis of the characteristics of the site or the method of operation (e.g., gas recovery), it may be necessary to incorporate special features for the control of the movement of gases and leachate from the landfill. Estimation of the capacity of a landfill is illustrated in Example 11-5.

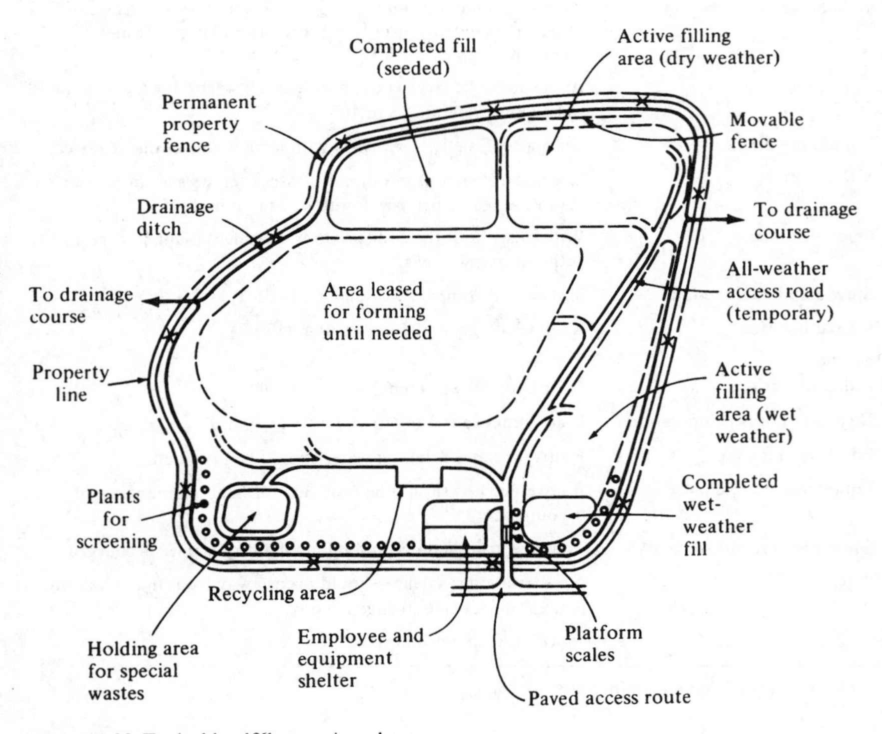

Figure 11-23 Typical landfill operation plan.

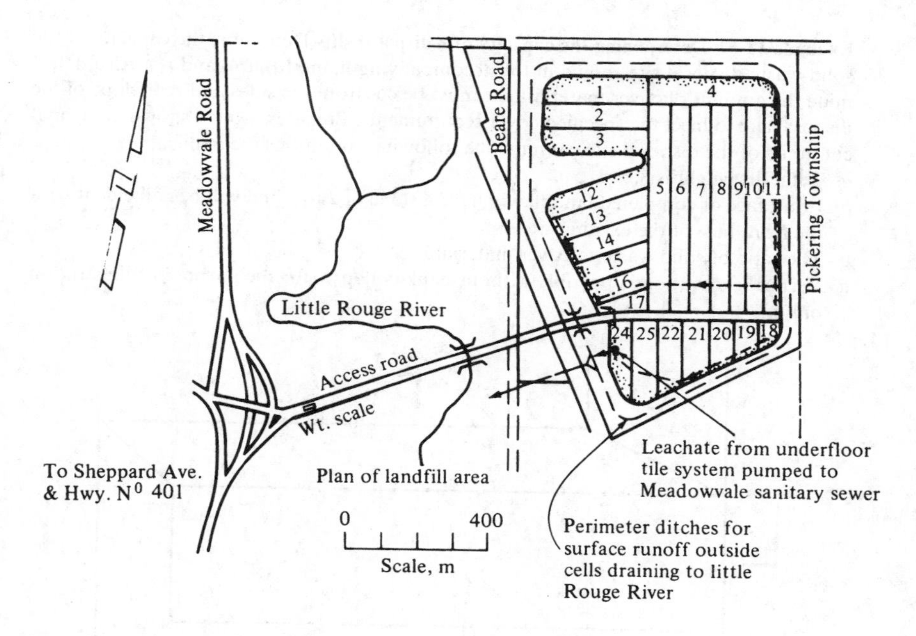

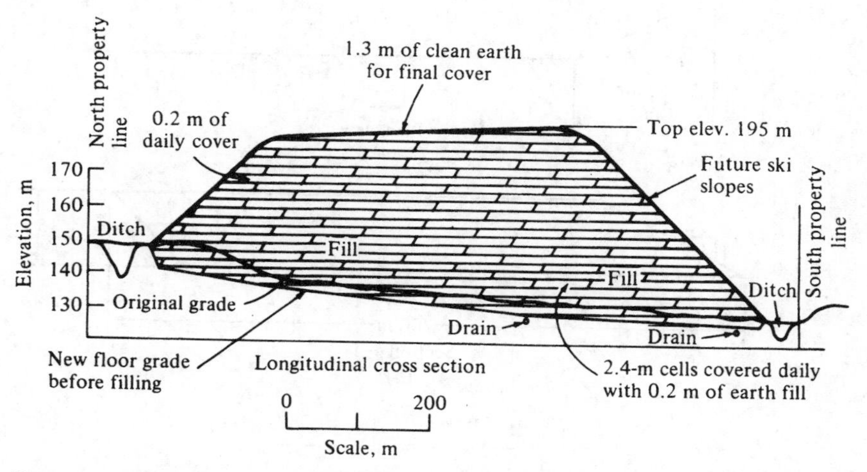

Figure 11-24 Typical plan for filling landfill. (*Courtesy of Municipality of Metropolitan Toronto, Department of Public Works.*)

Example 11-5: Determining the capacity of a disposal site Determine the capacity of the solid-waste disposal site shown in the accompanying figure (*a*), (*b*), and (*c*). Also determine the amount of cover material that must be cut from the sides and rear slope of the disposal site to meet the specified cover requirements. Prepare a sketch showing the final cut limits of the disposal site. Assume the following conditions are applicable:

a. Total lift height is 3.0 m.

b. Front face of completed landfill will have a slope of 2 to 1 and will be coincident with the front face of the existing site.

c. The ratio of solid waste to cover material is 5 to 1.

d. Vertical excavation is possible without bank collapse (for the purpose of illustration only).

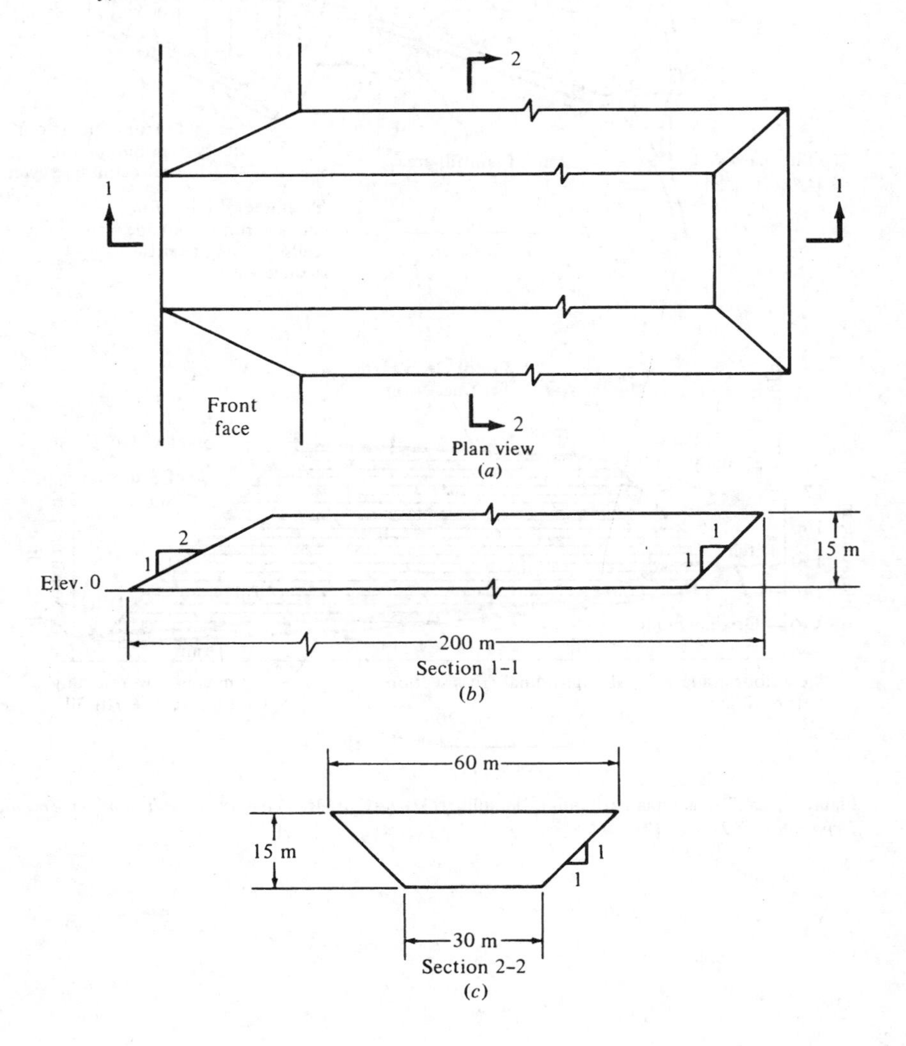

Plan view
(*a*)

Section 1-1
(*b*)

Section 2-2
(*c*)

SOLUTION

1. Determine the capacity of the disposal site and the amount of cover material required. Because the site will be excavated to obtain the necessary cover material, the capacity of the site is equal to the volume of the site. Because cover material will have to be excavated from the sides of the site, it will be useful to compute the volume of each lift separately. The necessary computations are summarized in the table.

Estimation of capacity of landfill

		Area, m^2			
Lift number	Elevation, m	At contour interval	Average between intervals	Capacity between contours, m^3	Cover material required, m^3
	0	5,550*			
1			6,015	18,045†	3,609‡
	3	6,480			
2			6,945	20,835	4,167
	6	7,410			
3			7,875	23,625	4,725
	9	8,340			
4			8,805	26,415	5,283
	12	9.270			
5			9,735	29,205	5,841
	15	10,200			
Total capacity, m^3				118,125	
Total cover, m^3					23,625

* Referring to the figure, the area at elevation zero is computed as 5550 m^2 = (200 m − 15 m) × 30 m.

† 18,045 m^3 = 6015 m^2 × 3 m (lift height).

‡ 3609 m^3 = 18,045 m^3 × (1/5), ratio of cover material to solid waste.

2. Determine the limits of excavation that will be needed to obtain the required cover material (see the table assuming that a triangular wedge of material will be excavated (see figure (*d*) all around the disposal site.

   a. Develop a relationship that can be used to determine the volume.

$$\begin{aligned}\text{Vol} &= 2\,(\text{volume of truncated excavated wedge sections})\\ &\quad + L\,(\text{area of continuous triangular wedge})\\ &= 2[\tfrac{1}{3}(2x \times x)x] + L(\tfrac{1}{2}x^2)\end{aligned}$$

where L = length of wedge measured along a line drawn on a horizontal plane of a distance $\frac{2}{3}x$ in from the bottom perimeter of the disposal site. Thus

$$\begin{aligned}L &= 2(155\text{ m}) + 30\text{ m} + 4(\tfrac{2}{3}x)\\ &= 340 + 2.67x\end{aligned}$$

Hence

$$\begin{aligned}\text{Vol} &= 1.33x^3 + (340 + 2.67x)(\tfrac{1}{2}x^2)\\ &= 170x^2 + 2.67x^3\end{aligned}$$

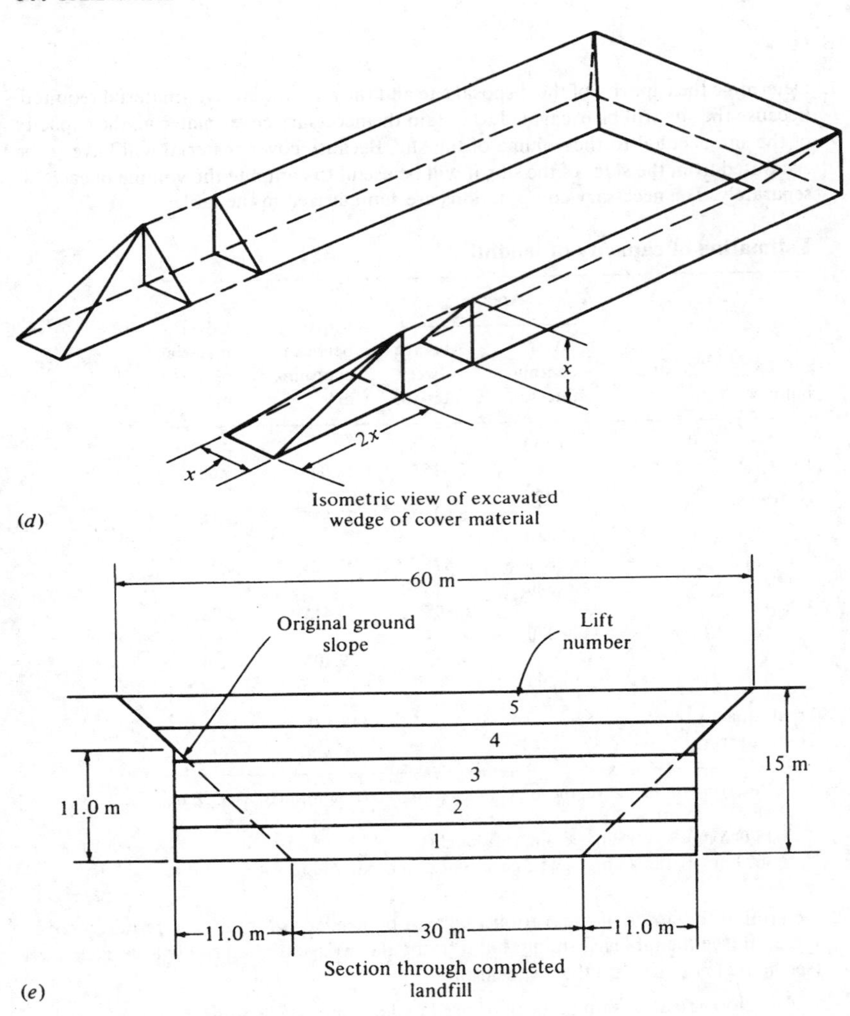

b. Equate above expression to require cover volume and solve for x by trial and error.
Try $x = 10$ m.

$$23{,}625 = 170(10^2) + 2.67(10^3)$$
$$23{,}625 \neq 19{,}670 \qquad \text{unacceptable}$$

Try $x = 11.0$ m

$$23{,}625 \approx 24{,}123 \qquad \text{O.K.}$$

3. Prepare a sketch showing the final cut limits of the disposal site. A section through completed landfill is shown in figure (*e*).

COMMENT An alternative solution would be to excavate the cover material in a stepped (staircase) fashion. In this situation, only enough material is excavated to meet the cover requirements for each lift (see the table). This solution is left to the reader as Prob. 11-15.

(a)

(b)

(d)

(c)

Figure 11-25 Typical equipment used at sanitary landfills.

Equipment Requirements

The types of equipment that have been used at sanitary landfills include both crawler and rubber-tired tractors, scrapers, compactors, draglines, and motor graders (see Fig. 11-25). The size and amount of equipment required will depend primarily on local site conditions, the size of the landfill operation, and the method of operation.

11-20 LANDFARMING

Landfarming is a waste-disposal method in which the biological, chemical, and physical processes that occur in the surface of the soil are used to treat biodegradable industrial wastes. Wastes to be treated are either applied on top of the land, which has been prepared to receive the wastes, or injected below the surface of the soil (see Fig. 11-26).

When organic wastes are added to the soil, they are subjected simultaneously to the following processes: (1) bacterial and chemical decomposition, (2) leaching of water-soluble components in the original wastes and from the decomposition products, and (3) volatilization of selected components in the original wastes and from the products of decomposition. Factors that must be considered in evaluating the biodegradability of organic wastes in a landfarming application include (1) composition of the waste; (2) compatibility of wastes and soil microflora; (3) environmental requirements including oxygen, temperature, pH, and inorganic nutrients, and (4) moisture content of solid-waste mixture.

Figure 11-26 Equipment used for the landfarming of liquid solid waste such as those from a wastewater treatment plant.

Landfarming is suitable for wastes that contain organic constituents that are biodegradable and are not subject to significant leaching while the bioconversion process is occurring. For example, petroleum wastes and oily sludges are ideally suited for disposal by landfarming. A variety of other organic wastes with similar characteristics are also suitable. Properly managed landfarming sites can be reused at frequent intervals with no adverse affects.

11-21 DEEP-WELL INJECTION

Deep-well injection for the disposal of liquid solid wastes involves injecting the wastes deep in the ground into permeable rock formations (typically limestone or dolomite) or underground caverns. The installation of deep wells for the injection of wastes closely follows the practices used for the drilling and completion of oil and gas wells. To isolate and protect potential water supply aquifers, the surface casing must be set well below such aquifers and cemented to the surface of the well (see Fig. 11-27). The drilling fluid should not be allowed to penetrate

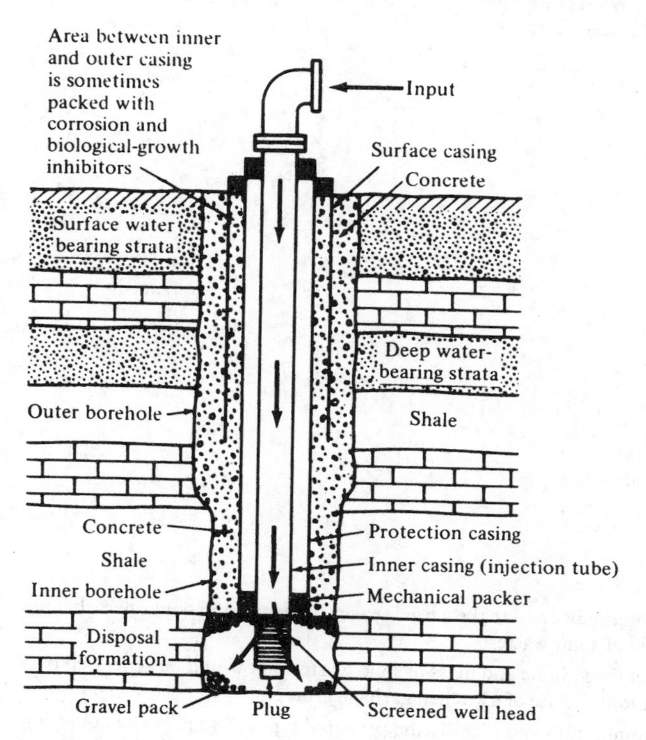

Figure 11-27 Well used for injection of liquid solid wastes.

the formation that is to be used for waste disposal. To prevent clogging of the formation, the drilling fluid is replaced with a compatible solution. Also, in some cases, it may be necessary to acid-treat the formation before injection of wastes is initiated.

Deep-well injection has been used principally for liquid wastes that are difficult to treat and dispose of by more conventional methods and for hazardous wastes. Chemical, petrochemical, and parmaceutical wastes are those most commonly disposed of with this method. The waste may be liquid, gases, or solids. The gases and solids are either dissolved in the liquid or are carried along with the liquid.

DISCUSSION TOPICS AND PROBLEMS

11-1 Obtain data on the solid-waste generation rates for your community. How do they compare with the values reported in the text? Explain any differences.

11-2 If you were asked to determine the solid-waste generation rates for your community, how would you assess the seasonal effects if the only data available had been collected during the month of December?

11-3 Plot a histogram of the following generation rates obtained at a commercial establishment over a period of a year. What conclusions can be drawn from histograms? What can you say about the histogram you have plotted?

Generation rate, m^3/wk	Frequency
100–199	1
200–299	3
300–399	7
400–499	10
500–599	6
600–699	2
700–799	1
800–899	1
900–999	0
1000–1099	1
1100–1199	4
1200–1299	7
1300–1399	5
1400–1499	3
1500–1599	1

11-4 Drive around your community and make a brief survey of the types of containers that are used for the outside storage of solid wastes.

11-5 Prepare an estimate of the volume and mass of solid wastes that would be generated by a family of four with and without the use of a kitchen garbage grinder.

11-6 Drive around your community and identify the principal types of solid-waste collection systems that are in use. Select two of the systems and obtain some field data on the times required for the various activities associated with the collection of solid wastes. How do the values compare with those reported in the text? Explain any differences in the values.

11-7 A new residential area composed of 600 single-family dwellings is being developed. Assuming that either two or three trips per day will be made to the disposal site, select an appropriate truck size and determine the number of trips that must be made.

1. Solid waste generation rate = 5.0 kg/residence · d
2. Containers per service = 2
3. Type of service = 80 percent curbside and 20 percent rear of house
4. Collection frequency = once per week
5. Collection vehicle compaction ratio = 2.5
6. Size of collection crew = two persons
7. Length of workday = 8 h
8. Off-route factor = 0.15
9. Round-trip haul distance = 34 km
10. Haul route constants: a = 0.016 h/trip; b = 0.011 h/km
11. At-site time per trip = 0.083 h/trip

11-8 Layout the collection routes for the commercial area shown in the accompanying figure. Assume the following data are applicable:

1. Collection system = stationary container
2. Container size = 4 m^3
3. Container utilization factor = 0.80
4. Collection frequency = as shown
5. Collection vehicle capacity = 26 m^3

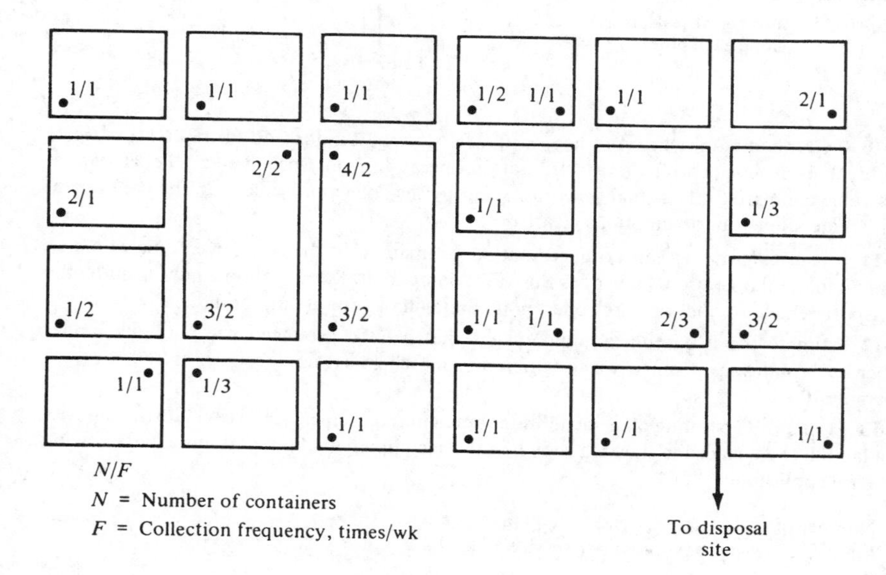

11-9 Lay out collection routes for the residential area shown in the accompanying figure. Assume the same conditions as specified in Example 11-3 are applicable.

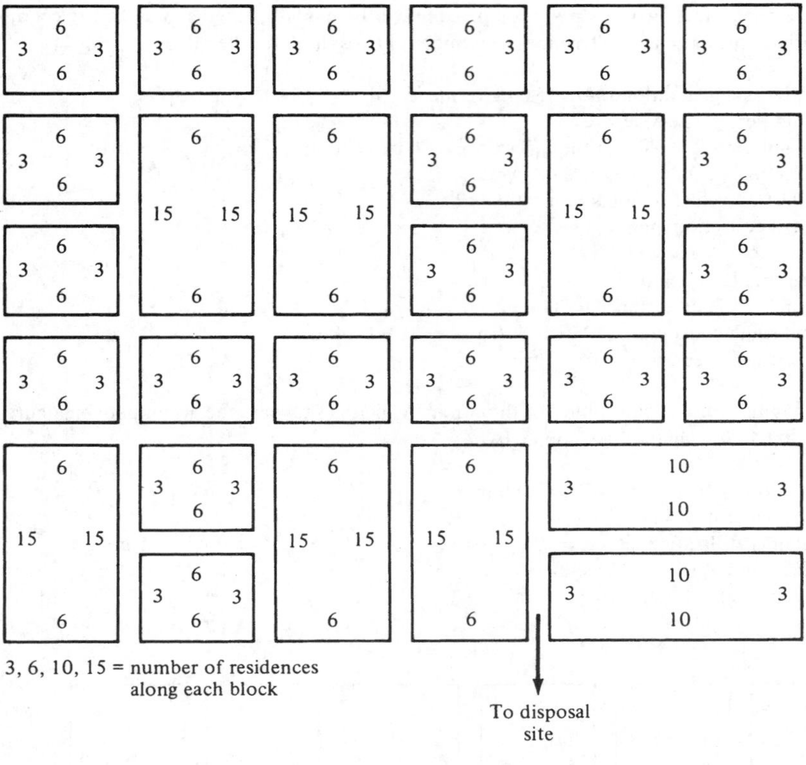

3, 6, 10, 15 = number of residences
along each block

To disposal
site

11-10 If your community does not have a transfer station (if it does, do Prob. 11-11) estimate the break-even time at which a transfer station operation would become feasible. How does this value compare to the actual time now spent by the collection vehicles in the haul operation? State all of your assumptions clearly.

11-11 If your community has a transfer station, estimate what the break-even time would be for a direct-haul operation. How does this value compare to the actual time now spent by the transport vehicles in the transport operation? State all of your assumptions clearly.

11-12 Estimate the maximum amount of gas that can be produced per kilogram, under anaerobic conditions, from a waste with the same chemical composition as the waste in Example 10-4.

11-13 Prepare an operating plan, including equipment requirements, and estimate the capacity of a landfill to be placed in the area shown in the accompanying figure. Assume the following data are applicable:

1. Number of collection services = 3000 (average over 20 yr)
2. Amount of wastes generated per service = 6.5 kg/d

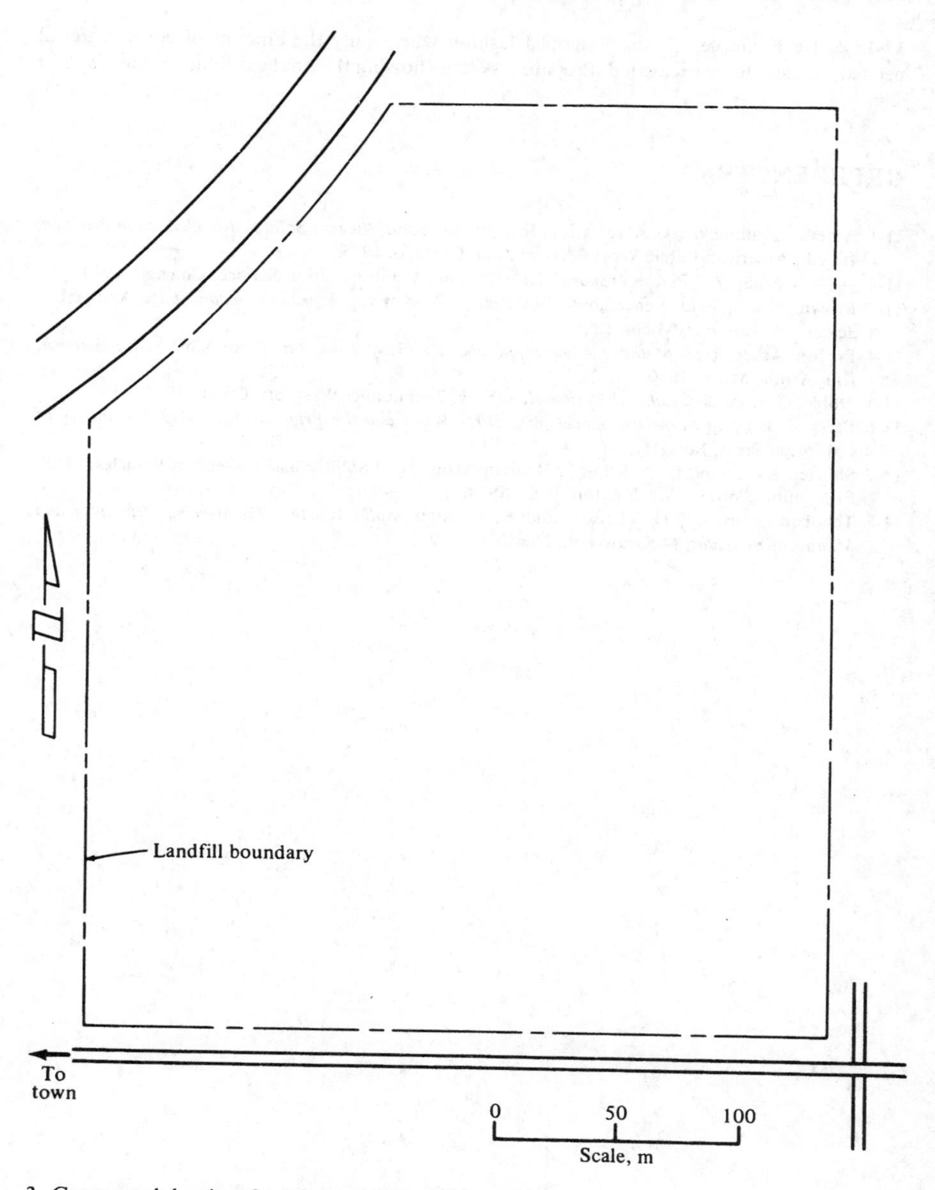

3. Compacted density of solid wastes in landfill = 475 kg/m^3
4. Maximum allowable finished grade above surrounding ground = 20 ft
5. Ratio of solid wastes to cover material = 6 to 1

11-14 Estimate the capacity of the landfill shown in Fig. 11-24. If the in-place density is 714 kg/m^3, how many tonnes have been placed in the landfill?

11-15 Solve Example 11-5 in a stepped fashion where only the amount of cover material needed for each lift is excavated. Prepare a sketch showing the final cut limits of the disposal site.

REFERENCES

11-1 American Public Works Association, Institute for Solid Wastes: *Solid Waste Collection Practice*, 4th ed., American Public Works Association, Chicago, 1975.
11-2 ———: *Municipal Refuse Disposal*, 3d ed., Public Administration Service, Chicago, 1970.
11-3 Brown, M. D., T. D. Vence, and T. C. Reilly: *Solid Waste Transfer Fundamentals*, Ann Arbor Science, Ann Arbor, Mich., 1981.
11-4 Emcon Associates: *Methane Generation and Recovery from Landfills*, Ann Arbor Science, Ann Arbor, Mich., 1980.
11-5 Noble, G.: *Sanitary Landfill Design Handbook*, Technomic, Westport, Conn., 1976.
11-6 Perry, S. E.: *San Francisco Scavengers: Dirty Work and the Price of Ownership*, University of California Press, Berkeley, 1978.
11-7 Shuster, K. A., and D. A. Schur: "Heuristic Routing of Solid Waste Collection Vehicles," U.S. EPA, publ. SW-113, Washington, D.C., 1974.
11-8 Tchobanoglous, G., H. Theisen, and R. Eliassen: *Solid Wastes: Engineering Principles and Management Issues*, McGraw-Hill, New York, 1977.

CHAPTER

TWELVE

ENGINEERED SYSTEMS FOR RESOURCE AND ENERGY RECOVERY

The purpose of this chapter is to introduce the reader to the techniques and methods used to recover materials, conversion products, and energy from solid wastes. Topics to be considered include (1) processing techniques, (2) materials-recovery systems, (3) recovery of biological conversion products, (4) recovery of chemical conversion products, (5) recovery of energy from conversion products, and (6) materials and energy recovery systems.

Because many of the techniques to be considered are in a state of flux with respect to application and design criteria, the objective here is only to introduce them to the reader. If these techniques are to be considered in the development of waste-management systems, current engineering design and performance data must be obtained from the records of operating installations, from field tests, from equipment manufacturers, and from the literature. References [12-3, 12-6, and 12-8] are recommended as a starting point.

Processing Techniques

Processing techniques, as noted previously in Chap. 11, are used in solid-waste management systems to improve the efficiency of solid-waste management systems, to recover resources (usable materials), and to prepare materials for the recovery of conversion products and energy. Processing techniques used to improve the efficiency of solid-waste systems and to recover materials manually were considered previously in Chap. 11. The more important techniques used for processing solid wastes to recover materials and to prepare the waste for subsequent processing are summarized in Table 12-1.

Table 12-1 Processing techniques used to recover materials and to prepare wastes for further processing

Processing technique	Function	Representative equipment and/or facilities and applications
Mechnical size and shape alteration	Alteration of the size and shape of solid-waste components	Equipment used to reduce the size of solid waste includes hammermills, shredders, roll crushers, grinders, chippers, jaw crushers, rasp mills, and hydropulpers.
Mechanical component separation	Separation of recoverable materials, usually at a processing facility	Trommels and vibrating screens are used for unprocessed and processed wastes; disk screens for processed wastes; zigzag, vibrating air, rotary air, and air knife classifiers for processed wastes. Jig, pneumatic, sink/float, inertial, inclined or shaking table flotation, and optical sorting are used to separate the light and heavy materials in solid wastes.
Magnetic and electro-mechanical separation	Separation of ferrous and nonferrous materials from processed solid wastes	Magnetic separation is used for ferrous materials; eddy current separation for aluminum; electrostatic separation for glass in wastes free of ferrous and aluminum scrap; magnetic fluid separation for nonferrous materials from processed wastes
Drying and dewatering	Removal of moisture from solid wastes	Convection, conduction, and radiation dryers have been used for solid wastes and sludge. Centrifugation and filtration are used to dewater treatment plant sludge.

12-1 MECHANICAL SIZE ALTERATION

The objective of size reduction is to obtain a final product that is reasonably uniform and considerably reduced in size in comparison with its original form. It is important to note that size reduction does not necessarily imply volume reduction. In some situations, the total volume of the material after size reduction may be greater than the original volume. A typical shredder used for size reduction in solid waste processing systems is illustrated in Fig. 12-1.

Figure 12-1 Horizontal-shaft reversible-type hammermill used for the size reduction of solid wastes: (*a*) schematic; (*b*) photograph of installation.

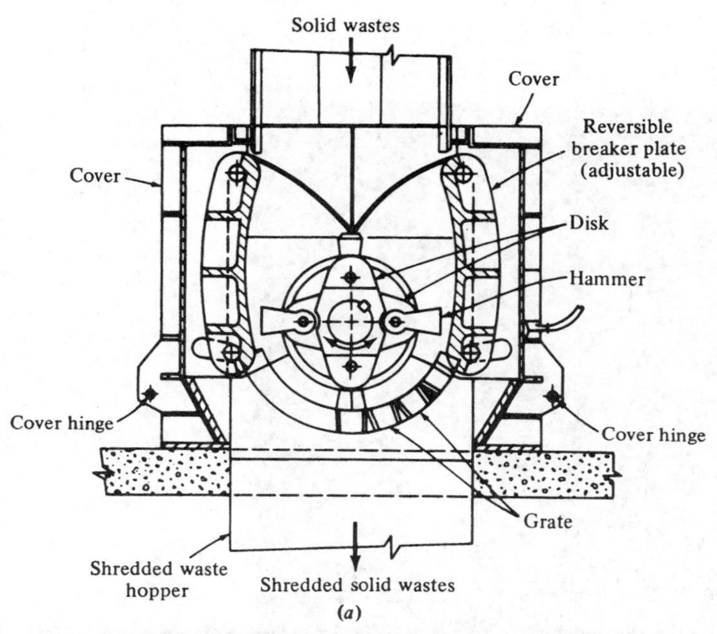
Solid wastes
Cover
Reversible
breaker plate
(adjustable)
Cover
Disk
Hammer
Cover hinge
Cover hinge
Grate
Shredded waste
hopper
Shredded solid wastes
(*a*)

(*b*)

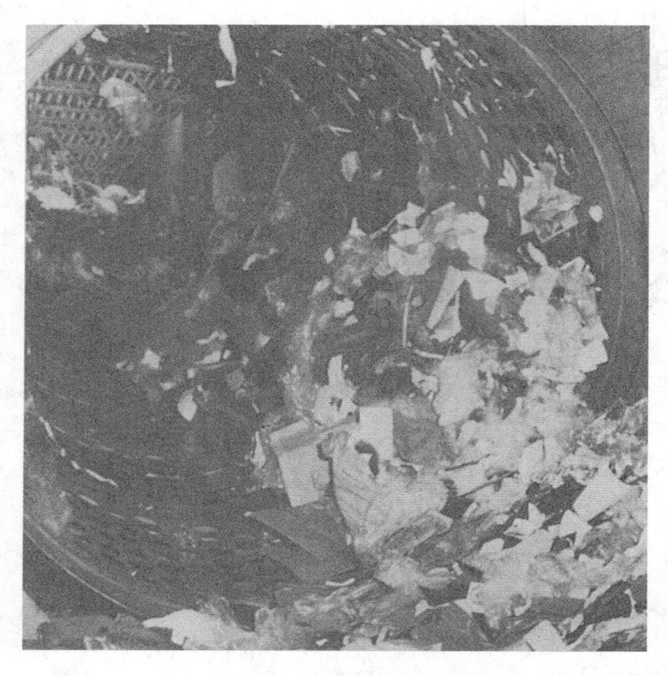

Figure 12-2 Trommel used for compost separation in waste processing systems.

12-2 MECHANICAL COMPONENT SEPARATION

Component separation is a necessary operation in the recovery of resources from solid wastes and where energy and conversion products are to be recovered from processed wastes. For example, trommels are now used routinely for the separation of unprocessed wastes (see Fig. 12-2). The use of trommels is a departure from what was done in the near past where shredding was the first step in the process flow sheet. Shredding of wastes as a first step has been replaced with screening because (1) shredding tends to shatter glass and entrap organic materials within tin cans, (2) it contaminates paper with liquids and putrescible organic materials, and (3) the operation has a high energy demand. [12-4]

12-3 MAGNETIC AND ELECTROMECHANICAL SEPARATION

Magnetic separation of ferrous materials, a well-established technique in the metals industry, is now used commonly for the removal of ferrous metals from solid wastes. More recently, a variety of electromechanical techniques have been developed for the removal of several nonferrous materials.

12-4 DRYING AND DEWATERING

In many solid-waste energy recovery and incineration systems, the shredded light fraction is predried to decrease weight. Although the energy requirements for drying wastes vary with local conditions, the required energy input can be estimated by using a value of about 4300 kJ/kg of water evaporated.

Materials Recovery Systems

Various types of processing techniques and equipment were discussed above. In this section the objective is to show how the individual processes can be combined in alternative flow sheets for the recovery of materials and the preparation of combustible wastes for subsequent processing.

12-5 MATERIALS SPECIFICATIONS

Paper, cardboard, plastics, glass, ferrous metals, and nonferrous metals are the principal recoverable materials contained in municipal solid wastes. In any given situation, the decision to recover any of or all these materials is usually based on an economic evaluation and on local considerations. In assessing the economics of materials recovery, the materials specifications will be a critical consideration.

12-6 PROCESSING AND RECOVERY SYSTEMS

Once a decision has been made to recover materials and/or energy, flow sheets must be developed for the removal of the desired components and for processing combustible materials, subject to predetermined materials specifications. Typical flow sheets for the recovery of waste components and the preparation of combustible materials for use as a fuel source are presented in Figs. 12-3 and 12-4, respectively. The light combustible materials are often identified as RDF (refuse-derived fuel).

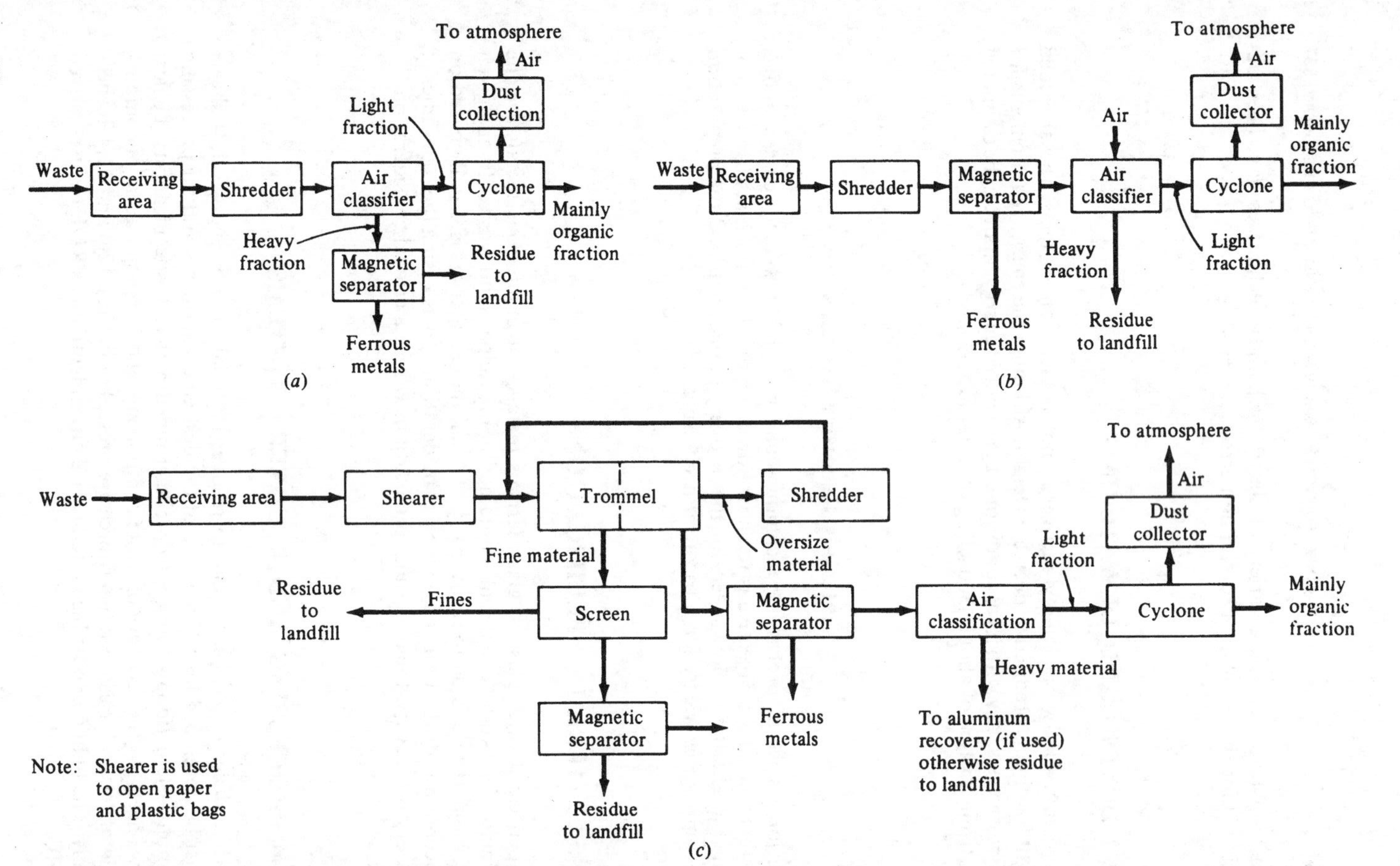

Figure 12-3 Flow sheet for the recovery of waste components from solid waste, (*a*) conventional; (*b*) conventional with shredder in different location; (*c*) trommel used to replace shredder in flow sheet (*a*). (*Adapted in part from Wilson* [*12-9*].)

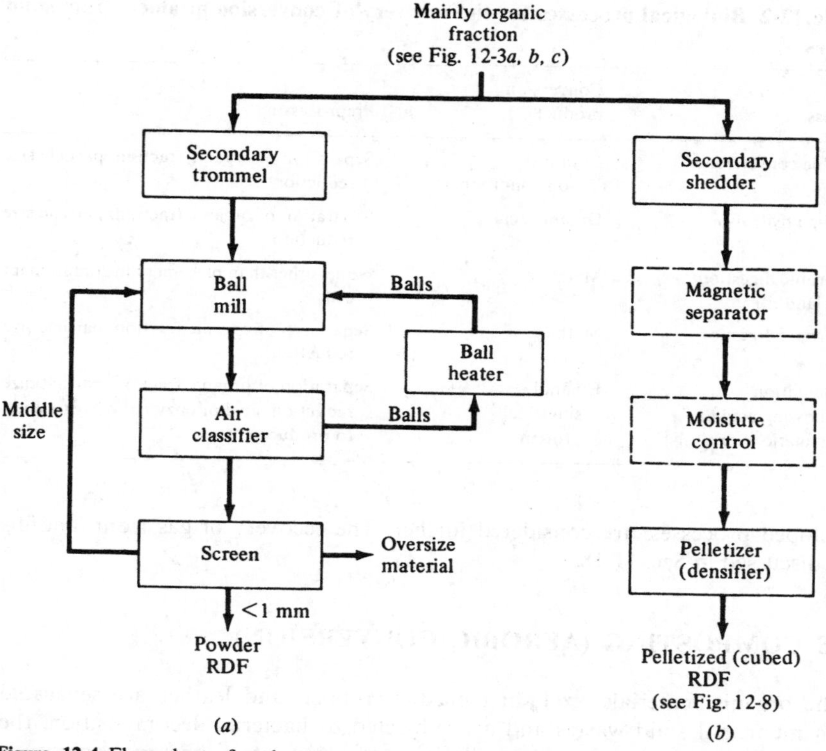

Figure 12-4 Flow sheets for the preparation of RDF (refuse-derived fuel): (*a*) powder RDF; (*b*) pelletzed (cubed) REF.

12-7 SYSTEM DESIGN AND LAYOUT

The design and layout of the physical facilities that make up the processing plant flow sheet are an important aspect in the implementation and successful operation of such systems. Important factors that must be considered in the design and layout of such systems include (1) process performance efficiency, (2) reliability and flexibility, (3) ease and economy of operation, (4) aesthetics, and (5) environmental controls.

Recovery of Biological Conversion Products

Biological conversion products that can be derived from solid wastes include compost, methane, various proteins and alcohols, and a variety of other intermediate organic compounds. The principal processes that have been used are reported in Table 12-2. Composting and anaerobic digestion, the two most

Table 12-2 Biological processes for the recovery of conversion products from solid wastes

Process	Conversion product	Preprocessing
Aerobic conversion	Compost (soil conditioner)	Separation of organic fraction, particle size reduction
Alkaline hydrolysis	Organic acids	Separation of organic fraction, particle size reduction
Anaerobic digestion (in landfill)	Methane	None, other than placement in containment cells
Anaerobic digestion	Methane	Separation of organic fraction, particle size reduction
Fermentation (following acid or enzymatic hydrolysis)	Ethanol, single-cell protein	Separation of organic fraction, particle size reduction, acid or enzymatic hydrolysis to produce glucose

developed processes, are considered further. The recovery of gas from landfills was discussed in Sec. 11-18.

12-8 COMPOSTING (AEROBIC CONVERSION)

If the organic materials, excluding plastics, rubber, and leather are separated from municipal solid wastes and are subjected to bacterial decomposition, the end product remaining after dissimilatory and assimilatory bacterial activity is called *compost*, or *humus*. The entire process involving both the separation and bacterial conversion of the organic solid wastes is known as *composting*. De-

Figure 12-5 Composting of garden and tree trimmings using the windrow method.

composition of the organic solid wastes may be accomplished either aerobically or anaerobically, depending on the availability of oxygen.

Most composting operations involve three basic steps: (1) preparation of the solid wastes, (2) decomposition of the solid wastes, and (3) product preparation and marketing. Receiving, sorting, separation, size reduction, and moisture and nutrient addition are part of the preparation step. Several techniques have been developed to accomplish the decomposition step (see Fig. 12-5). Once the solid wastes have been converted to a humus, they are ready for the third step, product

Table 12-3 Important design considerations for aerobic composting processes

Item	Comment
Particle size	For optimum results the size of solid wastes should be between 25 and 75 mm (1 and 3 in).
Seeding and mixing	Composting time can be reduced by seeding with partially decomposed solid wastes to the extent of about 1 to 5 percent by weight. Sewage sludge can also be added to prepared solid wastes. Where sludge is added, the final moisture content is the controlling variable.
Mixing/turning	To prevent drying, caking, and air channeling, material in the process of being composted should be mixed or turned on a regular schedule or as required. Frequency of mixing or turning will depend on the type of composting operation.
Air requirements	Air with at least 50 percent of the initial oxygen concentration remaining should reach all parts of the composting material for optimum results, especially in mechanical systems.
Total oxygen requirements	The theoretical quantity of oxygen required can be estimated.
Moisture content	Moisture content should be in the range between 50 and 60 percent during the composting process. The optimum value appears to be about 55 percent.
Temperature	For best results, temperature should be maintained between 323 and 328 K (50 and 55°C) for the first few days and between 328 and 333 K (55 and 60°C) for the remainder of the active composting period. If temperature goes beyond 339 K (66°C), biological activity is reduced significantly.
Carbon-nitrogen ratio	Initial carbon-nitrogen ratios (by mass) between 30 and 50 are optimum for aerobic composting. At lower ratios ammonia is given off. Biological activity is also impeded at lower ratios. At higher ratios nitrogen may be a limiting nutrient.
pH	To minimize the loss of nitrogen in the form of ammonia gas, pH should not rise above about 8.5.
Control of pathogens	If properly conducted, it is possible to kill all the pathogens, weeds, and seeds during the composting process. To do this, the temperature must be maintained between 333 and 343 K (60 and 70°C) for 24 h.

Source: Adapted from Tchobanoglous et al. [12-6]

preparation and marketing. This step may include fine grinding, blending with various additives, granulation, bagging, storage, shipping, and, in some cases, direct marketing. The principal design considerations associated with the biological decomposition of prepared solid wastes are presented in Table 12-3.

The amount of oxygen required for the complete aerobic stabilization of municipal solid wastes can be estimated by using the following equation:

$$C_aH_bO_cN_d + \frac{4a + b - 2c - 3d}{4} O_2 \longrightarrow aCO_2 + \frac{b - 3d}{2} H_2O + dNH_c \tag{12-1}$$

If the ammonia, NH_3, is to be oxidized to nitrate NO_3^-, the amount of oxygen required to accomplish this can be computed with the following equation.

$$NH_3 + 2O_2 \longrightarrow H_2O + HNO_3 \tag{12-2}$$

Computation of the amount of oxygen required for the stabilization of prepared solid wastes is illustrated in Example 12-1.

Example 12-1: Estimating air requirements Determine the amount of air required to oxidize completely 1 tonne of waste having the chemical equation $C_{50}H_{100}O_{40}N$.

SOLUTION

1. Determine the oxygen requirement for the given waste, using Eq. (12-1).

$$C_aH_bO_cN_d + \frac{4a + b - 2c - 3d}{4} O_2 \longrightarrow aCO_2 + \frac{b - 3d}{2} H_2O + dNH_3$$

a. The required coefficients are:

$a = 50$
$b = 100$
$c = 40$
$d = 1$

b. Using these coefficients the resulting equation is:

$$\underset{(1354)}{C_{50}H_{100}O_{40}N} + \underset{(1736)}{54.25O_2} \longrightarrow \underset{(2200)}{50CO_2} + \underset{(873)}{48.5H_2O} + \underset{(17)}{NH_3}$$

c. The oxygen required per tonne is:

$$O_2 = \tfrac{1736}{1354} \times 1000 = 1.282 \times 10^3 \text{ kg/tonne}$$

2. Using Eq. (12.2), determine the oxygen required to stabilize the ammonia.

$$\underset{(17)}{NH_3} + \underset{(64)}{2O_2} \longrightarrow \underset{(18)}{H_2O} + \underset{(63)}{HNO_3}$$

The oxygen required per tonne of waste is:

$$O_2 = \tfrac{17}{1354} \times \tfrac{64}{17} \times 1000 = 47.3 \text{ kg/tonne}$$

3. Determine the amount of air required. Assume air contains 23.15 percent oxygen by weight and that the density of air is 1.2928 kg/m³.
   a. The total amount of oxygen required is

$$O_{2,\,\text{total}} = (1.282 + 47.3)\ \text{kg/tonne} = 1329\ \text{kg/tonne}$$

   b. The mass of air required is

$$\text{Air}_{\text{mass}} = \frac{1329\ \text{kg/tonne}}{0.2315} = 5742\ \text{kg/tonne}$$

   c. The volume of air required is

$$V_{\text{air}} = \frac{5742\ \text{kg/tonne}}{1.2928\ \text{kg/m}^3} = 4442\ \text{m}^3\text{/tonne}$$

12-9 ANAEROBIC DIGESTION

Anaerobic digestion, or *anaerobic fermentation* as it is often called, is the process used for the production of methane from solid wastes. In most processes where methane is to be produced from solid wastes by anaerobic digestion, three basic

Table 12-4 Important design considerations for anaerobic digestion

Item	Comment
Size of material shredded	Wastes to be digested should be shredded to a size that will not interfere with the efficient functioning of pumping and mixing operations.
Mixing equipment	To achieve optimum results and to avoid scum buildup, mechanical mixing is recommended.
Percentage of solid wastes mixed with sludge	Although amounts of waste varying from 50 to 90+ percent have been used, 60 percent appears to be a reasonable compromise.
Hydraulic and mean cell-residence time, $\theta_h = \theta_c$	Washout time is in the range of 3 to 4 d. Use 7 to 10 d for design, or base design on results of pilot-plant studies.
Loading rate	0.6 to 1.6 kg/m³ · d (0.04 to 0.10 lb/ft³ · d). Not well defined at present time. Significantly higher rates have been reported.
Temperature	Between 328 and 333 K (55 and 60°C).
Destruction of volatile solid wastes	Varies from about 60 to 80 percent; 70 percent can be used for estimating purposes.*
Total solids destroyed	Varies from 40 to 60 percent, depending on amount of inert material present originally.
Gas production	0.5 to 0.75 m³/kg (8 to 12 ft³/lb) of volatile solids destroyed (CH_4 = 60 percent; CO_2 = 40 percent).

* Actual removal rates for volatile solids may be less, depending on the amount of material diverted to the scum layer.

Source: From Tchobanoglous et al. [12-6]

steps are involved. The first step involves preparation of the organic fraction of the solid wastes for anaerobic digestion and usually includes receiving, sorting, separation, and size reduction. The second step involves the addition of moisture and nutrients, blending, pH adjustment to about 6.7, heating of the slurry to between 328 and 333 K (55 and 60°C), and anaerobic digestion in a reactor with continuous flow, in which the contents are well mixed for a period of time varying from 5 to 10 d. The third step involves capture, storage, and, if necessary, separation of the gas components evolved during the digestion process (see Example 12-2). The disposal of the digested solids is an additional task that must be accomplished. Some important design considerations are reported in Table 12-4. Because of the variability of the results reported in the literature, it is recommended that pilot-plant studies be conducted if the digestion process is to be used for the conversion of solid wastes.

Example 12-2: Estimating methane production Estimate the theoretical volume of methane gas that would be expected from the anaerobic digestion of a tonne of a waste having the composition $C_{50}H_{100}O_{40}N$.

SOLUTION

1. Solve Eq. (11-13) for the given waste.

$$C_aH_bO_cN_d + \frac{4a - b - 2c + 3d}{4} H_2O \longrightarrow \frac{4a + b - 2c - 3d}{8} CH_4 + \frac{4a - b + 2c + 3d}{8} CO_2 + dNH_3$$

a. The coefficients are:

$a = 50$

$b = 100$

$c = 40$

$d = 1$

b. Using these coefficients the resulting equation is:

$$\underset{(1354)}{C_{50}H_{100}O_{40}N} + \underset{(103.5)}{5.75\,H_2O} \longrightarrow \underset{(434)}{27.125\,CH_4} + \underset{(1{,}006.5)}{22.875\,CO_s} + \underset{(17)}{NH_3}$$

2. Determine the mass of methane produced per tonne of waste.

$$\text{Methane} = \tfrac{434}{1354} \times 1000 \text{ kg/tonne} = 320.5 \text{ kg/tonne}$$

3. Using a density value for methane of 0.7167 kg/m^3, determine the volume of methane gas.

$$\text{Volume of methane gase} = \frac{320.5 \text{ kg/tonne}}{0.7167 \text{ kg/m}^3} = 447.2 \text{ m}^3\text{/tonne of waste}$$

COMMENT In practice, a portion of the waste would be used for the synthesis of cell tissue. The actual volume of gas would be about 0.85 times the value determined in step 3.

Table 12-5 Thermal processes for the recovery of conversion products from solid wastes

Process	Conversion product	Preprocessing
Combustion (incineration)	Energy in the form of steam or electricity	None in mass-fired incineratory; preparation of refuse-derived fuels for suspension or semisuspension firing in boilers
Gasification	Low-energy gas	Separation of the organic fraction, particle size reduction, preparation of fuel cubes or other RDF
Wet oxidation	Organic acids	Separation of the organic fraction, particle size reduction, preparation of fuel cubes or other RDF
Steam reforming	Medium-energy gas	Separation of the organic fraction, particle size reduction, preparation of fuel cubes or other RDF
Pyrolysis	Medium-energy gas, liquid fuel, solid fuel (char)	Separation of the organic fraction, particle size reduction, preparation of fuel cubes or other RDF
Hydrogasification/ hydrogenation	Medium-energy gas, liquid fuel	Separation of the organic fraction, particle size reduction, preparation of fuel cubes or other RDF

Recovery of Thermal Conversion Products

Thermal conversion products that can be derived from solid wastes include heat, gases, a variety of oils, and various related organic compounds. The principal thermal conversion processes that have been used for the recovery of usable conversion products from solid wastes are reported in Table 12-5. The more important processes in Table 12-5 are reviewed following an introductory discussion of combustion.

12-10 COMBUSTION OF WASTE MATERIALS

The principal elements of solid waste are carbon, hydrogen, oxygen, nitrogen, and sulfur. Under ideal conditions, when solid-waste materials are combusted (burned) the gaseous end products include CO_2 (carbon dioxide), H_2O (water), N_2 (nitrogen), and SO_2 (sulfur dioxide). In practice, a variety of other gaseous compounds are also formed, depending on the operating conditions under which the combustion process is occurring.

Combustion Calculations

In their simplest form all combustion calculations are based on the following fundamental laws. [12-2]

1. Conservation of mass: Mass can neither be created or destroyed.
2. Conservation of energy: Energy can neither be created or destroyed.
3. Gas law: The volume of gas is directly proportional to its absolute temperature and inversely proportional to its absolute pressure.
4. Law of combining masses: All substances combine in accordance with a definite, simple relationship with respect to relative masses.

The first and fourth of the above laws were illustrated in the computations in example 12-2.

Air Requirements For Combustion

To determine the amount of oxygen required for the complete combustion of solid wastes, it is necessary to compute the oxygen requirements for the oxidation of carbon, hydrogen, and sulfur contained in waste. The basic reactions are:

For carbon

$$\underset{(12)}{C} + \underset{(32)}{O_2} \longrightarrow \underset{(44)}{CO_2} \tag{12-3}$$

For hydrogen:

$$\underset{(4)}{2H_2} + \underset{(32)}{O_2} \longrightarrow \underset{(36)}{2H_2O} \tag{12-4}$$

For sulfur:

$$\underset{(32)}{S} + \underset{(32)}{O_2} \longrightarrow \underset{(64)}{SO_2} \tag{12-5}$$

If it is assumed that air contains 23.15 percent oxygen by mass, then the amount of air required for the complete oxidation of 1 kg of carbon would be equal to 11.52 kg [(32/12)(1/0.2315)]. The corresponding amounts for hydrogen and sulfur are 34.56 and 4.31 kg, respectively. In combustion computations, the oxygen requirements for the combustion of hydrogen usually are based on the net value of hydrogen available. The net value of hydrogen is computed by subtracting one-eighth of the percent oxygen from the total percentage of hydrogen present initially. This computation is based on the assumption that the oxygen in the sample will combine with the hydrogen in the waste to form water. Combustion computations are illustrated in Example 12-3.

Example 12-3: Determining air requirement for complete combustion Determine the air requirement expressed in kilograms per tonne of waste for the complete oxidation of a waste with the same composition as the waste in Example 12-1 ($C_{50}H_{100}O_{40}N$).

SOLUTION

1. Determine the molecular mass of the compound.

$$\text{Molecular mass} = \underset{\text{Carbon}}{(50 \times 12)} + \underset{\text{Hydrogen}}{(100 \times 1)} + \underset{\text{Oxygen}}{(40 \times 16)} + \underset{\text{Nitrogen}}{(1 \times 14)}$$

$$= 1.354$$

2. Determine the percentage distribution of the basic elements composing the waste.

Element	Percent by mass
Carbon	44.3*
Hydrogen	7.4
Oxygen	47.3
Nitrogen	1.0

* (50 × 12/1354)100

3. Compute the net available hydrogen not bound as water.

$$\text{Net available hydrogen} = 7.4\% - 47.3\%/8 = 1.49\%$$

4. Compute the air required per tonne of waste. Using the air requirements developed for carbon (11.52 kg/kg) and hydrogen (34.56 kg/kg), the required air is computed as follows:

Element	kg/tonne
Carbon (0.443 × 1000)11.52	5103
Hydrogen (0.0149 × 1000)34,56	515
Total	5618

COMMENT The reason there is a difference between the air requirement computed in step 4 of this problem and the corresponding value obtained in Example 12-1, step 1*c* (1282 ÷ 0.2315 = 5538) is because in Example 12-1 a portion of the hydrogen combines with the nitrogen to form ammonia.

The heat released from the combustion of solid wastes is partly stored in the combustion products (gases and ash) and partly transferred by convection, conduction, and radiation to the incinerator walls and to the incoming waste (see Table 12-6). The energy content of the waste can be estimated using the modified Dulong equation [Eq. (10-4)] or the heating value of the individual waste

Table 12-6 Heat losses in combustion of solid waste

Type of losses	Remarks
	The heating value of carbon is about 32,789 kJ/kg.
Reactor	
1. Unburned carbon	Typically, the grate residue is assumed to contain from 4 to 8 percent carbon.
2. Radiation	Heat lost through the reactor walls and other appurtenances to surroundings is estimated as 0.003–0.005 kJ/kg of furnace input.
Latent heat	
3. Inherent moisture	Water content of waste. The latent heat of vaporization for water is approximately 2420 kJ/kg.
4. Moisture in bound water	
5. Moisture from oxidation of net hydrogen	
Sensible heat	
6. Sensible heat in residue	Specific heat of residue is taken as 1047 J/kg · K (10.25 Btu/lb · °F)
7. Stack gases	

components. A heat balance for combustion of solid wastes is illustrated in Example 12-4.

Example 12-4: Determining combustion heat balance Determine the heat available in the exhaust gases from the combustion of 100 tonnes/d of solid waste with an energy content of 10,500 kJ/kg and the following composition. Assume the incinerator residue contains 5 percent carbon and that the temperatures of the entering air and residue from the grate are 25 and 425°C, respectively.

Element	Percent by mass
Carbon	28
Hydrogen	5
Oxygen	22
Nitrogen	4
Sulfur	1
Water	20
Inerts	20

SOLUTION

1. Compute carbon in residue.

Inerts = 20 tonnes/d
Total residue = 20/0.95 = 21.1
Carbon in residue = 1.1 tonnes/d

2. Compute net available hydrogen.

$$\text{Net available hydrogen} = 5\% - 22\%/8 = 2.75\%$$

3. Compute bound water.

$$\text{Hydrogen in bound water} = 5\% - 2.25\% = 2.75\%$$

$$\text{Bound water} = 22\% + 2.75\% = 24.75\%$$

4. Compute amount of water produced from combustion of hydrogen.

$$H_2O = \frac{9 \text{ kg } H_2O}{1 \text{ kg H}}(0.0225 \times 100 \text{ tonnes/d}) = 20.25 \text{ tonnes/d}$$

5. Prepare a heat balance for the combustion process. (*Note*: 100 tonnes/d = 10^5 kg/d)

Item	10^6 kg/d
Gross heat input 10^5 kg/d × 10,500 kJ/kg	1050
Heat lost in unburned carbon 0.011 × 10^5 kg/d × 32,789 kJ/kg	−36
Inherent moisture 0.2 × 10^5 kg/d × 2420 kj/kg	−48
Moisture in bound water 0.2475 × 10^5 kg/d × 2420 kJ/kg	−60
Moisture from oxidation of hydrogen 0.2025 × 10^5 kg/d × 2420 kJ/kg	−49
Radiation loss 10^5 kg/d × 0.005 kJ/kg	—
Sensible heat in residue 0.211 × 10^5 kg/d × 1.047 kJ/kg K × (698 K − 298 K)	−9
Sensible heat in hot gases	848

COMMENT If the boiler efficiency is 85 percent, the overall efficiency would be about 69 percent. This value is consistent with values obtained in modern incinerators.

12-11 Incineration with Heat Recovery

Heat contained in the gases produced from the incineration of solid wastes can be recovered by conversion to steam. The low-level heat remaining in the gases after heat recovery can also be used to preheat the combustion air, boiler makeup water, or solid waste fuel.

Existing Mass-Fired Incinerators

With existing mass-fired incinerators (see Fig. 12-6), waste-heat boilers can be installed to extract heat from the combustion gases without introducing excess amounts of air or moisture. Typically, incinerator gases will be cooled from a range of 1250 to 1375 K (1800 to 2000°F) to a range from 500 to 800 K (600 to 1000°F) before being discharged to the atmosphere. Apart from the production of steam, the use of a boiler system is beneficial in reducing the volume of gas to be processed in the air-pollution control equipment.

Water-Wall Incinerators

In these incinerators, the internal walls of the combustion chamber are lined with boiler tubes that are arranged vertically and welded together in continuous sections. When water-walls are used in place of refractory materials, they are not only useful for the recovery of steam, but also extremely effective in controlling furnace temperature without introducing excess air; however, they are subject to corrosion by the hydrochloric acid produced from the burning of some plastic compounds.

Figure 12-6 Section through typical mass-fired incinerator. (*Courtesy of M & E Engineers, Inc.*).

Figure 12-7 Densified fuel cubes prepared from source-separated waste paper (see Fig. 11-16). Length of ruler is 10 cm.

12-12 USE OF REFUSE-DERIVED FUELS (RDF)

Prepared RDF, typically in a powdered form, can also be fired directly in large industrial boilers that are now used for the production of power with pulverized coal or oil. RDF also can be fired in conjunction with coal or oil. Although the process is not well established with coal, it appears that about 15 to 20 percent of the heat input can be from prepared solid wastes. With oil as the fuel, about 10 percent of the heat input can be from solid wastes. Depending on the degree of processing, suspension, spreader-stoker and double-vortex firing systems have been used.

Densified RDF fuel is prepared using a modified agricultural cubing machine. The resulting fuel cubes (see Fig. 12-7) are suitable for use in a variety of thermal conversion processes including incineration, gasification, and pyrolysis.

12-13 GASIFICATION

The gasification process involves the partial combustion of a carbonaceous fuel to generate a combustible fuel gas rich in carbon monoxide and hydrogen. A gasifier is basically an incinerator operating under reducing conditions. Heat to sustain the process is derived from the exothermic reactions while the combustible components of the low-energy gas are primarily generated by the endothermic reactions. The reaction kinetics of the gasification process are quite complex and still the subject of considerable debate.

When a gasifier is operated at atmospheric pressure with air as the oxidant, the end products of the gasification process are a low-energy gas typically containing (by volume) 10% CO_2, 20% CO, 15% H_2, and 2% CH_4, with the balance being N_2 and a carbon-rich char. Because of the diluting effect of the nitrogen in the input air, the low-energy gas has an energy content in the range of 5.2 to 6.0 MJ/m^3. When pure oxygen is used as the oxidant, a medium-energy gas with an energy content in the range of 12.9 to 13.8 MJ/m^3 is produced.

12-14 PYROLYSIS

Of the many alternative chemical conversion processes that have been investigated, excluding incineration, pyrolysis has received the most attention. Depending on the type of reactor used, the physical form of the solid wastes to be pyrolyzed can vary from unshredded raw wastes to the finely ground portion of the wastes remaining after two stages of shredding and air classification. Upon heating in an oxygen-free atmosphere, most organic substances can be split through a combination of thermal cracking and condensation reactions into gaseous, liquid, and solid fractions. *Pyrolysis* is the term used to describe the process. In contrast to the combustion process, which is highly exothermic, the pyrolytic process is highly endothermic. For this reason, the term *destructive distillation* is often used as an alternative term for pyrolysis.

The characteristics of the three major component fractions resulting from the pyrolysis are: (1) a gas stream containing primarily hydrogen, methane, carbon monoxide, carbon dioxide, and various other gases, depending on the organic characteristics of the material being pyrolyzed; (2) a fraction that consists of a tar and/or oil stream that is liquid at room temperatures and has been found to contain chemicals such as acetic acid, acetone, and methanol; and (3) a char consisting of almost pure carbon plus any intert material that may have entered the process. It has been found that distribution of the product fractions varies with the temperature at which the pyrolysis is carried out. Under conditions of maximum gasification, the energy content of the resulting gas is about 26,100 kJ/m^3 (700 Btu/ft^3). The energy content of pyrolytic oils has been estimated to be about 23,240 kJ/kg (10,000 Btu/lb).

Recovery of Energy from Conversion Products

Once conversion products have been derived from solid wastes by one or more of the biological and thermal methods listed in Tables 12-2 and 12-5, the next step involves their storage and/or use. If energy is to be produced, then an additional conversion step is required.

12-15 ENERGY-RECOVERY SYSTEMS

The principal components involved in the recovery of energy from heat, steam, various gases and oils, and other conversion products are boilers for the production of steam, steam and gas turbines for motive power, and electric generators for the conversion of motive power into electricity.

Typical flow sheets for alternative energy-recovery systems are shown in Fig. 12-8. Perhaps the most common flow sheet for the production of electric energy involves the use of a steam turbine-generator combination (see Fig. 12-8*a*). As shown, when solid wastes are used as the basic fuel source, four operational modes are possible. A flow sheet using a gas turbine-generator combination is shown in Fig. 12-8*b*. The low-energy gas is compressed under high pressure so that it can be used more effectively in the gas turbine.

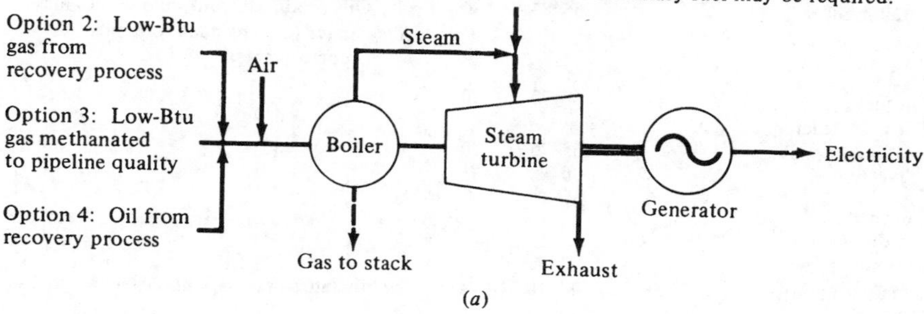

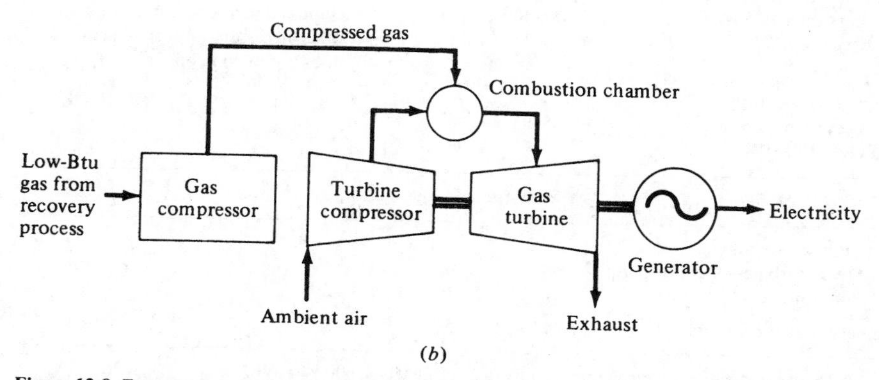

Figure 12-8 Energy-recovery systems: (*a*) options with steam turbine-generator; (*b*) options with gas compressor-gas turbine-generator.

12-16 EFFICIENCY FACTORS

Representative efficiency data for boilers, pyrolytic reactors, gas turbines, steam turbine-generator combinations, electric generators, and related plant use and loss factors are given in Table 12-7. In any installation where energy is being produced, allowance must be made for the station or process power needs and for

Table 12-7 Typical thermal efficiency and plant use and loss factors for individual components and processes used for the recovery of energy from solid wastes

Component	Efficiency*		Comment
	Range	Typical	
Incinerator-boiler	40–68	63	Mass-fired.
Boiler			
Solid fuel	60–75	72	Processed solid wastes (RDF).
Low-Btu gas	60–80	75	Burners must be modified.
Oil-fired	65–85	80	Oils produced from solid wastes may have to be blended to reduce corrosiveness.
Gasifier	60–70	70	
Pyrolysis reactor			
Conventional	65–75	70	
Purox	70–80	75	
Turbines			
Combustion gas			
Simple cycle	8–12	10	
Regenerative	20–26	24	Includes necessary appurtenances.
Expansion gas	30–50	40	
Steam turbine-generator system			
Less than 12.5 MW	24–40	29†‡	Includes condenser, heaters, and all other necessary appurtenances, but does not include boiler.
Over 10 MW	28–32	31.6†‡	
Electric generator			
Less than 10 MW	88–92	90	
Over 10 MW	94–98	96	
Plant use and loss factors			
Station service allowance			
Steam turbine-generator plant	4–8	6	
Purox process	18–24	21	
Unaccounted heat losses	2–8	5	

* Theoretical value for mechanical equivalent of heat = 3600 kJ/kWh.

† Efficiency varies with exhaust pressure. Typical value given is based on an exhaust pressure in the range of 50 to 100 mmHg.

‡ Heat rate = 11,395 kJ/kWh = 3600 kJ/kWh/0.316.

Table 12-8 Energy output and efficiency for a 1000-tonne/d steam boiler turbine-generator energy-recovery plant using unprocessed solid wastes with an energy content of 12,000 kJ/kg

Item	Value
Energy available in solid wastes, million kJ/h 1000 tonnes/d × 1000 kg/tonne × 12,000 kJ/kg)/(24 h/d × 10^6 kJ/million kJ)	500
Steam energy available, million kJ/h 500 million kJ/h × 0.7	350
Electric power generation, kW (350 million kJ/h)/(11,395 kJ/kWh)*	30.715
Station service allowance, kW 30,715(0.06)	−1842
Unaccounted heat losses, kW 30,715(0.05)	−1535
Net electric power for export, kW	27,338
Overall efficiency, percent (100)(27,338 kW)/[(5 × 10^8 kJ/h)/(3600 kJ/kWh)]	19.7

* 11,395 kJ/kWh = (3600 kJ/kWh)0.316.
Source: Adapted from Tchobanoglous et al. [12-6]

unaccounted process heat losses. Typically, the auxiliary power allowance varies from 4 to 8 percent of the power produced. Process heat losses usually will vary from 2 to 8 percent.

12-17 DETERMINATION OF ENERGY OUTPUT AND EFFICIENCY

An analysis of the amount of energy produced from a solid-waste energy-conversion system using an incinerator-boiler-steam turbine-electric generator combination with a capacity of 1000 tonnes/d is presented in Table 12-8. If it is assumed that 10 percent of the power generated is used for the front-end processing system (typical values vary from 8 to 14 percent), then the net power for export is 24,604 kW and the overall efficiency is 17.5 percent.

Materials- and Energy-Recovery Systems

During the past few years numerous systems have been proposed or built incorporating different types of processing and energy-conversion systems. Two typical examples are shown in Figs. 12-9 and 12-10. Unfortunately, few of the full-scale plants that have been built have proved to be successful. Although economics has been the major reason for their demise, some energy-conversion plants have failed

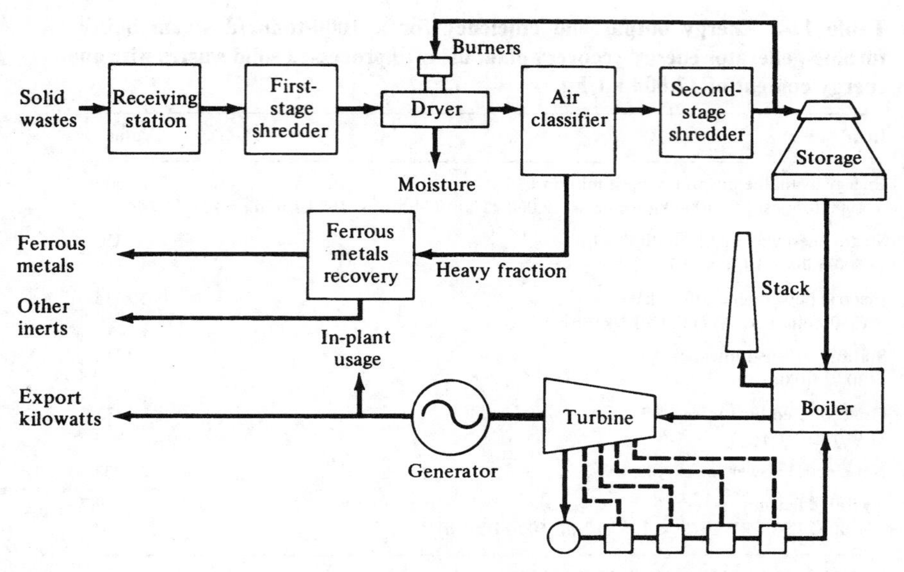

Figure 12-9 Flow sheet for the recovery of ferrous materials and energy from solid wastes. (*From Tchobanoglous et al.* [*12-6*].)

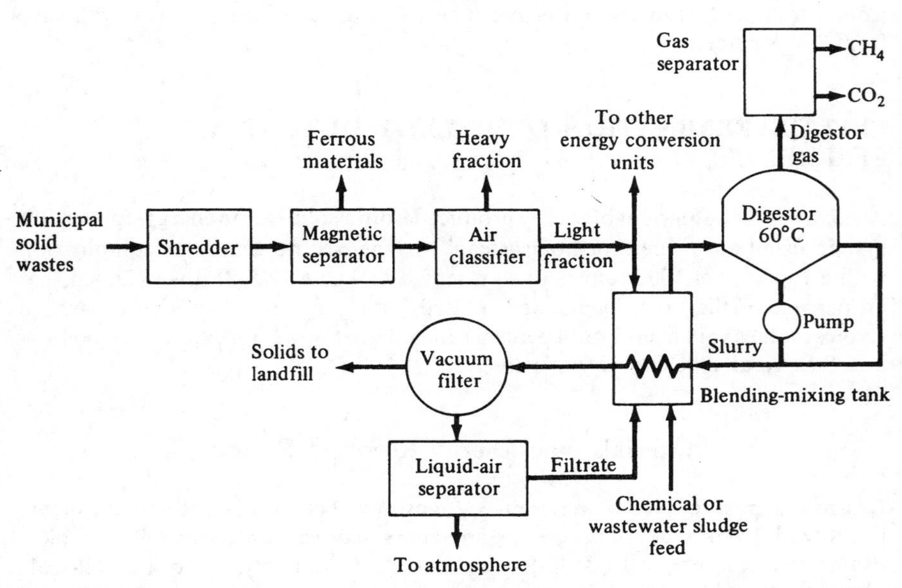

Figure 12-10 Flow sheet for the recovery of ferrous metal materials and digester gas from solid wastes. (*From Tchobanoglous et zl.* [*12-6*].)

because of technical difficulties. Thus, if the use of a materials- and energy-recovery system is contemplated, current operating systems should be visited and analyzed, and realistic cost estimates should be prepared.

DISCUSSION TOPICS AND PROBLEMS

12-1 Estimate the theoretical amount of air required to oxidize completely 1 tonne of waste having the composition $C_{60}H_{120}O_{30}N_3$.

12-2 Estimate the theoretical amount of air required to oxidize completely a waste having the composition given in Prob. 10-3.

12-3 Estimate the amount of compost that could be produced per tonne from a waste having the composition given in Prob. 10-3.

12-4 Estimate the heat that could be recovered per tonne from a waste with the composition given in Prob. 10-3.

12-5 If the overall efficiency of an energy-conversion process is 12.6 percent, estimate the original energy content of the solid waste using the following information:

a. Energy loss in the conversion process = 25 percent
b. Process fuel usage = 8 percent of energy in conversion gas
c. Gas-turbine efficiency = 24 percent
d. Electrical generator efficiency = 96 percent
e. In-plant power usage = 21 percent of the total power generated

REFERENCES

12-1 Golueke, C. G.: *Biological Reclamation of Solid Wastes*, Rodale Press, Emmaus, Penn., 1977.
12-2 Haslam, R. T., and R. P. Russell: *Fuels and Their Combustion*, 1st ed., McGraw-Hill, New York, 1926.
12-3 Haug, R. T.: *Compost Engineering: Principles and Practice*, Ann Arbor Science, Ann Arbor, Mich., 1980.
12-4 Homes, J. R.: *Refuse Recycling And Recovery*, Wiley, New York, 1981.
12-5 Price, E. C., and P. N. Cheremisinoff: *Biogas Production and Utilization*, Ann Arbor Science, Ann Arbor, Mich., 1981.
12-6 Tchobanoglous, G., H. Theisen, and R. Eliassen: *Solid Wastes: Engineering Principles and Management Issues*, McGraw-Hill, New York, 1977.
12-7 Vesilind, P. A., and A. E. Rimer: *Unit Operations In Resources Recovery Engineering*, Prentice-Hall, Englewood Cliffs, N.J., 1981.
12-8 Weinstein, N. J., and R. F. Toro: *Thermal Processes of Municipal Solid Waste for Resource and Energy Recovery*, Ann Arbor Science, Ann Arbor, Mich., 1976.
12-9 Wilson, D. C.: *Waste Management: Planning, Evaluation, Technologies*, Clarendon Press, Oxford, England, 1981.

APPENDIX

A

QUANTITIES AND UNITS

Table A-1 Atomic numbers and atomic masses*

Element	Symbol	Atomic number	Atomic mass
Actinium	Ac	89	227.0278
Aluminum	Al	13	26.98154
Americium	Am	95	(243)
Antimony	Sb	51	121.75
Argon	Ar	18	39.948
Arsenic	As	33	74.9216
Astatine	At	85	(210)
Barium	Ba	56	137.33
Berkelium	Bk	97	(247)
Beryllium	Be	4	9.01218
Bismuth	Bi	83	208.9804
Boron	B	5	10.81
Bromine	Br	35	79.904
Cadmium	Cd	48	112.41
Calcium	Ca	20	40.08
Californium	Cf	98	(251)
Carbon	C	6	12.011
Cerium	Ce	58	140.12
Cesium	Cs	55	132.9054
Chlorine	Cl	17	35.453
Chromium	Cr	24	51.996
Cobalt	Co	27	58.9332
Copper	Cu	29	63.546
Curium	Cm	96	(247)
Dysprosium	Dy	66	162.50
Einsteinium	Es	99	(254)
Erbium	Er	68	167.26
Europium	Eu	63	151.96
Fermium	Fm	100	(257)
Fluorine	F	9	18.99840
Francium	Fr	87	(223)
Gadolinium	Gd	64	157.25
Gallium	Ga	31	69.72
Germanium	Ge	32	72.59
Gold	Au	79	196.9665
Hafnium	Hf	72	178.49
Helium	He	2	4.00260
Holmium	Ho	67	164.9304
Hydrogen	H	1	1.0079
Indium	In	49	114.82
Iodine	I	53	126.9045
Iridium	Ir	77	192.22
Iron	Fe	26	55.847
Krypton	Kr	36	83.80
Lanthanum	La	57	138.9055
Lawrencium	Lr	103	(260)
Lead	Pb	82	207.2
Lithium	Li	3	6.941
Lutetium	Lu	71	174.97
Magnesium	Mg	12	24.305
Manganese	Mn	25	54.9380
Mendelevium	Md	101	(258)
Mercury	Hg	80	200.59
Molybdenum	Mo	42	95.94
Neodymium	Nd	60	144.24
Neon	Ne	10	20.179
Neptunium	Np	93	237.0482
Nickel	Ni	28	58.70
Niobium	Nb	41	92.9064
Nitrogen	N	7	14.0067
Nobelium	No	102	(259)
Osmium	Os	76	190.2
Oxygen	O	8	15.9994
Palladium	Pd	46	106.4
Phosphorus	P	15	30.97376
Platinum	Pt	78	195.09
Plutonium	Pu	94	(244)
Polonium	Po	84	(209)
Potassium	K	19	39.0983
Praseodymium	Pr	59	140.9077
Promethium	Pm	61	(145)
Protactinium	Pa	91	231.0359

Table A-1 (*continued*)

Radium	Ra	88	226.0254	Tellurium	Te	52	127.60
Radon	Rn	86	(222)	Terbium	Tb	65	158.9254
Rhenium	Re	75	186.207	Thallium	Tl	81	204.37
Rhodium	Rh	45	102.9055	Thorium	Th	90	232.0381
Rubidium	Rb	37	85.4678	Thulium	Tm	69	168.9342
Ruthenium	Ru	44	101.07	Tin	Sn	50	118.69
Samarium	Sm	62	150.4	Titanium	Ti	22	47.90
Scandium	Sc	21	44.9559	Tungsten	W	74	183.85
Selenium	Se	34	78.96	Uranium	U	92	238.029
Silicon	Si	14	28.0855	Vanadium	V	23	50.9414
Silver	Ag	47	107.868	Xenon	Xe	54	131.30
Sodium	Na	11	22.98977	Ytterbium	Yb	70	173.04
Strontium	Sr	38	87.62	Yttrium	Y	39	88.9059
Sulfur	S	16	32.06	Zinc	Zn	30	65.38
Tantalum	Ta	73	180.9479	Zirconium	Zr	40	91.22
Technetium	Tc	43	(97)				

* A value in parentheses is the mass number of the longest-lived isotope of the element.
Source: From *Pure Appl. Chem.*, **47**:75 (1976).

Table A-2 Base units in the international system of units (SI)

Quantity	Name	Symbol
Length	meter	m
Mass	kilogram	kg
Time	second	s
Electric current	ampere	A
Thermodynamic temperature	kelvin	K
Amount of substance	mole	mol
Luminous intensity	candela	cd
Plane angle*	radian	rad
Solid angle*	steradian	sr

* Supplementary units.

Table A-3 Derived SI units with special names

Quantity	SI unit symbol	Name	Units
Frequency	Hz	hertz	1/s
Force	N	newton	$kg \cdot m/s^2$
Pressure, stress	Pa	pascal	$kg/m \cdot s^2$ or N/m^2
Energy or work	J	joule	$kg \cdot m^2/s^2$ or $N \cdot m$
A quantity of heat	J	joule	$kg \cdot m^2/s^2$ or $N \cdot m$
Power, radiant flux	W	watt	$kg \cdot m^2/s^3$ or J/s
Electric charge	C	coulomb	$A \cdot s$
Electrical potential	V	volt	$kg \cdot m^2/s^3 \cdot A$ or W/A
Potential difference	V	volt	$kg \cdot m^2/s^3 \cdot A$ or W/A
Electromotive force	V	volt	$kg \cdot m^2/s^3 \cdot A$ or W/A
Capacitance	F	farad	$A^2 \cdot s^4/kg \cdot m^2$ or C/V
Electric resistance	Ω	ohm	$kg \cdot m^2/s^3 \cdot A^2$ or V/A
Conductance	S	siemens	$s^3 \cdot A^2/kg \cdot m^2$ or A/V
Magnetic flux	Wb	weber	$kg \cdot m/s^2 \cdot A$ or $V \cdot s$
Magnetic flux density	T	tesla	$kg/s^2 \cdot A$ or Wb/m^2
Inductance	H	henry	$kg \cdot m^2/s^2 \cdot A^2$ or Wb/A
Luminous flux	lm	lumen	$cd \cdot sr$
Illuminance	lx	lux	$cd \cdot sr/m^2$ or lm/m^2
Activity (radionuclides)	Bq	becquerel	l/s
Absorbed dose	Gy	gray	m^2/s^2 or J/kg

Table A-4 Derived SI units obtained by combining base units and units with special names

Quantity	Units	Quantity	Units
Acceleration	m/s^2	Molar entropy	$J/mot \cdot K$
Angular acceleration	rad/s^2	Molar heat capacity	$J/mot \cdot K$
Angular velocity	rad/s	Moment of force	$N \cdot m$
Area	m^2	Permeability	H/m
Concentration	mol/m^3	Permittivity	F/m
Current density	A/m^2	Radiance	$W/m^2 \cdot sr$
Density, mass	kg/m^3	Radiant intensity	W/sr
Electric charge density	C/m^3	Specific heat capacity	$J/kg \cdot K$
Electric field strength	V/m	Specific energy	J/kg
Electric flux density	C/m^2	Specific entropy	$J/kg \cdot K$
Energy density	J/m^3	Specific volume	m^3/kg
Entropy	J/K	Surface tension	N/m
Heat capacity	J/K	Thermal conductivity	$W/m \cdot K$
Heat flux density	W/m^2	Velocity	m/s
Irradiance	W/m^2	Viscosity, dynamic	$Pa \cdot s$
Luminance	cd/m^2	Viscosity, kinematic	m^2/s
Magnetic field strength	A/m	Volume	m^3
Molar energy	J/mol	Wavelength	m

Table A-5 Values of useful constants

Acceleration due to gravity, g = 9.807 m/s^2 (32.174 ft/s^2)
Standard atmosphere = 101.325 kN/m^2 (14.696 lbf/in^2)
= 101.325 kPA (1.013 bar)
1 bar = 10^5 N/m^2 (14.504 lbf/in^2)
Standard atmosphere = 10.333 m (33.899 ft) of water
1 meter head of water (20°C) = 9.790 M/m^2 (1.420 lbf/in^2)
= 0.00979 N/mm^2 (1.420 lbf/in^2)
= 9.790 kN/m^2 (1.420 lbf/in^2)

Table A-6 Standard prefixes

Factor	Symbol	Prefix
10^{-18}	a	atto
10^{-15}	f	femto
10^{-12}	p	pico
10^{-9}	n	nano
10^{-6}	μ	micro
10^{-3}	m	milli
10^{-2}	c	centi
10^{-1}	d	deci
10^{1}	da	deca or deka
10^{2}	h	hecto
10^{3}	k	kilo
10^{6}	M	mega
10^{9}	G	giga
10^{12}	T	tera

APPENDIX

B

CONVERSION FACTORS

Table B-1 Conversion factors for commonly used wastewater-treatment plant design parameters

	To convert, multiply in direction shown by arrows			
Parameter (in SI units)	SI units	→	←	U.S. units
Screening				
m^3 screenings/10^3 m^3 wastewater	$m^3/10^3\ m^3$	133.6806	7.4805×10^{-3}	ft^3/Mgal
Grit removal				
Air supply				
m^3 air/m of tank length · min	$m^3/m \cdot min$	10.7639	0.0929	$ft^3/ft \cdot min$
Grit removal				
g grit/m^3 wastewater	g/m^3	8.3454	0.1198	lb/Mgal
kg grit/m^3 wastewater	kg/m^3	8345.4	1.1983×10^{-4}	lb/Mgal
Surface overflow rate				
m^3 flow/m^2 surface area · h	$m^3/m^2 \cdot h$	589.0173	0.0017	$gal/ft^2 \cdot d$
m^3 flow/m^2 surface area · d	$m^3/m^2 \cdot d$	24.5424	0.0407	$gal/f^2 \cdot d$
Volume				
m^3 grit/10^3 m^3 wastewater	$m^3/10^3\ m^3$	133.6806	7.4805×10^{-3}	ft^3/Mgal
Flow equalization				
Air supply				
m^3 air/m^3 tank volume · min	$m^3/m^3 \cdot min$	133.6806	7.4805×10^{-2}	$ft^3/10^2\ gal \cdot min$
Mixing horsepower				
kW/m^3 tank volume	kW/m^3	5.0763	0.1970	$hp/10^2\ gal$
Sedimentation				
Particle settling rate				
m/h	m/h	3.2808	0.3048	ft/h
m/h	m/h	0.4090	2.4448	$gal/ft^2 \cdot min$
Sludge scraper speed				
m/h	m/h	0.0547	18.2880	ft/min

Table B-1 (*continued*)

Parameter (in SI units)	SI units	→	←	U.S. units
	To convert, multiply in direction shown by arrows			
Solids loading				
kg solids/m^2 surface area · d	kg/m^2 · d	0.2048	4.8824	lb/ft^2 · d
Surface overflow rate				
m^3 wastewater/m^2 surface area · d	m^3/m^2 · d	24.5424	0.0407	gal/ft^2 · d
m^3 wastewater/m^2 surface area · h	m^3/m^2 · h	589.0173	0.0017	gal/ft^2 · d
Volume of sludge				
m^3 sludge/10^3 m^3 wastewater	$m^3/10^3$ m^3	133.6806	7.481×10^{-3}	ft^3/Mgal
Weight of dry sludge solids				
g dry solids/m^3 wastewater	g/m^3	8.3454	0.1198	lb/Mgal
Weir overflow rate				
m^3 wastewater/m weir length · d	m^3/m · d	80.5196	0.0124	gal/ft · d
Activated sludge				
Aeration device mixing intensity, diffused aeration				
m^3 air/m^3 tank volume · min	m^3/m^3 · min	1000.0	0.001	$ft^3/10^3$ ft^3 · min
Aeration device mixing intensity, mechanical aeration				
kW/10^3 m^3 tank volume	kW/10^3 m^3	0.0380	26.3342	hp/10^3 ft^3
Air flow rate				
m^3 air/h	m^3/h	0.5886	1.6990	ft^3/min
Air requirements, organic removal				
m^3 air/kg BOD_5 removed	m^3/kg	16.0185	0.0624	ft^3/lb
Air requirements, volume of wastewater				
m^3 air/m^3 wastewater	m^3/m^3	0.1337	7.4805	ft^3/gal
Organic load				
kg BOD_5 applied/m^3 aeration-tank volume · d	kg/m^3 · d	62.4280	0.0160	lb/10^3 ft^3 · d
Oxygen requirements				
kg O_2/kg BOD_5 applied · d	kg/kg · d	1.0	1.0	lb/lb · d
Oxygen-transfer rate				
kg O_2 transferred/kW · h	kg/kW · h	1.6440	0.6083	lb/hp · h
kg O_2 transferred/m^3 wastewater · h	kg/m^3 · h	0.0624	16.0185	lb/ft^3 · h
Trickling filters and rotating biological contactors				
Hydraulic load				
m^3 wastewater/m^2 bulk surface area · d	m^3/m^2 · d	24.5424	0.0407	gal/ft^2 · d

m^3 wastewater/m^2 bulk surface area · h	$m^3/m^2 \cdot h$	589.0173	0.0017	$gal/ft^2 \cdot d$
m^3 wastewater/m^2 bulk surface area · d	$m^3/m^2 \cdot d$	1.0691	0.9354	Mgal/acre · d
L wastewater/m^2 bulk surface area · min	$L/m^2 \cdot min$	35.3420	0.0283	$gal/ft^2 \cdot d$
Organic load				
kg BOD_5/m^3 filter-medium volume · d	$kg/m^3 \cdot d$	62.4280	0.0160	$lb/10^3\ ft^3 \cdot d$
Specific surface loading, hydraulic				
m^3 wastewater/m^2 filter medium surface area · d	$m^3/m^2 \cdot d$	24.5424	0.0407	$gal/ft^2 \cdot d$
m^3 wastewater/m^2 filter medium surface area · d	$m^3/m^2 \cdot d$	0.0170	58.6740	$gal/ft^2 \cdot min$
m^3 wastewater/m^2 filter medium surface area · h	$m^3/m^2 \cdot h$	589.0173	0.0017	$gal/ft^2 \cdot d$
Specific surface loading, organic				
kg BOD_5/m^2 filter medium surface area · d	$kg/m^2 \cdot d$	0.2048	4.8824	$lb/ft^2 \cdot d$
Tank volume				
L/m^2 medium surface area (rotating biological reactor)	L/m^2	2.4542×10^{-2}	40.7458	gal/ft^2
Stabilization ponds and lagoons				
Organic loads				
kb BOD_5/ha surface area · d	kg/ha · d	0.8922	1.1209	lb/acre · d
Volumetric load				
kg BOD_5/m^3 basin volume · d	$kg/m^3 \cdot d$	62.4280	0.0160	$lb/10^3\ ft^3 \cdot d$
Chlorination				
Feed rate				
kg chlorine/d	kg/d	2.2046	0.4536	lb/d
Sludge thickening				
Sludge loading				
kg dry solids fed/m^2 surface area · d	$kg/m^2 \cdot d$	0.2048	4.8824	$lb/ft^2 \cdot d$
Surface overflow rate				
m^3 wastewater/m^2 surface area · d	$m^3/m^2 \cdot d$	24.5424	0.0407	$gal/ft^2 \cdot d$
m^3 wastewater/m^2 surface area · d	$m^3/m^2 \cdot d$	0.0170	58.6740	$gal/ft^2 \cdot min$
Sludge digestion				
Gas production				
m^3 gas/kg volatile solids fed	m^3/kg	16.0185	0.0624	ft^3/lb
m^3 gas/capita	m^3/capita	35.3147	0.0283	ft^3/capita
Loading rate				
kg BOD_5/m^3 digester volume · d	$kg/m^3 \cdot d$	62.4280	0.0160	$lb/10^3\ ft^3 \cdot d$

Table B-1 (*continued*)

Parameter (in SI units)	To convert, multiply in direction shown by arrows			
	SI units	→	←	U.S. units
Sludge heating				
W/m² surface area · °C	$W/m^2 \cdot °C$	0.1763	5.6735	$Btu/ft^2 \cdot °F \cdot h$
Volatile-solids loading				
kg volatile solids/m³ digester volume · d	$kg/m^3 \cdot d$	62.4280	0.0160	$lb/10^3\ ft^3 \cdot d$
Sludge drying beds				
Dry-solids loading				
kg dry solids/m² area · yr	$kg/m^2 \cdot yr$	0.2048	4.8824	$lb/ft^2 \cdot yr$
m² area/capita · yr	$m^2/capita \cdot yr$	10.7639	0.0929	$ft^2/capita \cdot yr$
Vacuum filtration				
Dry solids				
kg dry solids/m² surface area · h	$kg/m^2 \cdot h$	0.2048	4.8824	$lb/ft^2 \cdot h$
Pressure applied				
kPa (kN/m²) pressure	kPa	0.1450	6.8948	lbf/in^2 (gage)
Sludge feed				
m³ wet sludge/m² surface area · h	$m^3/m^2 \cdot h$	3.2808	0.3048	$ft^3/ft^2 \cdot h$
Vacuum applied				
kPa (kN/m²) vacuum	kPa (kN/m^2)	0.2961	3.3768	inHg (60°F)
Heat drying				
kJ heat energy required/kg water evaporated (sludge cake)	kJ/kg	0.4303	2.3241	Btu/lb
kg water evaporated/h	kg/h	2.2046	0.4536	lb/h
kg wet sludge/m² heating surface · h	$kg/m^2 \cdot h$	0.2048	4.8824	$lb/ft^2 \cdot h$
Incineration				
kJ heat energy/kg moisture evaporated	kJ/kg	0.4303	2.3241	Btu/lb
kg sludge/m² heating surface area	kg/m^2	0.2048	4.8824	$lb/ft^2 \cdot h$
kg sludge/m³ combustion chamber volume · h	$kg/m^3 \cdot h$	0.0624	16.0185	$lb/ft^3 \cdot h$
Land disposal				
kg mass/ha field area	kg/ha	0.8922	1.1208	lb/acre
bu yield/ha field area · yr	bu/ha · yr	0.4047	2.4711	bu/acre · yr
Mg loading/ha field area	Mg/ha	0.4461	2.2417	tons/acre
m³ wastewater/ha field area · d	$m^3/ha \cdot d$	106.9064	0.0094	gal/acre · d
Surface or in-depth filters				
L water (backwash)/m² surface area · min	$L/m^2 \cdot min$	0.0245	40.7458	$gal/ft^2 \cdot min$

Source: From Metcalf & Eddy, Inc.: *Wastewater Engineering: Collection and Pumping of Wastewater*, McGraw-Hill, New York, 1981.

Table B-2 Metric conversion factors (U.S. customary units to SI units)

Multiply the U.S. customary unit		by	To obtain the SI unit	
Name	Symbol		Symbol	Name
Acceleration				
feet per second squared	ft/s^2	0.3048*	m/s^2	meters per second squared
inches per second squared	in/s^2	0.0254*	m/s^2	meters per second squared
Area				
acre	acre	0.4047	ha	hectare
acre	acre	4.0469×10^{-3}	km^2	square kilometer
square foot	ft^2	9.2903×10^{-2}	m^2	square meter
square inch	in^2	6.4516*	cm^2	square centimeter
square mile	mi^2	2.5900	km^2	square kilometer
square yard	yd^2	0.8361	m^2	square meter
Energy				
British thermal unit	Btu	1.0551	kJ	kilojoule
foot-pound (force)	ft · lbf	1.3558	J	joule
horsepower-hour	hp · h	2.6845	MJ	megajoule
kilowatt-hour	kW · h	3600*	kJ	kilojoule
kilowatt-hour	kW · h	3.600×10^6*	J	joule
watt-hour	W · h	3.600*	kJ	kilojoule
watt-second	W · s	1.000*	J	joule
Flow rate				
cubic feet per second	ft^3/s	2.8317×10^{-2}	m^3/s	cubic meters per second
gallons per day	gal/d	4.3813×10^{-5}	L/s	liters per second
gallons per day	gal/d	3.7854×10^{-3}	m^3/d	cubic meters per day
gallons per minute	gal/min	6.3090×10^{-5}	m^3/s	cubic meters per second
gallons per minute	gal/min	6.3090×10^{-2}	L/s	liters per second
millions gallons per day	Mgal/d	43.8126	L/s	liters per second
million gallons per day	Mgal/d	3.7854×10^3	m^3/d	cubic meters per day
million gallons per day	Mgal/d	4.3813×10^{-2}	m^3/s	cubic meters per second

Table B-2 (*continued*)

Multiply the U.S. customary unit		by	To obtain the SI unit	
Name	Symbol		Symbol	Name
Force				
pound force	lbf	4.4482	N	newton
Length				
foot	ft	0.3048*	m	meter
inch	in	2.54*	cm	centimeter
inch	in	0.0254*	m	meter
inch	in	25.4*	mm	millimeter
mile	mi	1.6093	km	kilometer
yard	yd	0.9144*	m	meter
Mass				
ounce	oz	28.3495	g	gram
pound	lb	4.5359×10^2	g	gram
pound	lb	0.4536	kg	kilogram
ton (short: 2000 lb)	ton	0.9072	Mg (metric ton)	megagram (10^3 kilogram)
ton (long: 2240 lb)	ton	1.0160	Mg (metric ton)	megagram (10^3 kilogram)
Power				
British thermal units per second	Btu/s	1.0551	kW	kilowatt
foot-pounds (force) per second	ft · lbf/s	1.3558	W	watt
horsepower	hp	0.7457	kW	kilowatt
Pressure (force/area)				
atmosphere (standard)	atm	1.0133×10^2	kPa (kN/m^2)	kilopascal (kilonewtons per square meter)
inches of mercury (60°F)	inHg (60°F)	3.3768×10^3	Pa (N/m^2)	pascal (newtons per square meter)
inches of water (60°F)	inH_2O (60°F)	2.4884×10^2	Pa (N/m^2)	pascal (newtons per square meter)
pounds (force) per square foot	lbf/ft^2	47.8803	Pa (N/m^2)	pascal (newtons per square meter)
pounds (force) per square inch	lbf/in^2	6.8948×10^3	Pa (N/m^2)	pascal (newtons per square meter)
pounds (force) per square inch	lbf/in^2	6.8948	kPa (kN/m^2)	kilopascal (kilonewtons per square meter)

Temperature				
degrees Fahrenheit	°F	0.555(°F − 32)	°C	degrees Celsius (centigrade)
degrees Fahrenheit	°F	0.555(°F + 459.67)	K	degrees kelvin
Velocity				
feet per second	ft/s	0.3048*	m/s	meters per second
miles per hour	mi/h	4.4704×10^{-1}*	km/s	kilometers per second
Volume				
acre-foot	acre-ft	1.2335×10^{3}	m^3	cubic meter
cubic foot	ft^3	28.3168	L	liter
cubic foot	ft^3	2.8317×10^{-2}	m^3	cubic meter
cubic inch	in^3	16.3871	cm^3	cubic centimeter
cubic yard	yd^3	0.7646	m^3	cubic meter
gallon	gal	3.7854×10^{-3}	m^3	cubic meter
gallon	gal	3.7854	L	liter
ounce (U.S. fluid)	oz (U.S. fluid)	2.9573×10^{-2}	L	liter

* Indicates exact conversion.

Source: From Metcalf & Eddy, Inc.: *Wastewater Engineering: Collection and Pumping of Wastewater*, McGraw-Hill, New York, 1981.

Table B-3 Metric conversion factors (SI units to U.S. customary units)

Multiply the SI unit		by	To obtain the U.S. customary unit	
Name	Symbol		Symbol	Name
Acceleration				
meters per second squared	m/s^2	3.2808	ft/s^2	feet per second squared
meters per second squared	m/s^2	39.3701	in/s^2	inches per second squared
Area				
hectare (10,000 m^2)	ha	2.4711	acre	acre
square centimeter	cm^2	0.1550	in^2	square inch
square kilometer	km^2	0.3861	mi^2	square mile
square kilometer	km^2	247.1054	acre	acre
square meter	m^2	10.7639	ft^2	square foot
square meter	m^2	1.1960	yd^2	square yard
Energy				
kilojoule	kJ	0.9478	Btu	British thermal unit
joule	J	2.7778×10^{-7}	kW · h	kilowatt-hour
joule	J	0.7376	ft · lbf	foot-pound (force)
joule	J	1.0000	W · s	watt-second
joule	J	0.2388	cal	calorie
kilojoule	kJ	2.7778×10^{-4}	kW · h	kilowatt-hour
kilojoule	kJ	0.2778	W · h	watt-hour
megajoule	MJ	0.3725	hp · h	horsepower-hour
Flow rate				
cubic meters per day	m^3/d	264.1720	gal/d	gallons per day
cubic meters per day	m^3/d	2.6417×10^{-4}	Mgal/d	million gallons per day
cubic meters per second	m^3/s	35.3147	ft^3/s	cubic feet per second
cubic meters per second	m^3/s	22.8245	Mgal/d	million gallons per day
cubic meters per second	m^3/s	15,850.3	gal/min	gallons per minute
liters per second	L/s	22,824.5	gal/d	gallons per day
liters per second	L/s	0.0228	Mgal/d	million gallons per day
liters per second	L/s	15.8508	gal/min	gallons per minute

Force				
newton	N	0.2248	lbf	pound force
Length				
centimeter	cm	0.3937	in	inch
kilometer	km	0.6214	mi	mile
meter	m	39.3701	in	inch
meter	m	3.2808	ft	foot
meter	m	1.0936	yd	yard
millimeter	mm	0.03937	in	inch
Mass				
gram	g	0.0353	oz	ounce
gram	g	0.0022	lb	pound
kilogram	kg	2.2046	lb	pound
megagram (10^3 kg)	Mg	1.1023	ton	ton (short: 2000 lb)
megagram (10^3 kg)	Mg	0.9842	ton	ton (long: 2240 lb)
Power				
kilowatt	kW	0.9478	Btu/s	British thermal units per second
kilowatt	kW	1.3410	hp	horsepower
watt	W	0.7376	ft/lbf/s	foot-pounds (force) per second
Pressure (force/area)				
pascal (newtons per square meter)	Pa (N/m^2)	1.4504×10^{-4}	lbf/in^2	pounds (force) per square inch
pascal (newtons per square meter)	Pa (N/m^2)	2.0885×10^{-2}	lbf/ft^2	pounds (force) per square foot
pascal (newtons per square meter)	Pa (N/m^2)	2.9613×10^{-4}	inHg	inches of mercury (60°F)
pascal (newtons per square meter)	Pa (N/m^2)	4.0187×10^{-3}	inH_2O	inches of water (60°F)
kilopascal (kilonewtons per square meter)	kPa (kN/m^2)	0.1450	lbf/in^2	pounds (force) per square inch
kilopascal (kilonewtons per square meter)	kPa (kN/m^2)	0.0099	atm	atmosphere (standard)
Temperature				
degree Celsius (centigrade)	°C	1.8(°C) + 32	°F	degree Fahrenheit
degree kelvin	K	1.8(K) − 459.67	°F	degree Fahrenheit
Velocity				
kilometers per second	km/s	2.2369	mi/h	miles per hour
meters per second	m/s	3.2808	ft/s	feet per second

Table B-3 (*continued*)

Multiply the SI unit		by	To obtain the U.S. customary unit	
Name	Symbol		Symbol	Name
Volume				
cubic centimeter	cm^3	0.0610	in^3	cubic inch
cubic meter	m^3	35.3147	ft^3	cubic foot
cubic meter	m^3	1.3079	yd^3	cubic yard
cubic meter	m^3	264.1720	gal	gallon
cubic meter	m^3	8.1071×10^{-4}	acre · ft	acre · foot
liter	L	0.2642	gal	gallon
liter	L	0.0353	ft^3	cubic foot
liter	L	33.8150	oz	ounce (U.S. fluid)

Source: From Metcalf & Eddy, Inc.: *Wastewater Engineering: Collection and Pumping of Wastewater*, McGraw-Hill, New York, 1981.

APPENDIX

C

PROPERTIES OF WATER AND AIR

Table C-1 Physical properties of water (SI units)

Temperature, °C	Specific weight, γ, kN/m^3	Density, ρ, kg/m^3	Modulus of elasticity,* $E/10^6$, kN/m^2	Dynamic viscosity $\mu \times 10^3$, $N \cdot s/m^2$	Kinematic viscosity $\nu \times 10^6$, m^2/s	Surface tension,† σ, N/m	Vapor pressure, p_v, kN/m^2
0	9.805	999.8	1.98	1.781	1.785	0.0765	0.61
5	9.807	1000.0	2.05	1.518	1.519	0.0749	0.87
10	9.804	999.7	2.10	1.307	1.306	0.0742	1.23
15	9.798	999.1	2.15	1.139	1.139	0.0735	1.70
20	9.789	998.2	2.17	1.002	1.003	0.0728	2.34
25	9.777	997.0	2.22	0.890	0.893	0.0720	3.17
30	9.764	995.7	2.25	0.798	0.800	0.0712	4.24
40	9.730	992.2	2.28	0.653	0.658	0.0696	7.38
50	9.689	988.0	2.29	0.547	0.553	0.0679	12.33
60	9.642	983.2	2.28	0.466	0.474	0.0662	19.92
70	9.589	977.8	2.25	0.404	0.413	0.0644	31.16
80	9.530	971.8	2.20	0.354	0.364	0.0626	47.34
90	9.466	965.3	2.14	0.315	0.326	0.0608	70.10
100	9.399	958.4	2.07	0.282	0.294	0.0589	101.33

* At atmospheric pressure.
† In contact with air.

Table C-2 Henry's law coefficients for several gases that are slightly soluble in water

T, °C	$H \times 10^{-4}$, atm/mol fraction							
	Air	CO_2	CO	H_2	H_2S	CH_4	N_2	O_2
0	4.32	0.0728	3.52	5.79	0.0268	2.24	5.29	2.55
10	5.49	0.104	4.42	6.36	0.0367	2.97	6.68	3.27
20	6.64	0.142	5.36	6.83	0.0483	3.76	8.04	4.01
30	7.71	0.186	6.20	7.29	0.0609	4.49	9.24	4.75
40	8.70	0.233	6.96	7.51	0.0745	5.20	10.4	5.35
50	9.46	0.283	7.61	7.65	0.0884	5.77	11.3	5.88
60	10.1	0.341	8.21	7.65	0.103	6.26	12.0	6.29

Table C-3 Equilibrium concentrations (mg/L) of dissolved oxygen* as a function of temperature and chloride

Temperature, °C	Chloride concentration, mg/L				
	0	5,000	10,000	15,000	20,000
0	14.62	13.79	12.97	12.14	11.32
1	14.23	13.41	12.61	11.82	11.03
2	13.84	13.05	12.28	11.52	10.76
3	13.48	12.72	11.98	11.24	10.50
4	13.13	12.41	11.69	10.97	10.25
5	12.80	12.09	11.39	10.70	10.01
6	12.48	11.79	11.12	10.45	9.78
7	12.17	11.51	10.85	10.21	9.57
8	11.87	11.24	10.61	9.98	9.36
9	11.59	10.97	10.36	9.76	9.17
10	11.33	10.73	10.13	9.55	8.98
11	11.08	10.49	9.92	9.35	8.80
12	10.83	10.28	9.72	9.17	8.62
13	10.60	10.05	9.52	8.98	8.46
14	10.37	9.85	9.32	8.80	8.30
15	10.15	9.65	9.14	8.63	8.14
16	9.95	9.46	8.96	8.47	7.99
17	9.74	9.26	8.78	8.30	7.84
18	9.54	9.07	8.62	8.15	7.70
19	9.35	8.89	8.45	8.00	7.56
20	9.17	8.73	8.30	7.86	7.42
21	8.99	8.57	8.14	7.71	7.28
22	8.83	8.42	7.99	7.57	7.14
23	8.68	8.27	7.85	7.43	7.00
24	8.53	8.12	7.71	7.30	6.87
25	8.38	7.96	7.56	7.15	6.74
26	8.22	7.81	7.42	7.02	6.61
27	8.07	7.67	7.28	6.88	6.49
28	7.92	7.53	7.14	6.75	6.37
29	7.77	7.39	7.00	6.62	6.25
30	7.63	7.25	6.86	6.49	6.13

* Saturation values of dissolved oxygen in fresh water and sea water exposed to dry air containing 20.90 percent oxygen by volume under a total pressure of 760 mm of mercury.

Source: G. C. Whipple and M. C. Whipple: Solubility of Oxygen in Sea Water, *J. Am. Chem. Soc.*, vol. 33, p. 362, 1911. Calculated using data developed by C. J. J. Fox: On the Coefficients of Absorption of Nitrogen and Oxygen in Distilled Water and Sea Water and Atmospheric Carbonic Acid in Sea Water, *Trans. Faraday Soc.*, vol. 5, p. 68, 1909.

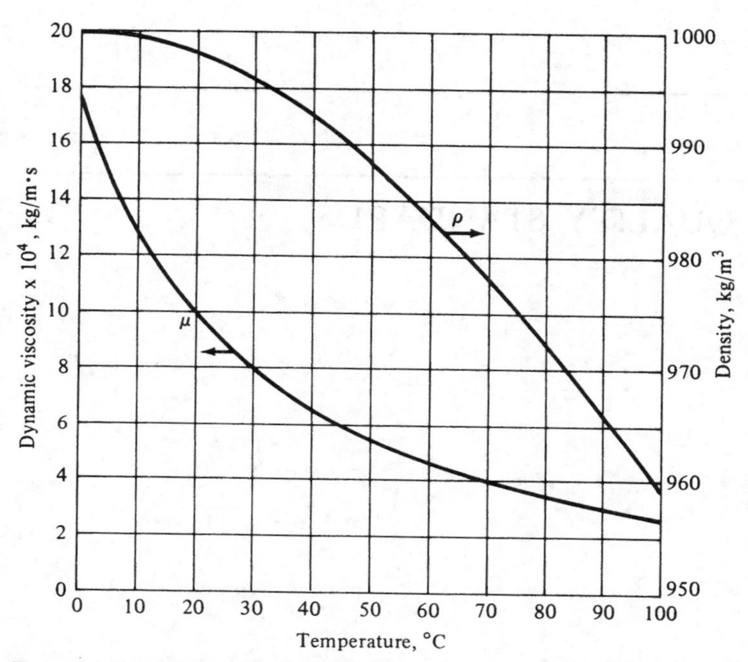

Figure C-1 Density and dynamic viscosity of liquid water as a function of temperature. (*From Martin Crawford: Air Pollution Control Theory, McGraw-Hill, New York, 1976.*)

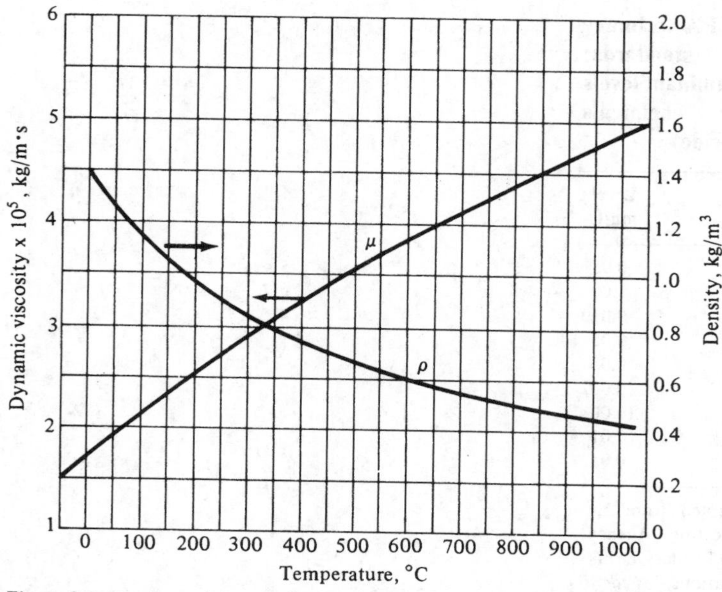

Figure C-2 Density and dynamic viscosity of pure air at 1.0 atm pressure as a function of temperature. (*Adapted from Martin Crawford: Air Pollution Control Theory, McGraw-Hill, New York, 1976.*)

APPENDIX

D

WATER QUALITY STANDARDS

Table D-1 EPA primary drinking-water standards: maximum contaminant levels for inorganic chemicals (other than fluoride)

Contaminant	Level mg/L
Arsenic	0.05
Barium	1.00
Cadmium	0.010
Chromium	0.05
Lead	0.05
Mercury	0.002
Nitrate (as N)	10.00
Selenium	0.01
Silver	0.05

Source: Adapted from Environmental Protection Agency: "National Interim Primary Drinking Water Regulations," *Federal Register*, part IV, December 24, 1975.

Table D-2 EPA primary drinking-water standards: maximum contaminant levels for fluoride

Annual average of maximum daily air temperatures of community in which water system is situated, °F	°C	Maximum contaminant levels, mg/L
53.7 and below	12.0 and below	2.4
53.8–58.3	12.1–14.6	2.2
58.4–63.8	14.7–17.6	2.0
63.9–70.6	17.7–21.4	1.8
70.7–79.2	21.5–26.2	1.6
79.3–90.5	26.3–32.5	1.4

Source: Adapted from Environmental Protection Agency: "National Interim Primary Drinking Water Regulations," *Federal Register*, part IV, December 24, 1975.

Table D-3 EPA primary drinking-water standards: maximum contaminant levels for organic chemicals

Chemical	Maximum contaminant level (MCL), mg/L
Chlorinated hydrocarbons:	
Endrin (1,2,3,4,10,10-hexachloro-6,7-epoxy-1,4,4a,5,6,7,8,8a-octo-hydro-1,4-endo,endo-5,8-dimethanonaphthalene)	0.0002
Lindane (1,2,3,4,5,6-hexachlorocyclohexane, gamma isomer)	0.004
Methoxychlor (1,1,1-trichloro-2,2-bis {p-methoxy-phenyl}ethane)	0.1
Toxaphene ($C_{10}H_{10}Cl_8$-technical chlorinated camphene, 67–69% chlorine)	0.005
Chlorophenoxys:	
2,4-D (2,4-dichlorophenoxyacetic acid)	0.1
2,4,5-TP silvex (2,4,5-trichlorophenoxypropionic acid)	0.01

Source: Adapted from Environmental Protection Agency: "National Interim Primary Drinking Water Regulations," *Federal Register*, part IV, December 24, 1975.

Table D-4 EPA primary drinking-water standards: maximum levels for turbidity*

Reading basis†	Maximum contaminant level (MCL), turbidity units
Turbidity reading based on monthly average	1 TU or up to 5 TUs if the water supplier can demonstrate to the state that the higher turbidity does not interfere with disinfection, maintenance of an effective disinfectant agent throughout the distribution system, or microbiological determinants
Turbidity reading based on average for 2 consecutive days	5 TUs

* As measured at representative point(s) in the distribution system.

† Failure to meet standards on either the monthly basis or the 2-consecutive-day basis constitutes a violation of the MCL.

Source: Adapted from Environmental Protection Agency: "National Interim Primary Drinking Water Regulations," *Federal Register*, part IV, December 24, 1975.

Table D-5 EPA primary drinking-water standards: maximum contaminant level (MCL) for microbiological contaminants

		Individual sample basis*	
Test method used	Monthly basis*	Fewer than 20 samples/mo†	More than 20 samples/mo
Membrane filter technique	1/100 mL average density	Number of coliform bacteria shall not exceed: 4/100 mL in more than one sample	4/100 mL in more than 5% of of samples
Fermentation tube method		Coliform bacteria shall not be present in:	
10-mL standard portions	More than 10% of the portions	Three or more portions in more than one sample	Three or more portions in more than 5% of samples
100-mL standard portions	More than 60% of the portions	Five portions in more than one sample	Five portions in more than 20% of the samples

* Failure to meet standards on either the monthly basis or the individual sample basis constitutes a violation of the MCL.

† For systems that are required to sample at a rate of less than four per month, compliance with the above regulations shall be based upon sampling during a 3-mo period, except that, at the discretion of the state, compliance may be based upon sampling during a 1-mo period.

Source: Adapted from Environmental Protection Agency: "National Interim Primary Drinking Water Regulations," *Federal Register*, part IV, December 24, 1975.

Table D-6 Proposed guidelines for secondary drinking-water standards

Parameter	Proposed standard
Chloride	250 mg/L
Color	15 CU (color units)
Copper	1 mg/L
Corrosivity	Noncorrosive
Foaming agents	0.5 mg/L
Hydrogen sulfide	0.05 mg/L
Iron	0.3 mg/L
Manganese	0.05 mg/L
Odor	< 3 TON
pH	6.5–8.5
Sulfate	250 mg/L
Total dissolved solids (TDS)	500 mg/L
Zinc	5 mg/L

Source: Adapted from E. W. Steele and T. J. McGhee: *Water Supply and Sewerage*, 5th ed., McGraw-Hill, New York, 1979.

Table D-7 Secondary treatment standards

Characteristic of discharge	Unit of measurement	Average monthly concentration	Average weekly concentration
BOD_5	mg/L	30*†	45†
Suspended solids‡	mg/L	30*†	45†
Hydrogen-ion concentration	pH units	6.0–9.0§	

* Or, in no case more than 15 percent of influent value.

† Arithmetic mean.

‡ Treatment plants with stabilization ponds and flows < 7570 m^3/d (2 Mgal/d) are exempt.

§ Continuous, only enforced if caused by industrial wastewater or in-plant treatment.

Source: From Metcalf & Eddy, Inc.: *Wastewater Engineering: Treatment, Disposal, Reuse*, McGraw-Hill, New York, 1979.

NAME INDEX

SUBJECT INDEX